A Handbook of Double Stars

For the Use of Amateurs

Edward Crossley
Joseph Gledhill
James M. Wilson

CAMBRIDGE
UNIVERSITY PRESS

CAMBRIDGE UNIVERSITY PRESS

Cambridge, New York, Melbourne, Madrid, Cape Town,
Singapore, São Paolo, Delhi, Tokyo, Mexico City

Published in the United States of America by Cambridge University Press, New York

www.cambridge.org
Information on this title: www.cambridge.org/9781108039772

© in this compilation Cambridge University Press 2011

This edition first published 1879
This digitally printed version 2011

ISBN 978-1-108-03977-2 Paperback

CAMBRIDGE LIBRARY COLLECTION

Books of enduring scholarly value

Physical Sciences

From ancient times, humans have tried to understand the workings of the world around them. The roots of modern physical science go back to the very earliest mechanical devices such as levers and rollers, the mixing of paints and dyes, and the importance of the heavenly bodies in early religious observance and navigation. The physical sciences as we know them today began to emerge as independent academic subjects during the early modern period, in the work of Newton and other 'natural philosophers', and numerous sub-disciplines developed during the centuries that followed. This part of the Cambridge Library Collection is devoted to landmark publications in this area which will be of interest to historians of science concerned with individual scientists, particular discoveries, and advances in scientific method, or with the establishment and development of scientific institutions around the world.

A Handbook of Double Stars

Used to describe both binary systems and optical doubles, the term 'double star' has been familiar to astronomers since the seventeenth century. This book, first published in 1879, outlines the history of their study, and describes the methods and equipment needed in order to observe the fascinating phenomenon. Written for non-specialists by Fellows of the Royal Society Edward Crossley (1841–1904), Joseph Gledhill (1837–1906) and James M. Wilson (1836–1931), the catalogue of over 1,200 double stars appears beside detailed notes and does not assume mathematical expertise. Also offered are a fully worked example of how to find the orbit of a binary star, and illustrations of telescopes, observatories, and even custom-made observation chairs. This reissue includes the supplement with corrections and notes published in 1880. A standard reference text in the late nineteenth century, the work remains a resources for students and scholars of the history of astronomy.

A

HANDBOOK

OF

DOUBLE STARS.

BERMERSIDE, HALIFAX.

A

HANDBOOK

OF

DOUBLE STARS,

WITH A

CATALOGUE OF TWELVE HUNDRED DOUBLE STARS AND
EXTENSIVE LISTS OF MEASURES.

With additional Notes bringing the Measures up to 1879.

FOR THE USE OF AMATEURS.

BY

EDWD. CROSSLEY, F.R.A.S.; JOSEPH GLEDHILL, F.R.A.S.,
AND JAMES M. WILSON, M.A., F.R.A.S.

"The subject has already proved so extensive, and still promises so rich a harvest to those
who are inclined to be diligent in the pursuit, that I cannot help inviting every lover of
astronomy to join with me in observations that must inevitably lead to new discoveries."—
SIR WM. HERSCHEL.

"Stellæ fixæ, quæ in cœlo conspiciuntur, sunt aut soles simplices, qualis sol noster, aut
systemata ex binis vel interdum pluribus solibus peculiari nexu physico inter se junctis composita.
Stellarum simplicium numerus est quidem major, at vero non nisi ter vel fortasse bis tantum
major quam systematum compositorum."—Σ.

London:
MACMILLAN & CO.
1879.

Hazell, Watson, and Viney, Printers, London and Aylesbury.

PREFACE.

THIS work has arisen out of our own wants as students of that branch of astronomy which deals with Double Stars, and it is on this account that we think it will be useful to others who are occupied in the same work. There does not exist any book which gives information sufficiently detailed to be of value to any one who seriously takes up this study. He must hunt through scores and hundreds of volumes if he wishes to get an accurate and complete list of the previous measures of any particular double star. These measures are scattered up and down the astronomical periodicals of all nations. If he wishes to know with what instruments, with what apertures, and what micrometers these measures were taken, a fresh research awaits him. And if he proceeds to attempt an orbit, he will fail, unless he is a tolerably expert mathematician, from want of sufficient guidance and detail in the various mathematical papers and pamphlets that have been devoted to this subject.

This branch of astronomy is peculiarly suitable to amateurs. It does not require long previous training; it does not demand unintermittent and severe work, nor the resources of a permanent observatory and staff. All it needs is a good telescope, a good eye, some patience, much conscientiousness, and—more than all—such an amount of guidance and co-operation as shall convince the amateur that his work is not useless, but that he is really contributing something, however small, to astronomical knowledge. And the construction of double-star orbits has always had a fascination for amateurs from the days of Admiral Smyth and γ Virginis to the present time; and it is perhaps the only branch of mathematical astronomy which is quite within the range of unprofessional mathematicians.

We venture to hope that this book will be of use in guiding amateurs in their work,—in pointing out what stars are of especial interest, what stars have had few or conflicting measures taken of them, at what times observations of certain stars are especially needful, and what stars have been so frequently and satisfactorily measured that for the present they need no attention. This sort of information has become a necessity owing to the extension of the subject and the number of observers. The Herschels, the elder Struve, and Mädler, might with equal advantage measure every double star they saw; but later observers must select their objects if they do not wish much of their work to be wasted. And so we find that Otto Struve, and Dawes, and Secchi, and others, have chosen stars that were certainly or probably of interest as subjects for their own work.

There has probably been no time in which so much work has been done in measuring double stars as during the last six or seven years. They have witnessed Burnham's lists of new double stars, which testify so highly to his telescope, his eye, his climate, and his industry; Otto Struve's two important volumes on his father's and his own double stars; Dembowski's lists in the *Astronomische Nachrichten;* Dunér's valuable volume of observations made at Lund; in America, the work of Hall, Stone, etc.; and in our own country, that of Knott and others.

The recalculation of orbits, also, is occupying much attention, both among foreign astronomers and at home; and every year will enable this to be done with greater accuracy, and to be attempted for a greater number of stars.

This work, then, consists of four parts. The first part is historical, and descriptive of instruments and methods; the second is mathematical; the third part contains lists of measures of the most interesting double and multiple stars, with historical notes on those which are of special interest; the fourth part is bibliographical.

In Part I., Chapter I. contains a historical introduction by Mr. Gledhill. Chapter II. is on the equatorial and the observatory, by Mr. Crossley; Chapter III. is an account of the equatorials which have been used by double-star observers, by Mr. Gledhill; Chapter IV. on micrometers, by Mr. Crossley; and Chapter V. on methods of observing, by Mr. Gledhill.

In Part II., Chapters I. and II. give a detailed account, with a fully worked example, of determining an orbit and an ephemeris by a purely graphical construction, founded on Herschel's and Thiele's methods, with some fresh extensions, by Mr. Wilson. Dr. Doberck, who has had very great experience in double-star calculations, has contributed Chapter III., giving an example of the application of analysis to a double-star orbit already approximately known by graphical methods, and shows how greater accuracy may be obtained by it; and Mr. Wilson gives Chapter IV. on the relative rectilinear motion of double stars; Chapter V. on the effects of proper motion and parallactic motion; and Chapter VI. on the mode of combining observations, and determining their weight.

Part III. contains a catalogue of double stars selected as of special interest, with a list of all accessible measures, and notes, etc., by Mr. Gledhill.

Finally, Part IV. contains the bibliography of the whole subject, and is due to Mr. Gledhill.

We may, perhaps, venture to say a word or two on the importance of this part of astronomy. It can scarcely fail to happen that accurate measures of double stars, especially when combined with a study of proper motion, will give in the future some sounder knowledge of the structure of the heavens. The calculation of double-star orbits, and the comparison of observed and calculated places, will bring out not only errors in the observations or of the computer, but the existence of forces that had been unsuspected. Resisting media and the laws of their condensation, unseen companions, and possibly new laws of force, may be discovered. And these investigations must throw light on the origin of these double and multiple systems, and thus indirectly on our own solar system.

Again ; if the difference of the linear velocities of the components of a binary system can be directly ascertained by the spectroscope, this fact, combined with a good knowledge of the orbit and of the period of revolution, and of the apparent mean angular distance, will lead to a knowledge of the parallax of the system, and therefore also to a knowledge of their mass.

At present we cannot see the significance of all that has been discovered : for example, the fact that the orbits hitherto computed are all elliptical, and very nearly all of large eccentricity, is too uniform to be an accident, and yet it is too isolated a fact to build theories on with safety. It does, however, seem to prove that these are genuine systems *ab initio*, and are not formed by the fortuitous approximation of single stars.

Nor, again, have we found the reason why the type of triple stars, such as μ Herculis, γ Andromedæ, ζ Cancri, μ Boötis— a bright primary and a faint binary companion—should be so common. When, further, we come to examine into the colours of binaries, we cannot yet see to what previous stage in their history is owing the absence of red stars in these systems, and the frequency of other colours which in their turn are rare in solitary stars. Spectroscopic observation will doubtless add some information on the point of fact, but will only remove the difficulty one stage further on. Again, the phenomenon of *variable* and *temporary* stars has always suggested the notion of a revolving dark companion. This may need further examination, and light may be thrown on the subject from tracing the gradual development of binary systems. In a word, the further study of binaries will help our successors to know what is the development-order of star systems and planetary systems.

The present work, therefore, is intended to facilitate the labours of future students of sidereal astronomy, by supplying the materials for the study of double stars in a convenient form, and as complete (so far as it is intended to go) as our utmost pains could make it.

The distribution of double stars has not been investigated, and it is perhaps at present premature to attempt it until more is known about them in both hemispheres; but there are already plain indications that it is not entirely fortuitous.

A knowledge of their distribution will scarcely fail to throw light on the great problem of the structure of the sidereal universe.

Similarly, it will be observed that we devote no chapters to the variability of colour or intensity in the components of double stars. We have been debarred from this branch of the subject by want of time, by the badness of our climate, and by the unsuitability of our instruments. It is to be hoped that this work will be taken up by some one else. Small telescopes, and especially small reflectors, are well suited to the examination of colour; but if possible a careful spectroscopic examination of each star should be made. We have, however, provided in the bibliographical part of the book some references to the chief works and papers on this subject.

We therefore commend this study to amateurs. They may be encouraged by the thought that, with few exceptions, all the great workers in this branch of astronomy have been amateurs; and be stimulated to exertion by the thought that observations made now will certainly be of value to their successors. The stars will not stand still. How can we be idle, and let slip the time for observations, which, if not made now, can never be made hereafter?

BERMERSIDE,
  September, 1879.

CONTENTS.

PART I.

HISTORICAL, AND DESCRIPTIVE OF INSTRUMENTS AND METHODS.

PART II.

ON THE CALCULATION OF THE ORBIT OF A BINARY STAR.

LIST OF ILLUSTRATIONS.

LIST OF PLATES.

DIAGRAMS.

ERRATA.

The plate facing p. 248, illustrating the looped path of ƺ Cancri $\left(\dfrac{A + B}{2}\text{ and C}\right)$, is taken from the *Observations de Poulkova*, vol. ix.

TEMPLE OBSERVATORY, RUGBY.

DOUBLE STARS.

PART I.

CHAPTER I.

HISTORICAL INTRODUCTION.

THE history of double-star astronomy begins with the year 1779, a year for ever memorable as that in which the greatest of observers began the investigations which created a new department of observational astronomy.

The results of the occasional attention of astronomers to this class of observation prior to the time of Herschel were small indeed. Riccioli, about the middle of the seventeenth century, saw that ζ Ursæ Majoris was double, and Kirsch also noted the same fact in 1700. Huyghens saw θ Orionis as a quadruple star in 1656; in 1664 Hooke first saw γ Arietis as a double star and α Centauri appears to have been the fourth double star which yielded to the power of the telescope, as Feuillée is said to have discovered it in 1709 at Lima. Bradley separated γ Virginis in 1718, and both Messier and Cassini watched the occultation of the components by the moon.* Castor was found to be a double star in 1719, 61 Cygni in 1753, β Cygni in 1755 ; then followed γ Andromedæ, ε Lyræ, 70 Ophiuchi, ζ Cancri, β Scorpii, ξ Ursæ Majoris, etc. Pigott discovered three in 1779.† Nor must

* See the *Histoire de l'Académie Royale des Sciences,* for the years 1678 1720, 1774.

† Phil. Trans., vol. lxxi.

the numerous wide pairs detected by Christian Mayer pass unnoticed. This industrious observer, working at Mannheim with an eight-feet mural quadrant by Bird and a power of about 60 to 80, observed and catalogued a considerable number of stars with Comites.* A short extract from his book † will give a good idea of the character of the objects and his mode of observation :—

1777.	Stella cum comite.	Gradus lucis.	Differentia Ascensionis rectæ.			Differentia Declinationis.		
Jan. 1	Comes Aldebaran	8·9	0°	2′	14″·2	0°	12′	29″
,, 6	Comes Electra	Teles.	0	0	8	0	0	32 ·5
,, 13	Comes Algol	8	0	2	49 ·5	0	9	9 ·5

At the end of the volume a table of the new pairs discovered by him (72 in number) is given ; among them are the following :—

	Mag.	Differentia in R.A.	Differentia Declinationis.	Dist.
		sec.	″	″
γ Andromedæ......... ...	2, 6	0·95	5·8	15·2
Castor	1, 6	0·7	3·8	11·0
ƺ Cancri	7, 8	0·0	7·7	7·7
γ Virginis...............	5, 5	0·5	6·3	9·9
α Herculis	3, 7	0·53	4·0	8·9
ε Lyræ...................	6, 8	0·2	3·0	4·2
β Cygni	3, 7	2·06	19·9	36·6

In 1777, Maskelyne, in a letter to Mayer, says that he saw a Herculis double in August 1777, magnitudes 3 and 6, the preceding star being the fainter, and that the distance of the centres was 7″. Mayer also wrote two other papers on this subject.‡

To return. It was in 1779 that Sir William Herschel began to direct his wonderful energy to the observation of double

---

* Mayer says that Flamsteed first used the word *comes* for the smaller star of a pair.

† See his work, *De novis in cælo sidereo Phænomenis*, etc., 1779.

‡ " *De centum stellarum fixarum comitibus, eorumque insigni usu ad determinandum motum proprium fixarum;* " and " *De miris fixarum comitumque mutationibus a me observatis a tempore cel. Flamsteedii.*"

stars ; and his famous paper is so interesting, and so fully exhibits the state of this department at the time he wrote, that a short account of it may here with propriety be given.

The great historical problem of finding stellar parallax had presented itself to him, and with his usual ardour he set himself the task of grappling with all its difficulties. After noticing Galileo's method, and the previous attempts to carry it out by Hooke, Flamsteed, Molineux, and Bradley, and pointing out the cause of their failure, he proceeds to describe his own method, viz., *to measure the position angle of two stars of unequal magnitudes at two opposite points of the earth's orbit.* He states the essential conditions to be, (1) that the stars be near each other ; (2) that their magnitudes be very unequal. He then criticises the attempt made by Dr. Long, and points out the causes of his want of success, viz., unsuitable double stars, and want of adequate optical power. (Dr. Long had chosen γ Arietis, Castor, γ Virginis, etc., and his magnifying power did not exceea 70.) His own method is then shown to be independent of refraction, nutation, precession, change of obliquity of the ecliptic, and aberration. The highest possible power is to be used ; and a figure showing a Lyræ under powers from 460 to 6450 is given. Having fully satisfied himself that the method was sound and practicable, the next step was the selection of suitable pairs of stars. And here his own noble words may fitly be quoted :—

" I resolved to examine every star in the heavens with the utmost attention, and a very high power, that I might collect such materials for this research as would enable me to fix my observations on those that would best answer my end. The subject has already proved so extensive, and still promises so rich a harvest to those who are inclined to be diligent in the pursuit, that I cannot help inviting every lover of astronomy to join with me in observations that must inevitably lead to new discoveries."—*Phil. Trans.,* vol. lxxii.

It was in this spirit, and with this glowing enthusiasm, that Herschel began those sweeps and measures which have added so much to our knowledge of the sidereal universe.

A full description of his method of finding the position angle and distance apart of the components of a double star, statements respecting the accuracy of his estimations and micrometric measures, etc., are then given. Then comes the catalogue of his discoveries. The pairs given number 269, and they are arranged in six classes, according to distance : Class I., close pairs requiring " indeed a very superior telescope, the utmost clearness of air," etc. II., those suitable for "very delicate measures of the micrometer." III., from 5″ to 15″. IV., from 15″ to 30″. V., from 30″ to 1′. VI., from 1′ to 2′.*

Of these 269 objects, 227 were new, 9 were known before Mayer's time, and 33 were known to Mayer and other observers. A single extract will show the form and character of the information given respecting these stars :—

" 16. η Coronæ borealis, Fl. 2.

" *Sept.* 9.—Double. A little unequal. They are whitish stars. They seem in contact with 227, and though I can see them with this power, I should certainly not have discovered them with it ; with 400, less than ¼ diameter; with 932, fairly separated, and the interval a little larger than with 460. I saw them also with 2010, but they are so close that this power is too much for them, at least when the altitude of the stars is not very considerable ; with 460 they are as fine a miniature of ε Bootis as that is of α Geminorum. Position 59° 19′ n following."†

In 1803 appeared Herschel's celebrated paper announcing the discovery of *binary* stars, and this was followed in 1822 by a list of 145 new double stars.

---

* Herschel's first measure of a double star is said to have been that of the trapezium in Orion.

† Phil. Trans., 1782.

During the first twenty years of this century, notwithstanding the splendour of the discoveries above described, double stars were but little observed. No doubt the principal cause was the want of instruments of suitable power and construction. In 1816 Sir John Herschel began to review the double stars discovered by his father, and was soon joined by Sir James South. For a list of his papers containing measures, etc., see List A, Part IV. For this distinguished observer, double-star measurement ever possessed a charm ; and from time to time, all through his long life, catalogues, measures, etc., were contributed by him to the Memoirs of the Royal Astronomical Society. Valuable results were also obtained during Sir John's stay at the Cape of Good Hope; and just before his lamented death he was busy at work on a general catalogue of double stars.

Two years before the reviews began at Slough, Friedrich Georg Wilhelm Struve, in the distant and ill-furnished observatory of Dorpat, was turning his attention in the same direction. Although an 8 feet transit by Dollond, and a 5 feet telescope by Troughton (power 126), were the only instruments at his command, he began to observe the positions, and occasionally to measure the position-angles and distances, of double stars. These results are to be found in the early volumes of the Dorpat observations. And in order to facilitate the study of this subject, he published in 1820 the places of double stars. In 1821 the fine Ertel Circle was received, and in 1824 the famous Fraunhofer refractor was added. Then began the great survey of the heavens between the pole and 15° of south declination, for the purpose of discovering new double stars, and the formation of a general catalogue of them. From 1824 to 1835 Struve and his assistants devoted themselves almost entirely to the execution of this noble scheme, and in 1837 appeared the results in the magnificent work entitled *Mensuræ Micrometricæ Stellarum duplicium et multiplicium.* Nor did double stars lose their attractiveness at the observatory of Dorpat after the conclusion of this vast undertaking. In 1839 the splendid observatory at Poulkova was established, and in 1861, on the resignation of his father, the directorship was placed in the hands of Otto Struve. From year to year careful and systematic measures have been made up to the present time, and the latest publication of the distinguished son of the great Struve is a noble series in two volumes of measures of the most important double stars.

Here, too, must be mentioned the labours of Admiral Smyth. With an 8 feet equatorial, this excellent observer measured 680 stars between 1830 and 1843, and the results were published in 1844, under the title *Cycle of Celestial Objects.* In 1860, the *Speculum Hartwellianum*, containing later measures, etc., was published.

Mädler, observing with the Dorpat refractor, measured a large number of double stars between the years 1834 and 1845, and published the results in 1847, in an elaborate work entitled *Untersuchungen über die Fixstern-systeme.* In this fine work are given extensive lists of double stars having probable direct motion, probable retrograde motion, and certain motion ; chapters dealing with the orbits of the most important binaries ; very complete lists of measures ; a chapter on the combinations of double stars to form "higher systems," etc., etc.

Between 1830 and 1868 Dawes communicated many important lists of measures and papers on double stars to the Royal Astronomical Society. His great catalogue was, however, not published till 1867. This work is enriched by the addition of valuable introductions, notes, and lists of measures made by previous observers.

Valuable measures were made at Lord Wrottesley's observatory between the years 1843 and 1860.

Powell and Jacob, at Madras, made many useful measures, the former from 1853 to 1862, and the latter from 1853 to 1857.

The Baron Dembowski began his fine series of measures in the year 1852 at Naples. He proposed to measure all the Dorpat "*lucidæ*" within the reach of his instrument. This important undertaking he successfully accomplished between the years 1852 and 1858 ; and a more valuable contribution to this department has rarely been made. In 1862 he resumed the examination of those Dorpat stars which exhibited changes in angle or distance; and the careful measurement of the great binaries has been continued up to the present time. The last review also included the measurement of a large number of the double stars discovered at Poulkova.

Secchi, in the years 1856 to 1859, paid considerable attention to double stars, and in 1860 appeared his *Catalogo di* 1321* *stelle Doppie misurate col grande equatoriale di Merz all' osservatorio del Collegio Romano.* Some years later he also published *Serie seconda delle misure micrométriche, fatte all' equatoriale di Merz del Collegio Romano, dal* 1863 *al* 1866 *inclusive, stelle doppie e Nebulose dal P. A. Secchi.*

In 1861, the late Rev. R. Main, Radcliffe Observer, began to observe a selected list of double stars. These observations have been published from year to year in the volumes issued by the observatory up to the present time. They have all been made with the Heliometer.

At Mr. Barclay's observatory the measurement of double stars has always held a prominent place in the work of the observers Mr. Romberg and Mr. Talmage.

Dunér, at the Lund Observatory, issued a volume of double star measures in 1876. It contains his results from 1867 to 1875, and is a valuable addition to the works on double-star astronomy.

Mr. O. Stone and his assistants at the Cincinnati Observatory have for some time paid special attention to double stars, and several lists of measures have already been published.

---

* The number is really 1221.

Mr. Burnham, of Chicago, has published no less than nine catalogues of double stars, his own discoveries, since 1871 : all these objects have also had their positions and distances either measured or estimated by this most industrious observer.

Dr. William Doberck, at Markree Observatory, has taken up this branch of astronomy with great spirit and success. For some of the results of his labours see List A.

Professor Pritchard, of the new Oxford University Observatory, assisted by Messrs. Plummer and Jenkins, is making careful measures of the principal binaries, and is also engaged in a re-investigation of their orbits, by a method possessing some new features, and which seems to yield good results.

M. Camille Flammarion has devoted himself with great ardour to double-star investigations : his catalogue of important objects, with lists of measures, will shortly be published.

This subject has always attracted the attention of patrons and wealthy amateurs, and the names of Lord Wrottesley, George Bishop, Esq., J. G. Barclay, Esq., Colonel Cooper, Edward Crossley, Esq., Isaac Fletcher, Esq., M.P., and G. Knott, Esq., must here be mentioned as deserving of special praise for the spirited manner in which they have established and supported observatories for the prosecution of this class of observation.

Lastly, compilers of useful catalogues of binary stars and the writers of handbooks must not be forgotten : among the former, Mr. A. Brothers, F.R.A.S., and among the latter the Rev. W. A. Darby, M.A., and, above all, the Rev. T. W. Webb, M.A., deserve especial mention.

Measures by the following observers and others have also been published : Auwers, Bessel, Bond, Brünnow, Challis, Dunlop, Ellery, Encke, Engelmann, Ferrari, Fletcher, Galle, Gledhill, Hall, Hind, Holden, Jacob, Kaiser, Knott, Lassell, Maclear, Miller, Mitchell, Morton, Newcomb, Nobile, Powell, Schiaparelli, Seabroke, Spörer, Waldo, Wilson.

THE BERMERSIDE EQUATORIAL.

CHAPTER II.

THE EQUATORIAL: ITS CONSTRUCTION AND ADJUSTMENTS.

In making a few remarks upon the instruments required by double-star observers, it is not our intention to give an exhaustive description, but rather to confine ourselves to a few points which may serve as some guide to the amateur who wishes to provide himself with these instruments, or who, being already equipped, desires to set to work with confidence.

It is first of all necessary to be furnished with a good refractor or reflector, equatorially mounted, of sufficient aperture, and driven by clockwork. And we do not hesitate to say that we much prefer a refractor, as being more stable in its adjustments, less disturbed by atmospheric conditions, and more durable in its optical surfaces,—conditions which seem to us to do more than counterbalance any advantages arising from the smallness of the star discs, and the absence of colour obtained from good reflecting telescopes.

We will assume that an equatorially mounted refractor is chosen. This should be of not less than six inches aperture, in order to be generally useful. An aperture of eight or nine inches would be a liberal and handsome provision. Good work may be done on some stars with smaller apertures, but we are afraid they would cause disappointment by their limited power.

To obtain a good instrument, it is best to secure the services of a first-class maker, who has made large equatorials his speciality. Among English makers it is hardly necessary to mention such names as those of Messrs. Troughton and

Simms of London, T. Cooke and Sons of York, and Mr.
Howard Grubb of Dublin, whose well-known achievements
speak for themselves.

We will now take up the different parts of the equatorial,
beginning with the object glass. This requires the greatest
possible amount of skill and patience in its construction, and
great care should be exercised in its selection by the employ-
ment of suitable tests.

After examining the lenses in their cell by transmitted light,
to discover any flaws of serious magnitude (for minute sand-
holes and bubbles are not serious), and then looking at its two
outer surfaces by reflected light to see if the polish is uniform
and good, replace the object glass in the tube and turn it
upon some elevated object, as a church spire or chimney
with a bright sky background. Focus carefully with a low
power, and if the outlines are sharply defined and free from
colour, the probability is that the glass is fairly achromatic.
To render this test more severe, Stokes recommended that
half the object glass should be covered by a semicircular
piece of cardboard.

For the next test the instrument must be directed to the sky
at night, and some patience and judgment will be needed in
selecting a night suitable for the work. Examine the moon or
any of the larger planets at an elevation of not less than 30°
above the horizon, the higher the better; and if there be sharp-
ness of outline, distinctness of detail, and absence of vibration,
the night is one suitable for the purpose. Now turn to stars
of different magnitudes, as near the zenith as may be, using a
high power; and if clean round discs are obtained, free from
wings and stray light, the result is so far satisfactory. Next
examine the rings which surround the central small disc, when
the eyepiece is moved a little within and without the focus. If
the rings are circular, and each of uniform brightness all round,
and sharply distinct from one another, the lens may be considered well centered and corrected. If the glass should fail

under this test, it must be carefully adjusted by the centering screws. It is, of course, best to have this done by the maker before the instrument leaves his workshop.

The central portion of the glass may now be covered with a disc of paper whose diameter is two-thirds of that of the aperture. Focus sharply on a star; remove the disc, and cover up the outer portion of the object glass with a diaphragm whose aperture is also two-thirds of that of the glass. If the focus remains unaltered, the figure is good.

The tests for separating and illuminating power may next be applied. Close double stars and minute points of light will supply the means. This can only be effectually done on the finest nights.

For lists of test objects and valuable information on these and other cognate matters, the excellent little book by the Rev. T. W. Webb, *Celestial Objects for Common Telescopes*, should be consulted.

It is scarcely necessary to discuss at length the merits of the different forms of mounting of the equatorial. The German form of mounting is now almost universally adopted, and with modern excellence of manufacture it may be considered quite equal in steadiness to the old English form.

The essential points are rigidity, strength, durability, and accuracy, facility, and permanence of adjustment. The tube is often made of sheet brass; sheet iron is lighter, cheaper, and more durable. The declination axis and polar axis should have plenty of bearing surface, and be of ample strength. The weight upon the polar axis should be relieved by friction rollers. The declination and hour circles should read by opposite verniers to 10″ or 20″ of arc, and 1 or 2 seconds of time respectively. The declination circle may be placed next to the telescope tube, so as to be read off conveniently by a reader from the eye-end, suitable illumination being provided. The hour circle should be moveable, and the telescope should have a clamp and slow

motion in declination.   The clockwork should be strong
and powerful, a weak clock being one of the commonest
defects of equatorials.   Slow motion is also required in
right ascension, and it is usually obtained by means of
differential wheels in connection with the driving clock, an
endless cord being brought to the eye-end.   The tangent
screw of the driving arc should be capable of perfect
adjustment, and should not have to be removed from the arc
for the purpose of releasing the telescope from the clockwork.
This should be done by a clamp on the polar axis.   The lamp
for illuminating the micrometer is best placed at the end of
the declination axis, away from the telescope, the axis being
perforated for the light to pass through into the tube, whence
it is reflected at right angles to the eye-end, either by a reflector just outside the cone of rays, or by a tiny reflector
say one-eighth of an inch in diameter, in the centre of the cone,
and carried by an arm in such a manner that it can be moved
to one side at pleasure.   The first plan is perhaps the least
objectionable ; the latter is, however, adopted by Mr. Grubb.
In Messrs. Cooke's form of mounting, the whole instrument is
carried upon a heavy central iron pillar, which takes up less
space in the observatory than any other form, and does not
interfere with the observing chair in any position of the
instrument.   The base of the pillar being turned true in the
lathe, is also easily bedded in the foundation-stone.

The following principal adjustments should be provided for,
viz., (1) the polar axis in altitude ; (2) the whole instrument in azimuth ; (3) the eye-end for collimation ; (4) the
verniers of both circles for index errors.

The declination axis is commonly set by the maker at
right angles to the polar axis.   When the bearing surfaces of
this axis are not equidistant from the polar axis, or bear
unequal weights, there may be a tendency to unequal wear,
and therefore to change of inclination, unless the bearing
surfaces are proportional to the weights they carry.

Before erecting the equatorial it will be well to see that the stand is carefully marked with a north and south point by the maker, and that a meridian line be drawn through the centre of the foundation-stone to the walls of the observatory. After preparing and levelling the stone, it is now easy to set the instrument to the meridian line approximately, or at least within the limits of the adjusting screws in azimuth.

We must now determine the following errors of the instrument, and make the necessary corrections :—

1. Error of altitude of the polar axis.
2. Index error of the declination circle.
3. Error of collimation, or deviation of perpendicularity of telescope to the declination axis.
4. Error of azimuth, or deviation of the polar axis from the plane of the meridian.
5. Index error of the hour circle.
6. Error of the declination axis from true perpendicularity to the polar axis.

No. 1 and No. 2 are determined by the same set of observations. Bring the telescope approximately into the plane of the meridian, say on the west side of the polar axis : put in the wire micrometer with a low power, and bring one of the moveable webs into the centre of the field of view approximately. Make a star run along the web by means of the slow motion in declination : move the micrometer through 180°, and if the star will not now run along the web from side to side of the field, bring the web half-way towards the star by turning the micrometer screw, and then set the star on the web by the slow motion in declination. Again, turn the micrometer through 180°, and if the star now travels along the web, the latter passes through the centre of the field of view, or the centre of rotation of the position circle of the micrometer.

Now set the centered web on a bright star south of the zenith near the meridian whose position is given in the

Nautical Almanac. Clamp in declination and read off the declination circle. Unclamp, swing the telescope over to the east side (being careful not to disturb the micrometer), set on the star again, clamp, and read off as before. If the star has north declination, the correction for refraction is subtracted from the readings; if the star be south of the equator, add the refraction correction. If the star has north declination, and half the sum of the two readings corrected for refraction be *greater* than the true declination as given in the Almanac, the north pole of the instrument is too high; if *less*, the pole is too low. If the star is south of the equator, and the result be too great, the pole is too low; if too small, the pole is too high.

Half the difference between the two readings in either case is the index error of the declination circle. The following example will illustrate this:—

Jan. 28, 1878.   Aldebaran was placed on the centered web.

Dec. Circle.

Telescope West, 16° 17′ 0″⎫
                16° 17′ 0″⎭ mean 16° 17′ 0″ N.

Telescope East, 16° 16′ 40″⎫
                16° 16′ 0″⎭ mean 16° 16′ 20″ N.

Sum	32° 33′ 20″
Half sum	16° 16′ 40″
Correction for refraction	− 44″
Observed declination	16° 15′ 56″
True declination	16° 15′ 53″
Error of altitude of polar axis, too high	3″

The *correction for refraction* is obtained thus:—

Colat. of place	36° 18′
North declination of star	16° 16′
Approximate altitude	52° 34′

Mean refraction, 44″.

The *mean* refraction is sufficiently correct for our purpose. A mean refraction table is to be found in all collections of mathematical tables, and in many astronomical handbooks.

3. If the polar axis is not far from the plane of the meridian, the error of collimation,—that is, the deviation of the telescope from perpendicularity to the declination axis,— can easily be determined as accurately as the hour circle will admit of. Thus: place the telescope on the west side, and near both the meridian and the equator. The micrometer having been undisturbed, turn it throug h90°: the centered web now points to the pole. Set the telescope a little in advance of the nearest bright star, and note by the sidereal clock the time of transit across the web. Read off the hour circle : throw the telescope over to the east side, transit the same star, and read off as before. If the difference between the transit times be greater than that of the hour circle readings, the angle formed by the telescope and the declination axis is *too great towards the eye-end*, and the eye-end must be moved towards the declination axis. If the difference of the transits is less, the angle is too small, and the eyepiece must be moved away from the declination axis. Half the difference between the interval by the clock and that by the circle is the error.

The following example will exhibit the method of proceeding in this case :—

Jan. 28, 1878. δ Orionis. Dec., 0° 23′ 28″.

	Clock.		Circle.	
	M.	S.	M.	S.
Telescope West ...	20	26	25	32
„ East ...	23	45	28	40
	3	19	3	8

Half the difference, 5·5 s. × cos. 23′ 28″ = error required.

As the clock interval is the greater, the eye-end must be moved towards the declination axis so as to diminish the angle between the telescope and the declination axis.

4. The error of azimuth is not so easily determined as the

2

previous errors, on account of the difficulty in correcting for the effect of refraction. This can be done by calculation, as is fully explained in Loomis's Astronomy, Arts. 32, 145 ; but it can also be done quite effectively, and much more readily, by the following method. Centre the web of the micrometer, set the telescope to the true declination of a Greenwich star about six hours east or west of the meridian, and from 30° to 60° in altitude. Sweep to the star in right ascension with the finder, and if the star is some distance from the centre of the field, move the telescope in azimuth until it passes a little below the centre of the field. Now take a small clinometer, (which can be readily constructed with a piece of hard wood, a semicircular protractor, and a small plumb-line,) and place it on the telescope ; read off the altitude to the nearest degree. Rotate the micrometer until the fixed wires are approximately in the vertical plane. Find the mean refraction for the observed altitude from the Table of Refractions. Now bring the web that is not centered below the centered one by a distance equal to the angle of refraction. Set in azimuth so that the star will pass through the intersection of the lower web and the fixed wires of the micrometer. Repeat the operation on a star in the opposite quarter of the heavens ; and if this star also comes to the corresponding intersection the polar axis is in the plane of the meridian.

If the micrometer screw have 100 threads to the inch, and the focal length of the object-glass be measured from its centre, the angular value of one revolution of the screw will be known well enough for the above purpose. (See the chapter on the Micrometer.)

5. The index error of the hour circle can only be determined by an independent observation for time, unless the declination axis is provided with a striding level for the purpose of rendering it horizontal, or truly east and west. In this latter case, all that is necessary after levelling is to set any division of the hour circle at the index point of the vernier which moves

with the telescope, then adjust the index point of the fixed vernier to the same division, and this will be the south reading. It is, however, still more convenient, when it can be done, to set the fixed vernier east or west according as the Observatory is west or east, by the difference in time between the longitude of the Observatory and Greenwich : this will save the trouble of always having to add or subtract this quantity from the right ascension of a star when setting the telescope by the circles. If the declination axis is not provided with a level, which is seldom the case, as it is not indeed necessary, then sidereal time must be obtained from occultations of stars by the moon, from Greenwich time when telegraphed to the nearest post-office or railway station, by Dent's Dipleidoscope ; or, best of all, from a small transit instrument of about two inches' aperture; for such an instrument will give the time to the tenth of a second, and help to make the Observatory complete and independent.

The telescope can now be brought into the meridian by a star at the time of transit, and the fixed vernier set as before.

6. The error of the declination axis from true perpendicularity to the polar axis should be so small as to fall within the error of the setting of the instrument. It is not usual to provide an adjustment for this error, as such would tend to weaken the construction of the instrument. It should, however, be determined by the following method :—

Set the telescope on a star of not less than 40° north declination, and near the meridian ; transit, read off the hour circle, and reverse the position of the telescope, as in the third adjustment. If there be no difference between the intervals, there is no error in the inclination of the declination axis to the polar axis : i.e., it is at right angles to it. If, however, the interval by the clock be greater than that on the hour circle, the declination axis towards the telescope is at too great an angle with the polar axis,—and vice versâ. Half the difference of the intervals (expressed in arc) divided by the tangent of the star's declination gives the error of inclination required.

The whole of these six adjustments should be repeated several times, and also from time to time, as they are liable to change.

As the errors mutually affect each other, the second set of observations will be more accurate than the first, and should be made with greater care.

Having completed the adjustments of our equatorial, we are now ready to set the telescope upon any object in the heavens which we may wish to observe, whose right ascension and declination are given in our catalogues. First, set the telescope in declination, and then set the moveable hour circle to the right ascension of the object by the fixed vernier (with no correction for longitude if the fixed vernier is put to the Greenwich meridian, as above recommended). Now sweep the telescope in right ascension until the upper vernier comes to sidereal time by the clock, and the object will be in the field of view.

It will now be desirable to determine, approximately, the focal length of the object-glass, the angular value of the field of view with each eyepiece, and the magnifying powers of the eyepieces. The makers usually furnish the first and last of these, but it is well for the observer to ascertain these values for himself with some care.

Firstly : to find the focal length of the object-glass. This is not a very easy matter, owing to the difficulty of finding the optical centre of the glass. According to Troughton, "the measure should commence from the interior part of the convex lens, at a distance from its exterior surface equal to one-fifth of the thickness of the double compound object-glass." (See Pearson, p. 19.) This point can of course be readily found by first ascertaining the thickness of the lens. A long, stout straight-edge, placed on the tube of the telescope and made level, will enable the observer to find the distance between the object end of the tube and the webs of the micrometer adjusted to stellar focus. A plumb-line gives the two points very quickly and accurately. If the telescope be not a large one, the following method will give good results : focus on a terrestrial object at a well-measured distance, and mark the draw-tube ; then focus on the sun, and again mark the tube ; then the formula

$$F = \frac{D. (F' - F)}{F}$$

where $F =$ the length of the solar focus required, F the length of the conjugate focus obtained from the terrestrial object, and D the distance of the object. Of course, the distance between the two marks on the draw-tube should be measured very carefully by means of a finely divided rule and a pair of compasses. The distance between the telescope and the terrestrial mark must be measured from the object-glass.

Again ; the focal length may be accurately determined as follows : find the value in arc of say 50 revolutions of the micrometer screw. This will of course be readily done by separating the webs 50 revolutions, transiting a star near the equator (or, better, a star not far from the pole), and reducing the observed interval by multiplying it by the cosine of the star's declination, and by 15. Next, measure with great accuracy the linear value of the space between the webs,* then the proportion

2 tan ½ the arc : radius :: linear value : focal length

will give the required quantity.

Secondly: to find the angular value of the field of view of the several eyepieces when in the telescope. This is easily done. Allow a star very near the equator to transit the field centrally, and convert the observed sidereal time into arc. If a chronometer or mean-time clock be used, the mean-

* The practical optician can do this with very great accuracy.

time interval must, of course, be converted into its equivalent
sidereal interval, and then the arcual value found from the
table. (See Loomis's Astronomy, p. 363.) Do this with each
eyepiece. The angular value of negative eyepieces may also
be found thus : as the field of view of a telescope depends
partly on the focal length of the object-glass, and partly on
the diameter of the diaphragm placed at its focus, the following formula will give it : F is the focal length of the
object-glass, and d the diameter of the diaphragm of the
eyepiece, both in inches :—

$$\frac{d}{F \sin. \, 1''}$$

This is Delambre's formula.*

Thirdly : the magnifying powers of the eyepieces have to
be found. One of the following methods may be chosen.

1. Measure the small illuminated circle seen in front of the
eyepiece (which is the image of the object-glass), by means
of the Dynameter. Then, the aperture of the object-glass is
to the diameter of its image at the focus seen through the
eyepiece in the ratio of the focal length of the object-glass to
that of the eyepiece. That is, the diameter of the object-
glass divided by that of the small image gives the magnifying power. The small image may, of course, be measured
without the aid of the Dynameter, by means of a finely
divided scale. Or the "Berthon Power-gauge"† may be
used.

2. In the case of small telescopes the powers may be conveniently found by means of a piece of white paper, say one
inch long, on a black ground, fixed at a known distance from

---

* To take Pearson's example : let the focal length of the object-glass
be 3·5 ft., and the diameter of the diaphragm of a negative eyepiece 0·3
in. : then 42 × ·000004848 = 000203616, and $\frac{0·3}{·000203616}$ = 1473″ = 24′ 33″.

† The Rev. T. W. Webb (*Celestial Objects*, p. 7) speaks highly of
this little instrument, which he says may be purchased for 7*s.* 6*d.* of
Mr. Tuck, watch-maker, Romsey.

the object-glass, a staff divided to inches being also placed near the paper. On looking through the telescope at the paper with one eye, and at the staff with the other at the same time, the number of inches on the latter covered by the paper will be seen, and the power at once found for that distance. From this terrestrial power, P′, the stellar power P is obtained from the following formula, F being the stellar focal length and F′ the terrestrial :—

$$P = \frac{F \times P'}{F'}.$$

3. The following method is convenient. Place a staff divided into feet and inches against a wall in a vertical position ; at a distance of three or four feet from the staff, hold the eyepiece to the eye, and, looking through it with one eye, and at the staff with the other eye, note how many feet and inches are contained in the diameter of the field of the eyepiece. For example, let the distance from the staff be 48 inches, and the observed diameter of the field 40 inches; then the tangent of half the angle $= \frac{20}{48} = 0.416$, and the angle is 45° 14′, or 162840 seconds of arc. Now if the angular aperture of the telescope with this eyepiece be 33 sidereal seconds (found by transiting, centrally, a star very near the equator), or 495 seconds of arc, we have

$$\text{Magnifying power} = \frac{\text{angular subtense}}{\text{angular diameter}} = \frac{162840}{495} = 329.$$

4. Valz's method is useful for small telescopes. Turn the telescope towards any celestial object of known angular magnitude, say the sun, whose angular diameter is given in the Nautical Almanac, page II, of each month. Let the image be received on a screen kept at right angles to the tube, and having a line nicely divided into inches and tenths marked on it. Observe the horizontal diameter in inches and tenths of the image on the screen. Then if a be the sun's true diameter, A the angular diameter of the image on the screen′

and D the distance between the middle of the eye-piece and the screen, then we have

$$\tan \tfrac{1}{2} A = \frac{\tfrac{1}{2} d}{D},$$

and the magnifying power $= \dfrac{\tan \tfrac{1}{2} A}{\tan \tfrac{1}{2} a} = \dfrac{d}{2 \, D \tan \tfrac{1}{2} a}$.

The measure of the image should be made when the sun is in the centre of the field of view.

The thickness of the webs of the micrometer may be found by bringing one up to a fixed web until the bright space between the two is estimated to be equal to the thickness of the web which is moved : read off the divided head, and then carry the web into contact with the fixed web. Read off again. Repeat five or ten times. Take the mean value, and convert it into arc.

The following information, drawn up in a tabular form, may, for convenient reference, be pasted inside the box containing the eyepieces : focal lengths of telescope and finder ; angular value, in arc, of the field of view of each eyepiece of telescope and finder ; magnifying powers of the eyepieces ; value in arc of one revolution of the micrometer screen, and a table for taking out at sight the arcual value of revolutions and parts ; the thickness, in arc, of the webs of the micrometer.

For fuller information on these and other matters, the following works may be consulted : Loomis's Practical Astronomy, published by Harper and Brothers, New York. (This work is essential.) Webb's Celestial Objects for Common Telescopes. Pearson's Practical Astronomy. Chauvenet's Practical and Spherical Astronomy (London, Trübner and Co.); and Brünnow's Spherical Astronomy (Asher and Co., London). The Nautical Almanac for the current year, a collection of mathematical tables (such as Hutton's or Chambers's), and a good Star Atlas, are of course necessary.

THE CLOCK.

A common well-made clock, if the pendulum be properly constructed and suspended, is all that is *necessary* for double-star observers. The piece supporting the pendulum should, of course, be very firm, and securely fastened to a good wall. The pendulum rod, 46 in. long, may be made of well-seasoned white deal soaked in melted paraffin, and $\frac{3}{8}$ in. in diameter; the bob should be of lead, and cylindrical, its length (for a seconds pendulum) being, say, 14·3 in., diameter $1\frac{3}{4}$ in. with a hole a little more than $\frac{3}{8}$ in. in diameter for the rod to pass through. The bob should be supported on the rod by means of a stout nut and screw, the latter having not more than thirty threads to the inch. A leaden bob of these dimensions would weigh about $13\frac{1}{2}$ lb, which is found in practice to be a suitable weight. Such a clock, beating seconds audibly, would keep its rate unchanged for a few hours, and would meet all the requirements of double-star work. The rate would be obtained with the aid of a small transit instrument, or the equatorial itself, if well adjusted; or the finder of the latter instrument might be used for this purpose. The rate should be small, and a *losing* rate, in order that the correction which becomes necessary from time to time may be made by putting the minute hand of the clock *forward.* If the clock be losing, say, ten or twenty seconds per day, the bob may be readily put near its true place by means of the nut under it, with the aid of the following formula :—

$$\text{Change in one day} = 43200 \frac{L}{l} \text{ seconds,}$$

where L is the breadth of one thread of the adjusting screw, and l is the length of the seconds pendulum; from this the effect of one turn of the nut on the clock's rate is obtained. Or, to put it in a still simpler way: if n be the number of turns of the screw in 1 inch, then $L = \frac{1}{n}$, $l = 39\cdot138$; and the change in seconds for one turn of the screw $= \frac{43200}{n \times 39\cdot138} = \frac{1103}{n}$.

Assuming that the losing rate has been reduced to, say, two seconds per day, and that it is desired to make it about half a second, either of the following methods may be adopted:—

(a) Place a small sliding metal collar on the rod, its weight being about $\frac{1}{1000}$th of that of the pendulum (bob and rod). At first this collar should be placed about 9 inches from the spring, and then gradually pushed downwards until the rate is what is desired.

(b) Let the sliding collar take the form of a cup into which small shot may be put, and let it be *fixed* to the rod at 19½ inches from the spring.

By trial the effect of one shot or of any number may be found, and the necessary change in the rate effected very readily.

The following extract from Baily's paper, in the *Memoirs of the Royal Astronomical Society*, vol. i., will be interesting in this relation.

Distance from axis in inches.	Variation in the rate per day. Sec.	Difference. Sec.
1	+ 1·08	+ 1·02
2	2·10	0·97
3	3·07	0·91
4	3·98	0·85
5	4·83	0·79
6	5·62	0·74
7	6·36	0·68
8	7·04	0·63
9	7·67	0·56
10	8·23	0·51
11	8·74	0·46
12	9·20	0·40
13	9·60	0·34
14	9·94	0·28
15	10·22	0·23
16	10·45	0·17
17	10·62	0·11
18	10·73	+0·06
19	10·79	0·00
20	10·79	−0·06
21	10·73	0·11
22	10·62	0·17
23	10·45	0·23
24	+ 10·22	

If the pendulum is found to go *slower* in *warm* weather and *faster* in *cold*, it is *under*-compensated, and more mercury should be put into the cylinder; if *faster* in *warm* and *slower* in *cold* weather, mercury must be taken away, the quantity in each case being found by trial.

Valuable information may be found in Baily's paper above referred to, in those by Bloxam ("Monthly Notices," vols. xiii. and xviii.), and in Denison's excellent "Clocks and Locks" (Adam and Charles Black, Edinburgh).

OBSERVING CHAIRS.

As the work of the double-star observer is laborious, and often protracted, it is essential that he should be in a comfortable position for his work.

Ordinary chairs and steps are quite insufficient for this purpose, though they often constitute the sole furniture of an observatory.

A special chair is required which will support the observer from head to foot, in any position of the telescope; such is Dawes's chair (see Figs. 1 and 2). We have used it for several years, and should not like to be without it. It consists of a horizontal wooden frame on castors, 6 feet by 2 feet 4 inches, well braced to an upper frame, and inclined at an angle of 35° from top to bottom; upon this upper frame is a sliding piece, carrying the seat which is nearly horizontal. The sliding piece is held at any point by a stout catch in a perforated iron plate on one side. The seat is 2 feet by 1 foot, and is padded; the back is also padded, and it is so hinged to the seat that it can be raised to any position by means of a handle on the left-hand side, and then clamped to an arc on the right-hand side of the observer: this padded back is 2 feet by 2 feet 9 inches. It may thus

FIG. 1.

FIG. 2.

be raised and clamped at any angle without leaving the chair. Dawes used a rack for supporting the back, but the clamp is more convenient. An arm is also attached to the chair on the right-hand side; this can be set at any angle by means of a notched arc, catch rod, and handle; and it makes an excellent rest for the right arm. An iron hook on the left-hand side of the chair carries a reading lamp.

FIG. 3. (A Chair for occasional use.)

THE OBSERVATORY.

The best form of Observatory is a square room with cylindrical dome. The corners of the room are always useful, if not necessary, for tables, shelves, chairs, etc.; and the cylindrical dome is manifestly more easily constructed than the spherical form. The shutters work horizontally, and are less liable to stick than curved shutters. Sufficient slope should be given to the roof to throw off a heavy fall of rain, and the top at least may be covered with thin sheet copper well painted. The conical form of roof is very effective, and also very cheap.

The Transit Instrument will require a small room, say 12 feet square, or rather less.

A Computing Room, on the north side of the Observatory, may be added, and this may be provided with a stove and chimney for heating the hot-water apparatus by means of which the observing rooms are kept dry in wet and cold

MR. EDWARD CROSSLEY'S OBSERVATORY, BERMERSIDE.

weather. The hot water must of course be turned off some time before the work of observation begins.

Four windows, north, east, south, and west, are of great use in ventilating the Observatory, and in rapidly reducing the temperature inside as nearly as possible to that outside, so as to avoid currents of heated air, which are so detrimental to optical definition.

CHAPTER III.

SOME ACCOUNT OF THE EQUATORIALS WHICH HAVE BEEN USED BY DOUBLE-STAR OBSERVERS.

AUWERS. (See KONIGSBERG.)

BARCLAY. (See LEYTON.)

BEDFORD.

The mounting of the 8½ ft. equatorial was by Dollond, the Sisson form being used. The object-glass had a diameter of 5·9 in., and was purchased in Paris by Sir James South. Tulley worked it. "It is considered by Captain Smyth to be the finest specimen of that eminent optician's skill, and will bear, with distinctness, a magnifying power of 1200."

The declination and hour circles had a diameter of 3 ft.: the former read to 10″. The negative powers were 22 to 1200, six of the highest being single convex lenses fitted in a polycratic wheel. The powers of the parallel-wire micrometer ranged from 62 to 850. The finder had an aperture of 1·6 in.

The driving clock was invented by Mr. Sheepshanks, and had a steam-engine governor and absorbing wheel. It worked very well.—*Monthly Notices, R. A. S.*, vol. i., and the *Celestial Cycle*.

Observer: Admiral Smyth.

BERLIN.

The refractor at this Observatory is similar to the famous Dorpat telescope in all essential respects.

Observers: Encke, Galle, Winnecke.

BERMERSIDE (Halifax).

Mr. Edward Crossley mounted his 9⅓ in. Cooke equatorial refractor in 1867. Its focal length is 148·5 in. The style of mounting is German. The diameter of the declination and hour circles are respectively 23½ in. and 12½ in., and they read to 10″ and 2 sec.

The lamp, which gives a bright field to the micrometer, swings at the end of the perforated declination axis.

The aperture, and amount and colour of the light for the bright field, are regulated from the eye-end by means of rods, and a rod and cords at the same end give the observer full control over the motion of the instrument in right ascension and declination.

The finder has an aperture of 2½ in., and a focal length of 2 ft. 4 in.

The negative eyepieces are ten in number: powers, 60 to 1000.

There are three micrometers, two filar and a double-image. The double-image and one of the filar micrometers are by Simms, and the other filar by Cooke. The eyepieces for these instruments are, in all, seventeen in number, and the powers range from 100 to 1200. The new filar micrometer by Simms is divided on the face: diameter of circle 4½ in.

The driving clock is by Grubb of Dublin.

Observers: Crossley and Gledhill.

BESSEL.   (See KONIGSBERG.)

BOND.   (See CAMBRIDGE, U.S.)

BONN.

The heliometer of this observatory has an aperture of 6 in.

The driving clock works "remarkably well," and its

construction is similar to that of the Poulkova refractor
—*Memoirs of R. A. S.*, vol. xx.

BRÜNNOW. (See DUNSINK.)

BURNHAM. (See CHICAGO.)

CAMBRIDGE (Northumberland equatorial).

English mounting: the tube is square, and of deal.
Object-glass by Cauchoix, $11\frac{1}{2}$ in. aperture, and $19\frac{1}{3}$ ft.
focal length; it was received in 1834. Hour circle $5\frac{1}{2}$ ft.
in diameter, and reads to 1 sec. The circles were graduated by Simms.—Main's *An Account of the Observatories
in and about London.*

Declination axis, 5 ft. $8\frac{1}{2}$ in. long. Finder, $2\frac{3}{4}$ in.
aperture, and $28\frac{1}{2}$ in. focal length. The declination is
obtained by means of divided rods. For a full account,
with elaborate drawings, see Airy's account of the instrument.—*Account of the Northumberland Equatorial and
Dome.*

Observer: Challis.

CAMBRIDGE (U. S.)

This instrument is of the same style of mounting, size,
and by the same maker, as the Poulkova refractor.
Focal length 22 ft. 8 in., aperture 15 in. "No colour
except a purple tinge round very bright objects, such
as the Moon and Venus."—*Monthly Notices of R. A. S.*,
vol. viii.

Observers: Bond and Waldo.

CAPE OF GOOD HOPE.

Prior to 1847 the equatorial was a 46 in. by Dollond,
aperture $3\frac{1}{2}$ in. There were four micrometers, viz., a
spider-line position, an annular, and two rock-crystal. A
flat-wire position micrometer was added subsequently.
In 1849 the equatorial by Merz was mounted; aperture
nearly 7 in., focal length $8\frac{1}{2}$ ft. The tube is of wood,
veneered with mahogany.

3

The declination circle is 12½ in. in diameter, and reads to 10″, and the hour circle has a diameter of 9·6 in., and reads to 4 sec. The Huyghenian eyepieces have powers 86, 128, 200, 302. and 458. Those of the micrometer, 123, 161, 273, 347, and 464. The power of the double annular micrometer is 64. The divided circle of the position micrometer is 4 in. in diameter, is divided to 15′, and reads to 1′: the total range of the screw is 60 revolutions. One head only is divided.

Observer: Maclear.

CHALLIS. (See CAMBRIDGE.)

CHICAGO.

Mr. Burnham has made most of his discoveries with his 6 in. refractor by Alvan Clark. He has also used the fine 18½ in. Clark refractor of the Dearborn Observatory, the 26 in. of the Washington Observatory, and the 9·4 inch of the Dartmouth College Observatory.

CINCINNATI. (U. S.)

The object-glass was purchased in 1842; it was begun by Fraunhofer, and finished by Merz and Mahler. Dr. Lamont pronounced it "one of the best ever manufactured." Aperture 11 in., focal length 17 ft. Diameter of hour circle 16 in., of the declination circle 26 in. The powers range from 100 to 1400. The stand is of iron, and is filled with sand. The driving clock is by Clark and Sons, and is good.—Loomis's *Recent Progress of Astronomy*, and the *Cincinnati Observations.*

Observers: Mitchell, Stone, Howe, and Upton.

CROSSLEY. (See BERMERSIDE.)

CUCKFIELD.

Mr. Knott's equatorial was mounted at Woodcroft, Cuckfield, and the measures lately published were made there between 1860 and 1873. The object-glass has a clear aperture of $7\frac{1}{8}$ in., a focal length of $110\frac{1}{2}$ in., and it was made by Messrs. Alvan Clark and Sons. The filar micrometer was made by Dollond; diameter of position circle $3\frac{1}{2}$ in.; it reads to tenths of a degree. The powers of the seven eyepieces range from 115 to 515.

DAWES (Rev. W. R.)

In 1831 this distinguished observer erected a 5 ft. achromatic at Ormskirk in Lancashire. It was by Dollond, and the mounting was like that of Smyth's refractor. The aperture was $3\frac{3}{4}$ in.; the circles 2 ft. in diameter; the powers used, 225, 285, and 625.—*Memoirs of the R. A. S.*, vols. iv. and v.

The Newtonian reflector, the mirrors of which were presented to Dawes by Sir John Herschel, was mounted by Dollond, and applied to the polar axis of the 5 ft. telescope. Focal length about 7 ft., aperture $6\frac{1}{4}$ in. This instrument was used between 1834 and 1839, but not much.— *Memoirs of the R. A. S.*, vol. xix.

In 1845 the Merz and Mahler equatorial was mounted at Cranbrook in Kent. The style of mounting was that of the great Dorpat refractor. The focal length was $8\frac{1}{2}$ ft., and the clear aperture $6\frac{1}{2}$ in. The object-glass was of first-rate quality. The hour circle read to 4 sec., and the declination circle to $10''$. Driving clock extremely steady and uniform.—*Memoirs of the R. A. S.*, vol. xvi.

In 1859 the equatorial by Alvan Clark and Sons (now at the Temple Observatory, Rugby), was mounted at Haddenham (Hopefield Observatory), in Bucks. The glass was cast by Chance and Co. Aperture $8\frac{1}{4}$ in., focal length 110 in. The figure is excellent to the circumference, and the dispersion "but a little over-corrected."

The finder has an aperture of 2 in. The micrometer was a parallel-wire by Dollond. Driving clock: this is

very good. Bond's spring governor renders the action very smooth.—*Memoirs of the R. A. S.*, vol. xx.

Dawes's micrometer by Merz and Son was made in 1846. It was a parallel-wire, and was used with the $8\frac{1}{2}$ ft. telescope. Powers 120, 155, 260, 322, 435, 572, and 690. His Amici micrometer was presented to him by Sir John Herschel : it was a double-image, and had but one power (1000 on the 20 ft. reflector). Dawes added three new eyepieces, which, on the $8\frac{1}{2}$ ft. refractor, were 212, 360, and 508.

DEMBOWSKI (Baron).

This eminent double-star observer used an excellent dialyte by Plössl 5 ft. focal length and 5 in. aperture equatorially mounted, from 1852 to 1862. The power generally used was about 300. It was not provided with a driving clock.

In 1862 the refractor by Merz was erected. Its aperture is $7\frac{1}{2}$ in. The object-glass is a fine one, and the powers range from 100 to 720. The driving clock is moderately good.—*Ast. Nachr.*, vols. xlii. and lxii.

DOBERCK. (See MARKREE.)

DORPAT.

This noble instrument was erected in 1825. It was the work of Fraunhofer. The tube was of deal overlaid with mahogany, and the framework of the stand was of oak inlaid with mahogany and polished. The polar axis was 39 in. long. Aperture of the object-glass 9·6 in. ; focal length 14 ft. The hour circle, with a diameter of 13 in., was divided to minutes, and read to 2 sec.; and the declination circle, with a diam. of 19 in., was divided to 10 min and read to 10 sec. Powers 86, 133, 198, 254, 420, 532, 682, 848, 1150, and 1500. The finder had an aperture of 2·4 in., and focal length of 30 in. The driving clock kept a star in the centre of the field when a power of

700 was used.—*Memoirs of R. A. S.*, vols. ii. and xxxvi.
Pearson's Astronomy.

Observers: *Σ.*, *O.Σ.*, and Mä.

DUNER.   (See LUND.)

DUNLOP.

Equatorial refractor, focal length 46 in.   Micrometers,
a parallel-wire and an Amici's double-image.

DUNSINK.

The object-glass is the work of Cauchoix : aperture
12 in. ; focal length 19 ft.   The mounting was by Thomas
Grubb.

ELCHIES.

The Elchies equatorial was mounted about 1850, by
Mr. J. W. Grant, at Elchies, in Scotland.   The German
form was adopted.   One portion of the stand weighed 11
tons.   Messrs. Ransome and May made the stand, and
the object-glass was by Ross.   The aperture was 11 in.,
and the focal length 16 ft.   The axes were 5 ft. long, and
6 in. in diameter.   The circles had a diameter of 30 in.,
and were 1 in. thick.   The eyepieces were twenty-three
in number.   The parallel-wire micrometer had two eyepieces, and one of the three finders had a focal length
of 5 ft.

ENCKÉ.   (See BERLIN.)

ENGELMANN.   (See LEIPSIC.)

FERRARI.   (See ROME.)

FLAMMARION.   (See PARIS.)

FLETCHER.   (See TARN BANK.)

GALLE.   (See BERLIN.)

GLEDHILL.   (See BERMERSIDE.)

GREENWICH.

In 1838 the Sheepshanks equatorial was mounted.
Grubb of Dublin supplied the stand, which was of the

German form. The object-glass was by Cauchoix : aperture 6·7 in. ; focal length, 8 ft. 2 in. Its definition was found to be good, the principal defects being outstanding colour, and a diffusion of light from brilliant objects. Negative eyepieces, a wire micrometer, a comet eyepiece, and a double-image micrometer were provided. The driving clock was regulated by governor balls at the ends of a horizontal arm on a vertical spindle. When a certain velocity had been acquired, projections on the balls rubbed against a fixed horizontal ring.

The mounting of the great equatorial is in the English style, and was executed by Simms. Messrs. Ransome and Sims made the engineers' work. The object-glass, by Merz and Son, has an aperture of 12½ in.,* and a focal length of 16 ft. 6 in., and it is a very fine one. The hour circle is 6 ft. in diameter, and the declination circle 5 ft. The driving clock is in the ground floor story, and the power is given by a flow of water acting through a turbine, the spindle of which passes up to the instrument. A Siemens' chronometric governor regulates the supply of water to the turbine. A Barker's mill drives the hour circle, and the regulation is obtained by a conical pendulum, Siemens' chronometric governor, and a spade dipping into a trough of water.—*Greenwich Observations,* 1864.

HALL. (See WASHINGTON.)

HARTNUP. (See LIVERPOOL.)

HERSCHEL (Sir William).

The gigantic reflector was erected in 1787, at Slough. Two concentric circles of brickwork, 42 ft. and 21 ft. in diameter, battened from a breadth of 2 ft. 3 in. at the bottom, to 1 ft. 2 in. at the top, and capped with

* In the "Monthly Notices" the aperture is always given 12¾ in. See vol. xxxvi.

paving-stones 12¼ in. wide and 3 in. thick, formed the foundation. A vertical beam 12¼ in. wide was fastened in the centre, and around this the whole framework had its circular motion in azimuth.

The tube was of iron, 39 ft. 4 in. long, and 4 ft. 10 in. in diameter. The speculum was of tin and copper; its weight 1050 lb., and diameter 4 ft. The power used seldom exceeded 200.—*Pearson's Astronomy*. See also *Phil. Trans.*, vol. lxxxv., for a full description.

HERSCHEL (Sir John).

The 20 ft. reflector was constructed in 1820, by Sir William and his son. The mirrors were fine, diameter 18 in., and focal length 20 ft. With the whole aperture, powers 150 to 160 were ordinarily used, the eyepiece being a single lens of 1½ in. focus.—*Memoirs R. A. S.*, vol. ii.

The reflector used at the Cape by Sir John was the 20 ft. The three mirrors were all fine; aperture 18¼ in. The 7 ft. refractor, aperture 5 in., was also used.—*Cape Observations*.

HIND. (See REGENT'S PARK.)

HOWE. (See CINCINNATI.)

JACOB. (See MADRAS.)

JENKINS. (See OXFORD UNIVERSITY.)

KAISER. (See LEYDEN.)

KONIGSBERG.

The famous heliometer of this Observatory is mounted like the great refractor of Poulkova. The focal length is 8 ft. 6 in., and the aperture 6¼ in., and a distance of 1° 52' can be measured. It was begun in 1824, by Fraunhofer, and mounted in 1829. The position circle at the object-glass has four verniers, and reads to minutes. For ordinary use there are five eyepieces: powers, 45

91, 115, 179, 290. A circle micrometer of the Fraunhofer kind has a power of 65. The ring micrometer and net micrometer have powers of 66, 92, and 165.—*Ast. Nachr.*, vol. viii.

Observers: Bessel, Anwers, Peters, Luther, and Schlüter.

LASSELL.

In 1841 the Newtonian reflector, 9 in. aperture and 112 in. focal length, was erected at Starfield, near Liverpool. The declination circle was divided to 15', and read to 30". The hour circle was of the same size, and read to 2 sec. The diameter of the circles was about 2 ft.

In 1848, the 20 ft. equatorial was mounted. The tube was of sheet iron, $\frac{1}{10}$ in. thick, and was 20 ft. long, and 25 in. diameter; its weight was 594 lb. The diameter of the speculum was 2 ft., and its weight 370 lb. The finder was a Newtonian reflector, aperture 4·2 in., focal length 42 in., power 27.—*Memoirs of the R. A. S.*, vols. xii., xviii., and xxxvi.

The two 4 ft. specula were constructed and mounted in 1859 and 1860; their focal lengths were 441·8 and 448·1 in.; length of tube 37 ft. The mounting was equatorial, and the motion in right ascension was given by an assistant.

LEIPSIC.

The mounting was by Pistor and Martins, and the optical part by Steinheil. Aperture, 8 Paris inches; focal length 12 ft.; powers, 72, 96, 144, 192, 288, 432, 576, and 720.

Observer: Engelmann.

LEYDEN.

The Leyden refractor is of Munich make. Aperture, 6 in.; focal length, 8 ft.

Observer: Kaiser.

LEYTON.

The 10 in. equatorial refractor, focal length 12 ft., by Cooke, was erected at Leyton in 1860, by J. Gurney Barclay, Esq. The mounting is in the German style. The polar axis is 4 ft. 2 in. long, and the declination axis 3 ft. 2 in. The declination circle is 2 ft. in diameter, and reads to 10″; and the hour circle is 13 in. in diameter, and reads to 2 sec.

The finder has an aperture of 3 in., and a focal length of 3 ft.

The driving clock is regulated by a double conical pendulum.

Observers: Romberg and Talmage.

LIVERPOOL.

This fine refractor was mounted in 1848. The mounting is a modified English form; the optical parts were by Simms, and the engineer's work by Messrs. Maudslay and Field. The object-glass, which is a very fine one, was by Merz; its aperture is 8½ in., and focal length 12 ft. The hour circle has a diameter of 4 ft., reads to 0·1 sec., and has two microscopes. The declination circle has the same diameter, and reads to 1″·0. There are six negative eyepieces (powers, 90 to 1100), and the two micrometers (filar and double-image) have powers 150 to 600. The driving clock was made by Simms, and drives fairly.

Observer: Hartnup.

LUND.

The instrument used by Dr. Dunér was mounted at the observatory of Lund in 1867. The tube and object-glass are by G. and S. Merz, of Munich. The rest of the mounting and the micrometer are by M. Emile Jünger of Copenhagen. The style of mounting is modified German. The object-glass is a very fine one; its aperture is 9·6 in., and the focal length 14 ft. The diameter of the declination circle is 21·2 in., and reads to 2″ ; and the hour circle, with a diameter of 19·6 in., reads to 0·2 sec., and, by microscopes, to 0·02 sec. The micrometer is a filar. The driving clock is a good one, the regulator being the invention of Professor Holten of Copenhagen.

MACLEAR. (See CAPE OF GOOD HOPE.)

MÄDLER. (See DORPAT.)

MADRAS.

The 4 in. equatorial was made by Simms, in the German style; focal length 63·2 in. The circles were for finding only, and read to minutes of space and seconds of time.

The micrometer was a parallel wire ; powers used 170 and 280. The spurious discs of stars were "sharp and round, but rather large."—*Memoirs of the R. A. S.*, vols. xxv. and xxxii.

The Lerebours and Sécretan equatorial had an aperture of 6·3 in., and a focal length of 89 in. A second object-glass was furnished by them in 1852, which proved good, but not perfect.—*Memoirs of the R. A. S.*, vol. xvii.

Observers : Jacob and Powell.

MAIN. (See OXFORD.)

MARKREE OBSERVATORY.

This equatorial was mounted in 1834, at Collooney, County Sligo, by the late Mr. E. J. Cooper. The German style was adopted, and the cast-iron stand was placed on limestone blocks.

The object-glass was the work of Cauchoix. It is not a very good one. Aperture 13½ in. ; focal length 25½ ft. The diameter of the declination circle is 1 ft. 9 in. ; it is divided to ¼°. The diameter of the hour circle is 30 in. ; it is divided to minutes.

The micrometer is of Munich make, and very good ;

powers, 100, 200, 300, 400, 500, 600, and 800. The position circle is 4½ in. in diameter, and reads to 1′. The driving clock is a rough machine.—See *Astr. Nachr.*, No. 2187.

Observer: Doberck.

MILAN.

The mounting is in the German style: both mounting and object-glass are the work of Merz and Mahler. The object-glass is a good one; its aperture is 9·5 in., and focal length 10 ft. 7·9 in. The diameter of the hour circle is 11 in., that of the declination circle 15·7 in. The negative eyepieces furnish the following powers: 67, 95, 155, 223, 322, 468. The filar micrometer was made by Merz: the powers are 87, 144, 210, 322, 417, 500, and 690; those generally used for double-star measurements are 322 to 690.

The driving clock, by Merz, is not a good one; it has a conical pendulum.—*Ast. Nachr.*, vol. lxxxix.

Observer: Schiaparelli.

MILLER. (See WHITEHAVEN.)

MITCHELL. (See NANTUCKET.)

MITCHELL. (See CINCINNATI.)

MORTON. (See WROTTESLEY.)

NANTUCKET (U. S.)

Miss Mitchell's telescope was a 5 in. refractor by Alvan Clark.

NAPLES.

Aperture 5¼ in.: focal length 7¾ ft.: powers used 268 and 362.

Observer: Nobile.

NEWCOMB. (See WASHINGTON.)

NOBILE. (See NAPLES.)

OXFORD (Radcliffe Observatory).

The mounting of the Oxford heliometer was designed and executed by Messrs. Repsold, and differs from the ordinary German equatorial. Aperture 7·5 in.; focal length 10·5 ft. The polar axis is 42½ in. long; diameter at upper pivot 4¾ in., and 3·85 in. at the lower. It is of steel, and the pivots turn in collars of bell-metal. It is perforated 2·1 in. throughout.

The declination axis is 43·4 in. long, 5 in. diameter in centre, 4·3 in. at the pivots. It is of steel, and perforated throughout, the bore being 1·9 in. The tube is of hammered brass; diameter at object-end 13 in., at the eye-end 9·2 in. The position circle is 22·7 in. in diameter. The hour circle is at the north end of the polar axis, has a diameter of 33·8 in., is graduated to 1 min., and reads to 0·2 sec. The declination circle has a diameter of 34·3 in., is graduated to 4′, and reads to 1″. The driving clock is governed by centrifugal balls, and the instrument is moved by a weight of about 30 lb.—*Radcliffe Obs.*, vol. xi.

Observer: Main.

OXFORD (University).

The equatorial refractor is by Grubb; aperture 12¼ in.; focal length 176 in. The declination circle has a diameter of 30 in. There are two filar micrometers, and a double-image. The driving clock is not faultless.—*Monthly Notices*, vol. xxxvi.

Observers: Plummer and Jenkins.

PARIS.

The instrument used by Flammarion is one of the equatorials of the Paris Observatory. The object-glass is by Lerebours, and has a diameter of about 15 in., and a focal length of 29 ft. It is not a very good one, and a diaphragm is therefore generally used. The hour circle has a diameter of 25 in., and reads to 1′. The declination circle is divided to 5′, and has a diameter of about 5 ft. The parallel-line micrometer is by Brunner, and the powers generally used are 300 and 400. The driving clock is also by Brunner, and has a Foucault regulator.

PLUMMER. (See OXFORD UNIVERSITY.)

POULKOVA.

A very fine instrument was mounted at this Observatory by Merz and Mahler. The weight of the instrument is 7000 lb. ; the clear aperture 15 in., and the focal length 22·5 ft. The driving clock is regulated by the friction of centrifugal balls against the interior of a conical box. There are 6 negative eyepieces, powers 152 to 1218 ; 21 positive eyepieces, powers up to 2000.

Observer : $O.\Sigma.$

POWELL. (See MADRAS.)

REGENT'S PARK.

In 1836 G. Bishop, Esq., erected an observatory in Regent's Park, London. The equatorial was by Dollond, and the mounting English in form. The tube was of brass, and painted. The aperture of the object-glass was 7 in., and its focal length $10\frac{3}{4}$ ft. The hour and declination circles were of brass, and 3 ft. in diameter, the former being divided to minutes and read off to seconds, and the latter divided to 10′ and read off to 10″.

The eyepieces gave the following powers : 45, 70, 108, 200, 320, 460, 700, and 800, and a polycratic wheel carried six of them.

The prismatic crystal micrometer was by Dollond, powers 185, 350, and 520 ; the parallel-wire was also by Dollond, powers 63, 105, 185, 320, 420, 600 ; also 190 and 300.

The driving clock was by Dollond : it was driven by a powerful spring, and regulated by two fans, and was found to work "extremely well."—*Bishop's Astr. Obs.*, 1852.

Observers : Dawes and Hind.

ROMBERG.   (See LEYTON.)

ROME.

This fine instrument is mounted like the great Dorpat refractor.

Aperture 9·6 in. ; focal length 14·2 ft.

Driving clock, very good.   " The rate of the regulating part of this instrument is controlled by the friction of two small brass balls against the sides of a conical box."— *Monthly Notices of the R. A. S.*, vol. xvi.

Observers : Secchi and Ferrari.

RUGBY.   (See DAWES.)

Observers : Wilson, Seabroke, and A. Percy Smith.

SCHIAPARELLI.   (See MILAN.)

SEABROKE.   (See RUGBY.)

SECCHI.   (See ROME.)

SMITH.   (See RUGBY.)

SMYTH.   (See BEDFORD.)

SOUTH.

The 5 ft. equatorial was erected in 1797 in London. "The whole scheme of its fabric was cast by the late Captain Huddart, many years a worthy Fellow of this Society.   All the tinned iron-work was made under the direction and inspection of the same able engineer." The brass-work was made by J. and E. Troughton, and the whole instrument was completed in 1797.   The excellent object-glass of 3¾ in. aperture was by P. and J. Dollond.   The powers used were 68, 116, 133, 240, 303, 381.   That most used was 133, the others being double eyepieces.   In some few cases a single lens, power 578, was used. The diameter of the declination circle was 4 ft.—*Phil. Trans.*, 1824, Part iii.

The 7 ft. equatorial had an aperture of 5 inches, and was, at the time it was made, the *chef-d'œuvre* of Tulley. "In distinctness under high magnifying powers, it is probably excelled by no refractor existing." The ordinary observing power was 179; occasionally, 105 and 273 were used.—*Phil. Trans.*, 1824, Part iii.

The 20 ft. refractor was mounted in 1829, at Kensington. The glass was by Cauchoix, and had a clear aperture of 11^3 in.—*Monthly Notices*, vol. i.

STONE. (See CINCINNATI.)

STRUVE and OTTO STRUVE. (See POULKOVA.)

TALMAGE. (See LEYTON.)

TARN BANK.

Mr. Fletcher's equatorial was erected at Tarn Bank in 1860. The optical part was by Cooke, and the stand was made under the direction of Mr. Fletcher. The Sisson polar axis was used in the mounting. The object-glass has a diameter of $9\frac{1}{2}$ in., and a focal length of $12\frac{1}{4}$ ft. The declination circle has a diameter of 42 in., and reads to 1"; the hour circle is of the same size. The driving clock had $22\frac{1}{2}$ lb. as a driving weight, and worked very well.

Mr. Fletcher's small equatorial, by Cooke, was mounted in the German style; aperture, 4·14 in.; focal length, 6 ft. This mounting was that used by Dollond, with a long polar axis. This axis was of mahogany, 9 ft. long, 9 in. square in the middle, and 7 in. square at the ends. The hour circle was 20 in. in diameter, read to 2 sec., and was loose on the polar axis. The declination circle had a diameter of 20 in. also, and read to 10". Powers, 50, 100, 160, 230, 300, 420, and 600, with the parallel-wire micrometer. The power generally used for double-star work was 300. The driving clock was a very elegant instrument and worked very well. The governor was like that used in steam engines.—*Monthly Notices of R. A. S.*, vols. x., xx., xxv. ; *Memoirs of the R. A. S.*, vol. xxii.

UPTON. (See CINCINNATI.)

WALDO. (See CAMBRIDGE, U.S.)

WASHINGTON. (The Great Refractor.)

This magnificent instrument has an aperture of 26 in. and a focal length of 390 in. The glass was by Chance, and Messrs. Alvan Clark and Sons were the makers of this noble lens. It was finished in 1872. The mounting is in the German style. The negative eyepieces are four in number, powers 155 to 1360. The positive eyepieces are sixteen in number, powers 173 to 1802. The tube is of steel, $\frac{1}{16}$ in. in thickness near the ends and $\frac{1}{8}$ in the middle. Length 32 ft. ; diameter of the middle one-third about 31 in. The object-glass is composed of an equi-convex front lens of crown-glass and a nearly plano-concave flint lens : thickness of the objective at the centre about 2·87 in. The glasses are free from all hurtful rings and striæ, and are of nearly perfect figure. There are three micrometers, two filar and one double-image. There are two finders, apertures 2 in. and 5 in. The driving clock was invented by Professor Newcomb : with careful attention to the oiling, etc., it works satisfactorily.—*Instruments and Publications of the United States Naval Observatory, Washington*, 1845-76.

The smaller instrument was made by Merz and Mahler. Aperture 9·6 in., focal length 14 ft. 3 in. The object-glass was under-corrected for colour, and in 1862 it was refigured by Messrs. Clark and Sons : the focal length was increased about one inch, and the glass corrected for defective achromatism ; the definition also was improved. The flint disc is not perfect. Hour circle 15 in., and declination circle 21 in. diameter. Finder 2·6 in. aperture, and 32 in. focal length. Micrometer, a repeating filar, by Fraunhofer. The driving clock is regulated by a Fraunhofer centrifugal pendulum, but it is scarcely powerful enough. There are eight eyepieces, powers 90 to 899.—*Washington Observations*, 1865.

Observers: Newcomb, Hall, and Holden.

WHITEHAVEN.

In 1850 Mr. J. F. Miller, of Whitehaven, began his double-star measurements. The instrument was a very good equatorial refractor by Cooke, the mounting in the German style, and of the same size as Mr. Fletcher's instrument. The micrometer was by Simms, and proved to be a very good one. Diameter of position circle 5 in. ; power generally used 300. The clock-work, too, was good.—*Memoirs of the R. A. S.*, vol. xxii. ; *Astr. Nachr.*, vol. xxxiii.

WROTTESLEY.

English mounting: polar axis of four mahogany planks 14 ft. 3 in. long and 10 in. square in the middle ; pivots of bell-metal. Focal length 10 ft. 9 in.; aperture 7¾ in. ; flint glass by Guinand ; crown by Dollond. Mounted at Wrottesley, Staffordshire, in 1843. Declination and hour circles each 3 ft. in diameter : verniers read to 10″ and 1 sec.

Parallel-wire micrometer: position circle 4 in. diameter, reads to 6′ ; powers used 450 and 320, and 600 and 820, occasionally. Driving clock not satisfactory.—*Memoirs R. A. S.*, vols. xxiii. and xxix.

Observer : Morton.

CHAPTER IV.

THE MICROMETER.

THE parallel-wire micrometer is *par excellence* the instrument of the double-star observer. Though used for many other purposes, it is specially adapted to his work, and has not been superseded by any other form of micrometer.*

It consists of the following parts: first, a stout brass tube or adapter fitting into the eyepiece end of the telescope, and carrying at its outer end a position circle divided from 0° to 360° in the direction contrary to the figures on a watch dial, and read off by two opposite verniers to tenths or twentieths of a degree; it is also provided with clamp and slow motion. The moveable vernier plate has attached to it the micrometer box, which is generally 5 to 6 inches long, 1½ to 2 inches wide, and ½ inch deep. The micrometer screws enter the box at each end, their divided and milled heads being outside. The screws, of a hundred threads to the inch, enter their respective frames, which fit nicely within the box, and move parallel to one another like two tuning-forks, one just small enough to work within the other. Across these frames, in the centre of the field, are stretched the fine webs at right angles to the direction of the screws. To prevent slack, the two frames are pushed towards one

---

* There are many other forms of micrometer, the most important being Airy's and Amici's, both double-image micrometers. The former consists of a positive eyepiece containing four lenses, the third from the eye being concave and divided into two halves, and each half carried by its own screw. Amici's double-image micrometer consists of two prisms, and has been used by Dawes and Doberck. It is considered the best of the kind.

another by spiral springs, thus bringing the inner heads of
the screws against the ends of the box. These heads are
often made square with the shaft of the screw; but they
are much better made spherical, so as to fit into conical
bearings at the ends of the box. A flat comb plate is
placed over the moveable frames across the open centre,
with a fine-toothed comb cut so as to form a chord to the
circle of the field of view at right angles to the moveable
webs. This comb plate carries two stout parallel wires
(called *position wires*), about 12" apart, across the centre
of the field, and at right angles to the moveable webs
and parallel to the comb. The eyepieces are attached

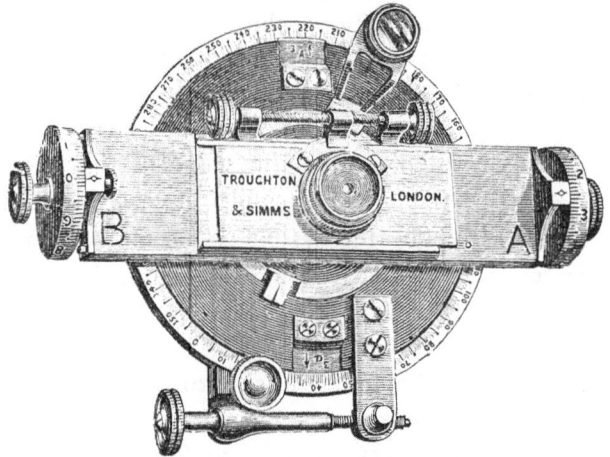

FIG. 3. (Parallel-wire Micrometer.) *

outside the box to a sliding-piece, moved by a screw for
centering over the webs in the direction of their motion.
The webs, position wires, and comb should be clearly defined
with a high power at the same time. The eyepieces should
as much as possible slide into the same adapter, to save
screwing and unscrewing.

* One reading lens is removed to show the slow-motion clamp.

It is usual to insert in the stout brass tube or adapter, close to the position circle, a thin ivory ring with openings all round through the adapter, to admit light for the purpose of giving dark ground illumination to the webs. English makers usually furnish both screws with heads divided into a hundred parts, and figured 0, 10, 20, etc., so as to give an increasing reading when the webs are moved towards the heads or against the spiral springs. Observations are always taken by setting the screw in this direction, as it is found in practice to give the best results. German makers divide only one of the heads, and simply use the other screw for setting in different parts of the field. It is desirable that both screws should have easy play through not less than fifty revolutions. A divided head to one of the screws is quite sufficient, and for distinction we will call this the micrometer screw, and the other the setting screw.*

We have now to determine the value of the revolutions of the micrometer-screw in seconds of arc, and for this purpose we can make the setting screw and its moveable frame an efficient auxiliary. Let the comb be divided in such a manner that every fifth notch is a longer one, and each tenth notch numbered by small holes—one, two, three, etc., counting from the notch nearest to the setting screw as Zero. Let the following webs be placed on the moveable frame of the setting-screw: No. 1, at Zero; No. 2, at 17·75 revolutions; No. 3, at 18·25; No. 4, at 19·0; No. 5, at 20·0; No. 6, at 25·0 (in the centre); and No. 7, at 50·0. On the micrometer screw but one fine web is needed, and it is placed in the centre of its moveable frame: let us call this web No. 8.

We are now in a position to *step* the micrometer screw throughout its whole length with great ease and accuracy,

* These are marked A and B, respectively, in Figure 3, and are held simply by friction, so as to admit of being set to any reading.

viz., at every five revolutions by webs No. 5 and No. 6; at every single revolution by webs No. 4 and No. 5; at every half revolution by webs No. 2 and No. 3; and also at every quarter revolution by webs No. 3 and No. 4.

It will probably suffice to test only the ten central revolutions for parts of a revolution. Use a high power and good illumination. The operation may be thus described. Bring No. 5 to Zero and No. 8 beyond Zero: the latter must now be brought carefully just into contact with No. 5, first on one side and then on the other, the head being read off to tenths of a division each time. No. 8 must now be brought up to No. 6 in precisely the same way, and this will complete the first *step* of five divisions. No. 5 must now be brought to five revolutions, and No. 8 set as before, first on No. 5 and then on No. 6; and this will be the second *step :* carry on this process throughout the fifty revolutions. Repeat this several times, and the mean readings of each step will give the comparative value of each five revolutions with great accuracy. Each group of five revolutions must now be tested in precisely the same way for *each single* revolution, by means of webs No. 4 and No. 5; and each of the ten central revolutions for *parts of a revolution* with webs Nos. 2, 3, and Nos. 3, 4. It is, of course, impossible for webs Nos. 1 to 7 to be placed absolutely at the distances named; but the exact distance will be determined by the observations and the proper allowances made in the computations.

Having thus obtained by the most accurate as well as the most convenient method the comparative value of the different parts of the screw, it now only remains to convert these values into seconds of arc. This is done by transits of a slow moving star from web No. 1 to web No. 7, the distance being fifty revolutions of the screw. The best stars for this purpose are α, β, and δ Ursæ Minoris, whose places are given in the Nautical Almanac.

If the telescope used has, say, 6 in. aperture and 9 ft. focal

length, the value of the fifty revolutions will be 954·93 ±
seconds of arc.   This, at the equator, is equal to 63·662
seconds of time, or 1″ = 0·066 seconds of time: but if we
multiply 0·066 by the secants of the declinations of β, δ,
and a Ursæ Minoris respectively, we get 0·2518, 1·127, and
2·859 seconds of time.   Now as it is difficult to take a
single transit with greater accuracy than 0·25 sec., the
advantage of a slow star is at once apparent.   If, for instance, the transit of δ Ursæ Minoris be taken to 0·5 sec.
by a single observation, the value of the screw will be
obtained with an accuracy of 1 in 2000; but as one observation cannot be relied on, a large number of transits of
different stars should be taken, and in this way an accuracy
of 1 in 5000, or even of 1 in 10,000, can be secured.

It is usual to express the value of the screw in seconds of
arc for one revolution ; and if an auxiliary table be constructed
giving the value of parts of a revolution, any measured distance can be readily converted into arc.

The effect of change of temperature on the screw is so small
that it may be entirely neglected.   The effect of refraction,
however, cannot be so disregarded when the above transits
are observed out of the meridian; and the following is a
simple and convenient mode of dealing with this, since it
enables the observer to transit, when away from the meridian, and to correct his results at once if the altitude be
not less than about 20°.   Find the altitude of the object to
the nearest degree or half-degree by the clinometer.   Observe
the transit as above and read off the position circle; then
bring the micrometer box into a vertical position by means
of the plumb-line of the clinometer.   Read off the position
circle, and the difference between the readings will give the
angle with the vertical, or the parallactic angle.   The full
effect of mean refraction on the position of the star, supposing the transit to be in a vertical plane, must now be
multiplied by the cosine of the angle with the vertical, and

this will give the correction for refraction in seconds of arc. It is always subtractive in the case of transits. The interval of transit must now be multiplied by the cosine of the declination to reduce it to the equatorial value, and then converted into seconds of arc. The correction for refraction must now be added. This method is also applicable to correct the measures of low wide double stars : in this case the correction is always additive.

The correction for curvature of path must be applied in observations of a and δ Ursæ Minoris, but for β it is insensible. Convert the observed interval into arc. Then twice the sine of half the arc thus obtained, divided by the arc expressed in terms of the radius, will give the factor by which the observed interval must be multiplied to reduce it to the true value. Dembowski preferred β to δ as requiring no correction for curvature, and taking less time to observe, and so lessening the chance of instrumental disturbance during transit.

The micrometer screw may also be tested by two terrestrial marks, and the angular value determined if the distance of the marks from the object-glass be ascertained ; but the definition so near the surface of the earth will rarely be found good enough for this kind of observation.

A powerful theodolite may also be used for this purpose, the two telescopes being turned towards each other, and the angular distance of the webs read off on the horizontal circle.

If the micrometer will include the sun's disc, its value may be obtained from the sun's diameter. In this case the *horizontal* diameter should be measured. If the *vertical* diameter be taken, the sun should have a considerable altitude, and the correction for refraction must, of course, be applied. The sun's semi-diameter for noon of each day will be found in the Nautical Almanac on page II of each month.

Some observers make use of the pairs of stars in the Pleiades whose places were determined by Bessel with the greatest care. The following pairs, consisting as they do of small stars of nearly the same magnitude, will be found very useful for this purpose; and to aid in their ready identification a rough map is also given.

Name.	Mag.	R. A. (1880).			Dec. (1880).		
		°	′	″	°	′	″
k (Asterope) .	7·8	54	41	21·19	24	10	46·84
l . . .	7·8		43	28·33		9	12·01
8 . . .	8·9	54	46	24·66	23	49	15·83
9 . . .	8·9		46	58·88		48	57·27
f (Atlas) .	4·5	55	30	22·16	23	41	11·53
h (Pleione) .	5·6			40·46		46	11·68
31 . . .	8	55	31	53·56	24	1	45·54
32 . . .	8		33	8·23		0	52·12
35 . . .	9	55	39	19·06	23	52	41·55
36 . . .	9		41	30·40		51	5·57

From the formula $r = \sqrt{(\varDelta\,\delta)^2 + (\varDelta\,a)^2 \cos^2 \text{mean } \delta}$
we find the following distances for the four pairs k l; 8, 9; 31, 32; 35, 36 :—

	Diff. of R. A. (Δ a).	Diff. of Dec. (Δ δ).	Distance.
k l	127·14″	94·83″	149·92″
8, 9	34·22	18·56	36·39
31, 32	74·67	53·42	86·64
35, 36	131·48	95·98	153·86

In order that the observer may be able to check the preceding results and also to select other pairs for special purposes, the following extract from Bessel's work (*Astronomische Untersuchungen, Erster Band*) is given :—

To face P. 56.

neb

Merope •

Atlas ● f

Alcyone η

Pleione ● h

b ● Electra

9
8

● Celæno
g

Maia

Taygeta e

● ι

Asterope

20
21

31
32

35
36

37

The Pleiades

(From Engelmann's Abhandlungen von F.W.Bessel.)

Name.	Mag.	R.A. 1840.	Precession Annual.	Precession Sec. change.	Proper Motion.	Dec. 1840.	Precession Annual.	Precession Sec. change.	Proper Motion.
Anon. 1	8	° ′ ″ 53 59 14·52	″ 53·119	″ + 0·270	″	° ′ ″ 23 31 42·30	″ 11·792	″ − 0·423	″
— 3	9	54 1 30·81	53·139	+ 0·270		34 36·65	11·782	− 0·424	
21 k (Asterope)	7·8	5 48·99	53·305	+ 0·274	+ 0·051	24 2 56·40	11·761	− 0·425	− 0·057
22 l	7·8	7 56·33	53·300	+ 0·274	+ 0·011	1 21·97	11·751	− 0·425	− 0·054
Anon. 8	8·9	10 56·98	53·192	+ 0·271		23 41 26·35	11·737	− 0·425	
— 9	8	11 31·24	53·191	+ 0·270		41 7·91	11·734	− 0·425	
— 18	8·9	27 20·05	53·198	+ 0·269	+ 0·011	38 16·98	11·659	− 0·426	
24 p	7·8	27 46·26	53·191	+ 0·269		36 55·12	11·657	− 0·427	
Anon. 21	8·9	28 41·90	53·377	+ 0·273		9 22·06	11·653	− 0·428	
— 23	8·9	29 36·40	53·046	+ 0·265		24 10 40·14	11·648	− 0·425	
25 η (Alcyone)	3·4	29 46·72	53·191	+ 0·268	+ 0·021	23 36 16·91	11·648	− 0·427	− 0·068
27 f (Atlas)	4·5	54 53·68	53·212	+ 0·266	+ 0·013	33 30·41	11·528	− 0·429	− 0·077
28 h (Pleione)	5·6	55 10·82	53·241	+ 0·267	+ 0·007	38 30·60	11·527	− 0·429	− 0·085
Anon. 31	8	56 20·32	53·331	+ 0·269		54 4·70	11·521	− 0·430	
— 32	8	57 35·11	53·328	+ 0·269		53 11·52	11·515	− 0·430	
— 35	9	55 3 47·46	53·290	+ 0·267		45 2·15	11·485	− 0·430	
— 36	9	5 59·14	53·285	+ 0·267		43 26·57	11·475	0·430	

The following table from Σ.'s *Mensuræ Micrometricæ* will give a good idea of the accuracy of the work done with the parallel wire micrometer [*e* is the probable error of a single distance, and *f* of a single measured angle].

A. TABLE of the probable ERRORS of single measures of Σ.'s *lucidæ*, i.e., those whose companions are not below the 8th magnitude.

Class.	Mean Distance.	No. of Stars.	No. of Measures.	e	f
I.	0·70	44	176	0·074	2 30·9
II.	1·48	111	447	·086	1 52·4
III.	3·08	128	563	·099	1 8·2
IV.	5·62	119	469	·116	0 48·9
V.	9·79	51	222	·127	0 30·2
VI.	13·94	46	199	·127	0 23·9
VII.	19·38	48	184	·145	0 18·3
VIII.	28·19	48	178	·156	0 14·9

B. TABLE of the probable ERRORS of single measures of Σ.'s *reliquæ*, i.e., those whose companions are between the 8th and 11th magnitudes.

Class.	Mean Distance.	No. of Stars.	No. of Measures.	e	f
I.	0·75	28	94	0·087	2 27·0
II.	1·54	186	642	·109	2 1·9
III.	2·93	383	1299	·122	1 29·5
IV.	5·82	426	1428	·156	1 7·1
V.	10·00	278	783	·184	0 47·1
VI.	13·88	161	455	·201	0 38·7
VII. & VIII.	22·60	383	1064	·207	0 27·0

C. TABLE of the probable ERRORS of single measures of Stars, the companions of which are below the 11th magnitude.

Class.	Mean Distance.	No. of Stars.	No. of Measures.	e	f
II. & III.	2·59	14	49	0·176	2 27·8
IV.	5·92	17	55	·221	2 2·1
V.	10·46	22	59	·362	1 20·7
VI.	14·19	11	37	·376	0 59·6
VII. & VIII.	21·93	12	35	·371	0 55·6

Dr. Dunér, of Lund, gives the following results for the value of his micrometer: they were obtained from transits of Polaris:—

	R.			R.
1867. Sept. 11	17·313	1868. Oct. 3	17·322
11	·303	12	·301
19	·313	21	·309
20	·308	22	·300
21	·309	25	·311
24	·326	26	·315
26	·336	Mean, r = 17″·313 ± 0″·002.	

The Baron Dembowski made a very elaborate investigation of his micrometer in 1873. He used star transits, terrestrial marks, and auxiliary webs or *types,* as he calls them, in the micrometer. The following extracts exhibit some of his results :—

> *Libres* means that all the transits taken on any given day are observed with the telescope in the same position with respect to the meridian, E. or W., the time of observation being any whatever within three hours of the meridian passage of the star.
>
> *Conditionnés* means transits observed with the instrument alternately E. and W. of the meridian, at the same culmination, the same number of observations being made on each side.

The values of the entire scale, and the probable errors are as follows :—

	Sets.		Inter-vals.		Probable error.	
β Ursæ Minoris	10	libres	84	50 rev. = 1054″·484	r ±0″·170	T centig.+28°·4
	7	,,	60	·874	·198	+ 0·4
	10	conditionnés	80	·384	·150	+30·4
	7	,,	84	·836	·292	− 0·8
	10	,,	80	·486	·209	+17·6
δ Ursæ Minoris	14	,,	28	·942	·311	+21·2
By Gauss' method	18	double sets	...	·375	·780	+12·7

And by the method of least squares he deduces the following results :—

> Value of the 50 rev. = 1054″·578 − (T − 19°·72). 0·01420.
> Probable error of the coefficient of (T − 19°·72) = 0·00295.

Hence it is inferred that the absolute value of the entire scale is known within the limits ± 0″·06.

The next table enables us to see the result of his examination of each 5 rev. of the scale, four different methods being used :—

Methods used.	Rev. 0 to 5.	Rev. 5 to 10.
Polaris : 13 transits (libres) ...	....	105″·662 r. = 0″·21
δ U. Minoris : 14 ,, (condit.) ...	105″·247 r. = 0″·16	105″·453 r. = ·21
Terrestrial marks, 14 measures	·382 r. = ·10	·444 r. = ·08
Types, 15 measures	·397 r. = ·06	·372 r. = ·07
Mean	105″·342	105″·423
Value of 1 rev.	21″·068	21″·085
The results from Polaris which are underlined in the tables are excluded from the means.		

Rev. 10 to 15.	Rev. 15 to 20.	Rev. 20 to 25.	Rev. 25 to 30.
105″·573 r. = 0″·29	105″·475 r. = 0″·28	105″·358 r. = 0″·29	105″·498 r. = 0″·35
105″·382 r. = ·15	·468 r. = ·18	·297 r. = ·26	·536 r. = ·18
·401 r. = ·12	·404 r. = ·10	·381 r. = ·09	·380 r. = ·08
·390 r. = ·03	·423 r. = ·07	·388 r. = ·07	·457 r. = ·04
105″·391	105′·442	105 ·356	105″·468
21″·078	21″·088	21″·071	21″·094

Rev. 30 to 35.	Rev. 35 to 40.	Rev. 40 to 45.	Rev. 45 to 50.
105″·459 r. = 0″·34	105″·607 r. = 0″·34	105″·826 r. = 0″·22	....
·348 r. = ·29	·536 r. = 0″·27	105″·690 r. = ·32	105″·655 r. = 0″·20
·436 r. = ·11	·484 r. = ·09	·670 r. = ·08	·632 r. = ·07
·438 r. = ·06	·499 r. = ·05	·655 r. = ·06	·590 r = ·05
105″·420	105″·506	105″·672	105″·626
21″·084	21″·101	21″·134	21″·125

These results present, on the whole, an increasing value from 0 to 50 revolutions ; a minimum value appears at 20 to 25, and the maximum is reached at 40 to 45. The probable error of one measure does not exceed 0″·07.

Then the value of each of the ten central revolutions (20 to 30) is given, by two different methods :—

Method.	Rev. 20 to 21.	Rev. 21 to 22.
Terrestrial mark : 50 measures	21″.078 r. = 0″·05	21″·070 r. = 0″·05
Types : 13 measures...	·063 r. = ·01	·079 r. = ·01
Mean	21″·070	21″·074

Rev. 22 to 23.	Rev. 23 to 24.	Rev. 24 to 25.	Rev. 25 to 26.
21″·079 r. = 0″·05 ·076 r. = ·01	21″·066 r. = 0″·05 ·082 r. = ·01	21″·085 r. = 0″·05 ·c81 r. = ·01	21″·083 r. = 0″·05 ·088 r. = ·01
21″·077	21″·074	21″·083	21″·085

Rev. 26 to 27.	Rev. 27 to 28.	Rev. 28 to 29.	Rev. 29 to 30.
21″·080 r. = ·05 ·090 r. = ·02	21″·086 r. = 0″·04 ·095 r. = .02	21″·099 r. = 0″·05 ·100 r. = ·02	21″·117 r. = 0″ 05 ·092 r. = ·02
21″·085	21″·090	21″·099	21″·104

Here, as in the preceding results, the mean values increase on the whole from 20 to 30; and De. finds that the probable error of one measure does not exceed 0″·05.

Résumé of the mean values of each quarter of the ten central revolutions in the seven different series, and the probable error of one measure :—

Series.	1st Quarter.		2nd Quarter.		3rd Quarter.		4th Quarter.	
I.	5″·005	0″·120	5″·175	0″·093	5″·513	0″·088	5″·390	0″·089
II.	·301	·107	·214	·059	·733	·099	·336	·088
III.	·111	·110	·185	·085	·356	·036	·432	·070
IV.	·143	·079	·195	·049	·277	·047	·469	·045
V.	·195	·064	·164	·090	·298	·076	·426	·108
VI.	·281	·097	·139	·043	·270	·063	·394	·062
VII.	·325	·008	·194	·024	·199	·004	·366	·026

Series.	4th Quarter.		3rd Quarter.		2nd Quarter.		1st Quarter.	
I.	4″·907	0″·144	5″·149	0″·158	5″·263	0″·141	5″·764	0″·180
II.	5 ·115	·126	·197	·116	·644	·104	·127	·053
III.	4 ·920	·076	·351	·089	·266	·081	·547	·165
IV.	·876	·049	·308	·030	·332	·069	·567	·069
V.	5 ·086	·114	·144	·059	·544	·060	·299	·174
VI.	·309	·121	·268	·093	·315	·048	·192	·146
VII.	·276	·022	·241	·026	·562	·029	·004	·024

The objects used in obtaining the series I. to VII. were as follows :—For I., II., double distances of 5 Lyræ; for III., IV., V., double distances of two terrestrial discs; for VI. double distances of μ Draconis; and for VII. the distance between two auxiliary webs in the micrometer.

Taking the mean of the values for each quarter of a revolution obtained by the positive and negative movements of the screw, the following results for each series are found :—

Mean of the values for each Quarter.			
I.	5″·384	5‴·219	5‴·331	5‴·148
II.	·214	·429	·215	·225
III.	·329	·225	·353	·176
IV.	·355	·263	·292	·172
V.	·247	·359	·221	·256
VI.	·236	·227	·269	·351
VII.	·164	·378	·220	·321

The means of these series for each quarter are 5″·276, 5″·300, 5″·272, 5″·233.

Difference between the mean measured value of a Quarter of a Revolution and the mean value 5‴·271.			
I.	+ 0″·113	− 0″·052	+ 0″·060	− 0″·123
II.	− ·057	+ ·158	− ·056	− ·046
III.	+ ·058	− ·046	+ ·082	− ·095
IV.	+ ·084	− ·008	+ ·021	− ·099
V.	− ·024	+ ·088	− ·050	− ·015
VI.	− ·035	− ·044	− ·002	+ ·080
VII.	− ·017	+ ·107	− ·051	+ ·050

In making the seven series of measures, the micrometer was removed from the telescope after each series.

Remarking on the whole investigation, De. is led to the following conclusions :—

1. The values of the four quarters of a revolution are not equal *inter se.*

2. Greater inequalities still are found between the + and − readings.

3. These inequalities do not depend on any defect in the division of the head.

The micrometer used at Bermerside Observatory (see the illustration, p. 51), was made by Mr. Simms last spring. It is a beautiful instrument, and a very careful examination of the screw by Dembowski's method (see p. 59) has shown that it may be regarded as perfect, at least for the purpose of double-star measurement.

From upwards of 200 transits of stars the value of 1 rev. was found to be 13″·8372, with a probable error ±0″·004.

The screw (marked A in the illustration) which is the one used in measuring double stars, was tested with the following satisfactory results :—

1. From ten careful settings of the micrometer web close to one of the fixed webs, it was found that the probable error of the mean was ±0″·003, and the probable error of one setting ±0″·014.

2. Careful stepping of the screw by 5 revolutions at a time showed the following differences from the mean value of eight sets of determinations : +0″·014, +0″·003, −0″·006, 0·0, −0″·008, +0″·004, +0″·004, −0″·001, +0″·005, −0″·005.

3. The ten central revolutions were then stepped singly, and the differences from the mean result were : −0″·014, +0″·001, +0″·001, −0″·005, −0″·004, −0″·007, −0″·005, +0″·021, −0″·012, +0″·007.

4. Each half revolution of the ten central ones was then measured five times, and the greatest difference from the mean result was +0″·04.

5. Lastly, each quarter of the six central revolutions was stepped four times ; the greatest difference between the mean of the whole and the means of the several quarters did not exceed 0″·02.

These results therefore show that there is no appreciable change of value in the different parts of the screw, and that there is no sensible eccentricity in its mounting.

The webs used for double-star work, No. 6 and No. 8, were measured, and the thicknesses found to be 0″·415 and 0″·372.

CHAPTER V.

METHODS OF OBSERVING, ETC.

IT is here proposed to give a somewhat full account of the methods of observing the positions and distances of double stars. The subject will be treated under the following heads :—

1. Methods of observing angles and distances.
    (*a*) The methods adopted by Sir Wm. and Sir John Herschel.
    (*b*) The methods used by Dawes and Dembowski in the measurement of angles : Dawes' prism.
    (*c*) Special methods for very close stars.
    (*d*) Methods which may be occasionally used.
2. The number of measures of angle and distance required to form a *set*, or complete observation, with an example.
3. Specimens of *Forms of Registry*.
4. Weights.
5. On contracted apertures.
6. Best time for observing : weather, etc.
7. Precautions to be used while observing.

(1) METHODS OF OBSERVING.

(*a*) The method Sir William Herschel adopted will be best given in his own words : " The distances of the stars are given several different ways. Those that are estimated by the diameter can hardly be liable to an error of so much as one quarter of a second ; but here must be remembered what I have before remarked on the comparative appearance of the diameters of stars in different instruments. Those that are measured by the micrometer, I fear, may be liable to an error of almost a whole second ; and if not measured with the utmost care, to near 2″. This is, however, to be understood only of single measures ; for the distance of many of them that have been measured very often in the course of two years' observations can hardly differ so much as half a second from truth, when a proper mean of all the measures is taken. As I always make the wires of my micrometer outward tangents to the apparent diameter of the stars, all the measures must be understood to include both their diameters ; so that we are to deduct the two semi-diameters of the stars if we would have the distance of their centres. What I have said concerns only the wire micrometers, for my last new micrometer is of such a construction that it immediately gives the distance of the centres ; and its measures, as far as in a few months I have been able to find out, may be relied on to about one-tenth of a second, when a mean of three observations is taken. When I have added *inaccurate*, we may expect an error of 3″ or 4″. *Exactly estimated* may be taken to be true to about one-eighth part of the whole distance ; but *only estimated*, or *about*, etc., is in some respect quite undetermined ; for it is hardly to be conceived how little we are able to judge of distances when, by constantly changing the powers of the instrument, we are, as it were, left without any guide at all. I should not forget to add that the measure of stars, when one is extremely small, must claim a greater indulgence than the rest, on account of the difficulty of seeing the wires when the field of view cannot be sufficiently enlightened.

"The angle of position of the stars I have only given with regard to the parallel of declination, to be reduced to that with the ecliptic as occasion may require. The measures

always suppose the large star to be the standard, and the
situation of the small one is described accordingly. Thus, in
Fig. 4, A B represents the apparent diurnal motion of a star
in the direction of the parallel of declination A B ; and the
small star is said to be south preceding at *m n*, north pre-

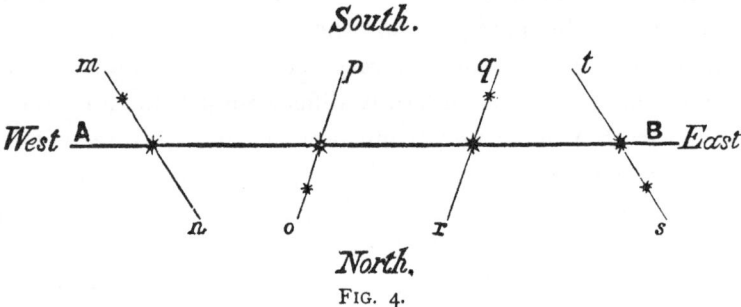

FIG. 4.

ceding at *op*, south following at *qr*, and north following at
st. The measure of these angles, I believe, may be relied on

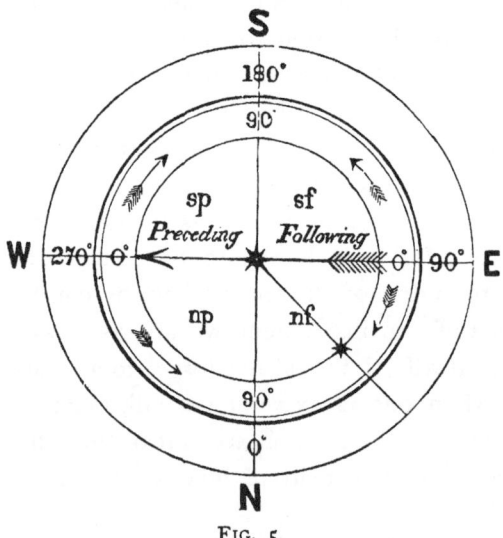

FIG. 5.

to 2°, or at most 3°, except when mentioned *inaccurate*, where
an error amounting to 5° may possibly take place. In mere

estimations of the angle without any wires at all, an error may amount to at least 10°, when the stars are near each other."*

The foregoing diagram will make this method of registering the position angles quite clear. The innermost circle represents the inverted field of view; the four quadrants are indicated by *n f, s f, s p, n p*, and the angle is given by the position circle: e.g., in the case supposed in the figure the position would be entered as 45° *n f.* The outer circles exhibit the method first suggested by Sir John Herschel, and now in universal use. In this the quadrants are dispensed with, the zero of the position circle is at the north point, and the circle is read all round to 360° in the direction N.E.S.W.; hence, according to this method, the above angle would be registered as 45° simply.

For distances, the methods used by Sir John Herschel and the later observers are identical.

(*b*) To measure accurately the position angle of a double star would seem at first sight to be a sufficiently simple process. Experience, however, has shown that in many cases it is most difficult. A glance at the measures of some double stars by different practised and eminent observers at the same epoch is quite enough to exhibit this fact in a striking way; and a comparison of the individual measures of the same star on the same night by one and the same observer and instrument, abundantly confirms it. Some of the disturbing causes are obvious enough; but even when the stars do not differ greatly in magnitude or brightness, and when the sky is clear and the air still, these discrepant measures still present themselves. And in the case of close and unequal pairs, "the eye, often at the very first glimpse, acquires a prejudiced bias." (H₂.) "When such stars are between the wires, the eye may unconsciously be directed to the edge of one wire rather than of

* Subsequent and more accurate measures show that Sir William's measures were liable to much greater errors than he here anticipates.

the other; "there is a tendency to place one of the double wires nearly in the direction of a tangent to the discs of moderately unequal stars." (H$_2$.) Further, we are told that there is a tendency in the eye to " accommodate its judgment to the position of the wires" before they are brought up to correct parallelism with the line joining the centres of the star discs.

Nor is this all. Not only have we to get rid of widely discrepant results, we must also be on our guard against accordant measures. This latter difficulty is often a very considerable one. However, as we are here rather concerned with the methods by which these tendencies are to be destroyed or counteracted, we proceed to describe those used by the most successful observers of double stars :—

1. By repeated small movements of the wires in the *same* direction till the eye is quite satisfied.

2. By bringing up the wires alternately from opposite sides of the true direction. If three or more measures be made both ways, the mean result will probably be near the truth.

3. By a succession of small movements of the wires, *the eye being removed from the telescope for a moment after each alteration.*

Whichever method be adopted, it will always be well to rest the eye a little, and to carry the webs some degrees away from the last position obtained, *after each reading.* *

When the stars are so faint that only very little arti-

* "It will occasionally happen that, after taking two or three very coincident angles, on recommencing after some slight interruption, a sudden difference of two or three degrees will occur, and a new set of angles, agreeing well *inter se*, but differing from the former, will be obtained. In such a case it is most probable that the one or other result has been affected by some bias of the kind above alluded to ; and, as it is highly necessary to ascertain which it is, the following method of trying such rival measures against each other will often be found serviceable. Suppose the two measures at issue were 63° and 65°, each being a mean of three or four pretty coincident

ficial light can be used, it is still possible to obtain useful angles by employing the method of *oblique vision*. The illumination is gradually increased until the webs are just well seen; and the eye is then directed, *not* to the star, but to another part of the field. "In this way, a faint star in the neighbourhood of a large one will often become very conspicuous." (H₂.)

Before concluding these remarks on the measurement of position angles, some account of Dawes's prism should be given. This distinguished observer, soon after he began to measure double stars in 1830, discovered a tendency in his own eye to "obtain a different result in position when the line joining the centres of the stars was nearly *parallel* to the line joining the centres of the eyes, from that which was obtained when these lines were nearly perpendicular to each other; and a still more decided difference was found to prevail when those lines formed a very *oblique* angle." He entirely overcame the difficulty by simply fixing a small prism to the eyepiece between it and the eye. By this means any double star can be placed in any desired position with respect to the horizon; and it was the uniform practice of this great observer to confine himself entirely to the vertical and horizontal positions. Dembowski and Struve always observed with the head vertical. *O.Σ.*, also, after accumulating a vast mass of measures, became aware of an error resulting from obliquity of position, and undertook a laborious series of measures of artificial double stars,

measures. As it is probable that one is decidedly right, and the other decidedly wrong, and as their difference is 2°, let the micrometer be set to 61° and 67°, one or the other of these being necessarily 4° in error, will be violently offensive, while the other will be affected only by an error which experience has already shown *may* be borne without detection in the particular star in question. Thus the false results will be made evident; and, in assigning weights to the measures, this must be taken into consideration as materially diminishing the influence due to it."—*Sir John Herschel*, in *Memoirs of the Royal Astronomical Society*, vol. v.

partly for the purpose of ascertaining the amount and law of this error; and in his measures lately published both the observed and corrected angles and distances are given. The objections to the prism on the score of loss of light and impaired definition were regarded by Dawes, after nearly forty years' use of it, as quite unfounded. It is obvious, too, that the comfort of the observer, and therefore, to some extent, the accuracy of the measures, will be considerably increased by this simple apparatus.

Of the extreme difficulty which attends attempts made to obtain accurate measures of distance of close and unequal double stars, nothing need here be said. So keenly was this felt by the late Sir John Herschel, that he devised a method of obtaining the elements of the orbit of certain double stars from the measured angles alone, the measured distances being used collectively for finding the value in seconds of space of the scale used in the construction. Extreme care, much practice, a good sky, patient repetition on different nights, the destruction of bias by removing the eye from the instrument for a few moments, and carrying the web far away from the last setting after each measure,—these and such like precautions naturally suggest themselves to the observer.

(c) In the case of close pairs, the following suggestions, if carefully carried out, will often be found of use :—

1. Place one star centrally over a web, and note the change of form which the disc undergoes, e.g., if it becomes elliptical in shape, place the other web so as to produce the same effect on the other star.

2. When the distance is less than one second, the two following methods will frequently give valuable results.

Place the inner edges of the webs at a distance apart as nearly as possible equal to that which separates the two stars, using a high power; bring the stars close up

to the webs, and compare the two spaces; correct, if necessary, and then read off the divided head of the micrometer. Repeat this from six to ten times; then, the reading
when the webs are just in contact, together with the readings given by the above settings, will furnish the means
for deducing the distance of the stars with considerable
accuracy.

A better method, however, is that of first placing the
threads a certain known distance apart, say 1″, bringing
the discs between them, and trying to estimate and express
in numbers the ratio between the distance of the discs
apart and the distance of 1″. Make several or many estimations; then, the distance between the threads being
known, the true distance of the discs is readily deduced
from the ratios. These two methods were used by Struve.
Baron Dembowski takes one measure by estimation, then
one with the webs, and places great confidence in the mean
of the two.

This will be a suitable place for a few words on the Barlow
lens. It was frequently used by Dawes, and he thus sums
up its advantages :—

1. The diameter of the micrometer threads subtends only
   about half the angle.
2. The moveable parallel threads are *both as nearly in focus*,
   with double the magnifying power.
3. The value of the micrometer divisions with the lens
   is only about half of its amount without it: hence a
   proportionably fine motion in the measurement of
   distance.
4. With any given power the threads are distinct to a
   much greater distance from the centre of the
   field.

(*d*) The method of *oblique transits* described by Sir John
Herschel may be noticed (see the " Cape Observations,"
p. 247). " If *p* be the polar distance of a double star; *θ* its

measured angle of position; a the angle of position of an oblique wire across which both stars are allowed to transit by their diurnal motion ; Δ the interval of their transits across it in seconds of time,—then will the distance of the stars from each other be given by the formula

$$\delta = \frac{15 \, \Delta \, \sin p \, \cos a}{\sin (a - \theta)}.$$

Convenient values of a are 100°, or 110°, or (on the other side of the vertical) 260° or 250°. The inclination of the oblique wire ought to be towards the opposite side of the meridian to that of the line joining the two stars. In situations not remote from the pole, a high degree of precision is attainable by this method."

Lastly, it is sometimes convenient, especially when the distance is very great, to measure *differences of declination*, and then to compute the distances of the components from them and the angles of position.

2. NUMBER OF MEASURES.

As regards the number of measures of position and distance which should be taken of a star on the same night, the practice of eminent observers differs. However, it is quite certain that at least three of the angle, and three double measures of the distance, should be taken. Six of angle and twelve of distance (six double measures) would be much better. On the other hand, it is better to measure the same object on two different nights, than to make a large number of measures on one night only. Of course the importance of the star and the quality of the night will also affect the number of measures taken. Sir John Herschel usually made ten of angle and ten of distance : Dembowski, four of angle and four double measures of distance: Wrottesley, ten of position and ten of distance. Dr. Doberck four of angle and one double distance.

The making of a complete observation of the position and distance of a double star may be thus described. After lighting the lamp which illuminates the field, and turning on the red or blue glass,* the micrometer is pushed into the tube, adjusted to distinct vision of webs and star, and the position circle set to zero. With eye on the star, the micrometer is then turned bodily until the star runs along one of the distance webs (which has been placed near the middle of the comb), from side to side of the field. The thick position webs, now coinciding with the meridian, are then moved until the stars lie between them. Then, if Dawes's practice be followed, let the webs be brought up to true parallelism with the imaginary line joining the centre of the stars by a succession of small changes, *the eye being removed from the telescope for a moment after each change.* Read off the circle, and repeat from four to ten times. If the method of Dembowski be preferred, the webs will be brought up alternately from opposite sides of the true direction, the *same* number of measures being made in each direction. The webs should be moved away some distance each time, so that the eye may be freed from any bias.

If the circle be not set to zero at the outset, the necessary correction must of course be applied to each reading when the set is complete.

It is well to examine the zero reading of the circle after the measurement of each star, to avoid errors from accidental derangement. To take an example, let the star run along

---

* "The colour I employ is that afforded by a brown-red glass of the Claude-Lorraine kind, which throws a strong sunshine glow over a landscape, almost verging to orange. A fuller red is even yet superior for distinct definition of wires."—*Herschel*, in *Memoirs of R. A. S.*, vol. v. Dembowski and Doberck prefer Cinnabar red glass.

the equatorial wire at 91° 30′ by the position circle : then will
— 1°·5 be zero correction.   If five readings be now taken, the
operation of reduction will be as follows :—

$$
\left.
\begin{array}{l}
\overset{\text{o}}{112\cdot7} \\
110\cdot5 \\
114\cdot2 \\
112\cdot3 \\
111\cdot8
\end{array}
\right\}
\quad
\begin{array}{ll}
\text{The sum} & = \underline{561\cdot5} \\
\text{Mean} & = \overline{112\cdot3} \\
\text{Correction} & - \ 1\cdot5 \\
& \overline{\underline{110\cdot8}}
\end{array}
$$

For the distance :—Let us suppose that the companion is
to the right of the principal star and the micrometer set to
measure position.   Fix one web, and place it on the centre
of the principal star : now move the free web to the right
until it bisects the companion star and read off the head.
Carry the free web to the left of the fixed one, and bring
the companion to the left until it is bisected by the latter ;
place the free web on the principal star ; and again read off.
Repeat this double measure, bringing the web up *in the same
direction as before*, i.e., from left to right, from four to ten
times, reading off the divided head each time.

To take an example as before : As the divided head is
held on the axis of the screw by friction only, it may be set
approximately to zero, when the moveable and fixed webs
are superposed.   Suppose this to be done, then the following
observation will illustrate the method :—

1875·7009 :—ε² Lyræ.

$$
\begin{array}{llll}
\text{d} & \text{d} & \text{d} & \text{d} \\
18\cdot5 & 100 - 81\cdot0 & = & 19\cdot0 \\
18\cdot5 & -81\cdot5 & = & 18\cdot5 \\
18\cdot0 & -81\cdot0 & = & 19\cdot0 \\
18\cdot7 & -82\cdot0 & = & 18\cdot0 \\
& & & \text{d}
\end{array}
$$

The sum of these eight readings is 148·2, and the mean is
d
18·52.

It still remains to convert these parts into seconds of arc.
This is most readily done by means of a short table from
which the values can be taken out at once.

Such a table may be thus constructed, supposing $13\overset{d}{\cdot}227$ = $1''$. The first column gives the divisions and the others the tenths.

Div.	·0	·1	·2	·3	·4	·5	·6	·7	·8	·9
0	·0	·007″	·015″	·022″	·030″	·037″	·045″	·052″	·060″	·067″
1	·075	·083	·090	·097	·105	·112	·120	·128	·135	·144
2	·151	·158	·166	·173	·181	·189	·196	·204	·211	·219
	etc.		etc.		etc.		etc.		etc.	

Here $2\overset{d}{\cdot}3 = 0''\cdot173$ at once from the table.

---

3. FORMS OF REGISTRY.

The importance of having ready a supply of forms for the entry of measures need not be here insisted on. Annexe are copies of those used by Sir John Herschel, Smyth, and Wilson.

Sir John Herschel's Form.

REGISTRY OF THE MICROMETRIC MEASURES OF DOUBLE STARS.

Number for Reference.	N.P.D.	Declination.	Right Ascension.
No.			
Instrument used.	Date.		Star's Name.
	$\begin{array}{c} 18 \\ = 18 \end{array}$ (Dec. of year.)		
Diagram.	Quadrant.		Magnitudes. Colours.
Face to	Micrometer reads		

FORM OF REGISTRY OF THE MICROMETRIC MEASURES OF DOUBLE STARS—*continued.*

Position.				Distance.						Remarks.
Power.	o	′	W	Power.	N.B. The + and − readings to be taken alternately.		Rev.	Pts.	Dec.	w
	.					+	.	.		
	.					+	.	.		
	.					+	.	.		
	.					+	.	.		
	.					+	.	.		
	.					−	.	.		
	.					−	.	.		
	.					−	.	.		
	.					−	.	.		
	.					−	.	.		

Position	Distance	Remarks
Mean . $Z =$ $=$ $=$ from n in direction $n f s p$	Mean $\left\{\begin{array}{l} + \quad . \quad . \\ - \quad . \quad . \end{array}\right.$ Div. by 2 . . Parts = Seconds = N.B.—When only positive readings are taken, a zero must be used, and the division by 2 omitted.	Sky Wind Steadiness Definition of Star Dist. from Merid. General ⎫ Pos, Judgment ⎬ of Obs. ⎭ Dist. Observer.

Zeros of Position and Distance.			Determination of Place.			
				H.	M.	S.
Star runs along the thread at ∴ Zero for position $Z =$			Clock (or clock + 24 h). — Hour Circle : + if East, — if West ; ⎫ if read on to 24 h. always ⎬ —. ⎭			
Threads close at	+	−	Instrumental correction True R. A.			
Mean ∴ Zero for distance $Z =$ To be used only in case opposite readings are not taken.	+	−	Declination Circle, ⎱ + North, — South ⎰ Instrumental correction True Declination			

Admiral Smyth's Form.

MICROMETRIC MEASURES OF DOUBLE STARS, AT BEDFORD, WITH THE 8½ FEET REFRACTOR.

Star's Name.	Right Ascension, 1830.	Declination, 1830.

Diagram.	Quadrant.	Magnitudes. Colours.
		A = B = C = D =

Position.			Distance.					Remarks.
Power.		W	Power.	N.B. The + and − readings to be taken alternately.	Rev. Pts. Dec.		w	
	·			+	· ·			
	·			+	· ·			
	·			+	· ·			
	·			+	· ·			
				+	· ·			
	·			−	· ·			
	·			−	· ·			
	·			−	· ·			
	·			−	· ·			
	·			−	· ·			
Mean	·		Mean {	+	· ·			Sky Wind Steadiness Definition of Star Face to Dist. from Merid.
Z =	·		{	−	· ·			
=			Div. by 2		· ·			General { Pos.
=			Parts =					Judgment {
from *n* in direction *n f s p*			Seconds =					of Obs. { Dist.

Zero of Position.	Date.
Star runs along the equatorial wire at } o I	18
∴ Zero for position Z =	= 18 , (Decade of the year.)

The Rugby Form.

TEMPLE OBSERVATORY.

No._____   _____ 187

DOUBLE STARS.

| R. A. | | DECL. |

Magnitudes.

POSITION.   DISTANCE

Zero.

Readings.   Direct.   Indirect.   ½ Diff.

Position =

Distance =

4. WEIGHTS.

Several practised observers have accustomed themselves to assign *weights* to every position and distance. Sir John Herschel, for example, gives the following account of his mode of doing this. " Although it is impracticable to estimate correctly in numbers the *goodness* of a measure, yet such is the powerful influence of atmospheric circumstances on this very delicate class of observations, as to render it imperatively necessary either to observe only on those rare nights when that cause of error does not exist, or to multiply observations on inferior nights, and reject, freely, all which exhibit great deviations, or which do not give satisfaction at the time. If this be not done, the greatest confusion will arise. The assignment of a weight to each measure, according to the best judgment the observer can form, offers a middle course, free from the objectionable point of arbitrary rejection, and admitting a multiplication of observations on different nights, which is, indeed, quite indispensable for coming at the truth in all the more difficult cases. The scale I have adopted is from 1 to 10; 1 applying to the worst possible measurement in the most unfavourable circumstances, and 10 to the most perfect which can be had in the most favourable." In casting up the mean of a set of measures, if the weights were pretty equal the arithmetical mean was adopted: if the weights differed much, the mean was found by the rule for finding the centre of gravity of a number of weights.—*Sir John Herschel*, in *Memoirs of the R. A. S.*, vol. v.

It will be understood that the assignment of the *weight* must *precede* the reading of the circle or divided head. Dembowski began to use Sir John Herschel's method in 1854.

Dawes followed Sir John Herschel's plan after 1831. He observes: " Scarcely any liberty has been taken in the rejection of observations considered tolerably satisfactory at the time. Occasionally the micrometer has been set to a suspected reading, and a re-examination instituted. If not found decidedly bad, it has been suffered to remain; if otherwise, another completely detached observation has been taken. If this last differed widely from the suspected one, and nearly coincided with the rest, it has been taken in its stead; if not, both the suspected measure and that taken to prove it have formed part of the set."

Wrottesley computed the probable errors and weights by the usual formula prior to 1857. After that year a more elaborate method was adopted: see *Proceedings of R. S.*, vol. x.

Secchi assigned weights (1 to 5) according to the agreement among the individual measures of the set; 5 was the highest and 1 denoted an approximate result.

For a fuller treatment of this subject, see page 144.

5. CONTRACTED APERTURES.

Sir John Herschel was probably the first observer who made constant use of these contrivances. In 1831 he has the following remark in the notes to his measures: "The action of a telescope is often surprisingly improved by stopping out the central rays, by a round disc from a fifth to a sixth of the diameter of the object-glass, which should be well sheltered." *

In 1834 Dawes wrote: "The use of a central disc on the object-glass having been suggested to me by Sir John Herschel, for the purpose of diminishing the images of the stars, I have frequently employed one from an inch to an inch and eight-tenths in diameter. The effect is decidedly good on the stars themselves, if not too faint to bear the loss of light. The *separating* power of the telescope is increased; but the concentric rings accompanying bright stars are multiplied, and rendered more luminous, and are also thrown further from the disc. Hence small stars may often be obscured or distorted by the ring passing through them." In the introduction to his last great catalogue, this eminent observer again takes up the use of apertures. His long experience enabled him to speak with much confidence, and the following is a summary of the contents of the chapter. He seldom used the central round disc before the object-glass, because it increased the number and brightness of the rings, and caused the rings to hide faint companions of bright stars and elongate the discs of nearly equal stars;—a *perforated whole aperture* was used with great advantage, and the "perforated cardboard used for making the Berlin-wool work is very suitable for bright stars." For fainter stars, a piece of cardboard covering the whole object-glass and pierced with holes in concentric circles may be used. These contrivances reduce the size of the discs and the brightness of the rings. The concentric prismatic rings produced are so distant as not usually to interfere with companion

* "Sheltered:" *i.e.*, provided with a dew-cap of ample length, blackened inside.

stars. *Angular* apertures were used by Sir John Herschel, especially the inscribed triangle for destroying the rings round bright stars ; but the rays often obliterate or distort the small companion star. Dawes recommends the inscribed *hexagon*. In order to destroy the tendency of discs to become triangular, " especially when the wind is in the east or south-east," he recommends " cutting off three equidistant segments from the whole aperture of the object-glass, the base of each of which is the chord of 60°. Then, the chords being placed so as to coincide in position with the angles of the telescopic inverted image, those angles will be reduced by the larger circular aperture between the segments, and a fairly round image will be substituted for the triangular one."

" A smaller aperture may sometimes show a very delicate and close companion to a bright star, when a larger aperture fails to do it."

The following table, from Dawes, may be of use in enabling the observer to form a correct estimate of the separating power of his object-glass : —

Aperture in inches.	Least separable distance.	Aperture in inches.	Least separable distance.	Aperture in inches.	Least separable distance.
1·0	4″·56	4·0	1″·14	8·5	0″·536
1·6	2·85	4·5	1·01	9·0	0·507
2·0	2·28	5·0	0·91	9·5	0·480
2·25	2·03	5·5	0·83	10·0	0·456
2·5	1·82	6·0	0·76	12·0	0·380
2·75	1·66	6·5	0·70	15·0	0·304
3·0	1·52	7·0	0·65	20·0	0·228
3·5	1·30	7·5	0·61	25·0	0·182
3·8	1·20	8·0	0·57	30·0	0·152

6. BEST TIME FOR OBSERVING, ETC.

The state of the atmosphere during double-star observation should always be described in the note-book. Secchi indicates the state of the sky by means of the initial letter of the words signifying *very fine, good, middling,* and *bad.* He considers the night *very fine* when distances under 1 can be

6

readily measured, the discs being sharp and clear ; *good*, when distances from 1″ to 2″ can be dealt with, the discs being less sharp than in the preceding; *middling*, when the discs are badly defined and unsteady ; *bad*, when discs 3″ apart cannot be clearly separated. Some observers express these conditions by numbers. From the experience of Dawes and Struve it would seem also to be worth while noting the direction of the wind. Both these practised observers frequently found that easterly winds were associated with *triangular* discs.

As regards the best time for observing, perhaps not much can be said, so much depends on local circumstances. Sir John Herschel, in the south of England, found that " the best time for astronomical observation, and especially for these measurements, is between midnight and sunrise. In the long nights of winter, it is true, distinct vision often comes on an hour or two before midnight, and in all seasons occasionally, of course, much earlier." He then notes the unsteadiness of the discs as morning twilight comes on, and uses the following descriptive terms in his notes : " twirling," " moulding," " convulsed," " twitchings," " wrinkled," " burred," " glimmering." " The rarest of all states of the atmosphere is that in which the rings are destroyed and the stars are seen perfectly round and tranquil."

In conclusion, Sir John's experience with respect to the action of dew and the use of the dew-cap is worthy of note. " The least dew on the object-glass must be most carefully avoided, as it produces a singular contortion in the stars, which I have usually termed *wrinkling ;* the discs are much diminished, the rings multiplied and rendered narrower, and are kept in constant motion ; and a material change of the apparent angle of position is often produced by the displacement of their centres." The remedy he found to be a tube of tinned iron about 20 in. long, bright without and blackened within, and fixed on the object end of the telescope. (This was for his 7 ft. refractor.)

7. PRECAUTIONS.

The following precautions and hints may be of use to amateur observers. They are drawn from the experience of such observers as Σ., H., Da., De., and Se. :—

1. At the outset it must be remarked that the observatory (doors, windows, slit, and ventilators) should be thrown open at least an hour before observation begins, in order to reduce the temperature of the room to that of the external air.

2. If the definition be bad and the motion great, it is useless to attempt the measurement of double stars. In short, if a power of at least 300 cannot be used, the results cannot generally be of any value.

3. Very bright stars should be measured in daylight or twilight.

4. The observations should be made near the meridian if possible.

5. The observer should be in an easy position,—the prism effectually secures this ; and the driving clock ought to go smoothly.

6. The bright-field should be used almost exclusively— red and blue colours are most in use.

7. Use the highest powers possible, and always the *same* powers.

8. A moderate number of measures of an object on each of two nights is better than a large number on one night.

9. Use printed forms.

10. Enter date, hour, weather, and distance from meridian, before observation begins.

11. Notes on definition, general impression as to the value of each measure or each set, etc., cannot well be too copious.

12. In all doubtful cases make a sketch, and add full description.

PART II.

CHAPTER I.

ON THE CALCULATION OF THE ORBIT OF A BINARY STAR.

INTRODUCTION.

IN his *Lettres Cosmologiques*, first published in 1761, the astronomer Lambert has the following remarkable words : " By observing the groups in which the stars are very much condensed, we may, perhaps, be enabled to ascertain whether there are not fixed stars which revolve in sufficiently short periods of time around their common centre of gravity." At the time these words were written, not more than from forty to fifty double stars were known to astronomers. In 1784 (see *Phil. Trans.*, vol. lxxiv., p. 477), about four years after Sir William Herschel began his famous discoveries of double stars, Michell wrote : " It is not improbable that a few years will inform us that amid the great number of double stars, triple stars, etc., observed by Herschel, there are some which form veritable systems of bodies revolving about one another." And again, in an earlier paper, (*Phil. Trans.*, 1767, vol. lii.,) Michell writes : " If, however, it should hereafter be found that any of the stars have others revolving about them, for no satellites shining by a borrowed light could possibly be visible, we should then have the means of discovering the proportion between the light of the sun and the light of those stars, relatively to their respective quantities of matter." Maupertuis, Cassini, and no doubt other thoughtful astronomers in the early part of the eighteenth century, speculated on the existence of siderial systems, but none with such clearness as did Christian Mayer of Mannheim. This diligent observer studied the proper motions of many bright stars by means of the small *Comites* he discovered near them, and speculated on binary systems, elliptical orbits, the origin of new stars, variables, a central sun (?), etc.* The actual discovery, however, of pairs of stars physically connected and in orbital motion was reserved for Sir William Herschel. In the year 1779 he began to sweep the northern heavens in search of double stars, and in his first catalogue, presented to the Royal Society in 1782, gave descriptions and measures of 269 of these objects. About twenty-five years after the conclusion of these sweeps for double stars, he carefully remeasured the angles and distances. The observed changes in angle and distance formed the subject of his great paper, " *Accounts of the changes that have happened during the last twenty-five years, in the relative position of double stars; with an investigation of the cause to which they are owing.*" (*Phil. Trans.*, 1803, Part ii.) In this paper he showed that " many of them are not merely double in appearance, but must be allowed to be real binary combinations of two stars, intimately held together by the bond of mutual attraction." And Castor is the star whose changes he first submits to examination. Indeed this splendid object seems to have commanded much of his attention for years before the publication of his famous discovery of binary stars ; for Sir John Herschel says of this star that its " unequivocal angular motion seems to have first impressed on my father's mind a full conviction of the reality of his long-cherished views on the subject of binary stars."—*Memoirs of R. A. S.*, vol. v., p. 196. Here too it is worth while noting that in 1798 Dr. Hornsby, reflecting on the well-marked proper motion of Castor, and the fact that the distance of the components had not changed for twenty years, drew the inference that both stars were moving with

* See his *Gruendliche Vertheidigung*, etc., 1777.

the same velocity and in the same direction, but quite failed to see that these facts supplied unequivocal evidence of physical connexion.[*]

In this way Sir William Herschel detected about fifty binaries. Since his time the list has been largely extended, and the researches of Struve, Mädler, and others brought the number up to about six hundred.

In the paper above referred to, rough guesses at the periods of revolution of some of the binaries were made by Herschel; *e.g.*, he assigned a period of about 342 years to Castor. It was reserved, however, for his distinguished son, Sir John, to grapple successfully with the interesting problem of finding by a graphical method the orbit which one star describes relatively to the other. If S represent the principal star, to which the motion of the companion is

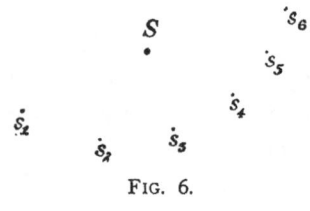

FIG. 6.

referred, and if at successive epochs the positions of the latter have been observed to be as in the figure, S_1, S_2, S_3, S_4, S_5, ... it is plain that, assuming that the observations are sufficiently numerous and accurate, a curve can be drawn through them which will represent the orbit. The positions thus marked down will not always form part of an ellipse; they may lie in a straight line. For instance, the charted positions of the companions of Vega, Σ 1263, and Σ 1516, appear to be well represented by straight lines; while γ Virginis, Castor, ξ Ursæ Majoris, certainly move in elliptic orbits. It is possible, too, that the path may be some other curve, the knowledge of which will in its turn throw light on the forces and conditions which obtain in these sidereal systems.

To describe a method by which the elements of the orbit of a binary star may be obtained without the aid of the higher departments of mathematics, is the object of the present section.

[*] See Grant's *History of Physical Astronomy*, p. 559.

Statement of the Problem.

From the observations of angle and distance at given epochs, to draw the apparent orbit which one star describes relatively to the other, and thence to determine the elements of the true orbit, and to construct an ephemeris.

The first part of the problem consists, then, in a careful study of the observations to determine their relative value, and in so arranging them as to obtain the *apparent* orbit. A little explanation will here be necessary. The orbit, or portion of it seen by us, is the *apparent* orbit: it is the *projection* on the background of the heavens of the *true* orbit, *i.e.*, the projection of the true orbit on a plane at right angles to the line of sight. Suppose, for example, the plane of the true orbit to be at right angles to the line of sight, then will the revolving star be seen to describe an elliptic path round the primary star in the focus, and the true and apparent orbits will coincide. If the plane of the orbit pass through the earth, and present its edge to the observer, the revolving star will appear to recede from, approach, occult, or be occulted by, and again recede from, the star in the focus of the ellipse. The plane of the orbit, again, may be but a little inclined to the line of sight, and then the companion will appear to pass a little below and above the principal star. In one word, the plane may have any inclination to the visual ray, and the projection will present corresponding phenomena. Hence, a circular or elliptic orbit, if its plane were oblique to the line of sight, would be projected into an ellipse; if the plane passed through the earth, the projection would be a straight line; and an elliptic orbit might be so situated as to have a circle for its projection.

The history of binary stars already furnishes us with illustrations on this point. Take the star ζ Herculis, discovered to be double by Herschel in July 1782. On looking at this object in October 1795, it was still seen double. Soon after the companion disappeared. During 1821, 1822, 1823, and

1825, the utmost endeavours of Herschel and Struve failed to elongate it. Encke caught it double in 1826. Of this phenomenon Herschel says: " My observations of this star furnish us with a phenomenon which is new in astronomy; it is the occultation of one star by another." Here then is an example of the orbital plane being in the line of sight. The period of this binary is about thirty-five years.

Once more : γ Virginis is already a famous binary. It was known as a double star in the seventeenth century. Herschel found the distance 5"·7 in 1780; in 1831 it was 2"·0. In 1836 Herschel wrote: "γ Virginis, at this time, is to all appearance a single star." About 1837 it again separated, and the distance is now nearly 5".

42 Comæ is a fine example of a binary, the plane of whose orbit coincides with the visual ray.*

Perhaps the accompanying figure will help to render this

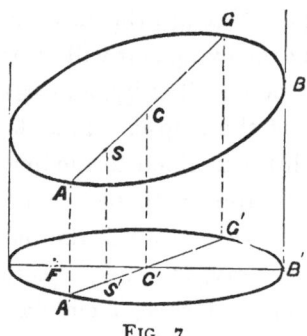

quite clear. Let C C' be the direction of the line of sight, A B G the real ellipse whose focus is S and centre C. Then will its projection on a plane at right angles to the line of sight be the ellipse A' B' G'. And it will be observed that S', the projection of S, does not coincide with the focus F. The principal star, therefore, will not in

FIG. 7.

general occupy the focus of the apparent ellipse, but will be displaced into some other position.

In many binary stars the observations do not yet extend over a sufficiently long period to enable us to compute any

* Examples.—In Σ 186, 1967, 2737 (A B) the plane of the apparent orbit coincides very nearly with the visual ray. The apparent orbit is nearly circular in Σ 1037, 1126, and λ Ophiuchi ; the orbit is extremely elongated in Σ 1516 (A C), 1909, and 2822. In Σ 1348, either the position of A in the apparent orbit is very eccentric, or the plane of the orbit is greatly inclined to the visual ray.

satisfactory orbit. In some, the portion of the orbit traversed since observations were commenced does not include any of the critical points; while in yet other cases complete revolutions have been made since the date of the discovery of the stars. Of this last class, ξ Ursæ Majoris, period about sixty-one years, η Coronæ Borealis, period about forty-two years, and ζ Herculis, period about thirty-four years, may be given as examples.*

The next part of the problem consists in determining the *real* from the *apparent* orbit, and the position occupied in the apparent orbit by the principal star. And when all this has been done satisfactorily we are in a position to put our orbit to the test by the construction of an ephemeris, *i.e.*, a series of computed positions and distances for the epochs of past and future observations. And if the computed quantities fairly agree with the measures made in past years, we must then proceed to compute positions and distances for future years at intervals of from a quarter of a year in the case of stars having rapid motion, to five or ten years in cases where the period extends over centuries.

That it is quite possible, however, for an ephemeris to represent all past observations in a satisfactory way, and yet to fail completely when it comes to be compared with future measures, will be evident on a little reflection. The subjoined table, however, will bring out the fact very clearly :—

Epoch.	Position.		Observer.	No. of Nights.
	Computed.	Observed.		
1848·0	239·0	249·16	Dawes.	7
1850·0	234·4		,,	
1852·0	227·3	246·39		1
1854·0	212·6	246·21	,,	7
1855·0	195·5			
1856·0	164·4	245·44	Dembowski.	7

The computed places are from an ephemeris for Castor con-

---

* In OΣ 208 and 298 it is probable that we shall soon be in a position to attempt the computation of the elements of the orbits.

structed by Sir John Herschel from an orbit which he published in 1832. The observations used by him extended from 1719 to 1831. The observed places are put by the side for comparison.

The small number of observations at the disposal of the computer, and the very small portion of the orbit dealt with, must, of course, be here remembered. Yet this orbit represented the previous measures very fairly indeed.

Even when a star has been measured by skilful observers during more than an entire revolution, it is not always an easy matter to obtain elements which will furnish materials for a good ephemeris. Take ξ Ursæ Majoris as an example. Its duplicity was discovered by Sir William Herschel in 1780; the companion was then not far from its apastron ; the periastron was reached in 1816, and again in 1876, and hence its period is about sixty years. Now in 1872 Dr. Ball gave a set of elements, and an ephemeris furnishing positions up to 1878·75. The subjoined table will show how far the predicted positions agree with recent measures.

Epoch.	Position.		
	Computed.	Measured.	
1872·50	22·4	19·39	in 1872·32, by Dembowski.
1872·75	17·5		
1873·00	12·5		
1873·25	7·3	358·91	in 1873·33, ,, ,,
1873·50	2·2		
1873·75	357·2		
1874·00	352·1		
1874·25	347·0	333·63	in 1874·35, ,, ,,
1874·50	342·2		
1874·75	337·6		
1875·00	333·2		
1875·25	329·0	317·56	in 1875·27, ,, ,,
1875·50	325·1		
1875·75	321·4		
1876·25	314·7	304·8	in 1876·30, ,, ,,
1877·25	303·2	294·9	in 1877·26, ,, ,,
1878·00	296·1		
1878·25	293·9		
1878·50	291·8	85·5	in 1878·45, by Wilson.

The agreement here is of course not satisfactory.

Methods of Solution adopted.

The first part of the problem—that is, the determination of the most probable *apparent* orbit—may be best solved by the methods given by Sir John Herschel ('Memoirs of the Royal Astronomical Society,' vols. v. and xviii.), with some slight additions. We shall give a brief explanation of it, but the method will be best understood by working through an example.

To pass from the apparent to the real orbit is a geometrical problem of considerable difficulty. Fine analytical solutions of it have been given by Savary ('Connaissance des Temps pour l'an 1830 et 1832'), Sir John Herschel ('Memoirs of the Royal Astronomical Society,' vol. v.), Encke ('Ueber die Berechnung der Bahnen der Doppelsterne, Berliner Astr. Jahrbuch für 1832'), Villarceau ('Méthode pour calculer les orbites relatives des étoiles doubles.' 'Connaissance des Temps pour l'an 1852 et 1877'), and Klinkerfues ('Ueber eine neue Methode die Bahnen der Doppelsterne zu berechnen.' Gottingen, 1855). Purely geometrical solutions have been given by Thiele ('Ast. Nachrichten,' No. 1227, vol. lii.), and by the writer ('Monthly Notices,' vol. xxxiii., p. 375). Of these, Thiele's is by far the most elegant, and it is the one we shall here adopt. The construction of an ephemeris, and the comparison of the observed with the calculated places, is essential for the completion of the problem. This will be effected in the present paper by a graphical method.

It must, however, be understood that the graphical method is only introductory, and that subsequent analytical methods are necessary in order to correct the elements, and attain the highest degree of accuracy that the observations permit of.

To Prepare the Observations for Use.

Some of the earlier measures of position and distance have

to be deduced from the differences of right ascension and

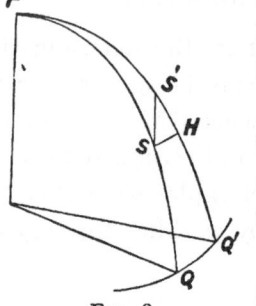

FIG. 8.

declination observed by Bradley, Piazzi, Lalande, and others. The process is as follows :—

Let S, S' be the two stars, whose right ascension and declination are a, δ, and $a + \Delta a$, $\delta + \Delta\delta$ respectively, S being the principal star.

Let Δa, $\Delta\delta$ be expressed in seconds of arc: then if P is the pole, P S Q, P S' Q' declination circles meeting the equator in Q, Q', and S H is an arc of a small circle parallel to Q Q', Q Q' = Δa, S' H = $\Delta\delta$.

Let P S S' = θ, S S' = ρ, the position and distance required to be calculated from Δa, $\Delta\delta$ observed.

Then $\tan \theta = \tan SS'H = \frac{SH}{S'H} = \frac{\Delta a \cos \delta}{\Delta \delta}$ (i.),

and $\rho = \Delta\delta \sec. \theta$,   (ii.),

from which θ and ρ may be obtained.

It must be observed that since θ is always measured in the direction n, f, s, p, if Δa is positive, that value of θ between $0°$ and $180°$, which satisfies (i.) must be chosen ; and if Δa is negative, the value between $180°$ and $360°$. ρ is always positive.

Further, if Δa, $\Delta\delta$ are observed with assigned limits of error, it is advisable to ascertain what are the corresponding limits of error in θ and ρ, by substituting in succession those values of Δa, $\Delta\delta$ which give the greatest and least values to θ and ρ.

Throughout the whole of the working of this problem it is advisable to have angles expressed in degrees and decimals of a degree.

REDUCTION TO A SELECTED EPOCH.

In all cases before observations made at different times can be combined, the effect of precession on the angle of position

must be eliminated. For it must be remembered that angles
of position are measured from the great circle which passes
through the star and the pole, and that in consequence of
precession the pole is constantly shifting its place, having a
slow retrograde motion round the pole of the ecliptic. Hence
the position that this circle occupies *at some selected epoch* must
be taken as the zero of position, and all observations must be
referred to it.

The subjoined figure will show how the effect of this motion
of the pole on the angle of position of any star can be computed.

Let E be the pole of the ecliptic; P, P′ positions of the
pole at an interval of a year; ♈ the intersection of equator
and ecliptic, from which the right ascension is reckoned;
S S′ a double star in right ascension $a°$ and declination $\delta°$.
Draw the circles P S, P′ S; then P S P′ is the $\varDelta\,\theta$ required.

Since ♈ E, ♈ P are quadrants, ♈ P E is a right angle, and
therefore P′ lies on ♈ P. Draw P′ p perpendicular to S P.

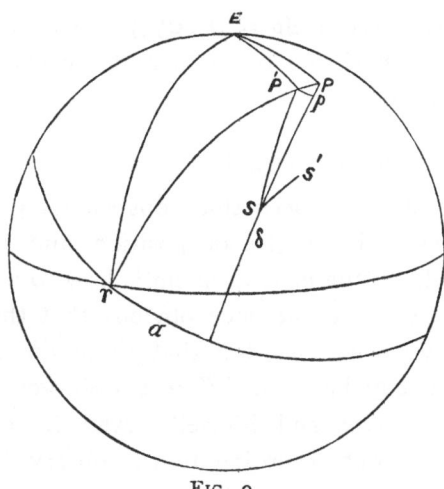

FIG. 9.

Then P P′ is known from the constant of precession to be for
the year 1850, and very approximately for any other year,
$20''{\cdot}0564 = 0°{\cdot}0055712$;

and $P'p = P\,P' \sin P'\,P\,p = P\,P' \sin a$;

also $P'p = \varDelta\,\theta \cos \delta$, as in the last section.

$$\therefore \quad \varDelta\,\theta \cos \delta = P'\,P \sin a\ ;$$

and $\varDelta\,\theta = 0°{\cdot}0055712 \sin a \sec \delta.$

The exact formula is $\{20''{\cdot}0564 - 0''{\cdot}000097(t - 1850)\} \sin a$ sec δ.

It appears further that the effect of precession is to increase the angle of position in the case chosen. Hence, in order to bring up to a certain date old observations of position taken t years before that date, we must add to those angles of position the quantity $0°{\cdot}0055712 \sin a \sec \delta \times t$. It is plain that this will be $+$ for values of a from $0°$ to $180°$, or from 0^h to 12^h and $-$ for values from $180°$ to $360°$, or from 12^h to 24^h.

EXAMPLE.—Dawes in the year $1831{\cdot}34$ observed the angle of position of η Coronæ Borealis in right ascension $15^\mathrm{h}\ 18^\mathrm{m}\ 14^\mathrm{s}$, and declination $30°\ 43'\ 31''$, to be $50°{\cdot}46$. Reduce this to the epoch 1880.

Converting the right ascension into degrees, it becomes $229°\ 35'\ 30''$. Hence $\varDelta\,\theta = 0°{\cdot}0055712 \sin 229°\ 35'\ 30'' \sec 30°\ 43'\ 31'' \times 48{\cdot}66 = -0°{\cdot}2292$; and the corrected angle is therefore $50°{\cdot}23$.

DRAWING OF THE INTERPOLATING CURVE.

When a table has been thus constructed, giving, for some selected epoch, the angles of position and distances at a number of dates, the next problem is how to use this mass of materials. It will be at once obvious that the observations are not very harmonious, but that there are serious discrepancies not only between different observers, but between the same observer and himself. And if the points were simply charted out according to the observed positions and distances, they would not lie on a curve, but on a broad irregular band.

Sir John Herschel was the first to suggest (*Mem. R. A. S.,* vol. v.) a graphical method of obtaining the positions at any selected epochs with a high degree of accuracy; a method which necessarily gives no weight to exceptionally bad observations, and makes use of all the good observations, both before and after any epoch, to determine the angle at that epoch.

Take a sheet of paper ruled in fine squares,—that called millimetre paper * is the best,—and let the divisions running horizontally, suppose, represent angles, each division standing for a tenth of a degree, and the divisions running vertically represent years, each division standing for a tenth of a year.

On this convention a dot on the chart represents a single observation.

The subjoined chart therefore represents the following table of observations,

$t.$	$\theta.$
1870·23	210·05
1870·38	209·95
1870·40	210·30
1870·25	212·38
1871·08	211·10
1871·34	212·08
1871·41	212·08
1872·10	214·62
1872·19	214·44
1872·20	213·21

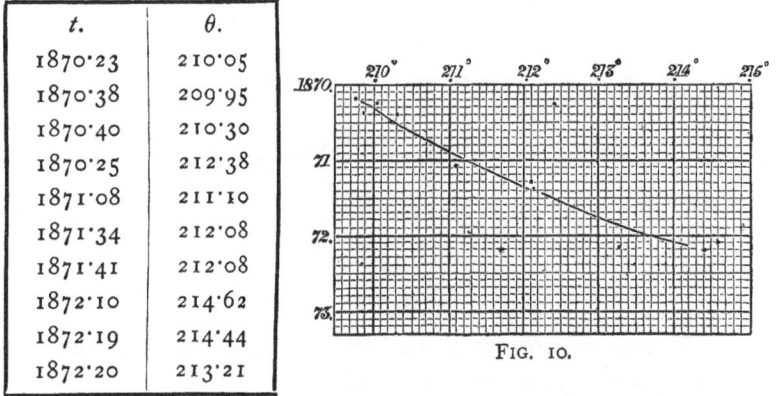

Fig. 10.

and the curve drawn among them cannot be very far from the truth, and is influenced by all the observations except the two outlying ones, which are obviously bad.

By this means we can obtain more accurate estimates of what the angle would be at any assigned date, or what is more used, of the date at which the angle would have an assigned value, than we can from the observations directly.

For example, from the diagram we see that in 1870·00 the

* Millimetre paper may be got at Messrs. Williams and Norgate's.

angle would have been 209·5, and that the angle was 213 at the time 1871·71.

All the measures, therefore, of position of the star must be charted, and the 'interpolating curve,' as Herschel calls it, must be drawn among them. This is a matter of the highest importance. The curve must be smooth and flowing. It may have points of contrary flexure, but it can have no abrupt changes of curvature.[*]

When the curve has been drawn, note the time indicated by it at which the angle had in succession a series of values, proceeding by some common difference, say of 2°, or of 5°, and construct a first table of interpolated angles and dates.

Let the subjoined table be a specimen :

$\theta°$.	t.
70	1837·34
75	1839·90
80	1842·12
85	1844·25
90	1846·08
95	1847·74

SMOOTHING THE CURVE.

The next process is to 'smooth' the curve by an arithmetical examination of this table.

Let Δt represent the result of subtracting any one number in column 2 from the number below it, and let the series of numbers so obtained be arranged in a column to the right of the column of t.

Similarly, let $\Delta^2 t$ be the differences between the numbers in the column of Δt, and be placed in a column to the right; and $\Delta^3 t$ be the differences of $\Delta^2 t$.

---

[*] A point of contrary flexure indicates a point where the line drawn to the principal star is normal to the apparent orbit of the star.

The table will then be as follows :

$\theta°$.	t.	Δt.	$\Delta^2 t$.	$\Delta^3 t$.
70	1837·34			
		2·56		
75	1839·90		− ·34	
		2·22		+ ·25
80	1842·12		− ·09	
		2·13		− ·21
85	1844·25		− ·30	
		1·83		+ ·13
90	1846·08		− ·17	
		1·66		
95	1847·74			

It is plain from this that the numbers are not quite right, that is, that the curve has not been drawn quite smoothly, or that some of the values of t have not been quite correctly estimated. For if they were, then the differences in each column ought to proceed regularly, and not show irregular and abrupt changes, as this series does, in the second and third differences.

It is necessary, therefore, to make slight changes in the second column such as will bring the difference columns into more perfect adjustment. To do this is not very easy, and requires patience. The following considerations may help in the process. The column of $\Delta^3 t$ is on the whole +, and therefore the column of $\Delta^2 t$ ought to have its terms, which are negative, continually decreasing in absolute magnitude. The ·09 is therefore, too small, and the ·30 too large. These can be changed in the right direction by increasing the 2·22 or diminishing the 2·13, and these in their turn make changes in the first column.

After successive attempts, we obtain the following result:

$\theta°$	t.	Δt.	$\Delta^2 t$.	$\Delta^3 t$.
70	1837·35			
		2·52		
75	1839·87		− ·23	
		2·29		·01
80	1842·16		− ·22	
		2·07		·01
85	1844·23		− ·21	
		1·86		·00
90	1846·09		− ·21	
		1·65		
95	1847·74			

7

By comparing this with the previous table, it will be seen that none of the dates have been altered more than ·04 of a year, which would be represented on the chart by an almost imperceptible space.

The values of t so obtained may therefore be regarded as a still closer approximation to the truth than those obtained directly from the graphical process, and *à fortiori* than those obtained by direct observation. All small errors arising from imperfect drawing of the curve, or wrong estimation of the decimals, have been got rid of. But it must not be forgotten that these values are still liable to be affected by serious errors of judgment in drawing the curve, or by errors of single observations when the curve depends on single observations. The curve may be smooth, and yet not the right curve. Errors of this kind cannot be detected at the present stage of the problem, but will be revealed later on.

EMPLOYMENT OF MEASURES OF DISTANCE.

In a precisely similar manner all the measures of distance should be charted on millimetre paper, and interpolated distances obtained, at equal intervals of time, and the distance curve 'smoothed.' The errors in observation of distance often bear a large ratio to the distance itself, and the interpolated distances are far more trustworthy than any individual measures.

If now a series of corresponding values of r, θ is found, and charted, these points will give a general indication of the nature of the curve. They will, for example, indicate whether the orbit is likely to be rectilinear or elliptical, and whether a sufficient portion of it has been described to make it worth while to attempt the computation. But in many cases it will be found that the points so obtained do not lie tolerably well on a curve, and that there will be liability to large error in attempting to draw a curve among them. This arises from the almost unavoidable error in the measurement of distances. Sir John Herschel, therefore, devised a method by which the *relative* distances could be obtained from the measurements of position alone, and this we now proceed to describe.

Determination of distance from the interpolating curve for angles of position.

If A C E is an ellipse, S the focus, it follows from Kepler's second law that equal areas are described in equal times, that the rate of change of angular position is much more rapid in some part of the orbit than in others.

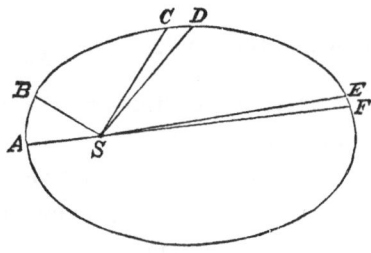

FIG. 11.

Let A S B, C S D, E S F, be equal areas ; then they would be described in the same time, and hence the change of position angle in that time would be A S B in one part of the orbit, C S D in another, and E S F in a third. And conversely, if the change of angle is greater at one part of the orbit than at another, it follows that the distance must be less, and less to such an amount as to make the areas described in equal times equal.

If r be the distance at any time, $\Delta \theta$ the small angle described in the time Δt, it follows that $\frac{1}{2} r^2 \Delta \theta$ is the area described in that time ; and therefore that the limit of $r^2 \frac{\Delta \theta}{\Delta t}$ must be constant at all parts of the orbit ; and therefore that r^2 varies as limit of $\frac{\Delta t}{\Delta \theta}$.

But from the table given above (p. 96) $\Delta \theta$ is constant, and Δt can be got by subtraction, and the limit of $\frac{\Delta t}{\Delta \theta}$ may be got either from the formula

$$\frac{\Delta t}{\Delta \theta} = \frac{1}{\Delta \theta} \left(\frac{\Delta t}{1} - \frac{\Delta t}{2} + \frac{\Delta^3 t}{3} - \cdots \right)$$

or, very approximately, by taking half the sum of the differences of the times that precede and follow the date seiected.

For example, referring to the previous table, $\Delta\theta = 5°$, and when $\theta = 80°$, by the first formula

$$r^2 \propto \frac{1}{5}\left(\frac{2\cdot07}{1} + \frac{\cdot21}{2} + \frac{\cdot00}{3}\right) = \cdot435,$$

and by the second formula

$$r^2 \propto \frac{\frac{1}{2}(2\cdot29 + 2\cdot07)}{5} = \cdot436;$$

when the angle is $70°$, $\frac{\Delta t}{\Delta\theta} = \frac{1}{5}\left(\frac{2\cdot52}{1} + \frac{\cdot23}{2} + \frac{\cdot01}{3}\right) = \cdot527$, and

therefore the values of r at $80°$ and $70°$ are as $\sqrt{436} : \sqrt{527}$, or as $2088 : 2295$.

In this manner relative values of r are obtained for all the values of θ in the previous table, at intervals of $5°$ or $10°$, and these will be in general more accurate than those obtained from direct measurement, as they depend on measures of position alone.

In order to compare with seconds of arc the unknown unit in terms of which these values of r are expressed, it will be necessary to take the whole series of values of r obtained in seconds at suitable points from its own interpolating curve, and the whole series obtained in the unknown scale from the formula above given, and compare the sums of the two series. Thus will be obtained the relative value of the two units to a high degree of approximation. Take the following values as an illustration :—

t.	r.	r' on scale.
1830	4·50	125·00
1835	4·62	127·60
1840	4·75	131·40
Sums ...	13·87	384·00

Here $13''\cdot87$ are equal to $384\cdot0$ scale divisions, and therefore 1 scale division corresponds to $0''\cdot03612$.

It will further be worth while to reduce to seconds each of the values of r, and chart them along with the interpolating curve which furnished the direct values of r, in order to see how far the calculated and observed values agree. A discrepancy, systematically recurring between them, may lead, as in Otto Struve's recent investigation of the orbit of the distant companion of ζ Cancri, to some novel and remarkable conclusions. (See *Observations de Poulkova*, vol. ix., and the *Comptes Rendus de l'Académie de Paris*, vol. lxxix., p. 1463.)

To Draw the Apparent Ellipse.

It may now be assumed that we have the values of r for a series of values of θ differing by 5°. Let these be converted into x and y by the formulæ $x = r \cos \theta$, $y = r \sin \theta$, and the points charted on the millimetre paper. They will be found to lie on a curve; and if a sufficient portion of the orbit has been described, the curve will be sensibly an ellipse. And here it may be observed that these points furnish the best possible test of the skill with which our final interpolating curve has been drawn; *for if any point or points lie out of the curve we must at once redraw that part of the interpolating curve.* Assuming that the correction has been made, the ellipse passing through the points may now be found either by the graphical or analytical methods. If the former be adopted, an ellipsograph, or a piece of string and two drawing pins, with a little patience, will suffice for this purpose. The line once drawn in pencil should be carefully inked in with a *fine* pen. This is the *apparent ellipse.* No care must here be spared in drawing the best possible ellipse, and drawing a fine line. With a pair of compasses we may now at once measure off the *maximum* and *minimum* apparent distances, and obtain directly the angles at which they occur. The larger star A occupies the *projected* focus of the *real* ellipse.

Determination of the Real Ellipse: Thiele's Method.

We must next proceed to the method of determining the *real* ellipse from the *apparent* one, and in doing this we shall follow Thiele's method, and give a geometrical proof of the elegant theorem he employs.

The problem is this :—Given an ellipse and a point in it which is not the focus, it is required to find the position and magnitude of the ellipse whose projection is the given ellipse, and the projection of its focus the given point.

The determination of the position and magnitude of the ellipse requires the determination of five elements, viz.,

(1) The angle ☊ that the line of intersection of the two planes, or line of nodes, makes with a fixed line.

(2) i the angle of inclination of the planes.

(3) e the eccentricity of the ellipse.

(4) a the semi-axis major of the ellipse.

(5) λ the angle between the line of nodes and the line of apsides, or the line to periastre.

The solution depends on the following geometrical property of the ellipse :—

Let P S Q be any focal chord of an ellipse ; M X N the

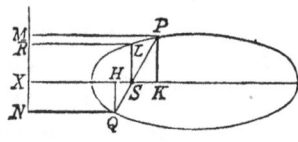

FIG. 12.

corresponding directrix; P M, Q N perpendiculars to the directrix ; P K, Q H perpendiculars to the axis major ; S L the semi latus rectum, and L R perpendicular to the directrix. Then, by similar triangles, H S : S K :: S Q : S P, and by the property of the ellipse S Q : S P :: Q N : P M ;

therefore Q N : P M :: H S : S K

:: L R − Q N : P M − L R ;

that is, Q N, L R, P M are in harmonic progression ; but Q N, L R, P M are respectively proportional to S Q, S L, S P ; therefore the harmonic mean of S Q and S P is constant.

And if along the chord P S Q a point Y be taken, so that S Y is the harmonic mean between S P, S Q, the locus of Y would be a circle of which S would be the centre, and S L the radius.

If now the ellipse and this harmonic circle (as it may be called) be projected on a plane inclined to their own, the circle will be projected into an ellipse, the direction of whose major axis gives the line of intersection of the two planes, and the ratio of whose semi-axes is the cosine of the inclination of the planes.

Conversely, if the harmonic ellipse be drawn, by taking, arithmetically or graphically, the harmonic means between the segments of a number of chords through the projected focus in the apparent ellipse, it follows that its major axis is equal to the latus rectum of the true ellipse; that its major axis is in the direction of the line of nodes; and that the ratio of its minor to its major axis is the cosine of the angle of inclination of the plane of the real ellipse to the plane of the apparent ellipse.

Further, if C is the centre (Fig. 13), S C A' is the projection of the major axis; and $\frac{CS}{CA} = e$, the eccentricity of the real ellipse, this ratio being unaltered by projection. Hence we find in succession ☊, i.e. the angle which the line of nodes makes with the axis of x, the meridian through the star; i, the inclination, from the condition $\cos i = \frac{Sb}{Sa}$, Sa and Sb being the major and minor axes of the harmonic ellipse; $e = \frac{CS}{AC}$; and $a = \frac{L}{1-e^2} = \frac{Sa}{1-e}$, L being the semi latus rectum.

Finally, λ, i.e. the angle the line to the periastron makes with the line of nodes, is found as follows :—

Let λ' be the angle X S C, $\lambda' - \Omega$ the angle A'S ☊ in the annexed figure where A' is the projected periastron, and therefore known. λ is the angle A S ☊ which is required.

Draw A' N, A N perpendicular to S ☊.

Then $\tan \lambda = \frac{AN}{SN} = \frac{AN}{A'N}$

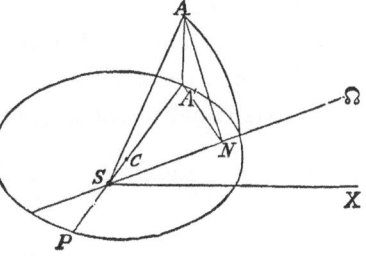

FIG. 13.

$\times \frac{'N}{SN} = \sec i \tan (\lambda - \text{☊})$, and therefore λ is known.

To construct the ephemeris graphically, it is necessary to divide the ellipse into equal sectorial areas by radii drawn from the focus. This may be accomplished as follows :—

Let A P A' be an ellipse (Fig. 14), P any point in it, S the focus, C the centre; A Q A' the auxiliary circle, Q P N an ordinate through P.

Let e be the eccentricity, a, b the semi-axes of the ellipse, T the periodic time for the whole orbit.   Then if t be the time taken in describing the area A S P from perihelion to the point P,

$$\frac{t}{T} = \frac{ASP}{\pi a b} = \frac{ASQ}{\pi a^2} = \frac{ACQ - SCQ}{\pi a^2};$$

and therefore if u is the circular measure of A C Q,

$$\frac{t}{T} = \frac{\frac{1}{2} u a^2 - \frac{1}{2} a e a \sin u}{\pi a^2} = \frac{u - e \sin u}{2 \pi}.$$

In order, therefore, to divide the area by focal radii into equal

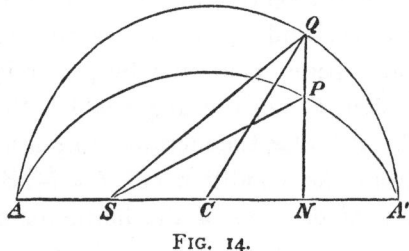

FIG. 14.

intervals, values must be given to $u - e \sin u$ in arithmetical progression.

Let A B A' be the semicircle described on the major axis of the ellipse as diameter, S the focus of the ellipse.

Divide the arc B A' into any number of equal parts, say of 10° each.   Draw the tangent at B, and mark off along it from B parts equal to the arcs of 10°, 20°, . . . 90°.

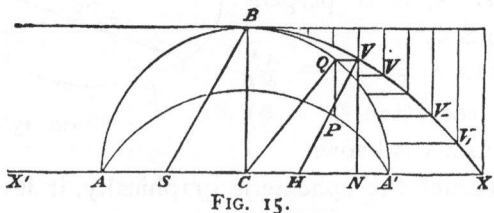

FIG. 15.

Through the points of division of the arc draw lines parallel to C X, and through the points of division of the tangent at B draw lines parallel to C B, thus determining a number of points $V_1 V_2$ . . ., and through these points draw a curve B V X. We will call this the ephemeris curve.

If now P is any point on the ellipse, Q the corresponding point on the auxiliary circle, V the corresponding point on the ephemeris curve, Q V being parallel to C X, C X′ equal to C X, A C Q = u. Then if V N is parallel to C B, X′ N = au. Join S B, draw V H parallel to S B, and join C Q. Since $\frac{HN}{VN} = \frac{SC}{CB} = e$, and $\frac{VN}{CQ} = \sin u$, therefore H N = $ae \sin u$, and therefore X′ H = $a(u - e \sin u)$.

Hence $2\frac{X'H}{X'X} = \frac{t}{T}$, and therefore the position P in the orbit can be at once found corresponding to any time t, and conversely the time t can be found corresponding to any position P in the orbit, by simply drawing parallel lines.*

Lastly, this method can be adapted to the further problem of dividing an ellipse into equal areas by lines from any point which is not the focus. To do this, instead of the auxiliary circle, an auxiliary *ellipse* must be taken, which will be similar and similarly situated to Thiele's harmonic ellipse.

The working of this will be readily understood from the example annexed.

* This problem can also be approximately solved with equal accuracy by *mechanical* means. The latest and best method is that given by Professor Bruhns in the *Vierteljahrschrift der Astronomischen Gesellschaft* 1875, Heft. 4. For an improved form of this apparatus, also by Professor Bruhns, see Heft. 4, 1877. Dr. Doberck, however, prefers to use the tables he has published in the *Ast· Nachrichten.*

CHAPTER II.

EXAMPLE OF AN ORBIT WORKED BY A GRAPHICAL
METHOD.

FOR this method we shall select Castor, as a double star of great historical interest, and sufficiently brilliant and widely separated to be within the reach of all telescopes that are likely to be used by amateurs. The orbit has been frequently computed before, both by graphical and analytical methods, and a comparison of the results arrived at is very instructive as showing the difficulty and uncertainty in problems of this nature, when the portion of the orbit described bears a small ratio to the whole.

Table I. gives in chronological order the observations arranged as follows. In column 1, headed t, is the date of the observation; in column 2, the observed angle, headed θ'; in column 3, the angle corrected for precession up to the year 1880, headed θ; in column 4, the number of nights of observation, an important element in estimating the weight to be assigned to an observation; in column 5, headed r, the observed distance; in column 6, the number of nights; and, lastly, in column 7 the initials of the observer.*

----

* For explanation of the initials see Part III.

Table I.—Castor. Angles and Distances : Angles reduced to 1880.
R. A. 1880.   7ʰ 26ᵐ 57ˢ = 111° 44′ 15″.
Dec.   32° 9′ 10″·
Correction = 0°·0055 sin α sec δ per annum = 0°·006.

t	θ'	θ	No. of Nights.	r	Observer
1719·84	355·88	356·85	...	...	Br. and P.
1759·80	323·78	324·50	...	...	Br. and M
1779·85	302·78	303·40	...	...	H₁.
1780·43	...	...	...	5·29	,,
1783·46	293·05	293·64	...	...	,,
1791·15	292·95	293·50			,,
1792·16	297·27	297·81			,,
1795·95	283·88	284·40			,,
1800·27	284·32	284·81			,,
1802·08	282·77	283·25			,,
1803·19	280·55	281·03			,,
1814·83	272·87	273·27	...	...	Σ.
1816·97	270·00	270·39	...	...	H₂.
1819·10	269·60	269·97	...	5·48	Σ.
1820·34	268·99	269·35	...	...	,,
1821·21	267·12	267·47	...	...	H₂ and So.
1822·01	266·81	267·16	...	...	Σ.
1822·10	...	...	...	5·36	H₂ and So.
1823·11	264·98	265·32	...	...	,,
1823·32	...	..	...·	4·71	Σ.
1825·24	263·30	263·63	...	4·77	So.
1826·22	262·54	262·87	5	4·40	Σ.
1827·28	262·32	262·64	4	4·42	,,
1828·69	261·87	262·18	...	4·64	H₂.
1828·89	261·10	261·41	...	4·36	Σ.
1829·88	260·97	261·28	...	4·52	H₂.
1830·52	259·02	259·32	...	4·68	,,
1831·06	259·38	259·68	...	4·73	Be.
1831·11	259·62	259·92	...	5·16	H₂.
1831·22	258·15	258·45	...	4·57	Da.
1831·31	259·58	259·88	...	4·46	Σ.
1831·91	259·35	259·64	8	4·74	H₂.
1832·12	258·42	258·71	14	4·71	Da.
1832·86	257·72	258·01	...	4·525	Σ.
1833·10	256·73	257·01	...	4·89	H₂.
1835·33	255·48	255·75	...	4·73	En. and Ga.
1836·88	256·12	256·38	...	5·28	Σ.
1838·34	254·40	254·65	...	4·81	Ga.
1839·35	253·73	253·98	...	5·20	Ka.
1840·06	253·97	254·21	...	4·71	Da.
1840·18	254·10	254·34	..	4·94	Mä.
1841·11	252·82	253·06	...	4·89	Da.
1842·25	252·38	252·61	...	4·91	,,
1843·15	251·71	251·94	...	4·87	Hi.
1845·93	249·80	250·01	...	...	Ja.
1846·34	250·38	250·58	...	5·89	Hi.
1846·73	249·46	249·66	4	...	Da.
1847·25	249·85	250·05	5	5·014	,,
1848·18	249·20	250·39	9	5·008	W. C. B.
1848·28	249·54	249·73	2	5·20	Da.
1849·32	248·97	249·16	4	5·027	Ft.

TABLE I.—*continued.*

t	θ'	θ	No. of Nights.	r	Observer.
1851·04	248·67	248·85	6	5·074	Da.
1851·21	248·11	248·29	10	5·068	Σ.
1851·88	247·65	247·82	...	5·044	Mi.
1852·04	247·97	248·14	6	5·075	Ft.
1852·20	246·39	246·56	1	5·070	Da.
1852·50	246·12	246·29	14	4·821	Mä.
1853·05	247·32	247·49	3	5·083	Ja.
1853·13	245·87	246·03	3	5·157	Da.
1853·34	246·26	246·42	9	4·931	Mä.
1854·23	246·21	246·37	7	5·098	Da.
1854·38	244·72	244·87	18	4·945	Mä.
1854·87	245·49	245·64	23	5·442	De.
1855·31	243·61	243·76	3	4·848	Mä.
1855·82	245·13	245·28	7	5·368	Se.
1856·20	245 44	245·58	7	5·145	De.
1856·35	243·78	243·92	6	4·875	Mä.
1856·73	245·51	245·65	4	5·172	Ja.
1857·34	244·25	244·39	4	5·382	Da.
1857·36	242·90	243·04	7	4·888	Mä.
1857·77	245·19	245·32	3	5·336	Ja.
1858·26	244·42	244·55	2	5·208	Mo.
1858·37	244·13	244·26	7	4·963	Mä.
1859·26	243·88	244·01	2	5·156	Mo.
1859·36	242·70	242·82	11	5·081	Mä.
1859·98	243·62	243·74	2	5·378	Mo.
1860·22	242·77	242·89	3	5·395	Da.
1863·02	242·75	242·87	11	5·537	Ro.
1863·03	241·66	241·78	14	5·381	De.
1864·60	241·53	241·88	10	5·59	Da.
1866·02	241·07	241·15	14	5·384	De.
1870·32	239·7	239·76	1	5·57	Gl.
1870·68	239·34	239·40	5	5·488	De.
1871·59	237·9	237·95	2	5·64	Gl.
1872·00	236·4	236·45	1	5·73	,,
1872·39	237·8	237·85	2	5·9	W. and S.
1873·24	237·9	237·94	1	5·6	,,
1873·29	236·3	236·34	1	5·62	Gl.
1873·78	236·92	236·96	8	5·557	De.
1874·10	236·6	236 63	7	5·7	Gl.
1874·13	236·9	236·93	2	5·6	W. and S.
1875·66	236·2	236·22	15	5·5	Gl.

These angles and distances are all charted on the millimetre paper as before described, and the result is shown in Plate I., in which each dot corresponds to an observation. A curve is then drawn as smoothly as may be among the points of observation. The first curve that was so drawn had to be abandoned, but the points at which it crossed the principal lines are shown by fine lines, which are in fact portions of the curve.

The first table of interpolated angles and epochs was as follows :—

TABLE II.—First Table of Interpolated Angles and Epochs.

θ	t	Δt	$\dfrac{\Delta t}{\Delta \theta}$	$r = 100 \times \sqrt{\dfrac{\Delta t}{\Delta}}$	x	y
355	1723·8					
350	1729·7	5·9 ⎫ 5·7 ⎭	1·16	107·7	106·00	18·70
345	1735·4					
340	1740·9	5·5 ⎫ 5·3 ⎭	1·08	103·7	97·65	35·54
335	1746·2					
330	1751·4	5·2 ⎫ 5·1 ⎭	1·03	101·5	87·90	50·74
325	1756·5					
320	1761·7	5·2 ⎫ 5·2 ⎭	1·04	102·0	78·11	65·54
315	1766·9					
310	1772·1	5·2 ⎫ 5·3 ⎭	1·05	102·5	65·86	78·49
305	1777·4					
300	1782·7	5·3 ⎫ 5·4 ⎭	1·07	103·4	51·72	89·58
295	1788·1					
290	1793·6	5·5 ⎫ 5·6 ⎭	1·11	105·4	36·04	99·00
285	1799·2					
280	1804·9	5·7 ⎫ 5·8 ⎭	1·15	107·2	18·63	105·60
275	1810·7					
270	1816·9	6·2 ⎫ 6·6 ⎭	1·28	113·1	...	113·10
265	1823·5					
260	1830·6	7·1 ⎫ 7·8 ⎭	1·49	122·1	21·76	123·40
255	1838·4					
250	1847·1	8·7 ⎫ 9·6 ⎭	1·83	135·3	46·27	127·10
245	1856·7					
240	1867·3	10·6 ⎫ 11·7 ⎭	2·23	149·4	74·67	130·90
235	1879·0					

From these values of r are obtained values of $r \cos \theta$ and $r \sin \theta$, and the corresponding points charted on millimetre paper, where they are indicated by the small crosses near the curve in Plate II., the values of x being taken horizontally, and those of y vertically.

It is at once seen that these points do not lie truly on any smooth curve, and hence it is inferred that the interpolating curve is wrong. It is necessary, therefore, to redraw the interpolating curve, and it is advisable, in order to save time and trouble, not to do this at random, but to ascertain from the errors of the points found on the erroneous curve, both the nature and as far as possible the amount of the modification required in the various parts of the interpolating curve. This may be done as follows.

If a curve be conceived as drawn through the extreme points and fairly among the others, it will leave the points corresponding to the angles 300°, 310°, 320° outside the curve; but those corresponding to 260° and 270° and 280° inside the curve. Hence the distance ought to be diminished in the neighbourhood of 310°, and increased in the neighbourhood of 270°. Also a simple measurement with compasses will show in what ratio the distances at these points ought to be respectively diminished and increased. But the distance varies as $\sqrt{\Delta t}$, and therefore the ratio in which Δt ought to be diminished or increased becomes known.

Hence the differences (Δt) in the neighbourhood of 300°, 310°, 320° were changed from 5·3, 5·3, 5·2, 5·2, to 5·2, 5·1, 5·1, 5·1; and those in the neighbourhood of 260, 270, 280 were changed from 7·1, 6·6, 6·2, 5·8, to 7·7, 6·8, 6·4, 6·0, and the whole table reconstructed as follows.

TABLE III.—SECOND INTERPOLATING CURVE.

	t	Δt	$\frac{\Delta t}{\Delta \theta}$	r	x	y
360	1717·1					
		6·2				
355	1723·3					
		5·9 }				
350	1729·2		1·16	...	106·00	18·70
		5·7 }				
345	1734·9					
		5·5 }				
340	1740·4		1·08	...	97·65	35·54
		5·3 }				
335	1745·7					
		5·2 }				
330	1750·9		1·04	...	88·32	50·98
		5·2 }				
325	1756·1					
		5·1 }				
320	1761·2		1·02	...	77·36	64·92
		5·1 }				
315	1766·3					
		5·1 }				
310	1771·4		1·02	...	64·92	77·36
		5·1 }				
305	1776·5					
		5·2 }				
300	1781·7		1·05	...	51·23	88·74
		5·3 }				
395	1787·0					
		5·4 }				
290	1792·4		1·10	...	35·87	98·55
		5·6 }				
285	1798·0					
		5·8 }				
280	1803·8		1·18	...	18·86	107·00
		6·0 }				
275	1809·8					
		6·4 }				
270	1816·2		1·32	...	...	114·90
		6·8 }				
265	1823·0					
		7·7 }				
260	1830·7		1·57	...	21·76	123·40
		8·0 }				
255	1838·7					
		8·7 }				
250	1847·4		1·83	...	46·38	127·40
		9·6 }				
245	1857·0					
		10·6 }				
240	1867·6		2·22	...	74·50	129·00
		11·6 }				
235	1879·2					

When these points are charted, they are found to lie satisfactorily on a curve. If they again failed to do so, a third interpolating curve would have had to be drawn. By proceeding to two decimals, and using the second column of differences, slightly more exact results could be obtained.

The next operation is to complete the ellipse of which the curve so found forms a part. This part of the problem requires much patience and some sagacity. Either an ellipsograph or a piece of string and two drawing pins may be used, and at last by methods of trial and error an ellipse is found which approximately passes through all the points. No pains should be spared here to make the ellipse pass as exactly as possible through the points. It must be remembered that a very slight alteration in the position of the foci and the length of the major axis will seriously affect the area of the curve, and hence the periodic time in the orbit we shall obtain. In cases like the orbit of Castor, when only a small portion of the orbit has been described, it is impossible to ascertain the apparent ellipse exactly, and hence the periods hitherto obtained by different computers differ seriously.

In the figure, Plate II., C is the centre of the apparent ellipse, and the part of it hitherto described is that part where the dots are seen and the dates are marked.

By inspection of this curve several facts are at once obtained. If A is the principal star, from axes through which the co-ordinates have been laid out, A must be the projection of the focus of the real ellipse; and C being the centre of the apparent ellipse, must also be the projection of the centre of the real ellipse. Hence A C produced both ways must be the projection of the major axis of the real ellipse.

If this cut the apparent ellipse in N, N must be the projection of the periastron, at an angle of about 338° 30',·which from the interpolating curve corresponds to a date of 1742·1.

Further, the ratio C A : C N being unaltered by the projection will give the eccentricity of the *real* ellipse. Measuring

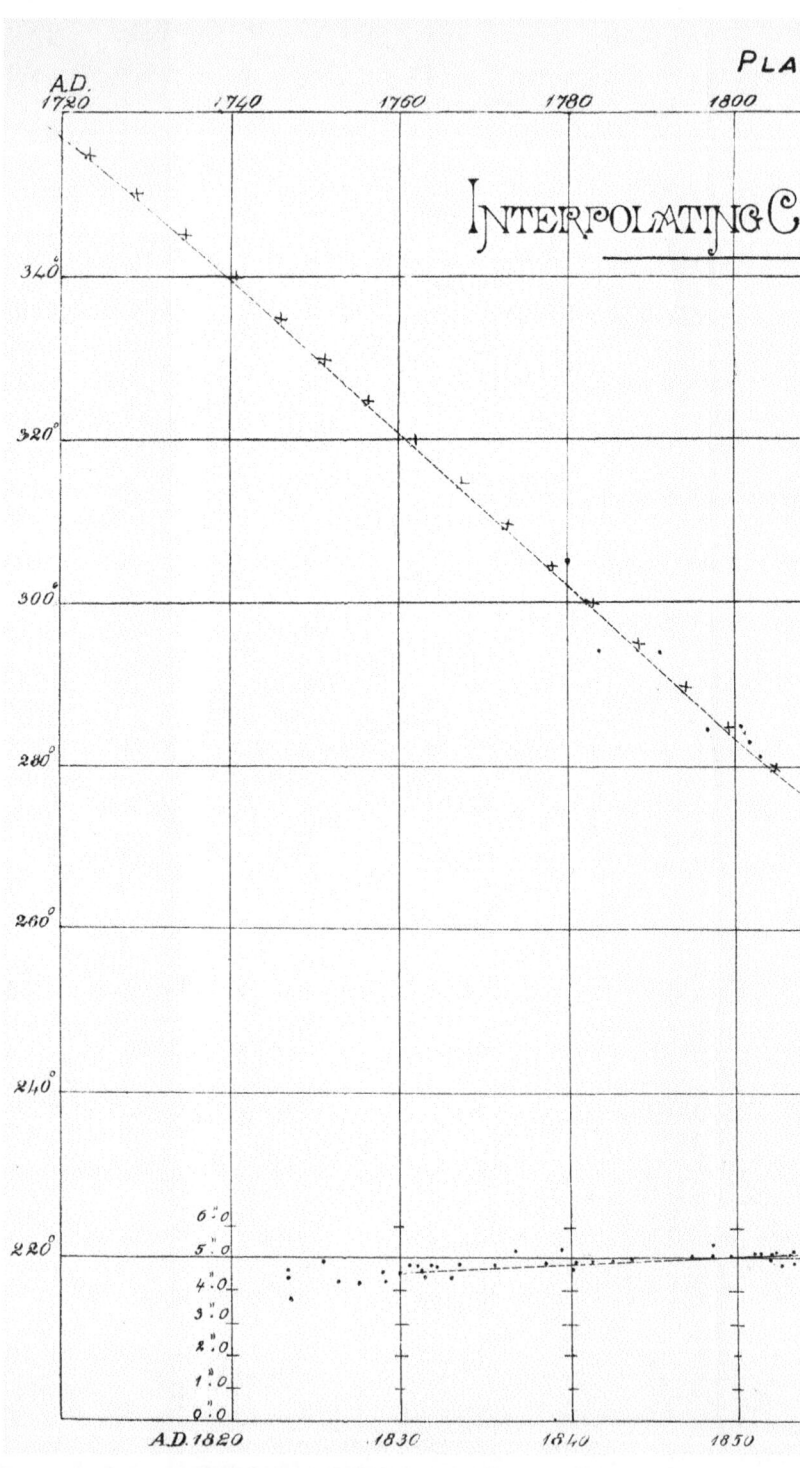

TE I.

1820　　　　1840　　　　1860　　　　1880

URVES OF CASTOR

6"·0
5"·0
4"·0
3"·0
2"·0
1"·0
0"·0

1860　　　　1870　　　　1880

these distances with the compasses and computing the ratio
it is found that $e = \cdot38 \ldots$

Again, it appears that the nearest approach of B to A was
at the angle 314, or at the time 1767·3 at the point U : this
distance on the millimetre scale is about 100. Similarly, the
greatest apparent distance on the same scale will be about
233·6.

In order to ascertain what these distances are in seconds of
arc, it will be necessary to make a table of the observed distances, obtained by interpolation at selected epochs from the
distance curve (Plate I.), and compare them with the distances
on the millimetre scale obtained from the apparent ellipse at
corresponding angles.

TABLE IV.

t.	r.	r' on scale.
1830	4·50	125·6
1835	4·62	127·6
1840	4·75	131·4
1845	4·89	134·5
1850	5·02	137·5
1855	5·14	140·5
1860	5·25	144·5
1865	5·38	147·2
1870	5·52	150·0
Sums ...	45″·07	1238·2 divisions.

Hence the least distance was 3″·64, and the greatest distance
will be 8″·50, at an angle of 174°, at the point V.

The next part of the problem consists in the construction
of Thiele's ellipse.

The axis of x is cut by the ellipse at distances 111·8 and
233·0 from A. The harmonic mean between these is 151·1.
Lay out this distance along this axis in both directions from
A, so obtaining the points 1, 2. Similarly, from the intercepts
on the axis of y obtain their harmonic mean, and the points
3, 4. Two more points can easily be obtained by drawing a

8

chord through A which is bisected in A. The extremities of this chord will plainly be points on Thiele's ellipse.

Construct an ellipse to pass accurately through these six points, A being of course the centre of this ellipse. Draw the axes A a, A b of this ellipse, and find its foci f, f'. Then by Thiele's theorem the ratio A $b : b f$ is cos γ, where γ is the inclination of the plane of the real orbit to the plane on which we see it projected—that is, to the plane perpendicular to the line of sight.

Hence γ is found to be 32° 15'.

The direction of the major axis of this ellipse is that of the line of nodes. This is found by a protractor or scale of chords. Hence ☊ = 28° 15'.

The elements of the orbit so far obtained are

$$
\begin{aligned}
e &= \cdot 38 \\
\gamma &= 32°\ 15' \\
☊ &= 28°\ 15' \\
T &= 1742\cdot 1.
\end{aligned}
$$

To obtain the period some further construction is required. Draw through C lines D C, E C parallel to the axes of Thiele's ellipse: these will be the directions of the axes of the ellipse which is the projection of the auxiliary circle. The ratio of the axes of this ellipse will be of course the same as that of the axes of Thiele's ellipse, and the magnitudes of the axes can be found from the consideration that C N and A M are radii drawn, one in each, in the same direction relative to the axes, and therefore have the same ratio as the axes. Hence if the proportions

A M : C N :: A a : C D

and A M : C N :: A b : C E

are worked out, C D and C E will be the semi-axes required.

Let this ellipse be drawn ; we will call it the auxiliary ellipse.

Draw C T, C T' parallel to the projection of the latus rectum of the real ellipse to meet the auxiliary ellipse in T, T', and draw through T, T' lines parallel to C M.

Measure off T Y, T'Y' along these lines, the length being found by the proportion $1 : \frac{1}{2}\pi :: C N : T Y$.

Divide T Y, T'Y' into nine equal parts, the points of division being numbered 1, 2 . . . 8 in the figure; and draw lines through them parallel to T C T'.

From a table of sines, and the known length of C T, compute C T sin 10°, and mark off this length along the line 1, 1, measuring from the central line C X on both sides of it, thus obtaining $G_1 K_1$, $G_1 K_1' = C T \sin 10°$.

Similarly, lay off $G_2 K_2 = C T \sin 20°$; $G_3 K_3 = C T \sin 30°$, etc. And through the points so determined draw the curve $X K_1 K_2 . . . T$. We will call this the ephemeris curve. This is the *projection* of the curve of sines.* Join A T, A T'.

Then, as was before shown, if through any point P on the apparent ellipse P Q be drawn parallel to C T to meet the auxiliary ellipse in Q, and Q O be drawn parallel to C M to meet the ephemeris curve in O ; O H be drawn parallel to A T to meet C M in H ; as P moves with its orbital motion in the apparent ellipse, H will move *uniformly* along the line X' C X.

Select two positions of P whose epoch is known, as at the first and last of the points interpolated, for which the times were 1867·6 and 1729·2 respectively, giving an interval of 138·4 years. Measure H X = 132, H' X = 15, X X' = 522. Then by the proportion

H X + H' X : 2 X X' :: 138·4 years : period,

we find the period to be 982·9 years.

To construct an ephemeris, divide C X into any number of equal portions, and determine as before the points on the apparent ellipse corresponding to each point of division.

To find the angle at any required date, say 1880, proceed as follows. Since 1880 − 1867·6 = 12·4 years ; and since 1044 divisions correspond to 982·9 years, 12·4 years correspond to

---

* The curve of sines was first suggested, we believe, by Professor Adams.

13·1 divisions. Take H h = 13·1, and determine by the same construction the point marked 1880. The angle is found by the protractor to be 234°·5. In the same manner the angle in the year 1890 is found to be 231·5. In the same manner the date of maximum distance will be found to be A.D. 2147·2.

We have still to determine the major axis (a) of the real ellipse, and the position of the periastron (λ) on the orbit.

Since the major axis of Thiele's ellipse is the latus rectum of the real ellipse, as before shown, and the eccentricity e of the real ellipse has been found,

$$a = \frac{l}{1 - e^2}, = \frac{157}{1 - (\cdot 38)^2} = 183\cdot 4 \text{ divisions,}$$

and this reduced to seconds by the equivalence in p. 113, gives us $a = 6''\cdot 67$.

Lastly, $\tan \lambda = \tan (\lambda' - \Omega) \sec i$, where λ' = the angle that the projection of the axis major makes with the initial line.

This gives $\lambda = 305° \ 10'$.

Hence our elements are as follows :—

Semi-axis major	a =	6″·67.
Eccentricity	e =	·38.
Position of node	☊ =	28° 15′.
Inclination	i =	32° 15′.
Position of periastron	λ =	305° 10′.
Period in years	P =	982·9 years.
Periastral passage A.D.	T =	1742·1.

It will be interesting to compare these with the elements obtained by a rigorous analytical investigation by Thiele, in *Ast. Nach.*, vol. lii., No. 1227.

THIELE'S ELEMENTS.

a =	7·5375.
e =	0·34382.
☊ =	31° 58′·0.
i =	42° 5·4.
λ =	294° 0·8.
P =	996·85.
T =	1750·326.

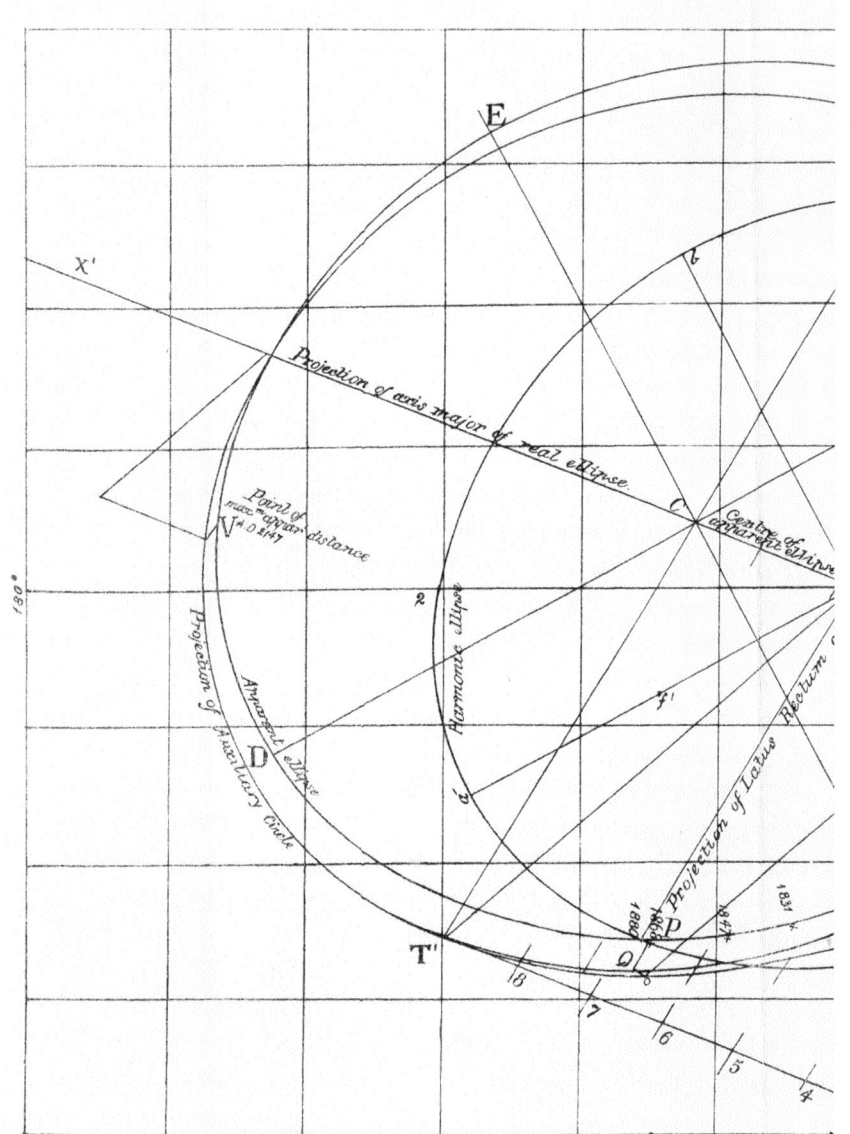

The Scale of this diagram is reduced from the original. The side of one of the squares represents 50 millimetres

90°

T

8

7

6

5

4

3

2

1

Y

a

f

K₃

real ellipse

A *Principal star*

K₂

I

1777

h H

1788

K₁

1748

0

Projection of Perastron

1751

M

H'

X

1761

1771

1788

Point of min'm Appar. dist. A.D. 1767.3

1792

1804

1816

6

5

0°

0°

270°

3

2

1

Y'

Since our graphical solution was finished in 1875, Dr. Doberck has also computed the orbit, and gives

DOBERCK'S ELEMENTS.

$$a = 7\overset{''}{\cdot}43.$$
$$e = 0\cdot329.$$
$$\Omega = 27°\ 46'.$$
$$\lambda = 297\cdot13.$$
$$P = 1001\cdot2.$$
$$T = 1749\cdot75.$$

CHAPTER III.

AN ORBIT WORKED BY ANALYTICAL METHODS.

THE following example, in which the orbit of σ Coronæ is worked out, will illustrate the application of analysis to the subject of double-star orbits. It possesses some independent interest on account of the discrepancy among the orbits hitherto published, which will be seen from the subjoined table. The method presupposes an orbit obtained approximately by graphical methods, and shows how greater exactness can be obtained in the elements. No further acquaintance with analysis is necessary than that of the elements of the differential calculus.

ON A DETERMINATION OF ELEMENTS OF σ CORONÆ (1877).

Herschel discovered in 1781 that σ Coronæ was double, and in 1802 he recognized its binary character. The motion is direct. The distance was small when first observed, but afterwards it increased rapidly, thus rendering the measures surer and easier. A re-determination of the elements, in which these later observations were taken into account, seemed to me likely to decide upon the question of the period, about which astronomers hitherto did not agree, as can be seen from the following table :—

T	Node.	λ	γ	P	a	e	Authority.
	° ′	° ′	° ′	yrs.	″		
1835·60	138 0	7 18	41 15	286·60	3·68	0·6112	J. Herschel.
1826·60	25 7	64 38	29 29	608·45	3·92	0·6998	Mädler.
1826·48	21 3	69 24	25 39	736·88	5·19	0·7256	Hind.
1829·70	3 8	96 53	45 6	240·00	2·94	0·3887	E. B. Powell.
1831·17	1 57	101 57	46 47	195·12	2·72	0·3088	Jacob.

I tried first to determine the elements by Sir J. Herschel's method (*Memoirs of the Royal Astronomical Society*, vol. v.) A first attempt with an ellipse corresponding to a moderate period failed to represent the observations; and I subsequently obtained the following orbit by the aid of ninety-eight annual means of angles and distances:—

FIRST ELEMENTS OF σ CORONÆ.

T	1828·91.
Node	6° 43'.
λ	89° 17',
γ	29° 40 .
φ	843·20 years.
a	6"·001.
e	0·7502.

The comparison of the angles of position, and the distances calculated from these elements with those given by the measures, has been published in the *Astronomische Nachrichten*, No. 2037.

I collected afterwards eighteen more annual means, partly in the library of the Royal Irish Academy in Dublin, partly they were communicated to me by Messrs. Wilson and Gledhill, and Dr. Dunér, of the Lund Observatory, Sweden. The comparison with all these measures proved the calculated angles to be a couple of degrees too small at the first epochs, about as much too large in 1830, and again too small at the present time. The corrections were graphically determined, and, when applied to the calculated angles, furnished new angles of position, from which the distances according to Herschel's method were deduced, and from these the second system of elements was calculated.

SECOND ELEMENTS OF σ CORONÆ.

T	1826·69.
Node	26° 10'.
λ	62° 14'.
γ	35° 8'.
φ	829·40 years.
e	0·7463.

The apparent ellipse was very like the former one, but this time it was possible to lay it nearly through all the points. I thought, therefore, that I had hit the right orbit this time; but the subsequent comparison with observation showed that the angles from 1825 to 1870 came several degrees short of the measures, though the agreement elsewhere was close. All the angles but for a short interval being represented, I thought that I had better correct the elements by *Klinkerfues's method*. This method requires six angles of position to be given, from which the six elements are deduced, the axis major being afterwards calculated from the observed distances. Sir W. Herschel's two epochs furnished the first two normal places; the angles measured between 1819 and 1828 the third; the unrepresented measures 1830—1839 the fourth. The fifth place was obtained from the measures 1839—1868 inclusive, as it was in this instance allowable to consider the deviation proportional to the time during this long interval, the difference between the observed and calculated angle being nearly constant throughout. The sixth normal place had been previously used to deduce epoch and period of the systems given above. It was determined on Gledhill's, Wilson's, Dunér's, Dembowski's, and Schiaparelli's measures only. Mädler's epochs of 1836·47 and 1842·73 were excluded, as also some of Talmage's measures and Copeland's for 1873·40.

SIX NORMAL PLACES FOR σ CORONAE.

I.	1781·79	$\theta^\circ = 347^\circ{\cdot}53$
II.	1802·74	$11^\circ{\cdot}40$
III.	1825·00	$77^\circ{\cdot}67$
IV.	1835·00	$128^\circ{\cdot}20$
V.	1855·00	$179^\circ{\cdot}73$
VI.	1872·11	$197^\circ{\cdot}37$

There is a well-known proposition which says that when a triangle is orthogonally projected on a plane, the area of the triangle in the projection is equal to the area of the real triangle multiplied by the cosine of the angle between the planes (cos γ). Now as the apparent orbit of a double star is the orthogonal projection of the real orbit, areas between the principal star and two places of the companion in the one orbit are in a constant ratio to the corresponding ones in the other, as,

$$\frac{r\,r^i \sin (v - v^i)}{r^i\,r^v \sin (v^i - v^v)} = \frac{\rho\,\rho^i \sin (\theta - \theta^i)}{\rho^i\,\rho^v \sin (\theta^i - \theta^v)}, \text{ and}$$

$$\frac{r\,r^{ii} \sin (v - v^{ii})}{r^{ii}\,r^v \sin (v^{ii} - v^v)} = \frac{\rho\,\rho^{ii} \sin (\theta - \theta^{ii})}{\rho^{ii}\,\rho^v \sin (\theta^{ii} - \theta^v)},$$

where r, r^i, etc., are the radii vectores corresponding to ρ, ρ, etc., the distances, and the angles of position θ, θ^i, etc., to the true anomalies v, v, etc.

Dividing the first equation by the second, we obtain

$$\frac{\sin (v^i - v) \sin (v^{ii} - v^v)}{\sin (v^{ii} - v) \sin (v^i - v^v)} = \frac{\sin (\theta^i - \theta) \sin (\theta^{ii}\, \theta^v)}{\sin (\theta^{ii} - \theta) \sin (\theta^i - \theta^v)}$$

If we write successively v^{iii}, θ^{iii} and v^{iv}, θ^{iv} in the place of v^{ii}, θ^{ii} in this equation, we obtain the two equations

$$\frac{\sin (v^i - v) \sin (v^{iii} - v^v)}{\sin (v^{iii} - v) \sin (v^i - v^v)} = \frac{\sin (\theta^i - \theta) \sin (\theta^{iii} - \theta^v)}{\sin (\theta^{iii} - \theta) \sin (\theta^i - \theta^v)}, \text{ and}$$

$$\frac{\sin (v^i - v) \sin (v^{iv} - v^v)}{\sin (v^{iv} - v) \sin (v^i - v^v)} = \frac{\sin (\theta^i - \theta) \sin (\theta^{iv} - \theta^v)}{\sin (\theta^{iv} - \theta) \sin (\theta^i - \theta^v)}.$$

The right side of the three equations contains nothing but the angles of the normal places; substituting their values, we obtain the equations as follows :—

$$\frac{\sin (v^i - v) \sin (v^{ii} - v^v)}{\sin (v^{ii} - v) \sin (v^i - v^v)} = a.$$

$$\frac{\sin (v^i - v) \sin (v^{iii} - v^v)}{\sin (v^{iii} - v) \sin (v^i - v^v)} = \beta.$$

$$\frac{\sin (v^i - v) \sin (v^{iv} - v^v)}{\sin (v^{iv} - v) \sin (v^i - v^v)} = \gamma.$$

The true anomalies being functions of the eccentricity, epoch, and period, it is theoretically possible to obtain these three elements from the three equations. The peculiarity of the method we shall follow is that it furnishes equations from which the elements fixing the plane of the orbit (node and inclination), and the position of the ellipse in the plane (λ), have been eliminated. It would, however, be very difficult directly to obtain the three elements—e, P, T,—from the above equations; but these equations are useful, when we, as in the present case, have already arrived at a very near approximation to the elements, which we want to advance further by representing by the orbit strictly the six angles of position. Instead of the elements e, P, and T, it is a slight improvement to substitute the annual mean motion for the period: $\mu = \frac{360°}{P}$, and the mean anomaly M_0 corresponding to the epoch of periastron-passage in the provisional orbit, e is obtained in degrees, from which its value in the usual form is computed by dividing by the number of degrees in the unit of circular measure—57°·296.

We calculate, firstly, the true anomalies, and hence a, β, and γ, with the provisional elements,—that is, with $M_0 = 0$. Secondly, we calculate the same quantities with the same e and μ, but $M_0 = + 1°$. Thirdly, with the same e, $M_0 = 0$, but adding to the mean motion a fifth of its value: $\mu' = 1·2\,\mu$. Fourthly, with $M_0 = 0$, the original mean annual motion μ, but with another eccentricity, $e' = e + 0·01$. By a comparison of the results obtained by the three last calculations with that from the first hypothesis, we get to know what influence any variation of the elements has on the three qualities a, β, and γ, that we are trying to represent,—that is, we learn their partial differential coefficients. The difference between M_0 in the two first hypotheses divided into the corresponding variations of a, β, and γ, give $\frac{da}{dM_0}$, $\frac{d\beta}{dM_0}$, and $\frac{d\gamma}{dM_0}$. The difference $\mu' - \mu$, divided into the corresponding variations of a, β, and γ, give $\frac{da}{d\mu}$, $\frac{d\beta}{d\mu}$, and $\frac{d\gamma}{d\mu}$. Finally, the difference $e' - e$ divided into the corresponding variations of a, β, and γ, give $\frac{da}{de}$, $\frac{d\beta}{de}$, and $\frac{d\gamma}{de}$.

If we now denote by a', β', and γ' the values corresponding to the first hypothesis, and by a, β, and γ the values calculated from the position angles, we obtain by Taylor's formula,—

$$a' + \frac{da}{dM_0} \Delta M_0 + \frac{da}{d\mu} \Delta\mu + \frac{da}{de} \Delta e = a.$$

$$\beta' + \frac{d\beta}{dM_0} \Delta M_0 + \frac{d\beta}{d\mu} \Delta\mu + \frac{d\beta}{de} \Delta e = \beta.$$

$$\gamma' + \frac{d\gamma}{dM_0} \Delta M_0 + \frac{d\gamma}{d\mu} \Delta\mu + \frac{d\gamma}{de} \Delta e = \gamma.$$

From which are easily obtained the corrections ΔM_0, $\Delta\mu$, and Δe, to be applied to the values in the provisional elements in order to be able to represent the six position angles of the normal places.

Professor Klinkerfues indicates further several processes, which must reduce the amount of work required in the computation of the single hypotheses. The true anomalies are calculated from the three elements by the following well-known formulæ,—

$$u - e \sin u = M_0 + \mu t,$$

$$\tan \tfrac{1}{2} v = \sqrt{\frac{1+e}{1-e}} \, \tan \tfrac{1}{2} u,$$

where u are the eccentric anomalies, t the time since the epoch, and e is expressed in the first equation in degrees in the second in absolute measure. From these are obtained the differential coefficients of the eccentric anomaly with respect to the mean anomaly,—

$$\frac{1}{1 - e \cos u^{\mathrm{i}}}, \ \frac{1}{1 - e \cos u^{\mathrm{ii}}}, \ \frac{1}{1 - e \cos u^{\mathrm{iii}}}, \ \text{etc.}$$

The eccentric anomalies of the second hypothesis are obtained from those of the first by multiplying the alteration of the mean anomaly, here $+ 1°$, by these differential coefficients. The eccentric anomalies of the third hypothesis are obtained by multiplication of the same coefficients by the corresponding alterations of the mean anomalies. Those of the fourth hypothesis are obtained by multiplying the variation of the eccentricity expressed in degrees by the differential coefficients of the eccentric anomaly with respect to the eccentricity:—

$$\frac{\sin u^{\mathrm{i}}}{1 - e \cos u^{\mathrm{i}}}, \ \frac{\sin u^{\mathrm{ii}}}{1 - e \cos u^{\mathrm{ii}}}, \ \frac{\sin u^{\mathrm{iii}}}{1 - e \cos u^{\mathrm{iii}}}, \ \frac{\sin u^{\mathrm{iv}}}{1 - e \cos^{\mathrm{iv}}}, \ \text{etc.}$$

The products are in all cases to be added to the eccentric anomalies of the first hypothesis. The true anomalies are then obtained from the eccentric, by the formula given above.

The results of these calculations in case of σ Coronæ were as follows :—

	Hypothesis I.	Hypothesis II.	Hypothesis III.	Hypothesis IV.
$M_0.$	$0°.$	$+ 1°.$	$0°.$	$0°.$
μ	$0·4340$	$0·4340$	$0·5208$	$0·4340$
e	$0·7463$	$0·7463$	$0·7463$	$0·7563$
u	$305°\ 52'$	$307°\ 38'$	$298°\ 56'$	$305°\ 3'$
u^i	$325\ 12$	$327\ 47$	$319\ 49$	$324\ 22$
u^{ii}	$357\ 9$	$1\ 5$	$356\ 34$	$357\ 2$
u^{iii}	$13\ 51$	$17\ 29$	$16\ 28$	$14\ 21$
u^{iv}	$39\ 25$	$41\ 47$	$45\ 12$	$40\ 16$
u^v	$54\ 33$	$56\ 19$	$61\ 31$	$55\ 22$
v	$253°\ 26'$	$255°\ 34'$	$245°\ 45'$	$251°\ 14'$
v^i	$281\ 8$	$285\ 42$	$272\ 20$	$278\ 25\frac{1}{2}$
v^{ii}	$352\ 32$	$2\ 50\frac{1}{2}$	$351\ 0$	$352\ 3$
v^{iii}	$35\ 21$	$43\ 56$	$41\ 35$	$37\ 20\frac{1}{2}$
v^{iv}	$86\ 27$	$90\ 4\frac{1}{2}$	$95\ 3$	$89\ 5\frac{1}{2}$
v^v	$107\ 3$	$109\ 6$	$114\ 43$	$109\ 15\frac{1}{2}$
α	$- 4·155$	$- 8·486$	$- 1·015$	$- 2·201$
β	$- 6·941$	$- 14·600$	$- 2·750$	$- 4·145$
γ	$+ 7·044$	$+ 10·990$	$+ 0·809$	$+ 2·733$

The constants calculated from the observed position angles are—

$$\alpha = - 3·382, \beta = - 5·739, \text{ and } \gamma = + 5·580.$$

Thus we obtain the equations

$$- 4·331\ \Delta M_0 + 3·140\ \Delta\mu + 1·954\ \Delta e = + 0·773$$
$$- 7·660\ \Delta M_0 + 4·191\ \Delta\mu + 2·796\ \Delta e = + 1·202$$
$$+ 3·950\ \Delta M_0 - 6·235\ \Delta\mu - 4·311\ \Delta e = - 1·464 ;$$

from which

$$\Delta M_0 = - 0°·051,\ \Delta\mu = - 0·0683 \times 0°·0868,\ \Delta e = 0·39 \times 0·01 ;$$

or $T = 1826·81,\ \mu = 0°·4281\qquad ,\ e = 0·7502 ;$

with which we get

$$u\ 305°\ 53',\ 325°\ 8',\ 356°\ 53',\ 13°\ 38',\ 39°\ 18',\ 54°\ 16'.$$
$$v\ 252°\ 59',\ 280°\ 33',\ 356°\ 27',\ 35°\ 6',\ 86°\ 46',\ 107°\ 12'.$$
$$\alpha = - 3·720,\ \beta = - 6·193,\ \gamma = + 5·805.$$

It has already been remarked that there hitherto existed some uncertainty as to the period of revolution of this system, and that I was by my first investigation led to consider the large period pretty well established. The possibility of fixing separately the eccentricity and period appears by an inspection of the three equations above, as the coefficients of $\Delta\mu$ and Δe, are not proportional; but they are not far from it, and if the measures employed embraced a shorter time than ninety-five years, they would be more nearly so, and it is therefore no wonder that we hitherto were in doubt about the orbit. In reality, a great number of different orbits, corresponding to every value of the eccentricity varying within wide limits, would have been obtained, if the elements had been represented as linear functions of the eccentricity.

The values of a, β, and γ, finally obtained, are far from those deduced from the observations. The differential coefficients employed in the calculation of the hypotheses may not be without influence hereupon, notwithstanding that the variations of the elements were small enough. But the disagreement arises principally from the circumstance that the higher orders of the differential coefficients of a, β, and γ with respect to the elements are not to be neglected, which has been supposed by confining ourselves to the first term of Taylor's series.

To represent a, β, and γ better, I extrapolated between the last values of the three elements and those of the first hypothesis. Thus I obtained

$$T = 1826\cdot85, \mu = 0°\cdot4264, e = 0\cdot7513.$$
$$u\ 305°\ 55', 325°\ 5',\quad 3°\ 10', 13°\ 32', 39°\ 12',\quad 54°\ 14'.$$
$$v\ 252°\ 52', 280°\ 18', 351°\ 37', 34°\ 57', 86°\ 45 , 107°\ 18'.$$
$$a = -3\cdot425, \beta = -5\cdot861, \gamma = +5\cdot530.$$

I substituted now these a, β, γ, for a', β', γ', in the equations, which then turn out as follows :—

$$-4\cdot33\ \Delta M_0 + 3\cdot14\ \Delta\mu + 1\cdot95\ \Delta e = +0\cdot043.$$
$$-7\cdot66\ ,,\quad +4\cdot19\ ,,\quad +2\cdot80\ ,, = +0\cdot122.$$
$$+3\cdot95\ ,,\quad -6\cdot24\ ,,\quad -4\cdot31\ ,, = +0\cdot050.$$

From which we obtain

$\Delta M = -0°.032$, $\Delta \mu = -0.049 \times 0°.0868$, $\Delta e = +0.02 \times 0.01$.

T = 1826·93, $\mu =$ 0°·4227 , $e =$ 0·7515.

u 306° 10', 325° 43', 3° 17', 13° 24', 38° 54', 53° 52'.

v 253° 9', 280° 31', 351° 18', 34° 38', 86° 18', 106° 53'.

$a = -3·778$, $\beta = -6·340$, $\gamma = +6·405$.

This very erroneous result shows that the coefficients, deduced for correction of the original elements, cannot be used strictly speaking for rectifying the new elements. The corrections which resulted from T and e are however so small, that no new approximation appears necessary. I therefore retained them, and calculated a, β, and γ with the former value of μ,—that is, $u = 0°·4264$.—The numbers obtained were the following :—

T = 1826·93, $\mu = 0°·4264$, $e = 0·7515$.

u 305° 51', 324° 59', 356° 43', 13° 27' , 39° 9', 54° 11'.

v 252° 46. 280° 5', 351° 18', 34° 45½', 86° 42', 107° 15½ .

$a = -3·341$, $\beta = -5·691$, $\gamma = +5·357$.

This result was then compared with the result from the same values of T and e, but $\mu = 0°·4227$. Denoting the constants calculated with $\mu = 0°·4264$ by a_1, β_1, and γ_1, those calculated with μ 0° 4227 by a_2, β_2, and γ_2, and those calculated from the observed position angles a, β, and γ, we obtain the following three equations for the determination of the correction to be applied to $u = 0°·4264$ in terms of the difference between the two mean motions :—

$(a_2 - a_1) \Delta \mu = a - a_1$, $(\beta_2 - \beta_1) \Delta \mu = \beta - \beta_1$ $(\gamma_2 - \gamma_1) \Delta \mu = \gamma - \gamma_1$.

These equations in the present case are—

437 $\Delta \mu$ − 41 = 0,   649 $\Delta \mu$ − 48 = 0,   1048 $\Delta \mu$ − 223 = 0 ;

from these it follows, according to the method of least squares, that $\Delta \mu = + 0°·22 \times -0°·0037 = -0°·00082$. This correction is applied to $\mu = 0°·4267$, and gives the following result. The errors still left behind correspond to errors of the normal places, which are far within their probable errors :—

T = 1826·93, $\mu = 0°·4256$, $e = 0·7515$.

u 305° 56', 325° 1', 356° 42½', 13° 27½', 39° 4', 54° 7 .

v 252° 52', 280° 10', 351° 17', 34° 47', 86° 34', 107° 12'.

$a = -3·406$, $\beta = -5·790$, $\gamma = +5·574$.

These, the so-called phoronomical elements, thus fixed, it remains to settle the position of the ellipse. Professor Klinkerfues uses three of the true anomalies and the corresponding observed position angles, applying at last a small correction to the node to represent better all the six normal places. He calculates the longitude of the node by eliminating λ and γ from three equations (which are obtained from a rectangular spherical triangle) of the form

$$tan\ (\theta - \Omega) - \cos \gamma\ tan\ (v + \lambda)\ ;$$

λ and γ are subsequently obtained from the equations, first γ, and then γ. The formulæ are however complicated, and differential equations to correct assumed values of the three elements now in question may be preferred to the direct solution of the above equations, whereby besides, by application of the method of least squares, more than three places may be used. The equations obtained from differentiation of the last equation are of the form—

$$\Delta\theta = \Lambda\Omega + \cos \gamma \frac{\cos^2 (\theta - \Omega)}{\cos^2 (v + \lambda)} \Delta\lambda - \tfrac{1}{2} tan\ \gamma \sin 2\ (\theta - \Omega)\ \Delta\gamma.$$

It is, however, still better to calculate the differential coefficients by variation of the elements in the equation. We assume values of Ω, γ, and λ, which are as exact as possible, and calculate with those from the true anomalies v, given by the previous investigations, the respective angles of position θ. Altering then γ with a suitable quantity, we again calculate the angles of position. The differences between the two values of these divided with the difference between the inclinations are the differential coefficients of the angles with respect to the inclination $\frac{d\theta}{d\gamma}$. Similarly, the differential coefficients of the angles with respect to λ : $\frac{d\lambda}{d\theta}$ are calculated, varying λ a certain quantity. We have, of course, $\frac{d\theta}{d\Omega} = 1$, and the equations of condition are as follows :—

$$\theta' + \frac{d\,\theta'}{d\,\Omega}\,\Delta\Omega + \frac{d\,\theta'}{d\,\gamma}\,\Delta\gamma + \frac{d\,\theta'}{d\,\lambda}\,\Delta\lambda = \theta,$$

$$\theta_1 + \frac{d\,\theta_1'}{d\,\Omega}\,\Delta\Omega + \frac{d\,\theta_1'}{d\,\gamma}\,\Delta\gamma + \frac{d\,\theta_1'}{d\,\lambda}\,\Delta\lambda = \theta_1,$$

. . . .

where θ, θ_1, etc., are given by the normal places. θ', θ'_1, etc. are obtained with the assumed Ω, λ, and γ.

The equations in the present instance are as follows :—

$$\Delta\Omega + 0.98\ \Delta\lambda + 0.31\ \Delta\gamma - 1°.18 = 0$$
$$,,\ + 0.86\ ,,\ + 0.15\ ,,\ - 0°.20 = 0$$
$$,,\ + 1.05\ ,,\ - 0.32\ ,,\ - 2°.75 = 0$$
$$,,\ + 1.18\ ,,\ + 0.12\ ,,\ - 3°.35 = 0$$
$$,,\ + 0.90\ ,,\ + 0.25\ ,,\ - 0°.50 = 0$$
$$,,\ + 0.85\ ,,\ + 0.08\ ,,\ - 0°.18 = 0$$

On further consideration, I however preferred to take the mean of the third and fourth equation, and combining this with the two last equations to deduce the three corrections. It must be remembered that Herschel's two epochs are not thereby excluded, for they helped to determine a, β, and γ, and in consequence the true anomalies used for calculating the position of the ellipse. Applying the resulting corrections to the assumed values of the three elements, we have the third system.

THIRD ELEMENTS OF σ CORONÆ.

T	1826·93.
Ω	16° 27'.
λ	73° 51'.
γ	31° 56'.
P	845^{m}·86.
a	5"·885.
e	0·7515.

The formulæ for calculating an ephemeris from these elements are :—

$$u - 43°.05 \sin u = 0°.4255\ (t - 1826.93).$$

$$tan\ \tfrac{1}{2}v = \sqrt{\frac{1+e}{1-e}}\ tan\ \tfrac{1}{2}u = \sqrt{\frac{1.7515}{0.2485}}\ tan\ \tfrac{1}{2}u = 2.655\ tan\ \tfrac{1}{2}u.$$

$$tan\ (\theta_c - 16°\ 27') = \cos\gamma\ tan\ (v + 73°51') = 0.8486\ tan\ (v + 73°51')$$

$$\rho = r\ \frac{\cos(v+\lambda)}{\cos(\theta-\Omega)} = a\ (1 - e\cos u)\ \frac{\cos(v+\lambda)}{\cos(\theta-\Omega)} = 5".885\ (1 - 0.7515$$

$$\cos u)\ \frac{\cos(v + 73°\ 51')}{\cos(\theta - 16°\ 27')}.$$

The half axis major was determined from the measured distances by dividing them by $(1 - e \cos u) \frac{\cos (v + \lambda)}{\cos (\theta - \Omega)}$.

The six normal angles of position were represented as follows,—always observation minus calculation :—

$$+ 12', + 2', - 8', + 5, + 5', + 2'.$$

Had a correction been applied to the node, the square sum of these errors could have been diminished, but as then the errors in the last places, which are the most certain, would have been increased, nothing would be gained. At any rate, the errors above are far below the errors of the normal places.

The above elements were now compared with all the observations which I had at my disposal. This comparison has been published in the *Astronomische Nachrichten*, vol. lxxxviii., No. 2103.

We have as yet seen but a small part of the ellipse described, but this part of the orbit has of course been so much the more observed, and so much the nearer are the measures lying to each other. Much more uncertainty must, however, always prevail about these slow-moving systems than about those of quicker revolution, apart from other considerations, at least because the angle changes so little in one observer's lifetime, that systematic corrections cannot so easily be expressed in laws. Engelmann has made extensive investigations on that part of the correction, which is constant for all the position angles measured by the same observer, in analogy with corrections to be applied to right ascensions and declinations in star catalogues. These corrections must, however, vary more or less with the time, as is the case with meridian observations. Exactly determined orbits of many double stars are wanted for the satisfactory solution of these different questions. Even before we may hope to lay the orbits down definitely, they will be of use in this respect.

The last-mentioned comparison showed large systematical errors in the angles and distances of σ Coronæ. Mädler's

angles are decidedly too large. His first angles, when the
position was very oblique, are much too large; his later angles,
when the position went through 180°, are about right, and
then the correction changes in sign. His distances are also
too large, but the correction is likewise diminishing, and disappears at the end of the series. Dunér's and Kaiser's
distances, on the other hand, are too small, which is a much
more remarkable feature. All these distances were excluded
in the above determination of the axis major.

When I had come so far in the calculation, I got for the first
time the series of measures at my disposal which has been
made by M. O. Struve with the large refractor in Pulkowa : the
comparison of these measures with the last elements showed
deviations similar to Mädler's. This shows that the corrections
O. Struve has applied to his observations, after measures made
on artificial objects, do not render his measures faultless.

It will be remembered that Mädler's evidently faulty angles
were introduced with as much weight in the derivation of the
normal places as all the others. This is a cause of the small
systematical deviations of the measures from the ephemeris
calculated after the last elements. I therefore now excluded
Mädler's, O. Struve's, Galle's, Main's, and Talmage's angles,
and Kaiser's first angles. The rest of the observations indicated that the normal place for 1835 should be diminished
about a degree ; the place for 1855 diminished a few minutes.
Such corrections were applied, and then ☊, λ and γ anew
calculated from the six normal places. Supposing node =
16° 27′ λ = 73° 51′, and γ = 31° 56′, and varying λ and γ a
degree respectively, the equations of condition are as
follows :—

$$\Delta\text{☊} + 0\cdot92,\ \Delta\lambda + 0\cdot27,\ \Delta\gamma - 0°\cdot20 = 0$$
$$,,\ + 0\cdot85\ \ ,,\ + 0\cdot05\ \ ,,\ - 0°\cdot03 = 0$$
$$,,\ + 1\cdot12\ \ ,,\ - 0\cdot27\ \ ,,\ + 0°\cdot13 = 0$$
$$,,\ + 1\cdot13\ \ ,,\ + 0\cdot22\ \ ,,\ + 0°\cdot92 = 0$$
$$,,\ + 0\cdot90\ \ ,,\ + 0\cdot15\ \ ,,\ + 0°\cdot07 = 0$$
$$,,\ + 0\cdot85\ \ ,,\ + 0\cdot00\ \ ,,\ - 0°\cdot04 = 0.$$

Allowing double weight to the two last equations, I obtained by the method of least squares the following normal equations:—

$$+ 8{\cdot}00 \, \Delta\Omega, + 7{\cdot}52 \, \Delta\lambda + 0{\cdot}57 \, \Delta\gamma + 0{\cdot}88 = 0$$
$$+ 7{\cdot}52 \,\,\,\, ,, \,\,\,\, + 7{\cdot}16 \,\,\,\, ,, \,\,\,\, + 0{\cdot}51 \,\,\,\, ,, \,\,\,\, + 1{\cdot}02 = 0$$
$$+ 0{\cdot}57 \,\,\,\, ,, \,\,\,\, + 0{\cdot}51 \,\,\,\, ,, \,\,\,\, + 0{\cdot}25 \,\,\,\, ,, \,\,\,\, + 0{\cdot}13 = 0.$$

After elimination of $\Delta\Omega$ from these equations, we obtain

$$\Delta\lambda = - 2^{\circ}{\cdot}25, \text{ and } \Delta\gamma = - 0^{\circ}{\cdot}57.$$

Substituting these values in the four last equations of condition, we obtain $\Delta\Omega = + 2^{\circ}{\cdot}14$, by taking the mean of the resulting four values of this quantity.—The final comparison of the elements, with all the measures (except those excluded), showed that the representation of the angles could be still further improved by diminishing the longitude of the node by $0^{\circ}{\cdot}24$.

These changes in the position of the ellipse were of no appreciable influence on the calculated distances, which came out about a hundredth of a second of arc larger in 1830: in 1835 there was no difference from those previously deduced. From 1840 to 1860 they were a hundredth of a second smaller than by the last orbit; afterwards there was no difference.

DEFINITIVE ELEMENTS OF σ CORONÆ BOREALIS.

T	1826·93.
Node	18 21 .
λ	71 36.
γ	31 22'.
P	845^{yrs}·86.
a	5 ·885.
e	0·7515

COMPARISON OF THE LAST ELEMENTS OF σ CORONÆ, WITH OBSERVATIONS.

Observer.	No	Epoch.	θ_0	θ_c	$\theta_0 - \theta_c$	ρ_0	ρ_c	$\rho_0 - \rho_c$
W. Herschel ..	1	1781·79	347·5	347·0	+ 0·5			
W. Herschel ..	2	1802·74	11·6	11·3	+ 0·3			
W. Struve ..	3	1819·62	48·0	53·8	− 5·8			
Herschel & South	4	1821·30	65·3	60·5	+ 4·8	$''$	$''$	$''$
Herschel & South	5	1823·47	72·9	70·1	+ 2·8	1·45	1·34	+ 0·11
South	6	1825·44	77·5	79·5	− 2·0	1·48	1·28	+ 0·20
W. Struve ..	7	1827·02	89·3	87·5	+ 1·8	1·31	1·27	+ 0·04
W. Struve ..	8	1828·20	96·5	93·7	+ 2·8			
J. Herschel ..	9	1828·50	92·1	95·2	− 3·1			
W. Struve ..	10	1830·11	104·9	103·5	+ 1·4	1·22	1·27	− 0·05
J. Herschel ..	11	1830·28	105·1	104·4	+ 0·7	1·22	1·27	− 0·05
Dawes	12	1830·52	107·3	105·6	+ 1·7			
Smyth	13	1830·76	107·6	106·9	+ 0·7	1·30	1·28	+ 0·02
Dawes	14	1831·34	111·5	109·9	+ 1·6	1·57	1·29	+ 0·28
J. Herschel ..	15	1831·36	108·8	110·1	− 1·3	1·38	1·29	+ 0·09
Smyth	16	1832·37	114·9	115·1	− 0·2	1·40	1·30	+ 0·10
J. Herschel ..	17	1832·52	113·6	115·9	− 2·3	1·07	1·31	− 0·24
Dawes	18	1832·55	115·4	116·0	− 0·6			
W. Struve ..	19	1832·99	118·8	118·2	+ 0·6	1·30	1·31	− 0·01
J. Herschel ..	20	1833·26	119·9	119·5	+ 0·4	1·33	1·32	+ 0·01
Dawes	21	1833·26	120·6	120·0	+ 0·6	1·30	1·32	− 0·02
Smyth	22	1833·58	120·7	121·0	− 0·3	1·20	1·34	− 0·14
Dawes	23	1834·55	125·6	125·5	+ 0·1			
Smyth	24	1835·50	130·9	129·7	+ 1·2	1·40	1·39	+ 0·01
W. Struve ..	25	1835·50	130·5	129·7	+ 0·8	1·31	1·39	− 0·08
Mädler ..	26	1836·47	138·5	133·7	+ 4·8			
W. Struve ..	27	1836·59	134·7	134·2	+ 0·5	1·43	1·42	+ 0·01
Dawes	28	1837·47	136·8	137·6	− 0·8			
W. Struve ..	29	1837·55	140·0	138·0	+ 2·0	1·42	1·45	− 0·03
W. Struve ..	30	1838·45	143·4	141·3	+ 2·1	1·48	1·49	− 0·01
Galle ..	31	1839·52	147·8	145·1	+ 2·7	1·55	1·55	0·00
Dawes	32	1839·53	144·3	145·1	− 0·8	1·60	1·55	+ 0·05
Smyth	33	1839·67	145·1	145·6	− 0·5	1·60	1·56	+ 0·04
Dawes	34	1840·57	147·9	148·5	− 0·6	1·66	1·61	+ 0·05
O. Struve ..	35	1840·82	150·2	149·3	+ 0·9	1·54	1·61	− 0·07
Dawes	36	1841·48	150·3	151·3	− 1·0	1·66	1·66	0·00
Mädler	37	1841·56	152·3	151·5	+ 0·8	1·60	1·66	− 0·06
Kaiser	38	1841·66	148·8	151·8	− 3·0	1·57	1·67	− 0·10
Mädler	39	1842·31	156·4	153·7	+ 2·7	1·81	1·70	+ 0·11
Dawes	40	1842·37	153·3	153·9	− 0·6			
Mädler	41	1842·73	157·6	154·9	+ 2·7	1·87	1·72	+ 0·15
Smyth	42	1843·35	155·9	156·6	− 0·7	1·80	1·75	+ 0·05
Dawes	43	1843·47	156·5	156·9	− 0·4	1·77	1·76	+ 0·01
Mädler	44	1843·51	157·3	157·0	+ 0·3	1·89	1·77	+ 0·12
Kaiser	45	1843·68	156·3	157·5	− 1·2	1·66	1·77	− 0·11
Mädler	46	1844·40	160·6	159·4	+ 1·2	2·05	1·81	+ 0·24
Main	47	1844·45	157·1	159·5	− 2·4			
Mädler	48	1845·51	163·0	162·0	+ 1·0	2·03	1·87	+ 0·16
Jacob	49	1846·21	162·0	163·7	− 1·7	2·25	1·91	+ 0·34
Hind	50	1846·32	162·8	163·9	− 1·1			
Mädler	51	1846·46	165·1	164·2	+ 0·9	2·07	1·92	+ 0·15
Smyth	52	1846·60	162·4	164·5	− 2·1	2·00	1·93	+ 0·07
O. Struve ..	53	1847·02	168·7	165·3	+ 3·4	1·74	1·95	− 0·21
Dawes	54	1847·44	166·0	166·4	− 0·4	1·88	1·97	− 0·09
Mädler	55	1847·44	166·6	166·4	+ 0·2	2·16	1·97	+ 0·19
Mädler	56	1848·41	168·4	168·3	+ 0·1	2·40	2·03	+ 0·37
Dawes	57	1848·53	168·6	168·6	0·0	1·99	2·04	− 0·05
Dawes	58	1849·45	170·1	170·4	− 0·3	2·09	2·09	0·00
O. Struve ..	59	1849·49	172·0	170·4	+ 1·6	1·95	2·10	− 0·15
O. Struve ..	60	1850·52	168·9	172·3	− 3·4	1·99	2·16	− 0·17
Mädler	61	1850·70	173·0	172·7	+ 0·3	2·23	2·17	+ 0·06
Fletcher ..	62	1851·22	174·4	173·6	+ 0·8	2·32	2·21	+ 0·11
Mädler	63	1851·25	175·5	173·7	+ 1·8	2·34	2·21	+ 0·13
Dawes	64	1851·42	173·8	174·0	− 0·2	2·26	2·22	+ 0·04
O. Struve ..	65	1851·63	173·4	174·3	− 0·9	2·06	2·23	− 0·17
Mädler	66	1851·76	176·2	174·5	+ 1·7	2·44	2·24	+ 0·20
Smyth	67	1852·25	176·8	175·4	+ 1·4	2·20	2·25	− 0·05
Miller	68	1852·31	176·5	175·5	+ 1·0	2·38	2·26	+ 0·12
Mädler	69	1852·60	177·5	175·9	+ 1·6	2·39	2·28	+ 0·11
O. Struve ..	70	1852·63	173·3	175·9	− 2·6	2·07	2·28	− 0·21
Jacob	71	1853·14	177·9	176·8	+ 1·1	2·18	2·30	− 0·12
Powell	72	1853·35	175·2	177·1	− 1·9			
Mädler	73	1853·38	177·7	177·2	+ 0·5	2·46	2·32	+ 0·14
Dawes	74	1853·63	177·9	177·6	+ 0·3	2·39	2·34	+ 0·05

COMPARISON OF THE LAST ELEMENTS, ETC.—*continued.*

Observer.	No.	Epoch.	θ_0	θ_c	$\theta_0-\theta$	ρ_0	ρ_c	$\rho_0-\rho_c$
O. Struve..	75	1853·66	175·6	177·6	− 2·0	2·17	2·34	− 0·17
Mädler	76	1853·77	178·8	177·9	+ 0·9	2·65	2·34	+ 0·31
Jacob	77	1854·05	177·9	178·3	− 0·4	2·22	2·35	− 0·13
Dawes	78	1854·56	178·5	179·0	− 0·5	2·26	2·38	− 0·12
O. Struve..	79	1854·66	179·0	179·1	− 0·1	2·24	2·38	− 0·14
Morton	80	1854·67	178·5	179·1	− 0·6	2·22	2·38	− 0·16
Mädler	81	1854·70	179·4	179·2	+ 0·2	2·51	2·38	+ 0·13
Dembowski	82	1854·86	179·8	179·4	+ 0·4	2·37	2·40	− 0·03
Dawes	83	1855·48	180·1	180·3	− 0·2	2·43	2·43	0·00
Winnecke..	84	1855·54	181·6	180·4	+ 1·2	2·49	2·43	+ 0·06
Secchi	85	1855·59	180·1	180·5	− 0·4	2·32	2·44	− 0·12
O. Struve..	86	1855·61	179·1	180·5	− 1·4	2·29	2·44	− 0·15
Mädler	87	1855·78	181·8	180·7	+ 1·1	2·64	2·45	+ 0·19
Winnecke..	88	1856·39	182·8	181·6	+ 1·2	2·52	2·48	+ 0·04
Dembowski	89	1856·42	181·8	181·6	+ 0·2	2·69	2·48	+ 0·21
Secchi	90	1856·43	182·4	181·6	+ 0·8	2·46	2·48	− 0·02
O. Struve..	91	1856·57	179·9	181·8	− 1·9	2·46	2·49	− 0·03
Jacob	92	1856·73	181·3	182·0	− 0·7	2·53	2·50	+ 0·03
Mädler	93	1857·39	183·3	182·9	+ 0·4	2·46	2·54	− 0·08
Secchi	94	1857·62	183·6	183·1	+ 0·5	2·43	2·55	− 0·12
Jacob	95	1857·66	183·1	183·2	− 0·1	2·53	2·55	− 0·02
O. Struve..	96	1858·01	181·9	183·6	− 1·7	2·51	2·58	− 0·07
Jacob	97	1858·20	184·0	183·9	+ 0·1	2·57	2·58	− 0·01
Dembowski	98	1858·29	183·2	184·0	− 0·8	2·66	2·58	+ 0·08
Mädler	99	1858·54	183·6	184·3	− 0·7	2·64	2·59	+ 0·05
Morton	100	1859·34	184·9	185·2	− 0·3	2·70	2·64	+ 0·06
O. Struve..	101	1859·94	186·1	185·9	+ 0·2	2·62	2·67	− 0·05
Dawes	102	1860·36	185·5	186·4	− 0·9	2·71	2·70	+ 0·01
O. Struve..	103	1861·58	187·4	187·7	− 0·3	2·69	2·76	− 0·07
O. Struve..	104	1862·76	189·1	189·0	+ 0·1	2·77	2·82	− 0·05
Dembowski	105	1863·09	190·1	189·3	+ 0·8	2·76	2·84	− 0·08
O. Struve..	106	1863·60	188·2	189·9	− 1·7	2·77	2·86	− 0·09
Engelmann	107	1864·45	190·5	190·7	− 0·2	3·11	2·91	+ 0·20
Dembowski	108	1864·95	191·2	191·2	0·0	2·79	2·93	− 0·14
O. Struve..	109	1865·36	191·9	191·5	+ 0·4	2·94	2·95	− 0·01
Dawes	110	1865·38	191·5	191·6	− 0·1	3·08	2·95	+ 0·13
Engelmann	111	1865·39	192·7	191·6	+ 1·1	2·96	2·95	+ 0·01
Talmage	112	1865·72	189·1	191·9	− 2·8			
Secchi	113	1865·81	192·5	192·0	+ 0·5	2·98	2·98	0·00
Talmage	114	1866·43	189·2	192·6	− 3·4	3·73	3·02	+ 0·71
O. Struve..	115	1866·63	193·0	192·7	+ 0·3	3·00	3·03	− 0·03
Kaiser	116	1866·68	193·9	192·8	+ 1·1	2·86	3·03	− 0·17
Dembowski	117	1866·92	193·2	193·0	+ 0·2	2·89	3·04	− 0·15
Main	118	1867·37	192·1	193·4	− 1·3	3·00	3·07	− 0·07
Dunér	119	1868·42	194·9	194·3	+ 0·6	3·07	3·11	− 0·04
Brünnow..	120	1868·55	194·1	194·4	− 0·3	3·11	3·12	− 0·01
O. Struve..	121	1868·58	194·7	194·4	+ 0·3	2·98	3·12	− 0·14
Dembowski	122	1868·88	195·7	194·7	+ 1·0	2·99	3·14	− 0·15
Talmage	123	1868·93	194·5	194·8	− 0·3	3·61	3·14	+ 0·47
Dunér	124	1869·63	195·0	195·3	− 0·3	3·00	3·17	− 0·17
Gledhill	125	1870·35	196·8	195·9	+ 0·9	3·35	3·20	+ 0·15
Dembowski	126	1870·95	197·1	196·4	+ 0·7	3·09	3·23	− 0·14
Dunér	127	1871·35	196·6	196·7	− 0·1	3·15	3·25	− 0·10
Gledhill	128	1871·45	196·5	196·8	− 0·3	3·28	3·26	+ 0·02
Wilson	129	1871·51	194·3	196·8	− 2·5	3·51	3·26	+ 0·25
Talmage	130	1871·86	197·3	197·1	+ 0·2	3·32	3·28	+ 0·04
Copeland	131	1872·28	196·5	197·5	− 1·0			
Wilson	132	1872·53	197·7	197·6	+ 0·1	3·25	3·31	− 0·06
O. Struve..	133	1872·57	195·3	197·7	− 2·4	3·26	3·31	− 0·05
Dembowski	134	1872·96	198·1	197·9	+ 0·2	3·12	3·33	− 0·21
Copeland	135	1873·40	196·7	198·3	− 1·6			
Wilson	136	1873·42	198·4	198·3	+ 0·1	3·14	3·35	− 0·21
O. Struve..	137	1873·56	197·6	198·4	− 0·8	3·15	3·36	− 0·21
Gledhill	138	1873·68	198·9	198·5	+ 0·4	3·40	3·37	+ 0·03
Copeland	139	1874·33	198·8	198·9	− 0·1			
Gledhill	140	1874·41	198·2	199·0	− 0·8	3·35	3·40	− 0·05
O. Struve..	141	1874·61	199·8	199·1	+ 0·7	3·41	3·41	0·00
Dembowski	142	1874·90	199·2	199·3	− 0·1	3·28	3·42	− 0·14
Schiaparelli	143	1875·46	198·6	199·7	− 0·9	3·34	3·44	− 0·10
Dunér	144	1875·54	199·7	199·8	− 0·1	3·29	3·45	− 0·16
Gledhill	145	1875·56	200·0	199·8	+ 0·2	3·28	3·45	− 0·17
Nobile	146	1875·65	200·6	199·9	+ 0·7	3·74	3·45	+ 0·29
Doberck	147	1876·29	199·3	200·3	− 1·0			
Gledhill	148	1876·48	200·6	200·4	+ 0·2	3·28	3·50	− 0·22

CHAPTER IV.

ON RELATIVE RECTILINEAR MOTION.

WHEN two stars happen to lie nearly in the same visual line, but one far behind the other, they are said to be optically double, and not to form a binary system. In this case, if one or both are affected with any proper motion, they will appear to change their relative position both in angle and distance, and the list of changing measures will resemble that of a binary system. But since it may be assumed that their proper motions are approximately uniform and rectilinear, it will follow that their relative motion will also be uniform and rectilinear; and hence that when a series of points is charted as before, they will lie approximately on a straight line.

No difficulty will be found in the graphical construction for this straight line, and it is not necessary to give an example. The processes will be as follows. Correct all the observed angles for precession, and chart them as before, thus obtaining interpolated angles for every five or ten degrees. Chart all the distances, and obtain interpolated distances corresponding to the same angles. It will be convenient to convert these coordinates r, θ into rectangular coordinates x, y by the usual formulæ $x = r \cos \theta, y = r \sin \theta$, referring to the adopted meridian as one of the axes, and then to draw a straight line passing among the points. By observing the times at which the star occupied certain points on this line, it is easy to ascertain what position it would occupy at any intermediate or any later time, and to ascertain by the protractor and compasses what its angle of position and distance would be at that time, and thus either to compare former observed positions with those which would have resulted from uniform movement in the straight line so found, or to construct an ephemeris for future years.

But the analysis of this problem is not at all beyond the reach of non-mathematical amateurs, and we shall therefore give a specimen of the more exact analytical handling of this problem in the case of 61 Cygni.*

In this case the problem is one of great interest. The two stars have very large proper motions nearly identical both in direction and in amount. That of one is given by W. Struve as $517'' \pm 10''$ per century in the direction $51° 16' \pm 1°$, and that of the other as $509'' \pm 10''$ in the direction $53° 38' \pm 1°$. The probability of a physical connection between these two stars is almost incalculably great. Struve has expressed it arithmetically, and illustrates it by saying that the physical connection between the components of 61 Cygni is more than a hundred thousand times more probable than that, after an experience of more than five thousand years, the sun will rise on the morrow.

But if they were physically connected, the relative motion would not be rectilinear, but orbital, from their mutual attraction ; and it becomes, therefore, a matter of some importance to examine accurately into their relative motion. What is certain is that hitherto the motion has deviated extremely little from a straight line.

The observations are charted on millimetre paper (see Plate III.), and curves drawn among the dots as before explained, and the interpolated angles and distances read off as in Table I. and converted into x and y.

---

* See *Monthly Notices*, vol. xxxv. p. 323 (1875). This chapter was written in 1876.

TABLE II.

t.	θ.	r.	r cos θ.	Δx.	r sin θ.	Δy
1820	82·7	15·15	1·9250		15·027	
1825	86·4	·35	+ 0·9638	·9612	15·320	·293
1830	89·9	·60	+ 0·0272	·9366	15·599	·279
1835	93·3	·90	− 0·9157	·9429	15·874	·275
1840	96·5	16·23	− 1·8373	·9216	16·125	·251
1845	99·6	·60	− 2·7683	·9310	16·367	·242
1850	102·6	17·00	− 3·7084	·9401	16·590	·223
1855	105·5	·43	− 4·6579	·9495	16·796	·206
1860	108·2	·89	− 5·5776	·9197	16·995	·199
1865	110·8	18·38	− 6·5268	·9492	17·182	·187
1870	113·3	·91	− 7·4797	·9529	17·368	·186
1875	115·7	19·49	− 8·4520	·9723	17·562	·194

In this case, however, it is plain from the columns of differences that these points do not lie on a straight line; for then the differences of the two columns would vary together. In fact, the differences in the column Δx are nearly uniform, while those in Δy steadily decrease; and thus indicate a curve slightly concave to the origin.

A little consideration, however, will show that this orbit cannot be elliptical. Taking the early observations into account, the companion has described about 80°, and yet has scarcely deviated from a rectilinear path. It is almost certain that the relative path is of a hyperbolic nature. It is given approximately, so far as at present described, in Fig. 16.

But if we wish to proceed with the problem in the case of any star, let us assume $p = r \cos (\theta - a)$ as the equation to the required line, p and a being the elements to be determined.

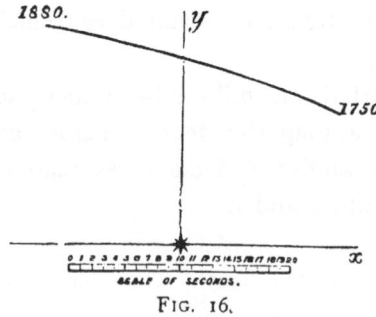

1880.

1750.

0 1 2 3 4 5 6 7 8 9 10 11 12 13 14 15 16 17 18 19 20

SCALE OF SECONDS.

FIG. 16.

No values of a and p can be found which will exactly satisfy the twelve derived equations, and the equations must therefore be combined by the method of least squares, so as to give the *most probable* straight line.

Let the equation be written

in the form $x \frac{\cos a}{p} + y \frac{\sin a}{p} - 1 = 0$. Then when the values of x and y for any of the selected points are substituted, we shall have $x_1 \frac{\cos a}{p} + y_1 \frac{\sin a}{p} - 1 = e_1$, where e_1 is the error.

Geometrically, e_1 is the intercept made on the line by parallels to the axes through any point P, and is therefore proportional to the perpendicular from P.

Let the series of twelve equations be thus formed; and let the first equation be multiplied by x_1, giving

$$x_1^2 \frac{\cos a}{p} + x_1 y_1 \frac{\sin a}{p} - x_1 = e_1 x_1;$$

and the second equation be multiplied by x_2, giving

$$x_2^2 \frac{\cos a}{p} + x_2 y_2 \frac{\sin a}{p} - x_2 = e_2 x_2;$$

and so on, and let all these equations be added together, giving

$$\Sigma (x_1^2) \frac{\cos a}{p} + \Sigma (x_1 y_1) \frac{\sin a}{p} - \Sigma (x_1) = \Sigma (e_1 x_1) \quad \text{(A)}.$$

Similarly, by multiplying the equations by y_1, y_2, etc., respectively, and adding, we shall obtain

$$\Sigma (x_1 y_1) \frac{\cos a}{p} + \Sigma (y_1^2) \frac{\sin a}{p} - \Sigma (y_1) = \Sigma (e_1 y_1) \quad \text{(B)}.$$

Now the method of least squares shows that on the assumption that all the twelve equations are to have equal weight, the most probable result will be obtained, or in this case that a line will be found such that the sum of the squares of the perpendiculars from the points on the line will be a minimum, by taking $\Sigma (e_1 x_1) = 0$, and $\Sigma (e_1 y_1) = 0$.

Hence our final equations for determining $\frac{\cos a}{p}, \frac{\sin a}{b}$ are

$$\Sigma (x_1)^2 \frac{\cos a}{p} - \Sigma (x_1 y_1) \frac{\sin a}{p} - \Sigma (x_1) = 0$$

$$\Sigma (x_1 y_1) \frac{\cos a}{p} - \Sigma (y_1^2) \frac{\sin a}{p} - \Sigma (y_1) = 0.$$

Solving these, we at once obtain

$$\tan a = \frac{\Sigma (x_1^2) \, \Sigma (y_1) - \Sigma (x_1 y_1) \, \Sigma (x_1)}{\Sigma (y_1)^2 \, \Sigma (x_1) - \Sigma (x_1 y_1) \, \Sigma (y_1)} = \frac{m}{n};$$

and $p^2 = \dfrac{\Sigma\,(x_1{}^2)\,\Sigma\,(y_1{}^2) - \{\Sigma\,(x_1\,y_1)\}^2}{m^2 + n^2} = \dfrac{\Sigma\,(x_1{}^2)\,\Sigma\,(y_1{}^2) - (\Sigma\,x_1\,y_1)^2}{n^2}\cos^2 a\;;$

and the equation to the straight line required is $p = r \cos (\theta - a)$.

Let the epoch at which the distance was a minimum be T. This can be approximately determined from the interpolating curve, by noting the time that corresponds to the angle a. But the correctness of this result would depend on the correctness of the curve at that point alone, and would not utilise other observations. We therefore proceed as follows. If θ_1 is the value of θ at the time t, $p \tan (\theta_1 - a) = m\,(t_1 - T)$; therefore $m\,(t - T) = c_1$ where $c_1 = p \tan (\theta_1 - a)$.

Obtain a series of equations of this type corresponding to all the points on the interpolating curve. It is required to combine them so as to obtain the most probable values of m and T.

For T write $T_1 + \tau$ where T_1 is an approximate value of T, and τ is small, and for $t_1 - T_1$ write k_1, and for $m\,\tau$ write z ; so that

$$m\,k_1 - z - c_1$$
$$m\,k_2 - z = c_2$$
$$\text{---}\quad \text{a series of } n \text{ equations.}$$

Treating these as before, we shall obtain the two equations for m and z

$$m\,\Sigma_1\,(k_1{}^2) - z\,\Sigma\,(k_1) = \Sigma_1\,(c_1\,k_1),$$
and
$$m\,\Sigma_1\,(k_1)\; - n\,z \;\;\;\;= \Sigma_1\,(c_1),$$

the solution of which only requires the formation of the quantities

$$\Sigma\,(k_1),\; \Sigma\,(c_1),\; \Sigma\,(c_1\,k_1) \text{ and } \Sigma\,(k_1{}^2).$$

PLA

INTERPOLATING CURVE

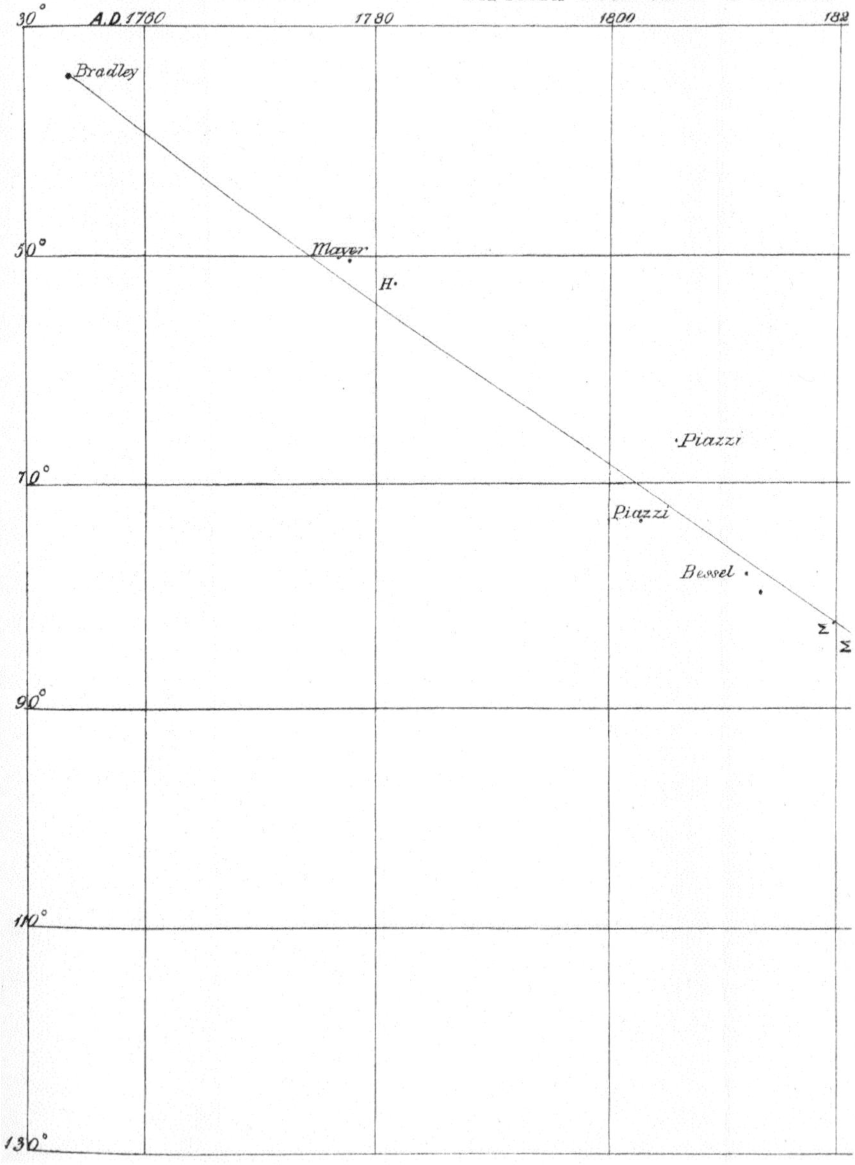

TE III
OF 61 CYGNI. (POSITIONS)

to face P.138.

0   1840   1860   1880

CHAPTER V.

ON THE EFFECT OF PROPER MOTION AND PARALLAX
ON THE OBSERVED POSITION ANGLES AND DIS-
TANCE OF AN OPTICALLY DOUBLE STAR.

IF a pair of stars is only optically double, and one is moving relatively to the other, it is plain that there will be a change in position angle and distance due to this cause.

If the near one is sufficiently near our system to have an appreciable parallax, it has long been seen that the circumstances were favourable for the determination of the parallax. The *proper motion* of either of the stars will of course complicate the result, and we proceed to show how in the first place by a preliminary examination the measures may be studied in order to see whether they show any trace of such parallax ; and then how they may be submitted to rigorous calculation for the purpose of ascertaining its amount.

It is well known that the annual motion of the earth would cause the nearer star to revolve, apparently, in an ellipse round its position as seen from the sun, the form of the ellipse being that which the earth's orbit would assume if seen from the star ; that the ratio of the axes would be the sine of the latitude of the star, and that the major axis would be the annual parallax.

Further, the proper motion of either or both stars would cause one to move relatively to the other in a straight line. If both causes are in operation, the motions of the stars will be combined.

Accidental errors being as far as possible got rid of by the graphical methods already described and illustrated, and the

position angles and distances for intervals of, say, twenty
days, having been obtained from the interpolating curve

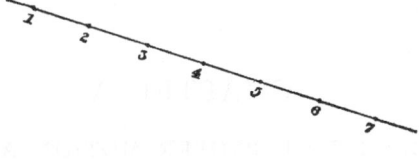

(supposed to embody the observations of some years, and
taken at all times of the year), the charted positions will

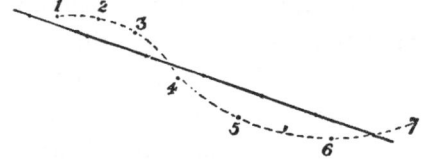

lie approximately in a straight line; and if the deviations
from it show no law, then it is not worth while proceeding
further. But if it is found that the points lie alternately
on one side and on the other of that straight line, and that
the total period is one year, then we have clear indications
of a measurable parallax.

The graphical proceeding will be as follows: by comparing the positions at intervals of as large an integral
number of years as may be, and using all the measures
available, first determine what the proper motion is. Then
chart over again the positions that would have been occupied by the star if proper motion had not affected it: the
resulting points ought to lie in an ellipse whose axis major
is in a position six hours distant from the longitude of the
star, and the ratio of whose axes is the sine of the latitude
of the star. The major axis itself is the parallax sought.
The preceding diagrams may help to make this clear:
No. 1 exhibits the effect of annual parallax only; No. 2
of proper motion only; and No. 3 the effect of both com-

bined. When this has been done graphically, it may be thought worth while to proceed further with the rigorous calculation as follows :

Let S be the principal star, σ the companion, at distance D″, and let S′ be the position of the principal star after the lapse of a year, in consequence of its proper motion, M S′ $= d\,\delta$ in declination, and S M $= d\,a \cos \delta$ in R. A, expressed in seconds of arc, δ being the declination.

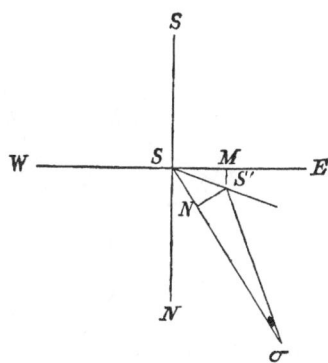

Then the change in position angle is $- S\,\sigma\,S'$.

FIG. 17.

Let the position angle N S σ be θ, and let E S S′ $= \phi$.

Here $\tan \phi = \frac{d\,\delta}{\delta\,a \cos \delta}$, which determines ϕ. ($< 180°$),

and $\quad S\,\sigma\,S' = \dfrac{180 \times S S' \sin S' S \sigma}{\pi \times \sigma S'}$ in degrees,

$\qquad\qquad = \dfrac{57\cdot 3 \times d\,a \cos \delta \sec \phi \cos (\theta + \phi)}{D}$ in degrees,

$\therefore d\,\theta = -\,m\,d\,a.$ \hfill (1)

and d D, the change in distance, $= - S N$,

$\qquad \therefore d\,D = -\,S S' \sin (\theta + \phi).$ \hfill (2)

These equations completely determine the change in angle and distance due to proper motion.

There will also be a change of position due to parallax, or rather to the difference of parallaxes of the two stars, and the investigation of this is of some importance, as it may lead to the determination of the parallaxes of some stars. It will be remembered that it was with this view that the examination of double stars was first entered upon by Sir William Herschel.

If S be the sun, Σ a star, E the earth moving in the ecliptic, then the apparent path of the star on the background of the heavens will be a small ellipse, whose major axis is parallel to the ecliptic, and whose minor axis is perpendicular to it, and parallel to the circle of latitude.

Now let C be the place of the star as viewed from the sun,

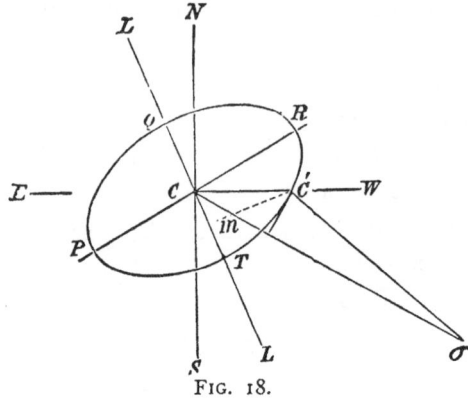

FIG. 18.

P Q R T the ellipse described in consequence of parallax, L L the circle of latitude, Q the position of the star when the longitudes of the earth and the star are the same, T where they differ by 180°, σ the smaller star supposed to have no parallax, C' the position of the star when the time is t, and the longitude of the earth is L ; then C σ C' is the change in the angle of position due to parallax.

Then we have, R being the radius factor of the earth at the time t, x the constant of parallax for the star at the earth's mean distance taken as unity, λ the latitude,

$$C R = R.x, C T = R.x \sin \lambda ;$$

therefore if Z' is the apparent position in degrees at the time t, Z the mean position, as viewed from C, D the distance of C σ in seconds, X the angle σ C C',

$$Z' = Z - \frac{57 \cdot 3 \, C \, C \, \sin X}{D}, \text{ C C' being expressed in seconds of}$$

arc, or $Z' = Z - \frac{57 \cdot 3 \, \rho . \, x. \, \sin X}{D}$ (3)

where ρ is the elliptic radius at the time t; and ρ and X are calculated as follows :—

If L, l are the longitudes of the earth and the star at the time t, I the angle L C C', A the angle L C N,

Then from the figure it is plain that

$$\cos I = \frac{C \, m}{C' \, m} = \frac{E \, M \sin \lambda}{S \, m} = \cot (L - l) \sin \lambda ; \quad (4)$$

and then $\rho = R \sin (L - l) \operatorname{cosec} I$,   (5)

and $X = Z + I - A$

      $= Z' + I - A$, very nearly,   (6)

    or $360 - (Z' + I + A)$,

and A is computed from the triangle E P C, E being the pole of the ecliptic, P the pole, C the star, by the formula

$$\cos A = \frac{\cos E\,P - \cos E\,C \cos P\,C}{\sin E\,C \sin P\,C}$$

$$= \frac{\cos \omega - \sin \lambda \sin \delta}{\cos \lambda \cos \delta} \qquad (7)$$

or by $\cos \dfrac{A}{2} = \dfrac{\sqrt{\sin S \sin (S - \omega)}}{\cos \lambda \cos \delta}$ (8)

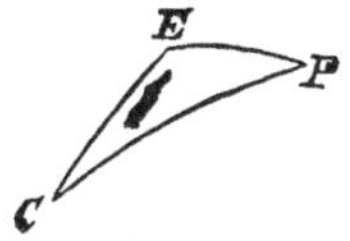

FIG. 19.

where $S = \frac{1}{2}$ (E C + C P + E P).

We compute therefore A by formula (8) or (7), which give no ambiguity, since A is $< 180°$;

I by (4), I being in the same quadrant as $L - l$;

ρ by (5), and X by (6) ;

and substituting in (3) we get an equation of 'the form $Z' = Z + n\,x$.

If a series of such observations is taken, they can be combined so as to give values of Z and x.

But in practice the proper motion will be involved with the parallactic motion, and the constants may be determined together as follows :—

Let $Z' = Z + n\,x - m\,(t - T)\,d\,a$ (9)

where t is the time,

T a fixed epoch,

m the constant determined above by (1),

x the unknown constant of parallax,

Z the position angle at the time T,

Z' the position angle at the time t.

It will be convenient to subtract from Z' the integral number of degrees z in it, and the same from Z, and put $Z - z = \zeta$, and if the weight of an observation or group of observations be w, we obtain a series of equations of the type

$$w\,\zeta + w\,n\,x - w\,\mu\,d\,a = w\,a,$$

which must be solved by the method of least squares for ζ, x, and $d\,a$.

This was done by Jacob for the star a Herculis, and he arrived at a parallax $0''\!\cdot\!06$, the proper motion being so small as to be neglected.

CHAPTER VI.

ON THE ERRORS OF OBSERVATION AND THE COMBINATION OF OBSERVATIONS.

FOR the general treatment of this subject we must refer the reader to Airy's *Theory of Errors of Observation ;* but an example or two may be given here of the application of the theory to the Observations of Double Stars.

Suppose the following series of measures of position was taken by one observer on one night : 211·8, 213·2, 209·9, 212·0, 212·5, 211·9, 210·8, 212·1. We require to know what is the most probable result of these measures, and within what limits it may be relied upon.

The most probable result is shown to be the arithmetical mean, which is easily found to be in this case 211·77.

Make a list of the separate errors of each of the observations from this mean, distinguishing between those in which the observation is in excess of the mean, from those in which it is less than the mean. In this case the errors are + ·03, + 1·43, − 1·87, + ·23, + ·73, + ·13, − 97, + ·33.

Take the mean of the + errors ·48, and the mean of the − errors, 1·42 ; and, finally, take the mean of these ·95.

This is a numerical quantity, without sign, and is a measure of the goodness of the measures. It is called the "mean error," and furnishes a ready means of comparing the value of the observations taken on one night, or with one instrument, from those taken on another night, or with a different instrument. It further gives a means of comparing the measures of one observer with those of another. Thus if another observer, on the same evening, with the same telescope, took the following six readings, 212·3, 212·7, 211·5, 211·2, 211·9, 212·1, it will be found that the mean of the errors is 42, and the most probable result is 211·95. This, however, does not show the *probable error*. This expression must not be taken to mean the error which is more probable than any other error, but the limit within which, on either side of the arithmetical mean, it is *probable* that the truth lies.

This is got by the formula, (Airy, § 60) probable error of the arithmetical mean $= 0·6745 \sqrt{\frac{\text{Sums of squares of apparent errors}}{n.(n-1)}}$, n being the number of the observations.

To take the first set of readings, the apparent errors of which were given above, the sum of their squares is 7·1952; and n is 8, whence the probable error of the mean $= ·24$. In the second case the probable error will be found to be ·15.

The next question is how to combine the observations made by these two observers so as to get the most probable result.

Let $\frac{1}{(\text{probable error of mean})^2}$ be called the "theoretical weight," or w.

Then in the first case $w_2 = \frac{1}{(.24)^2} = 17$, and in the second case $w_2 = \frac{1}{(·15)^2} = 44$; and the most probable result is shown to be

$$\frac{17 \times 211·77 + 44 \times 211·95}{17 + 44} = 211·90,$$

and the theoretical weight of the result is $17 + 44 = 61$, and probable error of result $= \frac{1}{\sqrt{61}} = ·13$.

Generally, if a, b, c, ... are successive results, whose theoretical weights are w_1, w_2, w_3 ... the most probable result is $\frac{w_1 a + w_2 b + w_3 c \dots}{w_1 + w_2 + w_3 \dots}$, with theoretical weight $w_1 + w_2 + w_3 + \dots$, and probable error $\frac{1}{\sqrt{w + w_2 + w_3 + \dots}}$.

It must be observed that this method assumes that the observations are really *independent* of one another and very numerous; and these conditions are not easily observed in double-star measures. No one who has long observed double

10

stars will have failed to notice that the readings taken on a single night tend to confirm one another, and yet may differ appreciably from those taken on another night. They may be taken with all honesty of purpose, yet the later readings are not strictly independent, but tend to confirm the early readings. Hence it is much more valuable work to take a moderate number of readings on several nights than to take very many readings on a single night. And it is not worth while to apply these methods of calculating the most probable result and the probable error to the observations of one night, but to the separate results of many nights, with the view of determining as accurately as may be one place for the year, and the weight to be attached to it.

The most useful form in which observations could be published would be to give the number of nights of observation, the resulting position, and the theoretical weight, the last number being thus not an arbitrary number assigned by guess,* but one which arises directly from the observations, and is referred to the same unit by all observers; a weight 1 assigned to an observation of position meaning a probable error of 1°, and generally a weight w indicating a probable error of $\frac{1}{\sqrt{w}}$ degrees.

So in determinations of distance the same elements should be given, and a theoretical weight w would indicate a probable error of $\frac{1}{\sqrt{w}}$ seconds.

It may be observed that the theoretical weight of a result varies inversely as the square of the probable error of that result: now it is also true † that the probable error of the arithmetical mean of a number of equally good observations varies inversely as the square root of the number of observations. Hence the theoretical weight of a number of *independent* equally good observations varies directly, in the case of any particular observer, as the number of observations.

This confirms what was said above of the importance of observing the same star on many nights.

Hence, finally, it is possible to determine exactly the weight to be assigned to a given series of measures of a star by an observer A. It will consist of the product of two numbers, one of which is the number of nights of observation of that star, and the other is his "theoretical weight," or the mean of as large a number as may be of the theoretical weights obtained, as above explained, from his observations on stars of similar magnitudes and distances.

It is necessary to say *similar magnitudes and distances*, as the probable errors of an observer in measuring such stars as γ^2 Andromedæ, δ Cygni, and Castor will be very different, and therefore the theoretical weights of such observations will be very different.

PART III.
THE CATALOGUE AND MEASURES.

THE CATALOGUE.

INTRODUCTION.

THIS Catalogue gives the places, etc., of the selected list of stars. Great care has been taken in the selection; and, for the most part, the stars will be found to be those which are either binary, probably binary, those in which certain change has taken place, or those deserving of at least occasional careful measurement on other grounds.

The R. A., Dec., and Magnitudes, are approximate. Column 6 gives Struve's and Otto Struve's numbers, the latter being placed in brackets. Column 7 gives the number in Herschel's great Catalogue (*Mem. R. A. S.*, vol. xl.) Column 8 gives, roughly, the apparent arc described by the star since its discovery. Column 9 the probable character of the object. In this column the following initial letters are used: B (binary); PB (probable binary); PM (proper motion); RM (rectilinear motion); CC (certain change); PC (probable change).

The stars are taken chiefly from the great works of Σ., $O.\Sigma$., H_1, and H_2. It was our original intention to include a large number of Mr. Burnham's discoveries; owing, however, to the difficulty in selection, the extreme faintness of many of these objects, and the consideration that, as a rule, well-known stars only are suitable for the amateur, we have omitted them. For the convenience, however, of those observers who, having sufficient skill, patience, and instrumental power, may desire to assist Mr. Burnham in obtaining thoroughly reliable measures of his recently discovered pairs, we have appended to our Catalogue a selection from his published lists.

In conclusion, we have to acknowledge much kind help in the form of suggestions, measures, and lists of stars in certain or probable motion; and our best thanks are offered to Mr. Burnham, Dr. Doberck, M. Flammarion, and Mr. Ormond Stone.

Ref. No.	Name of Star.	R. A. 1880.	Dec. 1880.		Σ's No.	H₂'s No.	Arc.	Char- acter.
		h. m.	° ′				c	
1		0 1·5	−5 12	8, 10	3063	10308	8	P B
2	α Andromedæ	2·2	28 36	2, 11·2			10	P M
3	Cephei 316 B	2.7	79 3	6, 6	2	2	300?	B
4	h. 1007	7·4	26 20	7, 8, 9	[2]	35	12	P B
5	Cephei 318 B	9·4	76 17	7, 7	13	48	24	B
6		10·4	35 48	7, 8	[4]	59	34	B
7		11·3	−0 21	7, 10	23	66	8	R M
8	Andr. 69 B	12·2	25 28	7, 8	24	72		P C
9	H₁ V. 85	13·1	37 28	7·4, 9·5			5	R M
10		15	66 20	7, 8, 9	[6]	93	3	P C
11		15	65 49	7, 8, 10	[7]	96		P C
12	42 Piscium	16·2	12 49	7, 11	27	103	7	R M
13	Cass. 49 B	21	49 20	7, 9	30	127	4	R M
14	49 Piscium	24·6	15 23	7, 11	32	156	3	R M
15	λ Cass.	25·1	53 52	6, 6	[12]	162	13	P B
16		25	36 16	8, 11	[13]	164		P C
17	51 Piscium	26·2	6 17	5, 9	36	176		P C
18		31·9	40 20	8, 9	44	215	6	C C
19	h. 1041	33	48 42	6, 11	[16]	218		P C
20		34·7	−7 53	7, 10	49	233		C C
21		36	3 32	7, 9	[18]	242	20	B
22		37	36 55	8, 11	[19]	249	3	P C
23		37·5	45 35	8, 9	52	251	10	B
24	P. O. 181	41·2	50 47	7, 8	59	281	8	B
25	η Cass.	41·8	57 11	4, 8	60	283	88	B
26		43·9	11 11	8, 11	63	295	26	R M
27		44·5	40 33	9, 10	64	297		P C
28		45·9	9 57	8, 9	67	305	12	P B
29	λ Toucanæ	47·8	−70 9	7, 8		318	10	P B
30		47·9	83 2	9, 10	69	307	11	C C
31	66 Piscium	48	18 32	6, 7	[20]	316	24	P B
32	Andr. 36 B	48	22 58	6, 7	73	319	50	B
33		49	8 44	8, 9	74	320		P C
34	So. 390	52·2	−16 20	7, 7		338	54	P B
35	P. O. 251	53·2	0 8	7, 8	80	344	17	C C
36		56	46 44	7, 8	[21]	354	4	P C
37		58·7	−6 7	8, 9	86	373	20	C C
38		59·1	14 44	8, 8	87	377	2	P C
39	ψ¹ Piscium	59	20 50	5, 5	88	378		B
40	Ceti 160 B	1 1	−2 24	7, 8	91	393	6	P B
41	φ Andr.	3	46 36	5, 6	[515]		42	B
42	ζ Piscium	7·5	6 56	4·2, 5·3	100	435	4	P B
43		8	80 16	7, 8	[28]	430	10	C C
44		7·9	−8 18	8, 10	101	439	10	C C
45		11	48 24	7, 8	102	453		C C
46		1 12	63 16	9, 10	109	460	2	P C

Ref. No.	Name of Star.	R. A. 1880.	Dec. 1880.	Mag.	Σ's No.	Hᵧ's No.	Arc.	Character.
		h. m.	° ′				°	
47	Polaris	1 13·7	88 40	2, 9	93	400		P B
48	42 Ceti	13·6	−1 9	6, 7	113	474	10	C C
49	h. 2036	14	−16 25	7, 7		478	27	B
50	ψ Cass.	17·4	67 30	4, 9	117	490	2	P B
51		20·6	82 44	8, 9	118	502	8	C C
52		20·7	2 55	8, 9	122	514	5	P C
53		20·8	−0 46	8, 10	125	515	43	R M
54		25·6	16 21	7, 10	132	542	34	P M
55		25·9	35 14	7, 10	133	543	6	P B
56	P. I. 107	27	7 36	7, 11	[31]	548		P C
57	100 Piscium	28·5	11 57	6·9, 8	136	560		P C
58		30	58 3	7, 8	[33]	564		P C
59	P. I. 123	29·7	7 2	7, 7	138	568	10	P B
60	P. I. 127	30·6	−30 31	6, 7		573	8	P C
61		33·5	14 38	8, 8	142	588	16	R M
62		35·8	−11 55	5, 7	147	611	4	B
63	6 Eridani	35·2	−56 49	6, 6		612	106	B
64		36	55 16	7, 10	[35]	606	6	C C
65		36	80 18	7, 7	[34]	592	5	P C
66		39·8	32 34	8, 9	158	637	18	P B
67		44·4	20 31	8, 9	175	677	48	R M
68	γ Arietis	47	18 42	4, 4	180	694	2	P C
69		48·3	28 13	7, 8	183	704	20	P B
70	P. I. 209	49·7	1 15	7, 7	186	714		B
71		51	74 55	7, 8	185	710	10	P B
72		52·9	20 26	9, 11	196	738		C C
73		54	34 42	7, 8	197	740		R M
74	α Piscium	55·8	2 11	3, 4	202	753	13	P B
75	γ Andromedæ	56·5	41 46	3, 5, 6	[38]	755	25	B
76	10 Arietis	56·8	25 22	6, 8	208	761	14	B
77		2 2·5	61 44	8, 8	216	789	2	P C
78		3	19 46	8, 9	221	799	4	P C
79	ι Trianguli	5·4	29 44	5, 6	227	814	9	P C
80	Andr. 259 B	6·4	46 55	7, 8	228	818	50	B
81	66 Ceti	7	−2 57	6, 8	231	821	4	B
82	Trianguli 28 B	7·7	29 50	7, 7	232	826	4	P C
83		8·7	60 48	8, 9	234	827	8	P B
84		9·3	51 55	8, 9	236	836	2	P C
85	Arietis 65 B	10	23 19	8, 8	240	852		P C
86	ο Ceti	13·3	−3 32	−, 9		876	48	R M
87		12·7	44 2	7, 9	249	876	2	P B
88		14	37 57	8, 9	[40]	880	3	P C
89		16·6	60 59	7, 8	257	892	26	B
90	ι Cass.	19·2	66 51	4, 7	262	906	25	B
91	P. II. 89	21·1	29 22	7, 10	269	925	20?	P B
92		2 28·3	68 47	8, 9	278	949	8	P B

Re No	Name of Star.	R. A. 1880.	Dec. 1880.		Mag.	Σ's No.	H₂'s No.	Arc.	Character.
		h. m.	°	′				°	
93		2 32	−11	54	8, 11	288	984	11	C C
94		34	26	6	7, 9	[43]	990	29	P B
95		35	42	10	8, 9	[44]	995	4	P C
96		35.5	56	31	9, 12	293	1003	13	C C
97	84 Ceti	35.1	−1	12	6, 9	295	09	10	B
98	θ Persei	35.9	48	43	4, 10	296	10	8	B
99	γ Ceti	37.1	2	44	3, 7	299	19	7	B
100		37.5	38	55	8, 8	300	20	14	P B
101	Arietis 114 B	40.6	18	52	7, 8	305	36	12	C C
102	π Arietis	42.2	16	58	5, 8, 10	311	47	18	C C
103		44.1	72	24	7, 8, 9	312	44	8	C C
104	Persei 85 B	44.3	52	30	7, 7	314	53	20	P B
105		48	26	24	8, 10	326	80		B
106		49	44	2	8, 9	328	84	4	R M
107	ε Arietis	52.3	20	51	6, 6	333	98	10	P B
108		53	6	10	8, 8	334	1104	8	P C
109	P. I. 230	54	17	32	7, 10	[49]	08	6	P C
110	Persei 104 B	54.1	31	56	6, 8	336	09		C C
111	52 Arietis	58	24	47	6, 6	346	1129		P C
112		3 0.7	71	6	7, 7	[50]	32	20	P B
113		1	7	56	9, 9	355	47	6	P B
114		4.6	83	34	8, 9	343	28		R M
115	12 Eridani	6.9	−29	27	4, 7		77		P B
116	P. III. 1	7	65	13	9, 10	[52]	67	20	P B
117		7.8	0	18	8, 8	367	79	35	P B
118		10	38	11	7, 8	[53]	88	20	P B
119		14	18	45	8, 9	377	1210	6	P C
120		15.3	8	20	8, 9	380	22	20	P B
121	P. III. 46	16.4	20	33	7, 9	381	24	6	P B
122		20	50	1	8, 9	388	40	103?	P B
123		20.5	58	57	7, 8	389	42	5	P C
124	h. 1135	21	67	11	7, 8	[54]	39		P C
125		24.3	19	22	8, 8	403	71	6	P B
126		24.7	−4	41	8, 8	408	79	7	C C
127		25.2	59	38	7, 8	400	70	13	P B
128	7 Tauri	27.3	24	4	7, 7	412	88	28	B
129		27.6	19	25	8, 8	414	91	5	P C
130	P. III. 98	30.6	0	12	6, 8	422	1308	15	B
131	H₁. II. 52	32.5	33	44	7, 7	425	18	2	P C
132		35.2	−13	0	7, 8	436	37	2	R M
133		36.1	38	0	7, 8	434	38	2	R M
134		36	7	31	7, 10	[61]	43		P B
135		40.1	37	58	8, 9	447	70	3	R M
136	Atlas Pleiad.	42	23	39	5, 8	453	81	78?	P B
137	P. III. 170	43.1	25	13	6, 7	[65]	92		B
138		3 43.6	29	17	8, 11	459	96	7	C C

Ref. No.	Name of Star.	R. A. 1880.		Dec. 1880.		Mag.	Σ's No.	H₂'s No.	Arc.	Character.
		h. m.		°	′				°	
139	32 Eridani	3	48·2	−3	19	4, 6	470	1436		P C
140	Camel. 9 Hev.		47	60	45	5, 8	[67]	09	4	P C
141	Cephei 49 Hev.		50	80	22	5, 6	460	06	37	B
142			56·2	39	8	8, 10	483	70		P B
143	h. 671		59	33	7	7, 9	[71]	83	4	P C
144	P. III. 242		59	37	46	7, 8	[531]	86		B
145	P. III. 249	4	1	17	1	6, 9	[72]	1500	2	P C
146	H₁ N. 17		1·7	22	47	8, 8	494	06		P C
147			6	9	20	8, 8	[74]	21		P B?
148			7·8	58	29	7, 8	511	28	27	B
149	40 Eridani		9·8	−7	47	4, 9	518	53	190	B
150			11·1	22	29	8, 8	520	56	2	P C
151	55 Tauri		13	16	14	7, 9	[79]	71	23	P B
152	P. IV. 46		15	42	9	7, 7	[80]	82	4	P C
153	Tauri 230 B		16·8	11	5	7, 8	535	1600	13	P B
154	56 Persei		17	33	40	6, 9	[81]	1595	3	P C
155			16·9	14	46	7, 9	[82]	1602	35	P B
156			20	−1	41	8, 11	547	31		?
157	80 Tauri		23·3	15	23	6, 9	554	48		B
158			28	48	10	7, 10	[85]	77	12	P B
159	α Tauri		29	16	16	1, 11·2				P M
160			29·5	19	14	8, 9	567	90	11	P B
161	2 Camel.		30·4	53	15	5, 7	566	87	15	B
162	Aurigæ 4 B		31·1	26	42	6, 6	572	1703	8	C C
163			34·1	37	17	8, 8	577	15	19	C C
164			38·5	5	4	8, 8	589	48	7	C C
165			48	8	24	7, 9	[90]	1819	2	P C
166	P. IV. 207		49	73	53	6,8	[89]	1803	200?	B
167			50	3	0	7,8	[91]	34	2	P C
168	P. IV. 258		51·9	1	28	8, 8	622	49	10	P B
169			52·1	50	5	9, 9	619	42	9	C C
170			53·1	73	25	8, 10	615	31	9	P C
171	5 Aurigæ		52	39	13	6, 10	[92]	44	11	P C
172			54·1	4	55	7, 9	[93]	62	3	P C
173	P. IV. 288		58·4	19	38	7, 7	[95]	97	9	P B
174			59·3	22	56	6, 8	[97]	1903		B
175	14 i Orionis	5	1·3	8	20	6, 7	[98]	23	49	B
176			2·1	37	9	7, 7	644	25	6	P C
177	Camel. 19 Hev.		2·8	79	4	5, 8	634	1892	9	R M
178			3	83	18	8, 11	629	71	15	C C
179	h. 693		3	8	1	7, 10	[100]	1941	6	C C
180			4·2	−7	14	8, 10	651	47	36	R M
181			6·7	−12	1	4, 10	655	62		P C
182	14 Aurigæ		7	32	33	5, 7	653	61		P C
183	λ Aurigæ		10·6	40	0	5, 9		91	21	R M
184		5	12·9	64	38	7, 8	676	2001	10	P B

Ref. No.	Name of Star.	R. A. 1800.	Dec. 1800.	Mag.	Σ's No.	H₂'s No.	Arc.	Character.
		h. m.	° ′				°	
185		5 13	63 16	8, 8	677	2005	17	P B
186		14	46 54	7, 11	[104]	24	4	C C
187		16·6	24 51	8, 8	694	50	7	P B
188	P. V. 70	16·8	−24 54	6, 10, 10		61		P C
189	111 Tauri	17·4	17 16	6, 9		60		R M
190	η Orionis	18·4	−2 31	4, 5		71	4	P B
191	115 Tauri	20	17 51	6, 11	[107]	86	3	P C
192		20.2	2 50	7, 9	712	91	13	P B
193		21·8	41 10	8, 9	715	97	3	B
194	118 Tauri	21·9	25 3	6, 7	716	2103	5	B
195		22·4	29 30	7, 9	719	05	7	C C
196		22	18 16	7, 10	[108]	09	5	P C
197	32 Orionis	24·4	5 51	5, 7	728	33	28	P B
198		25	44 42	8, 9	727	29		P C?
199		27	−6 35	8, 9	735	49		R M
200	Tauri 380 B	29·2	21 56	7, 8	742	65	23	P B
201	θ¹ Orionis	29	−5 30	7, 8, 5 6, 11, 11	748	78	10	P C
202		29·9	26 53	7, 7	749	82	17	P B
203		31·6	37 53	7, 8	[112]	90	6	P B
204	h. 3278	33	12 57	7, 11	[113]	2213	2	P C
205		34	16 10	7, 9	[114]	27	4	P C
206	ζ Orionis	34·7	−2 0	2, 6	774	35	4	P C
207		37	62 45	7, 8	3115	37	5	P C
208		37·6	15 2	7, 8	[115]	57	11	P B
209		40	30 31	7, 10	[117]	75		P C
210		41	7 57·	7, 8	[119]	90	13	P C
211	So. 503	49·1	13 56	7, 9, 8		2351	16	R M
212	θ Aurigæ	51·5	37 12	3, 11, 11		70	4	R M
213		52	12 49	6, 8	[124]	76	66	B
214		52	22 29	7, 9	[125]	79	4	P C
215		55·9	27 39	8, 9, 11	830	2401	5	P C
216	h. 3823	56	−31 4	9, 9		12	10	P B
217		59	36 16	7, 10	[131]	23	10	P C
218		59·8	10 48	9, 9	840	32	4	P C
219		6 0	37 59	7, 10	[132]	28	7	P B
220		1	21 19	7, 10	[133]	39		
221	Lacaille 2145	1·7	−48 27	8, 8		70	25	P B
222		2·5	11 41	8, 8	853	62	10	R M
223		3·2	5 40	8, 8	859		2	C C
224		4	30 46	8, 8, 8	861	75		C C
225		10·2	62 28	7, 11	878	2518	13	R M
226	4 Lyncis	11	59 26	6, 8	881	27	6	B
227	Monoc. 33 B	15·9	−11 42	6, 10	3116	88	5	C C
228		20	15 36	7, 10	[140]	2611	4	P C
229	11 Monoc.	6 23	−6 57	5, 6, 6	919	50	2	P C

Ref. No.	Name of Stars.	R. A. 1800.	Dec. 1800.		Mag.	Σ's No.	H₂'s No.	Arc.	Character.
		h. m.	°	′				°	
230		6 23	7	11	7, 11	[142]	2654		P C
231		24	17	1	7, 10	[143]	58		P C
232	Aurigæ 229 B	24·3	52	33	7, 8	918	47	9	P B
233		27	37	9	7, 11	[148]	86	5	P C
234		27·5	14	50	8, 8	932	95	7	β
235		29	52	24	8, 9	935	2700		P C
236		29	27	23	6, 9	[149]	7	34	B
237		30·2	41	41	7, 8	941	10	4	P B
238		30·3	23	19	8, 9	943	17	19	R M
239		32	41	5	7, 8	945	30	9	P B
240		34·4	10	0	6, 9, 11	950	55	4	P B
241		34	9	49	9, 9	3117	57	6	C C
242		35·4	−7	52	9, 9, 9	955	70	5	P B
243	12 Lyncis	35·6	59	34	5, 6, 7	948	49	45	B
244		36	40	46	7, 8	[154]	66	5	R M
245		38	24	48	7, 10	[155]	85	2	P C
246	Sirius	39·7	−16	32	1, 8		99	30	B
247		40·4	18	20	6, 7	[156]	2795	15	P B
248		41·5	0	29	7, 8	[157]	2811	13	P B
249	14 Lyncis	42·5	59	35	6, 7	963	02	16	P B
250	15 Lyncis	46·9	58	35	5, 6	[159]	51	34	B
251	38 Geminorum	47·3	13	20	5, 8	982	72	16	B
252	μ Canis Maj.	50·6	−13	53	5, 8	997	99		C C
253		53·4	54	21	7, 9, 9	1001	2907	6	P C
254		54	11	58	7, 8	[163]	26	3	P C
255	P. VI. 301	56·1	52	56	7, 7	1009	27	10	C C
256	Lacaille 2640	7 1·2	−59	0	6, 7		3003	5	P B
257	45 o Geminorum	1·5	16	8	5, 11	[165]	2985	41	P B
258		5·4	27	26	7, 7	1037	11	18	B
259		8	−8	43	8, 10	49	40	10	P B
260	P. VII. 52	11·2	9	31	7, 7	[170]	68	7	P B
261	δ Geminorum	13	22	12	3, 8	1066	84	11	P B
262		14·3	0	38	8, 8	74	3103	24	P B
263		14·4	45	14	8, 10	71	3092	10	P B
264		15	4	17	9, 9	76	3107		C C
265		17	21	41	8, 8	81	21	8	P B
266		19	31	52	7, 10	[171]	3138		P C
267		21	50	13	8, 9	1091	58	3	P C
268		21·1	50	14	8, 8	93	61	25	P B
269		24	−14	44	7, 8	1104	3214	22	P B
270	Castor	27	32	9	3, 4, 9	10	28	130	B
271		27	31	12	6, 7	[175]	34		P C
272		29·1	76	1	8, 10	1107	18	10	P B
273		32	0	47	7, 9	[176]	89	4	P C
274	Procyon	7 33	5	33	1, 11, 8, 9, 7		91		B

Ref. No.	Name of Star.	R. A 1800.	Dec. 1880	Mag.	Σ's No.	H₂'s No.	Arc.	Character.
		h. m.	° ′				°	
275	P. VII. 170	7 33·7	5 30	7, 7	1126	3297	27	B
276		34	37 44	7, 8	[177]	93	11	P B
277		36·2	−3 14	8, 9	1132	3315		R M
278	κ Geminorum	37·2	24 41	4, 8	[179]	21	7	C C
279	Pollux	37·2	28 19	1, 11, 12, 10		29	7	P M
280		42	65 13	7, 11	1136	40	4	C C
281		41·6	13 43	8, 10	42	54	19	R M
282		46	3 42	7, 7	[182]	3404	7	P C
283		48·5	−2 28	8, 8	1157	20	10	P B
284		51	1 27	7, 7	[185]	41		B
285		56	26 37	7, 8	[186]	82	2	C C
286		56·1	4 30	8, 9	1175	88	13	P B
287		56·5	33 24	7, 7	[187]	85	21	B
288		58·1	12 25	8, 8	1179	3501		R M
289	11 Cancri	8 1·5	27 50	7, 10	86	28		P B
290	Lyncis 85 B	2	32 36	7, 8	87	33	18	P B
291	ζ Cancri	5·3	18 1	5, 6, 5	96	57		B
292	γ Argûs	5·8	−46 58	2, 5, 8		74	6	C C
293	P. VIII. 13	7	11 13	8, 10	1202	72	10	P B
294		15·2	−1 13	7, 8	16	3646	42	B
295	φ² Cancri	20	27 19	6, 6	23	80		C C
296	υ¹ Cancri	19·5	24 56	6, 7	24	81	2	P B
297		19·5	−40 36	8, 8		78	12	P B
298	h. 447	21	33 57	7, 11	[193]	96		P C
299		37·1	42 8	7, 8	1263	3832	16	P M
300	ε Hydræ	40·4	6 52	4, 8	73	68	25	B
301		41·4	0 28	8, 9	81	77	11	R M
302		44·4	71 16	7, 7	80	79	9	P B
303		44·9	12 35	8, 10	87	3907	14	P B
304	σ² Cancri	47	31 2	6, 6	1291	20		P C
305	ι Ursæ Maj.	51	48 31	3, 10	[196]	43	9	B
306		51·8	35 25	8, 9	1296	47	5	P B
307		54·6	15 45	9, 9	1300	70	7	C C
308	σ² Ursæ Maj.	59·8	67 37	5, 8	06	89	38	B
309		9 1·9	−6 39	8, 11, 10	16	4021	10	C C
310		2·5	70 28	8, 9	13	08	2	P B
311		6·5	53 13	7, 7	21	46	7	R M
312		9·6	−0 44	8, 8	29	78	2	R M
313		10·7	29 7	7, 8	3121	83		B
314	Lalande 18289	11	35 52	7, 7	1333	84	3	P C
315	38 Lyncis	11·4	37 19	4, 7	34	87	6	P B
316	37 Lyncis	12	51 46	6, 10	[199]	95·		P C
317	Lyncis 157 B	13·4	38 42	7, 7	1338	4101	25	B
318		16·6	52 5	7, 8	[200]	23	3	P B
319		9 16·8	28 24	7, 9	[201]	28	4	P B

Ref. No.	Name of Star.	R. A. 1880.		Dec. 1880.		Mag.	Σ's No.	H₂'s No.	Arc.	Char- acter.
		h. m.		°	′				°	
320	21 Ursæ Maj.	9 17		54	32	7, 8	1346	4126		P C
321	Hydræ 116 B	18·2		6	52	7, 7	48	39	11	B
322	ω Leonis	22		9	35	6, 7	56	65	326	B
323	Hydræ 134 B	25		2	0	7, 8	65	90		C C
324	Leonis Min. 30 B	34		39	30	7, 8	74	4231	23	P B
325	P. IX. 161	37·2		3	11	8, 11	77	53	4	P B
326	υ Ursæ Maj.	42		59	36	4, 12	[521]	78		B
327		43·4		17	7	8, 11	1385	94	9	P B
328	φ Ursæ Maj.	44		54	37	5, 6	[208]	90	111	B
329		45·5		27	33	8, 9	1389	4305	13	P B
330		45·6		69	28	8, 8	86	4297	5	P B
331	8 Sextantis A. C. 5	46·6		−7	32	6, 6		4314	240	B
332		55·1		46	57	7, 8	[210]	59	3	P B
333		59		31	40	8, 9	1406	87	8	P B
334		10 6		28	1	8, 9	[213]	4429	2	P C
335	P. X. 23	9·7		18	20	7, 7	[215]	49	37	B
336	39 Leonis	10·6		23	42	6, 11	[523]	53	5	B
337		12·6		21	10	8, 9	1423	67	23	P B
338	γ Leonis	13·4		20	27	2, 3	24	69	40	B
339	Leonis 145 B	14·2		7	2	8, 8, 9	26	77	16	B
340		16·3		15	57	7, 10	[216]	86	17	C C
341		18·4		25	14	8, 8	1429	4501	10	P B
342	P. X. 58	18·4		53	14	7, 8	28	4497	4	P B
343		20		17	50	7, 8	[217]	4513	1	P C
344		21		4	10	7, 9	[218]	22		P C
345		22		51	36	7, 10	[219]	26	1	P C
346		23·5		21	25	8, 8	1439	36	9	P B
347		26·6		−0	15	9, 12	45	56	8	P B
348	49 Leonis	28·7		9	17	6, 9	50	75		P C
349		30		60	46	7, 11	[222]	84	5	P C
350		32·5		6	21	7, 8	1457	4606	27	P B
351	P. X. 128	33·4		9	28	7, 9	[224]	12	19	P B
352		33		19	52	7, 10	[225]	13		P C
353		35		11	21	7, 8	[227]	26	8	P C
354		36·2		45	15	8, 9	1465	28	10	P B
355		41		13	36	8, 8	72	69	1	P C
356		40·7		23	12	7, 8	[228]	71	18	C C
357		43·1		41	44	7, 7	[229]	90	12	P B
358		48		52	45	7, 9	1486	4714	2	C C
359		48		21	24	8, 11	[230]	17	7	P C
360	54 Leonis	49·1		25	23	5, 7	1487	19	5	C C
361		53·9		−2	51	8, 8	1500	54	13	P B
362	P. X. 229	59·2		4	17	7, 7	04	82	8	P B
363		11 4·1		66	46	8, 10	14	4820	10	P B
364	[539] = A C	7·4		74	7	7, 7, 10	16	33	3	B
365	P. XI. 9	7·4		20	47	7, 7	17	34	8	B

Ref. No.	Name of Star.	R. A. 1880.		Dec. 1880.		Mag.	Σ's No.	H₂'s No.	Arc.	Char- acter.
		h. m.		° ′					°	
366	P. XI. 14	11 8		38	14	7, 8	[232]	4839	4	P C
367		11		67	21	7, 10	[233]	57	3	P C
368	ξ Ursæ Maj.	11·8		32	13	4, 5	1523	60		B
369	Leonis 339 B	12·7		14	56	7, 8	27	65	5	P B
370		15·5		18	51	8, 11	34	85	6	C C
371	ι Leonis	17·6		11	12	4, 7	36	96	30	B
372	57 Ursæ Maj.	22·6		40	0	5, 8	43	4924	8	B
373	τ Leonis	21·8		3	31	5, 7		19	7	R M
374		24·3		41	58	7, 7	[234]	34	105	B
375		23·8		61	45	6, 7	[235]	42	107	B
376	90 Leonis	28·5		17	28	6, 7, 8	1552	70	4	R M
377		29		67	0	7, 11	[236]	73	4	P C
378		30		56	48	7, 8	1553	76	3	P C
379	P. XI. 111	30		28	27	6, 7	55	78	4	P B
380		32·5		41	48	7, 9	[237]	5000	15	P B
381		53·6		54	5	8, 9	[243]	5126	18	P B
382		56·1		73	2	8, 9	1588	41		C C
383		57·4		−1	47	8, 8	93	49	8	P B
384	2 Comæ Ber.	58		22	8	6, 7	96	53		P C
385		59·4		69	20	7, 7	3123	67		P B
386		12 1·1		69	45	7, 9	1602	77		C C
387	Virginis 59 B	3·2		−11	11	6, 9, 8	04	88	3	P B
388		4·7		40	34	6, 7	06	96	7	P B
389		5·5		36	45	8, 8	07	5205	6	C C
390		18·1		54	49	7, 8, 11	[249]	91	4	P C
391	Comæ 68 B	18·4		26	15	7, 8	1639	93	17	P B
392		18		43	45	8, 8	[250]	95	9	P C
393		18·6		38	24	10, 10	1641	96	11	R M
394	α Crucis	19·9		−62	34	1, 2, 6		98	2	B
395		21·2		27	42	8, 9	43	5305	17	P B
396		21·3		8	3	9, 9	44	07	2	C C
397		23		32	2	7, 9	[251]	13	28	P B
398	Virginis 191 B	24·5		10	23	7, 8	1647	19	16	B
399		29		8	6	8, 10, 8	58	41	9	P B
400		30		12	4	8, 8	61	50	15	C C
401		31·2		21	52	8, 9	63	54	8	P B
402		32·1		−10	51	8,9,11,11	64	58	20	R M
403	γ Centauri	34·9		−48	18	4, 4		70	17	B
404	Corvi 58 B	35		−12	21	6, 6	69	73		C C
405	γ Virginis	35·6		−0	47	3, 3	70	77	300	B
406		39·4		15	2	6, 7	78	5401	10	R M
407	35 Comæ Ber.	47·4		21	54	5, 8, 9	87	30	40	B
408		50·3		−0	18	7, 8	[256]	45	17	P B
409	h. 2625	51		46	16	7, 8	[257]	52		P C
410	Lalande 24180	53·1		8	33	8, 11	1703	64	2	C C
411		12 55·3		16	31	8, 10	07	77	3	C C

Ref. No.	Name of Star.	R. A. 1880.	Dec. 1880.	Mag.	Σ's No.	H₂'s No.	Arc.	Character.
		h. m.	° ′				°	
412		56·5	14 7	8, 9	1711	5483	7	P B
413		13 1·8	27 35	8, 8	[260]	5514	4	P C
414	Comæ Ber. 179 B	2	16 8	8, 9	1722	15	7	C C
415	42 Comæ Ber.	4·1	18 10	6, 6	28	23		B
416		6·4	32 43	7, 7	[261]	35	9	B
417		14·6	3 34	7, 8	1734	70	5	P B
418		18	2 2	7, 8	42	90	5	P C
419	ξ Ursæ Maj.	19·1	55 33	2, 4	44	96	5	B
420		22	10 5	8, 10	46	5608		P C
421		22·6	16 21	7, 8	[266]	10	9	P B
422		27	35 31	6, 7	[269]	35	39	P B
423	P. XIII. 127	28·2	0 18	8, 9	1757	39	55	B
424	25 Canum Ven.	32	36 54	6, 8	68	73	82	B
425	Smyth 488	32·3	28 56	9, 10		5674	4	P B
426		33·5	70 23	8, 8	71	88	6	P B
427		34·9	20 33	6, 9	72	91	12	C C
428		36·8	46 50	8, 8	76	5706		P C
429	o Virginis	37	4 9	6, 8	77	04	10	B
430		40·2	5 43	8, 8	81	26	20	B
431	τ Boötis	41·6	18 3	5, 11	[270]	37	6	B
432		43·6	27 35	7, 8	1785	54	34	P B
433	P. XIII. 238	48·6	−7 28	7, 8	88	89	20	B
434	P. XIII. 242	49	30 29	7, 10	[272]	97	7	P B
435		50	5 50	7, 8	[273]	5803	5	P B
436		14 4·7	27 10	8, 9	1808	80	8	P B
437	OΣ 277	7	29 17	8, 8, 9	12	94	10	P B
438		7	44 46	7, 8	[278]	97	26	P B
439		7	60 58	7, 11	[280]	5902	4	P C
440		7·4	5 58	8, 8	13	5895	3	P B
441	P. XIV. 20	8	12 34	7, 9	[279]	98	3	P C
442		8	29 40	7, 7	1816	5904	3	P B
443		9·1	55 53	8, 9	20	13	17	P B
444	κ Boötis	9·2	52 21	5, 7	21	12	4	B
445		9·3	3 41	8, 8	19	07	65	B
446	Boötis 121 B	10·7	20 41	7, 8	25	22	12	C C
447		11·9	57 14	8, 10	30	33	19	P B
448		13	4 27	9, 9	32	34	10	P C
449		14	9 8	7, 11	[281]	44	9	P C
450		15·9	49 3	7, 7	1834	54	9	R M
451	P. XIV. 70	18·2	−11 7	7, 9	37	64	17	B
452		21·6	4 14	9, 9	42	87	4	P B
453		22·2	−9 40	8, 10	47	94	8	R M
454		28	49 44	7, 11	[283]	6037	4	P C
455		29	36 6	7, 8	1858	40		P C
456	α Centauri	31·8	−60 20	1, 2		47		B
457		14 34	52 6	7, 7	63	62	15	P B

11

Ref. No.	Name of Star.	R. A. 1880.	Dec. 1880.	Mag.	Σ's No.	H₂'s No.	Arc.	Character
		h. m.	° ′				°	
458	π Boötis	14 35·1	16 56	5, 6	1864	6066	4	P B
459	ʓ Boötis	35·4	14 14	4, 4	65	69	9	P B
460		35·9	10 2	8, 8	66	72	3	P B
461		36·1	49 15	7, 11	[284]	77	4	C C
462		37·5	51 55	7, 7	1871	88	6	B
463		40	−6 53	8, 8	76	99	10	B
464	ε Boötis	39·7	27 35	3, 6	77	6101	24	B
465		40·4	10 10	8, 9	79	06		P B
466	P. XIV. 182	41	42 53	7, 8	[285]	15	19	P B
467		42·9	6 27	7, 7	1883	24	13	P B
468	ξ Boötis	45·8	19 36	5, 7	88	46	100	B
469		47·1	45 25	7, 7	[287]	59	27	B
470		47·7	16 12	6, 7	[288]	61	28	P B
471		50·2	29 58	8, 10	1893	81	10	C C
472		51	32 46	6, 10	[289]	77	8	C C
473	Boötis 342 B	56	31 51	8, 9	1901	6212	3	C C
474	i Boötis	59·8	48 7	5, 6	09	37	10	B
475	P. XIV. 279	15 1·8	9 41	7, 7	10	45	7	P B
476		10	−4 26	8, 8	3091	6302	4	P B
477		10	56 30	7, 11	[294]	07	4	P C
478		10·4	38 45	6, 8	1926	10	4	P B
479		10·4	37 16	7, 9	[295]	11	8	P B
480		10·5	−7 50	8, 9	1925	05	3	P B
481	5 Serpentis	13	2 14	5, 10	1930	27	3	B
482		13·2	44 14	8, 8	34	36	8	P B
483	Cor. Bor. 1 B	14·5	27 16	6, 6	32	31	30	P B
484		16·4	−1 6	8, 9	3093	48	3	C C
485	η Cor. Bor.	18·2	30 43	5, 6	1937	62		B
486	P. XV. 74	20	37 46	7, 7	38	71	220	B
487		21·8	6 31	7, 8	44	82	7	P B
488		22·2	44 26	7, 9	[296]	88	10	C C
489	δ Serpentis	29·1	10 56	3, 4	1954	6426	38	B
490		29	42 13	8, 9	56	30		
491		30	25 25	7, 11	[297]	32		C C
492		30·2	13 19	8, 9	1957	34	8	P B
493		30·3	43 57	9, 9	61	40	9	C C
494		31·7	40 13	7, 7, 7	[298]	46	130	B
495		32	64 15	7, 9	[299]	53	3	P C
496	ʓ Cor. Bor.	34·9	37 1	4, 5	1965	65		R M
497	γ Cor. Bor.	37·7	26 41	4, 7	67	69		B
498		46	35 51	9, 11	83	6523	4	P C
499	π² Ursæ Min.	46·2	80 22	7, 8	89	47		P B
500		48·1	53 16	6, 8	84	34	6	P C
501	H₁ II. 85	49·7	−1 49	7, 8	85	35	18	P B
502	H₁ V. 126	54	17 43	8, 8	93	66		C C
503		15 55·2	13 27	7, 8	[303]	75	23	P B

Ref. No.	Name of Star.	R. A. 1880.		Dec. 1880.		Mag.	Σ's No.	H 's No.	Arc.	Character.
		h. m.		°	′				°	
504	ξ Libræ	57·8		−11	2	5, 5, 7	1998	6582		B
505		16	0	13	39	7, 8	2007	99	2	C C
506	κ Herculis	2·6		17	22	5, 6	10	6610	5	P B
507		3·5		83	58	7, 8	34	63	3	P C
508	ν Scorpii	5		−19	9	4, 7, 7, 8				P B
509	49 Serpentis	7·7		13	51	7, 7	21	34	37	B
510		7		34	43	7, 9	[306]	35	5	P B
511		7·8		26	59	6, 10	2022	40	9	C C
512		8·6		5	50	8, 9	23	41	7	P B
513		8·9		·7	40	8, 9	26	45	22	P B
514	σ Cor. Bor.	10·2		34	10	5, 6	32	54	220	B
515		15		41	56	7, 8	[309]	81	5	P C
516		20		37	19	8, 8	2044	6702	2	P B
517		21		38	13	8, 10	[310]	09	4	P C
518	η Drac.	21		61	47	2, 8	[312]	24	2	P C
519	Drac. 99 B	22		61	58	6, 7	2054	23	7	C C
520	Antares	22		−26	10	1, 8		07	4	B
521		22·6		21	10	7, 10	[311]	16	7	R M
522		23		26	15	6, 7	2049	18	3	P C
523	Herc. 71 B	23·6		18	40	7, 7	52	22	8	B
524	λ Ophiuchi	24·9			215	4, 6	55	27	110	B
525		29		40	21	7, 8	[313]	53	10	P B
526	ʒ Herculis	36·8		31	49	3, 6	2084	99		B
527		39		23	44	7, 7, 11	94	6816	5	C C
528	(Dembowski)	40·3		43	42	8, 8			17	B
529		40·5		35	57	8, 9	97	23	5	P B
530	21 Ophiuchi	45·3		1	25	6, 8	[315]	40	10	P B
531		45·4		9	37	7, 8	2106	42	15	P B
532	Herc. 167 B	47·1		28	52	6, 8	07	47	62	B
533		49		44	36	7, 12	[317]	60	10	C C
534		51		14	18	7, 9	[318]	63	3	P C
535		52·5		4	10	8, 8	3107	67	5	P B
536		54		14	29	8, 9	[321]	79	4	B
537	20 Drac.	55·8		65	13	6, 7	2118	95		B
538	P. XVI. 270	56·2		8	37	6, 7	14	88	16	B
539	Herc. 210 B	17	0	28	15	6, 9	2120	6910	140	B
540		1·6		47	8	7, 10	[323]	24	5	C C
541	μ Drac.	2·9		54	38	5, 5	2130	35	60	P B
542		3		31	23	6, 11	[324]	33	2	P C
543		7		21	22	7, 8	2135	45	5	C C
544	36 Ophiuchi	8		−26	25	4, 6		46	23	B
545	α Herculis	9·1		14	32	3, 6	2140	58		
546	δ Herculis	10·1		24	59	3, 8	3127	68	20	P B
547		11·8		26	43	8, 9	2145	73	5	R M
548		14·8		49	26	8, 9	53	95	8	P B
549	ρ Herculis	17 19·5		37	15	4, 5	61	7016	11	B

Ref. No.	Name of Star.	R. A. 1880.	Dec. 1880.	Mag.	Σ's No.	H₂'s No.	Arc.	Char. acter.
		h. m.	° ′				°	
550	Herc. 281 B	17 21·6	29 34	7, 8	2165	7028	8	P B
551		22·6	−9 54	7, 7	71	32	5	B
552	Ophiuchi 221.B	24·7	−0 58	6, 6	73	40	6	B
553	P. XVII. 135	26	2 55	7, 9	[331]	53	8	P C
554		29	6 6	7, 10	2185	62	4	C C
555	P. XVII. 163	30·9	21 4	6, 9	90	76	10	C C
556	Herc. 315 B	35·4	29 18	7, 10	92	88	10	C C
557		36·4	55 49	7, 8	99	7104	15	P B
558		37·5	41 43	7, 8	2203	08	8	B
559	61 Ophiuchi	39	2 38	5, 6	02	10		
560		39·5	63 44	6, 8	18	37	7	B
561	Herc. 331 B	40·3	31 11	7, 8	13	31		
562		40·4	17 46	8, 9	05	28	10	B
563		40·6	17 45	6, 8	15	30	10	B
564	μ Herc.—A.C. 7	41·8	27 48	4, 9, 10	20	42	174	B
565	P. XVII. 260	44·8	7 17	7, 8	[337]	61	10	B
566		46·5	15 21	7, 7	[338]	77	17	B
567	τ Ophiuchi	56·5	−8 11	5, 6	2262	7245	150	B
568		57·7	52 51	7, 8	71	67	7	P B
569		57·8	40 11	8, 8	67	62	6	B
570		58	25 22	8, 9	68	64	2	C C
571	70 Ophiuchi	59·4	2 33	4, 6	72	73		B
572		59·3	39 21	9, 9	75	81	13	P B
573	Herc. 401 B	18 0	48 27	6, 8	77	88	3	C C
574		1	56 26	7, 7, 8	78	97		P C
575	72 Ophiuchi	1·6	9 33	4, 8	[342]	92		
576	73 Ophiuchi	3·6	3 58	6, 7	2281	7309	18	B
577	Herc. 417 B	4	49 41	7, 11	[344]	23	7	C C
578		4·8	16 27	6, 7	89	22	10	P B
579		7	5 47	7, 10	[345]	31		P C
580		7·3	27 37	8, 8	2292	35		P C
581		8·4	0 9	7, 8	94	40	7	P B
582		12·2	83 54	7, 8	[349]	7417	8	P B
583	L 33731	13·6	−8 2	7, 9	2303	70	14	P B
584		17	11 23	9, 10	11	88	6	C C
585		19	7 10	7, 11	[347]	98	8	C C
586	Herc. 452 B	20·2	27 20	7, 8	2315	7406	24	B
587		20·6	25 56	8, 10	18	12	8	C C
588	d Serpentis	21	0 7	6, 8	16	10		P C
589		22·1	48 42	7, 8	[351]	23	15	B
590	39 Drac.	22·1	58 44	5, 8	2323	25	9	C C
591	φ Drac.	24	71 17	5, 6	[353]	43	16	C C
592		25·7	13 6	7, 9	2330	44	4	C C
593	L 34438	29·6	4 50	6, 8	42	71	3	C C
594		30	11 37	7, 7	[357]	75	19	B
595		18 30·4	20 59	8, 10	2345	77	12	P B

Ref. No.	Name of Star.	R. A. 1880.	Dec. 1880.	Mag.	Σ's No.	H₂'s No.	Arc.	Char-acter.
		h.　m.	°　′				°	
596		18 30·5	16　54	7, 7	[358]	7479	24	P B
597		30·5	7　26	7, 9	2346	81	7	R M
598	P. XVIII. 132	31	23　31	7, 7	[359]	80	8	P B
599	α Lyræ	32·8	38　40	1, 10		7501	39	P M
600		33	4　45	6, 10	[360]	92		P C
601		33·6	28　36	8, 9	2356	05	9	B
602		35·9	30　11	7, 7, 8	67	23	10	P B
603		38·4	67　0	8, 8	84	63	25	B
604	Tauri Pon. 75 B	40	5　23	6, 7	75	51	3	P B
605	ε¹ Lyræ	40·4	39　33	5, 6	82	64	18	B
606	ε² Lyræ	40·4	39　29	5, 5	83	66	30	B
607		40·9	59　25	8, 9	98	99	10	R M
608		42·8	10　38	8, 11	96	93	82	R M
609		43	77　33	7, 8	[363]	36	1	P C
610		43·5	16　7	8, 11	2400	7604	69	C C
611		44·1	10　32	8, 8	02	09	18	B
612		46	13　23	8, 9	09	25	10	P C
613		48	25　12	7, 10	[364]	40		P B
614	o Drac.	49·4	59　14	5, 7	2420	60	21	R M
615	Lyræ 91 B	50	33　48	5, 10, 7	[525]	59	4	P B
616	OΣ 365	52	44　4	7, 8, 11	3130	7670		B
617		52·2	25　56	7, 8	2422	71	10	B
618	11 Aquilæ	53·5	13　28	6, 9	24	75	20	R M
619	H₁ I. 58	54	36　16	8, 10	29	89		P C
620	P. XVIII. 287	55·5	58　4	7, 8	38	7709	52	B
621	P. XVIII. 274	56·5	−0　53	9, 9, 10	34	02	27	B
622		56·6	19　0	8, 8	37	06	14	B
623		58·1	31　13	8, 9	41	23	10	P B
623		58	16　48	8, 9	42	21		C C
625		19　1·5	30　15	8, 9	54	52	33	P B
626		1·6	38　21	8, 8	56	56	6	R M
627	L 35821	1·8	21　59	7, 8	55	53	40	P B
628		3·6	11　41	8, 10	64	68	6	C C
629		5	7　56	8, 11	71	87	6	C C
630	Cygni 4 B	5·9	55　8	7, 8, 9	79	7806	20	B
631		7·1	38　35	8, 8, 9	81	10	36	B
632		9	18　52	7, 9	84	19	6	P B
633	Cygni 6 B	9	49　37	6, 6	86	28	3	B
634		10·5	15　57	7, 8	[368]	42	18	P B
635		11·4	28　4	8, 9	2491	54	16	P B
636	P. XIX. 108	15·6	62　59	7, 8	2509	7908	8	B
637		16·8	67　28	9, 11	14	22	39	C C
638		20	21　17	8, 9	15	26	4	R M
639		19·9	46　59	7, 9, 10	[372]	37	10	B
640	P. XIX. 128	21	19　39	5, 10	2521	46	4	C C
641		9 21·6	25　15	8, 8	24	54	3	C C

Ref. No.	Name of Star.	R. A. 1880.	Dec 1880.	Mag,	Σ's No	H₂'s No.	Arc.	Character.
		h. m.	° ′				°	
642	Cygni 22 B	19 21·6	27 5	7, 7	2525	7958	20	B
643		27	36 27	8, 8, 9	38	8006	3	C C
644	P. XIX. 185	30·2	−10 42	8, 10	41	24	13	C C
645		31	33 57	7, 10	[376]	35	5	P C
646		31·3	8 3	8, 10, 11	2544	37	13	P B
647		32	35 22	8, 8, 9	[377]	51	13	P B
648		32	40 44	7, 9	[378]	61	4	P C
649		34·3	21 59	7, 8	2556	79	25	P B
650	χ Aquilæ	36·9	11 33	6, 7	[380]	96	5	P C
651		38·8	40 26	7, 8	[383]	8123	2	P B
652		39·1	62 23	8, 8	2574	39	10	B
653		41	33 20	8, 8	76	46	22	B
654	δ Cygni	41·2	44 50	3, 8	79	53	96	B
655		42	40 16	7, 10	[385]	61	4	P C
656	π Aquilæ	43	11 31	6, 7	2583	68	2	P C
657		44	36 51	8, 8	[386]	77		P C
658		44·2	35 0	7, 8	[387]	79	106	B
659	α Aquilæ	44·9	8 33				23	P M
660		47	25 33	7, 7, 9	[388]	8205		P C
661	Aquilæ 192 B	48·5	14 59	7, 9	2596	19	10	P B
662	ε Drac.	48·6	69 58	4, 7	2603	40	30	P B
663	β Aquilæ	49	6 6	3, 11	[532]	28	8	B
664		53·9	32 57	7, 8	06	68	2	P B
665	Cygni 116 B	54	41 56	7, 9, 9	07	74	15	P B
666		54	44 4	7, 8	[393]	77	1	P C
667	16 h Vulpec.	56·9	24 36	6, 6	[395]	96	16	P B
668		57	47 56	8, 8, 12	2619	8313	6	P C
669		20 2	4 26	9, 11	27	50		P C
670		3·2	63 33	6, 10	40	86	7	C C
671	θ Sagittæ	4·6	20 33	6, 8, 7	37	82	1	B
672		5·3	−4 56	8, 9	36	88		C C
673		5·9	3 27	7, 11	41	99		P C
674		6·2	43 37	7, 8	[400]	8411	67	P B
675		7	61 43	7, 8	2652	39	7	P B
676		7·6	31 43	8, 9	49	28		C C
677	Aquilæ 241 B	8	−6 25	7, 9	46	25		C C
671		10	41 45	7, 7, 9	[403]	55		P B
679		10·5	52 45	7, 9, 10	2658	57		C C
680		14	32 52	8, 9	[405]	92	8	P B
681	P. XX. 177, 178	25·5	10 51	7, 7, 8	2690	8600	17	C C
682	Vulpec. 94 B	26·8	25 24	6, 8	2695	21		P C
683		27·5	5 2	8, 8	96	24	8	B
684		31·2	14 19	8, 8, 7	2703	56		C C
685	β Delphini	31·9	14 11	3, 4, 11	04	63	25	B
686	κ Delphini	33·3	9 40	5, 11	[533]	74	40	R M
687		20 34·1	38 13	7, 9	2708	91	30	R M

Ref. No.	Name of Star.	R. A. 1880.	Dec. 1880.	Mag.	Σ's No.	H₂'s No.	Arc.	Character.
		h. m.	° ′				°	
688		20 35·1	40 9	6, 7, 8	[410]	8703	5	P B
689		38·3	45 25	7, 10	[411]	40	18	C C
690		40·6	15 28	7, 8	2725	51	12	B
691	52 Cygni	41	30 17	4, 9	26	55	4	P C
692	γ Delphini	41·8	15 42	4, 5	27	57	9	B
693	λ Cygni	42·5	36 3	5, 6	[413]	73	36	B
694		43	41 59	7, 8	[414]	76		P C
695	4 Aquarii	45·1	−6 4	6, 7	2729	84	170	B
696		47·7	43 19	8, 8	[416]	8811	6	P B
697		48	28 41	7, 8, 9	[417]	10	4	P B
698		48·3	12 39	8, 9	2734	12	10	C C
699		49·9	32 15	7, 7	[418]	23	8	P B
700		50	40 15	7, 11	[420]	25	4	P C
701		51	44 42	7, 9	[422]	31	3	P C
702	ε Equulei	53·1	3 50	6, 6, 7	2737	39	10	B
703		54	15 6	7, 9	[424]	44		P B
704	P. XX. 429	54·6	50 0	6, 7	2741	50	10	B
705	P. XX. 440	56	48 13	7, 10, 11	[425]	61		P C
706		57	1 4	6, 7	2744	60	20	P B
707		57	38 47	8, 9	46	68	10	B
708		58·7	3 3	8, 9	49	76	23	B
709	61 Cygni	21 1·4	38 7	5, 6	58	98	81	B
710		1·9	33 39	7, 8	60	8902		P M
711	P. XXI. 1	4	29 43	6, 8	62	17	2	C C
712	δ Equulei	8·6	9 28	4, 5, 10	77	59	54	B
713		9·3	28 35	8, 8	79	65	4	C C
714		9·5	−1 44	8, 11	78	63	2	C C?
715	P. XXI. 50	9·7	40 39	7, 7	[432]	76	8	B
716	τ Cygni	10·0	37 32	6, 8			24	B
717	A. C. 19	11·4	63 57	7, 7		8998	12	B
718		15·4	2 23	7, 8	[435]	9016	7	P C
719		15·9	31 56	6, 7	[437]	21	13	P B
720		20·9	13 10	7, 8	2797	59	3	P C
721		22·1	79 50	7, 8	2801	87	10	B
722	Pegasi 20 B	23	10 34	7, 7	99	72	20	B
723		27	33 17	8, 8	02	9104	2	P C
724	Pegasi 29 B	27·1	20 11	7, 8	04	07	15	B
725	μ Cygni	38·9	28 12	4, 5	2822	9210	8	B
726	κ Pegasi	39·2	25 6	4, 11	24	13	7	C C
727		40·9	0 18	8, 8	25	26	11	B
728		42·9	82 23	8, 9	37	73	19	B
729		43·5	2 50	8, 9, 9	28	40	3	P R
730	Cephei 147 B	48	55 14	6, 7	40	94	3	P C
731		51	51 59	8, 8	[456]	9328	5	P B
732		52·1	19 40	8, 11	2849	33	10	P B
733		21 52·6	59 14	7, 9	[458]	42		P B

Ref. No.	Name of Star.	R. A. 1880.	Dec. 1880.	Mag.	Σ's No.	H₂'s No.	Arc.	Char- acter.
		h. m.	° ′				°	
734		59·4	60 16	8, 9	2860	9391	4	C C
735	ξ Cephei	22 0·3	64 2	5, 7	63	9403	6	B
736		1·1	69 38	8, 9	65	16	8	R M
737		4	13 9	7, 11	[463]	29	6	P C
738	P. XXII. 11, 12	4·5	58 42	8, 8, 8	2872	42	10	B
739		7	49 37	7, 11	[465]	61		P C
740	Pegasi 148 B	8·5	7 23	6, 8	2878	66	5	P C
741	P. XXII. 33	8·5	16 36	6, 10	77	69	40	R M
742		15·1	24 21	8. 10	95	9516	22	P B
743		15	34 31	7, 9	[469]	18	2	C C
744	33 Pegasi	17·9	20 14	6, 9, 8	2900	39	3	B
745	ζ Aquarii	22·6	−0 38	4, 4	09	80	45	B
746		22·5	22 55	8, 9	10	81	3	P C
747	37 Pegasi	23·9	3 49	6, 7	12	93	4	B
748		26·5	6 48	8, 9	2915	9614	10	C C
749		27·4	20 33	9, 10	19	20	6	C C
750		29·5	69 17	7, 7	24	46	10	B
751		33·1	−13 14	8, 8	28	70	8	C C
752		36·1	20 48	8, 9	34	9703	20	B
753		38	45 22	7, 11	[477]	20	26	R M
754		40·1	18 37	7, 10	2941	31	3	C C
755		41	38 51	7, 9	42	36	3	C C
756	τ′ Aquarii	41·3	−14 41	6, 9	43	40	5	C C
757	P. XXII. 219	41·6	−4 51	7, 7, 8	44	41	10	B
758		42	77 53	7, 9	[481]	57	2	P C
759		44·9	67 56	7, 7	2947	71	17	B
760		48	82 31	5, 10	[482]	9815		P C
761	H₁ N. 15	50·9	−3 53	6, 10	2959	18	5	C C
762		52·5	8 43	7, 7	[536]	32	180	B
763	Σ. 2966 rej.	52	72 12	7, 8, 11	[484]	43	28	B
764	52 Pegasi	53·2	11 5	6, 8	[483]	40	18	B
765		23 1·6	5 57	8, 10, 9	2976	9901	9	C C
766	π Cephei	4·1	74 44	5, 7	[489]	29	53	B
767		5	56 47	7, 9	[490]	33	7	P B
768	94 Aquarii	12·8	−14 7	5, 7	2998	82	6	B
769	o Cephei	13·7	67 27	5, 8	3001	93	22	B
770		15·4	34 47	8, 9	06	10004	10	P B
771		17	19 54	6, 9	07	15		P C
772	P. XXIII. 69	17·5	−9 7	7, 8	08	20	18	R M
773		19	56 52	7, 7	[495]	26		P C
774	P.XXIII. 100, 101	24	57 53	5, 7, 9, 10	[496]	69		C C
775		31·8	43 46	6, 7	[500]	117	22	B
776	H₁ II. 24 : So. 356	39·8	−19 21	6, 7		170	16	C C
777		40	59 48	7, 8, 9	3037			P C
778		40·4	61 59	9, 9	38		4	C C
779		23 41	27 45	7, 10	39	175	3	C C

Ref. No.	Name of Star.	R. A. 1880.	Dec. 1880.	Mag.	Σ's No.	H₂'s No.	Arc.	Character.
780		h. m. 23 43	64° 13′	7, 7, 8	[507]	10185	16°	B
781	h. 1911	45	41 25	7, 8, 9	[510]	200	13	B
782		50·2	−10 10	8, 8	3046	235	10	B
783	Andromedæ 37 B	53·4	33 4	6, 6	50	258	20	P B
784	B. A. C. 8350	55·9	26 27	6, 9			28	P M
785	L. 47206	58·5	33 36	7, 7, 9	56	291	7	B
786		59·9	57 46	7, 8	62	304	340	B

SUPPLEMENTARY LISTS.
A.

Ref. No.	Name of Star.	R. A. 1880.	Dec. 1880.	Mag.	Σ's No.	H₂'s No.	Arc.	Character.
787		h. m. 0 25·5	−2° 43′	9, 9	35	167	1°	C C
788		37·3	16 42	8, 9½	51	250	4	P B
789		1 37·4	39 21	8, 10	149	618	10	P B
790		42·6	−2 2	8½, 8½	171	666	2	C C
791		55·2	80 56	7, 9	[37]	734	9	P B
792		2 14·8	23 5	8½, 10	254	883	10	C C
793		48·2	33 59	8, 10	325	1078	27	C C
794		3 4·5	36 46	8, 8	360	1155	7	P B
795		10·3	46 35	8, 10	371	92	7	C C
796		44·2	−38 0	5, 5		1408	3	R M
797		4 1	39 51	8, 10	3114	94	11	P B
798		8·5	29 42	7, 9	[78]	1540	6	P B
799		14·1	55 22	7, 8	531	92	4	P B
800		16·2	−4 58	8, 9	536	97	11	P B
801		34·5	22 30	9, 11	579	1720	5	C C
802		40·2	−12 10	8, 10	596	64	7	C C
803		51·5	13 46	8, 9	620	1845	7	P B
804		5 20·8	69 34	7, 9½	704	2066	2	C C
805	Bu. 320	23·1	−20 51	3, 11			23	B
806		36·5	−0 1	8, 8	782	2249	1	R M
807		38·8	21 16	8, 8	787	64	20	P B
808		47·7	36 55	7, 8	[122]	2333		B
809		52·8	−1 20	8, 9	826	84	13	P B
810		6 8·6	30 10	9, 10	879		4	C C
811		18·3	22 31	7, 9	[139]	2600	10	P B
812	P. VI. 105	20·6	0 32	8, 9	910	27	5	P B
813		32	28 22	6, 8	[152]	2734	6	P B
814		44·8	39 1	7, 10	974	2832	8	C C
815		49·6	25 7	8, 9	991	84	6	P B
816		7 7·5	15 59	7, 10	1047	3033	3	C C
817		7·8	14 46	9, 12	46	36	3	C C

Ref. No.	Name of Star.	R. A. 1800.	Dec. 1800.	Mag.	Σ's No.	H₂'s No.	Arc.	Character.
818		h. m. 7 12·1	° ′ 73 19	6½, 8½	1051	3043	° 10	P B
819		53·8	23 55	6, 10½	1171	3464	8	P B
820		8 11·5	6 50	9, 11½	1213	3625	3	C C
821		21·6	17 15	8, 10	30	3704	2	C C
822		23·8	55 46	7, 8	34	18	2	C C
823		27·7	2 10	8, 10	43	59	4	P B
824		44·5	21 20	9, 9½	85	3899	2	C C
825		9 13·7	5 31	9, 9	1343	4109	2	C C
826		56·8	56 4	7, 8	1402	4370	2	C C
827		10 43·2	-3 23	7, 8	76	4693	8	P B
828	So. 621	11 3·9	66 40	7½, 7½			13	R M
829		26·3	25 0	8½, 9½	1549	4953		C C
830		57·3	42 3	9, 10½	94	5148	4	C C
831		12 9·9	6 18	9, 10	1621	5237	16	P B
832	γ Crucis	24·5	-56 27	2, 5		5317	2	C C
833	P. XII. 196	45·1	-9 41	7, 9	82	5416	2	R M
834		13 23·1	76 36	8, 8	[267]	5624		P B
835	h. 4649	14 0·6	-59 9	8½, 8½		5845	5	P C
836	H₁ N. 115	2·6	21 46	8, 9	1804	65	8	P B
837		3·1	37 19	8, 8, 11	[276]	68	8	P B
838	H₁ V. 9	11·9	51 55	5, 7½	3124	5932		B?
839		22	-1 41	5, 9½	1846	93	4	P B
840	So. 184	39·1	-24 56	5½, 9				C C
841	P. XIV. 212	50·5	-20 52	5½, 6½		6172	20	B
842	π Lupi	56·9	-46 35	5, 5		6210	6	P B
843		15 0·1	34 56	8, 9	1908	35	6	P B
844		36·2	-14 48	8, 10	3095	6468	12	C C
845	π' Ursæ min.	36·5	80 51	6, 7	1972	90	4	C C
846		51·1	12 50	7½, 8	88	6544		P B
847		58	59 15	7½, 9	2006	96	6	P B
848		16 6·6	14 52	8, 8	17	6627	3	P C
849		19·3	1 31	7, 10	41	97	3	C C
850		25·5	-6 47	8, 8	3105	6729	6	C C
851		34·4	38 34	8, 12	2080	92	1	C C
852		38·4	25 22	8, 11	89	6811	6	P B
853		41·1	2 17	6, 9	96	22	3	P C
854		17 17·8	-0 43	8, 9	2156	7005	4	P B
855	P. XVII. 94	19·1	15 43	5, 10	60	14	5	C C
856		19·5	42 16	9, 9	63	19	6	P C
857	A. C. 9	49·9	29 50	8, 9		7203	6	P C
858	h. 5014	59	-43 24	6, 6		57	66	B
859		18 2·2	19 38	7, 8	[524]		18	B
860		4·3	0 31	7, 10	2286	7315	7	P B
861	η Serpentis	15·1	-2 55	3, 12		81	32	R M
862		15·6	22 45	7, 10	2310	87	4	C C
863	A. C. 11	18·7	-1 39	7, 7		96	6	P B?
864	Bu. 134	22	46 49	8, 10			7	P B

Ref. No.	Name of Star	R. A. 1880.	Dec. 1880.	Mag.	Σ's No.	H₂'s No.	Arc.	Character.
		h. m.	° ′				°	
865	γ Cor. Aust.	18 58·3	-37 14	5½, 5½		7714	144	B
866	17 Lyræ	19 2·9	32 18	6, 10	2461	62	10	P B
867		10·2	19 49	8½, 10	88	7835	10	P B
868		11·1	27 14	7, 7	[371]	51	5	P B
869	h. 5113	17·5	-29 32	6, 9		7900	48	C C
870	h. 5114	18·1	-54 34	6, 11, 7		7897	129?	C C
871		29·3	17 51	7, 9	[375]	8019	24	P B
872		31·8	61 47	8, 9	2553	8065	13	P B
873		35·8	63 33	8, 10	64	8105	9	C C
874	Da. 10	42·8	23 57	8, 9		69	6	P B
875	ʒ Sagittæ	43·6	18 51	6, 9	85	75	8?	B
876	h. 2904	47·1	-24 14	6, 10		94	32	R M
877	A. C. 16	52·9	26 56	7, 8		8257	7	P B
878		53·9	41 56	7, 9, 9	[392]	74	20	B
879		55·5	6 36	8, 9	2612	82		C C
880		59	35 41	7, 8	24	8325	3	P B
881		59·5	30 12	8, 8	2626	28	10	P B
882		20 12·8	10 37	8, 11	62	8472	3	P B
883	Cygni 172 B	13·9	40 21	6½, 9	66	93	7	P B
884	Cygni 176 B	15·9	44 59	7, 8	[406]	8516	24	B
885		15·9	39 1	7, 9	2668	12	5	P B
886		17·1	12 57	8, 9	73	19	6	P B
887		17·2	12 57	9, 10	74	20		C C
888	Delphini 43 B	39·2	11 53	6, 8	2723	8742	7	P B
889	P. XX. 324	43·1	25 57	8, 10	28	75		C C
890	Bu. 269	58·6	7 19	8, 11			18	C C
891	Cephei 83 B	58·8	56 12	6, 7	51	8884	4	P B
892	Bu. 368	21 1	-8 43	7, 8			6	P B
893		2·4	4 40	7, 8	[527]		207	B
894	H₁ I. 47	5·7	-15 31	8, 8		8932	33	P B
895	θ Indi	11·3	-53 57	5, 9		74	14	C C
896		33·7	20 10	8, 8½	[445]	9158	6	P B
897	B. A. C. 7578	40·5	-47 51	6, 10		9217	6	C C
898		46	8 31	7, 10	2833	60	4	C C
899		47·9	63 28	8, 11	42	98	11	P B
900		50·1	45 13	8, 10	46	9308	7	C C
901		51·9	-4 4	8, 8	47	25	12	P B
902		22 9·1	28 57	7, 8	81	9474	8	P B
903	P. XXII. 93, 94	20	-17 21	6½, 6½		9560		C C
904		37·9	46 30	7, 7	[476]	9715	7	B
905	Cephei 241 B	46·7	61 4	6, 7	2950	88	7	P B
906		23 1·5	60 47	7, 11	77	9905	11	P B
907		7·2	19 20	8½, 10	89	44	6	P B
908		41·7	16 25	7, 8, 8	3041	10180	3	P B
909		51·8	56 43	9, 9, 12	47	247	8	P B
910		59·5	17 25	8½, 8½	60	297	5	P B

B.

MR. S. W. BURNHAM'S STARS.

Ref. No.	Bu.'s No.	R. A. 1880.		Dec. 1880.		Mag.	P.	D.	Date.	Remarks.
		h. m.		° ′			°	″	1800+	
911	394	0	24·2	46	52	8, 8	300	1	76	
912	233		49·1	−18	6	8, 9	90	1·2	74	
913	302		51·9	20	45	7, 8	94	0·7	74	
914	396		56·2	60	26	6, 11	85	1	76	Extremely difficult
915	397	1	0·9	46	12	8, 10½	160	10	76	
916	303		3·1	23	9	7½, 8	286	0·6	74	
917	398		4·9	47	10	8, 8	60	2	76	
918	83	2	39·9	−5	28	7½, 10	123	1·3	72	
919	400	3	5·1	−4	16	7, 11½	45	15	76	
920	308		32·1	−8	3	8½, 9½	320	1·2	74	
921	401		44·2	−1	52	7, 11	260	4	76	
922	87	4	15·3	20	32	5½, 9½	171	2·1	73	Very beautiful.
923	402		17	−1	33	8½, 10½	75	5	76	
924	403		19·3	−2	20	7, 8½	100·3	2·0	76	De.'s measures.
925	184		22·5	−21	46	7, 8	270	1·2	73	
926	186		40·2	−7	12	8, 10	180	1·7	73	
927	312		42·6	−21	1	8, 9½	330	2	75	
928	316		46·8	−5	29	8, 8	180	1·2	75	
929	404		49·8	8	58	9, 9½	113·4	1·5	76	De.'s measures.
930	319	5	21·2	−20	49	7½, 10½	225	4	74	
931	405		42·3	−13	34	8½, 11	150	10	76	
932	406		43	−13	28	9, 12	260	8	76	
933	16		55·7	−10	36	5½, 10	356	1·8	71	Kn.'s measures.
934	323	6	18·7	−1	41	8, 9	90	1·7	75	
935	97		18·5	−1	21	7½, 9	257	1·1	73	Exquisite.
936	194		28·1	38	6	8, 8½	283	1	74	
937	326		49·9	2	28	8, 8½	60	1·2	75	
938	329	7	4·1	−16	2	6, 10½	95	15	75	
939	197		7	−6	57	8, 10	150	2	74	
940	330		13·5	−0	41	8½, 9	220	1·2	75	
941	199		19·9	−20	56	7, 9	19	1·2	74	
942	198		20·6	−20	43	8, 11	212	3·5	74	
943	333		57·4	−22	1	7½, 9	40	2	75	A B.
						8½	60	30	75	A C.
944	203		57·6	−27	14	7, 10	246	5	74	
945	206	8	30·3	−24	42	8, 9	278	1·5	74	
946	207		33·7	−19	19	6½, 11	99	5	74	A splendid pair.
947	407		45·8	−6	20	8, 10	160	6	76	
948	408		48·9	63	54	7, 10	350	2	76	
949	409		54·9	−8	43	8, 10	180	10	76	
950	410	9	4·5	−25	19	7, 9	160	1·5	76	
951	212		10·2	−7	51	7, 9	218	1·5	74	
952	214		35·9	−17	56	7½, 11	264	2·5	74	
953	217	10	1·2	−24	8	7½, 7½	273	1·5	74	
954	218		1·6	−19	7	8, 8	109	1	74	
955	219		15·9	−21	56	7, 9	193	2	74	

Ref. No.	Bu.'s No.	R. A. 1880. h. m.	Dec. 1880. ° "	Mag.	P. °	D. "	Date. 1800+	Remarks.
956	411	10 30.4	−26 3	7, 9	310	1.3	76	
957	220	11 6.5	−17 51	6, 6	148	0.5	74	
958	412	12 2.2	−17 55	8, 9.5	160	1.5	76	
959	343	13 45.1	−31 1	6, 8½	120	1.2	75	
960	348	14 55.6	0 20	6, 6	130	0.5	75	Very fine.
961	350	15 8.5	−27 9	6½, 8	170	1.3	75	Very fine.
962	227	12.1	−23 50	7, 10½	184	1.7	74	[difficult.
963	32	14.4	1 11	5½, 13	30	3	72	Splendid, but very
964	36	45.8	−24 56	5½, 10	270	3	71	Very beautiful.
965	417	17 52.2	39 27	8, 9½	270	1.5	76	
966	418	18 1.5	64 26	8½, 11½	240	10	76	
967	419	25.7	−7 55	8, 9½	40	1.5	76	
968	56	19 58.2	−4 41	7½, 10	180	2	71	Fine pair. [cult.
969	57	59.4	15 8	7, 15	140	2	72	Fine, but very diffi-
970	63	20 24.1	10 28	6, 11	340	0.7	72	Very difficult in-
								[deed.
971	65	41.4	5 32	6, 10	195	1.2	71	
972	66	42.5	26 58	8, 8	160	2	72	[cult.
973	67	45.2	30 26	7, 11½	290	1.2	71	Beautiful, but diffi-
974	68	55.3	49 43	8½, 9	170	1.5	72	
975	69	56.7	21 7	8, 9	350	2	72	
976	472	57.2	61 23	8½, 8½	6	0.6	77	
977	70	58.4	11 31	8	235	70	71	A and B C.
				10, 10	110	2.0	71	B and C. .
978	473	21 1.4	−10 42	9, 10	115	1.7	77	
979	71	4	9 37	5, 16	10	25	71	A C.
980	72	23	−5 58	9, 12	50	2	72	
981	73	24.7	−6 8	3, 16	180	35	77	
982	74	29.2	20 49	6½, 10	315	1.5	71	
983	372	35.8	51 1	8, 10½	360	1.4	75	
984	75	49.2	10 16	8, 9	30	1.2	72	
985	474	22 1	60 25	8½, 12	360	10	76	
986	375	4.5	50 11	8, 9	330	1	75	
987	475	6.2	−8 36	7½, 11	240	1.5	76	
988	376	8.1	59 30	7½, 11½	150	3	75	
989	476	8.7	30 48	9½, 10	93	2.6	76	
990	477	10.5	30 49	9, 11	46	6.4	76	
991	377	11.4	54 4	8, 10	65	60	75	A B.
				10½	30	3	75	B C.
992	378	12.8	60 16	8½, 9	90	4	75	
993	379	16	53 13	8½, 9	330	1	75	
994	76	22.9	−0 52	8½, 12	335	1.5	72	
995	77	27.3	−2 27	8, 10	210	2	72	A B.
				12	225	20		A C.
996	381	23 27.4	32 47	8, 10½	210	1.2	75	
997	80	12.2	4 42	8½, 9	300	1	72	
998	81	28.5	−12 18	8, 12	20	1.5	72	Very difficult.
999	279	36.5	−15 12	5, 12	90	3	74	
1000	482	55.7	62 39	8½, 10	360	4	77	A B.
				11	150	10		A C.

ABBREVIATIONS USED IN THE MEASURES, ETC.

Auwers	.	Au.	Luther .	. .	Lu.
Bessel	. .	Be.	Mädler .	. .	Mä.
Brünnow	. .	Br.	Main .	. .	M.
Burnham	. .	Bu.	Miller .	. .	Mi.
			Mitchell (Prof.)	.	Mit.
Challis .	. .	Ch.	Morton .	. .	Mo.
Cincinnati Observations		C.O.			
			Nobile .	. .	No.
Dawes .	. .	Da.			
Dembowski	. .	De.	Otto Struve	. .	O.Σ.
Doberck	. .	Dob.			
Dunér .	. .	Du.	Plummer	. .	Pl.
Durham Observations .		D.O.	Powell .	. .	Po.
Ellery .	. .	El.	Romberg	. .	Ro.
Engelmann	. .	Eng.			
			Schiaparelli	. .	Schi.
Ferrari .	. .	Fer.	Seabroke	. .	S.
Flammarion	. .	Fl.	Secchi .	. .	Se.
Fletcher	. .	Flt.	Smyth .	. .	Sm.
			South .	. .	So.
Gledhill	. .	Gl.	Spörer .	. .	Sp.
Greenwich Observations		G.O.	Struve .	. .	Σ.
Herschel, Sir Wm.	.	H_1.	Talmage	. .	Ta.
Herschel, Sir John	.	H_2.			
Hind .	. .	Hi.	Vogel .	. .	Vo.
Jacob .	. .	Ja.	Wilson .	. .	W.
			Winnecke	. .	Wi.
Kaiser .	. .	Ka.	Washington Observations		W.O.
Knott .	. .	Kn.			

M.M. = Mensuræ Micrometricæ.
M. = Magnitude.
h. = H_2.
A. C. = Alvan Clark.
Mem. R. A. S. = Memoirs of the
   Royal Astronomical Society.

L. = Lalande.
P.M. = Positiones Mediæ.
C. = Colour.
B. A. C. = British Association Catalogue.
P = Piazzi.

MEASURES.

THE following measures have been compiled with great care, and the originals have been consulted where possible. Some, however, have been given on the authority of H², Mä., Da., and Fl.

The first column gives the position angle (P.); the second the number of observations or nights (*e.g.* 14 or 2n.); the third gives the distance, and the fourth the date.

Where the angles and distances are the result of two or more nights' work, they are the arithmetical means. In the case of O.Σ.'s measures the arithmetical means of the "corrected" angles and distances have been given.

The whole of the measures by any observer are given at once under the proper initials, and both these groups and the individual results are placed in chronological order. This arrangement has been found convenient in compilation, and it exhibits at a glance the whole of the work of each contributor.

The diagrams are not all drawn to one scale; but a scale of equal parts will at once show the value of 1″.

---

1   Σ. 3063.

R. A.	Dec.	M.
0ʰ 1·5ᵐ	5° 12′	8·3, 10·2

C. A, yellowish.

Slow retrograde motion. Probably a binary.

Σ.	232·9	3n.	1″.78	1831·50
Mä.	227·3	1n.	...	45·86
De.	223·7	3n.	1·85	64·84
	224·4		·84	5·55
C. O.	221·4	3n.	·80	77·86

---

2   α ANDROMEDÆ.

R. A.	Dec.	M.
0ʰ 2·2ᵐ	28° 36′	2, 11·2

C. A, white.

Rectilinear motion. The proper motion of α in R. A. is + 0ˢ·013, and in Dec + 0″·13.

	°		″	
H₁	280·6	1n.	55·7	1781·96
Da.	264·2		66·57	1830·68
Sm.	267·1		65·9	4·64
	266·9		64·8	7·74
Σ.	·8		·94	6·38
O.Σ.	269·4		66·92	51·93
De.	270·7		69·2	66·68
Gl.	269·8	1n.	...	76·07
Fl.	271·0	1n.	71·1	7·08

---

3   Σ. 2.

R. A.	Dec.	M.
0ʰ 2·7ᵐ	79° 3′	6·3, 6·6

C. A, yellow; B, deeper yellow.

This difficult double star was discovered by Σ. in 1828, and the steady change in angle and distance has secured for it the careful attention of observers. Σ., H$_2$, Da., Se., Demb., and others have measured it. H$_2$ says, " Charmingly divided with 320. The discs like two grains of mustard-seed separated by one-third of the diameter of either." In 1839 Dawes could not separate the pair, and in 1866 Secchi describes it as " ovale." Between 1828 and 1866 the change in angle amounted to about 20°, but owing to the extreme closeness of the stars it is difficult to detect in the measures the acceleration of angular velocity due to the decrease of distance. " If the measures in 1858 and 1869 are correct, the two stars have already passed their apparent peri-astre." (O.Σ. in 1877.)

Σ.	342·5	1n.	0·72	1828·22
	343·4	,,	·84	·27
	339·3	,,	·94	32·20
	337·5	,,	·70	·24
	344·8	,,	·85	3·34
H$_2$	339·7	,,	$\frac{2}{8}$	0·31
Da.	336·1	3	0·7	9·67
O.Σ.	338·4	3n.	·743	40·56
	334·9	6n.	·522	8·22
	329·3	10n.	·443 simple	58·50 69·17
Mä.	334·3	1n.	0·80	41·42
	336·6	,,	·55	·45
	337·9	,,	·64	·64
	332·7	,,	·65	2·45
	336·4	,,	·62	·81
	338·8	,,	·55	3·28
	343·5	,,	·65	·31
	331·7	,,	·60	5·12
	335·5	,,	·60	·14
Se.	324·9	2n.	·38	57·52
	136·8	,,	·25	66·95
De.		1n.	single	3·6
	295·5	,,	0·38	5·7
		,,	single	7·0
Ta.	295·6	8	0·30	5·76
Du.	325·0	1n.	elong^d.	9·02
	331·0	,,	,,	·75
	334·0	,,	,,	75·71
W.			,,	2·92
Fl.			,,	6·85
Dob.	315·8	3	...	7·82

4   O.Σ. 2.

R. A.   Dec.   M.
0^h 7·4^m 26° 20′   A 6·9, B 8·3, C 9·6

This is h. 1007. A slow retrograde movement in A B. Probably a binary.

A B.

O.Σ.	51·5	1n.	0·79	1844·83
	65·3	,,	·83	50·92
	57·0	,,	·82	·99
	53·8	,,	·78	2·67

O.Σ.	51·9	,,	0·79	1857·71
	43·8	,,	·88	74·71
Se.	51·4	2n.	·67	58·43
De.	47·4	3n.	·5	66·64
Du.	44·8	,,	·72	9·78

$\frac{A\ B}{2}$ and C.

O.Σ.	226·2	5n.	17·77	14·52
Ro.	224·2	1n.	·58	62·86
De.	225·3	3n.	·51	6·64
Du.	·3	2n.	·77	9·72

5   Σ. 13.

318 (B) CEPHEI.

R. A.   Dec.   M.
0^h 9·4^m 76° 17′   6·6, 7·1

C. Σ. yellowish white. Se. and De. white.

A very difficult object. In 1830, H$_2$ says, " With 320 and full aperture, both discs seen with a momentary hair-breadth separation." Struve calls it " oblonga, ex æqualibus." In 1828 Σ. could not divide it, but he did so in 1832.

" The diminution of the angle is evident. The positions of Σ. are probably subject to considerable systematic errors. A small increase in the distance appears probable." —(O.Σ. in 1877.)

Σ.	126·7	1n.	0·54	1828·22
	129·5	,,	·5	32·20
	125·7	,,	·54	2·24
	114·2	,,	·55	3·34
	124·8	1n.	·4	6·68
	116·9	,,	·5	·69
	117·6	,,	·4	·70
H$_2$.	311·8	,,	$\frac{1}{2}$	0·31
O.Σ.	125·5	3n.	0·64	40·58
	116·6	6n.	·57	8·22
	105·9	10n.	·70	58·50
	101·9	3n.	·73	71·22
Mä.	119·9	12n.	·54	43·20
De.	101·9	4n.	...	55·59
	105·9	3n.	...	8·56
	103·3	,,	...	62·76
	·9	2n.		3·35
	·0	1n.		4·69
	104·0	,,	0·5	5·93
	100·0	,,	·6	9·51
	96·2	,,	·47	74·82
	97·2	,,	·58	5·71
Se.	102·2	2n.	·69	57·52
	103·5	,,	·38	66·95
W. & S.	·1	9	·5	72·5
	0·7	7		3·3
Gl.	101·0	6	0·5	·91
Fer.	93·1		·47	4·82
Dob.	181·7	3	...	7·82

6   O.Σ. 4.

R. A.   Dec.   M.
$0^h\ 10\cdot4^m$ $35°\ 48'$ 7·4, 8·1

Certain retrograde motion. Probably a binary.

O.Σ.	206·7	2n.	0″·59	1845·26
	187·5	4n.	·55	54·01
	172·7	2n.	·56	61·66
Mä.	178·0	1n.	elong.d	51·75
	29·5	,,	0·25	·76
De.	184·7	3n.	elong.d	66·88
	358·8		,,	9·61

7   Σ. 23.

R. A.   Dec.   M.
$0^h\ 11\cdot3^m$ $-0°\ 21'$ 7·6, 9·9

C. yellowish. Both angle and distance have decreased.

The formulæ given in the M. M. by Σ. no longer satisfy the observations. From the observations by Σ., Da., O.Σ., and De., the following are deduced :—

$$\Delta A = -\ 0''\cdot48 - 0''\cdot030\ (T-1850\cdot0),$$
$$\Delta D = +\ 11'''\cdot40 - 0'''\cdot110\ (T-1850\cdot0),$$

and the comparison of the observed and computed quantities is very satisfactory.— (O.Σ. in 1877.)

Σ.	1·2	3n.	13·67	1828·52
	359·7	6n.	12·87	36·24
Da.	358·8	3n.	12·25	42·18
Ka.	356·9	1n.	10·98	54·00
	359·1	8n.	12·04	42·48
	1·2	9n.	11·86	3·98
Mä.	359·9	4n.	12·13	·91
	355·4	,,	10·72	58·00
O.Σ.	357·9	2n.	12·01	46·24
	·4	,,	11·00	54·35
	353·2	,,	9·44	67·88
Mo.	355·5	1n.	10·68	54·94
	356·7	,,	·87	6·96
De.	355·0	6n.	9·85	63·33
Ta.	354·5	4	...	5·70
	355·0	6	...	·76
	353·9	6	8·84	9·72
	351·6	6	·72	71·78
W. & S.	352·8	4	...	3·86
	353·9	2	...	·86
	·2	5	8·9	4·91
	352·8	9	6·96	6·95
Gl.	353·4	4	...	3·91
Dob.	348·7	2n.	...	6·89
Pl.	351 6	4n.	8·2	7·46
O.O.	350·3	3n.	7·92	·83

8   Σ. 24.

69 (B) **ANDROMEDÆ.**

R. A.   Dec.   M.
$0^h\ 12\cdot2^m$ $25°\ 28'$ 7, 8
C. white.

The angle is unchanged, but the distance slowly diminishes.

Σ.	248·4	4n.	5″·20	1831·11
Da.	246·2	3	...	46·76
De.	247·1	8n.	5·22	53·05
Se.	·5	4n.	·05	6·85
Mo.	·5	2n.	·06	8·41
Ta.	250·2	4	...	65·70
	248·5	6	5·21	·76
	·7		...	9·58
	247·3	5	5·98	·72
Du.	250·8	4n.	4·84	70·39
Fer.	249·1		5·22	2·95
Gl.	247·9	4	...	3·91

9   H₁ V. 85.

R. A.   Dec.   M.
$0^h\ 13\cdot1^m$ $37°\ 28'$ 7·4, 9·5
C. white.

Rapid increase of distance. Rectilinear motion.

H₁.	10·6	1	...	1783·04
			30·45	·63
So.	13·2		45·31	1824·91
O.Σ.	15·4		53·35	51·99
Fl.	·1	1n.	62·5	77·13

10   O.Σ. 6.

R. A.   Dec.   M.
$0^h\ 15^m$ $66°\ 20'$ A 7·2, B 8·2, C 9·5
Σ. 26 *rej.*

In A B the angle has probably diminished, and the distance between $\dfrac{A+B}{2}$ and C.

A B.

O.Σ.	143·9	4n.	0·77	1849·64
Mä.	135·3	1n.	·55	51·76
	133·9	,,	·5	·77
	140·4	,,	·7	2·21
De.	·6	3n.	·6	67·67

$$\dfrac{A+B}{2}\ \text{and}\ \mathbf{C.}$$

O.Σ.	114·8	4n.	13·49	49·64
Mä.	115·6	1n.	·46	51·76
	·8	,,	·56	·77
Ro.	·0	,,	·26	62·87
De.	114·1	3n.	·28	7·67

I 2

11   O.Σ. 7.

R. A.	Dec.	M.
0ʰ 15ᵐ	65° 49′	A 7·2, B 8, C 9·8

B C.

O.Σ.	97·6	ın.	0″46	1846·74
	116·8	,,	·46	7·91
Mä.	106·7	,,	·3	51·77
	100·4	,,	·55	2·21
De.	Too close for measurement in 1865.			

A B.

O.Σ.	76·3	2n.	52·44	47·23

12   Σ. 27.

R. A.	Dec.	M.
0ʰ 16·2ᵐ	12° 49′	6·3, 10·7

C. very yellow.

Rectilinear motion. The changes are probably due to the proper motion of the principal star. (See P. M., p. ccxxiv.)

Σ.	344·0	3n.	31·67	1829·50
	340·7	ın.	30·20	51·80
	·1	,,	·01	·89
Sm.	341·5		·5	33·95
Mä.	·8	4n.	31·84	44·26
De.	338·0		29·73	63·85
O.Σ.	·0	ın.	30·14	5·87
	·9	,,	·02	9·93
Gl.	·1	3	29·	73·89
Fl.	337·9	ın.	28·5	7·08

13   Σ. 30.

R. A.	Dec.	M.
0ʰ 21ᵐ	49° 20′	6·8, 8·7

C. A, white ; B, ash.
Change in angle and distance.

Σ.	295·8	3n.	21·23	1831·21
Mä.	296·6	ın.	20·42	45·08
W. & S.	299·3	4	18·72	·93

14   Σ. 32.

49 PISCIUM.

R. A.	Dec.	M.
0ʰ 24·6ᵐ	15° 23′	6·8, 10·6

The evident change is explained by the proper motion of the principal star.—(O.Σ. in 1877.)

Σ.	108·3	2n.	13″43	1829·24
	107·6	3n.	·84	32·90
	106·8	2n.	15·48	51·84
Sm.	109·5		·0	35·87
Mä.	107·2	ın.	14·87	44·01
De.	106·5	2n.	16·15	63·92
O.Σ.	105·8	3n.	·73	9·91
W. & S.	106·6	3	·6	73·93
Gl.	·8	4	·4	·94

15   O.Σ. 12.

R. A.	Dec.	M.
0ʰ 25·1ᵐ	53° 52′	5·6, 5·9

Certain direct motion. Proper motion of λ, +0ˢ·003 in R. A. and +0″·02 in P.D. Du.'s formulæ are—

$$1855·27 \quad \Delta = 0″·48.$$
$$P = 130°·7 + 0°·55 \ (t - 1860·0).$$

O.Σ.	299·2	ın.	...	1843·14
	303·7	,,	0·48	4·84
	295·5	,,	·54	5·16
	127·2	,,	·56	6·11
	305·2	,,	·49	7·13
	131·3	,,	·53	51·13
	129·1	,,	·44	·19
	310·0	,,	·52	4·67
	315·0	,,	·65	70·18
Mä.	122·3	2n.	·33	45·73
	·9	4n.	·29	51·99
	140·3	ın.	...	3·24
Da.	115·9	2n.	0·57	4·36
Se.	124·8	ın.	·25	9·01
De.	133·1	6n.	...	66·37
	134·3		0·38	9·39
	138·7	2n.	elong.	70·69
	133·9	2n.	0·5	1·28
	136·8	ın.	obl.	2·61
	324·1	ın.	elong.	3·69
	134·3	ın.	0·58	5·63
	318·7	ın.	·57	7·03
Du.	142·1	4n.	·49	5·80
W. & S.	140·7	7	·5	·92

16   O.Σ. 13.

R. A.	Dec.	M.
0ʰ 25ᵐ	36° 16′	7·8, 10·9

Probably a slight change in both angle and distance. There is a star of the 10·11 mag. at a distance of 41″ (De.)

A B.

O.Σ.	133·2	4n.	6·19	1850·06
De.	131·1	3n.	·39	66·63

A C.

De.	180·9	2n.	41·22	66·19

17 Σ. 36.

R. A.	Dec.	M.
0^h 26.2^m	6° 17′	5, 9

C. A, white; B, ash.

A difficult star to observe. The angle has probably decreased. Mädler gives the proper motion as + 1″.7 and + 1″.2, but Σ. has − 0″.2 and + 3″.4.

H₁	89°.4		22″.48	1783.63
Σ.	82.9	In.	...	1820.96
		In.	27.44	2.22
	.3	3n.	.42	33.20
H₂ & So.	.8		25.87	22.87
Mä.	81.0	2n.	28.25	52.86
	80.4	In.	.21	3.04
	81.9	In.	27.04	3.85
	.9	In.	.51	8.04
Se.	82.4	In.	.29	8.04
Eng.	81.8	In.	.47	64.94
W. & S.	82.7	5n.	28.4	73.86

18 Σ. 44.

R. A.	Dec.	M.
0^h 31.9^m	40° 20′	8, 9

C. yellowish.

Direct angular motion and increase of distance.

Du.'s formulæ are—

$$\Delta = 8''.34 + 0''.0237 (t - 1850.0).$$
$$P = 261°.6 + 0°.139 (t - 1850.0).$$

Σ.	258.8	3n.	7.86	1829.82
H₂.	260.4	2n.	10.35	30.16
Mä.	259.0	2n.	7.34	45.69
Se.	262.2	2n.	8.74	57.93
De.	263.2	6n.	.66	65.09
O.Σ.	262.3	1n.	.76	6.92
Du.	265.0	7n.	.80	70.09
W. & S.	264.2	4n.	9.1	2.64
	263.0		.6	4.86
	265.4	8	8.8	6.95
Gl.	264.6	4	.9	3.91

19 O.Σ. 16.

R. A.	Dec.	M.
0^h 33^m	48° 42′	6.3, 10.8

C. yellow.

The distance has probably diminished.

O.Σ.	25.6	3n.	14.76	1845.92
Mä.	.6	In.	...	49.28
	26.7	,,	15.39	52.84
De.	24.9	3n.	14.24	67.09

20 Σ. 49.

R. A.	Dec.	M.
0^h 34.7^m	−7° 53′	6.5, 10

C. yellowish white.

Σ.	321°.4	3n.	4″.49	1830.92
Mä.	320.4	In.	.93	44.05
De.	319.6	3n.	5.24	65.18
Ta.	304.4	5	.82	71.78
C.O.	320.5	4n.	.92	7.80

21 O.Σ. 18.

R. A.	Dec.	M.
0^h 36^m	3° 32′	7.4, 9.5

Probable direct motion.

O.Σ.	93.6	2n.	1.40	1845.70
	94.7	,,	.34	55.18
Se.	99.8	,,	.13	8.51
De.	106.2	3n.	.55	66.60

22 O.Σ. 19.

R. A.	Dec.	M.
0^h 37^m	36° 55′	7.8, 10.7

Probable change in angle.

O.Σ.	117.3	3n.	9.56	1847.22
Mä.	293.0		...	5.85
De.	114.4	3n.	9.74	66.60

23 Σ. 52.

R. A.	Dec.	M.
0^h 37.5^m	45° 35′	8, 9

C. yellowish.

Σ.	25.8	3n.	1.42	1831.44
Mä.	24.8	1	.70	45.08
De.	19.0	In.	...	63.97
	18.4	,,	1.36	.87
	19.3	,,	.06	8.88
	14.6	,,	.28	70.06
	17.9	,,	.33	7.87

24 Σ. 59.

R. A.	Dec.	M.
0^h 41.2^m	50° 47′	7, 8

C. very white.

Probably very slow orbital motion. Du. gives—

$$1855.42 \quad \Delta = 2'''.21.$$
$$P = 146°.8 + 0°.0860 (t + 1855.42).$$

H₁.	140·5	1n.	2±	1783·34
So.	147·6	3n.	2·57	1825·14
Da.	148·2		·32	30·78
Σ.	144·9	4n.	·19	2·33
Sm.	147·2		·3	·87
	146·8		·4	6·94
Mä.	·o	3n.	·20	45·34
	·5	,,	·24	51·52
	·9	,,	·05	6·70
	·5	2n.	·49	61·80
Mit.	·4	6	·25	47·63
De.	147·8	1n.	...	56·76
	·4		2·03	67·98
Se.	146·2	3n.	·12	58·58
Mo.	147·7	2n.	·17	·66
Du.	148·1	8n.	·21	71·94
W. & S.	149·0	4	·14	3·83
Gl.	148·5	4	...	·89

25   Σ. 60.

η CASSIOPEIÆ.

R. A.	Dec.	M.
0ʰ 41·8ᵐ	57° 11′	4, 7·6

C. Σ. A, yellow ; B, purple.

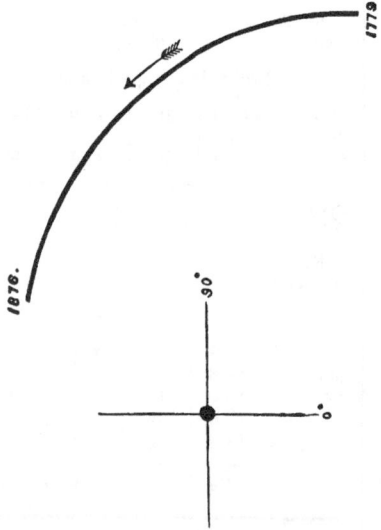

H₁. "3. η Cassiopeiæ, Fl. 24. In cingulo.
"Aug. 17, [1779].—Double. Very un-equal. L. fine W. ; S. fine garnet ; both beautiful colours. Distance 11″·275 mean measure. Position 27° 56′ n. following."
Again, he says (*Phil. Trans.*, 1804), "The situation of the two stars of this beautiful double star, June 14, 1782, was 27° 56′ north following ; and, Feb. 11,

1803, it was 19° 14′ ; which gives a change of 8° 42′ in 20 years and 242 days. This arises probably from a real motion of η in space ; for parallax would have had a con-trary effect.[7]
And H₂ (*Phil. Trans.*, 1824, part ii.,) re-marks that "The changes, both in position and distance, of this remarkable star, have been regularly progressive." He gives measures from 1779 to 1821, points out that Σ.'s position for 1814 is not reliable, and finds the angular motion to be 0·5133° per annum in the direction *n f s p*, and the period probably about 700 years. He ob-serves, "A connection between these stars cannot be doubted, as they have a common proper motion of nearly 2″ per annum. The distance having diminished almost 3″, the apparent orbit is evidently elliptic."
H₂ having predicted that the small star would probably be on the parallel in 1835, Sm. "carefully watched, both before and after, and saw the prediction verified." Sm. further remarks that "The lapse of 40 years after H.'s measure gives a mean velocity of + 0·45° per annum, and the 23 years since elapsed + 0·70°, while the distance may be regarded as but little altered."
Da. gives measures from 1831 to 1854 (*Mem. R.A.S.*, vol. xxxv.), and observes that the proper motions in R. A. and N. P. D. applied to the larger star would have *dimi-nished* the angle and distance, supposing the smaller star to be at rest.
The diminution in the distance, and the consequent increase in the angular velocity of this star, are well exhibited by the long list of measures. It may be observed, too, that the proper motion of η is unusually great, Argelander giving it as + 1·97″ in R. A. and − 0·495″ in D. (For some inte-resting remarks and results, see Mädler's *Die Fixstern-Systeme*.)
Mädler was of opinion that the bright-ness of the principal star is a constant source of error in the measures of distance, and was but little satisfied with his own results. He thought that the companion probably passed its aphelion between 1780 and 1803, and that the distance would sink to 2″ or 3″ about 1860.
Orbit.—The following are the more re-cent elements :—

Doberck, 1876, gives :	Grüber in 1876 :
T = 1909·24	T = 1901·25
☊ = 39° 57′	☊ = 33° 20′
π − ☊ = 223·20	π = 229·27
i = 53·50	i = 48·18
e = 0·5763	e = 0·6244
a = 9·83″	a = 8·639″
P = 222·435 yrs.	P = 195·235 yrs.

Grüber made use of all measures from

H.'s to those in 1875 : his normal positions are—

$$
\begin{aligned}
1780 \quad & 58°·90 \\
1800 \quad & 69·03 \\
1820 \quad & 80·25 \\
1840 \quad & 94·99 \\
1860 \quad & 117·66 \\
1880 \quad & 156·38
\end{aligned}
$$

And with O.Σ.'s parallax, 0″·154, he finds the mass of the system to be 4·63 times the sun's mass ; and the semi-major axis 56·10 times the sun's mean distance.

Dr. Dunér in 1876 obtained the following elements :—

$$
\begin{aligned}
T &= 1924·78 \\
\omega &= 245°·9 \\
\Omega &= 50·8 \ (\text{Equ. } 1850·0) \\
i &= 68·5 \\
e &= 0·6268 \\
\mu &= +2°·0411 \\
a &= 10″·68 \\
P &= 176·37 \text{ yrs.}
\end{aligned}
$$

He remarks that there still exists much uncertainty in the elements of this system.

	°		″	
H₁.		I	11·09	1779·8
		I	·46	80·5
	60·8	...		2·4
	70·6	...		1803·1
Σ.	81·1		10·68	20·16
	85·6	1n.	·25	7·21
	87·6	5n.	9·78	32·05
	91·2	3n.	·52	5·26
	92·1	4n.	·39	6·74
H₂ & So.	82·8	7	8·8	21·9
	83·1	42	9·90	5·78
	86·4	2	12·0	8·9
Be.	·2	5n.	10·06	30·75
	89·6		9·80	4·76
Sm.	87·8		·8	0·91
	88·2		·9	1·92
	·9		·9	3·74
	90·9		·7	5·20
	92·0		·4	6·81
	95·8		·4	43·19
	101·5		8·5	6·73
	110·6		7·7	54·17
Da.	88·6	2n.	9·74	32·87
	95·7	1n.	·33	41·80
	109·6	,,	7·91	54·00
Encke.	92·5	,,	9·64	37·62
Galle.	·6	,,	·47	8·68
Ka.	95·81	9n.	8·98	40·43
G.O.	96·4	31	·96	·44
O.Σ.	98·1	3n.	9·21	1·34
	101·7	5n.	8·48	7·40
	104·9	4n.	·26	9·66
	108·0	3n.	·03	51·84
	112·0	4n.	7·97	4·56
	114·1	2n.	·57	7·22
	119·8	,,	·17	60·68
	132·6	,,	6·44	6·22
	·9	3n.	·42	8·53
	136·2	2n.	·28	70·18

	°			
	140·8	,,	5·94	1872·18
	144·6	3n.	·68	3·53
	148·6	2n.	·58	5·15
Mä.	96·4	4n.	9·24	41·57
	98·3	2n.	8·75	2·39
	100·1	6n.	·58	4·56
	101·9	7n.	·46	5·39
	1·1		·58	6·67
	2·7		·26	7·42
	6·5		·02	50·80
	6·9	20n.	7·72	1·76
	8·6	15n.	·64	2·67
	110·1		·57	3·29
	112·7		·52	·90
	111·8		·60	4·80
	112·0		·60	5·51
	111·0		·77	5·87
	114·3		·07	8·52
	115·7		6·96	9·26
	119·0	14n.	7·00	61·91
D.O.	104·0		8·82	47·08
Mi.	101·4		·59	7·6
	106·9	27	·12	51·9
Mit.	101·3	6	·60	47·63
Ja.	105·5	26	·16	50·87
	106·4	15	·04	1·88
	107·9	10	7·98	2·75
	109·0	10	·91	3·13
	·7	22	8·01	·98
Peters.	·6		·17	·93
De.	111·2	4n.	7·87	4·77
	112·8	3n.	·83	5·08
	114·2	4n.	·40	6·58
	·5	1n.	·30	7·11
	115·9	3n.	·18	7·82
	·8	4n.	·26	8·46
	121·3	8n.	·04	62·74
	122·6	11n.	6·91	3·48
	124·5	2n.	·80	4·10
	·6	7n.	·78	·71
	126·3	10n.	·67	5·51
	129·3	13n.	·56	7·16
	132·4	5n.	·30	8·55
	134·1	,,	·19	9·68
	135·4	7n.	·17	70·52
	137·5	,·	·09	1·56
	139·1	6n.	5·97	2·62
	140·7	7n.	·77	3·65
	142·5	,,	·83	4·63
	146·2	6n.	·67	5·60
	149·9	13n.	·57	6·94
Mo.	111·0		8·12	54·95
	112·4		7·80	5·96
	117·3	2n.	·08	9·94
Wi.	110·9	2n.	·94	5·25
Se.	112·2		·90	5·79
	·8	3n.	·86	7·15
	127·7	4n.	6·79	66·86
Lu.	117·5		8·35	56·57
	123·6		7·12	63·18
Po.	109·4		·60	53·94
	111·5		·55	4·91
	112·5		·60	5·92
	116·6		·02	9·72

	°		″	
Po.	118.3		6.99	1860.97
	120.6		.7	1.95
Au.	119.8	5n.	7.37	.58
M.	118.1	1n.	6.44	.82
	129.9	,,	.31	7.65
	132.4	,,	.12	9.67
	143.9	,,	5.94	72.77
	146.1	,,	.78	5.78
Ro.	119.1	6	7.01	62.86
	121.7	6	.29	.90
	122.9	2	6.89	3.04
	.5	6	.87	.06
	121.0	4	7.00	.12
Kn.	125.3	10	6.73	5.69
	126.7	6	.74	.69
	125.2	6	.77	.70
	137.7	5	.10	72.65
	138.0	5	.13	.65
	.0	5	.03	.66
	137.7	5	.11	.66
Ta.	123.9	2n.	.43	65.73
	124.6	3n.	.38	6.63
	.3	1n.	.21	8.89
	.8	,,	.58	9.72
	.4	,,	.32	72.86
	141.2	,,	5.66	3.86
	149.3	,.	4.72	6.86
Du.	131.8	5n.	6.30	68.37
	135.2	4n.	.07	9.93
	140.5	7n.	.59	72.50
	144.9	1n.	5.72	4.22
	146.7	10n.	.67	5.51
Br.	131.5	3n.	6.35	68.84
Gl.	135.7	5	.15	70.65
	.8	5	.13	.70
	136.0	5	.0	.80
	137.6	5	.07	1.6
	138.3	5	.0	.8
	143.1	5	.1	3.51
	.8	6	5.8	.73
	144.3	7	...	.81
	147.5	4n.	...	5.69
	.9	5n.	5.6	6.37
	149.9	7n.	...	7.41
	153.3	6n.	...	8.67
W. & S.	140.9	8	...	1.93
			6.0	2.01
	142.3	7	...	3.06
	144.7	7	6.22	.83
			.64	.83
			.23	.83
	146.0	6	5.8	4.90
	153.5	14	.32	7.95
No.	143.6		...	3.98
Dob.	147.8	2n.	...	5.93
	150.2	5n.	5.7	7.76

26   Σ. 63.

R. A.   Dec.   M.
0h 43.9m   11° 11′   8.5, 11

C. yellow.

Rectilinear motion. The smaller star is at rest.

	°		″	
Σ.	195.2	4n.	11.43	1832.41
Mä.	199.9	,,	12.25	45.47
De.	214.8	,,	13.93	64.10
	218.6	,,	14.65	72.69
W. & S.	221.5	6	13.0	4.93

27   Σ. 64.

R. A.   Dec.   M.
0h 44.5m   40° 33′   9.2, 9.7

The distance has probably diminished.
Dunér's formulæ are—

$$\Delta = 3''.39 - 0''.11 \, (t - 1848.05).$$
$$1848.05 \quad P = 272°.8.$$

	°		″	
Σ.	270.7	1n.	3.64	1828.85
	272.5	2n.	.54	31.73
Mä.	274.2	,,	.31	45.16
	272.0	1n.	.45	8.07
Se.	273.2	,,	.58	58.89
Du.	.5	,,	.22	70.73
	272.5	,,	.14	.09

28   Σ. 67.

R. A.   Dec.   M.
0h 45.9m   9° 57′   8.3, 9

Change in angle.

	°		″	
Σ.	13.0	3n.	1.58	1830.91
Mä.	12.7	4	.82	43.35
De.	7.2	1n.	.9	63.88
	.5	,,	2.11	6.67
	.9	,,	1.76	7.63
	5.3	,,	.69	70.71
Fer.	1.6		.80	3.94

29   λ TOUCANÆ.

R. A.   Dec.   M.
0h 47.8m   − 70° 9′   7, 8

Probably binary.

	°			
Dunlop.	71.6		...	1826.80
H_2.	76.8		20	34.84
	78.5		.46	5.92
	80.8		22.25	6.73
	.6		20.94	7.74

30   Σ. 69.

R. A.   Dec.   M.
0h 47.9m   83° 2′   8.5, 9.7

Certain change in angle and distance, but the nature of it is uncertain.

Σ.	359·8	2n.	21·44	1832·23

Σ.	359·8	2n.	21·44	1832·23
Mä.	2·3		22·06	47·30
De.	6·6	,,	·42	64·02
Gl.	8·0	1	·4	74·90
W. & S.	10·0	3	·4	4·93

31   O.Σ. 20.

R. A.	Dec.	M.
0h 48m	18° 32'	5·9, 7

C. A, yellowish white; B, bluish white.

Direct motion.

O.Σ.	72·7	4n.	0·618	1847·33
	59·8	3n.	·667	60·34
Se.	78·8	1n.	elong^d.	57·84
	56·0	,,	0·35	8·00
	58·0	,,	con-tact.	9·01
De.	84·1	4n.	obl.	5·88
	48·7	3n.	...	66·85
	26·6	1n.	obl.	70·71
	50·0	,,	...	1·65
	36·9	,,	...	2·67
	45·1	,,	...	4·68
	31·1	,,	...	5·65
	15·2	,,	0·3?	7·87

32   Σ. 73.

R. A.	Dec.	M.
0h 48m	22° 58'	6·2, 6·9

C. Σ. golden.

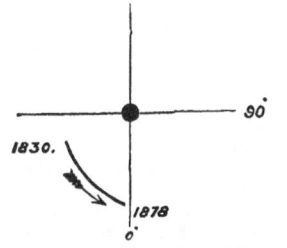

Discovered by H₁ in 1830, this star has been assiduously watched by observers, Dawes alone having measured it on no less than forty nights.

H₂ (*Mem. R. A. S.*, vol. v.) writes: "A miniature of η Coronæ. In glimpses, the two discs may be discerned in contact." "Very close; in contact; twirls much. Difficult measures." Smyth (*Cycle*, p. 21,) says: "This beautiful golden pair is very difficult." He used 600 with a central disc on the object-glass with advantage. From his own and H.'s measures he inferred that "there is a decided direct orbital motion."

The increase in the distance accords per-fectly with the manifest diminution in the angular movement. The distance appears already to have attained its maximum. (O.Σ. in 1877.)

Doberck gives the following elements:—

$$r = 1798·80$$
$$\text{Node} = 57° 54'$$
$$\lambda = 142\ 19$$
$$\gamma = 41\ 39$$
$$e = 0·6537$$
$$P = 349·1 \text{ yrs.}$$
$$a = 1'''·54.$$

Dr. Dunér has deduced the following formulæ:—

$$\Delta \sin. P = -0''·48 + 0''·0130\,(t-1854·42) + 0''·000235\,(t-1854·42)^2.$$
$$\Delta \cos. P = +1''·11 + 0''·0180\,(t-1854·42) - 0''·000361\,(t-1854·42)^2.$$

And on comparing these with the observations from 1830 to 1875, he finds very satisfactory agreement.

H₂.	305·0	2n.	0·849	1830·73
	308·6	4n.	·775	1·79
Σ.	307·8	3n.	·847	2·14
	320·4	,,	·937	6·90
Sm.	315·7		1·1	5·92
	318·5		·1	9·77
	332·9		·0	43·12
	335·8		·3	52·83
Da.	317·8	3n.	·092	39·79
	319·3	5n.	·080	40·98
	321·2	3n.	·102	1·87
	322·9	2n.	·007	2·94
	324·5	4n.	·117	3·88
	321·4	2n.	...	·99
	328·9	5n.	1·120	6·93
		1n.	·124	·96
	329·3	2n.	...	7·92
	·9	1n.	1·227	·93
	332·0	,,	·173	50·91
	334·5	2n.	·078	3·87
	336·5	,,	·218	·89
	334·4	,,	·170	·90
	335·8	1n.	·227	4·75
	340·2	2n.	·189	9·83
Mä.	324·7	6n.	·192	41·59
	325·8	2n.	·047	2·77
	329·0	1n.	·264	6·99
	·6	3n.	·219	7·90
	334·0	,,	·402	51·00
	336·5	,,	·280	3·87
	·5		·367	5·29
	340·3	5n.	·392	7·75
	·1	,,	·336	8·04
O.Σ.	324·2	3n.	·303	41·64
	328·6	1n.	·210	6·78
	335·9	3n.	·333	54·70
	344·4	1n.	·360	61·74
Ka.	323·4	3n.	0·99	42·34
	350·9		1·25	67·03
Mit.	330·2	6	1·05	47·70
Flt.	336·4	34	1·12	51·93
Ja.	338·0	15	1·26	3·96

	°		"	
Mo.	340·1	24	1·25	1854·91
De.	335·3	4n.	·2	5·59
	336·3	5n.	·16	6·46
	344·0	,,	·12	62·78
	·3	4n.	·14	3·82
	345·0	,,	·22	4·74
	·7	5n.	·21	5·64
	350·4	1n.	·31	8·65
	349·5	,,	·31	9·65
	350·1	2n.	·22	70·35
	352·4	,,	·40	1·62
	·5	,,	·34	2·65
	·5	,,	·35	3·68
	354·4	3n.	·26	4·80
	355·7	2n.	·25	5·62
	356·2	5n.	·34	7·27
Wi.	344·2	1n.	·30	56·09
Se.	339·2	3n.	·202	7·27
	349·5	1n.	·314	66·05
M.	329·0	,,	·10	1·80
Eng.	348·4	2n.	·62	5·11
Kn.	344·8	5	·323	·67
	·9	7	·393	·69
	·6	6	·322	·70
Ta.	347·0	3n.	·07	·73
	344·2	1n.	·38	6·83
	349·5	,,	...	9·72
	347·4	,,	1·24	72·86
	351·6	,,	·29	3·86
	·8	,,	·03	6·86
Br.	350·7	8	·57	68·76
Gl.	349·9	5	·2	70·14
	350·2	5	·3	·60
	352·7	5	·2	1·60
	354·0	5	·4	3·91
	356·3	3n.	·37	5·54
	354·8	1n.	...	6·07
	355·4	6n.	...	7·62
	357·2	2n.	...	8·45
W. & S.	352·3	4	1·36	2·03
	355·0	7	·14	·88
	354·1	4	·20	·88
	353·1	7	·34	3·81
	355·9	5	·43	4·93
	356·0	6	·38	·94
	358·6	4	·28	6·95
	·8	12	·33	7·94
Du.	356·1	5n.	·36	5·70
W.O.	3·1	1n.	·39	·97
	0·4	,,	·27	·98
	356·4	,,	·24	6·00
	355·7	,,	·46	·00
	357·0	,,	·12	·01
Dob.	359·1	5n.	...	·08
	354·9	4n.	1·15	·75
Sp.	355·9	,,	·28	7·02
Schi.	·8	1n.	·275	·01
Pl.	358·2	4n.	·49	·19

33   Σ. 74.

R. A.   Dec.   M.
$0^h\ 49^m$ 8° 44'   8, 9

C. white.

	°		"	
Σ.	301·2	2n.	2·96	1829·83
	303·3	1n.	3·19	32·86
Mä.	302·4	3n.	·15	43·59
Se.	304·9	2n.	·13	56·89

34   So. 390.

R. A.   Dec.   M.
$0^h\ 52·2^m$ − 16° 20'   7, 7·2

This pair was discovered by South in 1824. The angle has increased considerably, and the distance has diminished.

	°		"	
So.	32·3	16	7·78	1824·90
H₂.	·1		6·67	35·74
	·7		·65	7·80
Se.	86·8		·43	55·95
C.O.	214·9	3n.	·33	77·79

35   Σ. 80.

P. O. 251 PISCIUM.

R. A.   Dec.   M.
$0^h\ 53·2^m$ 0° 8'   7, 8·2

C. Σ. and De., yellow, blue. South, "small, blue."

A wide pair, first measured, probably, by South. Piazzi noted the duplicity of this star: "Duplex, comes 9ᵅ magnitudinis præcedit 1″ temporis parumper ad boream."

South measured it at Passy. "Double; 9th and 10th magnitudes; small, blue; 24° 43′ n. p. 19″·206 ; 5 obs. Oct. 25, 1824, extremely difficult."

Smith (*Cycle*, p. 23), "A neat double star bearing both illumination and high magnifying power." He observes that Piazzi assigns it to Pisces, but that it should be placed in the Whale ; and, from a comparison of his own measures with those of South, he infers a direct orbital motion of 0°·4 per annum.

O.Σ. finds that the observations from 1831 to 1868 are exactly represented by the formulæ

$$e = 18''·924 + 0''·040\ (T - 1850·0).$$
$$P = 305°·08 + 0°·31\ (T - 1850·0).$$

Engelmann's formulæ are

$$P = 299°·82 + 0°·3066\ (t - 1833·34).$$
$$\Delta = 18''·264 + 0''·04143\ (t - 1833·34).$$

And Dr. Dunér gives

$$\Delta = 19''·33 + 0''·0332\ (t - 1866·0).$$
$$P = 310°·0 + 0°·308\ (t - 1866·0).$$

	°		"	
Σ.	296·7		...	1822·29
	·9	1n.	...	4·99
	299·5	3n.	...	31·53
	300·6	,,	18·38	5·83

	°		″	
So.	296·5	3n.	18·87	1825·17
Sm.	299·8		·4	32·98
	301·8		·5	8·03
	305·1		·8	52·81
Mä.	303·8	2n.	17·87	42·78
	304·1	1n.	18·61	4·94
	305·5	2n.	·52	53·09
	306·3	1n.	19·05	8·01
O.Σ.	302·4	2n.	18·69	42·84
	305·5	,,	19·01	51·22
	311·0	,,	·60	68·42
Ka.	303·8	9n.	17·85	43·10
De.	307·3	1n.	18·80	55·99
	·1	,,	·63	6·03
	306·5	,,	·93	·62
	308·9	2n.	19·41	62·84
	309·1	,,	·39	3·80
Ja.	307·4		·07	57·95
Eng.	308·8		·78	62·97
	310·5	2n.	·54	5·03
Du.	·9	1n.	·62	8·84
	311·3	3n.	·69	9·71
	·5	1n.	·73	70·73
M.	310·1		·51	69·78
	311·4	1n.	20·10	70·77
W. & S.	312·9	8	·1	1·85
	311·6	4	19·7	·90
	·7	4	·9	2·00
	·8	7	18·4	·88
	·9	3	20·3	3·81
	312·9	6	·0	4·93
	·9	6	·31	·73
	313·7	3	...	6·95
Gl.	311·0	5		3·91
Fl.	313·3	1n.	20·9	7·06

36 O.Σ. 21.

R. A.	Dec.	M.
0h 56m	[46° 44′	7, 8

			°	
Mä.	45·1		0·97	1845·68
O.Σ.	177·1	4n.	·58	47·84
He.		Oblong?		64·7
				5·7

37 Σ. 86.

R. A.	Dec.	M.
0h 58·7m	−6° 7′	8, 8·7

C. white.

Σ. early recognized the angular change, and the measures since made confirm it. The distance may have increased slightly.

	°		″	
H1.	180·6	1	14·83	1783·08
Σ.	173·8	1n.	12·01	1822·03
	171·8	,,	·0	9·90
	·3	,,	·12	30·92
	0·0	,,	·25	5·85
	169·4	3n.	·11	6·58

	°		″	
So.	172·7	4n.	12·89	1824·89
Da.	167·6		·51	41·61
De.	·6	4n.	·51	3·59
	162·9	3n.	·64	63·47
Ka.	169·9	7n.	·22	43·59
Mä.	167·3	3n.	·36	·55
	164·0	2n.	·76	53·09
	162·2	1n.	...	8·01
O.Σ.	167·8	2n.	12·34	43·85
	163·0	1n.	·65	66·92
Ta.	·5	4	...	5·70
	·5	6	11·67	·78
	161·7	4	·82	6·84
	·8	5	...	8·84
	·6		...	9·72
	160·9	5	11·56	72·77
	161·8	1n.	12·38	3·86
	·5	1n.	11·71	6·89
W. & S.	163·4	4	12·6	3·93
	1·6	15	·5	6·53
Gl.	161·9	4	·6	3·93
C.O.	·4	1n.	·74	5·87
	160·3	5n.	·84	7·81
Dob.	158·9	2n.	·50	6·95
Pl.	160·1	3n.	·65	·94

38 Σ. 87.

R. A.	Dec.	M.
0h 59·1m	14° 74′	8·5, 8·5

C. yellowish.

	°		″	
Σ.	193·0	3n.	6·56	1829·85
H2.	195·1		9·14	30·33
Mä.	·2	3n.	6·55	43·89
	196·1		·88	7·95
	·9	1n.	7·10	51·00
	198·1	2n.	...	2·87
	·9	2n.	...	3·08
Se.	195·1	2n.	6·79	7·46
	198·7	1n.	·65	66·05
Gl.	195·8	3		74·03

39 Σ. 88.

ψ′ PISCIUM.

R. A.	Dec.	M.
0h 59m	20° 50′	4·9, 5

C. white.

A probable increase in distance.

Mädler gives the proper motion as + 5″·7 in R. A., and − 2″·3 in Dec.; Σ. gives +7″·6 and −3″·5.

This physical pair is easy of observation, and yet the measures are very discordant.

	°		″	
H1.			27·50	1779·82
Σ.	160·3	4n.	29·89	1832·11
	159·8	1n.	·61	51·80
Mä.	160·2	,,	30·26	36·20
	159·7	,,	29·81	44·01

DOUBLE STARS.

Mä.	160·3	,,	30″53	1845·08
	159·5	,,	29·83	6·74
	161·8	,,	...	50·96
	160·0	,,	29·50	·97
	·4	,,	30·29	·99
	159·0	·,	29·78	4·98
	·4	,,	30·34	6·98
	160·4	,,	29·69	7·90
	·8	,,	30·12	·95
Se.	159·8	,,	·02	8·04
Ro.	161·2	4	29·43	65·70
Du.	159·9	1n.	·95	9·08
	·5	,,	30·01	·09
Ta.	161·2	5	29·6	74·91
	159·7	6	·8	·91
	160·2	3	31·0	·94

40   Σ. 91.

R. A.	Dec.	M.
1ʰ 1ᵐ	− 2° 24′	6·7, 7·5

C. yellowish white.

Probably binary.

H₂	328·5		3·63	1830·67
Σ.	·8	3n.	·86	1·89
	325·5		·77	7·49
Mä.	324·0	2n	·35	42·85
	322·3	,,	·86	5·03
	·8	4n.	·65	53·08
	·9	1n.	...	8·00
De.	323·2	3n.	4·03	4·77
	322·0	1n.	3·69	63·88
	·3	,,	4·06	6·68
	321·5	,,	3·72	7·66
	324·2	,,	4·02	72·67
Wi.	·9	6	...	56·09
Se.	323·6	4n.	5·57	·72
	321·3	2n.	3·94	66·51
M.	323·2	1n.	·66	5·80
	320·5	,,	4·62	73·96
Fer.	322·8		3·98	·01
W. & S.	321·4	4	4·0	·81
	322·2	5	3·8	4·93
	·1	4	4·0	·93
Gl.	320·7	1n.	3·9	3·93
C.O.	324·0	3n.	...	5·97
	321·3	4n.	3·92	7·81

41   O.Σ. 515.

R. A.	Dec.	M.
1ʰ 3ᵐ	46° 36′	4·9, 6·5

C. A, yellow ; B, green.

A binary.

O.Σ.	309·9	4n.	0·535	1851·51
	303·9	3n.	·510	6·87
	302·6	2n.	·560	64·70
	267·2	1n.	obl.	75·14
De.			single	65·97
			single	7·52

42   Σ. 100.

ζ PISCIUM.

R. A.	Dec.	M.
1ʰ 7·5ᵐ	6° 56′	4·2, 5·3

C. white.

Probably a very slow orbital motion manifests itself. The proper motion, according to Mädler, is +12″·0, −7″·2; Σ. gives +14″·1, −5″·1 ; and Argelander +17″·1 and −8″·6. Mädler found the changes represented by

$$p = 64° \ 27'·3 - 4'·078 \ (t - 1825·19),$$
$$\Delta = 23''·450.$$

See the Dorpat Observations, ix. 63.

H₁.	67·4	1n.	22·17	1781·88
H₂ & So.	63·5	2n.	24·65	1821·92
Σ.	65·6	1n.	23·33	3·87
	63·7	5n.	·45	2·83
	64·0	1n.	·60	51·89
Mä.	63·0	2n.	·31	41·65
	64·1	1n.	22·99	2·94
	63·0	3n.	23·22	5·04
Flt.	·9	29	·32	51·66
Du.	·7	3n.	·33	71·67
W. & S.	·7	3	24·7	3·89

43   O.Σ. 28.

R. A.	Dec.	M.
1ʰ 8ᵐ	80° 16′	7, 8·5

O.Σ.	324·3	3n.	0·527	1847·5
De.	313·4	1n.	·7	65·93
	·2	,,	·92	7·61
	317·5	,,	wedge	8·65

44   Σ. 101.

R. A.	Dec.	M.
1ʰ 7·9ᵐ	−8° 18′	7·5, 9·8

C. yellow.

Probably an optical pair.

H₁.			19·50	1782·87
	333·6		·60	3·65
So.	337·9	2n.	·89	1825·30
H₂.	345·0		...	9·67
Σ.	339·3	3n.	21·33	32·22
	340·8	1n.	...	45·89
Mä.	·3	2n.	21·24	53·09
	332·2		48·32	65·80
De.	340·2		20·52	8·20
W. & S.	341·5	2n.	22·30	74·90
	340·8	1n.	·52	·93
Gl.	342·3	2n.	·3	·88

45   Σ. 102.

R. A.	Dec.	M.
1ʰ 11ᵐ	48° 24'	A 7, B 8·2, C 8·4, D 10·8

C. white.

A B and A C unchanged. Slight change in A D, both in angle and distance, appears certain; probably due to the proper motion of the triple system, D being fixed.

A B.

Σ.	309·1	4n.	0·57	1834·25
Mä.	303·9	1n.	·5	53·08
O.Σ.	310·9	4n.	·65	3·14
De.	305·7	4n.	wedged	64·89

$$\frac{A+B}{2} \text{ and } C.$$

Σ.	255·7	8n.	10·22	33·89
O.Σ.	224·6	4n.	·32	53·14
De.	·2	4n.	·09	64·89

$$\frac{A+B}{2} \text{ and } D.$$

Σ.	66·3	3n.	29·76	32·45
O.Σ.	64·9		28·99	54·62
	·4		·26	69·17
De.	65·0	4n.	·6	4·89

46   Σ. 109.

R. A.	Dec.	M.
1ʰ 12ᵐ	63° 16'	9, 10

Σ.	10·2	4n.	7·02	1832·72
Mä.	9·4	1n.	·62	45·64
Du.	8·6	3n.	6·96	68·82

47   Σ. 93.

α URSÆ MINORIS (POLARIS).

R. A.	Dec.	M.
1ʰ 13·7ᵐ	88° 40'	2, 9

C. A, yellow; B, white.

O.Σ., after applying the precession correction to the measures of Σ. in 1834·14, finds the angle 212°·24 for the mean epoch of his own observation. The differences −0"·220 and −1°·16 leaves it still uncertain whether any change has taken place in this system.

H₁.			19	1779·8
			17	81·6
	203·3		...	·9
	202·6		...	2·4
	208·3		...	1802·1
H₂ & So.	·8	6n.	18·7	23·06

Sm.	209·9		18·4	1830·78
Σ.	210·0	In.	·27	34·14
Mä.	209·3	,,	19·1	41·46
	210·5	2n.	18·67	2·33
	208·9	In.	·39	6·23
Se.	212·4	,,	·44	59·95
O.Σ.	·7	2n.	·56	61·33
	213·3	In.	·44	8·25
	214·6	,,	·55	72·19
	213·7	,,	·36	5·18
De.	211·6		·26	62·90
Du.	212·4	7n.	·54	70·90
Gl.	·9	In.	·7	3·94
Fl.	213·3	,,	·62	7·31

48   Σ. 113.

R. A.	Dec.	M.
1ʰ 13·6ᵐ	1° 9'	6·2, 7·2

C. Σ., De., and Se., white.

This star was measured by Σ. in 1829. It is a difficult object, and great discrepancies are found in the recorded observations. The positions given by Da. in 1841 and 1842 are 330°·77, 332°·32, 331°·43, 337°·23, 334°·32, 338°·30, 339°·06, 338°·02, each being the mean of five or six observations on different nights.

Sm. (*Cycle*, p. 34,) says: "A beautiful object, but very difficult to measure in distance. It seems to have a direct angular movement, to the amount of about 0°·7 per annum; but this requires verification."

Da. thought that the discrepancies were probably due to the closeness and oblique position of the stars, and remarks that it is "still uncertain whether any real change has occurred."

Sm. (*Spec. Hart.*, p. 220): "I think the angular motion in orbit is now clearly proved."

Se.: "The motion in angle appears certain, though slow." "A feeble angular movement" (O.Σ., 1877).

Σ.	333·6	4n.	1·245	1831·61
	334·3	3n.	·177	6·91
H₂.	325·6		...	1·81
Sm.	332·6		1·2	4·84
	344·6		·3	57·97
Da.	331·4	3n.	·014	42·64
	335·6	2n.	·185	3·86
	338·4	3n.	·164	54·51
O.Σ.	·1	2n.	·355	44·72
Mä.	343·9	2n.	·39	53·09
	345·7	1n.	·57	8·01
	340·0	2n.	·36	42·75
	338·9	,,	·16	5·04
Se.	339·7	4n.	·169	56·48
	346·9	In.	·48	66·07
Ta.	338·4	2n.	·14	5·73
	339·0	In.	·15	6·84
	342·3	,,	...	76·86

De.	343·3°	2n.	1"·1	1855·89
	·4	,,	·15	56·13
	·4	4n.	...	62·80
	342·3	2n.	·27	63·43
Ja.	340·5	18	·45	56·78
Fer.	346·5	1n.	·406	72·95
M.	357·6	,,	0·90	61·90
W. & S.	346·2	4	1·35	72·07
	9·5	4	·24	3·01
	350·7	3	·63	·81
	348·0	8	·45	4·93
Gl.	350·7	4	·5	4·94
W.O.	351·9	1n.	·24	6·06
	·0	,,	·06	·08
C.O.	349·7	1n.	·19	6·79
	347·5	4n.	·46	7·79
Schi.	348·7	,,	·38	·01
Sp.	·8	,,	·38	·02
Dob.	346·3	3n.	·29	6·87
Pl.	349·3	,,	·38	7·37

49   h. 2036.

R. A.	Dec.	M.
1ʰ 14ᵐ	− 16° 25′	7, 7

Rapid change in angle since 1870. Probably a binary.

H₂.	53·0	1n.	2±	1830·79
	45·0	,,	1·25	5·72
	38·1	,,	·82	6·96
Ja.	36·1	,,	·57	57·97
De.	26·5		·45	74·67
	·6		·63	5·62
C.O.	24·0	4n.	...	5·94
	26·4	1n.	1·64	6·78
	29·8	,,	·62	·79
	26·0	4n.	·39	7·76
Sp.	·6	1n.	·51	·01

50   Σ. 117.

R. A.	Dec.	M.
1ʰ 17·4ᵐ	67° 30′	A 4, B 9·5, C 10

C. A, very yellow.

A B probably an optical pair. B C probably binary. The magnitude of A is variously given: e.g., Se. 4; Σ. 4·4, 4·5; Heis 5·0; Fl. 4·7.

A B.

H₁.	100·2			33·4	1782·63
So.	101·3	1n.		·3	1823·20
Mä.	103·3			30·7	30·87
	104·4	2n.		·24	44·33
	·5			·36	5·53
	106·9	1n.		29·9	50·72
	105·7	2n.		30·29	2·84

Σ.	101·8°	5n.	32"·2	1831·04
Sm.	102·1		31·9	6·28
Da.	·7	1n.	...	9·74
	104·9	,,	30·55	54·07
Mit.	102·8	1n.	·32	47·67
Se.	104·9	2n.	29·6	58·82
M.	103·2	1n.	27·73	62·71
De.	105·1	3n.	29·74	5·50
W. & S.	·5	2	·8	73·83
Fer.	106·0	1n.	28·49	4·83
Dob.	105·4	2n.	...	5·92
	106·5	5n.	28·12	6·23
Fl.	105·7	1n.	29·5	7·12

B C.

Σ.	253·3	4n.	3·01	31·04
	252·0		2·93	2·28
Sm.	·6		·0	6·28
Da.	253·1	1n.	...	9·74
	255·4	,,	3·25	54·07
Mä.	251·8	1n.	·16	44·33
Mit.	253·3	1n.	·28	7·63
	·0	,,	·25	·67
Se.	256·4	2n.	2·25	58·82
De.	255·3	3n.	·82	65·50
W. & S.	·8	6	3·06	73·83
Fer.	257·2	1n.	·11	4·83
Dob.	·0	3n.	...	6·12
Fl.	256·1	1n.	2·9	7·34

A C.

De.	108·0	2n.	26·96	65·88

51   Σ. 118.

R. A.	Dec.	M.
1ʰ 20·6ᵐ	82° 44′	8·5, 9·4

Probable change in angle and distance.

H₂.	61.5		12	1830·00
Σ.	62·0	4n.	10·75	2·49
Mä.	60·8	2n.	·52	45·10
De.	69·9		11·18	63·71

52   Σ. 122.

R. A.	Dec.	M.
1ʰ 20·7ᵐ	2° 55′	8, 9

C. A, very white; B, blue.

Probable change in angle and distance.

H₂.	334·2		...	1831·81
Σ.	332·8	3n.	5·79	3·56
Se.	326·6	2n.	·95	57·97
	328·5	1n.	·60	66·05
Mä.	333·6	2n.	·97	43·49
	331·5	1n.	·44	58·00
W. & S.	329·2	2n.	6·2	73·87

53   Σ. 125.

R. A.   Dec.   M.
1^h 20·8^m −0° 46′   8, 10·3

C. white.

On reducing his observations for the effect of precession, converting the results into rectangular coordinates, and treating them by the method of least squares, O.Σ. obtains the following formulæ :—

$$\Delta A = +4''{\cdot}323 \pm 0''{\cdot}040 - (0''{\cdot}2910 \pm 0''{\cdot}0037)\,(T - 1850{\cdot}0).$$
$$\Delta D = +19{\cdot}994 \pm 0{\cdot}040 + (0{\cdot}3501 \pm 0{\cdot}0037)\,(T - 1850{\cdot}0).$$

The motion is thus rectilinear. The small star is at rest. (See the P. M., p. ccxxiv.)

	°		″	
Σ.	37·3	In.	15·82	1829·90
	36·3	,,	16·96	30·92
	35·9	,,	17·52	2·79
	30·2	,,	16·91	3·95
	·9	,,	17·09	5·85
	29·2	,,	·16	·96
	27·3	4n.	·20	6·62
	21·2	3n.	18·28	42·78
	18·7	In.	·92	4·05
	·7	,,	·89	5·04
O.Σ.	19·3	2n.	19·05	3·90
	8·6	,,	21·35	52·91
Mä.	9·8	In.	·30	3·09
De.	1·6	3n.	24·45	62·94
	0·7	4n.	·81	3·81
W. & S.	356·3	2	...	73·87
	354·8	3	30·3	4·93
	·8	3	·8	·95
	352·7	4	·34	6·95
Fl.	353·3	In.	28·8	7·08
C.O.	352·9	2n.	30·74	·84

54   Σ. 132.

R. A.   Dec.   M.
1^h 25·6^m 16° 21′   7, 10

C. A, yellow.

The proper motion of the principal star explains the observed changes. (See the P. M., p. ccxxv.)

H$_1$.	27·7	In.	16	1783·63
Σ.	5·4		24·25	1829·87
	359·2	In.	28·87	51·80
Mä.	0·2		·26	47·10
	359·7	,,	·52	51·01
	0·1	,,	·43	2·10
	358·2	2n.	30·89	8·11
O.Σ.	359·0	In.	28·88	1·82
	355·8	,,	32·43	68·91
De.	356·2		30·99	3·84
W. & S.	·0	In.	...	73·89
Fl.	353·6	,,	34·0	7·08

55   Σ. 133.

R. A.   Dec.
1^h 25·9^m 35° 14′
    M.
A 7, B 10·5, C 11·2, D 11·6

C. A, yellow.

In A B a small increase in angle. In A C and A D a decrease in distance.

A B.

	°		″	
Σ.	179·1	3n.	2·99	1833·04
Mä.	185·6	In.	2·6	43·97
	189·2	2n.	·7	5·65
	187·7	In.	...	7·07
	189·4		...	51·18
	190·0	2n.	2·86	2·84
	189·8	In.	...	3·78
De.	185·3	5n.	2·87	63·59
W. & S.	·0	In.	3·04	73·89
	182·0	,,	...	4·93

C D.

Σ.	346·2	3n.	4·76	33·04
Mä.	351·7	In.	5·25	45·18
	347·6	,,	·20	·64
	348·1	,,	·58	7·07
W. & S.	351·5	4	4·75	73·89
	·7	2	...	4·94

A C.

Σ.	199·5	3n.	29·08	33·04
Mä.	196·7	In.	...	47·07
De.	197·9	4n.	27·30	63·76
W. & S.	193·9	2	...	74·93

A D.

Σ.	193·3		33·8	33·04
De.	·8	4n.	32·3	63·76

A E.

W. & S.	1·0	1	...	73·89
	0·3	2	...	4·93

E F.

W. & S.	300·0	1	...	73·89
	·7	2	...	4·93

56   O.Σ. 31.

P. I. 107.

R. A.   Dec.   M.
1^h 27^m 7° 36′   7, 11

The change in angle and distance is very small.

O.Σ.	84·9	4n.	4·04	1850·02
De.	81·4	In.	·05	65·88
	83·8	,,	·13	6·64
	·0	,,	·12	7·63

57   Σ. 136.

100 PISCIUM.

R. A.	Dec.	M.
1ʰ 28·5ᵐ	11° 57′	6·9, 8

C. white.

The proper motion, according to Mädler, is $-4''\cdot2$, $-0''\cdot1$; and Σ. has $-4''\cdot4$, $+1''\cdot9$. The former observer gives the following formulæ:—

$$p = 78° 41''\cdot3 - 7'\cdot059 \ (t-1826\cdot54).$$
$$\Delta = 15''\cdot830.$$

H₁.	85·0	In.	15·87	1783·59
H₂ & So.	80·4	,,	16·02	1821·91
	79·9		15·79	3·00
Σ.	78·7	In.	16·16	8·82
	·9	,,	15·61	9·81
	·9	,,	16·31	30·92
	79·1	,,	·02	1·93
	78·6	,,	·04	5·85
Mä.	·4	3n.	15·15	41·70
	·6	,,	·33	2·82
	·8	In.	·36	3·81
	77·6	,,	·60	5·02
O.Σ.	79·1	5n.	16·17	4·58
Ta.	78·09	In.	·31	65·78
	79·60	,,	...	7·04
	·06	,,	15·48	72·77
	·20	,,	...	6·86
Du.	78·8	4n.	15·96	69·07
W. & S.	79·2	5	16·4	73·89

58   O.Σ. 33.

R. A.	Dec.	M.
1ʰ 30ᵐ	58° 3′	7·2, 8·3

A probable increase in the distance.

O.Σ.	74·4	3n.	24·26	1846·80
Mä.	·9	2n.	·28	51·76
De.	·9	In.	·49	65·57
	75·1	,,	·69	6·54
	·1	,,	·78	7·62

59   Σ. 138.

P. I. 123 PISCIUM.

R. A.	Dec.	M.
1ᵇ 29·7ᵐ	7° 2′	7·3, 7·5

C. A, white; B, yellowish.

H₁. (*Mem. R. A. S.*, vol. i., p. 166):—
"Oct. 21, 1792. Double, a pretty object, a little unequal, less than a diameter asunder."
"Oct. 5, 1801. A beautiful minute object with 400."
Da. (*Mem. R. A. S.*, vol. xxxv., p. 309): "Though the results of my observations of this star do not run very smoothly,

there can be no doubt of its binary character."
Sm. (*Spec. Hart.*, p. 221): "Though the above measures do not confirm the motion in this beautiful star, I have no doubt of its binarity."
Se. (*Catalogo di stelle doppie*, p. 22). The measures made by Secchi in 1857 and 1858 seem to him to indicate increase of angle.
The angle has increased (O.Σ., 1877).

A B.

H₁.	10°±		...	1801·94
Σ.	20·0	3n.	1·46	30·23
Sm.	19·8		·5	2·86
	26·9		·4	43·10
	·3		·5	53·91
Da.	24·1	13n.	·40	41·54
	29·3	3n.	...	53·81
		2n.	·26	4·09
O.Σ.	212·9	In.	·67	41·70
	31·5	,,	·63	5·73
	28·5	,,	·66	56·73
Mä.	24·6	2n.	·53	42·75
	23·1	,,	·50	3·50
	26·5	,,	·44	5·03
	25·7	In.	·53	50·99
	166·6	,,	...	8·04
De.	209·7	4n.	·3	5·89
	207·2	2n.	·2	6·74
	28·2	5n.	·66	62·87
	29·4	3n.	·51	3·96
	32·2	In.	·58	6·62
	212·7	,,	·55	70·71
	·2	,,	·53	2·69
	31·7	,,	·38	6·07
Se.	29·1	3n.	·46	57·89
	32·2	In.	·85	66·07
M.	30·4	,,	·24	2·02
Ta.	28·5	2n.	·53	5·74
	26·4	In.	·74	6·84
	·9	,,	...	7·04
	32·6	,,	·30	72·77
W. & S.	31·0	4	·42	2·07
	32·9	4	·14	3·01
	33·2	7	·36	·81
	29·9	7	·59	4·95
	30·2	4	·60	·95
Gl.	33·1	4	·3	3·94
W.O.	34·1	In.	·51	5·98
	32·1	,,	·38	6·00
	34·2	,,	·27	·09
Dob.	29·2	5n.	...	·39
	30·8	In.	·34	7·91
Schi.	212·0	,,	·46	·05
Sp.	32·1	In.	·46	·05
Pl.	30·1	3n.	·46	·32

A B and C.

W.O.	62·3	In.	22·5	75 93
	63·4	,,	·0	6·00

A C.

W. & S.	70·7	2	...	74·95
	·7	4	77·31	·95

60 P. I. 127.

R. A.	Dec.	M.
1ʰ 30·6ᵐ	— 30° 31'	6, 7

Perhaps the angle has increased a few degrees.

H₂.	75·8		3·65	1836·64
	·1		·04	7·80
Ja.	82·8	2n.	2·8	46·35

61 Σ. 142.

R. A.	Dec.	M.
1ʰ 33·5ᵐ	14° 38'	8·2, 8·4

C. white.

The relative movement has been in a straight line hitherto. Σ. gives the following formulæ :—

$$\Delta A = -(17''\cdot770\pm0''\cdot021)+(0''\cdot219\pm 0''\cdot003)\,(T-1840\cdot13).$$
$$\Delta D = +(17''\cdot055\pm0''\cdot021)-(0''\cdot039\pm 0''\cdot003)\,(T-1840\cdot13).$$

To the proper motion of the smaller star the changes are probably due. (See the P. M., p. ccxxv.)

Σ.	310·9	In.	26·86	1828·82
	311·1	,,	·88	29·81
	310·8	,,	·20	·93
	313·0	,,	25·23	35·85
	312·3	,,	·48	·96
	313·1	3n.	·29	6·90
	317·4	In.	22·54	51·88
O.Σ.	313·4	,,	24·84	39·95
	314·0	,,	·35	41·70
	315·7	,,	23·54	5·74
	324·6	,,	19·52	68·77
	325·2	,,	·46	·94
Ka.	314·1	,,	23·75	41·99
	323·0		19·44	67·05
Mä,	314·4	In.	23·57	42·78
	·9	2n.	·54	5·02
	318·2	In.	22·53	51·01
Se.	319·7	In.	21·46	8·04
De.	321·8		20·33	63·34
W. & S.	327·2	3	19·0	73·87
	·6	6	17·7	6·95
Fl.	326·8	In.	18·5	7·08

62 Σ. 147.

R. A.	Dec.	M.
1ʰ 35·8ᵐ	—11° 55'	5·3, 6·9

Probably a small change in the distance. The two stars have a common proper motion.

Σ. Gives +0ˢ·030 in R. A., and +0''·390 in P. D.

Σ.	86·0		3·53	1822·30
	87·2	5n.	4·01	31·90
	88·1	In.	·30	51·88
So.	89·6	2n.	·19	23·97
H₂.	86·1		·30	9·67
	89·0		6·0	30·80
	86·5		4·27	1·81
	·9		·65	7·80
Da.	87·5		3·95	6·97
Mä.	·2	In.	...	44·91
Se.	88·5	3n.	3·62	55·89
Mo.	89·6	10	·78	6·97
De.	88·2	In.	4·04	7·81
C.O.	86·5	3n.	·00	77·87

63 6 ERIDANI.

R. A.	Dec.	M.
1ʰ 35·2ᵐ	— 56° 49'	6, 6

Change in angle and distance.

Dr. Doberck has lately published the following elements :—

$$\Omega = 81°\ 42' \qquad P = 117\cdot51 \text{ years}$$
$$\lambda = 327\ \ 15 \qquad T = 1817\cdot51$$
$$\gamma = \ \ 44\ \ 40 \qquad a = 3''\cdot82$$
$$e = 0\cdot378$$

Dunlop	343·1		2·5	1825·96
H₂.	302·3		3·65	35·03
Ja.	276·0		4·16	45·88
	·5		·32	6·35
	270·0		...	9·82
	268·7		...	50·80
	266·4		4·30	1·79
	264·8		·14	2·76
	261·1	18	·70	6·09
	258·1	18	·49	7·96
Po.	263·2	9n.	...	3·96
	253·4	6n.	4·86	61·03
El.	237·3		5·0	77·03

64 O.Σ. 35.

R. A.	Dec.	M.
1ʰ 36ᵐ	55° 16'	7, 10

Retrograde motion in angle, and increase in distance, are pretty certain.

H₂.	114·1		9	1831·50
Sm.	120·0		10·0	5·74
O.Σ.	·1	In.	9·8	44·91
	112·0	,,	·81	7·59
	114·2	,,	·91	50·13
Mä.	118·9	4	·54	48·50
	111·4	2n.	·51	51·76
De.	109·2	3n.	10·24	66·58
	108·6		·29	9·32

65   O.Σ. 34.

R. A. 1ʰ 36ᵐ	Dec. 80° 18′	M. 7·3, 7·5

C. white.

Probably a small increase in the angle.

	°		″	
O.Σ.	113·7	3n.	0·603	1847·57
De.	115·4	1n.	obl.	65·93
	·7	,,	,,	6·61
	125·7	,,	,,	8·65

66   Σ. 158.

R. A. 1ʰ 39·8ᵐ	Dec. 32° 34′	M. 8·3, 8·8

Certain change in angle and distance. Probably a binary.

H₂.	239·5		1·0	1828·64
	246·7		·5	31·79
Σ.	·2	3n.	2·13	3·11
Mä	251·2	2n.	·19	45 11
	250·7	1n.	·18	50·71
	255·2	2n.	·11	1·17
	·5	1n.	...	·79
	253·6	1n.	2·93	5·86
Se.	252·7	2n.	·78	7·90
De.	254·5	1n.	·99	65·87
	256·5	,,	·15	6·57
	255·8	,,	·01	7·68
O.Σ.	260·8	,,	·19	9·95
W. & S.	257·5	4	·05	73·89
Gl.	256·7	4	1·9	·89

67   Σ. 175.

R. A. 1ʰ 44·4ᵐ	Dec. 20° 31′	M. 8, 9

C. white.

An instance of rapid rectilinear motion.

H₁.	293·2		4·8	1783·58
Σ.	327·9	4n.	10·43	1830·22
Mä.	334·0	2n.	11·81	44·06
	332·8	,,	·93	5·03
	336·3	,,	12·44	50·97
	·7	,,	13·27	3·08
De.	339·3		·26	63·89
Gl.	341·8		14 ±	74·01
W. & S.	343·2	6	·62	6·94

68   Σ. 180.

γ ARIETIS.

R. A. 1ʰ 47ᵐ	Dec. 18° 42′	M. 4·2, 4·4

Small change in angle and distance probable.

Dr. Dunér gives

$$\Delta = 8''·68 - 0''·01 \ (t-1850·0).$$
$$1848·91 \quad P=359°·1.$$

	°		″	
H₁.	175·0	2n.	...	1780·3
	179·2	1n.	...	1802·2
H₂.	177·4	,,	...	16·81
	178·7	3n.	9·11	22·88
Σ.	359·9	7n.	8·63	30·84
	358·5	1n.	·45	51·52
Be.	178·9	4n.	·96	30·93
Mä.	·3	,,	·82	41·78
	·5	13n.	9·11	3·4
	179·9	2n.	·53	50·97
	180·5	1n.	...	2·11
	358·7	3n.	·70	8·05
Da.	·8	6	8·98	46·95
	·2	8	·62	7·93
Mit.	357·0	3n.	·84	7·65
O.Σ.	356·4	1n.	·60	51·82
	179·4	,,	·45	70·18
De.	359·3	4n.	·71	53·47
Wi.	179·3	4	·34	7·87
	·3	4	·48	62·54
	·2	9	·49	4·80
Ro.	358·8	1n.	·79	2·94
	359·6	4n.	·62	3·49
Ta.	·0	2n.	·54	5·75
	358·9	,,	9·17	6·94
Du.	·9	6n.	8·41	71·47
Dob.	359·1	2n.	...	5·92
	358·6	3n.	8·32	7·89

69   Σ. 183.

R. A. 1ʰ 48·3ᵐ	Dec. 28° 13′	M. A 7·5, B 8·2, C 8·8

In A B there has been a diminution in the angle. A C seems unchanged. A B is probably binary.

A B.

Σ.	25·6	3n.	0·55	1833·12
O.Σ.	31·0	1n.	·70	41·70
	24·8	,,	·66	5·73
	12·7	,,	·73	56·73
	8·6	,,	·70	7·67
Mä.	26·5	,,	·6	44·91
De.	9·1	2n.	oval	64·04
Gl.	2±	1	0·55	74·02

$\dfrac{A\ B}{2}$ and C.

H₂.	163·4		5·07	28·75
Σ.	·7	5n.	·68	32·31
O.Σ.	·5	1n.	·86	41·70
	164·9	,,	·63	5·73
	·5	,,	·68	56·73
	162·2	,,	·58	7·67
Mä.	165·4	2n.	·78	44·48
De.	·7	,,	·67	64·04
Gl.	·7	2	·70	74·01

70 Σ. 186.

P. I. 209 PISCIUM.

R. A.	Dec.	M.
1ʰ 49·7ᵐ	1° 15′	7·2 7·2

C. white.

H₂ (*Mem. R. A. S.*, vol. v., p. 56) : "In contact. A division seen by glimpses. Like η Coronæ."

And in vol. viii., p. 39, he says : "Very clear and difficult, but less than η Coronæ. Well separated, and black divisions well seen."

Da. (*Mem. R. A. S.*, vol. xxxv., p. 473): "There can be no doubt that this double star, discovered by Σ., is a binary system." He remarks that although it was probably not single in 1851, it was probably so in 1863.

The distance has clearly diminished. The apparent orbit probably coincides very nearly with the visual ray. (O.Σ., 1877.)

The common proper motion is + 0″·09 in R. A., and + 0″·22 in P. D. (Σ.)

	°		″	
Σ.	64·7	4n.	1·23	1831·12
	61·3	1n.	0·97	41·70
			single	51·81
H₂.	56·9		1·23	30·66
	·4	2n.	0·96	1·80
Sm.	62·9		1·5	3·83
O.Σ.	242·8	1n.	1·11	41·70
	68·2	,,	0·82	6·11
De.	252·8	3n.	oblong	55·89
	258·8	1n.	,,	7·03
Se.	87·0		0·4	7·92
	83·0	In	contact	9·84
	85·0		0·3	63·85
	96·0		...	6·05
Da.	81·0		0·5	59·81
	85·1		0·3	63·85
W. & S.	Less than 0·5			73·93
Schi.	single			6·99

71 Σ. 185.

R. A.	Dec.	M.
1ʰ 51ᵐ	74° 55′	7, 8·5

Probable diminution of the angle and increase in the distance. Probably a binary.

Σ.	40·3	3n.	1·39	1831·95
	34·9	,,	·38	6·71
Mä.	36·2		1·31	39·24
	35·8	1n.	·25	44·33
	31·0	3n.	·26	52·50
Se.	30·7	2n.	1·48	7·94
O.Σ.	33·8	1n.	1·79	72·31
Gl.	20·0	1	1·5	4·00

72 Σ. 196.

، R. A.	Dec.	M.
1ᵇ 52·9ᵐ	20° 26′	A 9, B 11, C 10, D 6·5

A B is probably unchanged : the distance between A C has diminished, while that between A D has considerably increased.

A B.

	°		″	
Σ.	55·5	3n.	2·37	1832·42
Sm.	53·0		·5	4·99
Mä.	54·6	1n.	·4	45·04
Da.	53·7		·25	62·93
De.	60·7	1n.	...	4·73
	54·5	,,	2·24	7·70
	56·8	,,	·11	8·67
Kn.	50·0	,,	·5	2·95

A C.

Σ.	167·4	3n.	39·46	32·42
Sm.	165·0		40·0	4·99
Da.	·6		36·2	62·93
Kn.	166·3	1n.	37·36	2·95
De.	·5	,,	36·23	3·99
	·1	,,	35·01	7·70
	·3	,,	36·06	8·67

A D.

Sm.	359·2		165·0	34·99
Da.	361·6		182·5	62·93
Kn.	0·8	1n.	183·6	2·95

73 Σ. 197.

R. A.	Dec.	M.
1ʰ 54ᵐ	34° 42′	7·3, 8·3

C. white.

The changes are due to the proper motion of the larger star.

Σ.	233·6	3n.	18·33	1833·48
Sm.	·6		·3	·70
Mä.	242·9		20·07	47·12
	232·2	2n.	·48	51·17
	·4	1n.	·95	2·84
	·8	,,	·98	3·09
	236·6	,,	·40	·83
	234·1	,,	·53	5·83
De.	233·2	5n.	21·67	64·79
Gl.	·0	2	22·1	73·96

74 Σ. 202.

α PISCIUM.

R. A.	Dec.
1ʰ 55·8ᵐ	2° 11′

M.

2·8, 3·9 (Σ.) ; 4, 6 (Da.) ; De. 4, 5·3

C. Σ. A, greenish white ; B blue ; Da. very white, white.

Both stars probably vary both in colour and brightness.

H₁ (*Phil. Trans.*, vol. lxxii., p. 217): "α Piscium, Fl. ultima. In nodo duorum linorum. Oct. 19.—Double, cousiderably unequal. Both W. With 222, not quite 2 diameters of L; with 460, about 3 diameters of L. Distance 5″·123 mean measure. Position 67° 23′ n.p." This was in 1779.

H₁ (*Phil. Trans.*, 1804, part i., p. 384): "The position of the stars Oct. 19th, 1781, was 67° 23′ n.p.; and by a mean of 3 measures, taken Jan. 28 and Feb. 4, 1802, it was 63° 0′. This gives a change of 4° 23′ in 20 years and 105 days. The parallactic motion of α will account for the alteration, unless a proper motion should hereafter lead to a different conclusion, which, from the insulated situation of this double star, is not improbable."

H₂ and So. (*Phil. Trans.*, 1824, part iii., p. 47): "A beautiful double star; nearly equal. This star has undergone no appreciable change." H₂ thinks that Σ.'s measure, 70° 48′ n.p., 1819·9 (see *Additamenta*, p. 182,) is too large, and in quoting H₁'s first measure gives the date 1781·99.

Sm. (*Cycle*, p. 49). He observes that H₁ was led to suppose a retrograde motion, and adds, "All the subsequent observations, however, of H₂, So., Da., Σ., and myself prove the fixity of these stars."

Da. (*Mem. R. A. S.*, vol. xxxv., p. 310). He says that his later observations compared with his Ormskirk results indicate a slow diminution of angle, and that a slight diminution of distance is possible. He also remarks that the obliquity of position requires special care.

Se. (*Catalogo di stelle doppie*, p. 47) thinks the motion in angle is certain, but that in distance doubtful. The colours of the stars he suspects of change.

The movements proceed very slowly. It is, however, certain that both the angle and distance have changed. (O.Σ., 1877.) The proper motion in R. A. is + 0ˢ·009, and — 0″·01 in N. P. D.

Dunér (*Mesures Micrométriques*, p. 171): On making a graphical construction of the angles and distances, he found that both diminish, but the latter more rapidly and the former more slowly than according to the time; that this is contrary to nature, and most probably due to accidental errors of observation; and that the following formula represents the observations fairly well.

$$P = 330°\cdot3 - 0°\cdot290\,(t - 1850\cdot0) - 0°\cdot00113\,(t - 1850\cdot0)^2.$$

	°		″	
H₁	337·4		...	1781·8
	333·0		...	1802·8

	°		″	
H₂ & So.	335·8	5	5·401	1821·89
	·4	7	·448	·95
Σ.	336·9	3	3·94	·96
	335·7	5n.	·63	31·16
Be.	333·0		·77	0·93
	335·7		·64	1·16
	332·0		·76	4·85
	333·0		·64	40·03
Da.	332·1	5n.	·756	32·88
	330·1	1n.	·479	42·95
	·0	,,	...	6·92
	328·1	2n.	3·420	53·99
Sm.	334·7		·6	34·92
Mä.	331·9	3n.	·80	41·64
	332·0	5n.	·4	2·8
	331·2	4n.	·5	4·5
	·1	2n.	·4	5·0
	329·7	,,	·14	52·12
	328·4	4n.	·38	8·12
Ja.	333·0	3	4·04	42·94
	330·2	15	·20	5·84
	329·3		3·58	51·26
	333·2		·49	1·94
	328·4	15	·48	2·60
	327·8	3n.	·22	3·96
	·1	2n.	·20	6·45
	326·8	24	·29	7·94
	·9	19	·13	8·15
Mo.	331·1	30	·63	44·07
	326·8	12	·51	58·70
D.0.	329·5	1n.	·73	47·10
Mit.	·4	2n.	·08	·68
Flt.	·4	6n.	·40	51·75
	·5		·22	3·94
O.Σ.	330·4	3n.	·557	1·87
	325·0	1n.	2·980	70·18
De.	327·5	6n.	3·59	54·28
	329·4	2n.	·79	5·06
	328·4	1n.	·40	6·62
	327·1	,,	·71	7·61
	326·8	2n.	·51	8·63
	325·7	,,	·2	63·76
	326·6	3n.	·04	4·08
	325·7	2n.	·17	5·52
	·7	3n.	·11	6·56
	·1	1n.	2·96	7·61
	·5	,,	3·12	8·07
	·7	,,	·34	·65
	·7	,,	·12	9·66
	·2	,,	·12	70·76
	·0	,,	·21	1·64
	·7	,,	·23	2·67
	·7	,,	·15	4·68
	·0	,,	·04	5·08
Wi.	331·5	2n.	·41	56·09
Se.	327·8	3n.	·352	·16
Au.	·9		·56	61·18
M.	329·6	1n.	·15	·90
	323·2	,,	·06	8·84
	324·3	,,	·13	72·78
	323·2	,,	·09	4·79
	322·6	,,	·09	·85
	·7	4n.	·27	5·76
Ro.	324·9	2n.	·30	62·92
	325·3	,,	·17	3·54

	°		″	
Kn.	327·1	In.	3·077	1864·01
	325·7	3n.	·237	5·47
Eng.	328·5	,,	·35	·08
Ta.	·9	2n.	·48	·76
	·8	In.	·41	6·84
	325·6	,,	·38	70·86
	324·8	,,	·10	1·78
	325·7	,,	·06	2·77
Du.	·4	In.	·00	68·84
	326·1	2n.	·1	71·32
	325·6	,,	2·79	2·18
W. & S.	324·4	4	·3	·07
	·2	3	·16	3·93
	325·0	4	·11	4·01
Gl.	323·0	6	3·5	3·74
	324·5	3	·2	·93
	325·3	2	...	4·03
Dob.	327·2	In.	...	5·99
	325·2	6n.	3·77	6·09
	324·3	3n.	·23	7·83
W.O.	322·1	In.	·15	6·02
	·1	,,	·10	·03
Schi.	324·0	,,	·084	7·04
Sp.	·0		·08	·05
Pl.	325·9	5n.	·00	·18

75   Σ. 205 and O.Σ. 38.

R. A.	Dec.	M.
1^h 56·5^m	41° 46′	3, 5, 6*

C. golden, blue.†

The duplicity of B was discovered by O. Σ. in 1842, and the following is the account given by Sm. in the *Cycle*, p. 50. After expressing his conviction that γ_1 Andromedæ (Σ. 205; H₁, iii. 5) is not a binary system, he says : "Since the above was written, Mr. Baily put into my hand a letter which he had received from M. Struve in Oct., 1842, announcing the unlooked-for tidings that he had detected γ Andromedæ to be

* MAGNITUDES.—O.Σ., in the Pulkowa observations always make the n.p. star the smaller. Da. had never the slightest doubt about the s.f. star being the smaller, by at least half a magnitude. On the other hand, the discoverer says the more he looks at the star the more astonished he is that any one could place the smaller star in the n.p. quadrant.

† COLOURS.—H₁ gives γ₁ (H₁, iii. 5), "reddish white, and fine light sky-blue, inclining to green." H₂ and So. call the large star orange, and the smaller emerald green, while, in the *Mem. R. A. S.*, vol. vi., p. 8, the larger is called yellow, and the smaller pale blue, by H₂. De., golden and blue. Se., yellow and green. Sm., orange and emerald green. The components of the smaller star (γ₂) are registered pale yellow and small blue by Sm., clear blue by De., while Da. uses the following expressions: "both blue;" "both palegreen, precisely the same tint." "A, green; B, deeper green." Very pale yellow, blue;" "greenish yellow, bluish green."

triple, and that the companion is composed of two stars of equal size, separated by an interval of less than 0″·5." Sm. at once sent the news to Da., who readily elongated the star in the direction s.f. and n.f., "making it look like a dumpish egg." Sm. also received a letter from Challis, who could "easily recognise the small star as being double," and also thought the components unequal.

In 1843, Sm., at Hartwell, "fairly saw that the comes was not round, but elongated, in a direction n.p. and s.f. It was, however, so slightly oval, that, but for M. Struve's unexpected announcement, I must assuredly have overlooked it."

Da. (*Mem. R. A. S.*, vol. xxxv., p. 311) found that his measures indicated "on the whole a decrease in the angle."

O.Σ. (*Catalogue Revu et Corrigé*, 1850,) says that the Pulkowa observations from 1842 to 1850 indicate no perceptible change in the position of the close stars.

Se. gives no measures in his catalogue (1860), but in the second series he says that the companion was elongated, but not separated.

The progressive diminution of the angle is manifest, and an augmentation in the distance is probable. (Ō.Σ., 1877.)

A B.—No change in the relative position of this pair since its discovery in 1777 by C. Mayer. The proper motion of A is + 0ˢ·004 in R. A., and + 0″·04 in N. P. D.

It is worthy of passing notice that neither Messier in 1764 nor Mayer in 1776 observed the duplicity of A.

B C.—Dunér gives the following formulæ :—

$$\Delta = 0''·56 + 0''·07 \, (t - 1855·0.$$
$$P = 112°·5 - 0°·32 \, (t - 1860·0) + 0°·0038 \, (t - 1860·0)^2.$$

The distance formula rests entirely on O. Σ.'s measures; and Dunér thinks the following is better than the one given above.

$$= 0''·50 + 0''·07 \, (t - 1855).$$

He observes that the errors in the measured angles of position are enormous.

A and $\dfrac{B+C}{2}$

	°		″	
Σ.	62·4	6n.	10·33	1830·02
Da.	64·0	3n.	·62	2·94
	·2	In.	·47	63·86
Ch.	63·9	,,	·61	41·95
	·5	,,	·35	2·92
	62·9	,,	·49	3·12
Mä.	63·0	4n.	·08	·2
	62·4	In.	·02	5·1
	61·9	8n.	9·83	51·15
	62·6	3n.	·80	3·22

	°			
Mä.	62·1	1n.	9·94	1854·25
	·8	,,	·60	5·01
	·4	,,	·88	6·19
	·4	2n.	·63	7·23
	61·4	,,	·70	8·22
O.Σ.	63·0	1n.	10·57	42·72
	62·3	,,	·40	4·84
	64·1	,,	·58	8·13
	63·2	,,	·21	9·22
	·0	,,	·30	50·19
	62·8	,,	·42	4·20
	·9	,,	·43	7·23
	·7	,,	·62	9·10
	64·6	,,	·52	64·21
	·3	,,	·57	6·21
	63·7	,,	·41	8·20
	65·5	,,	·50	9·17
	63·8	,,	·53	70·18
	·3	,,	·11	2·18
	64·4	,,	·46	3·13
	63·8	,,	·41	5·14
Mit.	62·2	,,	...	46·65
D.0.	64·7	,,	10·20	·92
Ja.	61·4	5	·38	53·92
De.	63·2	3n.	·47	4·84
	·2	2n.	·50	5·09
	·5	3n.	·21	·89
	64·4	1n.	·45	6·56
	63·2	3n.	·43	62·80
	62·7	,,	·40	3·12
Wi.	63·4	2n.	·34	56 22
M.	62·4	1n.	9·58	61·85
	63·1	,,	10·21	2·71
	60·6	,,	·02	9·77
	63·6	2n.	·05	72·77
	·5	,,	9·86	·78
	66·4	,,	10·38	3·78
	64·9	5n.	·57	5·76
Ro.	62·6	1n.	·33	63·13
Eng.	63·8	2n.	·18	4·14
Kn.	·4	4n.	·36	5·67
Ta.	62·7	1n.	·47	·82
	61·1	,,	·56	6·53
	62·6	,,	·12	71·78
W. & S.	64·5	4	·45	2·03
	·4	2	·50	3·81
Gl.	63·9	4	·5	·94
	64·0	7	·3	4·93
	·8	4	·0	7·94
Du.	62·8	12n.	·09	1·33
W.0.	·4	,,	·56	5·96
Schi.	·6	,,	·10	7·11

B C.

	°			
Σ.	126·6	1n.	0·51	1842·72
Da.	125·8	1n.	·47	·83
	111·3	4n.	·62	7·82
	108·5	,,	·55	53·79
	112·0	1n.	·60	4·75
	108·7	5n.	·53	9·81
	107·7	1n.	·58	63·86
	106·9	,,	·60	5·68

	°			
Ch.	100·2	1n.	...	1842·84
	112·5	,,	0·15	·84
	115·1	,,	·31	3·31
Sm.	120·0	,,	·5	·33
O.Σ.	125·5	3n.	·47	·55
	117·9	5n.	·52	7·13
	114·9	4n.	·47	9·69
	113·0	3n.	·67	56·84
	109·9	,,	·70	66·21
	106·9	,,	·63	9·84
	105·4	5n.	·63	73·17
Mä.	116·9	4n.	·39	45·15
	·5	,,	·40	51·19
	115·3	5n.	·47	2·82
	116·5		·45	6·20
	114·9		·47	7·05
	115·7		·45	·24
	116·9		...	8·23
	115·2		·5	62·55
Mit.	111·3	7n.	·43	46·66
Ja.	·3	6	·5	52·78
	106·8	18	·4	3·94
	116·7	1n.	·5	6·12
	113·9	2n.	...	8·06
De.	92·7		oblong	4·81
	280·0		wedgᵈ.	·83
	270·8			·88
	274·0	2n.		5·09
	278·4	4n.		·85
	281·5	1n.		6·56
	107·4	3n.		62·77
	108·8	4n.		3·31
	109·1	1n.	...	4·57
	104·8	6n.	0·5	5·81
	106·6	1n.	wedgᵈ.	6·71
	107·2	,,	0·5?	7·60
Wi.	121·7	2n.	·41	56·21
Se.	109·7	6n.	·47	·90
	108·5	1n.	·45	8·99
Ro.	107·3	5n.	·6	63·49
	·3	1n.	·64	4·02
Kn.	·0	4n.	·59	5·67
	101·3	1n.	·6	72·67
Ta.	106·3	2n.	·58	65·76
	104·2	1n.	·64	6·84
Du.	103·0	1n.	·68	8·68
	112·0	4n.	·58	9·39
	106·0	2n.	·63	70·12
	111·3	,,	·63	1·19
	113·7	,,	·69	2·13
	110·1	,,	·65	·35
	109·2	1n.	·61	3·12
	108·9	,,	·62	4·17
Br.	101·9	6n.	·69	68·82
W. & S.	87·7	4	...	72·03
	95·9	10	·5	3·81
	105·7	6	...	6·79
Gl.	93·9	3	·46	3·94
	98·7	5	·56	4·93
	102·4	4	·84	7·94
W.0.	109·3	1n.	·53	4·00
	101·8	5n.	·4	7·10
Schi.	104·1	1n.	·48	·05
Dob.	103·9	,,	...	·71

76   Σ. 208.

R. A. 1ʰ 56·8ᵐ   Dec. 25° 22′   M. 6·2, 8·4

C. A, yellow ; B, ash.

The common proper motion is $+0^s\cdot013$ in R. A., and $+0''\cdot03$ in N. P. D.

Dunér's formulæ are

$$\Delta = 1''\cdot68 - 0''\cdot015\,(t-1850\cdot0).$$
$$P = 30°\cdot5 + 0°\cdot37\,(t-1850\cdot0) + 0°\cdot00325\,(t-1850\cdot0)^2.$$

	°		″	
H₂.	25·6		2·13	1830·79
	26·0		·00	2·80
Σ.	25·2	4n.	1·98	3·05
Da.	27·8		2·0	·36
	30·1	3n.	...	47·90
	...	1n.	1·73	9·00
	30·7	2n.	·84	53·92
Sm.	26·8		2·2	38·66
Mä.	30·2	3n.	1·60	42·77
	28·8	1n.	·75	4·12
	30·4	,,	·78	5·09
	32·6		·54	50·99
	34·7	1n.	·42	2·10
	31·6		·60	6·21
	36·1		·28	62·85
Ch.	32·1	1n.	·65	44·91
O.Σ.	29·6	,,	·60	51·76
De.	34·1	3n.	·3	6·72
	33·9	6n.	·43	63·07
	36·3	1n.	...	6·68
	41·2	,,	1·36	7·65
	·6	,,	·32	9·67
	44·1	,,	·45	71·59
	41·8	,,	·21	4·71
	46·1	,,	·29	6·54
Se.	34·4	3n.	·64	56·98
	38·1	1n.	·51	66·06
Eng.	40·6	,,	·90	5·01
Du.	39·0	5n.	·41	71·45
W. & S.	42·0	2n.	·44	2·46
	40·1	1n.	·32	3·93
Gl.	39·0	4	·4	4·93
Dob.	41·1	4n.	·27	7·86
	43·0	1n.	·16	8·08

77   Σ. 216.

R. A. 2ʰ 2·5ᵐ   Dec. 61° 44′   M. 7·7, 8·2

C. yellow.

Σ.	270·5	3n.	0·59	1831·23
Mä.	265·0	1n.	·75	41·50
O.Σ.	268·8	3n.	·7	51·05
Se.	262·4	,,	·43	7·63
De.	266·5		·6	67·36

78   Σ. 221.

R. A. 2ʰ 3ᵐ   Dec. 19° 46′   M. 7·7, 8·9, 12

C. yellowish.

A B.

	°		″	
H₁.	145·7		8·08	1783·13
Σ.	150·2		·64	1822·06
	145·7	4n.	·44	31·36
	·2	3n.	·38	6·91
H₂ & So.	150·1	1n.	·64	22·06
	148·8	,,	·95	4·87
Da.	143·5	7n.	·43	42·33
	·1	2n.	·15	3·41
	144·1	4n.	·46	54·50
Mä.	147·5		7·91	44·49
Mo.	144·6	50	8·34	55·45
Wi.	148·1	1n.	·95	6·09
Se.	143·1	2n.	·34	7·41
Ta.	144·4	,,	·51	65·76
	·4	1n.	·79	71·78

A C.

Wi.	226·2	1n.	61·0	56·09

79   Σ. 227.

ι TRIANGULI.

R. A. 2ʰ 5·4ᵐ   Dec. 29° 44′   M. 5, 6·4

C. A, yellow ; B, blue.

Dunér gives

$$1847\cdot21 \qquad \Delta = 3''\cdot58.$$
$$P = 77°\cdot4 - 0°\cdot0775\,(t - 1847\cdot12).$$

	°		″	
H₁.	85·6	1n.	...	1781·77
Σ.	79·1	1	3·02	1821·03
	77·9	5n.	·60	30·97
	80·5	3n.	·68	6·73
H₂ & So.	78·0	1n.	·88	21·94
H₂.	74·0		...	2·11
	77·9		·60	30·95
Sm.	78·1		·6	0·91
	77·9		·3	4·17
	78·8		·5	8·99
	·5		·5	57·95
Mä.	79·0		·60	32·84
	77·5	1n.	·57	41·77
	78·1	2n.	·44	7·55
	77·1	1n.	·51	52·25
	76·0		·55	6·83
	75·4		·41	61·87
Da.	79·0		·68	32·94
Ka.	77·1		·48	40·05
Mo.	78·8	40	·82	53·52
De.	76·9	6n.	·91	4·81
Se.	·9	3n.	·56	5·89
Ro.	77·6	3	4·38	62·90
	78·0	11	·05	3·38

Eng.	80°·2	1n.	4″·04	1865·12
Du.	76·9	2n.	3·25	70·55
Fer.	·2		·87	2·93
Dob.	75·6	5n.	~·86	7·85

80   Σ. 228.

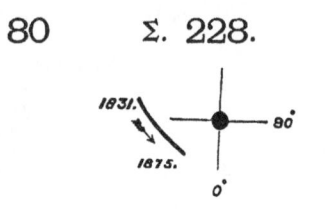

259 ANDROMEDÆ.

R. A.   Dec.   M.
2ʰ 6·4ᵐ   46° 55′   6·7, 7·6

C. Σ., De., and Se., white.

The angular motion is decided. The distance does not appear to have changed, unless we suppose that a maximum was reached about 1842, and that it has diminished since. (O. Σ., 1877.)

Dunér gives

$$\Lambda \sin . P = -1''·059 + 0''·01215\,(t - 1852·0) + 0''·000641\,(t - 1852·0)^2.$$
$$\Lambda \cos . P = +0''·173 + 4''·01243\,(t - 1852·0) - 0''·000169\,(t - 1852·0)^2.$$

The comparison with the measures is fairly satisfactory. From these formulæ it appears that the maximum distance was reached about 1847, that the star is rapidly hastening to its minimum distance, and that it will soon become excessively difficult. (Du.)

Σ.	264·0	1n.	1·08	1829·16
	262·2	,,	·22	31·75
	258·8	,,	·14	2·02
	263·3	,,	0·93	·20
	262·4	,,	1·03	·21
O. Σ.	274·7	2n.	·32	41·94
	299·6	1n.	0·86	70·18
Mä.	280·2	6	1·11	52·19
	284·1	1n.	0·9	7·21
De.	281·0	2n.	1·0	6·76
	·1	,,	·0	6·76
	286·4	,,	0·75	62·79
	·5	,,	1·1	3·11
	291·6	3n.	0·95	6·07
	304·4	4n.	·64	73·81
	314·9	,,	·52	6·89
Se.	286·1	3n.	·99	7·62
Du.	299·5	4n	·71	69·48
	311·4	3n.	·52	75·20
W. & S.	308·7	4n.		2·04
	309·0	3n.	·77	3·93
Gl.	307·2	2n.	·6	4·94

81   Σ. 231.

R. A.   Dec.   M.
2ʰ 7ᵐ   −2° 57′   6, 7·8

C. yellowish blue.

Rapid common proper motion.

Σ.	229·2	8n.	15″·47	1834·19
O. Σ.	·9	6n.	·55	52·20
Mä.	230·2	2n.	·09	3·09
	229·0	,,	14·73	8·11
De.	·2	5n.	15·36	4·83
Wi.	228·4		·35	7·87
Ta.	·8	2n.	·43	65·77
W. & S.	230·4	3	6·0	73·94
U. O.	·6	4n.	5·86	7·79
Dob.	229·8	2n.		6·07

82   Σ. 232.

R. A.   Dec.   M.
2ʰ 7·7ᵐ   29° 50′   7·5, 7·5

C. very white.

Dunér has the following formulæ :—
$$1848·69. \quad \Delta = 6·49.$$
$$P = 246°·8 + 0°·075\,(t - 1848·69).$$

Σ.	244·1		...	1821·00
	...		6·71	2·86
	245·5	3n.	·56	32·03
Mä.	246·4	,,	·45	43·67
O. Σ.	65·4	1n.	·6	51·82
Se.	246·8	3n.	·56	5·98
De	247·2	1n.	·49	6·83
Mo.	245·5	2n.	·30	7·96
Ta.	246·4	1n.	·27	65·84
Du.	248·4	2n.	·41	9·86

83   Σ. 234.

R. A.   Dec.   M.
2ʰ 8·7ᵐ   60° 48′   7·8, 8·7

C. white.

Σ.	239·2	3n.	0·84	1831·55
Mä.	232·7		·82	45·65
	·3	3n.	·87	52·50
O. Σ.	235·6	2n.	·83	46·98
Se.	231·4	3n.	·62	57·91
De.	·4	,,	·70	63·45
	220·4	5n.	·6	71·22

84   Σ. 236.

R. A.   Dec.   M.
2ʰ 9·3ᵐ   51° 55′   8·5, 9·3

Σ.	259·0	3n.	0·81	1831·87
Mä.	260·8	1n.	·76	45·23
O. Σ.	258·2	2n.	1·07	7·32
	257·9	1n.	·21	70·18
Se.	258·8	,,	0·5	57·92

85   Σ. 240.

R. A. 2^h 10^m	Dec. 23° 19'	M. 7.7, 8.2

C. white.

Dunér's formulæ are—

1854.50. Δ = 4".73.

P = 49°.7 + 0°.053 (t − 1853.50).

Σ.	48.0	3n.	4.71	1832.19
Mä.	50.7	1n.	5.03	44.04
Ta.	49.2	,,	4.73	65.84
Du.	51.1	5n.	.46	72.21

86   o CETI.

R. A. 2^h 13.3^m	Dec. 3° 32'	M. 2.5 to 9.5, 9.5

The period of A is about 331 days: maximum, Dec. 19, 1876; minimum, July 23, 1877.

The proper motion of A is − 0^s.003 in R. A., and + 0".23 in N. P. D.

Rectilinear motion.

Cassini.	130.7		119	1683
H$_1$.	...		110	1779
	...		110	80
	92.5		114	82*
Σ.	88.6		115	1819.96
	...		115	24.63
So.	88.6	...		1.90
Sm.	.9		116.0	31.03
O.Σ.	85.2		115.6	51.99
Po.	84.9	24	117.9	6.07
	.6	50	.1	9.96
Fl.	82.2	1n.	118.2	77.12

87   Σ. 249.

R. A. 2^b 12.7^m	Dec. 44° 2'	M. 7, 9

C. A, very white; B, ash.

Σ.	154.7	3n.	2.28	1831.11
Mä.	192.3		.39	5.74
	188.8	2n.	.55	45.17
	192.2	,,	.44	51.69
	189.4	1n.	...	5.87
	191.4	2n.	2.38	62.54
Se.	187.9	,,	.13	57.89
O.Σ.	192.2	,,	.47	69.08
Du.	191.9	7n.	.29	73.13

* Oct. 19, 1779. Distance, 1' 50'".468.
Dec. 5, 1779. Distance, 1' 52'".812, and 1' 50'".625.
Jan. 4, 1780. Distance, 1' 44'".687.
Sept. 8, 1780. Distance, 1' 50'".312, "with the utmost accuracy." Also 1' 50'".625.
Sept. 20, 1780. Distance, 1' 50''.
Aug. 17, 1781. Distance, 1' 47'" 54'".
Oct. 28, 1781. Distance, 1' 52'" 37'".
Aug. 25, 1782. Distance, 1' 54'" 36'".

88   O.Σ. 40.

R. A. 2^h 14^m	Dec. 37° 57'	M. 7.8, 8.6

O.Σ.	56.0	6n.	0".59	1850.64
Mä.	54.7	2	.25	48.75
De.	53.4	3n.	...	67.57

89   Σ. 257.

R. A. 2^h 16.6^m	Dec. 60° 59'	M. 7.2, 7.7

C. Σ. yellowish white.

Considerable direct movement. The distance has perhaps diminished a little, and this seems confirmed by the recent more rapid change in angle. (O.Σ., 1877.) Dunér observes that there are enormous discordances in the angles. He gives

Δ = 0".47 − 0".008 (t − 1850.0).

Σ.	170.8	1n.	0.60	1829.16
	162.1	,,	.51	30.22
	161.9	,,	.69	2.20
	169.5	2n.	.65	6.29
Mä.	171.6	1n.	.55	41.50
	186.6	6	.5	51.76
	183.9	4	.55	2.21
	180.0	1n.	.5	6.21
O.Σ.	172.3	2n.	.61	46.98
	190.8	1n.	.52	70.18
Se.	183.3	2n.	.40	57.48
De.	.5	,,	obl.	63.13
Du.	186.3	,,	0.28	9.37

90   Σ 262.

ι CASSIOPEÆ.

R. A. 2^h 19.2^m	Dec. 66° 51'	M. 4.2, 7.1, 8.1

C. Σ. A, yellow; B, blue; c. blue.

Dawes says, "This star is P. II. 72, and B. A. C. 744; but it is not Fl. 55 Cass., as H$_1$ supposed it to be."

H$_1$, *Phil. Trans.*, vol. lxxii., p. 219.— "Aug. 17, 1779. Double, extremely unequal, L.W.; S, bluish r. Distance 7".5 single measure. Position 10° 37' s.f." And he adds, in a note, "In a future collection this will be found as a treble star of the 1st class; the larger star having a small one preceding, easily seen with 460 and 932."

H$_1$, *Phil. Trans.*, vol. lxxv., p. 645.— "Treble. 20° 30' n.p. 1782 and 1783."

H$_1$, *Mem. R. A. S.*, vol. i., p. 173.— "No. 65, Nov. 4, 1788, double of the 2nd class. Very unequal."

So., *Phil. Trans.*, 1826, part i., p. 25.—"Dec. 9, 1823. Extremely difficult. Small star is decidedly blue, and bears only an indifferent illumination ; the large star may be suspected close double with 137 ; with 303 is seen such."

H₂ writes, "The position of the distant star C was stated in 1782 at 10° 37′ s.f., and in 1804 at 18° 57′ s.f. It is to be presumed that some mistake had been committed in the earlier measure."

Da. (*Mem. R. A. S.*, vol. xxxv., p. 313).—"His observations, compared with each other and Σ.'s, indicate a slow diminution of the angle in the close pair, while the distance may possibly have slightly increased."

The position and distance of the more distant star are unchanged.

Mä. (*Die Fixstern-Systeme*, p. 89).—He thinks both the close and the distant pair are in motion, and observes that in the projection the directions of the motions are opposite.

Se. (p. 22).—In A B the motion in angle is certain ; in A C there is no certain motion.

The common proper motion of the three stars is − 0ˢ·009 in R. A., and + 0″·02 in N. P. D.

A B.

	°		″	
Σ.	277·2	In.	1·73	1827·27
	280·9	,,	·85	28·27
	274·4	,,	·91	30·22
	273·5	,,	·88	1·26
	277·4	,,	·94	·27
Da.	·0	5n.	2·19	·13
	274·3	In.	...	7·04
	271·7	2n.	·23	40·91
	265·5	In.	·23	8·09
Mä.	274·6	4	2·49	1·50
	265·6	3n.	1·86	52·51
	269·5	In.	1·78	6·21
O.Σ.	267·6	7n.	2·08	45·35
	266·0	3n.	1·95	61·59
De.	263·6	In.	...	54·75
	262·5	,,	...	·81
	267·4	,,	...	·84
	266·8	,,	...	·91
	267·4	,,	...	5·09
	268·9	2n.	1·45	·90
	265·7	5n.	·90	6·49
	·3	2n.	·97	62·87
	266·4	,,	·88	3·11
Se.	·8	4n.	1·83	57·49
Ro.	268·8	2	1·72	63·95
Ta.	265·5	2n.	2·04	5·77
Br.	267·7	5n.	2·15	8·92
W. & S.	265·0	7	2·	72·04
	6·0	2	2·25	·92
	4·4	4	·04	3·06
Gl.	265·3	3	1·8	3·94
	6·7	2	2·0	·98
	5·7	6	1·9	4·93
Du.	265·7	7n.	1·94	0·91
Pl.	·7	,,	2·06	6·72
Dob.	264·0	In.	..	·22

A C.

	°		″	
H₁.	100·6		7·5	1779·6
	108·9		...	1804·43
So.	106·8		7·91	25·30
Σ.	107·1	In.	·57	8·22
	108·1	,,	·49	·27
	106·2	,,	·70	30·22
	107·3	,,	·74	1·26
	·8	,,	·63	·27
O.Σ.	111·1	,,	·55	41·21
	108·2	,,	·69	2·21
	111·2	,,	·87	6·20
	·7	,,	·75	8·21
	110·7	,,	·51	9·22
	108·1	,,	·98	55·25
	109·3	,,	·92	7·22
	·3	,,	·68	72·31
Mä.	108·5	4	8·45	41·50
	110·6	3n.	7·5	52·51
	·1	4	·52	6·21
De.	108·1	In.	8·05	4·75
	107·1	,,	·49	·81
	109·2	,,	·81	·84
	·9	,,	·73	·91
	·8	,,	·91	5·09
	·5	2n.	·91	·90
	108·4	5n.	·79	6·49
	107·5	2n.	·79	62·87
	·5	,,	·86	3·11
Se.	108·9	4n.	·84	57·49
Ro.	105·6	2	·76	63·95
Ta.	110·1	2n.	·31	5·77
M.	104·6	In.	·21	8·84
	108·1	,,	·00	75·78
Du.	109·1	4n.	7·47	0·29
W. & S.	·0	4	·81	2·92
Gl.	·0	5	·9	3·94
	7·5	2	·9	·98
Dob.	108·5	In.	...	6·22

91   Σ. 269.

R. A.	Dec.	M.
2ʰ 21·1ᵐ	29° 22′	7·5, 9·8

C. A, yellow ; B, ash.

H₁. "Oct. 8, 1781. A cluster of small stars in the Finder ; the middlemost and most north of them is. I believe, a very fine double star." Although H₁ examined this object a second time on the same night, and also on the 15th and 20th, he failed to satisfy himself of its duplicity. On the 22nd, however, he readily saw it double.

	°		″	
H₁.	325		2	1781·8
Σ.	340·4	3n.	1·90	1832·36
Sm.	342·1		2·3	4·11
Da.	340·3		1·93	40·96
	341·3		·92	8·63
O.Σ.	344·1	4n.	·65	6·39
Mä.	·1	2n.	·72	53·90
	342·5	,,	·72	·09

Mo.	345°2		2″04	1856·06
Se.	344·7	2n.	1·78	9·94
Ta.	334·8	,,	2·71	65·77
W. & S.	346·1	7	1·48	73·95
	345·6	4	·29	4·00
Pl.	342·2	2n.	·61	6·94

92   Σ. 278.

R. A.	Dec.	M.
2ʰ 28·3ᵐ	68° 47′	8·4, 8·7

C. Σ. and Se., white.

Σ., in 1827, found this object " oblong ; two discs in contract." He measured it in 1827, 1831, and 1833, and then Mädler took it up, giving measures in 1844, 1845. Se., in 1857, found the components "well separated," and thought there was "motion in angle." There has been no change in angle during the last quarter of a century. (O.Σ., 1877.)

Σ.	82·0	4n.	0·43	1830·77
H₂	75 ±		·4	1·80
Mä.	...	1n.	single	44·34
	90·1	,,	elongᵈ.	5·35
	102·5	,,	0·3	·65
	106·6	,,	·45	·74
Se.	67·7	2n.	·4	57·94
O.Σ.	71·4	4n.	·62	9·28
Da.	76·3		elongᵈ.	68·74

93   Σ. 288.

R. A.	Dec.	M.
2ʰ 32ᵐ	−11° 54′	8, 11

C. A, yellow.

H₁	224·8		10·8	1782·7
H₂	216·6		·0	1830·20
Σ.	213·6	3n.	11·92	1·20
De.	·7		·67	67·24
C.O.	215·1	2n.	12·48	77·95

94   O.Σ. 43.

R. A.	Dec.	M.
2ʰ 34ᵐ	26° 6′	7·2, 8·8

Certain increase in distance. Probably binary.

O.Σ.	85	1n.	obl.	1844·85
	95·9	,,	0·39	5·73
	90·1	,,	·53	51·72
De.	68·5	3n.	·98	67·30

95   O.Σ. 44.

R. A.	Dec.	M.
2ʰ 35ᵐ	42° 10′	7·8, 8·5

O.Σ.	58°6	4n.	1″47	1850·24
De.	54·2	3n.	·36	66·98
Du.	53·4	,,	·32	9·51

96   Σ. 293.

R. A.	Dec.	M.
2ʰ 35·5ᵐ	56° 31′	8·4, 11·7

C. A, yellow.

B is probably variable.

Σ.	57·5	4n.	6·61	1830·87
Mä.	68·6	2	...	52·65
De.	71·2	5n.	7·67	66·06
	B not seen			7·62

97   Σ. 295.

84 CETI.

R. A.	Dec.	M.
2ʰ 35·1ᵐ	− 1° 12′	6, 10

C. A, white ; B, lilac.

Common proper motion in R. A. − 0″·003, and + 0″·050 in N. P. D.

Dunér's formulæ are

$$1850·39. \quad \Delta = 4″·59.$$
$$P = 329°·3 − 0°·30 \; (t − 1850·0).$$

Σ.	335·1	2n.	4·6	1829·82
	334·2	,,	5·1	33·99
	331·6	1n.	4·84	51·91
Sm.	334·5		5·0	33·97
Mä.	329·2	,,	4·27	53·09
	330·1	1n.	5·16	8·11
O.Σ.	326·5	3n.	4·81	6·53
Se.	330·6	2n.	·57	8·03
De.	324·7	3n.	·63	63·97
Du.	323·3	2n.	·34	8·84
W. & S.	324·1	4	·7	72·08
Gl.	325·0	5	·7	3·94
C.O.	323·7	3n.	·49	7·84
Dob.	325·0	,,	·75	·84

98   Σ. 296.

θ PERSEI.

R. A.	Dec.	M.
2ʰ 35·9ᵐ	48° 43′	A 4, B 10, C 9

C. A, yellow ; B, violet ; C, grey.

Common proper motion of A B + 0ˢ·033 in R. A., and + 0″·14 in N. P. D.

The distant star is independent of the system of θ Persei.

H₁.	290·0	In.	13·52	1782·64
H₂.	292·4		16·0	1830·20
Σ.	294·6	3n.	15·40	2·20
Mä.	296·2	4	15·79	52·26
O.Σ.	296·8	3n.	16·05	3·09
	297·9	2n.	·32	72·06
De.	296·4	2n.	16·37	62·97
Da.	·9		16·34	4·36
W. & S.	·0	3	16·5	73·93
	215·3	3	A C	·93
	16·6	1	A D	·93

The distant star.

H₁, in 1782, observed a small star about 1′ south of θ Persei.

Sm.	219·0	27·0?	1833·65
O.Σ.	210·5	66·14	51·88
	·6	·08	2·71
	211·3	·11	4·67
	215·6	68·17	69·95

99   Σ. 299.

R. A.	Dec.	M.
2ʰ 37·1ᵐ	2° 44′	3·5, 7

C. A, yellow; B, blue.

Common proper motion, − 0ˢ·011 in R. A., and + 0″·19 in N. P. D.

Σ.	283·2		2·83	1825·43
	287·4	5n.	·59	32·48
	289·2	2n.	·67	6·74
	286·1	In.	·61	41·70
H₂.	280·7		2·58	28·69
	·7		3·74	31·79
Sm.	289·0		2·6	·85
	286·8		·72	5·89
	288·8		·8	8·92
	285·7		·6	43·16
	289·1		·9	55·09
Da.	286·1		·69	33·90
	287·6		·7	6·98
	·9		...	7·88
	288·9	11n.	2·72	41·65
	285·5	5n.	·65	2·99
	289·2	11n.	·77	9·08
Mä.	288·7		·61	38·90
	291·6	15	·84	41·37
	289·5	6	·77	7·12
	289·8	12	3·17	52·11
	292·0	3n.	·7	8·07
Flt.	289·2	35	·75	1·90
Ja.	291·0	In.	·7	3·06
O.Σ.	287·3	8n.	·92	·09
De.	287·5	6n.	2·9	5·06
Se.	288·3	5n.	·71	6·14
M.	289·1		·78	62·94
Ta.	290·6	6	...	5·84
	295·5	7	2·33	8·82
	297·9	5	...	72·86

Br.	291·9		2·91	68·80
Dob.	292·5	2n.	...	75·98
	291·4	6n.	3·40	6·08
Sp.	290·9		2·82	7·08
Pl.	291·1	3n.	·73	7·31

100   Σ. 300.

R. A.	Dec.	M.
2ʰ 37·5ᵐ	38° 55′	8, 8·4

C. white.

Probable binary.

Σ.	299·5	2n.	2·91	1832·80
Mä.	·8	3n.	3·11	43·62
Ta.	304·3	4	2·71	65·89
Pl.	299·9	3n.	·79	77·02

101   Σ. 305.

114 ARIETIS.

R. A.	Dec.	M.
2ʰ 40·6ᵐ	18° 52′	7·3, 8·2

C. Σ., "certainly yellow"; Da., "both white;" Se., "white."

H₂, in 1830, says: "Beautifully separated with the whole aperture [20 ft. reflector]. Measure excellent; taken with 320 and 12 inches. 329°·4, 1½″."

Da. (*Mem. R. A. S.*, vol. xxxv., p. 314): "The angle of position of these stars is slowly varying in a retrograde direction." A very decided increase of distance has also occurred.

Se. (p.23) says, "the motion appears secure and noteworthy."

An increase in the distance and a diminution in the angle are shown by the measures. (O.Σ., 1877.)

Σ.	331·8	In.	1·42	1829·81
	333·5	,,	·67	·90
	327·3	,,	·67	33·14
Da.	324·7	6n.	·96	41·94
	326·5	In.	·93	2·91
	325·7	,,	·91	3·01
	324·6	2n.	...	4·02
	·6	,,	2·19	7·48
	·8	In.	·27	8·94
	322·5	,,	·32	51·99
	·4	4n.	·33	3·90
	321·8	In.	·26	4·84
O.Σ.	325·1	,,	1·96	41·70
	324·3	,,	2·11	5·73
De.	322·1	3n.	1·9	56·77
	321·8	,,	2·53	62·83
	·8	2n.	·51	3·49
Se.	322·2	2n.	2·56	57·89
Ta.	321·3	4	2·46	65·89
	316·6	6	1·98	72·86

Br.	321.3		2".73	1868.78
W. & S.	319.4	4	.7	72.04
	320.1	4	.83	7.09
	.6	3	.6	3.94
Fer.	317.5		.77	3.94
Dob.	319.3	6n.	...	6.06
	317.8	1n.	3.59	7.88
Schi.	318.7	1n.	2.75	.07
Sp.	.8		.72	.08
Pl.	320.9	3n.	.77	.33

102   Σ. 311.

R. A.	Dec.	M.
$2^h 42.2^m$	16° 58'	4.9, 8.4, 10.2

Certain direct motion in A B.

A B.

H'.	109.1	8n.		1782.82
Σ.	119.6	1n.	3.06	1829.89
	118.3	,,	.40	31.11
	119.7	,,	.25	2.79
	118.5	,,	.40	.86
	120.4	,,	.31	4.95
Da.	.7	,,	3.24	48.97
Mä.	.6	6	2.84	50.76
	122.2	6	...	1.04
	119.3	18	3.26	2.11
	117.5	4	3.36	7.93
Se.	121.2	3n.	2.94	6.77
	128.6	,,	.66	73.13
O.Σ.	126.6	1n.	3.19	65.92
Gl.	121.2	5	.2	0.01
	2.5	3	.0	.02

A C.

H₁.	109.3			1782.82
Σ.	.9	1n.	24.43	1829.89
	110.8	,,	25.11	31.11
	109.8	,,	.69	2.79
	110.3	,,	.73	.86
	109.5	,,	.44	3.86
	110.5	,,	24.94	4.95
O.Σ.	.0	,,	25.12	65.92
	.6	,,	.57	73.13

103   Σ. 312.

R. A.	Dec.	M.
$2^h 44.1^m$	72° 24'	7, 8, 9

C. white.

Certain change in A B.

Dunér gives, for A B,

$$\Delta = 3".39 - 0".014\,(t - 1850.0),$$
$$P = 16°.3 + 0°.13\,(t - 1850.0);$$

and for A C,

$$1851.75 \quad \Delta = 42".37,$$
$$P = 127°.52 + 0°.026\,(t - 1851.75).$$

A B.

H₂.	10.3		4"	1830.10
Σ.	13.9	5n.	3.59	2.08
Mä.	15.6		.71	44.33
De.	16.1	1n.	.54	57.67
	17.5		.26	66.42
Se.	18.1	2n.	.32	57.95
M.	8.9		.37	62.23
O.Σ.	22.4	2n.	.3	72.31
Du.	19.5	7n.	.10	4.58

A C.

Σ.	127.0	2n.	42.32	31.75
Du.	128.0	3n.	.4	71.75
O.Σ.	128	,,	.32	3.48

104   Σ. 314.

PERSEI 85.

R. A.	Dec.	M.
$2^h 44.3^m$	52° 30'	7, 7.5

C. white.

Probably a binary.

H₁.	278.4	1n.	...	1782.63
	290.5	,,	...	1804.18
So.	291.1	,,	1.32	25.40
H₂.	292.0		.7	30.20
Σ.	.1	,,	.35	27.21
	294.5	,,	.51	30.22
	298.0	,,	.55	2.20
	297.2	,,	.42	.22
O.Σ.]	.0	2n.	.71	41.44
	296.5	1n.	.63	54.67
	300.2	,,	.45	72.18
Mä.	297.9	4	.56	52.26
De.	295.8	3n.	sep^d.	5.13
Se.	300.7	,,	1.46	7.62

105   Σ. 326.

R. A.	Dec.	M.
$2^h 48^m$	26° 24'	7.5, 9.7

Σ.	216.1	2n.	9.03	1831.46
Mä.	215.2	,,	.35	44.44
O. Σ.	217.4	,,	8.61	60.32

106   Σ. 328.

R. A.	Dec.	M.
$2^h 49^m$	44° 2'	8.5, 9

C. white.

Σ.	299.3	2n.	27.06	1832.18
Mä.	298.6	1n.	25.71	45.63

107   Σ. 333.

ε ARIETIS.

R. A.	Dec.	M.
2ʰ 52·3ᵐ	20° 51′	5·7, 6

C. Σ. and De., white.

Σ. first measured this object in 1827. He was led to think that the components are variable, but was struck by the fact that the difference of the magnitudes of the two stars always remained the same, viz., from 0 to 0·5 of the scale.

H₂ measured it in 1830. He says, " seen double with 320. Measured with 480 ; but measure not good, the illuminating lamp having gone out. 195° ¾″."

Mä. measured it from 1841 to 1845, and thought that the distance had probably increased.

Sm. (*Cycle*, p. 74). An increase of angle had, however, become so apparent to him in 1839, that he watched it at Hartwell, and was soon pretty sure that the companion had "a direct orbital motion" ; and taking this at 0°·85 per annum, he thought "its revolution may be made in four centuries, at most." He adds, "If we may place dependence on the observations as to the slight increase of distance, it will probably still widen for a few years longer, until the satellite shall have doubled the southern point of its course, which now seems to be on an ellipse shooting out from Σ in the micrometric direction of 210°, with a major axis about thrice the length of its minor."

Da. on examining his own observations, extending over 14 years, found "a decided increase of distance, with no perceptible variation of angle."

Dunér's formulæ are—

$$\Delta = 0''\cdot85 + 0''\cdot0136\,(t - 1850\cdot0).$$
$$P = 196°\cdot70 + 0°\cdot210\,(t - 1850\cdot0) - 0°\cdot0043\,(t - 1850\cdot0)^2 + 0°\cdot00007\,(t - 1850\cdot0)^3.$$

These were obtained by the graphical method, aided by calculation. The coefficients are still very uncertain.

	°		″	
Σ.	186·4	In.	0·51	1827·61
	191·1	,,	·57	29·21
	189·1	,,	·51	32·13
	188·9	,,	·60	·14
H₂.	195·0		·7	1·10
Sm.	193·5		·5	5·08
	195·7		·8	9·25
	199·6		·9	43·18
	200·1		1·0	53·08
Mä.	194·3		0·71	40·02
	196·2	3n.	·76	1·75
	·3		·78	3·61

	°		″	
Mä.	197·9	2n.	0·81	1847·12
	198·1		·84	9·96
	·7		·90	52·72
	200·1		·95	3·09
	198·2		·99	6·08
	201·2		1·02	·21
	198·3		·06	8·08
	200·5		·10	62·39
Da.	196·22	In.	...	40·09
	195·69	2n.	0·84	1·45
	194·47	3n.	·69	2·10
	200·43	In.	...	3·01
	198·30	,,	...	4·00
	195·97	,,	0·95	6·91
	·61	5n.	·84	7·62
	196·32	In.	·99	9·00
	195·55	,,	1·01	53·99
O.Σ.	203·9	In.	0·88	41·70
	196·1	2n.	·86	2·21
	·9	1n.	·75	6·17
	192·3	,,	·77	7·19
	195·0	2n.	·71	8·21
	201·0	In.	·86	9·14
Mit.	196·3	6	0·69	6·77
Ja.	·15	5n.	1·08	53·49
		4n.	·08	·68
De.	201·1	In.	wedge	4·81
	205·3	,,	...	·83
	203·3	,,	...	·92
	202·9	,,	...	·97
	203·2	,,	...	5·13
	206·1	2n.	1·0	·85
	197·3	3n.	1·0	6·26
	194·5	2n.	0·7	62·81
	·8	3n.	0·8	3·12
	196·2	In.	1·11	5·93
	197·4	,,	·04	7·14
	199·3	,,	·08	8·66
	198·3	,,	·13	70·76
Mo.	·3		0·98	56·03
Se.	196·73	3n.	0·87	·57
	201·1	In.	1·09	66·07
Ta.	199·0	6		5·71
	192·4	5	1·38	6·86
	198·8	7	·05	72·86
Kn.	198·3	4n.	1·052	66·69
Br.	196·7	3n.	1·34	8·78
Du.	199·0	In.	1·02	·68
	198·4	2n.	·03	9·18
	199·9	In.	·17	71·17
	198·7	2n.	·12	2·18
	199·8	,,	·09	4·17
	200·0	,,	·14	5·18
Gl.	198·0	In.	1·0	0·65
	9·7	2	·07	3·94
W. & S.	196·3	4	1·26	1·95
	200·5	5	·5	2·17
	198·7	12	·69	·86
	7·5	9	·10	·92
	200·5	4	·44	3·14
	1·2	7	·23	5·09
	0·7	6	·36	7·09
W.O.	200·7	2n.	1·52	6·06
	·6	,,	·18	·06

W.O.	201·8	2n.	1·14,	1876·06
	204·6	,,	·20	·12
Dob.	198·1	8n.	·52	·09
Schi.	197·6	1n.	·169	7·06
Sp.	·7		·17	·07

108 Σ. 334.

R. A. 2h 53m		Dec. 6° 10'		M. 7·7, 8·2
Σ.	322·8	3n.	1·59	1830·94
Mä.	325·2	1n.	·58	42·97
O.Σ.	314·7	2n.	·55	9·84
Se.	322·5	3n.	·55	57·02

109 O.Σ. 49.

R. A. 2h 54m		Dec. 17° 32'		M. 7, 10
O.Σ.	71·1	4n.	1·71	1846·80
Mä.	65·2	1	·2	52·1
Se.	·6	3n.	·76	66·24

110 Σ. 336

R. A. 2h 54·1m		Dec. 31° 56		M. 6·5, 8

Angle unchanged ; distance augmented.

Σ.	1·5	3n.	8·19	1831·17
Mä.	7·2	1n.	·61	44·95
Se.	·2	2n.	·35	58·03
De.	·8	,,	·34	·09
O.Σ.	·3	1n.	·83	68·77

111 Σ. 346.

R. A. 2h 58m		Dec. 24° 47'		M. 6, 6, 10·8

A B.

Σ.	265·8	6n.	0·73	1834·48
Mä.	266·6	2n.	·45	41·87
O.Σ.	84·2	3n.	·79	3·36
Kn.	270·2	2n.	·79	66·21
Du.	269·5	6n.	·64	72·26

$\dfrac{A+B}{2}$ and **C.**

Σ.	356·5	4n.	5·22	33·70
O.Σ.	355·0	1n.	·53	41·70
Du.	357·1	4n.	·14	72·68

112 O.Σ. 50.

R. A. 3h 0·7m	Dec. 71° 6'	M. 7·5, 7·5

C. white.

Probably binary.

O.Σ.	232·5	2n.	0·88	1847·22
	228·2	,,	·85	50·22
	216·1	1n.	·11	75·33
Mä.	56·4		0·85	42·30
	302·3	1n.	1·56	51·77
De.	217·3	3n.	1·10	67·40

113 Σ. 355.

R. A. 3h 1m		Dec. 7° 56'		M. 8·7, 9·5

C. white.

Σ.	148·8	5n.	2·75	1832·52
Mä.	149·6	2n.	·71	43·47
Se.	142·8	,,	·65	57·11

114 Σ. 343.

R. A. 3h 4·6m		Dec. 83° 34'		M. 8, 9

C. yellowish.

Rectilinear motion.

H_2.	326·2		20	1830·50
Σ.	325·4	3n.	22·66	2·60
Mä.	326·1	1n.	23·8	44·34
De.	325·2		24·95	65·00

115 h. 3555.

12 ERIDANI.

R. A. 3h 6·9m		Dec. − 29° 27'		M. 4, 7

H_2.	306·1	1n.	3	1835·86
Ja.	309·6		4·09	47·00
	310·0	21	3·31	56·16
Bu.	...		2·3	74·80
C.O.	316·9	3n.	·56	7·81

116 O.Σ. 52.

P. III. 1.

R. A. 3h 7m	Dec. 65° 13'	M. 6·4, 7

C. white.

Retrograde angular motion. Distance unchanged. Probably binary.

Mä.	157·4		0·38	1843·31
O.Σ.	150·6		·48	5·23
	154·2		·54	6·74
	155·7		·43	7·22
	153·3		·57	8·22
	145·6		·56	57·22
	·1		·62	62·23
	137·5		·51	75·33
De.	135·6	3n.	·5	66·29
	138·5		...	72·42

117   Σ. 367.

R. A.   Dec.   M.
3^h 7.8^m $0°$ $18°'$ 8, 8

Certain indirect motion: probably binary.

Σ.	281.2	In.	1".06	1829.90
	99.7	,,	0.87	32.11
	103.3	,,	.92	3.14
O. Σ.	273.9	,,	.91	41.70
Mä.	95.8	,,	.5	1.79
	98.3	2n.	.52	2.89
Se.	266.5	In.	.89	57.12
De.	257.1	6n.	...	64.01
W. & S.	246.7	6	0.73	72.51

118   O.Σ. 53.

R. A.   Dec.   M.
3^h 10^m $38°$ $11'$ 7.2, 8

O. Σ.	271.3	In.	0.72	1844.89
	274.9	,,	.64	6.09
	265.0	,,	...	64.21
Mä.	95.0	In.	0.7	51.77
De.	261.6	4n.	0.88	68.18
	257.9		.82	74.02

119   Σ. 377.

R. A.   Dec.   M.
3^h 14^m $18°$ $45'$ 8.3, 8.7

Σ.	115.4	3n.	0.823	1831.66
O. Σ.	121.5	,,	1.013	46.13
Mä.	120.6	,,	0.95	3.10
	117.5	In.	.7	7.91
De.	120.7		1.03	68.99

120   Σ. 380.

R. A.   Dec.   M.
3^h 15.3^m $8°$ $20'$ 8.3, 9.3

Probably binary.

Σ.	90.1	3n.	1.20	1831.62
Mä.	87.6	In.	0.8	43.97
	.5	,,	.6	4.13
	84.2	,,	1.05	5.04
	87.2	,,	.12	.08
	78.2	,,	.11	58.11
Se.	86.8	,,	.15	.03
De.	75.9		.2	64.00
W. & S.	73.5	In.	.26	73.93
	75.4	,,	...	4.00
	.3	,,	1.3	.00
	70.1	,,	...	7.09

121   Σ. 381.

R. A.   Dec.   M.
3^h 16.4^m $20°$ $33'$ 7, 8.7

Probably binary.

Σ.	93.7		0".75	1827.16
	91.0	4n.	.82	30.16
Sm.	87.6		.8	4.19
Mä.	88.0	In.	.5	41.79
	93.1	,,	.8	2.65
O. Σ.	91.1	2n.	.94	3.71

122   Σ. 388.

R. A.   Dec.   M.
3^h 20^m $50°$ $1'$ 8.2, 9.2

C. white.

Probably a binary.

Σ.	108.0	In.	2.93	1828.20
	110.8	,,	.78	32.17
	111.2	,,	3.05	5.18
Mä.	208.9	,,	2.82	45.64

123   Σ. 389.

R. A.   Dec.   M.
3^h 20.5^m $58°$ $57'$ 7, 8

C. A, white; B, purplish.

Dunér's formulæ are—

$$1852.52. \quad \Delta = 2''.71$$
$$P = 64°.1 + 0°.125 \, (t - 1850.0).$$

Σ.	61.8	4n.	2.81	1831.00
Da.	.7		.72	3.90
	62.7		.71	7.04
	.7		.82	54.75
Mä.	63.5	3n.	.78	43.77
	65.3	In.	...	52.22
	74.4		.81	61.23
De.	64.2	In.	.6	57.96
Mo.	.2		.77	9.86
Se.	67.3	2n.	.85	.95
Du.	66.9	4n.	.4	72.66
Gl.	63.3	In.	.7	3.94

124   O.Σ. 54.

R. A.   Dec.   M.
3^h 21^m $67°$ $11'$ 7.2, 8.5

H_2.	352.7		...	1829
O. Σ.	354.5	4n.	25.82	50.08
Mä.	174.6	In.	.71	52.26
De.	355.2	3n.	.413	66.74

125   Σ. 403.

R. A.   Dec.   M.
3^h 24.3^m $19°$ $22'$ 8.5, 8.5

C. white.

Probably a binary.

H₁.	172·8		...	1783·05
Σ.	181·7	3n.	2·91	1829·76
H₂.	178·5		2 ±	30·50
Mä.	180·9	2n.	3·26	43·14
Se.	176·7	,,	2·93	57·11
De.	178·0		·89	65·48
Gl.	·0	In.	·9	74·2

126   Σ. 408.

R. A.	Dec.	M.
3ʰ 24·7ᵐ	− 4° 41′	8, 8·2

Σ.	346·1	In.	1·47	1829·90
	348·5	,,	·38	32·86
	·0	,,	·26	3·14
O.Σ.	342·5	,,	·62	41·70
Mä.	346·7	,,	·15	4·13
Se.	338·2	2n.	·24	57·10
De.	·4	,,	·25	64·00
W. & S.	159·8	5	·44	73·93
	160·8	5	·43	4·00
Gl.	339·2	4	·4	3·93
C.O.	336·4	2n.	·34	7·86
Dob.	198·7	In.	0·95	6·13
	156·5	,,	...	7·91

127   Σ. 400.

R. A.	Dec.	M.
3ʰ 25·2ᵐ	59° 38′	7, 8

Certain change in angle and distance. Probably a binary.

Σ.	283·0	In.	1·64	1827·27
	281·7	,,	·50	31·25
	283·0	,,	·44	·30
H₂.	276·9		2·24	0·79
Mä.	284·5		1·35	6·14
	288·5	2n.	·08	45·45
O.Σ.	291·3	In.	·65	1·21
	285·3	,,	·35	8·21
	286·3	,,	·36	54·67
	293·4	,,	·00	62·23
Se.	286·2	2n.	·05	57·96
De.	293·6	3n.	·11	67·41
Gl.	295·0	In.	·2	73·96

128   Σ. 412.

7 TAURI.

R. A.	Dec.	M.
3ʰ 27·3ᵐ	24° 4′	6·6, 6·7, 10

C. Σ., A and B yellowish; Se., white; Sm., A white, B pale yellow, C bluish.

This is a triple star, but H₁ did not see that A was double. H₂ and So. measured A C in 1821, without detecting the duplicity of A, their attention being no doubt drawn away by the extreme faintness of C. Σ. in 1827 found that A was double. Mädler measured it from 1841 to 1845;

he remarked on its difficulty, and thought that after ten or fifteen years it would cease to be separable.

Smyth calls it a "fine and very difficult object," and says, "Now the first two epochs exhibited so great an orbital change, in less than forty years, as to excite much attention; but the accordance of those of Σ. and myself indicate some error of observation or entry. In this conclusion, however, Σ.'s angle for 1821·95, in the Dorpat observations, is rejected; since it must be deemed rather an essay than a conclusive measurement."

Dawes (*Mem. R. A. S.*, vol. xxxv., p. 316,) thinks that the decrease of angle continues in the close pair, and that the distance remains nearly the same.

Secchi in 1857 regarded the motion both in angle and distance as certain.

In A B there is decided change both in angle and distance. Σ. (*P. M.*, p. ccxxvi.) thought that the relative movement of C was explained by the proper motion of A B. This is probably not the fact. (O.Σ., 1877.)

A B.

Σ.	271·0	In.	0·03	1827·16
	272·8	,,	·64	9·21
	274·7	,,	·67	31·22
	266·2	,,	·84	2·14
	264·9	,,	·68	·19
	263·4	,,	·62	6·74
	266·4	,,	·57	7·05
H₂.	257·7		...	1·81
Sm.	265·0		·7	3·21
Mä.	·5		·55	9·70
	264·6	In.	·55	41·79
	254·6	,,	·4	6·84
	258·2	,,	·4	7·12
	257·6	,,	·4	50·96
	256·7	10n.	·4	1·16
	·1	In.	...	2·10
	252·9	3n.	·4	3·88
	255·5	In.	·4	4·85
	253·3	,,	·4	5·
	252·7	,,	·4	6·19
	259·1	,,	·3	7·06
	263·1	3n.	...	8·11
Da.	·4		·6	41·96
	259·9		·65	6·91
O.Σ.	262·3	In.	·76	1·70
	267·4	2n.	·74	2·21
	263·7	In.	·61	50·18
	241·1	,,	oblong	73·13
Se.	256·8	3n.	0·420	56·35
De.	72·0	In.	...	62·72
	71·9	2n.	...	3·40
Kn.	60·8	In.	·55	4·93
Ta.	261·9	,,	...	5·71
W. & S.	227·0	7	...	72·14
	239·7	15	·4	3·14
	254·1	4		4·01
Gl.	232·0	3	·4	3·94

$\dfrac{A\,B}{2}$ and C.

H₁.	66°8		20"0	1783·13
Σ.	63·5	In.	22·25	1827·16
	·2	,,	·38	32·16
	62·8	,,	·76	·18
	·6	,,	·24	·19
So.	56·1		21·05	21·97
Sm.	61·9		·8	33·21
Mä.	60·3	In.	22·50	41·79
O.Σ.	61·5	,,	·16	51·81
	60·7	,,	·07	69·10
	61·9	,,	·05	72·95
	59·5	,,	·12	3·13
Se.	60·1	2n.	...	55·99
De.	61·13	3n.	22·01	63·18
Kn.	·45	In.	·87	4·93
Ta.	64·8	,,	...	5·71
W. & S.	61·1	2	...	72·17
	60·5	4	3·2	·86
	·5	1	...	·92
	·9	3	...	3·14
	·3	5	2·3	·93
	·6	2	·9	4·00

---

129   Σ. 414.

R. A.	Dec.	M.
3ʰ 27·6ᵐ	19° 25′	8, 8

C. white.

Σ.	185·6	3n.	7·1	1829·76
Da.	·5		...	46·71
De.	184·6	In.	7·15	57·89
Mä.	·5	,,	·55	43·14
	185·2	,,	·8	58·04
Mo.	·3		·33	8·81
Se.	184·2	3n.	·37	6·68
Ta.	·4	In.	·18	65·92
M.	180·9		·25	6·13
Fer.	185·1		·17	73·06

---

130   Σ. 422.

P. III., 98 ERIDANI.

R. A.	Dec.	M.
3ʰ 30·6ᵐ	0° 12′	6, 8·2

C. Σ., deep yellow, blue ; Se., yellow, blue ; Sm., yellow, pale blue.

H₁, III. 45, *Phil. Trans.*, vol. lxxii., p. 220 : "In constellation Tauri. near Fl. 10. Oct. 22, 1781. Double. It is near the star *sub pede et scapula dextra*. Extremely unequal. L. pale r. ; S. d. Position, 35° 33′ s.p."

Smyth says, "This is 45°, H₁, III., who by measures in 1781·83 made the position angle 234° 27′; but H. informs us, that by a MS. note he finds it declared that the observation is too small by 6° or 8°. Hence the first measures for future reference must be those of So., No. 431.

225° 12′   5"·812   1824·02.

Mädler (*Die. Fixst. Sys.* p. 95) asks whether the companion passed its aphelion about 1833.

Secchi (p. 78) says, "Motion certain."

Direct motion certain.—(O.Σ., 1877.)

Dunér gives—

1860·00   Δ = 6"·23.

$$P = 239°·5 + 0°·27\ (t - 1860·0).$$

H₁.	227·5		...	1781·8
Σ.	226·2		5·4	22·08
	232·8	In.	6·18	32·14
	·6	,,	·12	·98
	231·3	,,	·09	3·14
So.	225·3		5·58	24·38
H₂.	231·2		7 ±	30·20
Sm.	·8		5·9	4·93
	255·9		6·0	45·81
Mä.	235·3	In.	5·64	2·94
	236·7	,,	...	5·09
	239·5		5·69	54·16
Da.	233·7		...	46·72
O.Σ.	238·8	In.	6·24	51·88
	239·8	,,	·26	65·92
Se.	237·3	3n.	·37	57·06
Ta.	240·6	In.	·45	65·92
De.	238·6	2n.	·02	57·87
	240·0	5n.	·26	64·86
M.	234·0		5·89	1·93
	237·2	In.	·71	6·10
Du.	239·5	2n.	6·28	8·84
	242·3	In.	·22	9·12
	243·9	,,	·63	73·15
	246·6	,,	·60	4·09
	·1	,,	·40	5·95
	243·1	,,	·43	6·09
W. & S.	241·3	4	5·9	1·95
	·2	5	6·2	·86
	·3	5	·3	3·93
	242·0	7	·41	7·09
Gl.	240·0	3	·4	3·94
Pl.	242·9	3n.	5·91	7·00
Dob.	244·5	2n.		·86

---

131   Σ. 425.

R. A.	Dec.	M.
3ʰ 32·5ᵐ	33° 44′	7·3, 7·3

C. very white.

Dunér's formulæ are—

$$\Delta = 2"·83 - 0"·0125\ (t - 1850).$$
$$P = 102°·5 - 0°·07\ (t - 1850).$$

H₁.	98·2		...	1783·00
So.	103·7		3·43	1823·98
H₂.	100 ±	2		6·90
	103·4		·89	30·97
	102·2		3·23	1·87

Σ.	104°6	3n.	2″87	1830·16
Da.	102·9		·99	·82
	·8		3·02	40·80
	101·7		2·81	8·13
	102·7		...	·66
	...		2·86	9·98
	102·0		·72	54·03
Mä.	103·7		3·10	37·09
	·9	1n.	·39	41·79
	102·6		·14	4·01
	101·9	2n.	·o	5·11
	102·7	3n.	2·93	51·37
	·5	1n.	·94	2·18
	101·8	,,	·70	7·21
	·0		·89	62·23
De.	102·5	2n.	·27	54·86
	100·8		·61	68·79
Se.	101·9	3n.	·8	57·64
Mo.	·3		·71	8·12
M.	103·9		·62	64·23
Ta.	101·9	1n.	·78	5·93
Gl.	98·7	,,	...	73·94
Du.	101·8	4n.	2·33	4·58
W. & S.	98·9	1n.	·52	7·09
	99·5	,,	·44	·15
	·3	,,	·61	·16

132   Σ. 436.

R. A.	Dec.	M.
3h 35·2	−13° 0′	7, 8·2

Rectilinear motion. Probably an optical pair.

Dunér has

$$\Delta = 31''\cdot70 + 0''\cdot096 \ (t - 1850\cdot0).$$
$$P = 233°\cdot1 + 0°\cdot043 \ (t - 1850\cdot0).$$

Σ.	232·4	4n.	30·22	1832·51
Mä.	·6	1n.	·58	43·14
De.	233·4		32·98	64·04
Du.	234·1	3n.	33·53	8·94
C.O.	·5	1n.	34·9	77·87
Fl.	·3	1n.	34·1	·08

133   Σ. 434.

R. A.	Dec.	M.
3h 36·1m	38° 0′	7, 7·8

Rectilinear motion. Probably an optical pair.

Dunér has

$$\Delta = 29''\cdot00 + 0''\cdot0335 \ (t - 1848\cdot0).$$
$$P = 87°\cdot70 - 0°\cdot029 \ (t - 1848\cdot0).$$

So.	88·4		28·43	1824·00
Σ.	·2	3n.	·34	30·59
De.	87·2	5n.	29·55	65·13
Du.	·1	,,	·71	8·69
W. & S.	·0	1n.	30·23	77·15
	·4	,,	·22	·16

134   O.Σ. 61.

R. A.	Dec.	M.
3h 36m	7° 31′	7, 10

In 1844, O.Σ. once suspected duplicity, magnitudes 7, 10; distance 1″·2; but on the whole was inclined to regard the star as single, and therefore rejected it from his list. De. in 1867 gave the distance 1″·93, and magnitudes 7·2, 10; and in 1875 Romberg readily saw the companion with the meridian circle at Poulkova.

O.Σ.	°		obl.? 1·2?	1842·93 4·91
De.	125·8	3n.	1·93	67·03

135   Σ. 447.

R. A.	Dec.	M.
3h 40·1m	37° 58′	7·8, 9

Change in angle and distance (O.Σ. 1877). Rectilinear motion.

Σ.	179·2		26·5	1828·15
	178·3	3n.	·46	30·59
	176·7	2n.	·62	6·11
Mä.	173·7	1n.	27·56	45·85
	174·8	,,	...	51·17
	175·8	,,	27·48	2·18
	177·0	,,	28·95	7·21
O.Σ.	175·0	,,	26·91	0·60
De.	173·2	,,	·91	62·97
	·1	,,	·97	3·11
	172·6	,,	27·35	7·83
	171·6	,,	·14	73·72

136   Σ. 453.

R. A.	Dec.	M.
3h 42m	23° 39′	5, 8

This star has not been seen double since 1830. Mä. looked for it more than twenty times in the years 1840 to 1857, but saw no trace of the companion. On the 11th of Jan., 1876, while observing an occultation of the Pleiades by the Moon, M. Hartwig, of the Strasbourg Observatory, noted that the disappearance of Atlas was not instantaneous.

Σ.	107·5	1n.	0·79	1827·16
	29·2	,,	·35	30·25
		,,	single	1·23
Mä.		,,	,,	41·79
		,,	,,	·99

137   O.Σ. 65.

R. A.	Dec.	M.
3h 43·1m	25° 13′	6·5, 6·8

14

A system probably in rapid change. An occultation in 1865? Special attention should be directed to this object.

O.Σ.	209·2	4n.	0"74	1846·16
Da.	202·9		·66	47·88
Mä.	204·3	2n.	·66	52·14
Se.	201·4	2n.	1·04	9·05
De.	195·		elongd.	65·93
			single	6·67
Du.			,,	71·18
	29·		elongd.	4·17

138   Σ. 459.

R. A.	Dec.	M.
3h 43·6m	29° 17′	7·8, 10·7

Σ.	318·3	3n.	12·84	1831·38
Mä.	·4	1n.	·88	45·00
De.	325·2	3n.	15·1	66·81

139   Σ. 470.

R. A.	Dec.	M.
3h 48·2m	−3° 19′	4, 6

C. A yellow, B blue.

H₁.	343·4		4·32	1781·81
Σ.	346·5	1n.	6·72	1833·14
	347·6	2n.	·68	·15
Mä.	348·1	1n.	·75	44·15
	·1	2n.	·62	5·06
Ro.	346·8	4	7·14	63·11
Ta.	345·5	1n.	5·99	5·92
	342·6	,,	·80	71·78
	346·2	,,	6·9	2·97
Du.	·7	2n.	·66	0·64
W. & S.	347·2	4	·8	3·93
C.O.	·3	2n.	·77	7·90

140   O.Σ. 67.

R. A.	Dec.	M.
3h 47m	60° 45′	5, 8·2

C. A, golden; B, green.

O.Σ.	39·3	3n.	1·72	1847·18
De.	43·8	,,	·94	66·24

141   Σ. 460.

49 CEPHEI.

R. A.	Dec.	M.
3h 50m	80° 22′	5·2, 6

C. Σ. A. yellow, B. bluish; Se. both white.

Secchi (p. 7) says that the motion in angle is certain. Dembowski, "Cephei 49. Couple toujours difficile."

Certain change in angle. (O.Σ., 1877.)

Σ.	348·5	1n.	0·88	1828·27
	350·6	,,	·80	·29
	354·1	,,	·94	32·28
	355·2	,,	·96	·29
	354·4	,,	·86	3·34
	355·8	3n.	·86	6·45
Mä.	356·7		·87	6·76
	0·1	1n.	·8	42·22
	1·6	,,	·85	4·33
	4·7	2n.	·90	5·35
O.Σ.	359·22	4n.	·74	0·26
	3·9	1n.	·73	6·30
	1·2	,,	·70	9·23
	5·3	,,	·72	50·26
	18·8	,,	·97	7·29
	22·4	,.	·76	64·43
	22·7	,,	·79	6·49
Se.	10·79	2n.	·72	57·90
De.	7·79	,,	...	5·19
	8·6	3n.	1·0	6·71
	15·6	4n.	0·8	62·85
	·5	3n.	·8	3·08
	19·2	1n.	1·00	5·97
	21·3	,,	0·85	7·68
	23·7	,.	·86	8·65
	25·4	,,	...	9·74
	26·6	,,	1·12	72·68
	·7	,,	0·93	3·90
	27·6	,,	·87	4·18
	30·8	,,	·89	·83
	29·9	,,	·68	5·71
W. & S.	24·4	6	·7	2·92
	25·7	7	·6	3·24
	27·4	4	·66	·25
	28·9	4	·75	·29
	25·1	5	·86	2·10
Gl.	27·3	3	·6	3·94
	29·3	3	·8	4·12
Dob.	27·5	4n.	...	6·12

142   Σ. 483.

R. A.	Dec.	M.
3h 56·2m	39° 8′	8, 9·5

Σ.	11·6	3n.	2·8	1830·52
H₂.	10·0		3·43	1·86
Mä.	5·6	2n.	·47	45·17
	4·0	1n.	2·5	53·23
De.	0·4	5n.	1·67	64·64

143   O.Σ. 71.

R. A.	Dec.	M.
3h 59m	33° 7′	7, 9

O.Σ.	206·4	2n.	0·98	1846·44
Mä.		,,	could not find it.	52·18
De.	202·8	3n.	1·08	67·05

144   O.Σ. 531.

R. A.	Dec.	M.
3ʰ 59ᵐ	37° 46′	6·5, 8·2

The small star partakes in the rapid proper motion of the principal star, the direction and amount of motion being almost identical with that of 50 Persei, which is about 12′ distant. These stars deserve special attention. (O.Σ.)

O.Σ.				
	148·5	2n.	3˝47	1851·16
	146·7	1n.	·09	3·18
	151·7	,,	·45	9·15
	142·4	,,	2·88	70·25

145   O.Σ. 72.

R. A.	Dec.	M.
4ʰ 1ᵐ	17° 1′	6, 9·2

Mä.	228·1	1n.	4·49	1845·96
O.Σ.	322·8	5n.	...	54·51
De.	325·3	3n.	4·37	67·44
Pl.	329·6	2n.	·97	76·81

146   Σ. 494.

R. A.	Dec.	M.
4ʰ 1·7ᵐ	22° 47′	8, 8

C. very white.

Dunér has,

$$1852·51 \quad \Delta = 5˝·11.$$
$$P = 187°·6 - 0°·100\,(t - 1852·51).$$

H₁.			4˝ to 8˝	1784·88
So.	185·9	1n.	6·30	1825·79
Σ.	191·0	,,	5·09	8·19
	190·2	,,	·06	31·18
	188·4	,,	·09	3·19
Da.	...		·11	41·98
	186·4		...	3·13
	·7		5·00	8·13
	187·0		·20	60·08
Mä.	188·2	1n.	·37	44·08
Hi.	187·9		·10	5·78
	186·8		...	6·74
Mo.	187·4		5·09	56·05
De.	186·9		·15	6·78
Se.	187·5		·19	7·39
Ta.	185·8	1n.	·65	65·95
	184·6	,,	·52	8·98
	187·9	,,	·51	9·13
	·6	,,	4·89	72·97
Du.	186·0	2n.	·83	0·64

147   O.Σ. 74.

R. A.	Dec.	M.
4ʰ 6ᵐ	9° 20′	8, 8·5

O.Σ.	270·1	1n.	0·53	1849·16
Se.			single	57·05
De.			,,	65

148   Σ. 511.

R. A.	Dec.	M.
4ʰ 7·8ᵐ	58° 29′	7·5, 8

C. Σ. both white. Se. A, white, yellow; B, white, bluish. De. both white.

Se. thought that the motion was certain. In 1863 De. found it "extremely difficult." Rapid angular motion. The distance seems unchanged; but the acceleration of the angular motion since 1845 leads to the conclusion that the distance has been much less than before. (O.Σ., 1877.)

Σ.	323·6	1n.	0˝55	1827·26
	319·2	,,	·46	8·29
	316·4	,,	·63	31·25
	320·8	,,	·54	·30
O.Σ.	317·3	,,	·68	41·21
	314·3	,,	·62	6·01
	316·9	,,	·65	8·21
	294·3	,,	·60	70·25
Mä.	320·7	3n.	·46	43·30
Se.	302·0	2n.	·4	58·02
De.	293·0		...	63·37
	296·2		...	4·08
W. & S.			failed	73·25
			,,	5·09

149   Σ. 518.

40 ERIDANI.

R. A.	Dec.
4ʰ 9·8ᵐ	—7° 47′

M.

A 4, B 9, C 10·8, D 12, E 11·4.

A remarkable ternary system : common proper motion − 2˝·17 in R. A., and + 3˝·45 in N. P. D.

A B. Σ. (see *M. M.*, p. 275) showed that A and B have a common proper motion, and that the distance was slowly diminishing. On reducing the observations from 1825 to 1871 to the equinox of 1850, and freeing them from the effects of refraction, O.Σ. finds a general confirmation of the diminution of distance, and that since 1855 the change has been almost inappreciable. This may be explained by the accidental errors in the measures, or, more probably, by the disturbing effect of the third star C : for B and C evidently revolve about A, the motion being retrograde.

B C. H₁ discovered C in 1783·13. It was seen by Σ. in 1825, but with great difficulty ;' and in 1835 and 1836 he could find no trace of it. In 1850, O.Σ. began to observe this system, and always saw C without difficulty. The distance appears to have attained its maximum since 1850, and we may therefore soon expect to find the object very difficult. Since their discovery, that is in 92 years, the two stars have described 190° about their common centre of gravity. An examination of the measures shows that the period is about 200 years, that the apparent orbit of C about B is very oblong, and that the passage between 1825 and 1850 was not a perfect occultation. The minimum distance was probably reached about 1835, and hence Σ.'s difficulty above mentioned. (O.Σ., 1877.)

The small stars D and E do not belong to the system.

A B.

	°		''	
H₁.	107·5		89±	1783·13
So.	·9		84·73	1824·90
Σ.	·5	2n.	85·32	5·05
	·3	4n.	83·48	36·04
	·6		·9	27·09
Sm.				
O.Σ.	106·3	2n.	82·23	50·94
	105·9	,,	·19	1·49
	106·0	3n.	81·93	3·64
	105·8	1n.	82·33	4·79
	·8	,,	81·72	6·80
	·8	,,	82·04	7·82
	·6	2n.	81·48	64·84
	·7	,,	·72	5·89
	·8	1n.	·83	9·10
	·9	,,	·67	72·18
	·8	,,	82·14	4·10
Se.	106·3	3n.	·22	56·38
De.	105·8		·17	63·47
Wi.	·6	2n.	81·48	4·85
Kn.	106·9	1n.	80·77	71·99
Fl.	104·7		81·5	7·12
W. & S.	106·7	2	82·5	·94

B C.

H₁.	326·7	1n.	4·08	1783·13
Σ.	287	,,	...	1825·12
O.Σ.	156·6	2n.	3·96	50·94
	155·0	,,	·87	1·49
	·4	3n.	·93	3·64
	·3	1n.	4·13	4·79
	152·9	,,	·51	6·80
	153·0	,,	·40	7·82
	147·6	2n.	·45	64·84
	143·8	,,	·26	5·89
	140·4	1n.	·46	9·10
	·6	2n.	·62	72·56
	133·9	1n.	·27	3·99
	135·7	,,	·99	4·10
	138·1	,,	3·80	5·14
Da.	160·0		3±	51·06
Wi.	147·6	2n.	4·45	64·80
C.O.	127·5	3n.	·36	77·84
Fl.	120·0		2±	·12

A D.

Wi.	185·0		75·85	64·84
Fl.	148·0		37·2	77·12

A E.

Wi.	312·5		89·45	64·84
Fl.	339·2		109·9	77·12

150   Σ. 520.

R. A.	Dec.	M.
4ʰ 11·1ᵐ	22° 29′	8, 8

C. white.

	°		''	
Σ.	101·8	5n.	0·88	1834·16
Da.	99·0		·87	41·52
	102·9		·97	54·13
Mä.	97·6	1n.	·45	41·79
	102·7	3n.	1·18	4·08
O.Σ.	99·4	2n.	0·96	1·96
Wi.	110·5	,,	·89	56·20
Se.	102·9	1n.	·63	7·11
De.	106·3		·9	68·08

151   O.Σ. 79.

R. A.	Dec.	M.
4ʰ 13ᵐ	16° 14′	7·9, 7, 8·8

C. A, white ; B, ashy.

Rapid change in angle.

O.Σ.	24·3	2n.	0·76	1846·06
Mä	25·2	,,	·32	8·56
	27·4	1n.	·35	52·09
Se.	32·5	,,	·98	9·05
De.	47·1	4n.	·64	67·25

152   O.Σ. 80.

R. A.	Dec.	M.
4ʰ 15ᵐ	42° 9′	6·5, 7

O.Σ.	188·6	5n.	0·52	1848·44
De.	184·6	4n.	·5	67·97

153   Σ. 535.

230 TAURI.

R. A.	Dec.	M.
4ʰ 16·8ᵐ	11° 5′	6·7, 8·2

C. Σ. A, yellowish ; B, bluish. Se. both white.

Dawes (*Mem. R. A. S.*, vol. xxxv., p. 319) sure that the angle has decreased, and that the distance is unchanged.

Se. (p. 24). The motion in angle is certain. The distance has diminished.

A retrograde motion is certain. (O.Σ., 1877.)

Σ.	355·0	2n.	1·96	1829·19
	353·4	,,	·92	32·58
	352·5	1n.	·98	3·14
	354·1	,,	...	1·86
H₂.				
Da.	...	5n.	1·95	41·41
	345·5	9n.	...	·83

Da.	344·6	2n.	1″·92	1847·04
	343·6	1n.	·85	8·13
	342·7	3n.	...	54·29
	...	2n.	2·06	·82
O.Σ.	348·0	1n.	·08	1·70
	346·8	,,	1·97	2·21
	340·7	,,	2·02	72·18
Mä.	351·3	,,	1·76	44·12
Se.	345·0	2n.	·54	56·52
Ta.	341·8	1n.	2·15	65·95
	343·6	,,	...	9·08
	344·8	,,	...	72·14
De.	343·0	2n	1·4	56·82
	342·2	3n. ·	·74	62·90
	341·8	2n	·73	3·11
Mo.	346·0	,,	·6	0·02
W. & S.	340·0	6	·95	72·08
	339·2	4	2·09	3·14
	340·1	4	1·71	·93
	339·7	9	·66	·13
	341·9	5	...	5·18
	339·3	5	·75	·19
	338·4	4	2·01	7·09
Gl.	339·7	5	1·8	3·94
W.O.	338·4	1n.	·71	6·11
	·4	,,	·68	·11
	340·8	,,	·78	·12
Sp.	335·6		·72	7·16
Schi.	·5	,,	·72	·16
Dob.	336·7	2n.	·56	·89
	·9	1n.	·65	8·08

154   O.Σ. 81.

R. A.	Dec.	M.
4ʰ 17ᵐ	33° 40′	6, 8·8

O.Σ.	53·0	4n.	4·49	1847·86
Mä.	235·1	1n.	·13	52·18
De.	50·4	3n.	·37	66·77

155   O.Σ. 82.

R. A.	Dec.	M.
4ʰ 16·9ᵐ	14° 46′	7, 9

Rapid change in angle.

O.Σ.	230·4	2n.	1·04	1848·66
Mä.	231·7	1n.	0·9	45·96
De.	195·9	3n.	0·94	66·73

156   Σ. 547.

R. A.	Dec.	M.
4ʰ 20ᵐ	−1° 41′	8·5, 11·5

Σ.	344·3	3n,	4·25	1831·39
Mä.	344·7	1n.	...	45·04

157   Σ. 554.

80 TAURI.

R. A.	Dec.	M.
4ʰ 23·3ᵐ	15° 23′	6·5, 9

C. yellow.

The distance has probably decreased. The common proper motion is + 0″·061 in R. A., and − 0″·003 in N. P. D. The measures are very discordant, but this may be explained by the faintness and closeness of the stars, and the nebulous character of the smaller.

Σ.	12·9	4n.	1″·73	1831·18
Sm.	13·9		·6	2·16
	11·0		·4	7·22
	13·9		·6	9·16
	15·2		·8	43·11
Da.	9·8		·5	36·96
	...		·66	40·10
	12·5		...	3·09
	10·6		1·41	59·15
Mä.	18·4	1n.	·45	44·17
	·6	2n.	·66	5·09
	21·2		·33	51·09
	24·3	1n.	·29	2·15
	22·2	,,	...	3·92
	20·6	,,	1·24	4·15
	21·3	,,	·37	5·20
	18·9	,,	...	7·95
	21·4	,,	1·31	8·21
O.Σ.	8·4	1n.	·61	1·85
	in contact.			72·18
Ja.	10·2		1·50	53·14
	6·9	10	·44	6·28
	7·7	11	·56	8·09
De.	10·5		·23	63·10
W. & S.	14·7	6	·29	70·07

158   O.Σ. 85.

R. A.	Dec.	M.
4ʰ 28ᵐ	48° 10′	7·5, 10

O.Σ.	23·65	2n.	1·07	1846·70
De.	35·75	4n.	·35	68·93

159   α TAURI.

R. A.	Dec.	M.
4ʰ 29ᵐ	16° 16′	1, 11·2

C. A, pale rose tint ; B, sky blue (Sm.)

This beautiful star has been observed for more than 2000 years. It has been called by various names, e.g., "The Hindmost," "Stella dominatrix," "The Bull's-eye," etc. Tycho considered it to be 125 times the size of our earth, while Ricciolus worked it up to 2810 times that magnitude." (Sm.) "Its ruddy aspect has long been noted, and old Leonard Digges, in his *Prognostication Everlasting*, 1555, pronounces that it is "ever a meate rodde." (Sm.) "I have repeatedly seen it apparently projected on the disc of the moon, even to an amount of nearly three seconds of time, at the instant of immersion, when occulted by that body. The phenomenon seems to be owing to the greater proportionate refrangibility of the white lunar light, than that of the red light of the star, elevating her apparent disc at the time and point of contact." (Sm.) The proper motion of this fine star has been variously estimated :—Piazzi, $+ 0''\cdot04$, $- 0''\cdot21$; Argelander, $+ 0''\cdot08$, $- 0''\cdot17$; Bessel, $+ 0''\cdot12$, $- 0''\cdot15$, in R. A. and Dec. respectively. The B. A. C. gives $+ 0''\cdot008$ in R. A. and $+ 0''\cdot15$ in N. P. D.

The position angle is about 36°, and the distance has changed from 95" to 114" since 1781.

Mr. Burnham has lately discovered with the 18½-inch refractor of the Dearborn Observatory an exceedingly faint companion to this bright star : distance about 30"·5.

		°		"	
H₁.		37·0		95	1781·96
		35·1		...	1802·10
So.		36·2		90	25·04
Σ.		·0		109	36·06
Sm.		35·9		107·9	·98
O.Σ.		·5		111·6	51·40
De.		34·1	.	112·7	63·37
Gl.		35·6	in.	110·9±	76·07
Fl.		·5	,,	114·5	7·06
Bu.		·2	,,	113·9	·9

Mr. Burnham gives the following measures of his new companion to this star:—

111·6	in.	30·16	1877·83
103·3	,,	·61	·86
112·1	,,	·27	·99

160   Σ. 567.

R. A.	Dec.	M.
4ʰ 29·5ᵐ	19° 14′	8·5, 9

C. yellow.

Σ.	302·9	3n.	1·43	1831·18
Mä.	·9	,,	·43	43·80
De.	313·7		·68	63·95

161   Σ. 566.

2 CAMELOPARDI.

R. A.	Dec.	M.
4ʰ 30·4ᵐ	53° 15′	5, 7·4

C. Σ. A, yellow ; B, bluish ; Se. A, yellow, B, blue ; Sm. A, yellow ; B, pale blue. De. A, blue ; B, ashy.

Dawes (*Mem. R. A. S.*, vol. viii , p. 78). "This star should be watched as it may be opening."

Smyth (p. 105). "Σ. recorded it 'Vicinæ'; but it is certainly wider and easier of measurement than those usually so classed by him."

Certain retrograde motion. Distance probably unchanged. (O.Σ., 1877.)

The common proper motion is $+ 0^s\cdot002$ in R. A., and $+ 0''\cdot11$ N. P. D.

	°			
Σ.	311·4	4n.	1·58	1829·79
H₂.	308·3		2·0	30·80
Mä.	309·2	in.	1·56	4·96
	305·3	2n.	·47	45·29
	306·9	2n.	·41	51·84
	302·7	,,	·50	2·26
	303·5	,,	·39	5·25
Sm.	307·9		·9	34·49
	308·7		·7	6·28
	307·2		·5	47·21
O.Σ.	304·6	3n.	·61	6·44
	302·4	in.	·63	51·29
	295·3	,,	·70	66·24
	296·7		·70	71·26
Ja.	303·7	10	·94	53·19
De.	301·9	3n.	...	4·95
	302·7	,,	...	5·09
	306·3	2n.	·75	·89
	304·2	in.	1·6	56·26
	298·8	2n.	·87	62·83
	300·9	in.	·50	4·07
Se.	·5	2n.	·73	58·92
W. & S.	294·3	5	·54	75·09

162   Σ. 572.

R. A.	Dec.	M.
4ʰ 31·1ᵐ	26° 42′	6·5, 6·5

C. yellowish.

Dunér's formulæ are

$$1852\cdot17 \quad \Delta = 3''\cdot42.$$
$$P = 207°\cdot8 - 0°\cdot146\ (t - 1850\cdot0).$$

	°			
Σ.	213·5		2·69	1822·27
	210·3	3n.	3·17	30·56
H₂ & So.	209·1		·92	23·97
H₂.	208·8		·4	32·30
Da.	210·6		·6	6·97
Mä.	·1	in.	·79	43·14
	206·5	,,	·62	58·23
De.	207·2		·60	6·75
	204·9		·35	68·29
Se.	206·0	3n.	·47	57·01
Mo.	·4	2n.	·43	·08
Du.	204·7	,,	·40	71·63
W. & S.	·7		·56	3·96
Gl.	·2		·4	4·04
	205·2		·6	5·01

163   Σ. 577.

R. A.	Dec.	M.
4^h 34·1^m	37° 17′	7·7, 7·7

C. Σ. white.　Se. white.　De. white.

Secchi (p. 24) says there is "a very small motion in angle."

Certain change in angle, retrograde ; distance unchanged. (O.Σ., 1877.)

	°		″	
Σ.	278·7	3n.	1·58	1829·57
H$_2$.	272·5		·5	32·60
O.Σ.	91·5	1n.	·68	41·70
	90·1	,,	·79	6·11
	79·3	,,	·53	70·25
	80·2	,,	·70	1·26
Se.	87·99	3n.	·63	57·66
Mä.	274·2		·64	35·81
	267·7	1n.	·83	45·17
	·3	2n.	·61	52·18
	·1	1n.	·67	1·04
	265·1	,,	·90	4·85
	266·5	,,	·66	5·20
	264·6	,,	·92	7·21
De.	85·1	,,	·5	6·93
	·3	2n.	·4	8·42
	84·6	3n.	·59	62·88
	·8	1n.	...	3·89
W. & S.	260·9	4	1·43	73·93
	258·6	4	·50	·95
	·9	4	·62	·99
	84·7	10	·47	2·16
	2·1	3	·24	5·09
	2·5	6	·36	·18
Gl.	260·9	1n.	·35	3·94
	259·7	,,	·4	5·90
Dob.	83·17	2n.	·42	7·88
	76·55	1n.	·65	8·08

164   Σ. 589.

R. A.	Dec.	M.
4^h 38·5^m	5° 4′	8, 8

C. yellowish white.

Σ.	310·9	3n.	4·47	1831·39
Mä.	·7	2n.	·41	43·09
Da.	311·3		...	6·73
De.	302·8		·39	63·04
M.	295·2		·65	5·05
Ta.	306·6	1n.	·82	6·04
Gl.	303·5	,,	·44	73·98

165   O.Σ. 90.

R. A.	Dec.	M.
4^h 48^m	8° 24′	7, 9

O.Σ.	343·9	2n.	2·06	1845·50
Mä.	352·9	,,	1·8	9·14
	355·3	1n.	·8	52·09
De.	345·6	4n.	·85	66·98

166   O.Σ. 89.

R. A.	Dec.	M.
4^h 49^m	73° 53′	6·2, 7·5

From the measures of O.Σ. and De. it appears probable that the periastron passage occurred about 1870.

	°		″	
O.Σ.	305·9	5n.	0·45	1848·28
De.	104·23	oblong		69·02

167   O.Σ. 91.

R. A.	Dec.	M.
4^h 50^m	3° 0′	7, 7·5

Perhaps one of the stars is a variable.

O.Σ.	62·8	3n.	0·77	1851·85
De.	240·9	,,	·7	66·61

168   Σ. 622.

R. A.	Dec.	M.
4^h 51·9	1° 28′	8·2, 8·2

H$_1$.	185·1		...	1783·06
Σ.	179·9	3n.	2·64	1832·09
H$_2$.	182·0		...	29·88
Sm.	180·4		2·4	33·92
Da.	175·9		·36	40·12
Mä.	181·1	,,	·94	2·50
	179·8	1n.	·93	3·14
	183·6	,,	...	5·11
	174·9	,,	2·77	58·10
O.Σ.	176·83	3n.	·78	46·02
Se.	·4	2n.	·41	58·08
De.	·0		·45	66·90
Gl.	173·2		·7	76·07

169   Σ. 619.

R. A.	Dec.	M.
4^h 52·1^m	50° 5′	8·7, 8·7

C. white.

Σ.	106·0	3n.	5·41	1830·23
Mä.	109·1	,,	·63	45·67
	110·7	1n.	·25	52·26
De.	115·1		·05	66·90

170   Σ. 615.

R. A.	Dec.	M.
4^h 53·1^m	73° 25′	8, 9·8

Probable increase in the angle. (O.Σ., 1877.)

Σ.	337·1	3n.	1·26	1831·9
Mä	·8	1n.	0·8	44·34
De.	345·9	3n.	1·40	66·81
O.Σ.	346·9	1n.	·41	73·35

171   O.Σ. 92.

R. A.	Dec.	M.
4ʰ 52ᵐ	39° 13′	6, 9·7

Da.	233·3		2·92	1847·08
O.Σ.	230·1	3n.	·78	49·09
Mä.	226·3	1n.	·55	52·18
	237·4	,,	·66	·26
De.	241·27	3n.	·82	67·39

172   O.Σ. 93.

R. A.	Dec.	M.
4ʰ 54·1ᵐ	4° 55′	7·5, 9

Mä.	72·9	1n.	0·82	1846·12
O.Σ.	65·6	2n.	1·37	47·18
De.	61·9	3n.	·07	66·29

173   O.Σ. 95.

R. A.	Dec.	M.
4ʰ 58·4ᵐ	19° 38′	6·6, 7·2

C. white.

Probable increase of distance.

O.Σ.	344·2	4n.	0·55	1845·96
Mä.	347·6	1n.	·5	52·09
	340·7	3n.	·77	63·54
De.	338·2	3n.	·5	6·97

174   O.Σ. 97.

R. A.	Dec.	M.
4ʰ 59·3ᵐ	22° 56′	6, 7·8

C A, yellow.

Between 1848 and 1861 O.Σ.'s observations show no trace of angular change. Yet, in 1846, the two stars were so close, that no separation could be effected by him. Probably one star occulted the other about 1844.

O.Σ.		round	1844·91
	248	oblong	6·19
	159·4	1n. | 0·53	61·20
De.		elongated	6
		single	9

175   O.Σ. 98.

14 i ORIONIS.

R. A.	Dec.	M
5ʰ 1·3ᵐ	8° 20′	6, 6·8

Rapid retrograde orbital motion.

Dunér gives

$$1859·08 \quad \Delta = 1'''·14.$$
$$P = 237°·1 - 1°·206 \ (t - 1860·0).$$

O.Σ.	250·83	3n.	1''·14	1844·53
	249·60	2n.	0·98	9·22
	237·80	,,	1·24	59·22
	224·13	3n.	·09	70·87
Mä.	258·18	1n.	...	44·05
	245·4	,,	...	52·15
Da.	...	,,	1·18	48·11
	240·9	,,	·29	54·82
De.	234·0	,,	·25	65·98
	232·1	,,	·28	7·15
	228·5	,,	·05	8·14
Du.	224·6	2n.	·00	9·19
	211·9	,,	·22	76·18
Sp.	209·9		0·98	7·18

176   Σ. 644.

R. A.	Dec.	M.
5ʰ 2·1ᵐ	37° 9′	6·7, 7

Probable change in angle.

Σ.	219·2	3n.	1·61	1828·60
Mä	223·7	2n.	·52	45·18
	·2	1n.	·73	51·04
	224·1	,,	·64	2·18
	240·7	,,	·50	7·24
O.Σ.	219·6	,,	...	41·22
	223·8	,,	1·79	·70
	227·7	,,	·92	2.21
	224·1	,,	·58	69.24
Du.	219·1	6n.	·46	71·98
W. & S.	41·6	4	·66	4·09
	220·5	4	·60	5·09
	221·9	4	·79	6·12
Dob	222·1	2n.	·68	7·90
	·1	1n.	·71	8·08

177   Σ. 634.

R. A.	Dec.	M.
5ʰ 2·8ᵐ	79° 4′	4·5, 7·9

O.Σ. finds that the following formulæ represent the observations :—

$$\Delta A = 4'''·120 \pm 0''·038 + (0''·1523 \pm 0''·0024)\ (T - 1850·0).$$
$$\Delta D = +28·583 \pm 0·040 - (0·3039 \pm 0·0026)\ (T - 1850·0).$$

Argelander (*Bonn Observations*, vol. vii.) gives the annual proper motion of the principal star, − 0ˢ·0365 in R. A. and + 0''·141 in Dec. The above formulæ assign the following values as the apparent proper motion of the smaller star, + 0ˢ·0170 in R.A. and − 0''·163 in Dec. These are nearly equal in amount, but in opposite directions. See also the *Bulletin de l'Académie de St. Pétersbourg*, vol. v. In vol. xix. of this work O. Σ. has the following remarks :—

"The distance will be 9″·2 in 1932 if there is no physical connection : if there is true orbital motion, it will be discovered in ten or twenty years if good observations are made."

So.	346·4		37·01	1825·10
Σ.	348·3	In.	34·50	31·30
	·3	,,	·64	2·18
	·0	,,	·46	·81
	349·2	,,	33·47	6·18
	348·9	,,	·46	·21
	·7	,,	·72	·22
Sm.	·8		34·1	3·16
	349·1		33·8	6·25
Mä.	350·1	2n.	30·24	45·35
De.	353·1	3n.	26·24	58·33
	355·0	5n.	24·63	63·15
	356·2	,,	23·65	6·12
O.Σ.	357·7	In.	·06	8·25
	358·0	,,	22·51	70·35
	359·4	,,	21·67	3·35
W. & S.	0·1	3	·30	5·09
Fl.	1·4	In.	20·29	·37

178   Σ. 629.

R. A.	Dec.	M.
5h 3m	83° 18'	8·2, 11·2

Certain change in angle and distance.

Σ.	342·5	In.	13·05	1832·29
	340·3	,,	·71	·30
	343·0	,,	·08	3·23
	342·7	,,	12·80	·25
Mä.	348·4	,,	·73	45·35
De.	355·5	3n.	·04	67·48
O.Σ.	357·6	In.	4·53	73·35
W. & S.	359·5	3	3·80	75·09

179   O.Σ. 100.

R. A.	Dec.	M.
5h 3m	8° 1'	7, 9·8

O.Σ.	247·0	In.	4·57	1845·17
	244·2	,,	·20	8·13
	250·5	,,	·20	52·22
	253·3	,,	·26	71·18
De.	249·8	3n.	·08	67·45

180   Σ. 651.

R. A.	Dec.	M.
5h 4·2m	—7° 14'	8, 10

Rectilinear motion.

Σ.	101·7	2n.	10·81	1829·67
H₂.	·0		10·5	30·30
Mä.	83·5	2n.	11·08	44·52
	82·6	In.	·49	5·19
	88·4	2n.	12·40	58·10
De.	64·7		14·1	65·28
W. & S.	56·8	6	16·92	77·94
C.O.	55·2	In.	·54	·95

181   Σ. 655.

R. A.	Dec.	M.
5h 6·7m	— 12° 1'	4·2, 10·5

C. A, greenish.

H₁.	359·3		12·33	1783·06
	360·0		...	85·08
Σ.	338·1		13·1	1829·05
	337·6	6n.	12·81	32·25
	·6		·81	52·25
Sm.	336·9		15·0	36·93
Mä.	337·7	2n.	12·60	3·56
Mo.	335·4	10	13·46	56·08
	337·7	6	12·76	7·08
Gl.	·2	In.	14 ±	76·07

182   Σ. 653.

R. A.	Dec.	M.
5h 7m	32° 33'	5, 7·2

C. A, greenish ; B, bluish white.

For A C, Dunér gives

$$\Delta = 12''\cdot16 - 0''\cdot020 \,(t - 1850\cdot0).$$
$$P = 345°\cdot11 + 0°\cdot14 \,(t - 1850\cdot0).$$

A B.

H₁.	232·6		16·13	1781·83
H₂ & So.	225·6		14·61	1822·09
Σ.	226·0	In.	·62	9·23
	224·7	,,	·67	30·25
	225·7	,,	·67	2·18
Sm.	224·5		13·5	·81
Mä.	225·0	2n.	·91	42·26
	·1	In.	·87	4·26
	224·6	,,	·55	5·21
	·1	3n.	·45	51·10
	·1	In.	·85	5·20
	·3	2n.	·56	7·58
	225·1	In.	·65	8·27
Da.	·8	2n.	14·91	47·78
Ro.	226·3	2n.	·53	63·09
Ta.	223·6	In.	·86	8·91
	226·9	5	·70	72·21
Du.	·2	3n.	·65	1·85
W. & S.	·7	2n.	15·08	5·13
	228·0	5	...	6·12
	226·3	4	14·94	7·15
	225·8	4	·99	·18
Pl.	·1	3n.	·8	6·69

A C.

Σ.	342°.4	3n.	12″.58	1830.55
Ta.	.4	1n.	13.31	66.04
Du.	348.1	2n.	11.79	72.17
W. & S.	349.4	1n.	13	5.18
	348.4	2n.	12	7.16

183　λ AURIGÆ.

R. A.	Dec.	M.
$5^h\ 10.6^m$	40° 0′	5.2, 8.7

The motion is rectilinear : proper motion, $+0^s.047$ in R. A., and $+0″.66$ in N. P. D.

So.	34.6		102.1	1825.10
Sm.	30.2		.8	35.88
Σ.	29.0		103.5	6.21
O.Σ.	22.7		109.7	52.14
Fl.	13.9	1n.	121.8	77.13

184　Σ. 676.

R. A.	Dec.	M.
$5^h\ 12.9^m$	64° 38′	7.5, 8.5

C. white.

Certain change in angle and distance.

Σ.	283.1	1n.	0.77	1831.30
	5.6	,,	.85	.31
	278.4	,,	.85	2.29
Mä.	.9	4n.	.81	42.84
O.Σ.	271.6	1n.	.88	5.32
	280.8	,,	.88	6.30
	278.6	,,	.87	7.34
	271.8	,,	.90	9.27
	274.0	,,	.91	51.27
	269.7	,,	1.10	71.30
	273.1	,,	.02	2.31
De.	276.0		.0	63.20

185　Σ. 677.

R. A.	Dec.	M.
$5^h\ 13^m$	63° 16′	7.7, 8

C. very white.

Σ.	279.8	1n.	1.83	1831.30
	281.0	,,	.48	.31
	282.1	,,	.81	.32
	274.7	,,	.83	3.14
Mä.	278.3	,,	2.03	44.34
O.Σ.	273.3	,,	1.88	5.32
	278.8	,,	.67	6.30
	268.2	,,	.81	9.27
	269.5	,,	.85	51.27
	262.7	,,	.72	71.30
	.1	,,	.67	2.31
De.	265.7	4n.	.77	63.13

186　O.Σ. 104.

R. A.	Dec.	M.
$5^h\ 14^m$	46° 54′	7, 11

O.Σ.	191.6	1n.	15″.73	1846.85
	189.8	,,	.83	7.20
	191.0	,,	16.29	51.27
Mä.	190.5	;.	...	2.26
De.	191.7	3n.	16.64	66.81

187　Σ. 694.

R. A.	Dec.	M.
$5^h\ 16.6^m$	24° 51′	8.2, 8.2

C. white.

Σ.	6.9	1n.	1.34	1827.16
	3.2	,,	.32	8.19
	2.4	,,	.37	33.19
Da.	357.7	3n.	.20	41.80
	358.5	2n.	.33	3.20
O.Σ.	359.9	1n.	.44	6.09
	3.9	,,	.41	7.16
Mä.	356.2	,,	.3	3.14
	358.2	,,	.4	4.91
	357.5	,,	.3	5.21
	6.0	2n.	.39	52.16
	.6	,,	.52	3.15
	5.3	1n.	.37	5.21
	.7	,,	.20	8.10
Wi.	185.8	2n.	.27	6.20
Se.	359.8	1n.	.30	7.12
Ta.	359.6		.59	66.04
	.6	6	...	72.14
Gl.	358.0	1n.	1.4	3.98
W. & S.	4.2	8	.24	4.10
Pl.	0.7	2n.	.24	7.08

188　h. 3752.

R. A.	Dec.	M.
$5^h\ 16.8^m$	−24° 54′	6, 10, 10

A B.

H₂.	100 ±		4	1835.05
C. O.	105.0	1n.	3.55	77.03

A C.

H₂.		30	35.05

189　III TAURI.

R. A.	Dec.	M.
$5^h\ 17.4^m$	17° 16′	6, 9

The proper motion of A is $+0^s.020$ in R. A., and $-0″.04$ in N. P. D.

H₁.	273.8	50.4	1783.16
So.	271.3	61.33	1825.06
Sm.	.2	63.0	32.95

Σ.	·1	65″70	1839·95
O.Σ.	·4	68·58	52·12
M.	272·9	72·91	62·11
Fl.	271·5	75·2	77·13

190 η ORIONIS.

R. A. 5^h 18·4^m Dec. −2° 31'   M. 4, 5

C. white, purplish white.

H$_1$, in *Phil. Trans.*, vol. lxxv., p. 225, has "vi. 67, Fl. 28, η Orionis, double 35° 12' n.f." H$_1$, therefore, did not observe the duplicity of the larger star. Dawes, on the 15th Jan., 1848, discovered that it was double, using 4¼ in. of his 6¾ in. refractor. He thinks the distance may have slightly increased since 1848. (*Mem. R. A. S.*, vol. xxxv., p. 323.)

The proper motion of the principal star is + 0^s·002 in R. A., and + 0″·02 in N. P. D.

H$_1$.		single		
Σ.		single		
Da.	88·7	In.	0·94	1848·11
	·6	9n.	0·93	·20
	86·2	4n.	1·08	51·69
Ja.	87·0	14	1·07	3·12
	83·7	10	0·75	·99
Kn.	87·6	5	0·98	63·12
	89·6	10	·90	·13
	88·4	5	1·08	6·06
	89·8	8	·02	·94
	·2	4	·03	71·99
Du.	88·0	In.	0·84	69·19
	·3	2n.	·89	71·23
	87·1	In.	·84	2·18
	83·8	,,	·85	3·22
	87·7	2n.	·97	4·17
	85·1	In.	·94	5·17
W. & S.	83·1	8	1·23	2·01
	87·6	8	·30	·04
	86·9	2	·3	·92
	85·5	4	·37	3·07
	·3	4	·30	·93
De.	84·7		·02	·69
Gl.	85·1	4	·34	·94
	86·5	5	·0	·99
	87·3	3	...	4·09
	88·5	3	1·0	·10
	84·0	7	...	·11
	85·8	6n.	...	·12
	88·0	3	1·25	·13
	86·3	3	...	·16
	·3	5	1·2	·17
	85·7	2n.	...	8·05
	82·4	In.	1·20	9·16
W.O.	83·8	,,	·11	6·13
	·8	,,	·11	·13
	85·7	,,	·02	·18
Schi.	81·8	,,	0·96	7·19
C.O.	82·6	3n.	·97	·94
Dob.	87·3	In.	1·12	8·08

191 O.Σ. 107.

R. A. 5^h 20^m Dec. 17° 51'   M. 6, 10·8

O.Σ. found a third star C closer and fainter than B. Angle B A C = 30°.

O.Σ.	304·1	In.	9″93	1847·25
	36·4	,,	10·59	9·16
	34·2	,,	9·92	52·22
De.	3·6	3n.	·89	67·93

192 Σ. 712.

R. A. 5^h 20·2^m Dec. 2° 50'   M. 7, 9

C. very white.

H$_1$.	40·3		2 to 4	1782·77
	46·6		,,	3·05
	45·9		,,	1802·06
So.	49·8		3·39	25·10
Σ.	45·4	3n.	·08	31·16
Mä.	·1	In.	2·85	44·12
	55·4	,,	·83	51·18
	54·2	2n.	·93	2·18
De.	·7		·89	64·14
W. & S.	53·6	5	3·33	74·10
Pl.	56·1	2n.	·17	7·08
Dob.	54·6	In.	·04	·91
	55·2	2n.	2·90	8·06

193 Σ. 715.

R. A. 5^h 21·8^m Dec. 41° 10'   M. 8·2, 8·9

C. very white.

Σ.	206·0	4n.	0·95	1831·47
Mä.	202·3		·86	45·49
O.Σ.	201·5	In.	·85	8·22
	208·5	,,	1·03	70·25
De.	200·6		0·91	67·78
Gl.	202·7	,,	1·1	76·07

194 Σ. 716.

R. A. 5^h 21·9^m Dec. 25° 3'   M. 5·8, 6·6

C. A, white; B, bluish white.

The common proper motion is + 0^s·005 in R. A., and + 0″·07 in N. P. D. (B.A.C.)

Dunér gives—

$$1856·22. \quad \Delta = 4″·92.$$
$$P = 198°·0 + 0°·079 (t - 1850·0).$$

H₁.	192.8		4.51	1783.74
H₂.	195.0		...	1817.20
So.	194.0		...	21.97
Σ.	196.8	5n.	4.89	9.63
Be.	195.4	4n.	5.1	30.81
Da.	196.3		.15	2.87
	.8		...	45.87
Sm.	195.5		5.3	33.78
	.9		.0	8.91
	197.4		.5	58.10
Mä.	.5	In.	4.98	43.14
	199.3	,,	5.16	5.21
	198.3		.07	4.17
	197.8	4n.	.19	51.52
	.5	2n.	.16	2.17
	.1	,,	.10	5.22
	.8		4.89	8.98
O.Σ.	.2	In.	.84	1.85
De.	200.5		.95	2.16
	197.5		.78	4.85
	198.7		.71	68.98
Se.	197.7	2n.	5.10	56.60
Mo.	.6	10	4.98	7.07
M.	196.3		.74	62.90
Ro.	.4	2n.	5.1	.91
	198.8	3n.	.06	3.11
Ta.	196.0	In.	4.72	6.17
Du.	199.8	6n.	.91	71.32
Gl.	201.4	In.	.8	3.98
W. & S.	200.1	4	5.0	4.10

195   Σ. 719.

R. A.   Dec.   M.
5ʰ 22.4ᵐ   29° 30'   A 7, B 9.5, C 8.9

C. A, very yellow.

In A B the distance has increased; in A C it has diminished; while the angle has probably increased in both pairs.

A B.

Σ.	326.5	4n.	0.68	1833.47
Da.	328.0	In.	.91	42.13
O.Σ.	.5	,,	.79	7.20
	331.4	.,	1.20	70.25
De.	329.6		.99	68.88

A C.

H₁.	344.9		16.02	1782.98
	345 ±		...	90.86
So.	351.9	2n.	15.45	1825.17
Σ.	.5	6n.	14.83	33.34
Da.	.8	In.	15.20	42.13
O.Σ.	.9	,,	14.96	7.20
	352.7	,,	15.21	70.25
De.	351.0		.07	67.12
Ta.	344.8	In.	.19	6.04
	352.1	,,	14.40	74.18

196   O.Σ. 108.

R. A.   Dec.   M.
5ʰ 22ᵐ   18° 16'   7, 10.5

O.Σ.	139.5	In.	3.64	1847.25
	138.1	,,	.61	9.16
	.5	,,	.52	52.22
De.	133.0	3n.	.43	68.01

197   Σ. 728.

R. A.   Dec.   M.
5ʰ 24.4ᵐ   5° 51'   5.2, 6.7

C. yellowish.

The observations are very discordant. In spite of this, however, a diminution in angle and distance is beyond a doubt.

Dunér has

$$\Delta = 0''.75 - 0''.0184\,(t - 1850.0).$$
$$P = 201°.5 - 0°.255\,(t - 1850.0) - 0°.00439\,(t - 1850.0)^2.$$

H₁	217.8	In.	1 to 2	1782.05
	216.5	2n.	,,	1802.06
So.	203.5		1.30	22.10
H₂.	214.5		.92	30.18
Σ.	200.0	In.	.04	0.21
	204.0	,,	0.99	1.21
	205.2	,,	1.22	.21
	.8	,,	0.91	.23
	207.7		1.04	3.96
Sm.	205.4		1.0	1.13
	206.2		.0	9.20
Mä.	205.0	In.	1.24	41.20
	203.6		0.98	3.96
	206.0	,,	...	4.11
	202.7		0.76	51.80
	203.1	2n.	.73	2.16
	202.3	In.	...	7.21
O.Σ.	218.1	2n.	0.96	41.22
	212.8	In.	.88	2.22
	219.2	,,	.74	5.23
	215.9	,,	.74	6.23
	205.8	,,	.75	8.21
	210.9	2n.	.77	9.23
	207.4	In.	.68	61.20
	.4	,,	.84	3.21
	185.3	,,	...	4.21
	195.2	,,	0.79	6.21
	189.6	,,	.77	8.21
	202.4	,,	.56	9.21
	196.2	,,	.76	70.21
	190.1	,,	.57	3.24
	188.3	,,	.62	5.19
Da.	205.1	3n.	0.89	44.94
Ja.	202.4	2n.	1.71	53.43
Se.	203.6	,,	.44	7.67
De.	192.2	5n.	...	63.33
Du.	193.6	5n.	0.3	73.41
W. & S.	198.5	2	...	4.10
Gl.	190.0	3	0.6	.14
Dob.	204.2	In.	...	6.24
Sp.	188.9		.44	7.19

198 Σ. 727.

R. A. Dec. M.
5^h 25^m 44° 42′ 8, 9·5

C. A, yellow.

Dunér has
1855·25. Δ = 2″·20.
P = 59°·5 + 0°·10 (t − 1850·0).

Σ.	56·7	3n.	2·18	1830·89
	61·7	1n.	·30	44·26
	62·2	,,	·08	5·20
Mä.	60·7	2n.	·55	52·26
Du.	·1	5n.	·12	71·49

199 Σ. 735.

R. A. Dec. M.
5^h 27^m −6° 35′ 8·2, 9

C. white.
Rectilinear motion.

Σ.	355·2	2n.	30·92	1831·15
Mä.	354·2		34·01	47·23
	·2	1n.	·51	51·20
De.	353·6		36·56	66·72
Fl.	354·3	1n.	38·05	77·13

200 Σ. 742.

R. A. Dec. M.
5^h 29·2^m 21° 56′ 7·2, 7·8

C. yellowish white.
Certain direct motion.

H$_1$.	233·6	1n.		1782·86
Σ.	246·3			1822·25
	247·1	1n.	3·18	28·19
	244·2	,,	·54	31·22
	247·4	,,	·22	·25
	251·1	2n.	·32	7·10
So.	248·3		2·97	26·10
H$_2$.	246·9		3·40	9·91
O.Σ.	251·0	1n.	·56	46·09
	250·8	,,	·56	7·16
	253·7	,,	·28	9·24
	252·7	,,	·41	50·19
	256·1	,,	·34	70·25
Da.	249·7	4n.	·27	42·01
	251·3	,,	·26	52·64
Se.	252·5	2n.	·21	6·50
Mä.	249·7	1n.	·47	41·22
	·7	,,	·27	4·91
	252·9	;,	80	5·21
	·4	4n.	·40	52·16
	·1	1n.	·33	5·21
	250·9	,,	·03	7·21
Mo.	·4	10	·51	8·45
M.	251·7		·46	63·23

Ta.	256·3	1n.	3·62	1866·09
	257·2	,,	·73	7·19
	258·1	,,	·93	8·96
	255·8	,,	·66	72·14
	256·6	,,	·62	·18
	257·1	,,	·35	4·18
De.	251·7	4n.	·67	55·16
W. & S.	255·9	4	·36	73·93
	·6	4	·31	·99
Gl.	·1	4	·4	·94
	256·6	4	·2	·98
Pl.	254·4	4n.	·16	7·11
Dob.	256·1	7n.	...	6·06

201 Σ. 748.

θ^1 ORIONIS.

R. A. Dec.
5^h 29^m −5° 30′
M.
A 7, B 8, C 4·7, D 6·3, E 11·3, F 11·3.

O.Σ. thinks that one of the two stars E, F is variable; and that E and F should be No. 10 of Σ.'s scale of magnitude. (See his Memoir on the Great Orion Nebula.)

After a very careful discussion of the measures of these stars, O.Σ. comes to the conclusion that probably no considerable changes have taken place since the earliest observations. He thinks that the changes in angle indicated by the measures of A E and A F are not real, but owe their existence to the difficulty of the objects. It is possible, however, that the angle and distance in A F have both increased.

"From the foregoing observations it may be gathered that in all probability not only the stars of the trapezium, but also many in the neighbourhood, are physically connected with the nebula. This is especially true of the groups, which, to the naked eye, form ι, θ, c, Orionis. For we see that each of these groups is accompanied by a nebula." (Bond.)

A B.

H$_1$.	...		8·78	1776·87
Σ.	30·8		9·08	1820·56
	31·6		8·49	31·18
	·6		·74	6·15
Mä.	32·3		·53	42·14
	31·7		·79	5·16
	35·6	1n.	...	53·21
	33·4	2n.	8·62	4·17
	31·0	1n.	·23	8·11
O.Σ.	32·9	1n.	·74	72·19
	30·8	,,	·48	5·19
C.O.	59·4	,,	...	7·95

A C.

	°		″	
H₁.	...		12.81	1776.87
Σ.	134.0		.62	1820.56
	131.0		13.08	31.18
	.5		.00	6.15
So.	130.8		.45	24.58
Mä.	132.5		12.99	42.14
	131.4		.88	5.16
	13.3	In.	...	53.21
	131.3	2n.	12.70	4.17
	130.7	In.	.75	.11
O.Σ.	132.5	,,	13.30	8.23
	133.2	,,	.31	66.19
	.2	,,	.44	9.21
	131.3	,,	.22	72.19
	.9	,,	12.99	5.19
C.O.	309.9	,,	...	7.95

A D.

	°		″	
Σ.	95.5		21.15	1820.56
	.7		.37	31.18
	.4		.41	6.15
Mä.	96.5		20.99	42.14
	95.3		21.23	5.16
	.0	In.	...	53.21
	.1	2n.	21.38	4.17
	96.9	In.	.16	.11
O.Σ.	95.1	In.	21.41	69.21
	.2	,,	.43	72.19
	.2	,,	.35	5.19
C.O.	94.5	In.	...	7.95*

B C.

	°		″	
Σ.	165.0		17.1	1820.56
	161.1		16.74	31.18
	162.1		.85	6.15
So.	165.0		.68	24.58
Mä.	162.5		.75	42.14
	.5		.78	5.16
	163.2	In.	...	53.21
	.3	2n.	16.65	4.17
	162.3	In.	.30	.11
O.Σ.	164.3	In.	17.10	8.23
	163.4	,,	.07	66.19
	164.4	,,	16.71	9.21
	162.8	,,	.79	72.19
	163.0	,,	.80	5.19
C.O.	342.2	,,	...	7.95†

D B.

	°		″	
H₁.	...		20.39	1776.87
Σ.	301.0		19.08	1820.56
	299.1		.26	31.18
	.3		.23	6.15
Mä.	300.5		18.90	42.14
	299.8		19.04	5.16
	298.6	,,	...	53.21
	299.6	2n.	19.04	4.17
	.3	In.	.18	8.11
O.Σ.	300.0	,,	.33	72.19
	.4	,,	.26	5.19
C.O.	299.6	,,	...	7.95

D C.

	°		″	
H₁.	...		15.21	1776.87
Σ.	240.5		13.70	1820.56
	.5		.09	31.18
	.3		.34	6.15
So.	.1		13.58	24.58
Mä.	.4		12.95	42.14
	.3		13.21	5.16
	.3	In.	...	53.21
	.7	2n.	13.41	4.17
	.9	In.	.37	8.11
O.Σ.	.3	,,	13.62	8.23
	243.2	,,	.40	66.19
	241.2	,,	.33	9.21
	.5	,,	.48	72.19
	242.6	,,	.20	5.19

A E.

	°		″	
Σ.	353.6	7n.	3.86	1832.53
Da.	352.5	2n.	3.82	41.92
Mä.	355.0	In.	...	2.18
	354.8	,,	3.68	4.91
Ja.	352.0	2n.	3.98	53.02
O.Σ.	353.1	In.	4.81	7.82
	351.0	,,	.10	8.23
	349.5	2n.	.13	61.21
	352.2	In.	.29	9.21
	347.6	,,	.17	72.19
Ta.	...	In.	3.32	66.09
	352.2	,,	.47	.16

B E.

	°		″	
Σ.	233.4	3n.		1832.53

C F.

	°		″	
Da.	127.3	3n.	2.79	1842.33
	124.5	In.	4.11	7.04
O.Σ.	127.0	,,	3.38	3.14
	124.8	,,	.19	50.18
	132.3	,,	.73	6.80
	128.0	,,	.93	7.21
	129.8	,,	4.43	.82
	125.8	,,	3.95	61.20
	128.2	,,	.31	.23
	131.0	,,	.71	9.21
	132.0	,,	.94	72.19
Ja.	123.4	2n.	3.26	53.02

A F.

	°		″	
H₂.	117.1	2n.	[3]	1836.50*
Da.	126.6	,,	3.12	43.51
O.Σ.	125.9	,,	3.28	46.66
	128.8	5n.	.92	58.85
	131.5	2n.	.82	70.70

* The C.O. observations also give
Aa 116°.7 1877.95.
Cc 351°.0 .95.

202   Σ. 749.

R. A.	Dec.	M.
5h 29.9m	26° 53′	7.1, 7.2

C. very white.

Direct motion.

* This is B C in the C.O. observations.
† This is A D in the C.O. observations.

Σ.	23·4 °	In.	0·70 "	1827·26
	·9	,,	·72	8·19
	21·5	,,	·60	31·23
	25·0	,,	·66	·26
Mä.	18·9		·77	44·04
O.Σ.	23·0	In.	·84	6·09
	16·3	,,	·61	·22
	19·8	,,	·70	7·20
	17·0	,.	·80	9·24
De.	191·8	3n.	·6	56·76
	186·4	5n.	·6	62·98
Se.	190·4	In.	·63	57·11
Ro.	186·9		·8	64·20

203   O.Σ. 112.

R. A.	Dec.	M.
5h 31·6m	37° 53'	7·3, 8

O.Σ.	80·8	In.	0·57	1846·19
	89·0	,,	·69	7·22
	85·8	,,	·67	52·26
Mä.	90·6	,,	·45	2·27
De.	79·8	3n.	elong^d.	67·43

204   O.Σ. 113.

R. A.	Dec.	M.
5h 33m	12° 57'	7, 10·7

O.Σ.	28·4	In.	10·13	1843·19
	27·7	,,	·16	9·22
	·3	,,	·15	50·19
De.	29·0	3n.	9·84	67·91

205   O.Σ. 114.

R. A.	Dec.	M.
5h 34m	16° 10'	7·3, 9·5

O.Σ.	273·9	In.	2·70	1844·90
	275·9	,,	·99	7·16
	276·4	,,	3·12	9·22
De.	279·5	3n.	2·79	67·95

206   Σ. 774.

R. A.	Dec.	M.
5h 34·7m	—2° 0'	2, 5·7, 10

C. A, yellow; B, reddish olive.

H$_1$, in 1792, did not see the faint star.
The proper motion of ζ is + 0^s·002 in R. A., and + 0"·03 in N. P. D.
The angles increase slowly both in A B and A C.   A C form an optical pair.

A B.

	°	B not seen.	"	1782
H$_1$.				
Σ.	147·8	...		1821·24
	151·3	6n.	2·35	31·22
	150·5	5n.	·47	4·93
	151·3	,,	·55	6·22
H$_2$.	149·8		·73	22·12
	·8		·62	32·11
Be.	148·3	In.	·76	0·93
	...	,,	·68	1·29
	148·2	4n.	·70	2·30
Da.	·4		3·00	2·56
	·5		...	5·27
	146·6		2·67	41·02
	148·4		·57	2·99
	·7		·63	7·84
	·7		·64	53·13
	149·2		·48	4·17
Sm.	148·8		·5	39·19
	149·4		·5	46·16
Mä.	·6	2n.	·39	1·24
	·8	In.	·47	3·42
	152·1	,,	·65	4·22
	148·3	3n.	·57	5·20
	·5	In.	·43	51·25
	·1		·46	·96
	147·8	3n.	·46	2·19
	149·6	2n.	·38	6·81
	·2	In.	·65	7·21
	150·0	2n.	·24	8·22
	146·5		·19	9·17
	151·7		·60	62·21
Flt.	149·6	2n.	·64	51·11
Ja.	152·1	,,	·64	·18
	149·9	4n.	·29	3·18
	151·6	,,	·32	·77
	148·9	2n.	·32	4·06
O.Σ.	154·6		·66	1·85
Mi.	149·0	16	·63	2·06
Mo.	151·0	30	3·06	4·15
De.	·9		2·45	4·56
Wi.	152·9	2n.	...	6·21
Se.	150·0	,,	2·45	7·10
Ta.	152·4	3n.	·86	66·13
	154·4	In.	...	8·98
	152·2	,,	...	9·00
	·8		3·35	77·07
Du.	153·8	9n.	2·28	2·12
W. & S.	·1	2n.	·51	4·15
Sp.	151·7		·56	5·24
	·5		·38	7·19
Dob.	157·2	In.	...	·91
	154·4	3n.	2·37	8·07

A C.

H$_1$.	7·0		60	1781·77
So.	·2		...	1822·61
Sm.	·8		56·0	39·19
W. & .	0·7	6	59·7	74·14
Fl.	9·3		60·3	7·17
Dob.	·2	2n.	...	8·07

207 Σ. 3115.

R. A.   Dec.   M.
5^h 37^m   62° 45′   6·7, 7·8

C. A, white; B, ashy white.

The distance and angle have diminished.

Σ.	35·6	3n.	1·68	1831·63
Mä.	34·5	2n.	·52	45·92
O.Σ.	30·7	1n.	·50	·32
	29·9	,,	·37	6·30
	31·6	,,	·48	9·27
	28·3	,,	·37	72·31
De.	·4	3n.	·48	66·83

208 O.Σ. 115.

R. A.   Dec.   M.
5^h 37·6^m   15° 2′   7, 8

C. A, yellow; B, olive.

Probably a binary.

O.Σ.	119·6	1n.	0·79	1844·90
	123·6	,,	·70	6·22
	127·7	,,	·72	9·23
	121·6	,,	·82	50·92
De.	123·0	3n.	·87	67·90

209 O.Σ. 117.

R. A.   Dec.   M.
5^h 40^m   30° 31′   7, 9·7

C. A, golden.

O.Σ.	28·9	1n.	11·98	1845·22
	30·0	,,	12·00	6·85
	29·0	,,	11·79	50·19
Mä.	30·2	,,	...	2·27
De.	29·9	3n.	11·51	67·25

210 O.Σ. 119.

R. A.   Dec.   M.
5^h 41^m   7° 57′   7·5, 8·3

O.Σ.	309·3	1n.	0·74	1845·22
	304·3	,,	·60	8·23
	298·1	,,	·57	52·22
De.	316·7	3n.	...	67·56

211 So. 503.

R. A.   Dec.   M.
5^h 49·1^m   13° 56′   A 7, B 9, C 8

Rapid rectilinear motion in A B.

A B

So.	134·1		39·95	1825·07
De.	120·3		8·23	73·79
	119·7		7·76	4·21
	118·8		·07	5·21
Fl.	115·3	In.	5·72	7·80

A C.

So.	337·3		201·76	1825·07
De.	335·8		230·04	75·21
Fl.	·7		231·6	7·80

212 θ AURIGÆ.

R. A.   Dec.   M.
5^h 51·5^m   37° 12′   3, 11, 11

B and C are fixed. The proper motion of A is

A B.

H_1.	286·0		35 ±	1782·68
Sm.	289·0		30·0	1832·64
O.Σ.	290·9		43·29	52·12

A C.

H_1.	150 ±		...	1780·74
So.	352·2		124·46	1823·17
O.Σ.	350·7		123·42	40·16
	·3		125·10	52·16

213 O.Σ. 124.

R. A.   Dec.   M.
5^h 52^m   12° 49′   6, 7·8

If the observation of 1873·25 (O.Σ.) be correct, no less than 66° of the apparent orbit have been described in 28 years.

O.Σ.	308·7	1n.	0·53	1845·22
	311·0	,,	·36	6·22
	242·2	,,	·66	73·25
De.	324·0	wedged		65

214 O.Σ. 125.

R. A.   Dec.   M.
5^h 52^m   22° 29′   7, 8·5

C. red.

O.Σ.	357·5	1n.	1·68	1844·90
	360·7	,,	·56	6·19
	353·5	,,	·39	52·22
Mä.	253·6	,,	0·3	44·21
De.	1·13	3n.	1·41	67·59

215 Σ. 830.

R. A.   Dec.   M.
5^h 55·9   27° 39′   8·5, 9, 10·5

C. A, yellow.

In A B there has been a slight increase in the angle, and the distance has probably diminished.

Dunér has

$$1851\cdot77. \quad \Delta = 12''\cdot65.$$
$$P = 251°\cdot4 + 0°\cdot10\,(t - 1850\cdot0).$$

A B.

Σ.	249°·6	3n.	12″·82	1830·54
Mä.	250·1	1n.	·37	45·00
Du.	253·2	3n.	·60	68·48
	254·2	1n.	·13	72·09
W. & S.	·8	4	·1	7·18

A C.

Σ.	187·7	3n.	25·21	1831·56
Du.	·8	2n.	·42	68·29
	·5	,,	·03	72·13
W. & S.	189·5	3	...	7·18

216   h. 3823.

R. A.	Dec.	M.
5^h 55·9^m	−31° 4′	9, 9

H₂.	131·7	6	3	1835·47
C.O.	122·7	1n.	3·85	77·13

217   O.Σ. 131.

R. A.	Dec.	M.
5^h 59^m	36° 16′	7, 10·2

C. A, blue.

O.Σ.	277·3	1n.	...	1846·19
	272·5	,,	...	8·21
De.	282·5	3n.	1·56	66·85

218   Σ. 840.

R. A.	Dec.	M.
5^h 59·8^m	10° 48′	A 6, B 8, C 9

C. A, yellow.

Probable angular change in B C.

A B.

Mä.	246·2	2n.	20·55	1843·10
O.Σ.	247·5		21·53	7·72
Se.	·1	1n.	...	57·11
De.	·4	3n.	21·28	66·68

B C.

Σ.	183·5	3n.	0·91	1830·89
Mä.	181·0	1n.	...	44·20
O.Σ.	179·4	2n.	0·92	7·72
Se.	181·5	,,	·55	57·12
De.	172·6	3n.	·97	66·73

219   O.Σ. 132.

R. A.	Dec.	M.
6^h 0^m	37° 59′	6·8, 10

C. A, white.

O.Σ.	313·95	2n.	1″·58	1847·20
De.	318·63	3n.	·64	67·61

220   O.Σ. 133.

R. A.	Dec.	M.
6^h 1^m	21° 19′	6·9, 10·1

O.Σ.	35·0	1n.	2·97	1844·90
	36·7	,,	3·06	6·22
	28·9	,,	...	52·26
	31·1	,,	3·20	70·25
De.	30·5	3n.	2·99	67·95

221   LACAILLE 2145.

R. A.	Dec.	M.
6^h 1·7^m	− 48° 27′	8, 8

The angle has increased, and the distance diminished.

Dp.	329·0		3·0	1826·00
H₂.	342·5		·86	35·02
	343·5		...	6·88
Ja.	348·5		3·22	46·94
	353·1		2·49	51·09
	350·7		·82	2·73
	351·5		·57	4·00
	354·1	2n.	·30	6·48
	355·1	,,	·19	7·54
	354·7	1n.	·18	8·17

222   Σ. 853.

R. A.	Dec.	M.
6^h 2·5^m	11° 41′	7·8, 8·3

Rectilinear motion.

Σ.	339·7	2n.	24·09	1829·19
	340·8	1n.	·01	33·19
Mä.	343·3		25·90	47·12
	345·0		·83	54·17
	346·2		26·01	8·11
Eng.	347·8		27·13	63·84
De.	346·9		26·15	4·51

223   Σ. 859.

R. A.	Dec.	M.
6^h 3·2^m	5° 40′	8, 8·5

Σ.	249·5	1n.	31·18	1828·20
	248·5	,,	·66	31·20
Mä.	·6	,,	32·05	45·19
Eng.	·4		34·01	63·17

224 Σ. 861.

R. A.	Dec.	M.
6ʰ 4ᵐ	30° 46′	A 7·8, B 8·2, C 8·2

Probably no change in B C. In A and $\frac{B + C}{2}$ the distance has diminished.

B C.

Σ.	318·2	4n.	1″·58	1830·95
Mä.	322·4	2n.	·59	44·28
O.Σ.	324·5	1n.	·92	1·23
	322·6	,,	·74	2·21
	·4	,,	·68	7·20
	321·8	2n.	·66	70·26

A and $\frac{B + C}{2}$.

Σ.	14·6	3n.	67·14	1831·18
O.Σ.	15·4	2n.	66·97	44·22
	·8	,,	·22	70·26

225 Σ. 878.

R. A.	Dec.	M.
6ʰ 10·2ᵐ	62° 28′	7·5, 11

C. A, yellow.
Rectilinear motion.

Σ.	311·7	2n.	16·19	1831·30
Mä.	317·2	3n.	17·07	45·30
De.	321·8		19·16	65·35

226 Σ. 881.

R. A.	Dec.	M.
6ʰ 11ᵐ	59° 26′	6·4, 7·6

C. white.

Σ.	88·9	4n.	0·81	1830·28
Mä.	89·9	3n.	·85	42·26
O.Σ.	95·6	14n.	·87	7·52

227 Σ. 3116.

R. A.	Dec.	M.
6ʰ 15·9ᵐ	−11° 42′	6·2, 10

Certain change in angle and distance.

Σ.	19·2	5n.	4·48	1831·16
De.	24·0		3·86	64·73

228 O.Σ. 140.

R. A.	Dec.	M.
6ʰ 20ᵐ	15° 36′	7, 9·5

C. A, blue.

O.Σ.	123·4	3n.	2·79	1847·22
De.	119·6	4n.	3·04	67·52

229 Σ. 919.

R. A.	Dec.	M.
6ʰ 23ᵐ	−6° 57′	A 5, B 5·5, C 6

C. white.

A B.

Σ.	130·0	3n.	7·25	1831·23
Mä.	131·3	2n.	·31	42·21
	·8	3n.	·18	3·12
	132·0	2n.	·44	4·41
	131·3	7n.	·51	5·16

A C.

Σ.	122·9	3n.	9·49	1831·23
Mä.	124·6	2n.	·67	42·21
	123·8	3n.	·97	3·12
	125·3	2n.	·43	4·41
	·0	7n.	·85	5·16

B C.

Σ.	101·7	3n.	2·46	1831·23
Mä.	103·7	2n.	·56	42·21
	·5	3n.	·73	3·12
	105·2	2n.	·47	4·41
	102·4	7n.	·49	5·16

230 O.Σ. 142.

R. A.	Dec.	M.
6ʰ 23ᵐ	7° 11′	7, 10·5

Σ.	352·2	2n.	8·56	1848·71
De.	353·5	3n.	·21	67·74

231 O.Σ. 143.

R. A.	Dec.	M.
6ʰ 24ᵐ	17° 1′	6·8, 9·9

C. A, yellow.

O.Σ.	105·7	1n.	7·30	1844·90
	104·3	,,	·29	7·23
	102·5	,,	·74	9·19
	105·2	,,	·86	68·21
De.	103·1	3n.	·88	7·29

232 Σ. 918.

AURIGÆ 229 (B).

R. A.	Dec.	M.
6ʰ 24·3ᵐ	2° 33′	6·7, 7·7

C. white.

Dunér gives

1850·80. Δ = 4″·48.

$P = 324°·3 + 0°·11 (t − 1850·0).$

Σ.	318.8	4n.	4″.18	1821.03
	319.9	2n.	5.09	2.27
	322.4	3n.	4.45	9.26
So.	319.4	,,	5.22	5.16
H₂.	324.7	2n.	4.45	9.99
Mä.	325.1	,,	.70	43.26
	324.6		.61	4.88
	323.3	,,	.54	5.29
	325.4	,,	.56	6.08
	323.9	3n.	.42	52.28
	.8	,,	.20	5.27
Da.	.5	2n.	.60	2.39
De.	325.6	,,	.56	6.15
Se.	.5	In.	.71	9.34
Mo.	324.5	2n.	.50	.93
Eng.	325.3	4n.	.84	65.46
Du.	327.1	7n.	.31	71.59
Gl.	328.5	In.	.5	4.11

233   O.Σ. 148.

R.A.	Dec.	M.
6h 27m	37° 9′	7.1, 10.8

C. A, golden.

O.Σ.	77.15	4n.	2.54	1849.24
De.	72.87	3n.	.63	67.95

234   Σ. 932.

R.A.	Dec.	M.
6h 27.5m	14° 50′	8.2, 8.3

C. Σ. and Se., white.

Dawes (*Mem. R. A. S.*, vol. xxxv., p. 329) thinks it is probably a binary. The measures from 1859 to 1863 confirm his opinion. Se. (p. 64) agrees with Dawes. O.Σ. Change very small, if any (1877).

Σ.	342.4	In.	2.52	1828.24
	341.8	,,	.44	30.22
	340.9	,,	.32	3.14
H₂.	346.1	...		0.18
	337.8		2 ±	2.20
Mä.	.3	2n.	.62	44.13
	.4	In.	.71	5.21
	336.8		.56	51.15
	335.9		.71	2.21
Da.	334.5		.43	48.19
	332.1		.56	59.15
	.3		.26	63.41
Se.	.4		.04	57.16
	333.2		.32	65.27
De.	.6	3n.	.28	3.18
	332.1	In.	.20	4.10
O.Σ.	334.5	,,	.36	8.21
W. & S.	331.0	5	.29	72.13
	.8	3	.14	3.93
	.0	8	.30	4.13
Gl.	.0	4	.25	3.94
	.6	4	.20	.94

Gl.	331.6	6	2″.26	1874.13
W.0.	332.6	In.	.11	6.09
	331.5	,,	.28	.05
	334.3	,,	.30	.06
Pl.	330.8	2n.	.07	7.15
Dob.	331.5	,,	.11	8.09

235   Σ. 935.

R.A.	Dec.	M.
6h 29m	52° 24′	8.2, 9

C. white.

Σ.	322.2	3n.	3.41	1829.58
Mä.	323.4	2n.	.55	44.24

236   O.Σ. 149.

R.A.	Dec.	M.
6h 29m	27° 23′	6.5, 9

A difficult object. Rapid angular motion.

O.Σ.	350.73	3n.	0.53	1848.23
De.	316.57	,,	...	68.33

237   Σ. 941.

R.A.	Dec.	M.
6h 30.2m	41° 41′	7, 8

C. A, bluish white; B, purplish white.

H₁.	76.0			1783.21
So.	85.1		1.66	1824.58
H₂.	77.5		...	9.88
Σ.	.6	4n.	1.95	30.29
Mä.	79.7	In.	.67	44.29
O.Σ.	82.3	5n.	2.12	9.66
Se.	80.7	2n.	1.93	57.25
	.9	In.	2.07	65.27
Gl.	.2	,,	.2	76.09

238   Σ. 943.

R.A.	Dec.	M.
6h 30.3m	23° 19′	8.5, 9

C. white.

Σ.	165.9	2n.	15.46	1829.74
Mä.	152.7	,,	16.40	44.27
	153.8	In.	...	5.22
De.	148.5		18.00	64.67

239   Σ. 945.

R.A.	Dec.	M.
6h 32m	41° 5′	7.1, 8

C. white.

The direction in 1841·23 appears to be 10° in error. It is probable that Σ.'s measures are also similarly erroneous.— (O.Σ.)

Dunér's formulæ are

$$1849\cdot45. \quad \Delta = 0''\cdot96.$$
$$P = 256°\cdot3 + 0°\cdot376 \ (t - 1856\cdot0).$$

Σ.	249·0	6n.	1·05	1830·77
Mä.	251·2		·0	5·38
	256·0	3n.	0·88	44·62
	254·2	1n.	·86	5·29
	260·8	2n.	1·09	51·75
	257·4	,,	0·95	5·76
O.Σ.	258·6	6n.	1·11	49·59
De.	256·7	2n.	0·85	56·88
	257·4	1n.	·82	65·38
Se.	258·9	2n.	·85	57·27
Du.	265·2	3n.	·84	72·87

240   Σ. 950.

R. A.   Dec.   M.
6ʰ 34·4ᵐ   10° 0′   6, 8·8, 11·2

C. A, green ; B, blue.

A B.

Σ.	208·6	6n.	2·77	1832·52
	·3	1n.	·81	6·15
Sm.	206·2		·5	5·13
Da.	209·3		3·07	42·19
Mä.	212·3	1n.	·13	3·14
	210·2	,,	2·63	4·22
	209·3	,,	·86	5·22
	212·6		·98	51·13
	·3		3·21	2·18
O.Σ.	211·6	1n.	·06	66·21
De.	·0		·02	8·74
Ta.	203·1		2·68	·99
	205·1		·62	72·14
W. & S.	210·8		·8	3·52
Gl.	·9		3·0	·53

A C.

Σ.	12·9	3n.	16·58	31·53
Sm.	15·0		15·0	5·13
Mä.	12·3	1n.	·89	43·14
O.Σ.	13·2	,,	16·67	66·21
De.	·9		·21	8·74

241   Σ. 3117.

R. A.   Dec.   M.
6ʰ 34ᵐ   9° 49′   8·9, 9·4

Σ.	93·7	4n.	0·65	1832·70
Mä.	88·3	1n.	·70	45·96
O.Σ.	87·5	5n.	1·01	63·22

242   Σ. 955.

R. A.   Dec.   M.
6ʰ 35·4ᵐ   −7° 52′   A 8·7, B 9, C 8·5

C. white.

A B.

Σ.	272·6	4n.	0·88	1830·65
	266·5	1n.	·89	6·19
Se.	276·3	,,	1·09	57·12
De.	267·4		·0	69·50

A C.

188±		11·5±	fixed.

243   Σ. 948.

12 LYNCIS.

R. A.   Dec.   M.
6ʰ 35·6ᵐ   59° 34′   5·2, 6·1, 7·4

C. Σ., A B, yellowish white ; C, bluish.

"This curious object, of which A and C are Piazzi's 185 and 184 of Hora VI., was discovered to be triple in 1780, and registered 6 H₁ I. and 22 H₁ III." (Smyth.)

H₁ writes: "Oct. 3, 1780. A curious treble star. Two nearest pretty unequal. L. w ; S. w, inclining to rose-colour. With 227, about ½ diameter ; with 460, full ¾ diameter of s. Position 88° 37′ s.p. The 1st and 3rd considerably unequal ; 2nd and 3rd pretty unequal. The 3rd pale red. Distance from 1st, 9″ 23‴ ; too difficult to be extremely exact. Position with regard to the 1st, 32° 33′ n.p." (*Phil. Trans.*, vol. lxxii., p. 215.) The 1st and 3rd are A C ; the close pair A B.

H₂ and So. (*Phil. Trans.*, 1824. pt. 2, p. 95.) "Triple ; A of 7th mag., B of 7½. C of 9th mag. The distant star C is decidedly blue."

"The position of the nearer star has sustained a remarkable change, while that of the more distant has scarcely altered. This star, therefore, deserves particular attention."

He then remarks that if the observed angular motion should continue uniform, "the lapse of 57 years will bring the three stars into one straight line, and in 646 years a complete revolution will have been performed." This was written in 1823.

H₂ (*Phil. Trans.*, pt. i., 1826, p. 318,) writes : "There is a considerable change in the position of the close star since the year 1823. At that time the angle was 68° 39′ s.f. Hence it appears that the small star has continued its motion in the direction there assigned to it; and, if we may confide sufficiently in both data, with an accelerated velocity, for the computed motion corresponding to an interval of 2·0 years, would be − 1°·148, whereas the observations make it − 4°·18 or − 4°·3. Meanwhile, the direction of the motion is as predicted, and we may therefore regard the reality of this star's rotation as fully confirmed."

Sm. (*Cycle*, p. 156) remarks the fixity of C, and says that a rough geometrical cast of the close pair gives "an annus magnus of nearly seven of our centuries."

Dawes (*Mem. R. A. S.*, vol. xxxv., p. 330) says, "evidently binary."

A B.

	°		"	
H₁.	181·3		...	1780·76
H₂ & So.	158·6	In.	2·59	1823·28
So.	154·3	4n.	·52	5·25
H₂.	157·2	3n.	1·76	30·24
	153·0	2n.	·67	1·19
Da.	·8	3n.	...	1·62
	·3	2n.	1·64	3·13
	149·5	In.	·75	6·97
	148·4	2n.	·72	41·20
	143·3	In.	·68	8·22
Σ.	153·7	5n.	·53	31·10
Sm.	154·3		·6	2·96
	149·5		·6	9·27
	143·7		·5	52·96
G.O.	151·9	12	·42	40·28
O.Σ.	152·7	3n.	·76	0·31
	153·6	In.	·58	1·20
	148·4	,,	·63	5·52
	149·2	,,	·66	6·30
	146·3	2n.	·57	7·34
	·2	3n.	·62	8·32
	·3	,,	·57	9·31
	147·8	2n.	·56	50·29
	146·7	,,	·72	1·28
	143·9	,,	·52	2·33
	·4	In.	·59	3·32
	142·3	,,	·52	5·32
	143·9	,,	·74	9·35
	·6	,,	·72	60·30
	138·8	,,	·84	7·31
	137·2	2n.	·75	8·31
	136·9	,,	·66	9·32
	132·6		·84	70·35
	135·2		·77	2·30
Ch.	148·6	3n.	·51	42·25
Mä.	·5	,,	·47	2·26
	149·4	2n.	·31	3·29
	147·5	4n.	·59	4·40
	145·4	2n.	·90	5·12
	146·7	10n.	·60	6·18
	·0	2n.	·54	7·25
	144·8	,,	·56	8·35
	142·6	3n.	·65	51·12
Ka.	147·0		·63	43·10
	141·5		·60	66·31
Mo,	142·0	30	·87	54·21
	140·3	12	·55	8·25

	°		"	
De.	141·4	In.	...	1854·91
	140·7	3n.	...	5·18
	142·5	2n.	1·8	·90
	143·0	,,	·55	6·44
	140·0	In.	·70	62·74
	138·1	3n.	·72	3·15
Se.	142·3		·68	57·20
Ta.	140·1	In.	2·04	66·09
	134·3	,,	1·34	74·18
M.	136·3	,,	·53	68·31
W. & S.	131·4	4	·84	72·08
	135·7	4	·27	3·19
	134·6	5	·56	·24
	135·8	2	...	·25
	133·3	3	...	·25
	·4	3	1·7	·29
Gl.	134·0	5	·41	4·13
Dob.	130·1	7n.	...	6·10

A C.

	°		"	
H₁.	302·5	In.	[9·38]	1782·34
Σ.	304·2	5n.	8·67	1831·10
O.Σ.	305·1	3n.	·83	40·31
	303·1	In.	·71	5·32
	305·8	2n.	·46	7·34
	·7	3n.	·55	8·32
	304·9	,,	·63	9·31
	305·3	2n.	·68	50·29
	·5	,,	·62	1·28
	304·8	,,	·59	2·33
	·8	In.	·75	5·32
	305·5	,,	·70	9·35
	306·4	,,	·76	60·30
	307·1	,,	·90	7·31
	306·1	2n.	·55	8·31
	305·9	,,	·61	9·33
	304·4	In.	·73	70·35
	306·0	,,	·50	2·30
G.O.	128·2	27	9·42	40·42
Ch.	304·0	3n.	8·74	2·25
De.	306·8	In.	·34	54·91
	·6	3n.	·38	5·18
	·6	2n.	·55	·90
	305·2	,,	·80	56·44
	·2	In.	·80	62·74
	·9	3n.	·63	3·15
M.	303·9	In.	·97	2·31
	305·2	,,	9·11	70·25
Ta.	307·0	,,	...	66·09
W. & S.	306·3	8	8·58	70·08
	307·9	4	7·2	3·19
	305·8	5	8·76	·24
	307·7	2	...	·25
	306·5	3	...	·25
	307·2	3	8·6	·29
Gl.	306·9	3	·5	4·10
	·4	3	·7	·13
Dob.	305·4	7n.	...	6·10

244 OΣ. 154.

R. A.	Dec.	M.
6h 36m	40° 46′	6·7, 8·4

C. A, golden; B, purple.

After reducing the angles to the equinox of 1850, and deducing the rectangular coordinates, O. Σ. finds the following formulæ for rectilinear motion :—

$$\Delta A = +20''\cdot046 \pm 0''\cdot015 + (0''\cdot0324 \pm 0''\cdot0010)\,(t-1850)\,;$$

$$\Delta D = -21''\cdot704 \pm 0'\cdot015 + (0'\cdot1421 \pm 0'\cdot0010)\,(t-1850)\,;$$

and these when compared with the observations are satisfactory.

O.Σ.			"	
	136°65	2n.	30'41	1846·76
	·35	,,	·28	8·76
	133·50	In.	29·28	61·26
	131·45	2n.	28·77	9·28
De.	·73	3n.	·77	7·91

245   O.Σ. 155.

R. A.	Dec.	M.
6h 38m	24° 48'	7, 9·9

C. golden.

O.Σ.				
	262·2	In.	15·20	1847·23
	·3	,,	14·82	8·21
	·1	,,	·86	9·24
	261·7	,,	·76	73·24
De.	260·4	3n.	15·36	67·86

246   α CANIS MAJORIS.

(SIRIUS.)

R. A.	Dec.	M.
6h 39·7m	− 16° 32'	1, 10

C. A, brilliant white ; B, deep yellow.

This magnificent star, the brightest in the heavens, has for thousands of years attracted the attention of mankind. Of all the stellar host Sirius stood first in the influences for good and for evil which these bodies were supposed to exercise over the earth and its inhabitants. A lively and interesting account of these and other such matters will be found in Smyth's *Celestial Cycle*.

Such being the brightness of this star, it is not surprising that it suggested to astronomers many speculations respecting the magnitude, distance, and relative brightness of the stars. Long before the days of accurate telescopic measures, attempts were made to estimate the apparent diameter of Sirius. Maginus made it 10', Kepler 4', Tycho 2',* Ricciolus 18". Passing by the curious results obtained from such estimates as these, and also from other erroneous assumptions,

we reach the times of Hevelius, who made the diameter of Sirius to be 6" 21''', of J. Cassini in 1717, who regarded 5" as the most correct value, and of Michell near the end of the 18th century, who considered that 0"·02 was too large for Sirius. Naturally the subject had a special attraction for our great observer Sir Wm. Herschel, and he did not fail to use the vast optical powers his genius had created in an attempt to solve this great question. But his success was not complete : in fact, the causes which determine the size of the telescopic disc of a star were far from being understood in Herschel's day. He found that α Lyræ had a diameter of 0"·3553, a value which, as he himself suspected, probably differs widely from the truth.*

The dazzling splendour of Sirius, too, early led speculative astronomers to attempt estimates of its distance, on the ground that the brightest star is most probably the nearest to the earth. Gregory, Huyghens, Chésaux, Lambert, Michell, Olbers, and others made attempts in this direction, the general result being that the parallax of Sirius was less than 0"·5. Wollaston by means of photometric methods deduced a parallax of 1"·8. Hooke was, however, the first who employed the telescope in observations for the purpose of detecting the annual parallax of the fixed stars. Then followed Bradley, Herschel, Piazzi (who found 4" as the value of the parallax of Sirius), Brinkley, Pond, Struve, Bessel, etc., etc.

Again, when exact meridional observations were made possible by the rapid progress of practical astronomy, the proper motion of the stars demanded the careful consideration of astronomers. Halley was the first to note the fact of stellar proper motion, and was led to it by a comparison of the places of Sirius and other stars in ancient and modern catalogues. J. Cassini, Bradley, Mayer, Herschel, Maskelyne, Bessel, Argelander, O.Σ., Henderson, Maclear, Main, Peters, and others have contributed to our knowledge of this subject. The following are some of the values of the proper motion assigned to Sirius :—

Piazzi	−0"·51 in R.A. and	−1"·44 in Dec.
Bessel	−0'·48 ,,	−1·23 ,,
Argelander	−0'·53 ,,	−1·23 ,,

A careful study of the path followed by Sirius led to the discovery of the fact that it was far from being a straight line ; that the apparent path was, in fact, an irregular sinuous line. Bessel found that the irregularity of the proper motion in R. A. was

---

* Tycho estimated the apparent diameters of stars of the second magnitude at 1' 30" ; those of the third at 1$\frac{1}{12}$' ; those of the fourth at $\frac{3}{4}$'.

* Chacornac in 1864, operating on the disc of Sirius by means of a prismatic telescope, found no perceptible diameter whatever.

very sensible between 1755 and 1844. The earliest suspicion of want of constancy was obtained in 1834. Recent observations have confirmed this, and the periodicity of the changes both in R. A. and N. P. D. has been established.—See *Monthly Notices*, vol. vi., p. 156, and vol. xx., p. 20. To account for this, Bessel in 1844 suggested the existence of an invisible perturbing body belonging to the system of Sirius, and in 1851 Peters, adopting this hypothesis, calculated the theoretical orbit which would satisfy the observations : he found

Passage through lower apsis ... 1791·431
Mean annual motion ... ... 7°·1875
Period... ... ... 50ʸʳˢ·01
Eccentricity ... ... 0·7994.*

In Sept. 1861 Safford sent to Brünnow an investigation of the perturbations of Sirius : in this paper he announced the angle of position of the centre of gravity with respect to the invisible mass : he gave for 1862·1, 83°·8 ; yearly diminution, 1°·4. Scarcely four months after this determination was arrived at, Mr. Alvan Clark, using his 18½ in. refractor, discovered a close companion to the bright star. The question now arose as to the identity of the new companion and Bessel's invisible disturbing body. Numerous and careful measures were made. Auwers computed the orbit, and gave the following table containing the values obtained from the elements for the quantities. D = distance of Sirius from the centre of gravity, *d* = distance of the hypothetical companion, assuming its mass to be in the ratio of 1 : 2·05 to that of Sirius, and P = the position angle of Sirius in its orbit + 180°.

	D.	*d*.	P.
	″	″	°
1861·0	3·159	9·64	87·86
2·0	·255	·93	85·81
3·0	·339	10·18	83·86
4·0	·412	·41	82·01
5·0	·475	·60	80·23
6·0	·525	·75	78·50
7·0	·567	·88	76·86

The last elements by this distinguished astronomer are as follows :—

T = 1843·275.
Ω = 61° 57′·8.
a = 18 54·5.
i = 47 8·7.
e = 0·6148.
a = 2″·331.
P = 49·399 years.

From these the minimum distance (2″·31,

In 1864 Auwers recomputed this orbit, and found the following results :—

Passage through lower apsis .. 1793·890
Annual motion .. .. .. 7°·28475
Period .. .. 49·418 yrs.
Eccentricity .. .. 0·6010

angle 302°·5 in 1841·84), the maximum distance (11″·23, angle 71°·7, in 1870·13), and the following ephemeris are obtained :—

1862·0	85·4	10·10
5·0	79·9	·78
8·0	75·0	11·15
71·0	70·3	·20
4·0	65·5	10·95
6·0	62·1	·59
8·0	58·4	·05
80·0	54·2	9·33

On comparing these with the measures observed since 1862, it will be seen that they do not agree at all well.

O.Σ., in 1864, communicated a paper to the *Monthly Notices*. He says, "According to Mr. Safford's computations, the hypothesis that the small star is in no physical connection with Sirius, and has for itself no sensible proper motion, demands for the same time

An annual change of distance + 0″·89
,, ,, ,, position − 5°·8

while the hypothesis that the small star was identical with Bessel's obscure body, would imply a feeble diminution of distance, and also a diminution (but only of 1°·4) in the angle of position for the same interval." The writer remarks that he does not regard the hypothesis of accidental juxtaposition as well established ; that the fact of Sir Wm. Herschel not having seen the companion strengthens his view ; that its light is probably variable, for in 1863 it was estimated as of the eighth magnitude, and in 1864 (March 28) it was easily seen a few minutes after sunset, when other stars of the ninth magnitude could only be seen with difficulty at greater altitudes. On the whole, he is disposed to attribute much of the uncertainty attending the measures to the existence of systematic errors in the observations.

Dunér gives the following formula for obtaining the corrections required by Auwers' ephemeris to bring the computed and observed angles into harmony :—

$$d\,P = -5°·0 - 0°·48\,(t - 1869·0) + 0°·03\,(t - 1869·0)^2.$$

He observes also that some of the measures are certainly faulty, and appear to have been made with bright wires in a dark field, a practice which he condemns in double-star measures of distance ; and he recommends that the observations of the star be made either just before or soon after sunset.

STARS NEAR SIRIUS.

PIAZZI at the end of the last century observed a small star near Sirius ; he wrote, "alia 8ᵃᵉ magnit. præcedit, 3″ temporis, 3′ ad Boream."

SMYTH (*Cycle*, p. 158) records a distant star of the tenth magnitude and of a deep yellow colour, distance 150″, angle 45°.

GOLDSCHMIDT (see *Monthly Notices*, vol. xxiii., 1863) in 1863 announced his discovery of five new stars near Sirius; the telescope used had an aperture of 4 inches, and all the stars lay between 15″ and 1′ from the bright star. Dawes readily saw the star d, but failed to detect the others.

ALVAN CLARK'S comes.

MARTH, at Malta, observing with Lassell's fine reflector, remarked, 1865, Jan. 13, a star considerably nearer to Sirius than d.

SECCHI in 1865 records having seen a faint star at a distance of about 44″.

In 1872 Messrs. Ellery, Le Sueur, and MacGeorge, observing with the Melbourne reflector, saw eight small stars near Sirius.

CLARK's Companion.

A B.

	°		″	
A.C.	85 ±		10 ±	1862·08
Bond	84·6		10·07	·19
	82·8		...	3·27
	76·0		9·0	5·26
Ch.	85·0		10·42	2·23
Lassell	83·9	In.	...	·28
	80·3	,,	9·53	4·15
	·1	,,	·67	·21
O.Σ.	82·5	2n.	10·14	3·21
	76·5	,,	·92	4·22
	77·2	,,	·60	5·20
	·2	,,	·60	·21
	75·2	,,	·93	6·20
	72·1	,,	·98	7·22
Chacornac	84·6		...	2·2
Rutherfrd.	85·0		...	·2
	81·2		...	3·1
Mit.	78·5		10·5	·08
	79·6		·9	·15
	·2		·4	·20
Da.	84·9	In.	·00	·23
Wi.	79·7	,,	...	·24
Marth	·3	3n.	10·60	·14
Se.	88·4	In.	7·62	·15
	75·5	9n.	9·59	5·22
	71·3	3n.	10·10	6·28
	65·9	,,	·75	71·16
Tietjen	76·8	,,	...	65·25
	73·8	In.	10·97	6·20
Bruhns	...	,,	·74	·20
	69·5	5n.	11·35	8·24
Br.	74·7	4n.	·26	9·10
Förster	77·9	2n.	10·78	5·22
	72·3	,,	...	7·24
Vo.	73·6	3n.	11·23	9·15
Eng	76·9	In.	9·0	5·26
	71·6	5n.	10·95	8·26

	°		″	
Σ.	73·7	In.	10·79	1865·24
	·3	,,	12·91	·70
Kn.	77·1	2n.	10·43	6·08
W.O.*	74·3	3n.	10·21	·23
	·3	2n.	·65	·25
	62·7	11n.	11·55	72·24
	58·0	,,	·39	4·17
	56·2	,,	·47	5·23
	52·8	5	·35	7·17
	53·4	5	10·95	·25
Tuttle	78·5	In.	·34	66·26
Du.	68·7	,,	11·17	9·20
	64·1	2n.	10·92	71·22
	59·8	,,	11·0	2·18
	60·8	4n.	10·57	3·22
	57·1	,,	·73	5·19
Pechüle	60·1	3n.	12·10	1·25
W. & S.	65·0	3	11·29	3·93
Bu.	53·2	5n.	10·71	7·93
	51·1	3n.	·06	8·03

SMYTH's Companion.

45·0	150	1835·80

GOLDSCHMIDT's Companion, d.

A D.

	°		″	
Marth	164·6	1	120 ±	1865·03
	163·9	3	...	·03
Bu.	158·9		104·24	77·87
	·5		102·99	·99
Pritchett	·4	5n.	103·1	8·21
Gl.	·0	In.	...	4·15

SECCHI's Companion.

169·8	In.	44·26	1865·06

MARTH's Companion.

A C.

	°		″	
Marth	126·6	1		1865·03
	127·0	2		·03
Hall	114·9	In.	72·09	77·16
Bu.	113·5	,,	...	·87
	115·1	,,	71·05	·99
	112·3	,,	·12	8·30
Pritchett	114·3	4n.	69·25	·21

---

247   O.Σ. 156.

R. A.	Dec.	M.
6ʰ 40·4ᵐ	18° 20′	6·5, 7.

	°		″	
O.Σ.	347·0	In.	0·38	1843·26
	339·3	,,	·33	4·26
	345·6	,,	·49	5·23
	338·2	,,	·49	7·22
	327·2	,,	·51	73·25
De.	324·2		...	67·35

By Messrs. Holden, Hall, Newcomb, Skinner, Eastman.

248   O.Σ. 157.

R. A.	Dec.	M.
6^h 41·5^m	0° 29'	7·5, 8

C. white.

O.Σ.	12·7	in.	0·76	1847·22
	2·4	,,	·66	48·25
	357·2	,,	·73	70·22
De.	354·8	3n.	...	68·14
W.O.	55·0	2n.	11·50	76·1
	51·1	,,	10·91	8·1
C.O.	53·1		11·20	7·1
Bu.	50·7	1on.	10·44	9·1

249   Σ. 963.

14 LYNCIS.

R. A.	Dec.	M.
6^h 42·5^m	59° 35'	5·9, 7·1

C. A, golden; B, purple.

Probably a slight change both in angle and distance.

H_1.	48·0		...	1830·18
Σ.	51·5	7n.	0·89	·88
Sm.	50·0		1·0	3·31
Mä.	53·2		0·86	8·41
	·3	3n.	·95	42·26
	54·6	2n.	·79	3·34
	55·6	3n.	·77	4·31
	·6	2n.	·75	5·19
O.Σ.	·2	14n.	·79	8·24
Se.	56·6	3n.	·76	57·20
De.	59·5		·70	63·44
W. & S.	62·3	in.	·63	73·24
Gl.	63·2	,,	·5	4·13
	64·1	,,	·7	5·10

250   O.Σ. 159.

15 LYNCIS.

R. A.	Dec.	M.
6^h 46·9^m	58° 35'	5·1, 6·2

C. A, yellow; B, golden; De., A, golden; B, blue.

So far the angular change has been very uniform. It may have slackened a little of late, and the distance appears to have increased since 1850. (O.Σ.)

In 1868 De. observed the partial superposition of the discs, the golden image of the larger star covering a portion of the azure blue disc of the smaller.

The common proper motion is + 0^s·004 in R. A., and +0"·18 in N. P. D.

O.Σ.	323·45	4n.	0·53	1844·04
	325·67	3n.	·46	6·32
	327·88	5n.	·43	8·72
	332·02	4n.	·45	50·79

O.Σ.	331·17	3n.	0·47	1852·66
	340·45	2n.	·45·	5·32
	341·57	3n.	·49	9·34
	344·40	2n.	·57	61·84
	348·67	3n.	·50	7·98
	354·97	,,	·58	9·67
	356·37	,,	·51	72·66
	357·03	2n.	·56	5·68
Mä.	336·6		·32	51·42
De.	354·9	in contact		66·87
		single		8·26

251   Σ. 982.

38 GEMINORUM.

R. A.	Dec.	M.
6^h 47·3^m	13° 20'	5·4, 7·7

C. Σ., A, yellowish; B, bluish.

"The colours so marked, that they cannot be entirely imputed to the illusory effect of contrast." (Smyth.)

This beautiful object was discovered by H_1.

He says (Phil. Trans., 1804, p. 384): "The position, Oct. 2, 1782, was 89° 54' s.f.; and April 6, 1802, it was 86° 6' s.p., which gives a change of 4° in 19 years and 186 days. This cannot be ascribed to parallactic motion."

H_2 and So. (Phil. Trans., 1824, part ii., p. 98): "Extremely unequal, large, white; small, bluish. The measures of this star would be attended with excessive difficulty, except in such a night as the present; it is one of rare occurrence. Moon nearly full. Small star appears a beautiful point; large one quite free from bur or flare." Again, he writes: "This star to-night admirably defined; the measures were gotten with a power of 133, with the greatest facility. With regard to the angle, a slight change may still be suspected, but the diminution of distance is not to be doubted, even should the rejected observations of March 19 [March 19, 1821, 86° 47' s.f., 6"·698] be the true ones."

Dawes (Mem. R. A. S., vol. viii., p. 70) writes: "The measures of this beautiful object point to a continued change in angle, though that in distance is not so strongly confirmed."

Smyth (Cycle, p. 165): "From a comparison of all the measures, a slight but constant diminution in the angle may be inferred." He also adds that the measures of H_1, H_2 and S., Σ., and Dawes, "suggest a retrograde slow motion of − 0°·16 per annum; and the distance appearing stationary, hints a period of upwards of 2000 years."

Dawes (Mem. R.A.S., vol. xxxv., p. 331) thinks that a slow diminution of angle is well established ; but that the diminution of distance is doubtful.

Secchi says : " The diminution in angle continues ; the distance increases."

O.Σ. H₁'s distance is probably much too great (1782, 7″·95). Retrograde motion. The distance appears to have increased since 1850.

The common proper motion is + 0″·04 in R. A., and + 0″·06 in N. P. D.

Dunér gives

$$\Delta = 6''03 + 0''01\ (t-1850\cdot0).$$
$$P = 169°\cdot5 - 0°\cdot225\ (t-1850\cdot0).$$

	°		″	
H₁.	179·9		7·80	1781·99
	176·1		...	1802·26
H₂ & So.	174·4	3n.	5·52	22·67
	·2		7 ±	32·20
Σ.	·8	5n.	5·73	29·24
Da.	172·4	1n.	·94	32·92
	171·9	,,	·79	6·17
	169·5	3n.	6·07	41·29
	·3	1n.	·09	3·15
	168·0	2n.	·00	51·45
Sm.	171·8		·0	36·10
	170·7		5·8	9·17
	171·8		6·0	43·20
	169·6		·0	8·22
Mä.	171·8	2n.	·50	1·27
	170·3	1n.	·16	2·21
	169·6	,,	·20	3·06
	171·2	2n.	·21	4·09
	169·2	,,	·38	5·21
Mo.	171·6		·16	·23
	168·4		·0	54·46
Ja.	169·8	,,	·22	46·27
	168·0	10	·00	51·10
	169·5	10	5·98	2·77
O.Σ.	166·2	1n.	·74	1·85
	168·0	,,	·78	64·30
	167·0	,,	6·17	8·21
	164·8	,,	...	9·23
	165·9	,,	6·26	·24
	164·2	,,	·08	70·22
	167·5	,,	·14	3·26
Po.	169·0	2n.	5·84	54·65
	165·3	5n.	·07	61·12
De.	168·3	7n.	6·07	54·46
	167·5	1n.	5·73	5·97
	·4	2n.	·83	6·51
	166·3	3n.	6·12	62·92
	·3	2n.	·14	3·15
Se.	169·3	2n.	·14	56·11
M.	167·8	,,	5·95	63·14
	166·1	,,	6·19	70·25
	·4	,,	·16	1·21
	163·7	,,	·00	2·14
Sch^i.	164·2	,,	·28	5·23
Ta.	165·0	,,	5·70	66·09
	·2	,,	·82	·16
	164·0	,,	6·81	70·35
	165·7	,,	·67	2·17
	·3	,,	·47	·98
	166·1	2n.	...	4·17

	°		″	
Du.	165·0	3n.	6·16	1870·12
W. & S.	·3	5	·42	2·13
	·7	6	·29	·08
	166·2	6	...	·12
	·3	2	5·7	3·16
	165·1	5	6·5	4·13
	·7	4	·31	·17
Ta.	159·4	1n.	·10	·07
Sp.	164·3		·28	5·24
Dob.	·1	8n.	...	6·09
Gl.	165·5	5	6·2	·10
W.O.	162·8	1n.	·42	·10
	164·3	,,	·34	·11
	165·3	,,	·42	·12
	162·8	,,	·32	·13
Pl.	·8	4n.	·37	·74

252   Σ. 997.

μ CANIS MAJORIS.

R. A.	Dec.	M.
6ʰ 50·6ᵐ	− 13° 53′	4·7, 8

C. A, yellow ; B, blue.

The proper motion of this star is 0ˢ·000 in R. A., and + 0″·01 in N. P. D.

	°		″	
Σ.	343·5	3n.	3·22	1831·20
Sm.	342·9		·5	4·15
	338·8		·0	50·79
Mä.	340·8	2n.	·13	44·17
Ja.	338·2		2·97	6·15
	335·9		·84	7·10
	338·1	11	·66	58·08
Flt.	·0	25	·95	2·60
	337·5		·86	6·24
Se.	338·9	3n.	·98	·47
De.	336·6		·96	7·94
	337·2		·76	64·09
M.	329·4		·90	3·16
Gl.	342·7	1n.	·28	4·13
W. & S.	343·5	,,	·4	2·14
	341·2	,,	·33	4·13
	343·9	,,	·75	5·19
C.O.	339·9	,,	3·14	77·19
Dob.	342·2	,,	2·53	8·08

253   Σ. 1001.

R. A.	Dec.	M.
6ʰ 53·4ᵐ	54° 21′	7·1, 8·7, 9

C. golden.

A B.

	°		″	
H₂.	58·2	1n.	10·04	1830·00
Σ.	63·9	5n.	8·9	1·48
Mä.	·2	1n.	·89	43·22
O.Σ.	64·6	2n.	9·20	58·29
Du.	65·3	,,	8·92	73·52

A C.

	°		ʺ	
Σ.	354.8	5n.	1.65	1831.48
Mä.	358.8	1n.	2.05	45.29
O.Σ.	0.3	2n.	1.87	58.29
Du.	359.4	,,	.66	73.45

254   O.Σ. 163.

R. A.	Dec.	M.
6h 54m	11° 58'	7.2, 8.5

O.Σ.	320.7	3n.	0.57	1848.57
De.	323.4	,,	...	67.40

255   Σ. 1009.

R. A.	Dec.	M.
6h 56.1m	52° 56'	6.7, 6.8

C. very white.

Dunér gives

$$1853.03. \quad \Delta = 3ʺ.21.$$
$$P = 157°.5 - 0°.055 \, (t - 1850.0).$$

H₁.	167.0	1n.	...	1782.86
So.	156.9	2n.	3.89	1824.59
H₂.	157.1	,,	.94	9.88
	159.9	,,	4.26	30.11
Σ.	.5	5n.	2.94	.34
Da.	160.6	3n.	3.32	1.15
Sm.	158.9	,,	.2	3.21
	159.4	,,	.0	43.30
Mä.	158.2	4n.	.02	2.27
	.5		.11	3.13
	156.1	3n.	.42	51.86
	.1	,,	.08	6.96
	154.9	1n.	.28	60.35
De.	157.8	3n.	.32	54.97
Se.	.0	2n.	.41	8.27
Mo.	156.2	,,	.13	9.21
M.	153.3		.21	62.31
Eng.	156.5	4n.	.59	5.21
Ka.	.0		.27	6.31
Du.	157.4	4n.	.18	70.70
Gl.	.0	,,	.0	4.11

256 LACAILLE 2640.

R. A.	Dec.	M.
7h 1.2m	-50°.0'	6, 7

H₂.	73.5	2.8	1835.03
	74.0	...	6.08
Ja.	78.1	2.06	38.11
	.4	.67	47.24

257   O.Σ. 165.

R. A.	Dec.	M.
7h 1.5m	16° 8'	5, 10.7

C. A golden.

Rapid change in angle and distance.

	°		ʺ	
O.Σ.	130.70	2n.	3.87	1847.22
	119.35	,,	.33	56.74
	89.70	,,	2.89	70.24

258   Σ. 1037.

R. A.	Dec.	M.
7h 5.4m	27° 26'	7.1, 7.1

The North Star is perhaps the smaller.

C. Σ. yellowish; Se. white; De. white.

Dawes (*Mem. R.A.S.*, vol. xxxv., p. 332) thinks there is evidence of slow diminution of angle, and that the distance is unchanged.

Mädler (*Die Fixst. Sys.*, p. 256,) after remarking the favourable position, brightness, etc., of this pair, and that they can be seen in bright twilight and even before sunset, proceeds to say that the observations indicate a double motion of the star, if the other be assumed to be at rest, and the existence of a third invisible star. From eight normal sets of observations he deduced a period of 16 years. As the point round which the star travels is invisible, he thinks that one of the stars may be found double, and that the year 1855 will probably be favourable for the discovery of the duplicity.

O.Σ. Retrograde motion : distance unchanged. The orbit is perhaps nearly circular.

Dunér gives

$$1855.76. \quad \Delta = 1ʺ.29.$$
$$P = 324°.7 - 0°.316 \, (t - 1850.0).$$

Σ.	337.8	2n.	1.24	1827.28
	332.6	6n.	.32	30.42
	327.4	3n.	.11	6.26
O.Σ.	148.8	,,	.21	40.27
	150.1	1n.	.41	5.22
	327.3	,,	.27	50.26
	323.1	,,	.23	67.24
	324.3	,,	.14	8.21
	140.9	,,	.11	9.24
	319.3	3n.	.32	70.24
Mä.	331.1		.33	41.80
	324.3		.37	52.36
	323.0		.29	5.51
	.0		.29	.66
	324.0		.37	9.22
	322.4		.45	60.55
Da.	326.8	1n.	.22	43.17
	324.6	,,	.32	8.17
Ka.	325.2	8n.	.35	3.20
	317.7		.07	67.21
De.	320.8	6n.	..	55.20
	.1	2n.	.1	6.15
	318.1	4n.	.22	63.20
Mo.	325.0	20	.35	0.11
M.	312.2	1n.	.52	.25
	305.1	,,	.35	1.25
W. & S.	316.9	7	.40	72.16

W. & S.	315°·5	3	...	1872·92
	317·6	6	1·31	5·19
Gl.	·6	5	·3	4·13
Du.	319·4	3n.	·35	1·92
	316·6	,,	·28	5·11
W.O.	308·8	1n.	·34	6·15
	314·9	,,	·18	·13
Pl.	312·3	3n.	·36	7·13

259   Σ. 1049.

R. A.	Dec.	M.
7h 8m	−8° 43′	8, 9·8

C. yellowish white.

Σ.	34·9	3n.	3·63	1830·53
De.	42·8		·50	67·71
Gl.	46·0	1n.	4·0	74·18

260   O.Σ. 170.

P. VII. 52.

R. A.	Dec.	M.
7h 11·2m	9° 31′	7·5, 7·5

C. yellow.

O.Σ.	133·0	2n.	0·96	1844·79
	132·0	,,	1·06	49·25
	120·6	1n.	·21	73·24
Mä.	134·1	,,	0·99	46·24
	127·8	2n.	1·08	52·72
De.	121·6	3n.	·29	67·13
Du.	·4	2n.	·05	72·70

261   Σ. 1066.

δ GEMINORUM.

R. A.	Dec.	M.
7h 13m	22° 12′	3·2, 8·2

Certain change in angle.

Dunér has

$$1854·53. \quad \Delta = 7''·15.$$
$$P = 199°·8 + 0°·155\,(t - 1850·0).$$

H1.	193·7	4n.	...	1797·53
So.	195·4	1n.	7·25	1822·14
Σ.	196·9	4n.	·14	9·72
Da.	·9		·13	31·02
	198·8		...	45·89
Sm.	·5		7·1	33·15
	196·8		·2	8·92
	199·8		·5	47·33
Mä.	·7	5n.	·46	4·42
	200·4	,,	·30	51·16
	·9	2n.	·31	2·74
	203·5		·08	4·03
	199·5		·21	5·24

Mä.	199°·4	12n.	7''·07	1856·07
	200·6		·00	7·21
	199·7		·27	8·21
	·4	3n.	·16	60·90
Ja.	200·6		·22	46·50
	201·2	18	·17	57·65
De.	203·5	9n.	7·08	4·03
	199·1	4n.	6·89	6·30
Se. .	200·0	3n.	7·16	6·11
Eng.	201·7	,,	·21	65·13
Ta.	·3	1n.	6·65	6·09
Du.	203·0	5n.	7·04	71.46
W. & S.	204·0	7	6·74	2·17
	·0	2	...	·18
	203·7	7	7·2	3·14
	204·3	4	·1	4·14
	202·9	4	·12	6·22
O.Σ.	204·7	1n.	7·14	3·26
	203·8	,,	·01	4·29
Gl.	204·0	2	6·9	·09
Sp.	202·9		·92	5·25
Dob.	205·4	1n.	...	·99
	201·8	5n.	...	6·05
	203·7	2n.	...	·22
	204·2	1n.	6·41	7·91
	·3	2n.	7·37	8·09

262   Σ. 1074.

R. A.	Dec.	M.
7h 14·3m	0° 38′	7·8, 8·2

Σ.	115·3	3n.	0·57	1831·54
O.Σ.	129·4	1n.	·61	52·25
	139·2	,,	·64	69·24
	138·6	,,	·62	70·22
	140·5	,,	·60	·24
De.	135·8	4n.	...	63·15
W. & S.	134·2	7	0·85	74·14
Gl.	·5	5	·87	·17

263   Σ. 1071.

R. A.	Dec.	M.
7h 14·4m	45° 14′	8·2, 10·2

Change in angle and distance.

Σ.	357·3	2n.	15·52	1829·73
De.	5·0	3n.	·87	67·30
O.Σ.	7·7	1n.	16·18	74·29

264   Σ. 1076.

R. A.	Dec.	M.
7h 15m	4° 17′	8·7, 8·7

C. white.

Σ.	105·8	1n.	2·74	1825·21
	106·3	,,	·83	8·17
	108·0	,,	·56	33·18
Mä.	110·6	2n.	·67	44·20
	109·2	1n.	·76	5·12

265 Σ. 1081.

R. A.	Dec.	M.
7ʰ 17ᵐ	21° 41′	7·5, 8·5
	C. white.	

Σ.	216·1	3n.	1·33	1828·93
Mä.	220·1		·34	36·76
Se.	222·9	In.	·58	56·11
De.	224·6		·40	67·83

266 O.Σ. 171.

R. A.	Dec.	M.
7ʰ 19ᵐ	31° 52′	7, 9·9
	C. yellow.	

O.Σ.	129·9	5n.	0·97	1851·25
De.	126·4	4n.	1·13	70·03

267 Σ. 1091.

R. A.	Dec.	M.
7ʰ 21ᵐ	50° 13′	8·2, 8·7

The distance has probably increased, and perhaps the angle has diminished.

Σ.	336·1	In.	28·69	1828·32
	335·6	,,	·49	30·25
O.Σ.	334·8	2n.	·74	43·31
	·1	,,	·57	9·76
	332·7	,,	·78	68·80

268 Σ. 1093.

R. A.	Dec.	M.
7ʰ 21·1ᵐ	50° 14′	8·2, 8·2
	Indirect motion.	

H₂.	94·1		1·0	1830·40
Σ.	96·4	3n.	0·57	1·94
O.Σ.	106·5	In.		40·32
	107·8	,,	0·88	2·32
	103·3	,,	·70	5·32
	105·1	,,	·67	6·33
	108·9	,,	·75	8·25
	107·9		·65	51·28
	121·8		·79	69·31
De.	110·0		·6	3·43

269 Σ. 1104.

R. A.	Dec.	M.
7ʰ 24ᵐ	−14° 44′	6·7, 8·3
	C. white.	

H₁.	...		2 ±	1795·22
Σ.	292·4	3n.	2·35	1834·88
De.	312·3		·21	64·50
W. & S.	314·0	In.	·55	74·17
Gl.	·2	,,	·5	·17

270 Σ. 1110.

CASTOR.

R. A.	Dec.	M.
7ʰ 27ᵐ	32° 1′	3, 3·5, 11*

C. H₁, both white ; Σ., both greenish ; Sm., A, bright white ; B, pale white ; C, dusky.

Of this beautiful object H₂ says, "The largest and finest of all the double stars in our hemisphere, and that whose unequivocal angular motion first impressed on my father's mind a full conviction of the reality of binary stars."

It is marked with a † in H₁'s catalogue, indicating that it had been observed by "different astronomers before Mr. Mayer."

EARLY HISTORY.

BRADLEY AND POUND'S OBSERVATIONS.

"1718. March 25.—The direction of the double star (*Castor*, or) α of *Gemini* was parallel to a line through Pollux (or β), which left κ to the westward, as also *g* tending to near the middle between *g* and *l* of *Gemini*.

"1719. March 30.—The direction of the double star α of *Gemini* was so nearly parallel to a line through κ and σ of *Gemini*, that, after many trials, we could scarce determine on which side of σ the line from κ parallel to the line of their direction tended ; if on either, it was towards β. This observation was made when the air was still, and with the 4½-inch eyeglass, which made the stars appear a good distance from each other.

"1722. October 1.—A line through the double star α of *Gemini* was parallel to another drawn through β and κ. The southernmost star is brightest." (Rigaud's *Miscellaneous Works of Bradley*, Oxford, 1836.)

These observations and the method adopted by Bradley are fully discussed by H₂ in the *Mem. R. A. S.*, vol. v.. p. 23, *et seq.*, and he shows that a correction amounting to 2° 43′ should be applied to the angles subtractively. The corrected angles then become 352°·28 in 1718·23, 355°·68 in 1719·24, 359°·88 in 1722·75.

Then comes the observation by Bradley and Maskelyne in 1759·80, giving as the angle 326°·50. (H₁, in *Phil. Trans.* 1802)

H₁ : "Feb. 28, 1781.—I saw with one eye the projection of the stars upon a wall at a distance of about six or seven feet, where they seemed to take up a space not less than four or five inches. I shall endeavour to construct a micrometer, from this

---

* Dawes observes that Se. has placed the smaller star in the n.f. quadrant five times, and that he suspects a variability of relative brilliancy.

hint, which may serve to measure such very small intervals exactly."

H$_1$ (*Phil. Trans.*, vol. lxxii., p 216). "April 8, 1778.—Double. A little unequal. Both W. The vacancy between the two stars, with a power of 146, is one diameter of S ; with 222, a little more than one diameter of S ; with 227, 1½ diameter of S ; with 460, near two diameters of L ; with 754, two diameters of L ; with 932, full two diameters of L ; with 1536, very fine and distinct, three diameters of L ; with 3168, the interval extremely large, and still pretty distinct. Distance by the micrometer 5″·156. Position 32° 47′ n.p. These are all a mean of the last two years' observations, except the first with 146."

In the *Phil. Trans.* for 1803, p. 339, H$_1$ announces his famous discovery of binary systems, and Castor is the one he first subjects to examination. He says, " I shall therefore now proceed to give an account of a series of observations on double stars, comprehending a period of about 25 years, which, if I am not mistaken, will go to prove that many of them are not merely double in appearance, but must be allowed to be real binary combinations of two stars, intimately held together by the bond of mutual attraction."

THE ORBIT.—As early as 1803 H$_1$ gave his speculations on this subject to the world. His results were, of course, merely intended as rough approximations. He found that between the years 1778 and 1803 the distance had not changed, but that the angle had diminished from 32° 47′ n.p. to 10° 53′ n.p. At great length he shows that orbital motion alone could account for this change. Taking the annual angular motion as 56′·18, he computes the position for the epochs of the observations, and an extract showing the results is here given :—

Times of observations.	Observed angles.	Calculated angles.
Nov. 5, 1779	32° 47′	32° 47′
Mar. 26, 1800	18 8	13 41
Jan. 10, 1802	10 53	12 1
Mar. 27, 1803	10 53	10 53

Using an observation of position by Dr. Bradley in 1759, a mean motion of 1° 3′·1 was obtained, and this was found to give a still closer agreement between the observed and computed positions. From the arc described in 43 years and 142 days, viz., 45° 39′, he inferred a period of about 342 years and 2 months.

H$_2$ and So. took up the subject in 1821, and H$_2$, after a careful study of all the observations up to 1822, found that the mean angular velocity was 0°·965. He used the observations of Bradley and Maskelyne in 1759·8, taking the angle as 56°·5 n.p., and gave equal weights to all. The

results he arrived at may be thus stated :— the orbit is elliptical, and nearly at right angles to the line of sight : there has been a sensible retardation of the angular velocity since 1780. (*Phil. Trans.* 1824, part ii., p. 103.)

Returning to the subject in 1825, H$_2$ found that the observations made since 1823 confirmed his previous speculations. (*Phil. Trans.* 1826, p. 320.)

But it was not till 1832 that this distinguished astronomer fairly grappled with the orbit of this star. In that year his famous paper " On the Investigations of the Orbits of Revolving Double Stars " appeared. (*Mem. R..A. S.*, vol. v., p. 171.)

His first example was γ Virginis, and Castor was the second. The sections on the latter may be thus summarized. The positions from the observations in 1759 and 1802 are perplexing : taking them as 320° 20′ and 383° 15′ respectively, the interpolating curve becomes a straight line, and the orbit a circle, with a uniform angular velocity of − 0°·8745 per annum. He decides at last on the following :— 1718·20, 160° 52′ ; 1756·00, 144° 22′ ; 1781·09, 130° 44′ ; 1803·20, 120° 19′ : taking up Σ.'s angles in 1819, 1820, 1822, and 1825. and also those by himself, South, and Dawes, down to 1831, he submits them to the graphical process. The final results are as follows :—

APPARENT ELLIPSE.

Major semi-axis	5″·34
Position of major axis	53° 53′
Minor semi-axis	2″·72
Farthest maximum of distance	6″·67
Position thereof	29° 5′
Nearest maximum of distance	5″·03
Position thereof	270° 30′
Farthest minimum of distance	4″·66
Position thereof	313° 5′
Nearest minimum of distance	0″·66
Position thereof	147° 20′

REAL ELLIPSE.

Major semi-axis	a = 8″·0861
Excentricity	e = 0·75820
Position of perihelion	π = 169° 10
Inclination	γ = 70° 3′
Position of node	Ω = 58° 6′
Distance of perihelion from node on orbit	λ = 262° 31′
Period in years	P = 252^m·66
Mean motion	n = − 1°·4248
Perihelion passage	T = 1855·83

On comparing the angles observed, up to 1833, with the computed angles a fair agreement was found. The following is the ephemeris from 1833 to 1856. For comparison, the observed angles and distances are given.

t	θ	ρ
1833·0	257° 10'	4″·82
36·0	254 22	·65
39·0	251 21	·37
42·0	248 1	·19
45·0	244 7	3 ·85
48·0	239 3	·37
50·0	234 25	2 ·91
52·0	227 19	·18
54·0	212 36	1 ·36
56·0	164 24	0 ·68

Observed Angle.	Distance.	Date.	Observer.
256°·73	4″·89	1833·10	H.
·12	5 ·28	36·88	E. & G.
253 ·73	·20	9·35	G.
252 ·38	4 ·91	42·25	Da.
249 ·80	...	5·93	H₁.
·20	5 ·008	8·18	Da.
248 ·11	·068	51·21	,,
246 ·39	·070	2·20	,,
·21	·098	4·23	,,
245 ·44	·145	6·20	De.

In 1842 (*Ast. Nach.*, No. 452, vol. xix.) appeared the following elements by Mädler:

a 7″·008.	λ 87° 37'.
e 0 ·79725.	P 232ʸʳˢ·124.
γ 70° 58'.	T 1913·90.
☊ 23 5'.	n −93″·054.

In 1845 Mr. Hind computed a set of elements. All known observations between 1718 and 1845 were used, and the method adopted was the graphical. In 1846 Captain Jacob obtained a set, and the two are here exhibited together:—

	Hind in 1845.	Jacob in 1846.
T	1699·26	1703·30
π	8° 15'	10° 0'
☊	11 24	10 0
λ	355 41	0 0
γ	43 14	43 17
e	0·2405	0·300
n	−34″·163	−0°·5512
a	6″·300	6″·30
P	632ʸʳˢ·27	653ʸʳˢ·1

On these Mr. Hind remarks, "The period of revolution of this star appears, therefore, to be very much longer than was formerly supposed, and the eccentricity, instead of being large, is possibly not greater than 0·25."

The next attempt to deal with this hitherto intractable star was by Mädler (see *Untersuchungen über die Fixstern-Systeme*, 1847, p. 233). The observations made use of extended from 1719·84 to 1847·19. The elements are,—

T	1688·28.
n	−41″·55654.
e	0·21938.
☊	10° 45′·6 (Æq. 1845·0).
λ	16 1′·7.
γ	41 46″·7.
P	519ʸʳˢ·77.

Mädler also gives an ephemeris, of which the following is an extract. Appended are the measured angles and distances for comparison:—

	Angle.	Distance.
1845	250° 0·1'	4″·849
50	246 54	·955
55	243 49	5 ·064
60	240 55·4	·177
65	238 9·2	·291
70	235 29·6	·406

MEASURED ANGLES AND DISTANCES.

Angle.	Distance.	Date.	Observer.
249·8	...	1845·95	H₁.
248·97	5″·027	49·32	Da.
243·61	4 ·848	55·31	M.
242·77	5 ·395	60·22	Da.
241·45	·678	65·31	,,

There is very satisfactory agreement between the computed and observed angles and distances all the way down from 1719 till the year 1845 is reached, when a divergence is manifest. In 1855·82 Secchi gives 245°·13, and in 1856·20 Dembowski has 245°·44. In 1870 the following measures may be given for comparison:—

De.	239°·34	5″·488
Ta.	240 ·51	5 ·650
Gl.	239 ·7	5 ·57

The next orbit was that deduced by Thiele, and given in the *Ast. Nach.*, vol. lii., No. 1227. His elements are,—

T	1750·326.
P	996ʸʳˢ·85.
n	−21′·6685.
e	0·34382.
a	7″·5375.
λ	294° 0′·8.
γ	42 5′·4.
☊	31 58″·0 (for 1850).

Thiele also gives, for comparison, the results for every two years from 1848 to 1880, as deduced from his own and other elements. Subjoined is an extract from this table:—

	Thiele's Orbit.	Mädler's.	Hind's.
1850	248·47 5·215	246·81 5·106	246·19 5·193
1856	245·45 ·388	243·17 ·237	242·74 ·347
1860	243·54 ·503	240·84 ·336	240·55 ·451
1866	240·83 ·674	237·50 ·477	237·41 ·607
1870	239·12 ·786	235·36 ·567	235·41 ·711
1876	236·67 ·953	232·26 ·700	232·55 ·867
1880	235·10 6·063	230·27 ·789	230·73 ·970

OBSERVED ANGLES AND DISTANCES.

Angle.	Distance.	Date.	Observer.
248°·29	5″·068	1851·21	Da.
245 ·58	·145	56·20	De.
242 ·89	·395	60·22	Da.
241 ·15	·384	66·02	De.
239 ·76	·57	70·32	Gl.
236 ·22	·5	75·66	,,
234 ·2	·58	76·70	Pl.
·9	...	77·31	Dob.
235	·35	78·11	,,

For further comparison, the *observed* angles, corrected for precession to 1880, from 1850 to 1875, are given in the last columns, together with the observed distances. The agreement between Thiele's angles and those observed is remarkably good, but the distances are not so accordant, and the difference is becoming greater.

A careful comparison of Thiele's elements with the observed angles and distances up to 1875 has led Du. to regard 7″·119 as the most probable value of *a*.

In 1877 Wilson obtained the following elements by the graphical method :—

$$a = 6''·67$$
$$e = 0 ·38$$
$$\Omega = 28° \ 15'$$
$$\gamma = 32 \ \ 15$$
$$\lambda = 305 \ \ 10$$
$$p = 982·9 \text{ years}$$
$$T = 1742·1.$$

Lastly, Doberck (see *Ast. Nach.*, vol. cxi., No. 8) gives these as provisional elements :

$$\Omega = 27° \ 46'$$
$$\lambda = 297 \ \ 13$$
$$\gamma = 44 \ \ 33$$
$$e = 0·3292$$
$$P = 1001^{yrs}·21$$
$$T = 1749·75$$
$$a = 7''·43 \ (\text{Æq. } 1850).$$

A B. The measures of the last six years appear to indicate that about 1872 the distance reached its maximum ; if so, we may expect it to diminish sensibly ere long.

A C. After reducing the angles for A C to those for $\frac{A + B}{2}$ and C, a change in distance to the amount of about 0″·2 and 1° in angle appears between 1835 and 1869. The measure by Σ. in 1829 was probably over-weighted by him when discussing the changes in this pair (see *P. M.*, p. ccxii) ; and it is probable too that an accumulation of accidental errors of considerable magnitude exists in the Pulkowa measures from 1851 to 1853. At present the real character of the changes cannot be ascertained. (O. Σ.)

The proper motion of Castor is − 0ˢ·013 in R. A., and + 0″·08 in N. P. D.

A B.

H₁.	302·78	1	5·31	1779·84
	293·05	1	·78	83·63
	292·95	1	4·69	91·14
	297·27	1	5·00	2·15
	283·88	1	·43	5·95
	288·13	1	·53*	1800·23
	280·51	1		·30
	277·97	1		1·99
	280·88	1		2·02
	·47	1		·06

* The dates of the distances are 1779·84, 1779·92, 1780 06, 1780·26, 1781·14, 1781·16.

H₁.	291·62	1		1802·15
	283·00	1		·16
	275·49	2		3·11
	277·88	1		·11
	284·59	3		·22
	280·88	2		·23
H₂ & So.	0·0		...	16·97
	267·12	24	...	21·21
	264·98	26	...	3·11
	·65	37	...	2·10
	263·3	42	4·76	5·23
Σ.	262·54	5n.	·40	6·22
	·32	4n.	·41	7·28
	261·10	5n.	·35	8·89
	259·58	,,	·46	31·31
	257·72	4n.	·52	2·86
	255·48	5n.	·73	5·33
	254·33	3n.	·78	8·34
H₂.	261·86		·64	28·69
	260·96		·52	9·88
	259·01		...	30·52
	...		4·68	·60
	259·61		...	1·11
	...		5·16	·19
	258·15		4·57	·22
Be.	259·7	4n.	·7	0·41
	260·0	1n.	·73	·76
	259·6	2n.	·54	1·30
	·5	1n.	·72	·40
Da.	258·42	14n.	...	2·12
	...	10n.	4·70	·17
	...	6n.	·78	3·14
	258·1	12n.	...	·15
	257·23	3n.	...	4·08
	...	2n.	4·85	·13
	255·73	3n.	·83	6·36
	254·95	4n.	·87	8·21
	·13	5n.	·93	40·20
	252·38	6n.	·91	2·01
	251·72	4n.	·87	3·16
	249·85	5n.	5·01	7·25
	·24	2n.	·14	8·13
	·16	7n.	4·97	·24
	248·97	4n.	5·02	9·32
	·11	10n.	·06	51·21
	246·39	1n.	·07	2·20
	245·87	3n.	·15	3·13
	246·2	7n.	·09	4·23
	244·25	4n.	·38	7·34
	242·77	3n.	·39	60·22
	·08	,,	·45	3·51
	·11	,,	·49	4·30
	241·49	2n.	·76	5·30
	·45	,,	·67	·31
Sm.	256·3		·7	34·24
	255·2		·8	6·31
	254·9		·8	8·33
	252·3		·9	43·13
Encke & Galle.	256·1	1n.	·28	36·88
Galle.	253·7	,,	·20	9·35
Ka.	·9	7n.	4·71	40·06
	255·0	6n.	·86	1·35
	253·8	8n.	·69	2·37

O.Σ.	°		″	
O.Σ.	254·9	7n.	5·07	1840·30
	253·8	4	4·99	2·78
	250·1	5	5·02	4·28
	249·4	4	·07	5·79
	·7	6	·06	7·93
	248·9	4	·18	9·28
	·2	3	·00	50·27
	247·0	4	·24	1·28
	246·2	3	·27	2·30
	245·3	3	·44	3·29
	246·0	3	·33	4·94
	244·7	3	·41	7·27
	242·9	3	·44	8·96
	243·6	3	·49	60·26
	241·9	4	·41	1·87
	242·4	3	·45	4·28
	241·0	4	·44	6·78
	238·9	3	·54	8·27
	·1	4	·55	9·76
	·2	3	·68	71·27
	236·3	5	·62	2·88
	238·0	3	·49	4·28
Ch.	250·5	3n.	·13	41·13
	249·0	1n.	4·92	4·27
Mä.	252·8	5n.	·88	1·11
	·1	6n.	·79	2·30
	245·8	9n.	·82	52·34
	246·3	5n.	·81	·66
	·2	9n.	·93	3·34
	244·7	18n.	·94	4·38
	243·6	3n.	·84	5·31
	·7	6n.	·87	6·35
	242·9	7n.	·88	7·36
	244·1	,,	·96	8·37
	242·7	11n.	5·08	9·36
Hi.	249·8	11n.	...	45·95
	·5	4n.	...	6·73
D.O.	248·0		...	6·40
	...		5·46	·91
	251·3		·38	7·07
Bond.	249·7		·1	8·30
	·3		·3	·26
	248·7		·2	·30
Johnson.	245·7	7	5·07	50·21
Ja.	248·3	20	...	·66
	247·9	30	5·11	1·74
	·3	15	·08	3·04
	·3	11	·24	4·03
Mi.	·6	32	·04	1·88
De.	244·6	5n.	·72	3·31
	245·4	,,	·60	4·21
	246·2	6n.	·36	5·14
	245·5	7n.	·16	6·19
	241·6	6n.	·37	62·74
	·5	8n.	·40	3·24
	·0	14n.	·38	6·02
	239·7	3n.	·45	70·30
	238·7	2n.	·54	1·25
	237·8	,,	·40	2·27
	·2	3n.	·67	3·29
Se.	63·8	1n.	·19	55·14
	·8	,,	·26	·14
	66·2	,,	·50	·28
	65·6	,,	·53	6·24

	°		″	
Se.	245·2	In.	5·39	1856·27
	244·5	,,	·36	·31
	246·4	,,	·31	·34
Schmidt.	...		·85	5
Mo.	244·4	12	·21	8
	243·9	20	·16	9
	·6	24	·38	9
Po.	·2	20	·15	9
	241·7	35	·29	61
M.	240·8	In.	·51	1·13
	236·4	,,	·31	·18
	248·1	,,	·53	·29
	241·8	,,	·06	·92
	242·2	,,	·46	2·31
	241·7	,,	·18	·94
	·5	,,	·28	3·13
	239·0	,,	·39	·14
	·2	,,	·58	64·23
	240·7	,,	·66	70·26
	·0	,,	·56	1·25
	239·5	,,	·40	·26
	238·4	,,	6·30	4·19
	237·4	2n.	5·85	5·16
Ro.	245·1	4	·27	62·80
	244·3	6	·47	·81
	243·4	6	·30	·86
	·8	6	·67	·86
	242·1	4	·66	·92
	243·0	6	·54	3·08
	241·8	4	·60	·10
	·8	4	·55	·13
	242·2	4	·70	·17
	239·9	4	·65	·25
	242·3	6	·46	·25
Mit.	241·0		...	·21
	242·7		5·6	·21
Eng.	·8	6n.	·52	4·16
Kn.	240·6	5	·48	·78
	239·7	10	·41	·90
	·6	7	·43	5·05
	238·8	7	·60	·87
	240·4	7	·72	6·03
	237·0	4	·70	71·99
	236·5	4	·75	2·02
Ta.	243·2	6	...	66·09
	...		·57	·14
	240·6	5	·32	·16
	...		·22	·20
	...		·44	·35
	242·9	5	·29	7·08
	·9	5	...	·27
	240·5	8	5·65	70·35
	·9	6	·16	1·33
	·7	6	·51	2·17
	235·9	2n.	...	4·16
Br.	238·9	,,	·71	69·16
Du.	239·2	8n.	·38	·38
	238·7	9n.	·49	70·79
	·4	6n.	·55	3·67
	237·0	1on.	·60	5·33
Gl.	239·7	4	·57	0·32
	·0	5	·6	1·21
	236·9	5	·69	·98
	·4	4	·73	2·00

16

Gl.	236°.3	12	5″.62	1873.29
	.0	3	6.0	4.03
	.8	3	5.8	.04
	237.2	2	.6	.05
	235.9	4	6.1	.06
	...		5.5	.14
	...		.5	.11
	237.0	2	.6	.29
W. & S.	.2	7	.8	2.20
	238.2	5	...	.26
	.9	4	...	.27
	237.7	4	6.1	.38
	.0	7	...	.86
	.9	5	5.6	3.24
	236.9	4	.5	4.12
	237.0	6	.7	.14
	239.0	4	...	3.28
	235.9	4	...	6.22
	234.2	4	...	.35
Fer.	237.5		5.86	3.14
De.	236.0	3n.	.54	5.25
Schi.	235.2	1n.	.58	.26
	234.6	,,	.53	7.17
Dob.	236.4	,,	...	5.99
	234.7	1 1n.	5.78	6.13
	.4	2n.	.59	7.30
	235.1		.55	8.11
Do.	234.8	12	...	6.12
	.9	2	...	7.31
Pl.	.2	7n.	5.58	6.70
Sp.	.6		.53	7

A C.

So.	161.6		70.2	1823
Σ.	162.4		72.8	29
	.5	7n.	.5	35
Sm.	.0		.9	0
	.8		73.1	7
	.2		72.4	8
	.6		73.0	43
O.Σ.	.5	5n.	72.59	1.87
	.6	,,	.52	8.86
	163.0	,,	.43	52.27
	.1	.,	.81	60.07
	.5	,,	.93	9.06
Flt.	.1		.1	52
Se.	161.3		73.0	8
De.	163.5		72.8	62
Fl.	.3		73.2	77

271　　O.Σ. 175.

R. A.	Dec.	M.
7ʰ 27ᵐ	31° 12'	6, 6.6

C. yellow.

The distance has probably increased.

O.Σ.	333.83	12n.	0.46	1847.60
De.	332.13	3n.	.81	67.92

272　　Σ. 1107.

R. A.	Dec.	M.
7ʰ 29.1ᵐ	76° 1'	8, 10

Σ.	200.5	3n.	1.27	1832.64
Mä.	201.8	1n.	.78	45.25

273　　O.Σ. 176.

R. A.	Dec.	M.
7ʰ 32ᵐ	0° 47'	7.3, 9.3

C. white.

O.Σ.	207.5	1n.	1.64	1848.24
	213.7	,,	.44	50.28
	210.0	,,	.53	67.24
De.	214.3	3n.	.66	8.07

274　　PROCYON.

R. A.	Dec.
7ʰ 33ᵐ	5° 33°

C. A, yellowish to white; B, orange.

Magnitudes.—Procyon is variously estimated: it was rated of the 1st by Hevelius, of the 2nd by Tycho, and at 1¼ by Sm. Smyth estimated the companion seen by him as an 8th magnitude star: Barclay's is of the 11th, Secchi's of the 7th, Flamsteed's of the 7th; those first measured by Powell about the 8th; and the three discovered at Washington are of the 10th magnitude.

Companions.—Flamsteed in 1692, and Christian Mayer in 1777, observed a star of the 7th magnitude. The distance was about 600″ at the latter date. Powell and Flammarion have made measures of it. It is G in list of measures; it is H₁ I. 23 and Σ. 1126.

Sm. in 1833 found a star of the 8th magnitude at a distance of 145″ and angle of 85°. "In 1848, Mr. Bond, of the Cambridge U.S. Observatory, announced that the small star was 'missing.' In 1850, I saw and measured the position of the companion with ease, and estimated it as of the 9th magnitude. My measures gave this result: 1850.17, position 84°.3. During the spring of this year I have looked most carefully for this small star with my 6-foot achromatic, but I have never obtained a trace of its existence." (Fletcher, in 1853.) Smyth himself and Dawes also failed to recover the missing star in 1858; but Dawes detected a minute star 48″± distant from Procyon, and having a position angle of 285° ±. This appears to have been the small star discovered by Mr. J. Gurney Barclay in 1856, of which he communicated an account to the R. A. Society in 1863. Mr. Barclay's star was measured by Mr. Romberg on the 17th of March, 1863 : position 295°·3, distance 45″·8.

Secchi in 1856 found a 7th magnitude star in the following place : position angle 83°·6, distance 33″·16. No other measures of this object are known to us.

In 1873 O.Σ. thought he had detected a close companion to this bright star. Familiar with the star (having observed it yearly for more than twenty years), second to no living astronomer as an observer of double stars, ever on the watch in such cases for false images, the utmost confidence was felt in the reality of this discovery. Special interest too attached to this new companion ; for it might prove to be the disturbing body Auwers and others supposed to be the cause of the irregularities in the proper motion of Procyon. Hence the extreme care used in testing the reality of the phenomena by changing eyepieces, reversing the telescope, placing the image in different parts of the field, and calling in the aid of assistants. Careful measures were made on every favourable opportunity, and transmitted to Dr. Auwers, who re-examined his computations, and predicted that the star, if really the disturbing body his theory required, would in March 1874 (when the star would again be visible), have a position angle of 97°. In this case Procyon would have to be regarded as having a mass eighty times that of the sun, and the companion itself would have a mass equal to seven times that of the sun. March 1874 proved very unsuitable for delicate astronomical measurements : however, one glimpse was obtained on the 21st both O.Σ. and his assistant saw the companion, and the position angle was 95°! Several confirmatory observations were made in April ; other Russian astronomers could see the companion ; Mr Talmage saw it and measured it. But, strange to say, the American astronomers at Washington, using the magnificent 26 inch refractor, although they examined the vicinity of the bright star on many fine nights during the years 1873, 1874, and 1876, could never obtain a glimpse of the new companion. In conclusion, the distinguished Poulkova astronomer himself announced that the point of light which he had taken for a star was an optical illusion or "ghost."

The American observers (Messrs. Holden, Clark, Watson, Peters, Newcomb, Hall, and Todd) did more than this ; during their scrutiny of the vicinity of Procyon they discovered at least three close com-

panions, A, B, C, and suspected the existence of one or two more (see the Measures). One was strongly suspected at a distance of 10″, and an angle of 320° to 330°.

Like Sirius, Procyon presents remarkable irregularities in its proper motion. Auwers investigated this case in 1861 : he found that a body moving round Procyon in a circular orbit situated at right angles to the line of sight, and having a distance of 1″·2 from the centre of gravity, would explain the observed phenomena ; and he gave the following elements of this orbit :—

Epoch of least distance in R. A. 1795·568
Annual motion - - - 9°·00634
Period - - - 39^{yrs}·972
Radius of orbit - - - 1″·0525.

This eminent astronomer on receiving O.Σ.'s measures of the supposed new star in 1873 proceeded to re-determine the proper motion of Procyon, but did not find that his results and the observations agreed well : these last elements were—

Epoch of minimum R. A. - 1795·629
Annual motion - - - 9°·02993
Period - - - - 39^{yrs}·866
Radius of orbit - - - 0″·9805.

The proper motion of Procyon is thus given :—

Piazzi :
—0″·71 in R. A., and —0″·98 in Dec.

Bessel :
—0″·63 in R. A., and —1″·05 in Dec.

Argelander :
—0″·69 in R. A., and —1″·05 in Dec.

In conclusion, it appears that none of the small stars hitherto seen near this fine star partake in its proper motion, those discovered at Washington excepted.

PROCYON and A.

W.O.	10°		6″	1876·03

PROCYON and B.

W.O.	36		8·8	1876·03

PROCYON and C.

W.O.	50		10	1876·03

SECCHI'S Companion.

83·6	1n.	33·16	1856·16

PROCYON and D.

Da.	285		48	1858·11
Ro.	294·9	30	45·8	63·22
Lassell.	296		44·6	64·17
Fl.	311·8		40	77·17

SMYTH'S Companion.

85		145	

PROCYON and E.

Po.	83.8		326".6	1855.91
	.1	35	332.2	60.81
Fl.	80.5	In.	346.5	77.17

(E is a close double star, 195° and 0".6.)

PROCYON and F.

Po.	282.1		384.3	1855.94
	.9	35	380.4	60.83
Fl.	286.4	In.	371.3	77.17

PROCYON and Flamsteed's Companion.

Fl.	116		588	1692
C. Mayer.	106		610	1777
Po.	99.7	25	643	1860
Fl.	96.8	In.	652	1877

275   Σ. 1126.

P. VII. 170.

R. A.	Dec.	M.
7^h 32.7^m	5° 30'	7.2, 7.5

The apparent orbit is probably nearly circular.

H$_1$: "Nov. 21, 1781.—The nearest of all double stars I have yet seen: in perfect contact with 460; nor can I get a glimpse of any separation. The morning not so fine as I could wish, therefore I still doubt the reality of this appearance till more confirmed.—Double. I saw it had changed the direction of position to the horizon in about an hour's time, as it should: this looks not like a deception of the telescope.

"Nov. 28, 1781.—Not open with 460. 12^h, the air very fine; with 278, $\frac{1}{8}$ of a diameter."

H$_1$.	117.3	In.	...	1781.91
So.	127.8		...	1823.13
H$_2$.	130.7		1.40	26.18
	123.0		.41	30.04
Σ.	132.0	I In.	.46	29.43
	.3		.47	30.09
	133.9		.23	50.26
Da.	.4	3n.	.35	32.10
Sm.	132.9		.4	3.22
O.Σ.	140.3	In.	.50	42.21
	137.1	,,	.37	6.29
	139.0	,,	.37	7.25
	136.6	,,	.27	8.24
	.5	,,	.03	9.27
	138.2	,,	.13	50.26
	137.8	,,	.41	1.26
	.2	,,	.27	2.25
	135.9	,,	.49	7.27
	143.9	,,	.51	60.25
	142.4	,,	.28	1.25
O.Σ.	136.8	In.	1".23	1864.26
	139.9	,.	.32	7.27
	144.2	,,	.25	70.24
	142.6	,,	.23	3.24
	145.8	,,	.24	.24
Mä.	138.2		.50	51.19
	140.8		.64	2.23
	138.8		.53	4.19
	141.1		.46	5.24
Mo.	136.9	12	.31	6.24
	140.0	18	.27	9.91
Se.	137.3	4n.	.27	6.64
Po.	138.4		.5	61.03
Ta.	144.1	In.	0.99	6.09
	.5	,,	...	7.09
	138.7	,,	0.92	74.23
Ka.	.4		1.15	67.21
W. & S.	140.2	4	.45	74.17
	139.9	4	.59	6.22
Gl.	143.6	6	.41	4.17
Sp.	139.8		.21	5.31

276   O.Σ. 177.

R. A.	Dec.	M.
7^h 34^m	37° 44'	7.5, 8.5

Rapid change in angle.

O.Σ.	149.9	3n.	0.57	1845.60
	138.9	2n.	.64	73.79
De.	127.4	3n.	...	68.65

277   Σ. 1132.

R. A.	Dec.	M.
7^h 36.2^m	—3° 14'	8.1, 8.7

C. white.

H$_1$.	246.0		18.32	1783.03
So.	238.1		19.88	1825.03
Σ.	.5	2n.	.32	.08
	237.4	,,	.19	33.72
	.2		.41	6.19
Mä.	236.4		.10	47.23
De	.5		.52	67.46
W. & S.	235.7	In.	20.6	74.18
Gl.	.9	,,	.88	.18

278   O.Σ. 179.

κ GEMINORUM.

R. A.	Dec.	M.
7^h 37.2^m	24° 41'	4, 8.5

De. suspects that the light of the companion is variable; but it is probable that Σ.'s estimate in 1828 was influenced by the bright field which he used.

H₂.	230± °		5± ″	1826·20
Σ.	229·62	1n.	6·19	28·27
Da.	228·9	2n.	·21	36·68
Sm.	231·9		·0	8·98
	232·3		5·8	51·21
O.Σ.	231·3	1n.	6·25	43·30
	·2	,,	·38	4·26
	232·8	,,	·11	5·25
	·0	,,	·07	8·24
	233·3	,,	·36	57·28
	237·3	1n.	·30	64·24
	234·8	,,	·21	73·26
Mä..	231·1		7·24	46·24
Ta.	233·1	1n.	6·22	66·59
De.	·10	3n.	·36	·73
Du.	235·7	7n.	·39	74·06
Gl.	232±	1n.	...	6·11

279   POLLUX.

R. A.   Dec.   M.
$7^h 37.2^m$ $28° 19'$ A 2, B 11, C 12, D 10.

C. A, yellow.

A perspective group.
H₂ and So. do not appear to have seen C. Is it a variable?

A B.

H₂.	65·5		116·7	1781·90
So.	66·4		132·3	1825·10
Sm.	·9		130·0	32·31
Fl.	72·1		175·0	77·08

A C.

Fl.	90·4		205·5	77·08

A D.

H₂.	74·1		160·7	1781·90
So.	72·7		198·0	1825·10
Sm.	73·6		202·7	32·31
Σ.	·6		203·8	6·26
O.Σ.	74·4		213·5	50·71
De.	·9		222·2	65·31
Fl.	75·2		228·9	77·08

280   Σ. 1136.

R. A.   Dec.   M.
$7^h 42^m$ $65° 13'$ 7·3, 11

Rapid change in distance.

Σ.	248·53	3n.	11·61	1830·65
O.Σ.	247·5	1n.	10·81	46·22
	246·4	,,	·55	7·34
	243·8	,,	9·58	68·30
	244·9	,,	·51	9·32

281   Σ. 1142.

R. A.   Dec.   M.
$7^h 41.6^m$ $13° 43'$ 8, 10·4

C. A, yellowish.

Rectilinear motion.

Σ.	275·8 °	4n.	24·36 ″	1829·47
H₂.	272·0		40·	32·30
Mä.	267·6		22·93	47·22
De.	262·0		·83	63·41
Gl.	259·1	1n.	23·4	74·17
W. & S.	258·1	,,	22·8	·17

282   O.Σ. 182.

R. A.   Dec.   M.
$7^h 46^m$ $3° 42'$ 7, 7·5

O.Σ.	47·5	2n.	1·08	1844·28
	51·7	1n.	·09	7·25
	46·6	,,	0·99	50·28
	44·6	,,	1·22	61·25
	·0	,,	·10	73·24
De.	39·83	3n.	·23	67·00

283   Σ. 1157.

R. A.   Dec.   M.
$7^h 48.5^m$ $- 2° 28'$ Σ. 8, 8

C. white.

Secchi (p. 26) says " certain retrograde motion."

Σ.	267·2	3n.	1·59	1831·20
Se.	256·4	,,	·30	56·47
	254·5	1n.	·41	65·87
De.	·4	3n.	·2	57·91
	76·4	1n.	...	62·93
	256·7	3n.	1·29	3·16
W. & S.	257·7	4	·00	72·17
	256·5	2	...	·14
Gl.	257·2	1n.	0·88	4·13

284   O.Σ. 185.

R. A.   Dec.   M.
$7^h 51'^m$ $1° 27'$ 6·8, 7

A mutual eclipse has probably taken place since 1855.

O.Σ.	15·3	1n.	oblong	1844·26
	23·6	,,	0·33	·30
	·4	,,	·46	50·28
	198	,,	obl⁸·?	61·25
	240	,,	,,	9·24
	...	,,	single	70·24
	...	,,	,,	3·24
Se.	265	,,	...	55·28
De.	...		single	66

285   O.Σ. 186.

R. A.	Dec.	M.
7ʰ 56ᵐ	26° 37′	7·5, 8·2

C. white.

	°		″	
O.Σ.	75·2	3n.	0·83	1844·28
	74·0	1n.	·72	9·26
	70·9	,,	·73	57·27
Mä.	81·0	,,	·55	1·33
De.	72·4	3n.	·81	67·93
Du.	78·7	2n.	·65	71·07

286   Σ. 1175.

R. A.	Dec.	M.
7ʰ 56·1ᵐ	4° 30′	7·8, 9·7

C. yellowish, bluish.

	°			
Σ.	204·6	5n.	2·37	1831·24
H₂.	206·7		1·5	·80
De.	217·8		2·07	66·83

287   O.Σ. 187.

R. A.	Dec.	M.
7ʰ 56·5ᵐ	33° 24′	6·7, 7·5

Secchi's measure in 1855 probably refers to some other star.

	°			
O.Σ.	306·9	4n.	0·46	1844·02
	299·2	,,	·35	48·82
	293·7	2n.	·43	58·28
	285·6	,,	·52	71·31
Se.	250·4	1n.	·85	55·28
De.	286·2	3n.	...	68·32

288   Σ. 1179.

R. A.	Dec.	M.
7ʰ 58·1ᵐ	12° 25′	8·5, 8·5

Rectilinear motion.

	°			
Σ.	205·5	1n.	17·86	1829·24
	204·8	,,	·96	30·22
H₂.	·8		18·0	2·20
Eng.	205·0		20·71	63·13

289   Σ. 1186.

11 CANCRI.

R. A.	Dec.	M.
8ʰ 1·5ᵐ	27° 50′	7·1, 10·4

	°			
Σ.	218·8	5n.	3·18	1828·26
Sm.	213·5		·2	39·70
Da.	211·5		·27	40·12
Ta.	206·4	1n.	...	66·10
	219·1	,,	3·05	74·23

290   Σ. 1187.

85 LYNCIS.

R. A.	Dec.	M.
8ʰ 2ᵐ	32° 36′	7·1, 8

C. white.

Certain retrograde motion.
Dunér gives

$$1855\cdot00. \quad \Delta = 1''\cdot79.$$
$$P = 62°\cdot6 - 0°\cdot432\,(t - 1850\cdot0).$$

	°		″	
Σ.	71·0	5n.	1·61	1829·50
H₂.	70·9	1n.	·45	30·18
	81·0	,,	2·09	·97
Mä.	68·7		1·62	7·69
	66·6	5n.	·63	43·55
	61·9	,,	·81	51·17
	62·6	1n.	2·23	2·27
	69·7	2n.	1·74	5·32
	74·7	1n.	·81	8·29
	59·1	,,	·20	9·14
	56·2	3n.	·96	60·32
Mo.	·7	2n.	·82	59·14
	55·5	,,	·83	60·28
O.Σ.	68·1	1n.	·87	42·30
	67·3	,,	·92	6·29
	56·7	,,	2·11	61·19
	55·5	,,	·00	4·30
	52·9	,,	·07	9·31
	55·6	,,	1·97	71·30
	52·4	,,	·98	3·31
De.	58·9	1n.	·8	55·99
	·8	3n.	·6	6·14
	59·4	1n.	·5	7·09
	57·1	,,	...	62·85
	56·1	5n.	·82	3·21
Se.	61·6	2n.	·75	58·21
Eng.	58·5	4n.	2·03	65·49
Du.	53·6	3n.	1·75	71·25
W. & S.	54·3	5	2·14	2·27
	3·5	4	1·95	·30
Gl.	52·7	5	·92	4·13
Schi.	50·8	1n.	2·22	5·30
Sp.	50·9		·22	5·30
Dob.	53·1	1n.	...	8·11

291   Σ. 1196.

ζ CANCRI.

R. A.	Dec.	M.
8ʰ 5·3ᵐ	18° 1′	A 5, B 5·7, C 5·5

C. H₁, A, pale red or red ; B, pale red or red ; C, pale red. Σ., A, yellower than C ; B, yellower than A and C. De. always saw all white till 1864. He remarked a change in the colour of C more than eight times in 1864 and 1865 ; he noted it as more or less "jaunâtre et olivâtre."

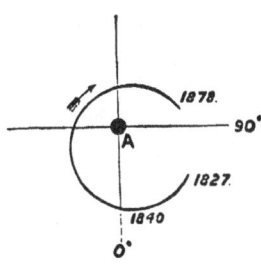

H₁, "Nov. 21, 1781.—If I do not see extremely ill this morning (4 a.m.), the large star consists of two." "Feb. 7, 1802.— After long looking, I cannot see the small star sufficiently well to measure its position."

H₁ (*Phil. Trans.*, vol. lxxii., p. 219), "ʃ Cancri, Fl. 16, Nov. 21, 1781.—A most minute treble star. It will at first sight appear as only a double star, but with proper attention, and under favourable circumstances, the preceding of them will be found to consist of two stars, which are considerably unequal. The largest of these is larger than the single star ; and the least of the two is less than the single star. The 1st and 2nd (in order of mag.) pretty unequal. The 2nd and 3rd pretty unequal. The two nearest both pale r. or r. With 278, but just separated ; with 460, ¼ diameter of S. Position 86° 32′ n.f."

Writing in the *Phil. Trans.* for 1804, H₁ observes, " The change is 9° 57′ in 20 years and 78 days ; and may be ascribed to a parallactic motion of the large star, which is in favour of the observed alteration."

H₁'s measure of A C, on the 5th of April, 1780, was "Distance, 8″·046 mean measure. Position, 88° 16′ sp." And the star is marked as one of those doubles known to astronomers before Mayer.

H₂ on the 21st of Feb. 1822, wrote, "Double ; pretty unequal ; is not to be seen triple, although beautifully defined and round." Ten measures of distance gave a mean value 6″·241. "In 40·25 years, then, the change of angle amounts to 23° 42′, which is at the mean rate of − 0°·5813 per annum, in the direction n.p.s.f.

or retrograde. The change of position has also been accompanied with a considerable diminution of distance ; and further observations must decide whether this is the result of rectilinear or orbital motion. If the former, the minimum distance will be attained in about 40 years from the present time, and the change during that period much less rapid than heretofore. On the other hand, an orbital motion will be indicated by the distance continuing to diminish beyond that limit, and probably too by an acceleration in the angular motion."

So. (*Phil. Trans.* 1826, part i., p. 332) gives measures of the close pair. "April 3, 1825, 31° 21′ n.f. 7 observations : 0″·887 5 observations. Difficult."

" I see the large star unquestionably elongated. At the time of perceiving the star elongated I was unaware that it had been observed by Sir W. Herschel as a close double star, as also that Mr. Herschel and myself, when we observed it in England as double of the 3rd class, had noted that ' it is not to be seen triple, although beautifully defined and round.'"

H₂ adds, " This star presents the hitherto unique combination of three individuals, forming, if not a system connected by the agency of attractive forces, at least one in which all the parts are in a state of relative motion." He then examines the measures in confirmation of these remarks ; and concludes, " If this be really a Ternary system connected by the mutual attraction of its parts, its perturbations will present one of the most intricate problems in physical astronomy. The difficulty will not be diminished by the circumstance of the rotations of the two small stars about the large one being (apparently at least) performed in opposite directions, being the reverse of what obtains in our planetary system, or by that of the deviations of the relative angular velocities from Kepler's law, being such as to indicate either great masses in all the three bodies, great excentricities in their orbits, or a different law of gravity from what obtains in our system."

In the *Mem. R. A. S.*, vol. v., p. 30, H₂ returned to this subject. Finding that his latest measures indicated a retrograde movement of − 6°·505 per annum, (that given in 1826 being + 1°·254 per annum, and direct,) he collects the measures by Σ. and Dawes, which also show that the motion was retrograde. Hence he is led " to assign the end of March 1873 as the time when it will have completed an entire revolution since the epoch of my father's first observations, in a periodic time of 55·34 years."

He adds, " ʃ Ursæ has hitherto afforded the only example of a double star of which the bimestral motion can be distinctly perceived and measured. It is now no longer a solitary instance." H₂ also points out "the remarkable difficulties in the way of any fair statement of the history of the position of this star [*i.e.* A C]. Most of the measures have been taken from the point of bisection of the close double star, seen as one or elongated ; some few directly from the larger."

On the whole, H₂ is greatly puzzled, and fears that "mistakes have been committed which it is now impossible to rectify or allow for."

Dawes (*Mem. R. A. S.*, vol. v., p. 136) gives his measures in 1831. He says, after examining the previous observations, that it "would appear as if the motion has been performed in a direct sense, or n.f.s.p., for perhaps 30 or 40 years ; and that the star B had then come to a stand, or appeared to do so, faced about, and is now proceeding in the opposite direction." The only explanation which offers itself to his mind is that B has performed almost an entire revolution in a retrograde sense, in the 49 years elapsed since H₁'s measures. The stars, too, differing but little in size, might lead to an error in placing the f. one as the p.

As to C, his observations corroborate the presumed motion of this star in direction, and indicate a considerable acceleration since 1825.

H₂ (*Mem. R. A. S.*, vol. vi., p. 27). "The motion of this star is steadily continued, and its binary nature and rapid retrograde motion must henceforth be considered as established beyond all possibility of doubt." [A B.]

Dawes in 1831-2-3, speaking of his measures, says, "just separated," "neatly separated." "A C somewhat difficult from the position of B with respect to A."

And in the *Mem. R. A. S.*, vol. xxxv., p. 337, after giving measures from 1831 to 1854, he writes that more than a complete revolution has been made by B since 1781, that the motion has been accelerated within the last ten years, and that the distance has diminished. He also says that the motion of C is orbital, but incomparably slower than that of B, and that its period is from 600 to 700 years.

Mädler in 1848 computed an orbit :—

$$P = 58\text{·}2708 \text{ years}$$
$$P. \text{ passage} = 1816\text{·}687$$
$$P. \text{ from node} = 133°\ 0'\text{·}7$$
$$\Omega = 33\ 34\text{·}3$$
$$i = 24\ 0\text{·}4$$
$$e = 0''\text{·}44385.$$

In 1855 Winnecke published the following elements of A B :—

P. passage = 1815·53
$$P = 58\text{·}94 \text{ years}$$
$$\Omega = 18°\ 23'\ (1855\text{·}0)$$
$$\pi - \Omega = 141\ 54$$
$$i = 48\ 36$$
$$\phi = 14\ 50$$
$$a = 1''\text{·}030.$$

O.Σ. has also computed the orbit of this pair : he gives—

$$T_0 = 1869\text{·}3$$
$$\omega = 199°\text{·}0$$
$$\Omega = 109\text{·}0$$
$$i = 20\text{·}7$$
$$e = 0\text{·}353$$
$$\mu = -5\text{·}77$$
$$P = 62\text{·}4 \text{ years}$$
$$a = 0''\text{·}908.$$

On these Du. observes that the distances are in general too large.

Du. has compared the elements with the latest observations. The following selection will show the excellence of O.Σ.'s orbit :—

	"	°	"	°	
1868·20	0·5	211·5	− 0·06	− 2·6	De.
9·32	·5	198·4	− 0·05	− 0·3	O.Σ.
70·30	·41	188·3	− 0·15	+3·0	Du.
1·31	·57	171·3	− 0·00	− 0·7	O.Σ.
2·21	·70	167·8	+0·10	+6·8	W.&S.
3·28	·61	152·0	− 0·02	+3·1	O.Σ.
4·13	·45	140·1	− 0·20	− 0·1	Gl.
5·33	·57	129·5	− 0·12	+0·2	Du.

For A C the following formulæ were deduced by O.Σ. :—

$$P = 155°\text{·}00 - 0°\text{·}50\ (t - 1831\text{·}3)$$
$$- 3°\text{·}04 \sin 18°\ (t - 1831\text{·}3).$$
$$\Delta = 5''\text{·}50 + 0''\text{·}20 \cos. 18°\ (t - 1831\text{·}3).$$

In 1871 Mr. W. E. Plummer made a new determination of the orbit, employing the elements given by Dr. Winnecke :—

Periastron passage	1872·44
Period	58·23 years
Ω	150° 17'·4
Long. of periastron	171 46·8
Inclination	36 14·4
Excentricity	0·30230, $\phi = 17°35'\text{·}8$
Mean motion	2'·56930
Mean distance	0''·908

Mr. Plummer also computed the following ephemeris :—

Epoch.	Angle.	Distance.
1870·0	196° 44'	0''·65
1·0	187 22	·64
2·0	177 30	·62
3·0	167 8	·60
4·0	156 5	·58
5·0	144 7	·56

A B. The maximum distance was reached about 1871. The influence of the third

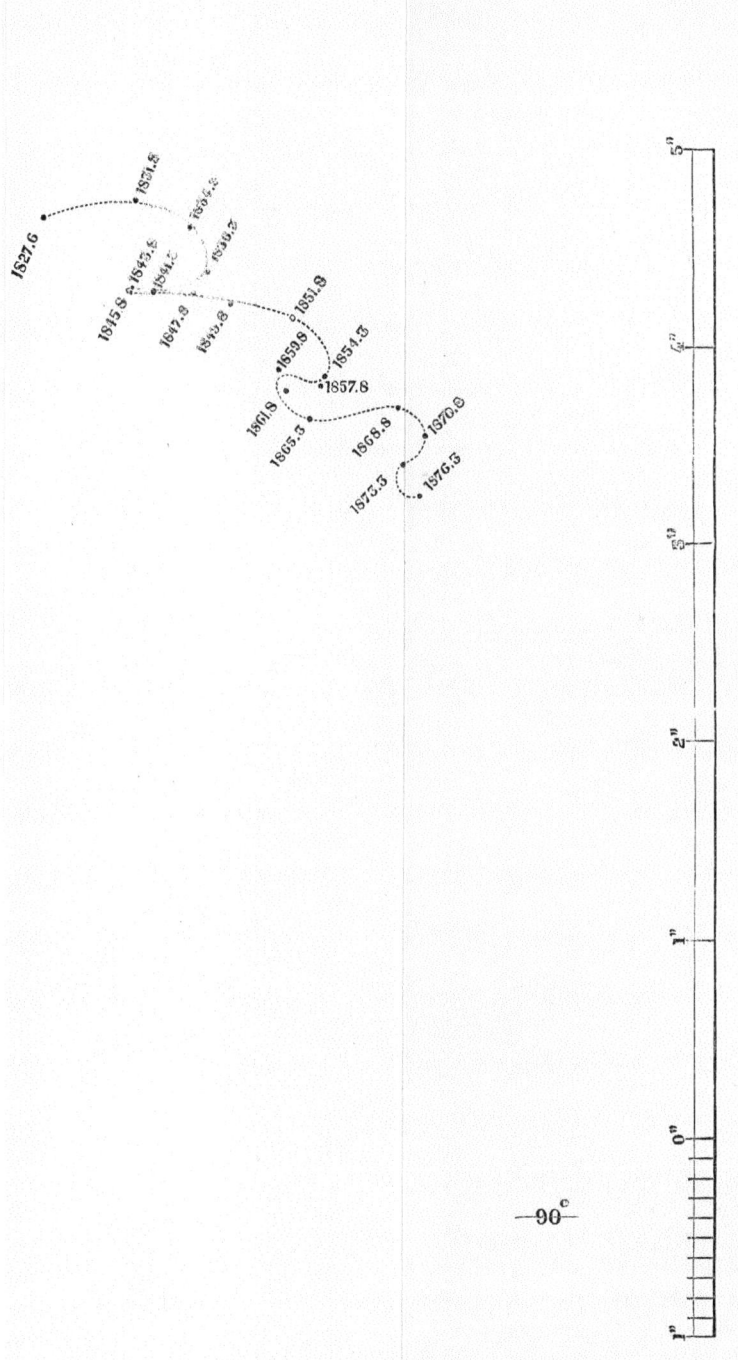

1527.6
1851.8
1854.2
1845.8
1845.8 &1843.2
1844.2
1851.3
1847.8
1849.8
1854.5
1859.8
1857.8
1861.8
1865.3
1868.8
1870.0
1873.3
1876.3

—90°—

5"

4"

3"

2"

1"

0"

1"

star is at once revealed by an examination of the measures of A B at different periods. (O.Σ.)

$\dfrac{A + B}{2}$ and C. On computing a table of the annual means from 1840 to 1874, great anomalies are found both in the angles and distances—anomalies which it is impossible to attribute solely to errors of observation. They present themselves also when the table is extended so as to embrace the Dorpat observations. We have here an example of the problem of the three bodies for which analysis does not yet furnish the means of solution. (O.Σ.) See also *Comptes Rendus de l'Académie de Paris*, vol. lxxix., p. 1463. A diagram showing the sinuous and looped path of the star is given in the 9th vol. of the Poulkova Observations.

A B.

	°		"	
H₁.	3·4		...	1781·9
So.	57·9		·1·0	1825·27
Σ.	·6	3n.	·14	6·22
	38·4	2n.	·04	8·80
	29·8	6n.	·04	31·28
	27·5	4n.	·15	2·28
	22·1	3n.	·14	3·27
	18·4	2n.	·13	5·27
	20·2	5n.	·13	·31
	15·3	3n.	·19	6·27
H₂.	35·6		·01	0·45
	27·5		·5	2·25
Da	30·7	3n.	·09	1·30
	27·0	7	...	2·12
	26·2	9	1·19	3·21
	16·1	4	...	6·68
	4·3	8	1·19	40·20
	0·8	5	·17	1·16
	356·2	6	·18	2·22
	355·0	8	·12	3·18
	338·5	1	·05	8·13
	·1	6	·06	·24
	334·2	5	·11	9·29
	327·9	7	·01	51·25
	324·4	3	·05	2·23
	315·3	3	0·97	4·20
	281·0	1	·70	60·26
	262·5	1	·66	3·25
	253·1	2n.	·70	4·29
	243·4	3	·63	5·30
Sm.	28·3		·3	32·23
	12·8		·2	7·11
	5·2		·3	9·32
	355·1		·2	43·11
Ka.	6·1		·25	0·15
	...		·27	3·12
O.Σ.	7·5	7n.	·99	0·29
	359·3	4	1·29	2·29
	354·2	3	·16	3·30
	350·3	4	·16	4·28
	347·9	3	0·97	5·31
	344·8	3	·95	6·29

	°		"	
O.Σ.	342·2	5	0·96	1847·33
	337·6	5	·91	8·30
	336·1	4	·80	9·32
	332·8	3	·94	50·29
	327·2	3	1·02	1·28
	321·7	2	0·89	2·32
	319·8	2	·97	3·30
	310·2	3	·91	5·31
	298·4	3	·97	7·27
	295·5	1	·98	8·28
	286·5	2	·91	9·30
	281·3	2	·84	60·27
	275·3	3	·87	1·27
	267·5	2	·74	2·31
	253·3	2	·72	4·30
	237·8	1	·70	6·27
	214·7	2	·72	8·28
	198·4	2	·62	9·32
	186·3	4	·66	70·28
	171·3	3	·59	1·31
	162·9	3	·58	2·31
	152·0	3	·61	3·28
	144·4	3	·63	4·28
Ch.	150·1	3n.	5·23	41·22
	146·8	1n.	4·80	2·15
	145·5	,,	·82	4·09
Mä.	1·0	6n.	1·05	1·31
	358·5	,,	·07	2·26
	325·8	,,	·06	52·25
	324·7	8n.	·05	3·25
	318·6	10n.	·07	4·27
	310·6	4n.	·06	5·26
	307·5	2n.	0·99	6·28
	304·5	3n.	·96	7·29
	297·5	,,	1·09	8·20
	294·9	8n.	0·97	9·26
Ja.	349·4		1·02	45·83
	346·5		·20	6·00
	322·0	15	·22	53·19
	317·2	10	·15	·94
	306·3		·21	6·20
	299·7		·14	7·88
Bond.	342·7		·0	48·25
Flt.	333·5	24	·1	51·18
	329·0	12	·0	2·16
	321·0	12	·1	3·30
De.	308·9	1n.	oblong	4·81
	309·0	,,	sepᵈ·	·88
	308·5	5n.	...	5·19
	306·4	1n.	1·0	·87
	302·2	8n.	·0	6·28
	299·2	1n.	·0	7·09
	293·8	,,	...	·84
	294·2	6n.	...	8·23
	265·7	3n.	...	62·85
	261·0	14n.	0·68	3·30
	254·4	3n.	·6	4·11
	256·9	2n.	·5	3·93
	255·1	9n.	·5	4·29
	245·4	11n.	·5	5·22
	238·4	9n.	·5	6·19
	224·4	7n.	...	7·22
	211·4	,,	0·5	8·20
	185·5	9n.	...	70·25

	°		″	
De.	175·2	6n.	...	1801·19
	162·8	7n.	...	2·23
	150·2	10n.	0·5	3·19
	144·5	2n.	·74	·86
	139·6	6n.	·75	4·29
	128·9	7n.	·72	5·18
Mo.	309·4	12	1·12	56·24
Kn.	268·1	1n.	0·69	63·13
	240·8	3n.	·63	5·36
	234·0	4n.	·79	6·26
	228·2	1n.	·65	·94
	166·7	2n.	·6	72·11
Ro.	267·3	1n.	·95	63·25
Ta.	231·5	,,	·72	6·38
Du.	203·6	4n.	·46	9·37
	188·2	3n.	·41	70·30
	178·2	,,	·53	1·29
	163·3	2n.	·66	2·33
	142·8	,,	·59	4·29
	129·4	5n.	·57	5·33
Gl.	185·1	6	·2	0·32
	177·0	5	...	·80
	176·3	5	0·2	1·21
	173·9	5	·2	·32
	160·7	5	·5	2·00
	·8	6	·5	·10
	154·7	6	·5	3·32
	144·0	5	·6	·94
	141·1	4	·4	4·10
	139·2	4	·5	·16
Br.	177·7		elong^d·	1·20
	175·0		,,	2·16
	159·5		0·71	3·14
W. & S.	168·6	4	·78	2·18
	169·3	3	·63	·19
	165·5	6	...	·25
	150·9	25	0·5	3·22
	141·2	5	·5	4·17
	142·2	4	·67	·17
	140·6	3	...	·20
	133·7	4	·80	5·26
	·0	9	·75	·32
Sp.	130·4		·69	·26
	108·2		·82	7·18
Dob.	122·1	8n.	...	6·21
	108·4	3n.	0·99	7·24
	104·1	1n.	·73	8·08
Pl.	110·3	3n.	·81	7·23
Schi.	130·3	1n.	·69	·25
	108·1	,,	·81	·18

A C.

	°		″	
T. Mayer	205·4		3·3	1756
C. Mayer	180·0		7·7	1778
H₁.	181·7		8·0	1781
H₂ & So.	171·8		...	1802·11
	158·3	12	6·24	22·14
	159·7	15	·19	4·22
	157·9	27	5·43	5·27
Σ.	154·6	3n.	·30	26·22
	151·4	,,	·31	8·99
	148·6	6n.	·40	31·28
	·6	4n.	·52	2·28
	147·6	3n.	·47	3·27

	°		″	
Σ.	145·1	2n.	5·35	1855·27
	150·1	5n.	·66	·31
	148·9	3n.	·62	36·27
Da.	150·3	1n.	·59	31·30
	148·8	4n.	·59	2·18
	147·1	1n.	·44	·87
	145·6	6n.	4·96	41·07
	146·7	2n.	·94	3·22
	·3		·88	8·14
	140·4	1n.	5·04	54·07
Sm.	149·4		·4	32·23
	147·1		4·8	4·36
	148·3		5·2	5·28
	146·9		·4	7·11
O.Σ.	147·0	7n.	4·92	40·29
	·1	4n.	·68	2·29
	149·4	3n.	·77	3·29
	·1	4n.	·87	4·27
	150·3	3n.	·81	5·30
	149·2	,,	·92	6·29
	148·4	5n.	·98	7·54
	147·1	,,	·97	8·29
	145·9	4n.	·93	9·31
	146·6	3n.	·92	50·29
	143·6	,,	5·07	1·27
	·1	2n.	4·95	2·32
	140·6	,,	·99	3·30
	141·2	3n.	·99	5·31
	·1	,,	5·04	7·27
	142·6	1n.	·02	8·28
	144·6	2n.	·06	9·19
	·1	,,	·06	60·27
	145·0	3n.	·14	1·37
	144·2	1n.	4·95	2·33
	143·9	,,	5·11	4·30
	141·4	,,	...	6·27
	138·4	2n.	...	8·28
	137·1	4n.	...	70·32
Ch.	2·1	2n.	1·07	41·22
	1·1	1n.	·08	2·15
	354·0	,,	0·97	4·09
Ka.	144·8		4·72	2·35
	146·0		·81	3·33
	137·9		5·41	66·28
Ja.	148·0		4·85	45·83
	147·5		·85	6·00
	142·1	15	·89	53·19
	140·0	8	·94	·95
	141·2		·95	6·20
	·9		·94	7·88
Bond.	149·3		·6	48·25
Flt.	143·7	20	·8	52·49
Mo.	·2	18	5·23	3·22
Po.	141·9	68	...	4·27
De.	139·9	1n.	·26	·88
	140·8	4n.	·56	5·17
	·0	1n.	·63	·87
	·3	5n.	·36	6·43
	141·0	1n.	·21	7·81
	139·2	4n.	·20	8·23
	140·2	,,	·59	62·88
	·8	5n.	·38	3·18
	·1	1n.	·35	4·99
	139·6	4n.	·49	5·20

	°		″	
De.	138·3	7n.	5·58	1866·84
	137·0	4n.	·54	8·22
	134·6	3n.	·61	70·21
	·1	,,	·61	1·17
	133·2	,,	·46	2·23
	132·8	,,	·40	3·22
	·7	1n.	·70	·88
	·6	2n.	·51	4·19
	131·6	4n.	·40	5·17
Se.	·2	3n.	4·93	55·19
	·7	2n.	·85	6·25
	142·8	5n.	·99	7·29
	·9	2n.	5·47	65·23
	140·7	,,	·62	6·28
M.	141·8	1n.	5·41	1·30
	·4	,,	·39	72·11
Ro.	·6	2n.	·62	3·18
W.O.	137·4	5	4·59	63
Kn.	143·9	,,	·23	·13
Ta.	·0	,,	·53	7·08
	138·9	1n.	·43	72·17
	135·8	,,	...	4·23
Du.	137·7	2n.	·52	69·37
	132·1	1n.	·78	75·33
W. & S.	133·8	1	·87	2·18
	·3	1	·63	·19
	131·2	2	·85	·25
	132·3	14	·43	3·22
	·0	2	...	4·17
	·2	2	...	·17
	131·5	4	·72	5·26
	·7	4	·53	·27
Gl.	132·7	6	5·1	4·10
	133·0	5	·7	·16
Schi.	130·4	1n.	.38	5·25
	·6	,,	.26	7·18
Pl.	·7	2n.	4·95	·20

B C.

	°		″	
O.Σ.	153·5	3n.	...	1844·28
	152·9	2n.	...	5·30
	149·0	5n.	6·15	8·29
	148·1	4n.	·19	9·31
	147·2	3n.	·16	50·29
	144·3	,,	·39	1·27
	142·2	2n.	·17	2·32
	139·8	,,	·13	3·30
	·4	3n.	·09	5·31
	138·1	,,	5·97	7·27
	·5	1n.	·98	8·28
	139·8	2n.	·82	9·19
	·9	,,	·78	60·27
	·7	3n.	·74	1·37
	137·8	1n.	·66	2·33
	138·1	,,	·31	4·30
	134·9	,,	...	6·27
	133·8	2n.	...	8·28
	132·4	4n.	...	70·32

$$\frac{A + B}{2}\ \text{and C.}$$

	°			
So.	161·2		...	1820·29
	160·1		...	1·07
	163·2		...	5·27

	°			
H₂.	158·2		...	1822·14
	159·4		...	4·49
	·2		...	30·21
Σ.	154·5		12	2·20
	159·0	3n.	5·40	26·26
	156·3	,,	·54	8·99
	153·2	6n.	·67	31·28
	·4	4n.	·84	2·28
	152·2	3n.	·82	3·27
	150·1	2n.	·67	5·31
	148·9	3n.	·63	6·29
	150·5		·31	40·29
	·7		·48	2·29
	152·0		·31	3·30
	151·3		·42	4·28
	·7		·30	5·31
O.Σ.	·5	7n.	·30	0·29
	·7	4n.	·48	2·29
	152·0	3n.	·31	3·30
	151·3	4n.	·42	4·28
	·7	3n.	·29	5·31
	150·6	,,	·38	6·29
	149·6	5n.	·42	7·33
	148·0	,,	·56	8·30
	147·0	4n.	·56	9·32
	149·9	3n.	·54	50·29
	143·9	,,	·73	1·28
	142·7	2n.	·56	2·32
	140·4	,,	·56	3·30
	·3	3n.	·54	5·31
	139·6	,,	·50	7·27
	140·5	1n.	·50	8·28
	142·2	2n.	·43	9·30
	·0	,,	·42	60·27
	·3	3n.	·44	1·27
	141·0	1n.	·30	2·31
	·6	2n.	·30	4·30
	138·1	1n.	·56	6·24
	136·1	2n.	·69	8·28
	·8	,,	·61	9·32
	134·7	4n.	·69	70·28
	·3	3n.	·61	1·31
	133·5	,,	·63	2·31
	135·0	,,	·39	3·28
	133·8	,,	·43	4·28
Sm.	148·2		·1	39·32
	147·2		·0	47·11
	·4		·0	7·28
Ja.	·8		...	6·29
De.	140·7		5·39	55·18
	·3		·35	6·42
	139·6		·15	8·15
	140·6	9n.	·48	63·05
	139·7	5n.	·47	5·17
	137·8		·58	7·26
	134·2		·61	70·21
	·1		·60	1·53
	133·2		·46	2·23
	132·8		·40	3·23
	·7		·57	4·09
	·8		·4	·13
M.	141·8		·41	61·30
Eng.	·3		·49	4·31

Du.	135·1	8n.	5·47	1870·70
	133·2	7n.	·46	5·02
Br.	132·7		·54	1·19
	133·3		·46	2·16
W. & S.	132·0		·64	3·70
	131·6		·63	5·27
Gl.	·4	in.	·39	5·17
	127·7	,,	·2	6·22
Sp.	130·4		·38	5·25
	·6		·27	7·18
Dob.	131·5	in.	...	5·99
	·0	6n.	...	6·20
	127·2	4n.	5·62	7·26
	132·0	in.	·48	8·08

De.	327·4	2n.	2·5	1863·11
O.Σ.	328·4	in.	·64	9·24
	324·6	,,	·35	75·27
Du.	328·1	4n.	·34	3·19
W. & S.	326·5	3	·1	·22
	·6	2	·09	·24
	324·2	6	·09	4·14
	328·9	9	1·9	2·28
	324·9	5	2·15	5·28
Gl.	325·5	5	·1	4·10

292　　γ ARGÛS.

R. A.	Dec.	M.
$8^h\ 5\cdot8^m$	$-46°\ 58'$	A 2, B 5, C 8

Certain change in the angle of A B.

A B.

H₂.	220·7		41·19	1835·03
	219·6		·12	·18
El.	214·8		42·5	77·03

A C.

H₂.	151·6		62·4	35·10
El.	·1		·6	77·03

293　　Σ. 1202.

P. VIII. 13.

R. A.	Dec.	M.
$8^h\ 7^m$	$11°\ 13'$	7·7, 9·8

C. white.

Σ. discovered this double star, and was led to think it a binary from the results of his own measures. A subsequent set, however, seemed not to confirm this opinion.

Dawes (*Mem. R. A. S.*, vol. xxxv., p. 339) thinks that the obliquity of direction may partly account for the discrepancies in the measures.

O.Σ. thinks that a retrograde movement is very probable.

Dunér gives

$$1854\cdot83.\quad \Delta = 2''\cdot32.$$
$$P = 331°\cdot9 - 0°\cdot220\ (t - 1850\cdot0).$$

Σ.	335·9	3n.	2·35	1829·55
	337·4	in.	·28	36·19
Sm.	338·0		·5	2·27
Mä.	333·6	2n.	·57	44·21
	332·2	3n.	·36	8·31
	·1	2n.	·27	51·22
	329·2	,,	·52	2·27
Da.	328·5	in.	·21	48·24
Se.	325·5	,,	·07	56·17
	331·9	in.	·10	65·28

294　　Σ. 1216.

R. A.	Dec.	M.
$8^h\ 15\cdot2^m$	$-1°\ 13'$	7·5, 8·2

Rapid change in angle.

Σ.	109·5	in.	0·53	1825·20
	115·2	4n.	·48	31·24
O.Σ.	142·0	in.	·54	51·27
	136·9	,,	·44	·28
Se.	149·9	,,	...	7·34
De.	151·1	7n.	...	63·35
W. & S.	170·0	est.	··	73·19
	168·7	2	0·4	·24
	166·2	3	...	5·27
	164·8	4	...	·28
Gl.	167·0	6	0·5	4·18

295　　Σ. 1223.

φ² CANCRI.

R. A.	Dec.	M.
$8^h\ 20^m$	$27°\ 19'$	6, 6·5

Dunér's formulæ are

$$1851\cdot02.\quad \Delta = 4''\cdot75.$$
$$P = 213°\cdot7 - 0°\cdot07\ (t - 1850\cdot0).$$

H₁.	33·3	in.	...	1783·06
Σ.	...	4n.	4·58	1820·17
	207·8	in.	3·89	2·23
	212·0	7n.	4·56	9·45
	214·3	2n.	·82	38·34
H₂ & So.	211·2	4n.	5·51	22·48
Sm.	212·5		4·8	33·25
Da.	·6	in.	·95	40·15
Mä.	214·4	3n.	·99	53·59
Mo.	213·3	,,	·88	4·24
De.	215·4	7n.	·90	·46
Se.	214·3	3n.	·73	6·20
Ka.	·1	5n.	·71	66·24
Du.	215·1	,,	·75	73·10
O.Σ.	35·0	in.	5·01	4·27

296　　Σ. 1224.

ν¹ CANCRI.

R. A.	Dec.	M.
$8^h\ 19\cdot5^m$	$24°\ 56'$	6, 7·1

Dunér gives

1849·52. Δ = 5"·89.
P = 39°·6 + 0°·10 (t − 1850·0).

H₁.	57·2		5·5	1783·07
Σ.	34·5	7n.	6·28	1820·60
	37·4	4n.	5·78	2·18
	·3	9n.	·84	30·76
	·5		·91	40·24
So.	·8	1n.	6·05	22·12
	·5	5n.	·74	5·26
H₂.	38·4	3n.	·47	30·18
	·6	2n.	·04	1·07
Sm.	37·9		·0	·17
	38·6		5·7	7·26
	40·1		·8	43·18
Da.	38·3	4n.	6·24	0·88
	39·8	2n.	5·87	3·17
	·9	4n.	6·03	9·54
O.Σ.	40·9	3n.	·27	0·24
	36·8	1n.	·11	4·28
	37·2	,,	5·98	6·30
	38·0	,,	·91	9·32
	·7	,,	·99	51·28
	·3	,,	6·05	3·30
	39·0	,,	·02	7·26
	41·0		·02	62·29
Mä.	39·5		5·93	42·32
	·1	1on.	·83	4·03
	·4		·60	5·98
	40·2	5n.	·73	51·15
	·2	3n.	·95	2·95
	39·5		·63	5·24
	40·7		·59	6·27
	38·3		·76	7·29
	39·2		·58	8·10
	·5	2n.	6·20	61·33
Mo.	40·6	,,	5·99	52·25
	·2	3n.	6·06	4·16
De.	·0	5n.	5·81	4·89
Se.	41·2	4n.	·95	6·20
M.	40·2	1n.	6·0	62·11
Ro.	38·6	,,	5·89	3·14
Eng.	41·0	5n.	·95	5·41
Ta.	39·4	1n.	6·04	6·10
	40·5	,,	·13	9·38
Ka.	·0	5n.	5·65	6·24
Du.	42·1	3n.	·74	9·54
	43·2	2n.	·79	74·17
Gl.	41·2	1n.	...	·18
W. & S.	·6	,,	·9	·18

297   h. 4087.

R. A.	Dec.	M.
8ʰ 19·5ᵐ	− 40° 36′	7·8, 8

Probably a binary.

H₂	146·6		1	1837·15
	147·5		0·83	8·08
Ja.	134·9	12	1·45	58·20

298   O.Σ. 193.

R. A.	Dec.	M.
8ʰ 21ᵐ	33° 57ᵏ	7, 11
	C. yellowish.	

O.Σ.	297·1	1n.	14·14	1844·30
	293·9	,,	·36	8·25
	294·6	,,	·17	50·26
	·9	,,	·12	61·26
De.	295·6	3n.	13·64	8·11

299   Σ. 1263.

R. A.	Dec.	M.
8ʰ 37·1ᵐ	42° 8′	7·6, 8·2

On reducing the angles of position to the equinox of 1850, O.Σ. finds twelve relations, and on converting them into rectangular coordinates and treating them by the method of least squares, he obtains

$$\Delta A = (+ 5"\!\cdot\!503 \pm 0"\!\cdot\!032) + (0"\!\cdot\!2646 \pm 0"\!\cdot\!0024)\,(t - 1850\!\cdot\!0).$$
$$\Delta D = (+ 19\!\cdot\!054 \pm 0\!\cdot\!016) + (0\!\cdot\!6554 \pm 0\!\cdot\!0012)\,(t - 1850).$$

Uniform rectilinear motion perfectly satisfies the Δ D. In the Δ A the differences are less satisfactory. He thinks it probable that Σ.'s measures contain systematic errors.

Σ.	359·0	1n.	4·86	1828·36
	4·1	,,	5·43	9·46
	·9	,,	7·08	31·31
	7·3	,,	·46	2·33
	8·0	,,	·97	3·29
	·4	,,	8·93	4·36
	9·3	,,	9·59	5·35
	10·3	,,	10·34	6·42
	·3	,,	·47	7·06
	11·8	,,	11·63	8·34
	12·4	,,	12·88	40·27
O.Σ.	·3	4n.	13·05	·28
	14·8	,,	16·50	5·08
	15·7	,,	18·89	8·80
	·6	,,	20·56	51·05
	16·7	,,	22·64	4·07
	17·8	2n.	27·06	60·27
	18·9	3n.	34·35	70·63
Da.	13·2		14·28	41·33
	17·1		22·47	54·13
Ka.	15·5		14·59	42·69
Flt.	16·6	22	20·84	52·15
	17·1	22	21·22	3·19
Mo.	16·4	30	22·50	4·22
	18·3	20	26·74	60·08
De.	17·0		23·04	55·37
	18·2		29·12	63·38
M.	16·8	1n.	·72	4·30
Ta.	19·0	,,	31·24	6·10
	21·6	,,	...	70·03
	19·6	,,	36·01	6·26
Du.	18·7	3n.	35·12	1·91

	°		"	
W. & S.	18·4	4	...	1872·25
	·2	4	35·6	·30
	·1	3	...	3·19
	·8	5	37·15	5·27
Gl.	19·0	1	...	4·72
Sp.	18·6		37·44	5·27
Dob.	·6	2n.	...	6·13
Pl.	·9	,,	38·57	·36
Fl.	·8		·9	7·25

300   Σ. 1273.

ε HYDRÆ.

R. A.	Dec.	M.
8ʰ 40·4ᵐ	6° 52'	3·8, 7·8

C. A, yellow; B, blue.

It is extraordinary that this beautiful star should have escaped the scrutinizing eye of Sir Wm. Herschel.

Σ. (*M. M.*, p. 4) had no doubt about the motion.

Sm., at the request of Dawes, examined this object and thought that the angular motion was about + 0°·8 per annum, "or a circuit of 4½ centuries."

The later observations only partially support his former impression of a diminution of distance. (Dawes.)

Secchi says the motion is orbital.

O.Σ. in 1860 and 1864 suspected that A was oblong in the vertical direction.

The common proper motion is − 0ˢ·013 in R. A., and + 0″·04 in N. P. D.

Dunér gives 1853·39 Δ = 3″·34 for the distance; and for the angle, P = 206°·5 + 0°·543 (t − 1850·0).

	°		"	
Σ.	192·4	3n.	3·31	1825·23
	195·9	,,	·14	31·29
	198·3	,,	·16	5·28
	·6	,,	·20	6·27
	200·8	2n.	·39	40·30
Da.	195·2	1n.	4·34	31·13
	197·6	2n.	·26	2·20
	199·1	1n.	3·65	4·00
	197·9	,,	...	7·23
	201·6	16n.	3·50	40·95
	203·5	3n.	·42	3·21
	205·7	1n.	·42	8·14
	206·7	3n.	·50	·83
	208·5	1n.	·43	51·32
Sm.	198·4		·4	37·11
	199·1		·5	9·22
	203·2		·6	43·14
Ch.	197·8	1n.	·44	1·20
	202·7	,,	·82	3·17
	200·0	,,	2·83	4·16
Mä.	203·0	7n.	3·32	2·64
	209·1	4n.	·37	52·30
	208·3	,,	·28	6·25
	212·4	2n.	·04	7·29
	210·7	11n.	·39	60·28

	°		"	
Ka.	203·0	7n.	3·13	1842·36
O.Σ.	203·5	2n.	·50	3·31
	208·5	3n.	·27	8·97
	211·0	,,	·43	61·62
	217·5	,,	·44	8·88
Ja.	203·6		·74	46·20
	209·1	9	·33	53·24
	·6	10	·26	·99
	212·7	3n.	·21	7·39
Mo.	208·0	1n.	·69	2·27
	·9	3n.	·04	4·28
	210·6	2n.	·42	8·30
De.	211·3	1n.	·74	4·92
	·9	5n.	·44	5·21
	210·9	2n.	·24	6·15
	211·3	1n.	·65	7·08
	213·0	,,	·53	62·92
	212·8	3n.	·45	3·20
	216·3	9n.	·44	6·01
	217·0	1n.	·36	8·25
	218·4	2n.	·26	70·27
	217·5	1n.	·32	1·17
	·8	,,	·45	3·23
	218·5	,,	·40	5·28
Se.	210·0	4n.	·33	56·19
	215·6	1n.	·48	65·27
Po.	213·2	15	·06	1·27
W.O.	199·3	2	·32	63
M.	200·9	1n.	·40	·17
	210·1	,,	...	70·26
	216·5	,,	3·59	1·22
	215·6	,,	·37	·23
Eng.	216·8	3n.	·41	65·18
Ta.	204·7	3n.	·87	6·10
	207·2	1n.	·68	7·20
	216·8	,,	2·94	74·23
	217·6	,,	...	6·34
W. & S.	216·2	5	3·28	2·19
	219·3	5	·18	3·19
	·3	1	...	·19
	216·7	6	3·33	4·18
	221·5	7	·47	5·32
	216·2	3	·4	4·09
Du.	219·3	5n.	·20	5·28
Schi.	217·9	1n.	·31	·29
Sp.	·9		·31	·29
Dob.	219·2	1n.	...	·99
	·7	9n.	...	6·20
Pl.	216·7	2n.	3·67	·73

301   Σ. 1281.

R. A.	Dec.	M.
8ʰ 41·4ᵐ	0° 28'	7·3, 8·3

The motion appears to be rectilinear hitherto.

	°		"	
Σ.	329·6	5n.	25·02	1833·48
H_2	328·6		25±	2·30
Mä.	326·7		27·15	47·23
De.	323·8		29·47	64·50
Fl.	321·9	1n.	31·10	77·31

302 Σ. 1280.

R. A. 8ʰ 44·4ᵐ — Dec. 71° 16' — M. 7·5, 7·6

C. yellowish.

The angle has increased and the distance has diminished.

H₂ & So.	31·2	2n.	8·75	1823·33
Σ.	34·0	4n.	7·43	31·90
Mä.	33·9	1n.	·42	43·05
Mo.	36·0	3n.	6·61	52·13
	37·5		·51	64·71
De.	36·2	3n.	·67	56·27
	37·5	,,	·51	64·71
Du.	40·1	6n.	·14	73·83

303 Σ. 1287.

R. A. 8ʰ 44·9ᵐ — Dec. 12° 35' — M. 8, 10·3

Probably a binary.

Σ.	109·4	3n.	1·41	1830·60
De.	95·0		·86	63·20

304 Σ. 1291.

σ² CANCRI.

R. A. 8ʰ 47ᵐ — Dec. 31° 2' — M. 5·9, 6·4

Dunér's formulæ are

$$1850·50 \quad \Delta = 1''·42.$$
$$P = 334°·1 - 0°·06 \,(t - 1850·0).$$

H₁.	338·2	1n.	...	1782·28
H₂ & So.	340·2	,,	1·89	1822·14
Σ.	333·3	5n.	·51	9·71
Mä.	335·4	,,	·47	42·90
O.Σ.	332·8	2n.	·41	5·28
	·2	1n.	·28	7·36
	·1	,,	·13	8·30
	·0	,,	·24	9·32
	·5	,,	·53	53·30
	335·2	,,	·42	9·30
	334·0	,,	·50	60·28
	·0	,,	·47	·29
De.	331·0	2n.	·2	56·19
Se.	333·8	3n.	·34	·27
Mo.	336·6	2n.	·40	7·29
	334·9	3n.	·29	60·15
Du.	332·5	4n.	·43	71·02

305 O.Σ. 196.

ι URSÆ MAJORIS.

R. A. 8ʰ 51ᵐ — Dec. 48° 31' — M. 3·1, 10·3

Orbital motion has distinctly shown itself. The common proper motion of this system is no less than − 0 ·047 in R. A., and + 0"·28 in N. P. D.

H₂.	348·8		...	1831·71
Sm.	·0		12·0	9·12
Ch.	350·0	1n.	10·68	41·19
O.Σ.	351·8	4n.	·69	5·27
	·8	5n.	·54	51·68
	355·0	1n.	·18	61·24
	356·9	2n.	9·78	71·80
Mä.	350·7		10·14	52·27
De.	357·1		9·72	69·38

306 Σ. 1296.

R. A. 8ʰ 51·8ᵐ — Dec. 35° 25' — M. 8·5, 9

Probably a binary.
Dunér has

$$\Delta = 2''·59 - 0''·0128 \,(t - 1850·0).$$
$$P = 73°·7 + 0°·136 \,(t - 1850·0).$$

Σ.	71·2	3n.	2·83	1831·59
Mä.	72·6	2n.	·67	44·27
Du.	76·6	,,	·31	71·26

307 Σ. 1300.

R. A. 8ʰ 54·6ᵐ — Dec. 15° 45' — M. 8·7, 8·8

C. yellow.

Σ.	210·0	3n.	4·11	1830·19
H₂.	211·0		2	2·20
Se.	204·2	3n.	4·67	56·58
	·5		·98	65·27
O.Σ.	24·6	1n	·68	6·28
	203·3	,,	·86	8·29
	202·4	,,	·79	70·28
W. & S.	203·4	4	·83	4·18
	·4	6	·5	·18
Gl.	204·0	4	·6	·18

308 Σ. 1306.

R. A. 8ʰ 59·8ᵐ — Dec. 67° 37' — M. 5, 8·2

The diminution in distance and increase in angular movement have been marked since 1850.

The apparent orbit is probably considerably elongated.

The proper motion of A is − 0ˢ·005 in R. A., and +0"·11 in N. P. D.

H₁.	283·0	1n.	[7·93]	1783·68
H₂.	267·3		5·0	1832·10
Σ.	253·2	5n.	4·59	·99
Sm.	262·4		5·0	5·27

O.Σ.	260.3	2n.	4.50	1840.34
	.6	1n.	.24	6.37
	257.5	,,	3.89	51.39
	.5	,,	.71	4.37
	249.5	,,	.17	66.42
	246.8	,,	.16	72.41
Da.	262.8	2n.	4.46	41.20
	258.2	,,	3.92	51.28
Ka.	261.6		4.36	42.49
Mo.	258.3	30	3.61	54.26
Se.	257.2	3n.	.41	6.34
De.	253.5	8n.	.25	63.19
	252.6	,,	.22	5.81
	247.5	3n.	2.88	71.52
	245.2	2n.	.68	5.21
Ta.	261.8		3.51	66.10
	258.1		2.76	71.39
W. & S.	249.5	4	.85	2.28
	246.7	4	3.07	.30
	249.8	4	.20	3.24
	246.8	4	.20	.29
Gl.	247.2	5	2.9	4.18

309　　　Σ. 1316.

R. A.	Dec.	M.
$9^h 1.9^m$	$-6° 39'$	A 8.2, B 11.5, C 10.5

C. white.

Certain change, but of uncertain nature.

A B.

Σ.	146.3	3n.	6.78	1832.88
Se.	139.6	1n.	5.79	57.26
	138.9	,,	6.94	65.19
De.	.4		.74	4.84
W. & S.	139.7	1n.	.18	74.17
C.O.	.8		7.38	7.18

A C.

Σ.	153.1	3n.	13.05	32.88
Se.	156.2	1n.	11.28	57.26
	153.9	,,	10.05	65.19
De.	158.7		.08	4.84
W. & S.	157.8	1n.	9.5	74.17
Gl.	155.0	,,	...	4.18
C.O.	163.5	,,	9.06	7.18

B C.

C.O.	28.7	1n.	4.20	77.18

310　　　Σ. 1313.

R. A.	Dec.	M.
$9^h 2.5^m$	$70° 28'$	8.5, 8.7

The angle has probably increased.

Σ.	240.8	3n.	0.84	1832.39
O.Σ.	242.2	4n.	.87	45.84
De.	50.5		1.0	66.74

311　　　Σ. 1321.

R. A.	Dec.	M.
$9^h 6.5^m$	$53° 13'$	7.4, 7.4

The distance is probably unchanged.
The proper motion of A is probably large.

Dunér has

$$\Delta = 19''.87 - 0''.01\ (t - 1850.0).$$
$$P = 52°.4 + 0°.24\ (t - 1850.0).$$

Σ.	43.8	1n.	...	1820.92
	...	11n.	21.12	2.07
	48.1	3n.	20.14	31.35
	.9	,,	19.93	5.73
So.	45.8	5n.	20.80	24.46
H₂.	51.5		.0	31.40
O.Σ.	50.5	3n.	.17	40.32
	52.2	4n.	.00	8.57
Da.	50.8	5n.	.22	2.89
	51.8		19.87	8.29
Mä.	.4	2n.	20.10	6.29
	53.0	1n.	19.47	51.09
	.0	2n.	.23	2.42
	54.9	1n.	.91	8.38
	52.3	,,	.38	61.67
De.	55.7	5n.	.74	3.12
M.	53.4		.33	.25
Ta.	...	1n.	20.23	6.35
	56.2	,,	19.63	7.23
	58.0	2n.	20.17	71.39
	55.0	1n.	19.92	2.18
	57.0	,,	20.29	4.23
Gl.	.0	,,	.0	.22
Dob.	58.4	,,	...	6.07

312　　　Σ. 1329.

R. A.	Dec.	M.
$9^h 9.6^m$	$-0° 44'$	8.3, 8.5

C. white.

Σ.	65.5	4n.	27.68	1831.21
	.7		.19	4.26
Eng.	67.4		24.56	63.25
De.	.3		23.73	4.76

313　　　Σ. 3121.

R. A.	Dec.	M.
$9^h 10.7^m$	$29° 7'$	7.5, 7.8

The orbit of this system was computed in 1866 by Fritsche from the observations between 1831 and 1864. The predicted places for 1874 are, however, far from representing the observations. O.Σ. explains this by saying that the distances were not corrected for systematic errors, and he predicts that the star will be single in 1877. The period is about 39 years.

Fritsche's latest elements are—

	(1)	(2)
$\Omega =$	19° 56'.4	23°.5
$\omega =$	143 17'.2	141 .6
$e =$	0.3471	0.3725
$\mu =$	+ 9.188	+ 8.862
$\pi =$	52° 23'.0	54.11
$a =$	0".696	0".715
$P =$	39^yrs.18	40.62
Epoch	1850.0	1850.0

The elements in (2) are based on the same observations with O.Σ.'s corrections applied. According to this system of elements Dunér finds that the distance in 1878 would be 0".49; but the star was oblong in 1874.

Σ.	20.0	In.	0".85	1832.31
	239.1	3n.	elong^d.	40.32
	190.2	In.	,,	4.28
	5.0	,,	1.25	.30
	18.8	,,	0.48	6.29
	210.7	,,	.45	7.34
	29.6	,,	.44	8.25
	34.4	,,	.41	9.32
O.Σ.	250	In.	oblong?	0.28
	246	,,	ob.&sep.	.32
	243.5	,,	0.40	.35
	198.4	,,	oblong	4.26
	8.5	,,	0.33	.30
	27.6	,,	.55	6.29
	214.2	,,	.54	7.34
	33.0	,,	.53	8.25
	43.3	,,	.48	9.32
	228.6	,,	.42	50.30
	59.7	,,	.33	1.26
	Certainly double		...	61.29
	8.9	,,	0.67	.30
	13.0	,,	.71	4.30
	29.5	,,	.78	8.29
	23.8	,,	.85	.30
	26.1	,,	.88	9.31
	35.3	,,	.82	71.28
	215.4	,,	.76	.31
	36.4	,,	.68	2.31
	40.4	,,	.64	4.28
	53.0	,,	.43	.28
	250.1	,,	oblong	5.29
De.	14.8		0.7	63.11
	19.6		.68	6.22
	21.3		.7	7.26
	.8		.7	8.25
	27.6		...	9.85
	32.6		0.6	71.21
	210.5		elong^d.	2.23
	213.9		,,	3.38
	214.9		,,	4.21
	251.9		,,	5.31
Gl.	210.4	5	0".5	0.44

	°		"	
Gl.	212.7	4	0.5	1871.21
	214.5	4	.5	3.70
Du.	206.9	2n.	.65	0.33
	208.2	3n.	.75	1.27
	209.3	In.	.68	2.09
	220.0	2n.	.3	4.24
	225.0	In.	.2	5.20
Sp.	245.2		.3	.29
	279.9		.35	7.18

314   Σ. 1333.

R. A.	Dec.	M.
9h 11m	35° 52'	6.6, 6.9

Dunér gives

1853.82.   Δ=1".49.
P=41°.7 + 0°.10 (t−1850.0).

H_1.	38.6		...	1782.86
Σ.	39.4	4n.	1.42	1828.59
H_2.	40.5	3n.	.38	31.12
Mä.	42.6	5n.	.46	45.51
	.5	6n.	.57	53.13
	.9	2n.	.38	9.83
O.Σ.	.8	6n.	.633	0.13
Se.	43.6	2n.	.43	6.32
	.4	In.	.78	65.19
De.	41.5	,,	.3	57.92
	43.0		.51	66.46
Mo.	39.2	3n.	.44	59.27
Eng.	45.3	4n.	.65	65.55
Du.	41.4	,,	.45	72.24

315   Σ. 1334.

R. A.	Dec.	M.
9h 11.4m	37° 19'	4, 6.7

C. A, white; B, blue.

The common proper motion is −0s.007 in R. A., and +0".04 in N. P. D.

Dunér has

Δ=2".80.
P=240°.4 − 0°.05 (t−1850.0).

H_1.	244.2	In.	...	1780.90
H_2 & So.	242.7	4n.	2.89	1822.46
Σ.	240.2	6n.	.69	9.17
	.7	In.	.70	52.00
Sm.	241.6		.8	32.35
H_2.	239.2		.5	2.40
Mä.	240.8	5n.	.97	43.30
	.9		.47	7.19
	241.7		.62	54.30
	240.6		.46	6.44
	.4		...	7.43
	239.5	2n.	.84	9.83
De.	241.5	3n.	3.0	5.19
	239.1		2.80	66.56
Se.	241.5	4n.	.84	57.36
Mo.	238.3	2n.	.91	9.16

Ro.	239·0	1n.	2″90	1862·13
Ta.	·0	,,	·89	6·16
Du.	240·8	2n.	·49	71·25
W. & S.	238·8	1n.	·82	3·19
Gl.	·6	,,	·92	4·18

316　　O.Σ. 199.

R. A.	Dec.	M.
9ʰ 12ᵐ	51° 46′	6·1, 10·2

The distance has probably diminished.

O.Σ.	116·78	4n.	5·74	1847·02
De.	117·17	3n.	·32	68·11

317　　Σ. 1338.

157 LYNCIS.

R. A.	Dec.	M.
9ʰ 13·4ᵐ	38° 42′	7, 7·2
	C. white.	

From the measures between 1827 and 1833, Σ. suspected a slow direct motion.

Se. (p. 27) says the motion is certain.

O.Σ. A gradual angular change, the distance remaining as in 1829; the apparent orbit is, therefore, probably nearly circular.

Dunér's formulæ are

$$1854 \cdot 50. \quad \Delta = 1'' \cdot 67.$$
$$P = 133° \cdot 1 + 0° \cdot 6257 \, (t - 1850 \cdot 0).$$

Σ.	121·1	5n.	1·76	1829·53
	119·7	2n.	2·51	30·25
	116·4	1n.	1·42	2·19
Mä.	127·1		·70	8·10
	130·3	6n.	·66	43·42
	132·7	2n.	·80	51·04
	134·8	5n.	·70	2·78
	135·5	,,	·71	4·78
	140·5	2n.	·61	9·83
O.Σ.	128·1	4n.	·82	40·33
	130·2	1n.	·89	2·30
	128·2	,,	·58	7·36
	134·5	,,	·66	54·25
	135·4	,,	·82	·26
	142·4	,,	·84	68·34
	149·8	,,	·72	73·31
Da.	125·9	3n.	·75	41·23
	127·3	,,	·72	2·23
	·7	1n.	·66	3·17
	131·8	3n.	·66	8·17
	·9	2n.	·69	50·12
	134·6	5n.	·65	4·20
Mo.	137·0	4n.	·92	·24
De.	135·7	3n.	...	5·20
	·6	1n.	1·2	·91
	138·6	,,	·5	6·04
	·8	,,	·73	62·85
	141·3	3n.	·51	3·15
	·0	1n.	·63	4·07

Se.	137·1	2n.	1″63	1856·31
Eng.	143·5	3n.	2·00	65·31
Kn.	142·5	2n.	1·71	·47
Ta.	143·4	,,	·66	6·67
	142·6	1n.	·58	71·40
	141·2	,,	...	2·18
	145·2	,,	1·24	4·23
	142·9	,,	...	5·18
	150·0	2n.	1·26	6·34
Du.	149·0		·66	1·24
	151·8	4n.	·63	6·29
W. & S.	145·8	4	·87	2·27
	147·7	8	·46	3·19
	·6	2	...	·23
	·8	6	1·57	4·17
	150·0	4	·73	5·27
	·5	4	·55	·29
	·0	5	·78	6·26
Gl.	147·0	4	·7	4·18
Sp.	149·3		·76	5·33
Schi.	·3	1n.	·76	·33
Dob.	·6	5n.	2·13	6·21

318　　O.Σ. 200.

R. A.	Dec.	M.
9ʰ 16·6ᵐ	52° 5′	6·7, 8·4

Slow direct motion.

Mä.	334·4	4n.	1·39	1845·64
	337·2	2n.	·52	51·34
O.Σ.	335·2	5n.	·41	47·09
Da.	338·5	1n.	·56	8·28
	333·7	,,	...	60·25
De.	338·3	3n.	1·34	7·60
Du.	·5	2n.	·54	71·36

319　　O.Σ. 201.

R. A.	Dec.	M.
9ʰ 16·8ᵐ	28° 24′	7·5, 9

Mä.	236·3		1·24	1843·25
	230·3		·33	6·32
	229·9		·47	7·41
O.Σ.	233·4	6n.	·45	52·43
De.	229·2	3n.	·48	67·72

320　　Σ. 1346.

21 URSÆ MAJORIS.

R. A.	Dec.	M.
9ʰ 17ᵐ	54° 32′	7, 8

C. A, white; B, bluish.

H₁.	306·7		...	1782·87
	317·6		...	1802·39
H₂ & So.	309·0	1n.	6·47	22·12
Σ.	·0	,,	...	·12
	310·9	5n.	5·69	30·99

Mä.	310·2	7n.	5·65	1842·57
	·3	2n.	·95	51·11
	·8	,,	·94	6·29
De.	311·9	1n.	·48	5·99
Se.	310·3	3n.	·80	6·98
Mo.	·7	,,	·81	8·18
Du.	312·7	2n.	·51	71·69
Gl.	·0	1n.	·3	4·26

321   Σ. 1348.

116 (B) HYDRÆ.

R. A.	Dec.	M.
$9^h\ 18\cdot2^m$	6° 52′	7·5, 7·6

C. white.

Slow angular change. The distance has probably increased.

Σ.	334·3	4n.	1·09	1831·02
Mä.	331·3		·27	40·38
O.Σ.	·1	1n.	·07	5·31
	325·4	,,	·21	8·25
	·8	,,	·64	64·30
	149·3	,,	·67	8·29
Se.	327·7	2n.	·41	56·74
De.	328·1	,,	·66	63·15
W. & S.	323·8	5	·78	72·19
	325·3	4	·61	·26
	326·1	4	·70	3·22
	·1	2	·7	·24
	·2	11	·69	4·16
	324·3	4	·80	6·28
	·9	4	...	·29
Gl.	326·0	1n.	1·6	4·18
Sp.	323·2		·70	7·19
Dob.	325·3	2n.	·45	·31

322   Σ. 1356.

ω LEONIS.

R. A.	Dec.	M.
$9^h\ 22^m$	9° 35′	6·2, 7

C. A, yellow; B, yellower.

This very difficult double star was discovered by H_1 in 1781, and he early suspected that these two stars were receding from each other, and subsequent observations confirmed his suspicion. On the 21st of April, 1795, they were ½ diam. of the small star asunder. Feb. 5. 1804, with a power of 527, the vacancy between them was nearly 1 diam. of the small one." Between Nov. 13, 1782, and Feb. 4, 1802, the angle had changed 19° 59′, "probably owing to a real motion of ω Leonis, for the effect of a parallactic motion would have shown itself in a contrary alteration of the angle of position." (*Phil. Trans.* 1804.)

So. (*Phil. Trans.* 1826, p. 154). A power of 420 with refractor by Lerebours, 8·4 inches aperture and 11 ft. focus, at the Royal Observatory, Paris, separated the small star "⅓ a diam. of the large star; with 560, ¾ of a diam.; with each power the stars are admirably defined, and as round as possible." This was on March 15, 1825. H_2 adds, "There can be little doubt, therefore, that this very curious double star is entitled to a place among revolving stars or Binary systems."

Neither H_2 nor So. could get measures of distance; they could only wedge it.

Dawes in 1831 says, "decidedly elongated."

O.Σ. It is evident that the distances in 1840 and 1842 were estimated by me much too great.

Mädler's elements, from observations to 1846, are

$$T = 1843\cdot408$$
$$\Omega = 159° \ 50′\cdot5$$
$$\lambda = 120 \ 27 \cdot5$$
$$i = \ 50 \ 38 \cdot2$$
$$e = 0\cdot62564 = \sin. \ 38° \ 43′\cdot8$$
$$m = 183′\cdot711$$
$$a = 0″\cdot8505$$
$$\pi' = 0 \ \cdot03544$$
$$P = 117\cdot577 \text{ years;}$$

while those from observations extending to 1841 are

Perihelion passage	1849·76
Mean annual motion	+261 ·72
Node	135° 11′
Perihelion from node	185 27
Inclination	46 34
Excentricity	0·64338
Semi-axis major	0″·857
Period	82·533 years.

Klinkerfues, in 1858, gave the following elements :—

	Node.	γ	λ	ε	P	T	α
I.	111° 51′	57° 14′	217° 22′	0·3605	$133^{yrs}\cdot35$	1876·44	0″·703
II.	169 12	60 13	84 10	·7225	227 ·77	41·40	1 ·307
III.	162 13	54 25	107 9	·6286	142 ·41	43·39	1 ·092

Doberck in 1876 published the following as "definitive until further observations under the now more favourable circumstances have been taken" :—

$$
\begin{array}{ll}
\text{Node} & 148°\ 46' \\
\gamma & 64\ \ 5 \\
\lambda & 121\ \ 4 \\
e & 0\cdot5360 \\
P & 110\cdot82\ \text{years} \\
T & 1841\cdot81.
\end{array}
$$

On these Dunér remarks that they are probably in better agreement with the observations than any preceding elements.

	°	n.	"	
H₂.	110·9		...	1783·26
	130·9		...	1804·09
Σ.	153·9	5n.	0·97	25·21
	163·4	2n.	·51	32·25
	172·8	3n.	·44	3·29
	173·9	,,	...	5·34
	358·7	,,	...	6·28
	180·0		...	8·33
So.	154·2	6	...	26·11
Sm.	180·0		0·5	32·11
		round		4·25
	355·0	elongated		9·33
	193·0		0·3	43·14
O.Σ.	247·5	2n.	·49	0·29
	302·3	4n.	·41	2·31
	316·8	2n.	·37	3·30
	320·9	3n.	·48	4·29
	321·0	,,	·44	5·31
	322·9	2n.	·45	6·30
	328·8	,,	·53	7·33
	332·1	4n.	·43	8·32
	331·8	3n.	·43	9·32
	335·8	,,	·48	50·63
	339·0	,,	·46	2·66
	348·7	2n.	·47	5·32
	358·1	1n.	·52	7·28
	6·7	2n.	·60	9·30
	10·2	,,	·62	60·28
	11·9	,,	·56	1·28
	29·2	1n.	·52	4·30
	44·2	3n.	·55	8·63
	53·6	2n.	·58	70·28
	56·7	3n.	·57	1·30
	58·8	2n.	·52	2·31
	63·6	3n.	·59	3·96
Da.	354·5	1	...	41·18
	300·6	3	...	2·26
	299·0	2	0·45	3·17
	347·1	3	·6	54·17
	345·7	3	·53	·26
Mä.	280·2		·85	43·14
	330·0		...	6·28
	338·4		...	7·24
	337·1		...	8·33
	342·5		...	51·24
	350·0	4n.	0·47	2·30
	346·5		·35	3·36
	351·9		...	4·31
	359·3		...	5·28
	1·0		0·36	6·42

	°	n.	"	
Ja.	343·3	2n.	0·45	1853·18
	350·0	,,	·4	·96
	2·3		·4	6·22
	5·5		·5	7·98
	4·6		·4	8·10
Se.	0·0	5n.	...	5·29
	2·3	,,	0·35	7·86
	32·9	1n.	·30	66·30
Wi.	6·2		...	55·34
Mo.	355	elongated		7·29
		round		8·29
De.	30·0	1n.	wedg^d·	65·25
	25·7	,,	,,	8·12
	52·4	,,	,,	70·15
	51·5	,,	,,	1·13
	53·9	,,	,,	2·20
	60·2	4n.	in contact	3·42
	64·6	5n.	0·46	5·25
Eng.	23·0	2n.	·50	65·67
	33·7		·57	7·34
Du.	37·9	2n.	·27	70·33
	42·7	1n.	·3	1·31
	66·7	5n.	·42	5·31
W. & S.	65·3	4	·57	2·18
	67·3	5	·4	·19
	53·8	2	...	3·23
	58·7	2	...	·23
	74·7	5	0·5	6·26
	72·3	4	·6	·29
Gl.	57·0	3	·4	3·29
Schi.	62·7	1n.	·49	5·25
Sp.	·7		·49	·26
Dob.	52·6	3n.	...	6·23
	73·0	,,	0·51	7·21
Pl.	71·2	5n.	·54	·21

323   Σ. 1365.

134 (B) HYDRÆ.

R. A.	Dec.	M.
9ʰ 25ᵐ	2° 0'	7, 8

C. A, yellowish ; B, bluish white.

		n.		
So.	164·3		3·76	1825·11
Σ.	163·0	1n.	·17	·28
	164·3	,,	·03	8·27
	162·0	,,	·00	31·29
	·0	,,	·13	5·26
Mä.	161·5	2n.	·52	42·28
	·7	,,	·33	3·22
	163·3	1n.	·41	5·13

324   Σ. 1374.

30 (B) LEONIS MINORIS.

R. A.	Dec.	M.
9ʰ 34ᵐ	39° 30'	7, 8·3

C. A, yellowish ; B, very blue.

Dunér gives

$$1855\cdot72. \quad \Delta = 3''\cdot37.$$
$$P = 277°\cdot8 + 0°\cdot25\,(t - 1850\cdot0).$$

H₁.	261·5	1n.	...	1783·06
Σ.	274·7	3n.	3·31	1828·34
H₂.	273·5	„	·79	30·21
Mä.	274·9	1n.	·77	44·27
Mo.	275·0	2n.	·35	52·28
	277·7	6n.	·54	5·30
	279·0	2n.	·43	6·12
Du.	284·1	7n.	·11	72·81

325   Σ. 1377.

P. IX. 161 SEXTANTIS.

R. A.	Dec.	M.
9ʰ 37·2ᵐ	3° 11′	7·9, 11

C. A, yellowish; B, blue.

The change in angle between 1830 and 1868 amounts to about 4°, that in distance to about 0″·3. Secchi's measure in 1856 appears to be so seriously in error that one is led to suppose it refers to another system. (O.Σ.)

Σ.	142·2	4n.	3·31	1830·24
Mä.	140·6		·37	6·41
	137·8		·22	47·30
Se.	129·4	1n.	·12	56·28
O.Σ.	145·9	„	·75	68·29
W. & S.	136·8	„	...	73·24

326   O.Σ. 521.

υ URSÆ MAJORIS.

R. A.	Dec.	M.
9ʰ 42ᵐ	59° 36′	4·2, 11·8

C. yellowish.

The two stars have a considerable common proper motion.

O.Σ.   295·3 | 7n. | 11·32 | 1855·58

327   Σ. 1385.

R. A.	Dec.	M.
9ʰ 43·4ᵐ	17° 7′	8·5, 10·7

Σ.	0·2	3n.	1·23	1829·94
Mä.	356·6	2n.	·10	42·18
De.	351·0		·10	63·53

328   O.Σ. 208.

φ URSÆ MAJORIS.

R. A.	Dec.
9ʰ 44ᵐ	54° 37′

Magnitudes.—O.Σ., 5, 5·6; Mädler, 5, 5;

Dawes, 5½, 5¾. Secchi "estimated the diameters as 4 : 5."

Dawes was sure that the star in the n.f. quadrant was the smaller.

One of O.Σ.'s discoveries. Mädler, with the observations from 1845 to 1851 before him, thought that a direct motion had been maintained, and that the distance had decreased since 1843. Dawes, too, was of opinion that there was a slow increase of angle; and after he had received all O.Σ.'s measures, he was convinced of the binarity of the star. O.Σ., writing in 1875, suspects a feeble increase in the distance between 1873 and 1875, and observes that if this be so the periastre has been passed, and the elements of the orbit may soon be calculated with success.

O.Σ.	8·0	4n.	0·48	1843·11
	10·5	„	·36	7·65
	14·9	3n.	·33	50·39
	·9	4n.	·32	1·90
	18·3	„	·36	3·64
	36·7	5n.	·38	8·80
	47·9	3n.	·37	61·74
	48·3	2n.	·25	5·42
	77·6		·23	72·42
	96·6	3n.	oblong	3·45
	115·0	2n.	„	5·47
Da.	25·9	5	0·4	54·28
Mä.	193·8	3n.	·4	46·01
	196·8	2n.	·3	7·41
	207·2	4n.	·31	51·39
	209·7	„	·24	2·40
Se.	30·6	1n.	·3	7·34
Kn.	45·9		·24	66·40
Du.	46	oblong		9·37
	44	„		·43
	80	„		70·42
	83	oblong?		·43
W. & S.	...	single		3·24

329   Σ. 1389.

R. A.	Dec.	M.
9ʰ 45·5ᵐ	27° 33′	8, 9

C. yellowish.

Σ.	329·2	3n.	1·67	1830·61
Mä.	327·2		·64	43·19
De.	316·7		·99	63·66

330   Σ. 1386.

R. A.	Dec.	M.
9ʰ 45·6ᵐ	69° 28′	8·2, 8·2

This star is in the Nebula Messier 81. Very slow retrograde motion.

H₂.	302.3°		1.5″	1831.10
Σ.	296.0	3n.	.98	32.11
Mä.	293.1	2n.	.84	42.69
	291.3	In.	.22	5.31
O.Σ.	115.8	5n.	2.01	.93
Kn.	295.7	In.	1.61	64.10
De.	294.2		.89	9.15
Gl.	291.4	In.	...	70.12

331   8 SEXTANTIS.

R. A.	Dec.	M.
9ʰ 46.6ᵐ	−7° 32′	6, 6.5

A star first seen double by Mr. Alvan Clark when observing with one of his earliest glasses 4¾ in. aperture in 1854. Dawes had a strong impression that this star would prove a binary.

Da.	50.1	5	0.6	1854.17
	51.2	3	.5	.26
	38.2	5	.5	60.34
W. & S.	...	single?		72.19
De.	173.8	elongated		3.26
	169.0	,,		5.30

332   O.Σ. 210.

R. A.	Dec.	M.
9ʰ 55.1ᵐ	46° 57′	7.5, 8.3

Very slow retrograde motion.

O.Σ.	270.6	3n.	0.93	1845.27
Mä.	278.1		.8	3.31
	269.5		.7	5.43
	272.0		.8	6.32
	268.3		.8	7.41
	267.7	8n.	.75	8.38
De.	.2	3n.	.80	68.57
Du.	271.9	2n.	.84	70.80

333   Σ. 1406.

R. A.	Dec.	M.
9ʰ 59ᵐ	31° 40′	8, 8.7

Σ.	228.2	3n.	1.14	1830.27
Mä.	231.6	2n.	0.96	44.28
O.Σ.	236.0	3n.	1.21	5.60

334   O.Σ. 213.

R. A.	Dec.	M.
10ʰ 6ᵐ	28° 1′	7.8, 9.5

O.Σ.	117.7	In.	0.87	1843.30
	121.6	,,	.96	4.26
	115.3	,,	1.02	8.25
	114.7	,,	.19	60.29
	107.0	,,	0.93	71.28
	115.1	,,	.99	4.28
De.	113.2	3n.	1.11	67.83

335   O.Σ. 215.

R. A.	Dec.	M.
10ʰ 9.7ᵐ	18° 20′	7, 7.2

The distance has increased, and this has been accompanied by a considerable diminution in the angular movement.

O.Σ.	266.52	4n.	0.47	1844.54
	258.50	2n.	.45	8.32
	254.52	4n.	.48	51.34
	243.70	2n.	.60	60.30
	233.60	3n.	.74	7.20
	231.50	2n.	.68	9.78
	229.15	,,	.82	75.81
Mä.	257.8		.30	49.04
Se.	243.5	In.	.47	57.34
De.	233.6	3n.	.74	67.20
Sp.	223.4		.63	75.32

336   O.Σ. 523.

39 LEONIS.

R. A.	Dec.	M.
10ʰ 10.6ᵐ	23° 42′	5.8, 11.4

The companion is probably variable. The two stars have a large common proper motion : it amounts to −0″.44 in R. A., and +0″.08 in N. P. D.

O.Σ.	295.65	4n.	6.73	1851.26
	298.05	2n.	.96	61.24
Da.	295.55	In.	...	54.28
De.	300.33	3n.	6.69	66.86

337   Σ. 1423.

R. A.	Dec.	M.
10ʰ 12.6ᵐ	21° 10′	8.6, 9.3

C. yellowish.

Probably a binary.

Σ.	99.3	6n.	1.12	1830.94
Se.	76.8	In.	0.40	56.28
De.	.8		1.27	65.23

338   Σ. 1424.

γ LEONIS.

R. A.	Dec.	M.
10ʰ 13.4ᵐ	20° 27′	2, 3.5

C. Σ., A, golden ; B, greenish red ; H₁, "white, white with a little pale red" ; H₂ and So., "both reddish" ; Sm., "bright orange, greenish yellow."

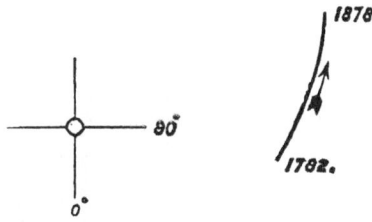

siderably, with a corresponding diminution of the angular movement ; for the change in angle per annum between 1782 and 1828 was 0°·41, and but 0°·28 between 1828 and 1872.

For some account of the distant star see Lists of Measures ; it is of the 7th magnitude. Whether or not it forms with γ a Ternary system the measures are insufficient to show.

The proper motion is thus given :—

Piazzi, R.A. + 0″35 Dec. − 0″20
Bessel   + 0·35   − 0·15
Argelander + 0·30   − 0·14

Doberck has published the following elements :—

	1876.	1879.
T	1741·11	1741·00
Node	111° 50′	111° 34′
λ	194 22	195 22
γ	43 49	43 6
e	0·7390	0·7327
P	402·62 yrs.	407·04 yrs.
a	2″·00.	1″·98.

This beautiful double star was discovered by H_1 in 1782 In his famous paper (*Phil. Trans.* 1803) he examines the motion at length : he finds that the change in 21 years and 38 days amount to 13° 58′, and thence infers a rough period of 1200 years, and that the changes must be ascribed to orbital motion. "The result of a great number of observations on the vacancy between the two stars made with the magnifying powers of 278, 460, 651, 840, 932, 1504, 2010, 2589, 3168, 4294, 5489, and 6652, is that with the standard power and aperture of the 7 feet telescope, the interval in 1782 was ¼ of a diameter of the small star, and is now ¾. With the same telescope, and a power of 2010, it was formerly ½ of a diameter of the small star, and is now full one diameter. In the years 1795, 1796, and 1798 the interval was found to have gradually increased, and all observations conspire to prove that the stars are ½ a diameter of the small one farther asunder than they were formerly. The proportion of the diameter of γ to that of x [the companion] I have, by many observations, estimated as 5 to 4."

H_2 wrote, in 1824 : "There can be no doubt of the motion of γ Leonis, though it is probably less rapid than supposed by Sir W. Herschel. That no mistake in the quadrant (n.f. for s.f.) was made in the observations made in the years 1782-3 is proved by the diagrams made at the time." "The mean annual motion from the most distant observations comes out + 0°·30, direct, or in the direction n.f.s.p." In 1826 H_2 adds, "The present observations, therefore, confirm this motion fully in point of reality and direction, but indicate an acceleration which (considering the number of observations) may have some claim to probability. The distances disagree more than might have been expected."

Mädler paid much attention to this fine star, and was strongly impressed with the idea that measures of it made after sunset were very likely to be erroneous. He strongly recommended that this star should always be observed in full sunshine.

The slow increase in the angle was noted by all the great observers, Σ., Sm., Da., Mä., etc.

O.Σ. The distance has augmented con-

H_1. Feb. 16, 1782 : 7° 37′ n.f.
April 18, 1783 : 5 24 n.f.
Jan. 24, 1800 : 3 15 ·75.
Feb. 19, 1800 : 5 33 ·45 ; the measure is too open :—3° 22′·5 ; this is better, but still open enough.
Mch. 26, 1800 : 3° 46′·8.
Jan. 22, 1802 : 6 4 s.f.
Feb. 10, 1803 : 3 33 s.f.
Mch. 22, 1803 : 6 34 s.f., and 6° 31′ s.f.

A B.

			°	″	
H_2 & So.	98·4	3n.	3·24	1822.24
	101·3	6n.	2·71	5·30
	102·4	,,	3·03	30·28
	104·2	1n.	2·65	3·22
Σ.	102·0	6n.	·45	28·14
	103·2	5n.	·48	31·34
	·4	,,	·50	2·75
	104·9	,,	·56	5·16
Da.	101·8	3n.	·54	0·39
	102·8	5n.	·52	1·33
	·9	8n.	·64	2·31
	103·7	3n.	·64	3·18
	105·8	,,	·84	40·29
	106·2	2n.	·83	1·23
	·5	,,	·72	2·33
	107·4	,,	·85	3·26
	·8	,,	·80	7·28
	108·1	5n.	·82	8·46
	·7	2n.	·80	51·87
	109·7	3n.	·84	4·37
	·2	2n.	3·12	9·37
	110·3	5n.	·09	60·37
	·1	1n.	·01	4·50
	·3	3n.	·17	5·37
Sm.	103·2		2·6	31·36
	104·9		·5	6·42

	°		″	
Sm.	106·0		2·6	1839·23
	107·2		·8	43·18
Encke.	100·6	2	3·26	37·19
	104·2	4	·54	8·33
	105·8	12	2·90	9·34
Ga.	·8		·90	9·36
Ka.	107·6	6n.	·89	40·15
	105·2	,,	·96	1·35
	107·1	8n.	·72	2·37
	109·7	7n.	·97	3·33
	110·3		3·11	66·28
O.Σ.	107·5	5n.	2·83	40·35
	·2	,,	·81	1·40
	106·9	1n.	·51	2·43
	107·1	2n.	·74	4·31
	109·6	1n.	·81	5·35
	108·5	2n.	·79	6·34
	·0	3n.	·81	7·35
	107·5	4n.	·66	8·36
	106·8	3n.	·79	9·35
	107·3	1n.	·88	50·32
	108·7	,,	·74	2·37
	107·7	,,	·85	5·32
	109·0	,,	3·02	7·28
	112·1	,,	2·96	8·38
	110·7	,,	3·15	9·35
	109·3	,,	·24	60·33
	110·8	3n.	·04	1·36
	111·2	2n.	·05	2·36
	110·2	1n.	·35	6·36
	113·0	2n.	·15	8·36
	111·2	1n.	·12	70·35
	114·2	2n.	·26	1·34
	·3	1n.	·27	4·42
Ch.	102·9	,,	2·76	41·20
	105·4	,,	·85	2·28
	106·1	,,	·87	4·25
Mä.	105·1	8n.	·78	1·26
	·9	4n.	·77	2·23
	107·1	5n.	·78	6·27
	·7	10n.	·64	8·39
	108·0	8n.	·74	51·28
	·9	3n.	·81	3·82
	107·9	16n.	·78	4·48
	108·7	9n.	·88	6·21
	·7	,,	·67	7·34
	·7	12n.	·94	8·38
	·9	9n.	·92	9·34
Ja.	105·6		·90	45·80
	107·8	10	·91	53·22
	108·4	10	3·07	·96
	109·6	3n.	2·92	6·79
	·0	,,	3·09	7·76
Hi.	107·4		...	45·89
D.O.	101·1		3·40	7·22
	111·5		2·96	·26
Bond.	108·3		·9	8·27
Flt.	·1	25	·84	50·91
	·4	38	3·00	3·21
Mo.	105·6	20	·16	2·29
	108·8	30	·05	5·34
	110·1	20	·07	60·12
De.	108·1	6n.	2·94	54·36
	109·6	5n.	3·03	5·23

	°		″	
De.	109·9	5n.	3·14	1856·19
	·3	2n.	2·91	62·78
	·3	9n.	·84	3·35
	110·3	7n.	·99	6·90
	111·2	2n.	3·00	8·37
	110·6	3n.	·12	70·27
	·2	4n.	·13	1·28
	·8	3n.	·15	2·34
	111·2	2n.	·29	3·25
	·6	4n.	·14	5·27
Se.	108·1	,,	·05	55·35
	110·3	5n.	2·97	6·95
	108·1	4n.	3·05	8·87
	110·3	3n.	·18	65·04
Wi.	111·1		·07	55·29
	·6		2·87	6·29
Au.	109·6		3·35	61·32
Po.	108·7	30	·22	54·13
	109·8	43	...	5·10
	108·8	60	3·32	61·13
M.	107·3	1n.	·10	2·35
	110·0	,,	·16	7·34
	111·9	,,	·55	8·39
	109·6	,,	·53	71·23
	113·6	,,	·56	·44
	112·3	,,	·69	4·32
	111·8	3n.	·50	5·16
Ro.	109·6	2n.	·24	63·21
	·7		·25	·21
Eng.	112·9	3n.	·39	4·31
	111·5		·24	5·42
Kn.	110·5	3n.	·21	6·21
	112·7	,,	·03	71·38
Ta.	109·1	2n.	·38	66·27
	108·4	1n.	·17	7·23
	111·6	,,	·16	8·18
	110·4	2n.	·97	70·32
	108·6	3n.	4·53	1·38
	109·2	1n.	2·62	4·32
	112·9		·93	6·34
Du.	111·2	11n.	·98	69·39
	·9	4n.	3·10	70·38
	112·3	8n.	2·98	1·44
	·4	,,	3·14	2·44
	113·2	3n.	·06	4·12
	·4	6n.	·10	5·46
Gl.	110·7	5	·11	0·30
	113·0	4	·1	1·32
	110·0	6	·0	·41
	110·9	5	...	3·30
	112·6	5	3·7	4·12
W. & S.	113·0	4	·78	1·48
	112·6	4	·36	2·19
	·5	5	·50	3·23
	·6	5	·43	4·20
	110·2	4	·04	3·23
	111·8	8	...	·25
	112·9	5	...	6·25
Sp.	110·9		3·38	5·29
Schi.	·8	1n.	·38	·28
Dob.	112·7	7n.	·84	6·16
	111·1	8n.	·63	7·23
Pl.	·8	2n.	·51	6·45
Fl.	112·0	1n.	·30	7·41

A C.

H₁.	295°·2		111″	1782
	300·0		...	83
Be.	294·8		196·5	1825
Se.	293·6	In.	215·0	56
	·0	,,	217·8	59
Po.	·5	10	·1	61
Fl.	292·8		229·3	77*

339 Σ. 1426.

145 (B) LEONIS.

R. A. 10ʰ 14·2ᵐ Dec. 7° 2′ M. A 7·8, B 8·3, C 9·3

C. A and B, yellowish.

Σ. discovered the duplicity of the larger star.

Se. says "the motion in angle appears certain"; but Dawes, writing in 1867, observes, "The measures at different epochs scarcely decide the question of relative motion in the close pair; the discordances being rather unusually large even for so difficult an object." He also says, "There seems to be no doubt of the fixity of the small distant star with respect to the close pair."

A B. The distance appears to have increased about 0″·1, and the angle about 4°, between 1833 and 1847. (O.Σ.)

$\frac{A + B}{2}$ and C. Here also there has been an increase in the distance.

A B.

Σ.	256·7	3n.	0·62	1832·26
	267·2	In.	·8	6·28
O.Σ.	262·6	3n.	·77	40·30
	263·7	In.	·88	68·29
Mä.	262·0	3n.	·55	42·25
Da.	257·7		·73	·30
	263·3		·88	54·16
Se.	271·8	3n.	·65	6·25
De.	269·7		·78	69·15
W. & S.	single ?			
	could not divide it			74·21
	278·3	2	...	6·26
W.O.	276·3	In.	0·72	·35
	277·6	,,	·60	·36
Dob.	274·0	3n.	...	·26

A C.

H₁	5·0		...	1782·13
H₂ & So.	9·8	7	6·72	1821·10
Σ.	·1	3n.	7·43	32·22
	8·5	In.	·29	6·28

* This star was also observed by Flamsteed in 1691, T. Mayer in 1755, and C. Mayer in 1777, the differences of R. A. being, respectively, 2, 4, 4·75 seconds.

O.Σ.	8·3	3n.	7·83	1840·30
	12·0	In	·88	68·29
Da.	9·4		...	52·33
Se.	4·8	2n.	7·68	6·25
De.	9·7		·57	67·17
Ta.	11·3	In.	·79	71·36
	8·58	,,	·88	6·34
Gl.	11·0	2	8·0	4·22
W.O.	9·3	In.	7·81	6·35
	10·5	,,	8·03	·36
W. & S.	·9	3	·25	·26
	·7	2	7·78	·29
Dob.	7·8	4n.	9·17	·24

340 O.Σ. 216.

R. A. 10ʰ 16·3ᵐ Dec. 15° 57′ M. 7, 10·5

O.Σ.	167·9	3n.	2·06	1845·62
	150·6	,,	...	73·29
De.	151·1		1·66	66·89

341 Σ. 1429.

R. A. 10ʰ 18·4ᵐ Dec. 25° 14′ M. 8·3, 8·3

Σ.	272·2	3n.	1·48	1827·29
	267·4		·58	33·26
H₂.	270·0	I		2·30
Da.	265·8		·37	49·76
De.	263·2		·09	66 55
W. & S.	·3	In.	0·95	73·24
	elongated			4·21
Gl.	262·5	In.	...	3·30

342 Σ. 1428.

R. A. 10ʰ 18·4ᵐ Dec. 53° 14′ M. 7·5, 7·8

C. white.

Dunér's formulæ are

$$\Delta = 3''\cdot68 - 0''\cdot015 \, (t - 1850\cdot0).$$
$$P = 85°\cdot7 + 0°\cdot102 \, (t - 1850\cdot0) + 0°\cdot0004 \, (t - 1850\cdot0)^{2}.$$

H₂.	83·7	2n.	4·10	1830·60
Σ.	84·3	3n.	3·84	1·69
Sm.	85·0		·6	2·49
Mä.	86·7	2n.	·99	44·21
De.	85·5	In.	·69	58·00
Se.	86·6	2n.	·75	·44
Mo.	·2	,,	·42	9·27
Du.	88·2	4n.	·36	71·32

343   OΣ. 217.

R. A. 10ʰ 20ᵐ	Dec. 17° 50′		M. 7·3, 7·8	
OΣ.	150·4	2n.	0″49	1844·27
	143·5	1n.	·55	·31
	149·3	,,	·50	8·33
	151·8	,,	·60	75·33
De.	148·1	3n.	·82	67·24

344   OΣ. 218.

R. A. 10ʰ 21ᵐ	Dec. 4° 10′		M. 7·3, 9·2	
OΣ.	66·7	2n.	1·24	1844·29
	63·1	1n.	·21	8·31
	60·6	,,	·26	61·24
	61·8	,,	·23	4·31
	59·1	,,	·08	8·29
Se.	65·2	,,	0·91	57·34
De.	·9	3n.	·98	67·28

345   OΣ. 219.

R. A. 10ʰ 22ᵐ	Dec. 51° 36′		M. 7, 10·3	
OΣ.	299·2	1n.	13·22	1844·31
	297·6	,,	·15	8·25
	·9	,,	·25	50·39
De.	·3	3n.	12·78	67·93

346   Σ. 1439.

R. A. 10ʰ 23·5ᵐ	Dec. 21° 25′		M. 8, 8·5	
Σ.	131·4	3n.	2·02	1829·26
H₂.	129·2		1	31·30
Da.	·0	1n.	...	40·29
Se.	123·4	2n.	1·98	56·78
Ta.	124·3	1n.	...	66·37
	122·3	,,	2·33	71·36
	121·0	,,	1·25	4·32
OΣ.	123·9	,,	·89	68·29
De.	122·5	4n.	·81	·72
Gl.	121·0	1n.	2·0	74·22
W. & S.	122·3	2	1·96	5·27
	1·0	4	·82	·30

347   Σ. 1445.

R. A. 10ʰ 26·6ᵐ	Dec. −0° 15′		M. 8·8, 11·8	

C. A, yellowish.

Σ.	167·4	3n.	2·42	1827·58
De.	159·4		·95	64·87

348   Σ. 1450.

49 LEONIS.

R. A. 10ʰ 28·7ᵐ	Dec. 9° 17′		M. 6, 8·7	
Σ.	161·0	2n.	2″43	1825·31
	·1		·39	30·76
	·1		·37	3·50
	160·5	2n.	·49	5·31
Mä.	159·5		·54	7·47
	158·9		·59	42·29
	·4		·73	51·26
	·6		·75	2·26
Sm.	158·1		·5	38·37
	159·0		·8	55·29
OΣ.	·0		·97	2·09
Da.	155·3		·60	4·28
Se.	157·1	2n.	·3	6·74
M.	169·8		·53	63·19
Ta.	154·9	1n.	...	6·14
	160·2	,,	1·97	7·23
	...	,,	2·29	76·34
Gl.	156·6	,,	·8	4·70
W. & S.	·0	,,	·54	·20

349   Σ. 222.

R. A. 10ʰ 30ᵐ	Dec. 60° 46′		M. 6·7, 10·7	
OΣ.	340·3	3n.	4·56	1847·72
De.	345·5	4n.	·58	68·70

350   Σ. 1457.

R. A. 10ʰ 32·5ᵐ	Dec. 6° 21′		M. 7·4, 8·4	

C. whitish yellow.

Se. says " the motion in angle is beyond doubt ; " and Dawes observes that there is but little doubt of its binarity.

Σ.	287·8	4n.	0·71	1829·55
OΣ.	302·0	3n.	·75	40·29
	310·1	1n.	1·01	64·30
	316·1	,,	·11	8·29
	315·0	,,	·15	71·31
Mä.	304·9	5n.	0·69	42·24
	305·9	1n.	·66	6·30
	312·4	3n.	·84	51·27
	311·2	4n.	·99	2·29
	...	2n.	·97	3·29
	310·1	4n.	...	·34
Da.	302·7	2n.	0·92	0·78
Se.	307·5	5n.	·76	6·24
De.	304·6	4n.	1·0	8·30
	309·8	3n.	0·91	63·2₀
W. & S.	312·3	5	·81	72·2₈
	·7	3	...	3·2₃
	316·1	4	1·26	5·28

W. & S.	313.6	5	1".20	1875.28

W. & S.	313.6	5	1".20	1875.28
	315.3	7	.12	.32
	314.2	5	.17	6.34
Gl.	311.9	4	.0	3.20
Schi.	.9	1n.	.18	5.36
Sp.	312.0		.18	.37
Dob.	316.6	2n.	...	6.31
	314.5	,,	1.38	7.21

351 O.Σ. 224.

R. A. 10h 33.4m   Dec. 9° 28'   M. 7.2, 9.2

Retrograde motion. The distance may have increased.

O.Σ.	20	1n.	wedg^d.	1844.31
	352.6	,,	0.48	51.27
	348.8	,,	.59	61.26
	328.4	,,	.59	71.31
	336.8	,,	.55	2.31
Mä.	15.6	,,	.22	48.29
Se.	13.6	1n.	elong^d.	57.34
De.	339.3		...	67.32

352 O.Σ. 225.

R. A. 10h 33m   Dec. 19° 52'   M. 7.5, 9.8

O.Σ.	351.3	2n.	6.57	1844.30
	350.8	1n.	.62	5.28
	349.2	,,	.64	8.31
	350.3	,,	.51	9.32
	351.2	,,	.41	75.33
De.	.7	3n.	.08	67.26

353 O.Σ. 227.

R. A. 10h 35m   Dec. 11° 21'   M. 7.5, 8.5

O.Σ.	326.4	2n.	0.54	1844.30
	.7	1n.	.51	8.33
De.	334.1	3n.	...	67.38

354 Σ. 1465.

R. A. 10h 36.2m   Dec. 45° 15'   M. 8.5, 8.8

Dunér gives

1855.76. Δ = 2".23.
P = 10.6 - 0°.1 (t - 1850.0).

Σ.	14.4	3n.	2.24	1829.32
Mä.	7.3	,,	.15	44.59
Du.	11.2	6n.	.39	71.13

355 Σ. 1472.

R. A. 10h 41m   Dec. 13° 36'   M. 7.8, 8.5

The distance has increased considerably.

Σ.	39.5	3n.	33".74	1828.55
O.Σ.	.5	2n.	34.36	40.33
	.1	1n.	.62	2.32
	38.8	,,	.88	6.33
	39.1	,,	.11	9.32
De.	38.6	6n.	35.81	64.57

356 O.Σ. 228.

R. A. 10h 40.7m   Dec. 23° 12'   M. 7.2, 8.1

O.Σ.	203.3	2n.	0.53	1844.30
	179.6	1n.	.43	8.24
	192.4		.33	50.38
	201.7	1n.	.63	71.31
De.	199.1	3n.	...	67.20
Sp.	13.2		.37	75.35

357 O.Σ. 229.

R. A. 10h 43.1m   Dec. 41° 44'   M. 6.7, 7.1

C. white.

O.Σ.	347.0	5n.	0.68	1846.65
	344.2	4n.	.78	59.84
Mä.	350.2	2n.	.80	45.42
Da.	347.3	1n.	.92	9.27
De.	338.3	3n.	.78	66.95
Du.	.7	4n.	.78	72.05

358 Σ. 1486.

R. A. 10h 48m   Dec. 52° 45'   M. 7.5, 8.8

Σ.	102.8	2n.	28.32	1831.38
Mä.	104.5	1n.	.66	40.40

359 O.Σ. 230.

R. A. 10h 48m   Dec. 21° 24'   M. 7.7, 11.2

O.Σ.	5.9	1n.	8.90	1844.30
	3.5	,,	.59	5.28
	4.7	,,	.45	51.27
De.	11.4	3n.	.30	67.27

360   Σ. 1487.

54 LEONIS.

R. A.	Dec.	M.
10ʰ 49.1ᵐ	25° 23′	5, 7

Slow increase in angle and distance.
The proper motion of A is — 0ˢ·002 in R. A., and 0″·00 in N. P. D.
Dunér has

$$1850.02. \quad \Delta = 6''.20.$$
$$P = 103°.7 + 0°.064 \, (t - 1850.0).$$

	°		″	
H₁.	99.2	1n.	...	1782.13
	100.6	,,	...	1802.10
H₂ & So.	98.3	2n.	7.02	21.68
Σ.	103.5	4n.	...	.17
	...	14n.	6.20	.60
	102.8	4n.	.17	30.35
Sm.	.5		.5	2.26
	.7		.2	9.33
Da.	103.8	2n.	.22	40.30
	.8	4n.	.26	50.19
O.Σ.	.5	,,	.34	40.61
	104.3	6n.	.21	7.68
	.6	5n.	.40	60.34
Mä.	.3	,,	.02	43.78
	103.9	,,	.11	51.27
	.0	10n.	.34	3.85
	104.4	4n.	.03	61.14
De.	103.3	,,	.24	52.25
	.8	1n.	5.78	5.94
Mo.	102.5	3n.	6.34	4.27
Se.	104.3	,,	.33	6.59
Ro.	103.8	1n.	.35	63.15
Ta.	...	2n.	.43	6.33
	103.6	1n.	.28	7.23
	105.4	,,	.81	8.21
	103.5	2n.	7.05	70.32
Du.	106.1	,,	5.90	69.27
W. & S.	105.4	4	6.3	73.24
	.0	6	.0	4.20
Gl.	104.6	1n.	.5	.21
Sp.	.5		.25	5.32
Dob.	.9	4n.	.50	7.32

361   Σ. 1500.

R. A.	Dec.	M.
10ʰ 53.9ᵐ	− 2° 51′	7.6, 8.2

C. white.
The distance has increased, and the angle has diminished.

Σ.	330.9	2n.	1.06	1825.22
	321.4	4n.	0.96	32.09
O.Σ.	320.2	2n.	1.11	40.30
	317.3	1n.	.53	71.31

	°		″	
Da.	317.1		0.91	1841.20
	315.8		1.15	60.34
Mä.	322.9		.06	42.24
Se.	318.3	3n.	.05	56.28
Ta.	314.4	1n.	.46	67.23
	319.5	,,	2.09	76.34
W. & S.	313.8	3	1.37	3.23
	315.4	5	.39	.24
	314.1	1	.40	4.22
	317.1	4	.27	5.28
	316.9	7	...	.29
	315.6	5	.34	6.34
	316.2	5	.31	.36
Gl.	314.2	5	.42	4.22
Sp.	313.2		.41	5.37
Dob.	324.1	2n.	.76	6.31

362   Σ. 1504.

R. A.	Dec.	M.
10ʰ 59.2ᵐ	4° 17′	7.5, 7.6

The relative brightness of the two stars is probably variable. Other observers have always noted the following star as either of equal or of greater magnitude than the preceding. Our measures between 1845 and 1848 are decisive as to the superiority of the preceding star. (O.Σ.)

Σ.	275.6	5n.	1.07	1829.13
Sm.	280.0		.3	36.29
Mä.	279.0		0.95	42.27
O.Σ.	278.1	3n.	1.07	54.99
De.	283.4	5n.	.11	66.67
W. & S.	284.5	7	.19	74.23
	.7	10	.14	.23
	286.4	4	.08	5.28
	285.5	4	.12	.29
	286.7	4	...	6.34
Gl.	283.9	4	1.2	4.22
Sp.	286.3		.16	5.37

363   Σ. 1514.

R. A.	Dec.	M.
11ʰ 4.1ᵐ	66° 46′	8.5, 10

Σ.	334.9	4n.	1.15	1832.92
Mä.	336.6	2n.	.02	45.55
De.	344.0		.15	66.70

364   Σ. 1516.

R. A.	Dec.	M.
11ʰ 7.4ᵐ	74° 7′	7, 7.5, 10

C. Σ., yellowish, ashy yellow; Sc. and De., white.

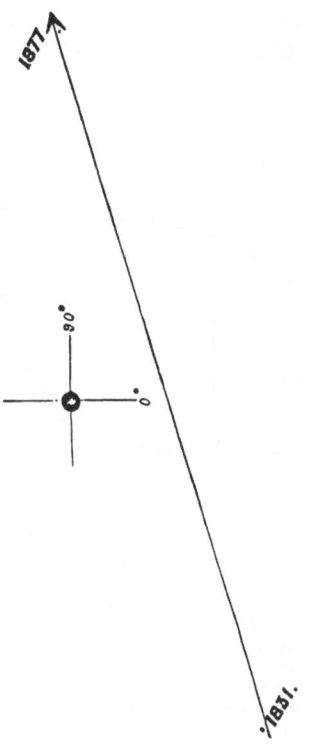

A B.

Σ.	298.6	∓15°	29.26	1790.21*
...			14.22	1823.92†
	298.7	2n.	9.93	31.54
	299.3	,,	.56	2.84
	.7	,,	.25	3.46
	300.9	,,	8.94	4.43
	301.6	4n.	.42	5.56
	302.6	8n.	.13	6.64
	304.0	3n.	7.78	7.61
So.	296.2		12.48	24.28
H₂.	301.0		12	31.40
	300.0		9.85	3.26
O.Σ.	308.3	3n.	6.66	40.45
	310.6	2n.	.17	1.92
	322.8	,,	4.17	6.94
	329.6	,,	3.37	8.94
	341.7	,,	2.97	50.92
	22.5	1n.	.32	5.47
	48.0	3n.	.96	8.87
	64.4	2n.	3.70	61.33
	76.9	,,	5.18	5.43
	81.2	,,	6.30	8.58
	86.0	3n.	7.90	72.54
Ka.	312.6		5.23	43.65
De.	8.3	6n.	2.70	54.55
	16.1	,,	.81	5.14
	23.7	1n.	...	.99
	26.0	3n.	2.66	6.16
	44.2	5n.	.87	8.29
	68.6	1n.	4.03	62.95
	70.5	3n.	.18	3.48
Mo.	6.8	30	2.49	54.26
	56.5	26	3.25	60.21
Se.	29.5		2.61	56.29
M.	71.4	1n.	4.43	64.44
	256.9	,,	5.78	7.27
	258.6	,,	6.18	8.40
Eng.	80.3	3n.	5.29	6.29
Du.	78.5	1n.	.58	7.78
	82.9	4n.	6.49	9.51
	86.8	2n.	7.02	70.46
	87.3	1n.	.32	1.49
	90.5	3n.	8.81	5.54
W. & S.	269.1	4	7.6	2.30
	.0	5	.8	3.25
	89.6	4	8.81	5.28
Gl.	270.5	4	9.1	4.13
Fl.	91.0	1n.	.5	7.37

A C.

O.Σ.	294.1	3n.	8.18	1858.87
	297.0	2n.	.06	61.33
	.2	3n.	7.73	6.49
	.7		.60	72.54
Du.	299.4	2n.	.48	5.54

Σ. states that the first observation of this star is found in the *Mem. Acad. Parisiensis* 1790, p. 389. He thinks that South's distance is probably not very accurate, and that the motion is probably orbital.

Se. found that the graphical construction gives a straight line for the apparent orbit with minimum distance about 1853.

O.Σ. in 1858 discovered a third star near A. Finding that the latest measures of A B depart widely from the rectilinear path deduced by Σ. (see *P. M.*, p. ccxxviii.), the investigation was repeated with the following results :—

$$\Delta A = -1''{\cdot}318 + 0''{\cdot}4070 \, (T - 1850{\cdot}0).$$
$$\Delta D = -2''{\cdot}914 - 0''{\cdot}1077 \, (T - 1850{\cdot}0).$$

According to these the minimum distance 2″·48 was reached in 1854·8, and the angle was 14° 50′, and on the whole the observations are well represented. The difference between the formulæ of Σ. and O.Σ. is probably due to application of the systematic corrections. The star B is therefore fixed, and has no physical relation with A. The star C, on the contrary, participates in the large proper motion of A.

* Lalandius.
† Σ. from six observations with the transit instrument.

365   Σ. 1517.

R. A.	Dec.
11ʰ 7ˑ4ᵐ	20° 47′

Magnitudes.—7ˑ3, 7ˑ3. Se. 7ˑ5, 7ˑ7.

The variability of the relative brightness was suspected by Σ., and the observations of his distinguished son confirm the suspicion. Σ. and Se. generally noted the following star as the brighter ; O.Σ., on the contrary, has invariably regarded it as the fainter of the two.

The common proper motion is − 0″·377 in R.A., and + 0″·16 in N. P. D.

Dunér has obtained the formulæ

$$\Delta = 0''·89 - 0''·013 \ (t - 1850·0).$$
$$1849·94. \quad P = 287°·3.$$

Σ.	108°4	8n.	1″08	1832·20
	286·4	1n.	·06	52·22
H₂.	288·0	,,	·19	30·24
	289·3	,,	0·81	1·07
Sm.	288·6	,,	1·2	3·31
Da.	·9	,,	·09	40·30
	283·6	1n.	0·91	54·17
O.Σ.	108·8	,,	·93	45·31
	104·8	,, .	·81	·32
	100·3	,,	·88	71·25
Se.	287·4	3n.	·78	56·98
De.	285·0	2n.	elongᵈ·	7·97
Du.	287·3	5n.	0·59	69·39
	288·8	1n.	·64	72·40
	286·8	,,	·58	5·20
Ta.	284·3	,,	...	2·40
Gl.	283·5	4	...	4·22
W. & S.	105·1	4	0·82	5·28
	6·0	4	·75	6·34
Sp.	284·0		·67	5·39
Dob.	104·5	1n.	...	6·30
	·8	3n.	0·61	7·26

366   O.Σ. 232.

R. A.	Dec.	M.
11ʰ 8ᵐ	38° 14′	7, 7·8

O.Σ.	238·1	5n.	0·72	1849·93
Mä.	237·0	4n.	·55	47·05
De.	234·0	3n.	·6	67·66
Du.	235·5	6n.	·56	74·49

367   O.Σ. 233.

R. A.	Dec.	M.
11ʰ 11ᵐ	67° 21′	6·9, 9·8

O.Σ.	334·72	4n.	4·98	1849·87
De.	337·67	3n.	·93	68·59

368   Σ. 1523.

ξ URSÆ MAJORIS.

R. A.	Dec.	M.
11ʰ 11·8ᵐ	32° 13′	7·3, 8·2

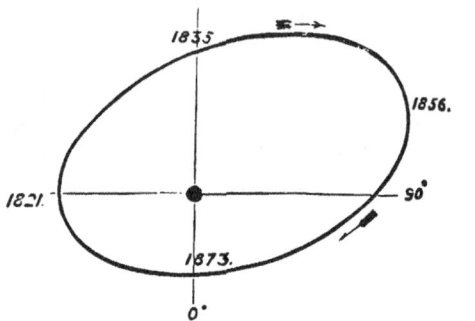

This remarkable pair was discovered by H₁ in 1780. He writes, "1780, May 2nd. A fine double star, nearly of equal magnitudes, and ⅔ of a diameter asunder ; exactly estimated. May 21. Unequal stars ; very bright ; one diameter of the large star asunder. But the air is rather tremulous. A little wind. Feb. 4, 1802. 7° 31′ s.f. ; very accurate. Jan. 29, 1804. 2° 38′ s.f."

In his review (*Phil. Trans.* 1804, p. 363) H₁ says : "This double star has undergone a very extraordinary change in the angle of position. Dec. 29, 1781, the smallest of the two stars was 53° 47′ s.f. ; Feb. 4, 1802, it was 7° 31′ ; and, Jan. 29, 1804, the position was only 2° 38′. This gives a motion of 51° 9′ for 22 years 41 days, and amounts to 2° 19′ per year." And he proceeds to point out the possible causes of these changes.

H₂ (*Phil. Trans.* 1824, p. 146) writes, "The position and dates here given (11° 33′ s.p., 1823·29 ; 2″·809, 1823·19,—means of 58 and 20 measures respectively), as well as the distance, are all derived on the supposition of each measure being independent of all the rest, and all equally good. The angle thus obtained from no less than 58 measures, with its corresponding mean date, will serve for an epoch in which the computer, at some future period, may rely with confidence in any investigation relative to the orbit of this star. A double star in which the two stars are nearly equal, connected undoubtedly in a binary system by their mutual gravitation, and revolving round their common centre of gravity, with a motion so rapid as to admit of being traced and measured from *month* to *month*, must be allowed to be a phenomenon of no common interest, and deserving every attention, both from the practical and theoretical astronomer." And he further observes that the observations "indicate a remarkable alteration in its velocity, which can only be accounted for by supposing the relative orbit to be one of great ellipticity." And in the *Memoirs of the R. A. S.*, vol. v., p. 34, he adds, " In the interval from 1839 to 1841 we may now securely predict that this star will have completed a full revolution from the epoch of the first measurement of its position in 1781, having occupied therein a periodic time of about 59 years."

In 1830 Savary's elements appeared in the *Connaissance des Tems;* they are as follows :—

$$a = 3''\!\cdot\!857$$
$$e = 0\cdot4164$$
$$\pi = 304^\circ\ 58'$$
$$\Omega = 95\ 22$$
$$\gamma = 59\ 40$$
$$\lambda = 131\ 38$$
$$P = 58\cdot2625 \text{ years}$$
$$n = -6^\circ\cdot1786$$
$$\tau = 1817\cdot25.$$

H_2 published the following in 1832 ; they were obtained by means of his graphical process :—

APPARENT ELLIPSE.

Major semi-axis	... ...	$3''\!\cdot\!169$
Position thereof ...	... ...	$281^\circ\ 20'$
Minor semi-axis...	... ...	$1''\!\cdot\!756$
Greatest maximum of distance ...		$4\cdot101$
Position thereof ...	... ...	$110^\circ\ 0'$
Least minimum ...	... ...	$2''\!\cdot\!338$
Position thereof ...	... ...	$257^\circ\ 35'$
Greatest minimum	... ...	$2''\!\cdot\!119$
Position thereof ...	... ...	$206^\circ\ 0'$
Least minimum ...	... ...	$1''\!\cdot\!059$
Position thereof ...	... ...	$0^\circ\ 40'.$

REAL ELLIPSE.

Major semi-axis	$a = 3''\!\cdot\!278$	
Excentricity ...	$e = 0\cdot3777$	

Position of perihelion $\quad\pi = 307^\circ\ 29'$
,, node ... $\Omega = 97\ 47$
Inclination ... ... $\gamma = 56\ 6$
Angle between major
axis and line of nodes $\lambda = 134^\circ\ 22'$
Period ... ... $P = 60\cdot72$ years
Mean motion... ... $n = -5^\circ\cdot9289$
Perihelion passage ... $\tau = 1816\cdot73.$

Savary's orbit represented the observations very well ; Sir John's not so well, as he himself points out and explains.

Mädler, making use of the observations to 1847, arrived at the following elements :

$$\text{Period} = 61\cdot30 \text{ years}$$
$$\Omega = 96^\circ\ 21'\!\cdot\!9$$
$$i = 50\ 55\cdot4$$
$$\lambda = 132\ 28\cdot7$$
$$\phi = 23\ 48\cdot7$$
$$\tau = 1817\cdot102.*$$

These also satisfied the observed angles and distances very well on the whole.

Villarceau computed an orbit for this star and obtained the following results :—

$$a = 2''\!\cdot\!439$$
$$e = 0\cdot43148$$
$$\Omega = 95^\circ\cdot83$$
$$\gamma = 52\cdot82$$
$$\lambda = 128\cdot95$$
$$P = 61\cdot576 \text{ years}$$
$$\epsilon = 1816\cdot86.$$

And Captain Jacob :—

$$\epsilon = 1816\cdot66$$
$$\Omega = 96^\circ\ 6'$$
$$\lambda = 129\ 28$$
$$\gamma = 53\ 52$$
$$e = 0\cdot4116$$
$$P = 61\cdot175 \text{ years}$$
$$a = 2''\!\cdot\!82.$$

The following extract from Dr. Ball's paper will show the relative value of some of the above orbits :—

Epoch.	Observed position.	Savary.		H_2.		Mädler.		Villarceau.	
		Calc^d position.	Differ- ence.	Calc^d position.	Differ- ence.	Calc^d position.	Differ- ence.	Calc^d position.	Differ- ence.
1781·97	143·78	143·89	+0·11	140·08	−3·70	14·85	−0·93	144·12	+0·34
1840·29	150·85	143·65	−7·20	148·92	−1·93	153·80	+2·95	155·48	+4·63
52·13	122·28	112·84	−9·44	116·22	−6·06	121·37	−0·91	122·22	−0·06
63·23	96·66	90·92	−5·74	92·85	−3·81	98·54	+1·88	98·48	+1·82
68·30	77·50	72·02	−5·48	75·27	−2·23	83·02	+5·52	84·02	+6·52
72·28	24·19	20·57	−3·62	41·62	+17·43	57·19	+33·00	62·10	+37·91

* Mädler also published the following elements : $a = 2''\!\cdot\!417$, $e = 0\cdot41359$, $\Omega = 98^\circ\cdot87$, $\gamma = 54^\circ\cdot93$, $\lambda = 130^\circ\cdot80$, $P = 61\cdot464$ years, $\epsilon = 1816\cdot44$: these were obtained from the measures made up to 1843.

This eminent astronomer, using the measures made up to 1836, computed the following elements :— $\epsilon = 1816\cdot95$, $\Omega = 95^\circ$, $\lambda = 129^\circ\ 41'$, $\gamma = 52^\circ\ 16'$, $e = 0\cdot40368$, $P = 60\cdot4596$, $a = 2\cdot290.$

On this Dr. Ball observes that Savary's elements represent the observations up to 1825 very well, and then begin to fail. Sir John's. although but a first approximation, present no violent differences till 1872,—a point where both Mädler's and Villarceau's also fail.

Dr. Ball's elements are,

$$a = 2''\!\cdot\!591$$
$$e = 0\cdot3786$$

$$\Omega = 103°\!\cdot\!6$$
$$\gamma = 53\ \cdot1$$
$$\lambda = 135\ \cdot3$$
$$P = 59\cdot88 \text{ years}$$
$$n = 6°\!\cdot\!012$$
$$\epsilon = 1816\cdot405.$$

A short extract from his Table VI. will exhibit the results when these elements and the observations are compared :—

Epoch.	Observed position.	Computed position.	Difference.	Observer.
1781·97	143°·78	147°·37	+ 3°·59	H₁.
1802·09	97·52	98·53	+ 1·01	,,
23·29	258·45	259·20	+ 0·75	H₂ and So.
32·27	196·72	195·13	− 1·59	Da.
40·29	150·85	152·95	+ 1·10	,,
52·13	122·28	120·87	− 1·41	Mi.
63·23	96·66	94·87	− 1·79	De.
72·28	24·19	26·47	+ 2·28	Br.

Dr. Ball also gives an ephemeris showing the position angle at intervals of three months from 1872·50 to 1878·75, and says, "The greatest velocity of change in the angular position occurs about 1873·25. At this date the rate will be fully 20° per annum. The periastron passage takes place about 1876·28; thus the period included in the ephemeris contains the most critical part of the entire orbit." A portion of the ephemeris is here given, together with the observed positions by De. up to 1877 :—

Epoch.	Computed position.	Observed position.		Epoch.	Computed position.	Observed position.	
		Angle.	Date.			Angle.	Date.
			1800+				1800+
1872·50	22·4°	19·39	72·32	1875·75	321·4°		
·75	17·5			76·00	317·9		
73·00	12·5			·25	314·7	304·8	76·30
·25	7·3	358·9	73·33	·50	311·6		
·50	2·2			·75	308·6		
·75	357·2			77·00	305·8		
74·00	352·1			·25	303·2	294·9	77·26
·25	347·0	333·6	74·35	·50	300·7		
·50	342·2			·75	298·4		
·75	337·6			78·00	296·1		
75·00	333·2			·25	293·9		
·25	329·0	317·5	75·27	·50	291·8		
·50	325·1			·75	289·7		

Dunér, in 1876, computed a set of elements : he gives

$$\tau = 1875\cdot29$$
$$\omega = 234°\ 3'$$
$$\Omega = 101\ \cdot5 \text{ (Equ. 1850·0)}$$
$$i = 56\ \cdot9$$
$$e = 0\cdot3952$$
$$\mu = -5\cdot9215$$

$$a = 2''\!\cdot\!549$$
$$P = 60\cdot79 \text{ years.}$$

A table is also given comparing the observed and computed quantities from 1781·96 to 1876·48. How well the agreement is maintained all through this period the following selection will show :—

Date.	Δ	P	d Δ	d P	Observer.
1781·96		143·8°		− 1·4°	H₁.
1820·13	″	276·4	″	+ 4·6	Σ.
31·25	1·90	201·1	+ 0·21	− 2·6	H₂.
40·25	2·08	152·4	− 0·13	− 1·8	Ka.
50·30	3·38	124·3	+ 0·43	− 1·0	Ja.
60·08	2·84	105·3	− 0·06	+ 0·7	Mo.
70·33	1·35	57·2	+ 0·09	0·0	Gl.
72·24	1·06	22·1	+ 0·13	− 2·6	W. and S.
73·33	0·98	358·9	+ 0·09	+ 0·5	De.
73·42	0·85	358·4	− 0·04	+ 2·2	Du.
74·13	1·00	338·4	+ 0·05	− 1·5	Gl.
74·35	1·02	333·6	+ 0·04	− 1·7	De.
75·27	1·09	317·6	− 0·04	− 1·4	,,
75·45	1·08	316·4	− 0·08	+ 0·1	Du.
76·48	1·31	303·8	− 0·04	+ 0·2	,,

Dunér observes that his elements are intended merely to represent tolerably well the existing observations, and that we must wait till about 1880 before correct elements can be obtained.

The most recent elements of this interesting star are those published in No. I. of the observations of the *University Observatory*, Oxford ; they are as follows :—

$$T = 1875\cdot26$$
$$P = 60\cdot80 \text{ years}$$
$$\Omega = 100° \ 13'$$
$$\gamma = 56 \ 40$$
$$\lambda = 235 \ 0$$
$$e = 0\cdot41590$$
$$a = 2''\cdot580.$$

	°		″	
H₁.	143·7	1n.	...	1781·97
	97·5	,,	...	1802·09
	92·6	,,	...	4·08
H₂ & So.	258·4	58	2·81	23·29
	244·5	55	·44	5·22
	224·5		·0	8·39
	212·3		·0	30·20
	201·1	4n.	1·90	1·25
	189·8		2·06	3·14
Σ.	238·7	3n.	1·74	26·20
	228·2	4n.	·71	7·27
	213·5	7n.	·67	9·35
	203·8	5n.	·70	31·44
	195·9	,,	·75	2·41
	188·4	,,	·76	3·84
	180·1	,,	·76	5·41
	171·2	4n.	·97	6·44
	165·3	3n.	·92	7·47
	160·3	9n.	2·26	8·43
	...	1n.	·29	40·44
Be.	203·1		1·85	30·86
	199·0		·93	1·39
Da.	201·9	17n.	·98	1·34
	196·7	10n.	·76	2·27
	189·8	4n,	·98	3·23
	171·3	1n.	·92	6·28

	°		″	
Da.	150·8	5n.	2·44	1840·29
	147·9	4n.	·40	1·21
	144·7	,,	·44	2·27
	142·1	7n.	·48	3·28
	131·6	1n.	·57	7·30
	·4	,,	·75	·34
	129·5	,,	·70	8·13
	·3	3n.	·94	·19
	126·6	5n.	3·01	9·30
	122·9	2n.	2·98	51·31
	119·9	1n.	...	2·38
	115·8	3n.	2·95	4·36
Sm.	196·9		1·9	32·29
	180·2		·9	5·37
	170·9		·8	6·33
	165·5		·8	7·28
	160·7		2·1	8·48
	156·9		·0	9·23
	143·2		·3	43·16
Encke.	168·5	9	·48	37·31
	166·8	17	·56	·49
	157·9	4	1·89	9·46
Galle.	·9		·89	·47
Ka.	152·2	7n.	2·08	40·25
	145·1		·7	2·50
	140·2		·55	3·60
	87·8		·08	66·45
	29·7		1·00	72·09
O.Σ.	153·6	6n.	2·28	40·40
	150·5	,,	·22	1·40
	147·5	4n.	·34	2·40
	140·4	3n.	·45	4·34
	138·1	2n.	·51	5·46
	137·2	4n.	·56	6·37
	133·1	3n.	·61	7·41
	130·0	5n.	·66	8·41
	127·6	4n.	·78	9·37
	124·1	,,	·67	50·39
	122·9	5n.	·80	1·41
	120·6	4n.	·75	2·40
	119·0	,,	·88	3·40
	115·9	,,	·90	4·38
	·2	3n.	·85	5·44

18

	°		"			°		"
O.Σ.	110·2	3n.	2·96	1857·46	Se.	114·3	In.	2·96	1855·29
	108·9	,,	·96	8·39		113·9	4n.	3·13	6·26
	104·9	5n.	·84	9·57		109·7	2n.	·11	7·36
	101·1	4n.	·69	61·40		89·9	In.	2·53	65·51
	99·3	,,	·62	2·39		86·5	,,	·26	6·31
	95·7	2n.	·55	3·46	Au.	100·4		3·03	1·56
	94·2	3n.	·33	4·42	Ro.	95·5	In.	2·79	3·14
	85·4	,,	·11	6·40		93·3	,,	·59	·50
	81·0	2n.	1·91	7·47	M.	87·2	In.	·60	·19
	72·6	4n.	·63	8·42		91·6	,,	·62	·21
	59·2	,,	·32	70·18		77·0	,,	1·77	8·39
	45·7	2n.	·12	1·40	Eng.	95·8	6n.	2·58	4·16
	17·8	3n.	0·96	2·41		91·4	19n.	·44	5·12
	358·4	5n.	·96	3·43	Ta.	93·0	5n.	·72	6·23
	338·1	3n.	1·03	4·41		82·2	In.	...	7·23
Ch.	150·6	2n.	2·45	41·19		79·1	2n.	2·49	8·23
	148·2	,,	·71	2·30		70·8	In.	...	70·35
	139·3	In.	·53	4·25		66·2	,,	...	1·39
Mä.	150·2	7n.	·44	1·29		68·0	2n.	1·28	2·35
	146·9	4n.	·41	2·24		335·6	,,	...	4·29
	122·1	9n.	3·02	51·78		334·5	In.	1·64	6·34
	120·8	6n.	2·75	2·35	Du.	68·6	11n.	·29	69·40
	118·8	13n.	·93	3·32		53·8	9n.	·16	70·43
	116·3	14n.	·89	4·37		40·0	11n.	0·98	1·47
	115·7	2n.	·87	5·44		16·6	14n.	·91	2·46
	112·7	13n.	·97	6·42		358·4	In.	·85	3·42
	109·7	8n.	·75	7·43		335·1	4n.	·93	4·45
	108·8	5n.	·92	8·42		316·4	14n.	1·08	5·45
	106·1	3n.	·97	9·37	Gl.	59·3	5	·4	0·22
	99·9	4n.	·90	62·35		55·1	6	·3	·44
D.O.	131·1		3·12	47·11		50·8	5	·3	1·13
	127·3		2·78	·12		44·6	5	·1	·50
Bond.	128·6		·7	8·45		32·0	5	·1	2·00
	129·7		3·1	·45		29·4	5	·0	·10
Ja.	124·2	10	·37	50·30		356·0	5	·0	3·50
	120·9	5	·01	2·29		342·2	3	0·8	·93
	119·4	10	·01	3·19		343·0	3	·7	·93
	117·0	10	·11	·93		338·0	5	1·0	4·12
Flt.	123·1	6n.	2·83	1·19		339·0	5	0·8	·13
	119·8	,,	·92	2·20		338·3	0	1·2	·13
	118·9	,,	·98	3·23	W. & S.	43·9	4	·1	1·48
Mi.	122·3	56	·89	2·13		23·1	7	·04	2·13
	118·8	32	3·01	3·19		22·9	14	0·97	·17
Mo.	117·7	3n.	2·90	2·34		23·0	6	1·19	·20
	108·1		·85	8·20		·3	4	0·91	·20
	105·3	2n.	·84	60·08		20·3	3	...	·24
De.	116·5	7n.	3·16	54·64		19·7	4	1·18	·33
	115·4	5n.	·2	5·21		22·2	4	·07	·38
	112·3	7n.	·19	6·34		23·9	23	0·90	3·22
	108·6	In.	...	7·89		359·6	4	...	·36
	·0	5n.	3·14	8·25		338·0	9	0·92	4·17
	98·1	3n.	2·81	62·85		334·5	2	...	·23
	96·4	16n.	·50	3·29		317·6	4	1·30	5·28
	93·6	10n.	·27	4·37		316·9	5	·22	·29
	90·1	9n.	·18	5·32		318·0	5	·31	·32
	86·7	10n.	·06	6·30		316·3	10	...	·43
	82·2	8n.	1·90	7·31		304·9	4	1·59	6·34
	77·5	,,	·73	8·30		306·4	9	·45	·36
	57·7	9n.	·39	70·24		305·2	6	·30	·37
	47·7	8n.	·20	1·22		294·1	6	·47	7·39
	19·3	10n.	·07	2·32		·7	7	·59	·40
	358·9	,,	0·97	3·33	Fer.	295·0	7	·50	·41
	333·6	6n.	1·01	4·35		15·4		0·97	2·47
	317·5	8n.	·09	5·27		337·0		1·47	4·20

Kn.	29°6	3n.	1″08	1872·08
	16·8	,,	·11	·44
Schi.	317·5	In.	·31	5·30
Sp.	·4		·31	·31
Dob.	311·7	In.	...	·99
	306·3	13n.	1·73	6·27
	304·2	10n.	·74	7·26
Pl.	301·2	5n.	·52	6·46
	297·0	7n.	·57	7·20

369 Σ. 1527.

LEONIS 339 (B).

R. A.	Dec.	M.
11ʰ 12·7ᵐ	14° 56′	7, 8

C. yellow, blue.

A appears to be variable : its magnitude is thus given : South, 8 ; Dawes, 8, 7·2 ; Σ., 8, 6·9 ; Du., 6·5, 6.
Certain change.
Dunér has the following formulæ :

$$1851·01. \quad \Delta = 3''·79.$$
$$P = 11°·8 + 0°·09 (t - 1850·0).$$

Σ.	9·7	In.	3·73	1822·20
	10·2	,,	·88	9·30
So.	·4	3n.	4·93	4·60
Da.	·2	In.	·00	40·60
	11·5	,,	3·90	54·27
Mä.	10·4	,.	4·10	44·27
Mo.	11·6	,,	3·93	55·70
Se.	12·2	2n.	·74	6·70
De.	13·3	3n.	·99	8·16
Ta.	12·8	2n.	·68	66·28
	15·3	In.	·24	7·24
	14·2	,,	·05	72·40
	10·9	,,	·16	4·32
	14·0	,,	·08	6·36
Du.	·4	4n.	·43	5·30
Pl.	11·4	3n.	·48	6·57

370 Σ. 1534.

R. A.	Dec.	M.
11ʰ 15·5ᵐ	18° 51′	8, 11·2

Σ.	342·2	2n.	4·79	1828·24
	339·0	,,	·88	33·28
Se.	332·2	In.	·08	56·30
De.	330·6	4n.	·74	64·76
O.Σ.	·9	In.	5·13	70·30
W. & S.	335·8	5	3·61	4·23
	3·6	5	...	·24
	4·5	3	...	6·35
Gl.	336·2	2	4·2	4·13

371 Σ. 1536.

ι LEONIS.

R. A.	Dec.
11ʰ 17·6ᵐ	11° 12′

Magnitudes.—Σ. 3·9, 7·1. Sm. 4, 7·5. Se. 4·2, 8·5. De. 4·8, 7·9.

C. Σ., A, yellowish ; B, blue.

One of Σ.'s discoveries. He found the angular motion indirect, and by the method of least squares obtained for the angle

$$\omega = 92°·38 - 0°·834 (t - 1832·01).$$

Smyth and Dawes assert its binary character.

Of late years the angle has not changed, and this accords with the increase in the distance. (O.Σ.)

The common proper motion is $+ 0''·133$ in R. H., and $+ 0''·028$ in N. P. D. (Σ.) Main gives $+ 0''·007$, $+ 0''·07$ in R. A. and N. P. D. respectively.

Dunér has the following formulæ :

$$\Delta = 2''·45 + 0''·01 (t - 1855·0).$$
$$P = 78°·9 - 0°·526 (t - 1855·0) + 0°·0021 (t - 1855·0).$$

Σ.	97·0	2n.	2·29	1827·81
	93·0	4n.	1·99	30·62
	90·4	3n.	2·17	3·34
	·3	,,	·40	5·33
	·1	In.	·41	7·39
Da.	91·8	,,	·44	4·00
	87·6	3n.	·44	40·29
	86·8	7n.	·52	1·29
	85·3	4n.	·45	2·27
	·3	,,	·63	3·27
	83·6	2n.	·47	7·72
	81·6	,,	·64	9·29
	80·6	3n.	·61	51·55
	79·5	2n.	·55	4·38
	76·0	In.	·68	60·29
	72·1	,,	·80	5·40
Sm.	90·5		·4	36·40
	87·7		·4	9·32
	86·0		·5	43·38
	81·3		·5	53·29
O.Σ.	91·0	3n.	·67	40·59
	92·3	In.	·22	1·40
	87·5	,,	·49	2·34
	86·3	2n.	·29	7·36
	83·1	In.	·23	9·36
	·6	,,	·42	51·37
	78·9	,,	·40	2·37
	80·8	,,	·70	8·38
	75·3	,,	·58	61·42
	76·6	,,	·58	2·39
	·0	,,	·80	6·36
	·3	,,	·66	8·36
Ch.	86·0	2n.	·41	41·23
	89·3	In.	·61	2·32
	87·4	2n.	·84	4·28

Mä.	86.6	5n.	2.29	1841.32
	.3	4n.	.27	2.22
	82.8	5n.	.31	6.31
	81.3	8n.	.35	7.35
	80.0	4n.	.47	51.28
	79.0	7n.	.42	2.38
	78.9	8n.	.70	3.34
	.8	6n.	.53	4.37
	76.1	7n.	.48	6.37
	.0	4n.	.38	7.37
	75.1	2n.	.46	8.35
Ka.	88.2	5n.	.30	42.59
	71.5	,,	.75	66.32
Ja.	81.2		.84	48.30
	79.7	10	.44	53.20
	78.7	11	.63	.96
	76.6		.64	8.21
Mo.	83.2	2n.	.71	3.35
Flt.	81.7		.09	5.27
De.	80.4	1n.	.2	.95
	78.6	2n.	.48	6.25
	79.4	1n.	.5	7.08
	76.7	3n.	.6	8.34
	.7	7n.	.51	63.23
	74.9	9n.	.56	6.08
	73.3	2n.	.55	8.24
	71.7	,,	.53	70.26
	.7	,,	.54	1.24
	70.6	1n.	.54	2.27
	.1	,,	.66	3.22
	71.1	,,	.57	4.22
	70.1	2n.	.54	5.19
Se.	76.4	5n.	.26	56.26
Po.	74.5		...	61.18
M.	73.9	1n.	.72	2.25
	72.8	,,	.75	7.27
	.8	,,	.84	8.40
	75.9	,,	.88	71.27
	67.6	,,	3.08	5.36
Eng.	76.8	5n.	2.92	65.70
Ta.	75.8	3n.	.91	6.28
	76.8	1n.	.91	7.24
	.8	,,	3.06	8.21
	78.9	,,	2.85	9.19
	.2	2n.	...	71.37
	77.0	,,	2.71	2.41
	76.6	1n.	.46	4.32
	69.2	,,	.46	6.36
Br.	74.0	2	.73	69.24
Gl.	71.7	4	.6	70.44
	72.0	4	.5	1.32
	71.2	2	.7	3.29
	67.0	5	.7	4.10
	.5	3	.7	.12
	71.0	6	.7	.13
	68.0	5	.69	.17
W. & S.	70.3	5	3.2	2.27
	71.0	7	...	.35
	70.1	4	2.57	3.19
	.5	4	.7	.23
	68.8	6	.81	.25
	69.8	5	.71	4.22
	68.5	4	.77	5.28
	70.5	8	.69	6.35
	.2	6	.62	.36

Fer.	73.2		2.01	1873.28
Du.	70.6	4n.	.58	5.31
Sp.	68.1		.73	.32
Schi.	.1	1n.	.73	.32
Dob.	69.6	,,	...	.99
	65.7	8n.	2.88	6.27
	64.5	4n.	.82	7.23
W.O.	70.3	,,	.73	6.29
	69.4	,,	.81	.31
Pl.	67.7	3n.	.86	7.10

372   Σ. 1543.

57 URSÆ MAJORIS.

R. A.	Dec.	M.
11h 22.6m	40° 0'	5.2, 8.2

B is probably variable. Dawes notes it as of the 9th mag., South as of the 10th, and H$_1$ saw it as a mere point of light.

The common secular proper motion is −7″.2 and + 2″.4. (Σ.)

Dunér has the following formulæ:

$$1851.77. \quad \Delta = 5''.40.$$
$$P = 10°.0 - 0°.11 \ (t - 1839.0).$$

H$_1$.	14.4	1n.	...	1783.10
So.	10.3	2n.	5.86	1825.25
Σ.	.7	6n.	.37	31.91
	7.2		.43	48.24
Sm.	9.9		.9	35.42
	8.3		.5	46.38
Ja.	.3		.5	.32
	6.8	2n.	.26	53.24
Da.	7.2		.52	48.70
Mä.	.6	1n.	.61	51.27
	9.5	,,	4.89	8.43
Mo.	6.0	,,	5.38	7.28
Se.	.5	2n.	.16	.89
De.	5.5	,,	.42	8.16
M.	355.6	1n.	4.72	64.43
Du.	5.8	5n.	5.46	73.05
Gl.	6.5	1n.	...	4.29
Sp.	5.3		5.62	5.33
Fl.	.2	1n.	.80	7.41

373   τ LEONIS.

R. A.	Dec.	M.
11h 21.8m	3° 31'	5, 7

The proper motion of τ is − 0s.001 in R. A., and + 0″.02 in N. P. D.

H$_1$.	165.3		90	1782.28
So.	169.8		95.2	1823
Be.	166.9		96.9	25
Σ.	169.6		94.7	34.94
Se.	252.2		.42	59.21
Eng.	171.7		93.4	63.26
Fl.	172.2	1n.	92.2	77.42

374   O.Σ. 234.

R. A.	Dec.	M.
11^h 24.3^m	$41°$ $58'$	7, 7.4

C. white.

Very distinct orbital motion.

O.Σ.	177.4	3n.	0″.42	1844.66
	188.9	,,	.37	8.66
	200.3	,,	.31	52.09
	243	2n.	oblong	8.88
	257	3n.	,,	61.35
		1n.	simple	6.49
	282	,,	oblong	70.46
De.			elongated	66.20

375   O.Σ. 235.

R. A.	Dec.	M.
11^h 23.8^m	$61°$ $45'$	6, 7.3

C. A, yellow; B, red.

Since 1856 the distance has increased considerably, and the angular motion has diminished in a corresponding degree.

O.Σ.	293.0	2n.	0.60	1844.90
	311.3	,,	.55	6.94
	318.6	,,	.52	9.89
	327.9	,,	.54	51.42
	331.5	,,	.55	2.94
	348.7	3n.	.52	6.51
	358.7	2n.	.68	8.92
	15.6	3n.	.68	61.74
	29.3	2n.	.81	5.46
	40.2	,,	.99	71.53
De.	38.2		.84	68.59

376   Σ. 1552.

R. A.	Dec.	M.
11^h 28.5^m	$17°$ $28'$	6, 7.3, 8.5

A B, probably binary. In A C, rectilinear motion.

A B.

H₁.	208.8		...	1782.28
H₁.	210.0		...	1802.18
So.	208.9		4.45	22.27
Σ.	209.4		3.01	9.94
H₂.	207.3	5n.	3	32.40
Mä.	210.9		.13	4.54
	211.0		.18	54.20
	.3		2.89	6.36
	210.6		3.22	7.37
Sm.	209.1		.5	35.38
Mo.	.5	20	.10	46.40
	208.9	30	.10	8.32
	.0	20	.18	54.37
	209.6	60	.46	5.31

Da.	210.8		3.03	1851.30
M.	208.8	In.	.28	61.33
Ro.	212.3	,,	.31	3.17
Se.	214.1		.55	5.33
Ta.	213.1	In.	4.14	8.21
	210.8	,,	3.58	70.35
Gl.	211.7	,,	.2	4.24
W. & S.	.1		.25	.24

A C.

H₁.	234.9		...	1782.28
	.2		53.72	3.29
So.	233.3		60.75	1822.27
Sm.	.9		58.8	35.38
Se.	234.5		63.33	65.33
W. & S.	235.4	In.	...	74.23
Gl.	234.0	,,	...	5.20

377   O.Σ. 236.

R. A.	Dec.	M.
11^h 29^m	$67°$ $0'$	7.5, 11

O.Σ.	209.2	3n.	2.33	1847.00
De.	213.4	In.	.71	68.13

378   Σ. 1553.

R. A.	Dec.	M.
11^h 30^m	$56°$ $48'$	7.3, 7.8

Probable change.   Dunér gives

$$1851.37.\quad \Delta = 5''.40.$$
$$P = 170°.7 - 0°.044\ (t - 1850.0).$$

Σ.	171.5	3n.	5.34	1832.58
Mä.	170.2	2n.	.45	44.38
Da.	171.3	In.	.38	51.29
De.	170.4	,,	.38	8.01
Mo.	.4	2n.	.30	9.25
O.Σ.	168.3	,,	.56	65.86
Du.	169.5	4n.	.51	75.30
Pl.	.1	2n.	.66	6.94

379   Σ. 1555.

R. A.	Dec.	M
11^h 30^m	$28°$ $27'$	6.4, 6.8

A B, probably binary. Of A C no other measures but Smyth's are known. Dunér gives

$$\Delta = 0''.98 - 0''.115\ (t - 1850.0).$$
$$P = 341°.7 + 0°.15\ (t - 1850.0).$$

A B.

Σ.	339.3	5n.	1.24	1829.12
H₂.	338.0	In.	...	30.26
Da.	340.3	,,	.45	2.24
Sm.	.1		.4	4.31

O.Σ.	341.4	In.	0.93	1841.41
	338.1	,,	1.05	2.34
	.8	,,	0.86	6.37
	337.1	,,	.79	9.36
	343.3	,,	.70	66.42
	342.0	,,	.67	8.36
Mä.	339.0	5n.	.94	42.96
Se.	.0	3n.	.80	55.95
De.	338.6	In.	1.0	6.09
Mo.	342.8	2n.	.14	9.35
Eng.	343.7	5n.	0.92	65.75
Du.	345.9	6n.	.78	70.06
Gl.	344.0	2	1.0	4.29
W. & S.	343.8	7	0.75	5.28
	342.8	4	.74	.30
	.8	3	.75	6.35
Sp.	344.0		.74	5.37
Dob.	337.1	In.	.71	7.33

A C.

Sm.	145.0			17.0	34.31

380   O.Σ. 237.

R. A.	Dec.	M.
11h 32.5m	41° 48'	7.4, 9

If Mädler's angle is correct, the angular change has amounted to 164° in 28 years.

O.Σ.	287.0	4n.	0.74	1845.82
	274.8	3n.	.92	61.68
Mä.	113.2		.64	47.40
De.	272.0	3n.	1.08	67.94
Sp.	277.0		.02	75.36

381   O.Σ. 243.

R. A.	Dec.	M.
11h 53.6m	54° 5'	7.8, 8.8

If Mädler's angle be correct, 18° of the apparent orbit have been described in 21 years.

O.Σ.	10.9	3n.	0.71	1846.04
Mä.	26.6		.42	.41
De.	8.9	3n.	.90	67.96

382   Σ. 1588.

R. A.	Dec.	M.
11h 56.1m	73° 2'	8.5, 8.7

Σ.	60.7	2n.	16.49	1831.59
Mä.	.1	,,	.33	45.55
De.	57.6		15.30	63.56

383   Σ. 1593.

R. A.	Dec.	M.
11h 57.4m .	—1° 47'	8 8.3

Σ.	18.2	3n.	1.43	1829.26
Mä.	24.1		.54	37.45
	28.5		.89	47.30
Se.	26.6	In.	.08	56.39

384   Σ. 1596.

R. A.	Dec.	M.
11h 58m	22° 8'	6, 7.5

Motion doubtful. Dunér gives these formulæ :—

$$1854.79. \quad \Delta = 3''.65.$$
$$P = 239.7 - 0°.025 \, (t - 1850.0).$$

H₁.	242.3	In.	...	1782.30
Σ.	239.9	,,	3.82	1827.28
	242.6	,,	.60	8.23
	240.2	,,	.77	9.30
	239.8	,,	.73	33.37
Mä.	239.1	,,	.76	41.32
	.6	2n.	.61	2.23
	238.9	,,	.47	3.33
	239.6	,,	.89	4.38
Mo.	.5	3n.	.92	55.33
	238.5	9n.	.72	9.27
Se.	239.5	3n.	.76	6.96
De.	240.0	In.	.84	8.07
Du.	239.5	3n.	.54	71.27
W. & S.	.2	2n.	.77	4.23
Gl.	.8	In.	.6	.30

385   Σ. 3123.

R. A.	Dec.	M.
11h 59.4m	69° 20'	7, 7

A very difficult object in 1832. Since 1851 there has been no trace of the companion.

Σ.	289.7	4n.	0.3	1832.20
O.Σ.	79	In.	oblong	40.42
	271	,,	,,	.45
	88.7	,,	0.44	1.41
	231		oblong	51.44
			single	8.44
			,,	61.26
			,,	2.39
			,,	8.56
De.			,,	2.95

386   Σ. 1602.

R. A.	Dec.	M.
12h 1.1m	69° 45'	7.5, 9

Σ.	179.8	2n.	13.00	1831.56
Mä.	178.7	,,	.70	45.54

387   Σ. 1604.

VIRGINIS 59 (B).

R. A.	Dec.	M.
12ʰ 3·2ᵐ	− 11° 11′	A 6·5, B 9, C 8

In A B the distance may have decreased slightly since 1831 : the angle appears to diminish very slowly.

C is in motion, rectilinear and uniform ; and Dunér gives the following formulæ :

$$\Delta \sin P = + 52''\cdot05 - 0''\cdot3074\,(t - 1850\cdot0) ;$$
$$\Delta \cos P = - 5''\cdot20 + 3''\cdot0995\,(t - 1850\cdot0) ;$$

whence it appears that the minimum distance, 10″, will be reached in A.D. 2008.

A B.

Σ.	93·3	3n.	11·98	1821·95
Mä.	94·8	In.	·01	44·35
Se.	92·8	,,	·75	56·40
De.	92·7		·16	64·19
Du.	91·6	2n.	·46	9·85
Fl.	91·5	In.	·6	77·40

A C.

Σ.	96·9	3n.	58·00	31·95
Se.	95·2	In.	50·38	56·40
De.	94·8		47·85	64·19
Du.	·0	2n.	46·04	9·85
Fl.	93·1	In.	41·9	77·40

388   Σ. 1606.

R. A.	Dec.	M.
12ʰ 4·7ᵐ	40° 34′	6·3, 7

The angle diminishes slowly. Probably binary.

Dunér gives the following formulæ :

$$1853\cdot01. \quad \Delta = 1''\cdot29.$$
$$P = 346°\cdot8 - 0°\cdot1\,(t - 1850\cdot0).$$

H₂.	348·4	3n.	0·89	1830·87
Σ.	·6	,,	1·39	31·48
Mä.	349·3	,,	·43	43·21
De.	246·8	In.	·1	56·46
Se.	344·7	2n.	·23	7·36
Du.	·8	5n.	·21	69·38

389   Σ. 1607.

R. A.	Dec.	M.
12ʰ 5·5ᵐ	36° 45′	7·8, 8·3

Considerable change both in angle and distance.

Σ.	350·3	3n.	33·07	1830·99
O.Σ.	352·5	In.	32·43	45·35
	·3	,,	·42	6·42
	356·1	,,	31·25	68·36

Mä.	352·7		32·91	1847·27
De.	355·0	2n.	31·35	63·31
Gl.	·0	2	32·0	74·30
W. & S.	357·2	6	30·77	6·48

390   O.Σ. 249.

R. A.	Dec.	M.
12ʰ 18·1ᵐ	54° 49′	7·2, 8, 11·2

A B.

O.Σ.	315·1	5n.	0·53	1853·19
De.	311·4	3n.	·5	68·04
	308·0		·5	72·46

$\frac{A+B}{2}$ and C.

O.Σ.	149·7	2n.	13·23	55·86

391   Σ. 1639.

COMÆ 68 (B).

R. A.	Dec.	M.
12ʰ 18·4ᵐ	26° 15′	6·7, 7·9

The distance has diminished considerably. Probably binary.

Σ.	290·9	6n.	1·18	1831·40
O.Σ.	293·2	In.	0·98	41·39
	289·8	2n.	1·13	2·36
	288·7	In.	·20	4·34
	·0	,,	0·93	55·32
	279·8	,,	·73	70·31
Se.	285·8	2n.	·85	56·90
Sp.	273·1		·4	75·39

392   O.Σ. 250.

R. A.	Dec.	M.
12ʰ 18ᵐ	43° 45′	7·7, 8

O.Σ.	330·7	3n.	0·44	1845·98
De.	321·3	In.	...	68·16

393   Σ. 1641.

R. A.	Dec.	M.
12ʰ 18·6ᵐ	38° 24′	10, 10

Rectilinear motion.

H₂.	53·4		4	1830·40
Σ.	50·4	2n.	6·14	1·38
De.	42·3		7·73	67·59

394   α CRUCIS.

R. A.	Dec.	M.
12ʰ 19·9ᵐ	−62° 34′	1·5, 2, 6

Proper motion of A_1 − 0ˢ·009 in R. A., and + 0″·02 in N. P. D.

A B form a binary system, while A C are probably an optical double-star.

A B.

H_2.	121·6		5·26	1834·39
	·0		·75	5·20
	120·8		·61	6·19
	·0		·55	7·18
	·0		·96	8·08
Ja.	·4		·74	47·10
	117·6	11	4·77	58·20
Po.	120		5·7	5·15
	118·5	40	4·98	61·18

A C.

H_2.	201·5		92·4	35·27
	·9		89·9	7·30
Ja.	202·0		89	47·25
	·2	2	90	58·20

395 Σ. 1643.

R. A. Dec. M.
$12^h 21·2^m$ 27° 42′ 8, 8·3

Probably binary.

Σ.	71·2	5n.	1·95	1830·36
H_2.	66·2		2·0	2·35
De.	54·4		1·79	64·75

396 Σ. 1644.

R. A. Dec. M.
$12^h 21·3^m$ 8° 3′ 8·7, 9·2

The distance has probably diminished.

Σ.	248·6	3n.	21·82	1827·55
De.	247·0		·08	67·89
	·0		20·88	70·31

397 O.Σ. 251.

R. A. Dec. M.
$12^h 23^m$ 32° 2′ 7·4, 9·1

Extraordinary discrepancies are presented by the measures of this difficult star.

O.Σ.	128·35	2n.	0·42	1843·77
	132·05	,,	·33	9·88
	156·55	,,	·49	52·42
De.	...		single	67
	149?		obl;·?	8

398 Σ. 1647.

191 (B) VIRGINIS.

R. A. Dec.
$12^h 24·5^m$ 10° 23′

Magnitudes.—Σ., 7·5, 7·8. Se., 7·5, 7·6. De., 7·5, 8·2. "The relative brightness is undoubtedly variable." (Σ.)

C. Σ. and Se., "white."

Σ. discovered the duplicity of this star, and also, from five years' observations, suspected direct motion. In 1836, however, he saw cause for changing his opinion.

Dawes' observations in 1840 "showed that the variation of angle continued in the same direction, accompanied possibly by a slight increase of distance."

Secchi says that "direct motion is undoubted."

With an increase of distance there has probably been a diminution in the angular motion. Secchi's distance is too small. (O.Σ.)

Σ.	198·6	2n.	1·25	1828·36
	202·8	3n.	·13	9·37
	203·5	1n.	·19	32·34
	205·2	,,	·21	3·34
	204·1	2n.	·24	6·32
H_2.	198·6		...	0·34
Mä.	204·2		1·20	5·06
	214·3		·35	51·27
	212·3		·26	2·31
O.Σ.	213·5	2n.	·50	40·32
	·9	1n.	·46	6·37
	217·6	,,	·57	61·24
	·6		·65	74·28
Da.	207·0	1n.	·27	40·31
	212·1	2n.	·17	8·43
	210·9	1n.	·36	54·37
De.	214·2		·2	5·81
	212·9	3n.	·39	63·24
Se.	211·6	2n.	·19	56·36
Eng.	218·0		·58	64·31
W. & S.	216·1	5	·42	73·36
	214·2	6	·15	4·29
	217·2	5	·28	5·29
	216·9	4	·08	·30
Gl.	215·7	3	·2	4·34
Fer.	209·1		·44	·23
Sp.	216·2		·30	5·31
W.O.	214·1	1n.	·33	6·35
	215·8	,,	·19	·39
	220·3	,,	·28	·39
Dob.	216·3	3n.	...	6·24
	214·4	2n.	1·55	7·22

399 Σ. 1658.

R. A. Dec. M.
$12^h 29^m$ 8° 6′ A 8, B 9·8, c 8

Probably binary.

A B.

Σ.	341·5	3n.	2·02	1830·64
Se.	348·8	2n.	1·90	56·90
De.	349·1		2·24	69·08
O.Σ.	350·6	1n.	·37	70·31
W. & S.	352·0	8	1·97	4·29
	·4	5	2·18	·30
Gl.	·0	3	·1	·34
Dob.	340·2	1n.	·27	7·26

A C.

O.Σ.	257·6	1n.	10·88	70·31

400   Σ. 1661.

R. A. 12h 30m	Dec. 12° 4'		M. 8·5, 8·5	
Σ.	226·0		2"·56	1828·67
Mä.	228·2		·63	43·33
	221·1		·42	4·24
Se.	227·3	2n.	·62	56·85
De.	234·4		·41	66·84
O.Σ.	232·6		·7	70·30

401   Σ. 1663.

R. A. 12° 31·2m	Dec. 21° 52'		M. 7·8, 8·7	
Σ.	117·5	3n.	0·81	1830·38
Mä.	123·3		·55	42·33
	119·7		·64	4·32
O.Σ.	124·1	4n.	·72	·26
Da.	112·4		·91	52·22
Se.	118·0		·40	7·34
De.	110·7		·77	68·55
W. & S.	100·3	In.	·8	74·31
	111·1	,,	·7	5·83
Gl.	100·8		...	4·36
Dob.	95·2	In.	...	7·30

402   Σ. 1664.

R. A. 12h 32·1m   Dec. − 10° 51'   M. 7·7, 8·8, 11, 11

C. A, yellow; B, blue.

Rectilinear motion.

Σ.	271·6	3n.	17·10	1830·23
De.	254·7		19·44	65·25
Gl.	253·2	In.	...	74·26
W. & S.	252·7	,,	...	·36

403   γ CENTAURI.

R. A. 12h 34·9m   Dec. −48° 18'   M. 4, 4

A binary system.

Common proper motion − 0"·022 in R. A., and + 0"·03 in N. P. D.

H2.	351·6		0·8	1835·38
	357·4		·8	6·38
	1·9		1·0	7·14
Ja.	20·6		0·7	56·20
	13·7	15	1·1	7·97
Po.	12·8	27	...	60·68
El.	8·5	70	1·3	76·63

404   Σ. 1669.

CORVI 58 (B).

R. A. 12h 35m   Dec. − 12° 21'   M. 6, 6·5

Σ.	298·9	3n.	5·44	1828·66
H2.	301·4		6·50	30·26
	302·6		9·2	1·30
	·3		7·38	7·31
Sm.	298·9		5·4	5·50
Se.	302·4	4n.	·78	56·53
M.	301·5		·95	63·30

405   Σ. 1670.

γ VIRGINIS.

R. A. 12h 35·6m   Dec. − 0° 47'   M. 3, 3

C. yellowish.

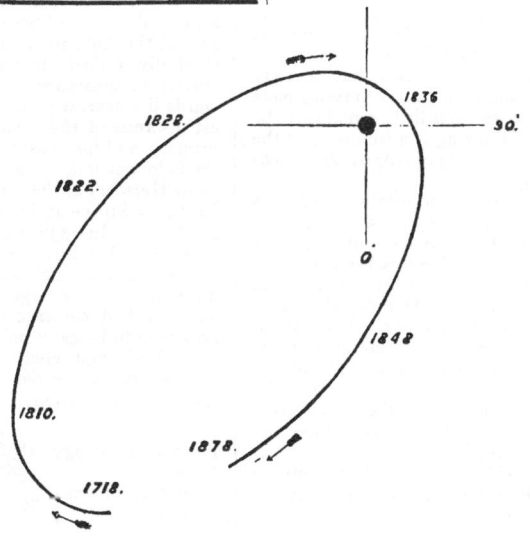

The variability in the relative brightness of the two stars has long been observed. In 1851 and 1852 O.Σ. paid special attention to this point, and the following results made when the stars passed the meridian will be read with much interest :—A is the south star, and B that to the north.

1851. April 17. A > B by 0·7 mag.
    28. ,,   ,, 0·5 ,,
  May 15 and 22. Very nearly equal. Perhaps A a little larger than the others.
  June 3. A > B by 0·2 mag.
    4. Equal, perfectly.
    5. A a little larger than the others.

1852. March 10. Equal, perfectly.
  April 3. A > B by 0·5 mag.
    6. A the larger, perhaps.
    8. Very nearly equal. A a little the larger, perhaps.
    21. A > B by 0·2 mag.
    29. Perfectly equal.
  May 4. A > B by 0·2 mag.
    19. ,,   ,,   ,,
    25. Very nearly equal. A may be a little the larger.
  June 7. Very nearly equal.
    14. Perfectly equal.

And he then observes that it is very remarkable that in the seven years 1825—1831 B was certainly the predominating star, while his observations almost give the superiority to A. His conclusions are that the amount of variability is about $0^m\cdot7$; that it is impossible to say how far each star participates in the changes ; and that, owing to the hopeless nature of such observations in the climate of Poulkova, he has discontinued the observations.

The Story of γ Virginis has already been written, and to Admiral Smyth's most interesting "story" but little remains to be added. The following is a summary of the statements in his paper (*Spec. Hartwellianum*, p. 335) :—

Although various occultations of this double star by the moon have been recorded, allusion to the two components has rarely been made. Nine observers watched an occultation in 1780 (March 20), yet at Paris only is mention made of one star being occulted 10^s before the other. Four astronomers watched the occultation on Jan. 21, 1794, but no mention is made of the duplicity. Yet Cassini, in 1720, saw the two stars, and noted that "the western disappeared 30" before the other behind the moon's dark limb."

The Orbit.—H₁ was the first to compute a set of elements for this system : they are as follows :—

$$a = 11''\cdot830$$
$$\pi = 17^\circ\ 51'$$
$$e = 0\cdot88717$$
$$\tau = 1834\cdot01$$
$$P = 513\cdot28$$
$$n = -0^\circ\cdot70137$$
$$\gamma = 67^\circ\ 59'$$
$$\Omega = 87\ 50.$$

"If they be correct," says Sir John, "the latter end of the year 1833, or the beginning of the year 1834, will witness one of the most striking phenomena which siderial astronomy has yet afforded, viz., the perihelion passage of one star round another, with the immense angular velocity of between 60° and 70° per annum, that is to say, of a degree in five days. As the two stars will then, however, be within little more than half a second of each other, and as they are both large and nearly equal, none but the very finest telescopes will have any chance of showing this magnificent phenomenon." (*Mem. R. A. S.*, vol. v.) In 1833, however, the measures were found to deviate materially from the ephemeris, and Sir John recalculated the orbit, with the following results :—

$$a = 12''\cdot090$$
$$e = 0\ \cdot8335$$
$$\pi = 36^\circ\ 40'$$
$$\lambda = 282\ 21$$
$$\gamma = 67\ 2$$
$$\Omega = 97\ 23$$
$$P = 628\cdot90$$
$$n = -0^\circ\cdot57242$$
$$\tau = 1834\cdot63.$$
$$(\textit{Mem. R. A. S.}, \text{vol. vi.})$$

"From the extreme delicacy of so novel a case, all the conditions were not yet met, so that this bold prediction was not circumstantially verified, although it was admirably correct in substance. Whilst rushing towards the nearest point of contact, or shortest distance of the revolving star from its primary, and the proximity became extreme, the field was left, so far as I know, to Sir John Herschel at the Cape of Good Hope, Professor Struve at Dorpat, and myself at Bedford." In 1836 Smyth was the first to observe γ Virginis a single star. "The companion now took such a movement as quite to confute a large predictive diagram which I had constructed." In fact, was now seen to be extremely elongated.

Mädler's first elements are as follows : (2) are the corrected elements of 1841 (*Dorpat Observations*) :—

	(1)	(2)
Perihelion passage	1836·103	1836·313
Node	58° 23'	60° 38'
Perihelion from } node ... }	266 0	78 22

Inclination	...	$35°\ 48'$	$24°\ 39'$
Excentricity	...	0·86805	0·86815
Mean annual motion	137′·0886	$-148'·453$

Period ... ... 157·562   145·453 yrs.
Semi-axis major ... $3''·638$ $3''·402$.

(See *Ast. Nachr.*, No. 363 ; *Dorpat Observations*, 1841.)

In 1836 Sir John was convinced that Bradley's observation in 1718 was wrong, and that it had misled "Mädler and all of us." He rejected it, and considered that the period was about 143 years.

In the Cape Observations he gives his final results :—

$$e = 0·87952$$
$$\gamma = 23°\ 35'\ 40''$$
$$\Omega = 5\ 33$$
$$\lambda = 313\ 45$$
$$P = 182·12 \text{ years}$$
$$\tau = 1836·43.$$

Henderson's elements, published in 1843, are—

Perihelion passage ..	1836·29
Mean annual motion	$2°\ 30'·59$
Excentricity ...	0·8590
Perihelion on orbit...	$319°\ 23'$
Inclination... ...	23 5
Node	70 48
Period	143·44 years.

(*Spec. Hartw.*, p. 345.)

In his *Untersuchungeu über die Fixsterne-System*, Mädler gives elements deduced from observations up to 1847 : they are—

Perihelion passage ...	1836·279.
Angle between perihelion and node	$79°\ 4'$.
Node	62 9
Inclination	25 25
Excentricity	0·88064
Period	169·445 yrs.

Mr. Hind, in 1845, computed the orbit of this star : his elements are—

Perihelion passage ...	1836·228
Perihelion on the orbit ...	$319°\ 46'·1$
Node	78 28 ·4
Inclination	25 14 ·1
Excentricity	0·85661
Period	141·297 years.

(*Mem. R. A. S.*, vol. xvi.)

Lastly, Thiele published the following set (*Ast. Nachr.*, vol. xviii.) :—

$$T = 1836·68$$
$$\omega = 283°·7$$
$$\Omega = 35·6 \text{ (Equ. 1850·0)}$$
$$i = 35·1$$
$$e = 0·896$$
$$\mu = -1°·9459$$
$$a = 3''·97$$
$$P = 185·0 \text{ years}.$$

And the following comparisons by Dunér will show that Thiele's elements satisfy the observed angles, but not the distances :—

1864·76	$4''·13$	$164°·4$	$-0''·14$	$-1°·1$	De.
66·46	·01	165·9	$-0·38$	$+1·4$	Ka.
68·28	·31	163·5	$-0·21$	0·0	De.
70·72	·63	162·3	$-0·06$	$+0·1$	,,
71·05	·72	162·1	$+0·01$	$+0·1$	Gl.
72·12	·47	161·2	$-0·32$	0·0	Du.
73·95	·97	161·4	$+0·07$	$+1·0$	W. & S.
75·14	·55	159·1	$-0·42$	$-0·6$	Du.

Cassini.		$319°·1$	$7''·49$	1720·31
Tobias Mayer.		324·4	6·50	56·6
H_1.		130·7	...	81·89
		120·3	...	1803·20
Σ.		103·0	2·86	22·00
		97·9	6n.	·37	5·32
		91·5	1n.	·07	8·38
		88·3	5n.	1·78	9·39
		80·9	,,	·49	31·36
		73·5	4n.	·26	2·52
		65·5	7n.	·05	3·37
		51·6	5n.	0·91	4·38
		33·6	1n.	...	4·84
		331·5	3n.	0·25	6·41
		257·9	6n.	·58	7·41
		231·1	11n.	·80	8·43
H_2 & So.		103·4	2n.	3·79	22·25
		96·8	4n.	·26	5·32
		87·7	2n.	1·79	9·19
		82·8	5n.	2·21	30·24
		77·4	10n.	1·73	1·32
		70·2	18n.	·21	2·26
		61·7	12n.	·41	3·20
		43·1	8n.	·51	4·37
		34·9	7n.	...	·54
		21·5	8n.	...	5·11
		237·4	1n.	...	8·08
Amici.		...		3·30	23·19
Be.		262·1		1·59	30·59
Da.		78·4	6n.	·98	1·29
		69·9	9n.	·33	2·31
		60·1	8n.	·13	3·36
		47·3	,,	...	4·29
		351·6	1n.	...	6·27
		347·4	,,	...	·29
		233·4	,,	...	8·32
		214·6	2n.	1·26	9·32
		205·7	11n.	·24	40·38
		200·0	7n.	·57	1·34
		194·9	9n.	·73	2·38
		192·7	6n.	·82	3·37
		191·5	4n.	...	·46
		183·7	2n.	2·45	6·90
		182·8	,,	·32	7·29
		·1	6n.	·48	·41
		180·4	2n.	·63	8·36
		·6	7n.	·60	·38
		179·0	5n.	·85	9·37
		176·5	,,	·99	51·40
		175·2	2n.	3·01	2·32
		174·1	3n.	·06	3·36
		172·5	7n.	·19	4·39

284 — DOUBLE STARS.

	°		″	
Da.	172·8	In.	3·22	1854·40
	171·2	4n.	·36	5·33
	·1	,,	·30	·46
	170·1	7n.	·58	7·35
	169·9	6n.	·56	·42
	168·8	8n.	·68	8·45
	·2	5n.	·77	9·46
	166·5	,,	·94	62·03
	165·4	4n.	4·10	4·44
	164·0	7n.	·37	5·42
Sm	74·9		1·6	31·38
	71·4		·2	2·40
	62·7		·3	3·44
	45·5		0·8	4·39
	15·0		·5	5·40
	round		...	6·06
	,,		...	·15
	blotty		...	·25
	350·9		...	·30
	348·6		...	·39
	265·4		0·6	7·21
	235·7		·8	8·28
	192·8		1·9	43·08
	191·6		·9	·33
	185·4		2·1	5·34
	181·8		·6	7·41
	179·5		·8	8·36
	175·5		3·2	52·42
	173·9		·2	3·35
	171·6		·4	5·40
	170·6		·5	7·41
	169·9		·8	8·39
Encke.	113·9	In.	...	36·59
	117·5	,,	...	7·19
	83·8	3n.	0·77	·38
	74·3	10n.	·65	·48
	49·0	4n.	·70	8·46
	30·3	In.	·93	9·24
	36·1	4n.	1·37	·36
Galle. Ka.	35·5		·29	39·35
	27·9		·30	40·26
	14·5		·76	2·82
	345·9	5n.	4·01	66·46
O.Σ.	211·6	5n.	1·42	40·45
	202·4	4	·63	1·41
	197·1	4	·86	2·41
	184·5	2	2·23	5·46
	182·9	2	·35	6·38
	·5	3	·39	7·42
	179·1	3	·54	8·43
	172·9	2	·64	9·41
	175·2	4	·73	50·39
	173·0	3	·87	1·41
	·0	3	·99	2·43
	172·0	4	3·13	3·40
	171·6	4	·36	5·18
	170·2	2	·63	7·44
	169·2	2	·67	8·44
	167·9	3	·76	9·38
	166·9	4	·93	61·15
	165·9	2	·97	2·40
	167·3	2	·90	3·46
	165·0	3	4·05	4·42
	164·0	2	·29	6·42
	163·2	2	·30	8·44

	°		″	
O.Σ.	163·3	3	4·44	1870·77
	159·9	3	·64	2·41
	160·9	3	·54	3·43
	·4	3	·86	4·41
Mä.	200·1	12n.	1·72	41·35
	196·6	10n.	·58	2·21
	176·4	4n.	3·30	51·96
	174·6	2n.	·16	2·43
	·2	6n.	·25	3·39
	172·0	8n.	·44	4·39
	174·0	2n.	·41	5·45
	171·7	6n.	·59	6·38
	170·2	9n.	·59	7·42
	169·8	2n.	4·00	8·37
	·1	9n.	3·88	9·37
Ch.	200·9	2n.	1·42	41·19
	192·2	,,	·85	2·35
	180·7	In.	2·05	3·30
	189·0	,,	·63	4·33
G.O.	197·8	In.	1·62	2·33
	199·2	,,	·64	·34
	197·6	,,	·83	·34
	195·0	,,	·74	·35
	193·6	,,	2·08	3·39
	182·9	,,	·20	4·34
	184·7	,,	·30	6·36
	191·0	,,	·28	·38
	183·3	,,	·18	·41
	190·8	,,	·38	7·41
	195·2	,,	·36	·41
	180·2	,,	·65	8·42
	...	,,	·70	·42
	180·7	,,	·55	·52
	174·4	,,	·90	9·40
	176·6	,,	·96	·41
	180·0	,,	·90	·43
	179·9	,,	3·02	50·46
	182·9	,,	2·82	·47
	178·1	,,	·98	·48
	177·8	,,	·96	·50
	179·1	,,	·98	1·34
	175·3	,,	3·09	·36
	174·3	,,	·05	·37
	179·1	1	·19	2·42
	186·1	1	·16	·50
	173·8	1	·24	·50
	175·8	1	·32	3·38
	178·9	1	·29	·38
	174	10	·40	5·37
	173	10	·49	·39
	169	12	·54	6·28
	171	10	·81	·93
	174	10	·59	·94
	174	10	·56	·95
	173	10	·61	·98
	168	10	·90	8·46
	168	10	·79	·47
	170	25	4·18	9·39
Ja.	179·9	In.	2·88	47·94
	175·4	15	3·12	52·24
	173·2	10	·12	3·24
	·0	10	·06	·91
	170·5	4n.	·44	6·10
	170·6	5n.	·50	57·96
	178·0	9	2·9	60·30

	°		"	
Ja.	177·7	10	3·12	1861·19
Bond.	179·2		2·5	48·45
	181·6		·7	·45
	179·8		3·0	9·45
	·7		·0	·45
M.	180·5	2n.	2·60	8·48
	177·0	3n.	·92	9·42
	179·6	4n.	·94	50·48
	176·3	3n.	3·04	1·36
	179·7	,,	·19	2·48
	177·3	2n.	·30	3·38
	166·2	1n.	4·11	61·26
	169·0	,,	·06	·27
	167·7	,,	3·69	·29
	168·2	,,	4·09	·29
	166·5	4n.	·0	2·38
	·9	1n.	·12	3·21
	168·4	,,	·02	·25
	164·6	,,	·05	·30
	167·3	,,	·30	4·39
	164·1	,,	·24	·41
	165·2	4n.	·28	5·36
	·2	2n.	·34	6·44
	161·4	6n.	·42	7·38
	160·8	7n.	·63	8·42
	·0	1n.	·85	9·49
	159·3	,,	·49	·49
	160·1	,,	·89	·50
	·2	3n.	·70	70·36
	·5	,,	·79	·40
	·3	1n.	·44	1·33
	161·5	,,	·56	·34
	160·8	,,	·59	·34
	·9	,,	·49	·36
	161·0	,,	·60	·37
	159·9	2n.	·72	2·39
	160·9	1n.	·91	·45
	·2	5n.	·87	3·40
	158·4	6n.	5·33	4·33
	159·7	,,	·09	5·29
Flt.	176·6	,,	2·94	50·36
	175·9	,,	3·04	1·40
	·4	5n.	·14	2·42
	174·5	6n.	·18	3·32
Hartnup.	176·4		2·91	0·30
	178·6		·88	·30
Mi.	175·9	15	3·04	1·47
	·2	48	·12	2·26
De.	174·9	56	·10	3·27
	353·6	7n.	·26	54·46
	171·2	4n.	·55	5·19
	170·6	5n.	·58	6·25
	169·5	1n.	·79	7·09
	348·5	6n.	·79	8·34
	345·9	18n.	4·08	63·33
	164·9	8n.	·09	4·28
	344·5	9n.	·18	5·26
	163·6	13n.	·23	7·05
	·4	6n.	·31	8·28
	162·6	,,	·61	70·25
	161·7	5n.	·64	1·26
	·3	,,	·54	2·37
	160·2	,,	·64	73·34
	159·9	3n.	·65	4·35
	158·9	5n.	·84	5·28

	°		"	
Se.	172·5	5n.	3·37	1855·39
	171·6	6n.	·54	6·38
	170·7	7n.	·73	7·39
	172·0	3n.	·61	8·40
	169·4	,,	·91	9·44
Baxendell	172·9		·58	7·40
Kn.	169·2	2n.	4·05	60·44
	165·3	,,	·27	4·44
	164·3	3n.	·33	5·45
	159·7	,,	·49	71·38
	161·4	1n.	·82	2·40
Mit.	171·9	1n.	...	61·44
	167·6	,,	...	·46
	170·1	,,	...	2·45
Ro.	165·1	2n.	·33	3·27
Eng.	346·3	9n.	·01	5·14
	166·3	·,,	·01	·14
Ta.	...	1n.	·07	6·21
	...	,,	·37	·31
	163·4	,,	5·15	·48
	162·8	,,	·28	7·24
	160·5	,,	·05	8·26
	159·5	,,	·33	9·25
	158·6	,,	...	70·39
	163·1	2n.	4·76	1·38
	158·6	1n.	·80	2·37
	159·3	,,	5·39	4·32
	·9	,,	...	6·36
Du.	341·8	17n.	4·43	69·98
	·1		·58	72·12
	339·1	14n.	·66	5·14
Br.	164·9	2	·77	69·22
Gl.	163·2	7	·7	70·22
	162·0	5	·6	·44
	·3	5	·7	1·33
	·0	5	·6	·41
	161·0	5	5·0	·85
	159·7	5	4·4	2·30
	·4	5	·7	3·32
	160·0	3	·6	·50
	·0	3	5·0	4·00
	·9	10	·16	·53
	158·5	4n.	4·86	5·22
	·5	13n.	·78	6·27
	·5	2n.	...	7·07
	·6	1n.	5·00	9·35
W. & S.	160·1	6	·30	1·37
	159·2	6	·40	·40
	162·5	5	·59	2·30
	161·6	10	...	·33
	162·6	8	...	·38
	161·9	8	4·9	3·23
	·8	7	5·0	4·30
	160·4	6	·02	·31
	340·5	5	4·97	5·30
Lindstedt.	·5	3n.	·96	3·69
Schi.	339·5	1n.	·91	5·37
Sp.	·6		·91	·37
	·0		·84	6·45
C.O.	159·8	4n.	5·30	·38
	158·1	8n.	·19	7·30
W.O.	160·0	1n.	·17	6·39
	159·9	,,	·24	·41
	160·8	,,	·08	·41
	159·8	,,	·12	·41

Dob.	336·2	5n.	5·34	1876·26
	335·8	4n.	·04	7·28
Fl.	338·4	1n.	4·96	·43
Pl.	160·0	5n.	·65	·24
Goldney.	157·1	3n.	5·06	8·37

406   Σ. 1678.

R. A.	Dec.	M.
12^h 39·4^m	15° 2′	6·3, 7

Rectilinear motion.

The angle has diminished, but the distance has changed very little, if at all.

Dunér gives

$$1852·57. \quad \Delta = 32''·34.$$
$$P = 207°·4 + 0°·24 \; (t - 1850·0).$$

So.	213·4	2n.	33·36	1825·30
Σ.	212·5	3n.	2·73	8·29
	210·7	,,	·45	36·25
H₂.	·4		30·	2·28
O.Σ.	209·6	1n.	32·83	40·29
	·1	,,	·87	2·41
	·4	,,	·70	·42
	208·0	,,	·54	45·35
	202·7	,,	·31	68·36
Mä.	208·2	,,	·01	45·29
	207·7	3n.	·39	51·27
	·1	1n.	31·76	2·32
	206·4		·90	4·38
	·0	5n.	32·09	5·57
	205·7		31·99	6·36
	204·8	1n.	32·75	8·36
	·7	,,	·96	61·41
Da.	207·6	,,	·06	51·29
De.	205·5	3n.	·45	8·35
	204·0	4n.	·17	63·23
M.	201·9		35·06	6·43
Du.	202·7	2n.	32·41	9·94
W. & S.	·0	4	·2	73·35
	201·3	2	·4	4·30
Pl.	·1	4n.	...	6·95
	...	3n.	32·28	7·16
Dob.	200·4	4n.	31·9	·29
Fl.	·3	1n.	·1	·35

407   Σ. 1687.

35 COMÆ BERENICES.

R. A.	Dec.	M.
12^h 47·4^m	21° 54′	5, 7·8, 9

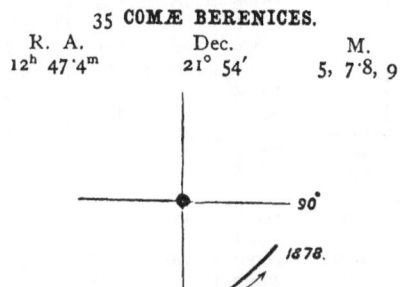

C. Σ., A, yellowish; B, blue.

Σ. discovered that the larger star was double.

H₂ and So., "Double: a small star, extremely faint; so much so, that it has been overlooked in former observations."

A.C. says there is no good ground for thinking there is anything but a small relative motion, and that the earlier observations may be faulty owing to the faintness of the smaller star.

Dawes writes: "My measures at Mr. Bishop's observatory in 1842 left no doubt of the close pair having an orbital motion." "There is no evidence of change in the more distant star."

The distance may have diminished of late, but the decrease in the angular change is opposed to this view. A C still unchanged. (O.Σ.)

A B.

Σ.	25·3	5n.	1·43	1829·99
	28·4	1n.	·38	33·37
Sm.	30·0		·0	4·28
	42·0		·5	43·32
Da.	36·6	5n.	·50	2·39
	38·9	4n.	·41	3·34
	39·2	3n.	...	8·12
	...	2n.	·57	·45
	40·9	,,	·55	9·33
	43·8	3n.	·61	53·38
	·6	,,	·50	4·41
	44·7	1n.	·59	7·45
	47·7	,,	·44	60·34
O.Σ.	39·6	,,	·58	42·39
	·6	,,	·53	5·31
	41·2	,,	·53	·31
	52·8	,,	·17	66·42
	51·8	,,	·26	74·40
Mit.	40·4		·32	47·57
Mä.	43·1	4n.	·23	51·00
	·9	3n.	·23	2·32
	40·5	1n.	·17	4·38
	44·5	,,	·33	5·42
	42·2	4n.	·26	6·39
	43·3	,,	·23	8·12
Se.	41·4	5n.	·31	6·41
De.	46 1	,,	·2	·48
	·6	3n.	·2	7·66
	42·8	4n.	·3	8·44
	54·3	1n.	...	62·95
	49·6	6n.	·26	3·31
	53·2	7n.	·23	5·94
	54·3	3n.	·27	8·32
	57·4	1n.	·16	70·15
	·5	,,	·23	1·33
	56·5	3n.	·40	2·43
	55·2	,,	·50	3·41
	57·4	2n.	·29	4·31
	58·2	,,	·33	5·31
Mo.	45·4	12	·44	57·28
Kn.	52·8	1n.	·31	65·31
W. & S.	57·0	5	·28	73·24

W. & S.	58°7	6	1"33	1873·35
	59·1	4	...	4·26
	56·8	5	1·44	·30
	57·7	4	·40	5·30
	58·4	7	·27	·32
	61·1	7	...	·39
	56·2	5	...	·43
	58·5		1·40	6·36
	57·1	3	...	·36
Gl.	59·1	4	1·32	4·34
Schi.	61·3	1n.	·07	5·31
Sp.	·3		·07	·32
Dob.	66·0	3n.	...	6·34
	61·5	2n.	1·40	7·29
Pl.	·3	2n.	·34	·00

A C.

Σ.	124·7	4n.	28·60	30·13
O.Σ.	·9	1n.	·56	45·31
Mo.	·2	12	·16	57·29
Kn.	·8	,,	·32	65·31
Ta.	125·5	,,	27·94	71·39
	61·6	,,	...	6·36
W. & S.	125·3	4	8·69	5·30
Dob.	124·9	2n.	...	6·33
	·1	4n.	8·68	7·27

408   O.Σ. 256.

R. A.	Dec.	M.
12h 50·3m	-0° 18'	7·2, 7·6

One of the two stars is probably a variable.
Probably binary.

O.Σ.	57·2	6n.	0·65	1848·70
De.	242·1	3n.	·50	67·37
Sp.	244·1		·71	75·36

409   O.Σ. 257.

R. A.	Dec.	M.
12h 51m	46° 16'	7·5, 8·2

O.Σ.	353·5	3n.	13·08	1846·73
De.	·8	,,	12·78	67·22

410   Σ. 1703.

R. A.	Dec.	M.
12h 53·1m	8° 33'	8, 11

C. A, yellowish.

Σ.	283·1	2n.	22·65	1829·27
Mä.	·2	1n.	·78	44·31
De.	·0		19·71	65·30

411   Σ. 1707.

R. A.	Dec.	M.
12h 55·3m	16° 31'	8·5, 10·3

Σ.	30·8	3n.	10·22	1828·94
Mä.	·3	1n.	·76	44·25
De.	33·0	—	9·22	65·30

412   Σ. 1711.

R. A.	Dec.	M.
12h 56·5m	14° 7'	8·7, 9·5

Probably binary.

Σ.	355·9	2n.	1"43	1829·35
De.	348·4		·28	63·24
W. & S.	352·3	1n.	·13	76·41

413   O.Σ. 260.

R. A.	Dec.	M.
13h 1·8m	27° 35'	7·9, 8·3

Mä.	120·0		0·75	1843·30
	107·0		·50	6·28
O.Σ.	111·3	5n.	·75	5·75
De.	115·2	3n.	·78	67·39

414   Σ. 1722.

R. A.	Dec.	M.
13h 2m	16° 8'	7·8, 8·8

C. A, yellowish; B, bluish.
Slight retrograde movement.

Σ.	343·9	2n.	3·54	1829·30
Se.	339·8	,,	·30	56·40
O.Σ.	336·8	1n.	·36	68·36
W. & S.	335·0	2	...	74·26
	339·8	5	...	·30
	...		3·41	·30
	...		·07	·30
	341·5	4	·09	·41
	339·2	4	·39	5·30
	340·6	5	·37	·31
Gl.	341·9	6	·4	4·34
Dob.	335·1	5	·35	7·31

415   Σ. 1728.

42 COMÆ BERENICES.

R. A.	Dec.	M.
13h 4·1m	18° 10'	6, 6

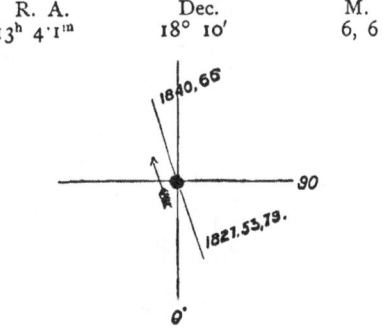

C. A, yellow; B, yellower than A; De., both white.

Σ., the discoverer, writes thus in the *M. M.*: "This star is worthy of all attention; for there is a suspicion that its period is smaller than that of ξ Ursæ. It seems certain that in five years the angle has changed 180°, and that the minimum distance fell between 1829 and 1833, and nearer to 1833."

In the *P. M.* he adds, "Between 1829 and 1851 the star has twice become single, first in the years 1833 and 1834, and then in the years 1845 and 1846."

Sm. found it round in 1832, and in 1839 he "could not palpably notch them."

Dawes writes, "One of the closest of Σ.'s discoveries. It requires the finest and largest telescopes."

In 1874 O.Σ. discussed the observations made at Dorpat and Pulkowa, and found a period of 25·71 years. In 1866, under less favourable circumstances, the period deduced was 25·5 years.

He remarks, "During the last forty years [the star] has presented three times more the rare phenomenon of an occultation of one star by another." His elements are

$$T = 1859 \cdot 92 \pm 0^y \cdot 080$$
$$\lambda = 99° \ 11' \pm 0°.45^s \cdot 6$$
$$a = 0'' \cdot 657 \pm 0'' \cdot 0126$$
$$e = 0 \cdot 480 \pm 0 \cdot 0239$$
$$m = 14° \ 0' \cdot 2 \pm 2' \cdot 75, \text{ or revolution}$$
$$= 25^y \cdot 71 \pm 0^y \cdot 084.$$

(See *Bull. de Acad. Imp. de St. Pétersbourg*, t. iii. and v).

The common proper motion is − 0″·433 in R. A., and − 0″·18 in N. P. D.

Obs.	°	n.	″	Year
Σ.	9·5	2n.	...	1827·83
	11·6	3n.	0·64	29·40
	...	2n.	...	33·37
	...	1n.	...	4·42
	11·2 }	4n.	...	5·39
	191·2 }			
	10·2 }	3n.	0·30	6·41
	190·2 }			
	10·9	6n.	·39	7·40
	11·5	3n.	·35	8·41
Sm.	round		...	2·38
	10·0		...	9·41
	5·0		0·3	42·50
Da.	198·5		·42	0·74
	single		...	2·53
	,,		...	3·45
	14·2		0·62	53·09
	12·7		·55	4·39
	183·5		·2	60·34
	191·0		·5	3·25
	193·4		·45	4·43
O.Σ.	195·7	3n.	·55	40·45
	194·5	2n.	·49	1·41
	193·9	3n.	·31	2·40
	...	single		5·47

Obs.	°	n.	″	Year
O.Σ.	66·8	oblong? ,,		1846·40
	15·5	1n.	0·20	7·42
	12·7	3n.	·26	8·42
	8·6	,,	·42	9·42
	11·4	,,	·48	50·39
	7·0	4n.	·48	1·42
	10·9	3n.	·56	2·43
	·8	,,	·57	3·40
	14·1	1n.	·60	4·38
	9·1	2n.	·62	5·44
	7·7	,,	·43	7·49
	8·5	,,	·38	8·44
	...	single		9·37
	185·6	2n.	0·43	61·42
	191·6	,,	·54	2·40
	189·3	1n.	·55	3·44
	192·5	3n.	·51	4·42
	188·5	,,	·40	6·44
	193·0	2n.	·36	7·47
	195·8	,,	·21	8·44
	195	1n.	obl.?	9·47
	...	single		70·44
	...	,,		1·43
	20	1n.	oblong	2·42
	9	2n.	0·20	3·46
	9·2	,,	·30	4·41
Mä.	4·7	11n.	·32	41·40
	15·5	4n.	...	2·45
	194·5	3n.	0·46	51·96
	190·9	6n.	·52	2·42
	194·0	14n.	·62	3·35
	193·6	8n.	·68	4·40
	198·7	2n.	·57	5·38
	192·7	5n.	·58	6·40
	188·2	2n.	·50	7·40
	196·2	6n.	·40	8·40
	215·8	3n.	·2	9·36
O.Σ. De.	3·8	4n.	·42	1·42
De.	189·1	7n.	...	63·23
	197·1		...	8·13
	194·9		...	71·19
	190·4	5n.	0·50	5·43
Ta.	191·6	1n.	...	69·25
Du.	19·0	3n.	0·13	9
	16·5	4n.	·1	70
	18·5	2n.	...	2
	11·5	6n.	0·32	5·53
W. & S.	...	round		3·36
	192·5	1	·5	5·30
Schi.	186·4	4	·5	6·36
	12·2	1n.	·39	5·43
W.O.	13·1	,,	·47	·44
	190·2	,,	·38	6·37
	194·3	,,	·42	·39
	193·9	,,	·40	·39
	195·2	,,	·42	·41

416   O.Σ. 261.

R. A.	Dec.	M.
13h 6·4m	32° 43′	6·9, 7·4

C. yellowish.

Probable change in angle and distance. The increase in distance has been accompanied by a retardation in the angular movement. (O.Σ.)

Dunér gives

$$\Delta = 0°\cdot81 + 0''\cdot025\ (t - 1860\cdot0).$$

	°		''	
O.Σ.	359·2	2n.	0·63	1843·80
	356·5	4n.	·55	7·17
	352·5	3n.	·91	57·76
	350·4	2n.	1·08	66·86
Mä.	366·0		0·48	45·86
	362·7		·35	8·00
	386·2		·55	51·27
De.	350·7		·99	66·99
Du.	351·3	10n.	1·05	70·04
Sp.	349·6		·16	5·35
Pt.	353·5	3n.	·11	7·20
W. & S.	350·9	5	·14	·45
	348·4	9	·22	·46

417 Σ. 1734.

R. A.	Dec.	M.
13h 14·6m	3° 34'	7·2, 7·9

C. white.

Very slow change. Probably a binary.

	°		''	
Σ.	198·1	4n.	0·73	1830·35
Mä.	200·0		1·1	41·37
	202·5		0·85	2·46
	203·0		·96	3·30
	196·1		·96	51·35
	200·4		1·02	2·39
O.Σ.	204·0	5n.	0·96	47·16
Se.	198·2	1n.	·84	56·31
	·8	,,	·79	·38
Gl.	192·6	,,	1·1	74·32
W. & S.	191·3	2n.	·06	·32
	195·0	3n.	·06	5·35
Sp.	193·2		·24	·37

418 Σ. 1742.

R. A.	Dec.	M.
13h 18m	2° 2'	7·4, 7·9

C. yellowish white.

	°		''	
Σ.	351·1	4n.	1·29	1831·85
Mä.	·0	2n.	·39	43·30
O.Σ.	346·1	3n.	·18	50·99

419 Σ. 1744.

MIZAR.

R. A.	Dec.	M.
13h 19·1m	55° 33'	2·1, 4·2

C. greenish white.

A very fine object, and probably the first star which was observed to be double. Riccioli discovered it in 1650; it was seen double by Kirch in 1700, and first measured by Bradley in 1755. The common proper motion of the pair is $+ 0^s\cdot017$ in R. A. and $+ 0''\cdot04$ in N. P. D. Alcor also seems to have the same proper motion as Mizar.

Between these two stars Einmart in 1691 discovered one of the 8th magnitude: its position in 1839 was 102°·6, distance 8' 45''. (Sm.)

In 1857 Bond tried some experiments in Stellar photography. Mizar and Alcor were the objects chosen: the distance between these two stars was found to be 707''·8, and that between Mizar and its companion 14''·6. The following results were obtained from an examination of eighty-six photographs: "the probable error of the distance of the centres of the photographs of Mizar and its companion is ± 0''·072 for a single pair of images: the probable error of a single micrometer measurement of a double star of this class, taken in the ordinary way, is ± 0''·127; so that the relative value of the photograph is $\left(\frac{0\cdot127}{0\cdot072}\right)$; or the photograph is worth three times as much as a single direct measure."

Dunér gives the formulæ

$$1852\cdot16.\quad \Delta = 14\cdot29.$$
$$P = 147°\cdot7 + 0°\cdot025\ (t - 1850\cdot0).$$

	°		''	
Bradley.	143·1		13·88	1755·00
H1.	...	3n.	12·3	79·76
	...	7n.	14·0	80·40
	146·8	2n.	·3	1·88
	141·2	1n.	...	1802·75
Σ.	145·6	,,	...	21·80
	...	2n.	14·74	2·76
	147·6	6n.	·36	30·63
H2 & So.	·8	2n.	·45	22·44
	·8	3n.	·21	30·44
Sm.	·0		·6	·85
	·4		·4	9·32
	148·1		·2	54·72
Be.	147·7	4n.	·43	30·79
Encke.	146·8	2n.	·70	7·63
Ga.	147·2	3n.	·65	8·62
Mä.	...		·42	9·59
	147·7		·58	41·55
	148·2	8n.	·53	2·80
	·1	16n.	·22	51·18
	·4		·20	·86
	·3		·10	3·32
	·6		·15	4·56
	·2	14n.	·13	6·07
	·6		·03	7·39
	·8	8n.	·12	61·99
O.Σ.	147·8	9n.	·39	46·55
Da.	148·0	1n.	·16	8·49
De.	·0	9n.	·24	52·14
	147·9	1n.	·52	8·54

	°		"	
Se.	148.1	2n.	14.6	1855.30
	.5	4n.	.4	7.70
M.	146.8	1n.	13.90	62.44
	148.8	,,	.04	9.57
Ro.	.1	2n.	14.48	3.27
Eng.	.4	,,	.51	4.33
Ta.	...	1n.	15.06	5.57
	148.1	,,	13.90	71.39
Du.	.6	10n.	14.52	69.50
W. & S.	.0	2n.	.5	73.28
Gl.	147.9	1n.	.7	4.22
Fl.	148.7	,,	.55	7.50

420   Σ. 1746.

R. A.	Dec.	M.
13h 22m	10° 5'	7.7, 10.3

C. A, yellowish.

	°		"	
Σ.	250.3	1n.	29.32	1828.31
	251.4	,,	30.08	9.36
	250.8	,,	29.47	31.21
Mä.	248.7	,,	28.72	47.27
	249.4	,,	29.11	.28

421   O.Σ. 266.

R. A.	Dec.	M.
13h 22.6m	16° 21'	7.3, 7.8

C. white.

The angle has increased, and the distance may have increased also.

	°		"	
O.Σ.	324.2	4n.	1.15	1846.10
Da.	325.3		...	7.34
	326.2		1.07	50.80
	327.2		.08	4.27
Mä.	.7		.18	49.27
Se.	333.5	1n.	.71	56.44
De.	.4	3n.	.34	67.30
W. & S.	336.2	5	.27	77.45
	.1	6	.22	.46

422   O.Σ. 269.

R. A.	Dec.	M.
13h 27m	35° 31'	6.5, 7

Direct motion.

	°		"	
O.Σ.	218.0	1n.	0.33	1844.31
	222.3	,,	...	46.37
	240.4	,,	...	.38
	230.5	,,	.39	.39
	...		oblong	9.47
	228.9	1n.	0.33	51.39
	223.6	,,	.27	5.47
	242.8	,,	.33	61.26
	257.1	,,	oblong	72.47
De.	45?		obl.?	65
	...		simple	68

423   Σ. 1757.

R. A.	Dec.	M.
13h 28.2m	0° 18'	7.8, 8.9

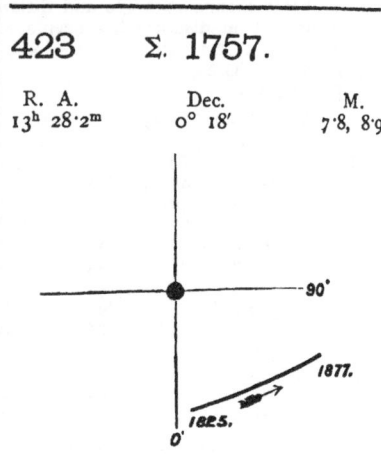

C. Se., A, white; B, bluish. Sm., A, pale white; B, yellowish.

From his measures made between 1825 and 1835, Σ. inferred direct motion.

Sm., from his own and Σ.'s measures, found that the angular progress was at first 2° per annum, that it then diminished to 1°, and that "it is now on the increase, amounting to 1¼°." Hence he concludes that the object was then seen "full face," and that its period is about 240 years."

Dawes, having the measures up to 1863 before him, considered that the increase in angle was established, and that the distance remained unchanged.

O.Σ. The distance has increased, and the angular change has diminished.

	°		"	
Σ.	10.0	1n.	1.60	1825.37
	19.5	2n.	.44	9.82
	23.9	,,	.54	33.38
	25.5	3n.	.66	5.37
	29.4	2n.	.64	6.42
Sm.	31.0		.7	8.48
	37.9		.7	42.52
	51.7		2.0	52.38
Mä.	36.0	4n.	1.74	41.38
	40.8	2n.	2.02	5.88
	52.2	3n.	.05	53.09
	50.1	5n.	.16	4.37
	54.7	2n.	1.91	8.37
	53.7	,,	.82	9.36
Da.	37.4	,,	.67	42.39
	38.8	,,	...	3.45
	54.3	1n.	2.31	60.34
	53.4	,,	.08	.35
Ka.	40.9		...	43.51
O.Σ.	43.7	3n.	1.89	4.72
	48.8	,,	.85	50.38
	60.8	,,	2.34	66.01
Ja.	48.0	15	.14	53.73
	52.8		1.76	8.08

	°		″	
De.	51.3	4n.	1.7	1855.31
	.8	2n.	.5	6.32
	59.0	5n.	2.01	63.32
	60.4	,,	.09	5.97
	62.7	3n.	.04	8.30
	63.5	1n.	.03	70.15
	.4	,,	.13	1.19
Mo.	51.9	12	.01	56.42
	54.2	10	.00	7.29
Se.	52.9	2n.	1.84	6.88
Ta.	63.7	1n.	2.60	67.27
	64.3	,,	...	9.24
	.1	,,	2.00	70.37
	67.3	,,	...	2.37
	69.8	,,	1.08	4.32
Br.	63.4	2	2.59	69.22
W. & S.	64.8	8	.30	72.33
	.0	5	.05	3.23
	65.0	5	.00	.35
	.5	7	.15	4.32
	.9	4	...	.32
	64.2	5	2.16	.41
	66.5	6	.21	6.41
	67.2	6	.15	.36
Schi.	66.6	1n.	.00	5.31
Sp.	.6		.00	.31
Dob.	61.1	3n.	...	6.32
	64.2	2n.	2.33	7.23

	°		″	
Mä.	44.7	4n.	0.31	1852.33
	36.2	1n.	.35	3.32
	26.7	2n.	.2	8.65
O.Σ.	72.6	4n.	1.01	41.17
	69.8	3n.	.71	6.80
	65.6	,,	.65	9.77
	...	4n.	single	59.72
Da.	67.7	3n.	0.99	42.35
	36.2	,,	.35	54.43
	10° or 15° round	1n.	.15	60.36
	...			5.44
Se.	25.7	1n.	elong.d	56.49
	...	,,	,,	7.59
De.	180	,,	,,	62.95
	315	,,	,,	3.15
	...		single	.50
Du.	178?	elongated		9.40
	186	,,		70.43
	47?			1.45
	...		round	5.49
W. & S.			,,	2.38
W.O.		single		5.36
Gl.		round		.40
Schi.	161.3	1n.	0.42	6.44
Sp.	.4		.42	.45

424 Σ. 1768.

25 CANUM VENATICUM.

R. A.	Dec.	M.
13h 32.1m	36° 54′	5, 7.6

C. Σ., A, white ; B, blue.

Σ., from his measures between 1827 and 1836, suspected orbital motion, and subsequent observations proved the correctness of his suspicion.

O.Σ. says, "The feeble angular motion from 1833 to 1841 indicates that the satellite was in aphelio in that period. The apparent ellipse is evidently very narrow. We shall probably see the companion emerge from the rays of the principal star under an angle of position between 180° and 90°. If so, the period of revolution does not greatly exceed a century."

Dr. Doberck, in 1877, found the following elements for this pair:—

$$\Omega = 82°.0$$
$$\lambda = 202.0$$
$$\gamma = 51.5$$
$$e = 0.66$$
$$P = 124.50 \text{ years}$$
$$T = 1862.98.$$

Σ.	79.5	5n.	1.05	1829.89
	72.4	,,	.09	33.12
	71.7	3n.	.07	6.50
Mä.	70.8	4n.	0.99	41.39
	56.5	6n.	.39	51.28

425 SM. 488.

R. A.	Dec.	M.
13h 32.3m	28° 56′	9.5, 10.5

C. white.

This close pair was discovered by Smyth in 1835, while looking through H₁'s 20 ft. reflector at Slough. It was afterwards elongated by Smyth, Challis, and Dawes. Smyth also noticed "a small blue telescopic companion in the n.f. quadrant."

Sm.	191.5		1.0	1835.48
	...	elongated		43
	195.0		1.0	51.37
Challis.	193.0		.6	42.47
Da.	196.5		.8	8.42

426 Σ. 1771.

R. A.	Dec.	M.
13h 33.5m	70° 23′	7.8, 8.5

Σ.	69.9	1n.	1.81	1829.81
	71.0	2n.	.67	31.73
Mä.	74.6		.76	45.56
	73.5		.82	52.67
	75.1		.71	4.21

427 Σ. 1772.

R. A.	Dec.	M.
13h 34.9m	20° 33′	6.2, 9.1

C. A, bluish white ; B, very blue.

	°		″	
H₂.	140		6·0	1826·00
	147·7		4·88	8·33
	146·8		5	31·00
	145·0		6	2·00
Σ.	150·4	2n.	4·71	28·30
	147·6	3n.	·92	33·74
	...	1n.	5·18	52·16
Sm.	147·1		4·9	32·23
Mä.	149·2		·79	44·34
	144·8		5·22	8·35
O.Σ.	...	1n.	·18	52·16
	145·6	,,	4·81	68·36
	149·4	2n.	·82	70·33
	·7	1n.	·95	4·28
Se.	144·1		·60	56·93
M.	140·3		·63	62·44
Fl.	137·9	1n.	·68	77·43

428   Σ. 1776.

R. A.	Dec.	M.
13ʰ 36·8ᵐ	46° 50′	8, 8

C. white.

Dunér's formulæ are

$$\Delta = 7''\cdot23 - 0''\cdot006\ (t - 1850\cdot0).$$
$$1849\cdot09. \quad P = 199°\cdot3.$$

	°		″	
H₂.	199·9	1n.	8·58	1830·32
Σ.	200·2	3n.	7·32	2·09
Mä.	199·4	2n.	·55	43·57
Se.	158·5	,,	6·93	57·55
Mo.	·2	,,	7·16	9·22
Du.	199·5	,,	·10	70·90
Ta.	201·0		6·90	3·37

429   Σ. 1777.

R. A.	Dec.	M.
13ʰ 37ᵐ	4° 9′	5·8, 8·2

C. A, yellow ; B, very blue.

The colour of B probably changes.
A beautiful pair: binary. Common proper motion − 0ˢ·023 in R. A., and + 0″·05 in N. P. D.
Dunér's formulæ are

$$1853\cdot01. \quad \Delta = 3''\cdot45.$$
$$P = 233°\cdot3 - 0°\cdot09\ (t - 1850\cdot0).$$

	°		″	
H₁.	240·9	1n.	...	1782·10
	239·8	,,	...	1802·31
Σ.	234·1	5n.	4·23	21·30
	235·3	,,	3·39	8·77
	234·0	1n.	·67	52·22
H₂ & So.	229·8	,,	·91	21·37
	232·9		4·06	30·20
	235·3		...	1·28
	231·8		2·5	2·00
	228·8		...	3·26

	°		″	
Sm.	232·9		3·7	1831·19
	231·8		·6	6·35
	233·4		·5	9·27
Mä.	234·8		·42	6·06
	233·1		...	42·40
	234·5		3·51	3·31
	233·9		·44	4·30
	232·5	5n.	·50	51·56
	233·3	6n.	·18	4·96
	232·7		2·89	8·38
	235·0		...	61·41
Da.	233·4	3n.	3·72	41·24
	231·8	1n.	·61	3·34
	234·1		·61	·59
	233·6	1n.	·60	60·38
Mit.	231·9	,,	·49	47·57
De.	233·4	,,	·33	56·41
	235·0		·50	67·80
Se.	231·6	3n.	·26	57·03
Mo.	230·8	2n.	·35	8·28
Eng.	231·9	4n.	·79	65·35
Ta.	227·8	2n.	·87	6·40
	·7	1n.	·86	9·25
O.Σ.	236·6	,,	4·03	70·35
Gl.	231·7	,,	3·52	1·32
Du.	·7	4n.	·39	2·53
W. & S.	·5	2n.	·54	3·35
	232·5	1n.	·22	4·32
	230·0	,,	·60	6·46
Sp.	231·7		·54	5·44
Dob.	232·3	4n.	...	6·34
Fl.	229·4		3·58	7·45
Pl.	230·1	3n.	·25	·51

430   Σ. 1781.

R. A.	Dec.	M.
13ʰ 40·2ᵐ	5° 43′	7·8, 8·2

C. yellowish white.

A binary.

	°		″	
Σ.	240·3	3n.	1·35	1830·31
H₂.	235·7		·5	2·00
O.Σ.	244·2	2n.	·40	41·91
Da.	238·6	,,	·20	2·31
Mä.	242·2		·09	3·34
Se.	246·5	2n.	0·99	56·39
De.	249·8	3n.	1·2	8·07
	251·7	,,	·15	64·75
Ta.	255·6	1n.	...	9·25
Gl.	256·0	5n.	1·1	71·32
W. & S.	257·5	4	0·95	2·38
	256·4	5	1·20	3·23
	261·4	5	·23	·34
	260·7	4	·02	·35
	256·0	8	·23	4·23
	·5	9	·3	·32
	261·7	5	·20	6·36
	262·5	6	·20	·45
Sp.	259·5		·21	5·37
Dob.	256·3	3n.	·03	7·36

431 O.Σ. 270.

R. A.   Dec.   M.
13^h $41\cdot6^m$ $18°$ $3'$ $4\cdot8$, $11\cdot4$

C. greenish yellow.

These stars have a common proper motion.

O.Σ.	347·8	5n.	10·26	1849·54
De.	348·9	3n.	9·03	67·36

432 Σ. 1785.

R. A.   Dec.   M.
13^h $43\cdot6^m$ $27°$ $35'$ $7\cdot2$, $7\cdot5$

Change in both angle and distance. The measures by Dembowski differ considerably from those by O.Σ. It is probable, however, that those by the former observer are the less accurate of the two, seeing that the angular movement has not augmented in so great a degree as the diminution in the distance demanded. (O.Σ.)

The common proper motion is − 0″·50 in R. A., and + 0″·003 in N. P. D.

Dunér has the following formulæ:

$$\Delta \sin. P = - 0''\cdot038 - 0''\cdot0437\ (t - 1852\cdot50)$$
$$- 0''\cdot000033\ (t - 1852\cdot50).$$
$$\Delta \cos. P = - 3''\cdot240 + 0''\cdot0267\ (t - 1852\cdot50)$$
$$+ 0''\cdot000857\ (t - 1852\cdot50).$$

So.	160·4		5·07	1823·40
Σ.	164·0	2n.	3·44	9·41
	165·3	1n.	·57	31·53
H2.	164·6		4·62	0·20
	166·2	2n.	...	1·34
Mä.	172·1		3·47	40·85
	174·6		·39	3·48
	·9	8n.	·47	4·88
	178·7	2n.	·48	51·28
	183·7	3n.	·03	5·66
	191·1	1n.	·52	61·59
Po.	176·2	4n.	·20	46·41
O.Σ.	·0	2n.	·18	51·41
	194·1	1n.	2·66	66·42
	198·2	,,	·96	8·38
	200·5	,,	·68	70·31
	198·3	,,	·91	·35
Se.	185·9	2n.	3·24	56·36
De.	184·7	5n.	·12	8·38
	190·7	7n.	2·69	63·27
	191·9	4n.	·60	4·39
	192·7	6n.	·60	5·34
	194·5	9n.	·56	6·81
	196·8	4n.	·51	8·34
	199·1	,,	·46	70·32
	200·1	,,	·40	1·30
	201·5	,,	·32	2·38
	202·4	,,	·32	3·39
	204·4	3n.	·21	4·46
	205·8	4n.	·16	5·30

Mo.	185·4	2n.	2·89	1859·30
M.	192·8	1n.	·69	63·31
	199·3	,,	·65	71·27
	198·9	,,	·32	·44
Eng.	193·5		·88	64·47
	·8	7n.	·87	5·42
Du.	198·5	5n.	·46	70·19
	201·7	2n.	·59	2·43
	206·4	4n.	·39	5·24
Gl.	204·1	5	·46	1·32
	199·0	4	·6	0·32
	·0	4	·4	·55
	·8	11	·7	1·32
Kn.	·2	3n.	·51	·38
W. & S.	200·2	5	·47	2·38
	203·2	4	·47	3·23
	201·0	4	·46	·35
	204·5	9	·41	4·32
	206·1	5	·46	5·35
	207·9	7	·58	·39
	206·7	5	...	·41
	208·4	6	·28	6·41
	208·8	5	·14	7·47
Lindstedt	200·2	2n.	·41	3·42
Schi.	205·3	1n.	·34	5·32
Sp.	·4		·34	·33
	206·9		·15	6·45
Pl.	208·9	2n.	·61	·42
Dob.	206·8	6n.	·21	7·32

433 Σ. 1788.

R. A.   Dec.   M.
13^h $48\cdot6^m$ $- 7°$ $28'$ $6\cdot7$, $7\cdot9$

C. white.

The common proper motion is − 0″·137 in R. A.

So.	51·7		2·76	1825·39
H2.	49·6		·57	30·27
	50·6		·68	1·44
Σ.	54·0	5n.	·36	1·38
	64·9	2n.	·66	52·23
Sm.	55·0		·5	34·29
Mä.	60·4		·49	44·35
	61·0		·44	54·38
	63·3		·47	8·37
Se.	62·6	2n.	·46	6·39
M.	64·2	1n.	·38	62·32
De.	67·7		·46	4·85
Gl.	69·6	1n.	·58	71·32
W. & S.	75·4	,,	·55	2·38
	70·0	3n.	·64	3·63
Sp.	·0		·45	5·40
C.O.	·2	2n.	·62	7·39
Dob.	67·5	4n.	·68	7·31

434 O.Σ. 272.

R. A.   Dec.   M.
13^h 49^m $30°$ $29'$ 7, $9\cdot9$

C. A, white.

Mä.	22°2		1″52	1843·33
	25·5		·86	9·34
O.Σ.	23·4	4n.	·89	·56
De.	17·8	3n.	·78	66·71

435 O.Σ. 273.

R. A. 13^h 50^m	Dec. 5° 50′	M. 7·5, 8

O.Σ.	106·1	3n.	0·74	1845·99
De.	111·2	,,	0·98	67·73

436 Σ. 1808.

R. A. 14^h 4·7^m	Dec. 27° 10′	M. 8, 9

C. white.

Probably binary.
Dunér has

$$\Delta = 2''·70 - 0''·007 \ (t - 1850·0).$$
$$P = 72°·1 + 0°·187 \ (t - 1850·0).$$

Σ.	68·8	3n.	2·82	1832·31
Mä.	71·0	2n.	·76	44·39
Du.	76·1	,,	·54	71·32

437 Σ. 1812.

R. A. 14^h 7^m	Dec. 29° 17′	M. A 7·8, в 8, C 9·3

A B probably binary.

A B.

O.Σ	333·7	4n.	0·42	1845·85
	344·3	3n.	·47	65·42
Se.	342·7	1n.	·4	56·44

$\dfrac{A\,B}{2}$ and **C.**

Σ.	108·2	3n.	14·18	32·37
O.Σ.	·1	7n.	·02	54·24
Pl.	107·4	2n.	13·89	77·51

438 O.Σ. 278.

R. A. 14^h 7^m	Dec. 44° 46′	M. 7·5, 7·7

Probably a binary.

O.Σ.	146·0	3n.	0·40	1846·03
	145·1	2n.	·45	54·00
	124·2	1n.	·53	75·48
De.	128·2	3n.	...	67·48
Du.	·3	,,	0·32	9·48

439 O.Σ. 280.

R. A. 14^h 7^m	Dec. 60° 58′	M. 7, 11·2

C. A, golden.

O.Σ.	20·5	3n.	7″20	1848·61
De.	24·5	,,	6·95	66·67

440 Σ. 1813.

R. A. 14^h 7·4^m	Dec. 5° 58′	′M. 8, 8·1

C. white.

Probably binary.
Dunér gives

$$1850·61. \quad \Delta = 4''·95.$$
$$P = 192°·6 + 0°·04 \ (t - 1850·0).$$

H$_1$.	180·0		...	1793·36
H$_2$ & So.	190·7	1n.	6·06	1823·34
	192·5	,,	4·5	31·00
Σ.	191·0	4n.	·76	29·81
Mä.	193·9		5·34	41·37
	192·9	4n.	·21	3·07
	191·5		·24	4·30
	194·7		...	6·25
	193·1	1n.	5·15	51·28
	·1	,,	4·83	8·38
Da.	192·2	7n.	·95	42·27
	194·0	2n.	·84	3·35
De.	192·9		·92	55·30
Se.	193·9	3n.	·82	7·05
M.	199·8	1n.	·86	63·31
Eng.	192·5	4n.	·98	5·32
Ta.	193·2	2n.	·67	6·40
	·0	,,	...	7·37
	194·5	1n.	3·83	73·48
	·8	,,	4·21	4·33
Du.	192·4	3n.	·88	9·38
Gl.	·7	1n.	5·2	1·32
W. & S.	193·1	,,	·0	3·36
Pl.	192·7	5n.	·0	7·27

441 O.Σ. 279.

R. A. 14^h 8^m	Dec. 12° 34′	M. 6·8, 9

C. A, yellow.

O.Σ.	248·4	3n.	2·27	1845·68
De.	251·8	,,	·11	66·71

442 Σ. 1861.

R. A. 14^h 8^m	Dec. 29° 40′	M. 7, 7·1

C. yellowish.

Very slow change. Probably binary.
Dunér gives

$$\Delta = 1''\!\cdot\!68 - 0''\!\cdot\!0122\,(t - 1850\!\cdot\!0).$$
$$P = 80°\!\cdot\!3 + 0°\!\cdot\!087\,(t - 1850\!\cdot\!0) + 0°\!\cdot\!0006\,(t - 1850\!\cdot\!0)^2.$$

Σ.	80°·1	5n.	1·86	1831·33
H₂.	76·3	In.	·92	3·26
Da.	79·4	3n.	·69	41·88
	78·4	,,	·84	3·34
Mä.	79·8	2n.	·73	2·88
O.Σ.	·5	6n.	·79	53·43
De.	80·9	2n.	...	5·27
	81·8	3n.	1·57	66·58
	82·5	In.	·57	74·23
Se.	80·4	2n.	·49	56·41
Du.	84·7	4n.	·32	71·16
Gl.	79·8	In.	·6	·32
W. & S.	·6	2n.	·55	3·84

			″	
Σ.	233°·6	In.	...	1821·78
	237·7	7n.	12·60	32·50
	·7	In.	·50	7·70
So.	238·7	4n.	13·14	22·62
H₂.	·8		12·79	30·48
Sm.	237·9		·5	·93
	238·1		·7	8·78
Mä.	237·0	8n.	·76	43·42
	236·4		·63	4·90
	237·0	5n.	·65	52·37
	236·1	4n.	·49	5·37
	237·3	2n.	·66	61·57
Mo.	·1	,,	·66	54·46
	238·6	3n.	·75	5·46
De.	·1	2n.	·46	·73
Du.	236·3	5n.	·92	72·90
W. & S.	242·8	In.	·99	6·46

443   Σ. 1820.

R. A.	Dec.	M.
14ʰ 9·1ᵐ	55° 53′	8·2, 8·5

Direct motion. Dembowski's distance in 1866·75 is probably too small; it is probably explained by the note that the observation was made in haste.

Dunér has

$$1850\!\cdot\!57. \quad \Delta = 2''\!\cdot\!35.$$
$$P = 54°\!\cdot\!0 + 0°\!\cdot\!422\,(t - 1850\!\cdot\!0).$$

Σ.	47·3	6n.	2·40	1834·14
Mä.	52·0	2n.	·50	45·47
	50·3	In.	·17	51·27
	63·1	,,	·35	4·21
De.	60·5	3n.	·11	66·75
Du.	63·2	5n.	·27	71·45
O.Σ.	68·0	In.	·64	4·70

444   Σ. 1821.

R. A.	Dec.
14ʰ 9·2ᵐ	52° 21′

Magnitudes.—Σ. 5·1, 7·2. The estimations of the magnitudes differ considerably. Du. has 3·5, 6·5; 4, 7; 5, 7·5; 4, 6.

C. Σ., A, yellowish; B, bluish. The colours also are variously given.

The proper motion of κ is + 0ˢ·009 in R. A. and + 0″·02 in N. P. D., and in this the companion probably shares.

Dunér gives

$$1847\!\cdot\!34. \quad \Delta = 12''\!\cdot\!68.$$
$$P = 237°\!\cdot\!2 - 0°\!\cdot\!05\,(t - 1850\!\cdot\!0).$$

H₁.	240	In.	...	1779·75
	...	,,	12·08	80·56
	...	,,	14·33	1·70
	242·5	,,	...	2·29
	...	,,	11·09	97·75
	240·7	In.	...	1802·66

445   Σ. 1819.

VIRGINIS.

R. A.	Dec.	M.
14ʰ 9·3ᵐ	3° 41′	7·9, 8

C. Σ. yellowish; Se. white; De. white.

Σ. discovered this double star, and in 1836 pointed out that it was a binary.

O.Σ. says, "The increase in the distance is certain, but slow; it is confirmed by the diminution in the angular motion."

Σ.	88·0	2n.	0·86	1828·35
	81·7	,,	1·10	32·42
	76·1	3n.	·12	6·43
H₂.	83·3		·0	2·00
Mä.	65·2	In.	0·95	41·35
	63·2	3n.	·86	2·40
	57·1	In.	1·04	5·39
	54·1	5n.	·16	7·38
	49·6	,,	·26	51·30
	44·4	4n.	·14	4·40
O.Σ	66·4	2n.	·07	41·93
	52·9	,,	·19	9·36
	36·5	In.	·43	66·42
Da.	60·5	8n.	·08	42·81
	61·9	In.	·02	3·34
Ka.	62·8		...	·24
Se.	43·7	2n.	0·98	56·39
	34·5		1·17	64·41
De.	44·0	3n.	·1	56·45
	40·8	7n.	·0	8·41
	211·4	4n.	·32	62·47
	32·8	7n.	·28	3·31
	31·6	6n.	·23	5·85
	27·9	2n.	·17	8·41
	·0	,,	·25	70·35
	25·5	,,	·34	1·22
M.	38·2		·0	58·38
Ta.	31·9	In.	·94	67·28
Gl.	25·7	4	·2	70·32
	26·7	4	·27	·34
	26·3	6	·22	·43
	26·2	5	·32	1·24

Gl.	27·0°	3	1·4″	1871·32
	25·1	4	·34	·36
	24·7	6	·4	·42
W. & S.	23·9	8	·25	2·38
	25·7	10	·13	·39
	27·4	3	·17	3·23
	25·0	4	·35	·36
	23·5	6	·33	·36
	·2	6	·33	4·41
	22·2	4	·40	5·35
	26·5	3	....	·39
	23·6	7	...	·41
	·2	5	1·37	6·41
Schi.	21.5	In.	·46	5·36
Sp.	·6		·47	·36
W.O.	201·7	In.	·25	6·39
	199·3	,,	·37	·39
	·4	,,	·15	·41
Dob.	17·2	2n.	...	7·33
Pl.	...	4n.	1·23	·27
	18·9	3n.	...	·47

446   Σ. 1825.

R. A.	Dec.	M.
14h 10·7m	20° 41′	6·8, 8·5

H₂.	186·5		4·0	1830·00
	185·5		2·5	2·00
Σ.	·7	3n.	3·45	0·66
Mä.	184·5		4·05	41·52
	·3		...	2·40
	183·8		3·89	3·31
Se.	182·2	3n.	·74	57·77
De.	178·8		·90	64·47
Gl.	180·1	1n.	4·2	71·22
W. & S.	178·0	3n.	·1	3·36
	177·5	1n.	3·93	4·93
Dob.	174·7	4n.	4·02	7·39
Pl.	179·1	3n.	3·67	·51

447   Σ. 1830.

R. A.	Dec.	M.
14h 11·9m	57° 14′	8·5, 9·8

C. Σ. A, yellowish ; Se. A, white ; B, blue.

Certain change both in angle and distance.
Dunér gives

$$\Delta = 5''\cdot30 + 0''\cdot019 \ (t - 1850\cdot0).$$
$$P = 273°\cdot9 + 0°\cdot454 \ (t - 1850\cdot0) - 0°\cdot00167 \ (t - 1850)^2.$$

Σ.	263·0	2n.	4·86	1829·71
	266·0	1n.	·79	33·26
Mä.	267·6		5·12	38·19
	271·3	3n.	·40	45·48
	275·3	1n.	·30	51·27
	276·2	2n.	·48	2·69
	277·4	,,	·67	6·46
	276·9		·71	8·72
Se.	278·2	2n.	·31	60·06
Du.	279·9	3n.	·65	71·50

Gl.	286·4°	2	5·7″	1871·32
O.Σ.	282·4	In.	·76	2·54
	285·3	,,	·60	4·70
W. & S.	283·9	4	·5	3·25
	285·9	4	...	·29

448   Σ. 1832.

R. A.	Dec.	M.
14h 13m	4° 27′	9, 9

Probable change in angle.

Σ.	118·3	3n.	0·47	1830·28
O.Σ.	132·6	In.	·66	47·40
	122·1	,,	·58	9·37
	131·9	,,	·51	53·41
Se.	120·6	2n.	·41	6·41
W. & S.	120·5	2	·4	76·46

449   O.Σ. 281.

R. A.	Dec.	M.
14h 14m	9° 8′	7·3, 10·8

O.Σ.	161·4	3n.	1·25	1847·72
De.	152·3	,,	·59	67·33

450   Σ. 1834.

R. A.	Dec.	M.
14h 15·9m	49° 3′	7·1, 7·2

Rectilinear motion.
Dunér has

$$\Delta = 1''\cdot04 - 0''\cdot0175 \ (t - 1850\cdot0).$$
$$P = 113°\cdot8 + 0°\cdot05 \ (t - 1850\cdot0).$$

H₂.	104·0	2n.	1·09	1830·24
	108·3		·20	1·37
	115·0		...	3·26
Σ.	113·7	4n.	1·36	1·20
Da.	111·8	3n.	·14	40·51
	112·7	2n.	·04	8·50
	111·1		·10	9·48
Mä.	113·9	2n.	·37	3·23
De.	114·0	In.	...	57·51
Se.	·8	2n.	0·92	7·57
Ta.	110·9	In.	·87	66·49
Du.	115·5	4n.	·66	71·21
Gl.	·3	In.	·6	1·53
W. & S.	113·7	,,	·6	4·52

451   Σ. 1837.

R. A.	Dec.	M.
14h 18·2m	− 11° 7′	7·1, 8·7

A physical pair. Common proper motion.

Σ.	326·9	4n.	1.41	1829·83
Sm.	325·8		·6	33·36
H₂.	321·5		·3	7·48

Mä.	323.4°		1".55	1848.38
Mit.	324.5	1n.	.50	.45
Se.	312.3	2n.	.40	56.47
M.	348.8	1n.	.05	62.37
De.	314.1		.34	5.07
Gl.	313.0	1n.	.41	71.40
W. & S.	311.6	,,	.33	3.36
Sp.	309.8		.26	5.87
C.O.	307.0	2n.	.45	7.42

452   Σ. 1842.

R. A.	Dec.	M.
14h 21.6m	4° 14'	8.7, 8.7

C. white.

Probably a binary.

Σ.	10.9	4n.	2.84	1828.86
H₂	9.4		0.96	30.34
Mä.	11.9		3.06	4.18
	13.8		2.90	44.35
Se.	15.8	3n.	.80	56.77
Gl.	13.7		.9	71.40
W. & S.	14.5	2n.	.75	4.42
Dob.	12.7	,,	.97	7.38
Pl.	.9	,,	.86	.51

453   Σ. 1847.

R. A.	Dec.	M.
14h 22.2m	-9° 40'	8.5, 9.8

Σ.	248.4	4n.	18.73	1829.81
Mä.	251.5		.22	44.34
Mit.	253.1		20.17	8.45
De.	256.0		21.67	65.36

454   O.Σ. 283.

R. A.	Dec.	M.
14h 28m	49° 44'	7.3, 11.2

O.Σ.	134.6	3n.	4.93	1848.19
De.	130.1	,,	5.11	66.79

455   Σ. 1858.

R. A.	Dec.	M.
14h 29m	36° 6'	7.2, 8

C. white.

H₂	30.7	2n.	1.44	1830.78
Σ.	35.2	3n.	2.20	1.84
	33.0	1n.	.73	51.80
Mä.	35.6	1n.	.34	43.46
Se.	.8	2n.	.36	56.89
De.	34.7	1n.	...	8.27
Mo.	31.5	3n.	2.51	9.33
O.Σ.	30.9	1n.	.73	70.45

Du.	34.2°	5n.	2".45	1870.99
Dob.	.4	4n.	.59	7.32
Pl.	35.6	2n.	.61	.51

456   α CENTAURI.

R. A.	Dec.	M.
14h 31.8m	-60° 20'	1, 2

C. "Both strong reddish yellow" (Dunlop). "Both yellowish;" "A, yellow; B, greenish yellow" (Jacob). "Both yellow" (H₂).

A fine double star discovered by Feuillée in 1709. He wrote: "Je trouvai cette étoile composée de deux, dont l'une est de la troisième grandeur, et l'autre de la quatrième. Celle de la quatrième est la plus occidentale, et leur distance est égale au diamètre de cette étoile." *
Richer was probably the first to examine this fine star with a telescope: this was in 1673, at Cayenne. Halley observed it at St. Helena in 1677, but neither observer records it as a double star.
La Condamine observed it while in Peru. See *Phil. Trans.* for 1749.
In 1709 the distance was probably about 7"; in 1751 Lacaille observed it, 22".5; in 1761 Maskelyne found it 15" or 16" (see *Phil. Trans.* for 1764); and in 1825 Dunlop made it 23" (*Mem. R. A. S.*, vol. iii.)
In 1848 Captain Jacob computed the orbit: his elements are—

$$\pi = 26° 24'$$
$$\gamma = 47 \quad 46$$
$$☊ = 86 \quad 7$$
$$\lambda = 291 \quad 22$$
$$e = 0.950$$
$$\tau = 1851.50$$
$$P = 77.0 \text{ years}$$
$$n = 4°.675$$
$$a - 15''.5.$$

Maximum distance, 21.85 at 207°.5.
Minimum   ,,   0.50 at 5°.0.
Greatest daily motion = 2° 40'.
Mass = ¾ of solar mass.

Mr. Hind in 1851 published the following elements:—
Perihelion passage 1859.42.
$$☊ = 16° 42'$$
$$\lambda = 26 \quad 2$$
$$e = 0.7752$$
$$\gamma = 62° 53'$$
$$a = 13''.57$$
$$P = 80.94 \text{ years.}$$

* See Feuillée's *Journal des Observations Physiques*, etc., tome i., p. 425. Paris, 1714. The telescope used was one of 18 ft. focal length.

In 1854, Powell published the following elements :—

	(1)	(2)
$\tau =$	$1857 \cdot 012$	$1858 \cdot 012$
$\pi =$	$30^\circ\ 14'$	$29^\circ\ 33'$
$\Omega =$	$2\ 35$	$177\ 50$
$\gamma =$	$77\ 19\frac{1}{2}$	$77\ 50$
$e =$	$0 \cdot 96887$	$0 \cdot 966$
$n =$	$4^\circ \cdot 35882$	$4^\circ \cdot 78$
$P =$	$82 \cdot 59$ yrs.	$75 \cdot 3$ yrs.
$a =$	$31'' \cdot 7574$	$30''.$

Powell thought that the correct elements lay between the two sets given ; that the next periastral passage would occur between $1857 \cdot 5$ and $1858 \cdot 5$; that the semi-major axis of the orbit is a little greater than $30''$; that the sum of the masses of the two stars is between six and six-and-a-half times the mass of the sun ; and that the orbit is something like a magnified image of the path of Halley's comet.

In 1877 Mr. Hind computed an orbit (see *Monthly Notices*, vol. xxxvii., p. 96). In this the observations made by Lord Lindsay in $1874 \cdot 85$ were used :—

Periastron passage, $1874 \cdot 85$.
$$\Omega = 21^\circ\ 48' \cdot 0.$$

Node to periastron on orbit　$59^\circ\ 32' \cdot 1$.
Inclination... ... ...　$82\ 18\ \cdot 4$.

$$e = 0 \cdot 6673$$
$$a = 21'' \cdot 797$$
$$P = 85 \cdot 042 \text{ years.}$$

And the comparison of the elements with the observations from $1752 \cdot 2$ to $1874 \cdot 85$ shows a very satisfactory agreement.

Mr. Hind remarks that " Lord Lindsay's measures fall exactly at the computed time of nearest approach of the component stars in the real orbit." " If, for the annual parallax, a mean of Henderson's value, as corrected by Peters, and that of Moesta, be taken, giving $0'' \cdot 928$, we find the mass of this system $= 1 \cdot 79 \times$ the sun's mass, and for the semi-axis major of the orbit $23 \cdot 49$."

Mr. Maxwell Hall has measured this star with great care. His results are—

$$1878 \cdot 38 \qquad 139^\circ \cdot 1 \qquad 2'' \cdot 4.$$

See *Nature*, vol. xviii., p. 225.

Lastly, Dr. Doberck in 1879 obtained the following elements :—

$$\Omega = 25^\circ\ 14'$$
$$\lambda = 45\ 58$$
$$\gamma = 79\ 24$$
$$e = 0 \cdot 5332$$
$$P = 88 \cdot 536 \text{ yrs.}$$
$$T = 1875 \cdot 12$$
$$a = 18'' \cdot 45.$$

Feuillée	...		6—9	1709
Lacaille	$218 \cdot 7$		$20 \cdot 51$	$52 \cdot 2$

			$''$	
Maskelyne ...			15—16	1761
Fallows	$209 \cdot 6$		$28 \cdot 75$	$1822 \cdot 00$
Brisbane	$215 \cdot 4$		$22 \cdot 45$	$4 \cdot 00$
Dunlop	$213 \cdot 2$		$\cdot 45$	$6 \cdot 01$
Johnson	$215 \cdot 0$		$19 \cdot 95$	$30 \cdot 01$
Taylor	$\cdot 9$		$22 \cdot 56$	$1 \cdot 00$
	$216 \cdot 4$		$19 \cdot 85$	$2 \cdot 16$
H_2	$217 \cdot 5$		$18 \cdot 67$	$3 \cdot 00$
	$218 \cdot 5$		$17 \cdot 4$	$4 \cdot 79$
	$219 \cdot 6$		$16 \cdot 52$	$6 \cdot 30$
	$220 \cdot 7$		$\cdot 11$	$7 \cdot 34$
Maclear	$223 \cdot 2$		$14 \cdot 74$	$40 \cdot 00$
	$262 \cdot 8$		$5 \cdot 03$	$52 \cdot 56$
Ja.	$232 \cdot 4$		$10 \cdot 96$	$46 \cdot 21$
	$234 \cdot 3$		$9 \cdot 82$	$\cdot 87$
	$235 \cdot 1$		$\cdot 45$	$7 \cdot 09$
	$238 \cdot 0$		$8 \cdot 05$	$8 \cdot 02$
	$250 \cdot 7$		$5 \cdot 97$	$50 \cdot 96$
	$251 \cdot 2$		$\cdot 90$	$1 \cdot 05$
	$267 \cdot 6$		$4 \cdot 55$	$3 \cdot 05$
	$276 \cdot 3$		$\cdot 21$	$4 \cdot 00$
	$283 \cdot 5$		...	$\cdot 63$
	$307 \cdot 7$	6n.	$3 \cdot 92$	$6 \cdot 38$
	$320 \cdot 0$	,,	$4 \cdot 01$	$7 \cdot 29$
	$329 \cdot 2$	2n.	$\cdot 29$	$8 \cdot 12$
Po.	$270 \cdot 1$	22	...	$3 \cdot 59$
	$277 \cdot 0$	155	...	$4 \cdot 06$
	$281 \cdot 1$	25	...	$\cdot 38$
	$283 \cdot 5$	38	...	$\cdot 63$
	$289 \cdot 0$	128	...	$5 \cdot 04$
	$293 \cdot 6$	140	$4 \cdot 07$	$\cdot 32$
	$294 \cdot 9$	26	...	$\cdot 50$
	$301 \cdot 1$	140	$3 \cdot 95$	$6 \cdot 02$
	$340 \cdot 0$	125	$5 \cdot 12$	$9 \cdot 38$
	$345 \cdot 3$	110	$\cdot 68$	$60 \cdot 11$
	$348 \cdot 7$	10	$\cdot 6$	$\cdot 48$
	$351 \cdot 0$	130	$6 \cdot 08$	$1 \cdot 05$
	$353 \cdot 2$	60	$\cdot 2$	$\cdot 30$
	$354 \cdot 3$	30	$\cdot 29$	$\cdot 58$
	$358 \cdot 0$	80	$\cdot 79$	$2 \cdot 20$
	$1 \cdot 4$		$7 \cdot 2$	$3 \cdot 03$
	$5 \cdot 7$		$\cdot 85$	$4 \cdot 11$
	$20 \cdot 4$		$10 \cdot 24$	$70 \cdot 00$
El.	$0 \cdot 0$		$\cdot 0$	62
	$5 \cdot 2$		$8 \cdot 5$	$3 \cdot 75$
	...		$\cdot 1$	$4 \cdot 72$
	...	2n.	$9 \cdot 4$	$8 \cdot 18$
	...	1n.	$10 \cdot 2$	$70 \cdot 65$
	...	,,	$8 \cdot 3$	$3 \cdot 16$
	$30 \cdot 5$	,,	$\cdot 0$	$4 \cdot 15$
	$50 \cdot 6$	3n.	$3 \cdot 9$	$6 \cdot 72$
	$69 \cdot 1$	5n.	$\cdot 1$	$7 \cdot 25$
Russell	$22 \cdot 3$	3n.	$10 \cdot 46$	$0 \cdot 74$
	$\cdot 9$	2n.	$\cdot 12$	$1 \cdot 47$
	$25 \cdot 3$	,,	$9 \cdot 73$	$2 \cdot 47$
	$28 \cdot 1$	1n.	$\cdot 5$	$3 \cdot 33$
	$30 \cdot 0$	2n.	$7 \cdot 96$	$4 \cdot 47$
	$47 \cdot 0$	,,	$4 \cdot 35$	$6 \cdot 41$
	$76 \cdot 1$	5n.	$2 \cdot 23$	$7 \cdot 54$
Lindsay	$34 \cdot 2$		...	$74 \cdot 85$
Gill	$69 \cdot 4$	1n.	...	$7 \cdot 55$
	$80 \cdot 6$	,,	...	$\cdot 56$
	$75 \cdot 3$	,,	...	$\cdot 57$
	$80 \cdot 5$	,,	...	$\cdot 59$
	$\cdot 7$	,,	...	$\cdot 61$

457   Σ. 1863.

R. A.   Dec.   M.
$14^h 34^m$ $52° 6'$ 7·1, 7·4

C. yellowish white.

Certain change in both angle and distance.
Dunér gives

1855·72. $\Delta = 0''·61$.
$P = 101°·5 - 0°·25 (t - 1850·0)$.

Σ.	109·7	4n.	0·65	1830·14
Mä.	104·1		·60	8·94
	101·6		·6	41·56
	·4	7n.	·55	3·32
	98·7	2n.	·68	51·27
	97·6	,,	·77	2·67
	100·2	,,	·58	4·50
	91·2	,,	·67	8·69
O. Σ.	107·3	3n.	·65	41·21
	105·0	2n.	·67	50·14
	94·6	1n.	·88	72·54
De.	97·3		elong.ᵈ	56·03
	95·2	3n.	,,	64·37
Se.	101·5	1n.	0·77	59·52
Eng.	108·1	2n.	·76	65·80
Du.	101·7	5n.	·50	9·49
	99·6	1n.	·58	75·52
W. & S.	·5	4	·6	2·41
	89·5	4	·6	3·25
	94·3	4	·57	·30
	·1	1	·65	·36
Gl.	93·4	3	·5	4·36

458   Σ. 1864.

π BOÖTIS.

R. A.   Dec.   M.
$14^h 35·1^m$ $16° 56'$ 4·9, 6

C. very white.

H_1. "Sept. 20, 1879. The Rev. Mr. Hornsby told me it was a double star, and I found it so. He observed that this had been found to have changed its place 16"."

H_1. (*Phil. Trans.*, vol. lxxii., p. 219).
"π Boötis, Fl. 29. Sept. 20 [1781] Double. Pretty unequal. L, w ; S, w inclining to r. Distance, 6"·17. Position, 6° 28' s.f."

H_2 and So. (*Phil. Trans.* 1824, p. 199). "Nearly equal ; large, white ; the smaller perhaps inclines to blue."

Sm. (*Cycle*, p. 323). From the words used by Piazzi, and the measures of H_1 and H_2 and So., he infers a slight direct orbital motion. "This suspicion," he adds, "would have been confirmed by my observations, but that Σ. found the angle 9° 50' s.f. in 1819·61 ; and ten years after-

wards he concluded 9° 12' s.f. to be the mean position."

Dunér gives the following formulæ :—

1852·21. $\Delta = 5''·87$.
$P = 100°·3 + 0°·065 (t - 1850·0)$.

H_1.	96·5	1n.	6·17	1781·83
	97·6	2n.	...	1803·19
So.	·9	13	6·88	22·05
Σ.	98·7	1n.	·08	7·28
	99·3	3n.	5·93	9·35
	·2	5n.	·71	31·50
Da.	·5		6·28	3·19
	100·3		5·50	45·39
Mä.	·2	10n.	·89	3·33
	101·0	4n.	6·01	52·36
	100·8	14n.	5·85	6·08
	·4	6n.	·96	60·73
Po.	98·8	3n.	6·08	46·43
D.O.	100·7		·90	·40
	97·3		·76	7·31
De.	101·1	5n.	5·76	54·46
	102·3	1n.	...	5·21
Se.	100·9	3n.	5·97	6·79
Mo.	·6	2n.	6·14	7·34
Ro.	101·8		·01	63·27
Ka.	100·6	6n.	5·73	6·45
Ta.	101·6	1n.	6·35	·49
M.	98·9	,,	5·94	7·34
	100·4	,,	6·14	8·40
	102·5	,,	...	70·30
	100·5	,,	6·22	·39
	·8	,,	5·92	1·37
	101·6	,,	6·18	·44
	100·0	3n.	5·99	3·40
	102·7	4n.	6·39	4·37
	·2	5n.	·40	5·33
Du.	101·2	4n.	5·89	69·47
	102·0	2n.	·51	71·47
	·6	,,	·82	2·47
W. & S.	101·5	6	...	·38
	100·8	4	5·90	3·36
	101·3	6	6·11	·37
	102·2	7	5·96	4·41
	·5	2	·86	·42
	103·7	4	·91	7·46
	·0	2	6·22	·46
	102·5	8	5·95	·47
Gl.	101·2	2	6·2	4·36
Schi.	·0	1n.	5·87	5·47
Pl.	·9	4n.	·84	6·96
Dob.	·0	6n.	6·16	7·31
Fl.	103·3	1n.	·11	·47

459   Σ. 1865.

ζ BOÖTIS.

R. A.   Dec.   M.
$14^h 35·4^m$ $14° 14'$ 3·5, 3·9

C. white.

The angle was unchanged from 1796 to

1841 ; a slow retrograde movement then began, accompanied by a diminution in distance.

Dunér has the formulæ

$$\Delta = 1''{\cdot}03 - 0''{\cdot}010\,(t - 1850{\cdot}0) + 0''{\cdot}00014\,(t - 1850{\cdot}0)^2.$$

$$P = 309°{\cdot}3 - 0°{\cdot}1244\,(t - 1835{\cdot}0) - 0°{\cdot}0015\,(t - 1835{\cdot}0)^2.$$

	°		″	
H₁.	312·0	In.	...	1796·59
So.	307·0	2n.	1·68	1823·27
H₂.	312·6	In.	·58	30·34
	308·5	,,	·15	1·39
	309·1	,,	·0	3·24
Σ.	·2	1In.	·19	0·47
	312·2		·16	3·42
	305·0		·17	44·40
Be.	310·7		·29	31·18
	307·5		·33	2·34
Da.	308·3		·32	·47
	311·2	In.	·20	3·30
	310·1		...	4·43
	307·0	In.	1·04	43·32
	306·5	4n.	·08	8·11
	·6		·03	8·43
Sm.	309·9		·3	33·39
	308·6		·3	8·45
	307·3		·2	42·43
	308·2		·0	52·38
Galle.	309·8		·20	38·66
O.Σ.	310·4	6n.	·24	41·16
	307·5	8	·00	7·72
	303·9	7	·00	61·12
	304·5	11	0·99	2·95
	303·1	6	·88	8·68
	301·5	4	·83	73·01
Mä.	310·0		1·31	41·39
	311·0		·16	2·36
	309·4	16n.	·14	·85
	308·4		·05	3·40
	309·7		·19	5·26
	·2	1on.	·23	6·88
	308·7		·23	7·65
	310·0	In.	·21	8·36
	307·8	6n.	·04	52·54
	306·7	In.	·23	5·94
	305·1		·34	7·43
	308·1		·02	8·44
	307·8		·07	9·38
	308·0		·26	61·42
	306·5		·24	2·63
Ch.	·7	2n.	·5	42·50
Ja.	307·0		·2	6·19
	306·2		·24	53·49
	·6		·35	5·44
Mit.	307·2	2n.	·11	46·67
	308·4	In.	0·93	7·57
Flt.	305·8	12	1·1	51·75
Mi.	307·2	32	·19	3·31
Se.	305·7	4n.	0·99	5·70
De.	306·1	9n.	1·0	·83
	303·2	6n.	·02	64·78
Mo.	304·7	2n.	·18	59·34
Eng.	306·0	In.	0·87	64·50

	°		″	
Ta.	307·2	In.	1·10	1866·49
Du.	303·1	15n.	0·75	9·16
	·1	21n.	·75	71·37
	301·4	9n.	·72	5·42
W. & S.	298·9	3	...	1·50
	301·6	10	0·93	2·38
	300·2	5	·88	3·36
	302·2	7	·92	4·41
	301·0	6	·98	·42
Gl.	308·2	5	1·38	·36
Kn.	303·1	In.	0·97	0·47
Sp.	299·4		·91	5·40
	298·5		·88	6·50
Dob.	302·0	4n.	·8	6·32
	300·6	5n.	·88	7·33

460   Σ. 1866.

R. A.	Dec.	M.
14ʰ 35·9ᵐ	10° 2′	8·2, 8·2

C. yellowish.

	°		″	
Σ.	19·1	In.	0·86	1827·27
	·2	,,	·99	8·32
	·3	,,	·90	33·22
Mä.	27·0		·65	42·36
	25·8		·77	3·35
	29·6		·89	7·27
	33·5		·80	52·43
	32·8		·79	4·42
	27·5		·75	7·43
	38·4		·70	8·39
De.	19·9		·89	65·08
W. & S.	23·5	In.	·93	74·42
	22·9		·77	5·41
Gl.	·4	In.	1·0	4·53

461   O.Σ. 284.

R. A.	Dec.	M.
14ʰ 36·1ᵐ	49° 15′	7·2, 11·2

If De.'s angle be correct, a great change took place between 1852 and 1866.

	°		″	
O.Σ.	106·3	3n.	6·98	1848·19
Mä.	143·5		...	6·29
	141·6		...	52·69
De.	102·3	3n.	6·79	66·69

462   Σ. 1871.

R. A.	Dec.	M.
14ʰ 37·5ᵐ	51° 55′	7, 7

C. white.

Probably a binary.
Dunér has

$$1844{\cdot}80. \quad \Delta = 6''{\cdot}80.$$
$$P = 286°{\cdot}1 - 0°{\cdot}20\,(t - 1850{\cdot}0).$$

Σ.	283°.2	3n.	1".82	1829.10
	.0	,,	.80	36.18
H₂.	279.0	5n.	.55	0.47
Mä.	285.1	6n.	.93	43.61
	286.5	2n.	.82	52.67
De.	287.2	1n.	...	7.11
Se.	.9	2n.	1.68	.57
Du.	289.8	,,	.61	71.42

463 Σ. 1876.

R. A. 14^h 40^m Dec. −6° 53'   M. 8.1, 8.6

C. Σ., yellowish; Se. and De., white.

Probably a binary in rapid motion.

Σ.	51.7	1n.	0.89	1829.31
	39.6	,,	1.19	31.37
	49.2	,,	0.94	2.34
	55.4	4n.	1.31	3.32
Mä.	57.2		.24	44.34
Mit.	59.8	1n.	0.95	8.45
Se.	|60.8		1.0	56.88
	63.8		elong^d·	65.48
De.	64.2	1n.	1.1	56.49
	61.8	,,	.0	7.49
	57.5	,,	...	8.42
	65.8	3n.	1.2	63.39
W. & S.	68.6	7	.17	72.38
	69.7	7	.27	3.36
	56.0	2	...	4.42
Gl.	69.1	2	1.3	.54
C.O.	67.0	1n.	.40	7.41

464 Σ. 1877.

ε BOÖTIS.

R. A. 14^h 39.7^m Dec. 27° 35'   M. 3, 6.3

C. H₁, A, reddish; B, blue, or rather a faint lilac. H₂ and So. A, yellow; B, blue-green. Σ. A, decided yellow; B, decided green. Sm. A, pale orange; B, sea-green. Se. A, yellow; B, blue.

H₁ (Phil. Trans., vol. lxxii., p. 213): "1. ε Boötis. Flamst. 36. Sept. 9, 1779. —Double. Very unequal. L, reddish; S, blue, or rather a faint lilac. A very beautiful object. The vacancy or black division between them, with 227, is ⅔ diameter of S; with 460, 1¼ diameter of L; with 932, near 2 diameters of L; with 1159, still further; with 2010, extremely distant, 2¾ diameters of L. These quantities are a mean of two years' observations. Position 31° 34' n.p."

In his paper read June 9, 1803, H₁ says: "This beautiful double star, on account of the different colours of the stars of which it is composed, has much the appearance of a planet and its satellite, both shining with innate but differently coloured light. There has been a very gradual change in the distance of the two stars; and the result of more than 200 observations, with different powers, is, that with the standard magnifier, 460, and the aperture of 6.3 inches, the vacancy between the two stars in the year 1781 was 1½ diameter of the large star, and that it now is 1¼."

He found, from many observations, that the proportion of the diameters of the two stars was 3 : 2.

H₂ and So. (Phil. Trans. 1824, p. 204): "Large, yellow; small, blue-green; a very marked contrast of colours.

"Nothing can be more unsatisfactory than the measures of this very difficult star, especially in position, the difference between the greatest and least among the single measures amounting to the enormous quantity of 16° 10', and even among the mean results of the whole sets of observations extending to 10° or 11°." H₂ then remarks on the difficulty of accounting for this, and rejects bias of eye, error of judgment, refraction, imperfection of vision, closeness, and difference of size and colour, as insufficient. "The angular motion is indisputable."

In 1826 (Phil. Trans. 1826, p. 337) he writes: "The motion of this star is therefore satisfactorily confirmed."

In vol. v., p. 46, of Mem. of R. A. S., he observes, "After a long and obstinate contest with ε Boötis, which is certainly one of the most difficult double stars to measure correctly, Rigel itself excepted, I remain unconvinced of its motion. My father's measure in 1796 differs only 3° from Σ.'s in 1826; yet this might arise from the conspiring effect of extreme errors. But, again, the mean of my measures for 1830, which I believe to be the truth, tallies within 0° 26' with the joint result of Sir James South and myself in 1822, which rests on upwards of sixty individual measurements."

Σ. (M. M., p. 49), referring to the discrepancies in his own measures, says:— "These probably arose from neglecting the position of the eye and head." He thinks that a slow direct motion is beyond doubt, and that H₂ and So.'s angles for 1822.55 and 1825.34 are too large.

Sm. (Cycle, p. 325), after referring to the observations and conjectures of H₁ and H₂, submits the following details and epoch which have led him to consider the question to be, as yet, unestablished:—

H₁.	301° 34'	Dist. 4".00±	1779.67
H₂ & So.	322 59	3 .93	1822.55
Σ.	320 58	2 .64	1829.39
D.	321 35	...	1831.35

Dawes (*Mem. R. A. S.*, vol. xxxv., p. 374): "Recent measures seem to confirm the idea that this is really a binary system."

Writing in 1865 in *Astronomical Register*, he says, "All my measures of this star point to a slow increase of angle."

The changes are very slowly produced in this system. The angle has increased, but the distance has not changed probably. The more rapid change in angle of late years leads us to expect that a diminution of distance will manifest itself. (O.Σ.)

Dunér has the formulæ

$$1847 \cdot 40. \quad \Delta = 2'' \cdot 67.$$
$$P = 323° \cdot 9 + 0° \cdot 165 \ (t - 1850 \cdot 0).$$

	°		''	
H₁.	32° 19′			Aug. 31, 1780.
	30 21			Mch. 13, 81.
	33 1			May 10, 81.
	38 26			Feb. 17, 82.*
	45 32·4			Aug. 18, 96.
	49 18			Jan. 28, 1802.†
	43 55			}
	42 42			} Mch. 23, 1803.
	44 33			}
	40 29 n.p.			}
	44 33			} Mch. 26, 1803.
	44 52 ‡			}
	305·1	6	...	1781·73
	314·6	8	...	1803·01
Amici.	...		2·35	16·04
H₂ & So.	322·9	5n.	3·93	22·55
	324·3	6n.	·35	5·43
	322·4	5n.	·87	30·27
Σ.	318·2		...	22·39
	317·9	2n.	2·69	6·79
	323·5	1n.	·55	7·27
	321·4	9n.	·68	8·56
	·2	1n.	·61	9·58
	320·0	3n.	·59	31·56
	321·8	2n.	·55	3·41
Da.	·5	1n.	...	1·36
	320·8	2n.	2·90	41·43
	322·1	5n.	·67	8·19
	·9	3n.	·68	54·52
	323·6	1n.	·63	5·53
	324·4	3n.	·83	60·05
	325·5	,,	·91	5·48
Sm.	321·6		3·2	31·46
	323·8		·8	3·53
	321·2		2·9	8·68
	322·1		·8	48·54
Be.	316·2		·96	31·56
Encke.	321·3		3·37	7·44
Galle.	·9		2·88	8·67
	324·9		·66	9·45
Ka.	320·0	6n.	·80	40·05
	319·8	9n.	·74	2·37
	325·3	5n.	·67	66·48
Ch.	321·4	1n.	·88	41·38
	316·6	,,	·34	2·43
	311·3	,,	0·86	3·33

* Very exact.
† Very accurately taken.
‡ The best.

	°		''	
Mä.	323·8		2·70	1843·93
	325·5	7n.	·61	50·74
	·7	6n.	·57	4·41
	326·0	,,	·63	7·03
	320·0	1n.	·90	7·42
	326·9	7n.	·61	8·47
	·5	,,	·77	9·38
G.O.	319·3		·75	42·35
	321·6		·75	4·44
O.Σ.	·6	12n.	·67	·49
	323·4	5n.	·56	55·33
	324·8	23n.	·70	64·14
	329·1	4n.	·68	73·26
Ja.	323·3		3·50	46·28
	·1	10	2·63	53·20
Mit.	320·8	2n.	·57	46·66
Bond.	326·7		3·2	8·38
	322·0		·2	·38
	·1		·1	·45
Flt.	·7	36	2·77	50·95
Mi.	321·2	25	·83	2·94
De.	·8	5n.	·64	4·48
	324·0	2n.	·24	5·22
	·1	3n.	3·03	6·39
	·3	1n.	·02	·49
	323·0	,,	2·80	7·54
	322·9	5n.	·88	8·49
	324·0	2n.	·78	63·50
	325·0	4n.	·69	5·46
Se.	323·6	,,	·61	55·37
	322·9	3n.	·59	6·50
	324·7	1n.	3·29	65·49
Mo.	327·8	12	2·95	57·44
	325·0	22	·72	8·46
M.	324·4	1n.	·60	62·39
	148·2	,,	·88	7·36
	140·4	,,	·81	9·55
	141·2	,,	·74	·55
	...	,,	·32	·56
	...	,,	3·16	70·39
	145·7	,,	2·97	3·40
	324·0	2n.	·93	4·36
	328·0	,,	·80	5·31
Ro.	323·5	1n.	·79	63·53
Eng.	326·5	3n.	·92	5·40
Ta.	323·1	,,	·97	6·40
	321·3	1n.	·87	9·62
	324·9	,,	3·08	70·46
	...	,,	·05	1·41
Kn.	323·9	,,	2·73	67·34
Du.	326·8	2n.	·52	·72
	327·6	4n.	·75	8·56
	326·1	8n.	·64	9·55
	327·6	7n.	·67	70·49
	·3	6n.	·56	1·56
	328·3	,,	·77	2·52
	327·6	7n.	·63	5·55
Br.	332·4	2	3·04	68·46
W. & S.	327·2	4	2·97	71·56
	326·9	3	3·06	2·38
	·9	5	·00	3·36
	327·6	6	·00	4·42
	326·1	8	2·9	·44
	327·1	6	3·19	5·39

Gl.	327°·0	4	2"·9	1874·54
Schi.	328·2	1n.	·80	5·41
	327·6	,,	·75	6·49
Sp.	328·3	1n.	·80	5·41
	327·6		·75	6·49
W.O.	328·6	1n.	·94	·41
	330·4	,,	·90	·43
	327·2	,,	3·09	·43
Dob.	324·5	1 1n.	·13	·42
	·7	2n.	2·95	7·23
Pl.	329·1	6n.	·92	6·70

465   Σ. 1879.

R. A.	Dec.	M.
14^h 40·4^m	10° 10'	7·8, 8·8

C. yellowish.

Probably binary.

Σ.	67·8	2n.	1·17	1827·80
	66·3	1n.	·21	34·39
Mä.	59·2		0·79	42·42
Dᴇ.		single		63·51
		elongated		5·31
Du.		single		9·38
		perfectly round		·40

466   O.Σ. 285.

R. A.	Dec.	M.
14^h 41^m	42° 53'	7·1, 7·6

C. white.

O.Σ.	72·1	3n.	0·60	1845·80
	57·8	4n.	·50	52·74
	53·9	3n.	·51	5·84
De.	36 ?		obl. ?	65·53

467   Σ. 1883.

R. A.	Dec.	M.
14^h 42·9^m	6° 27'	7, 7

C. yellowish.

H₂.	266·7		1·20	1830·23
	271·2		...	1·37
	270·0		·25	2·00
Σ.	272·0	3n.	·24	0·27
Mä.	269·8		·07	8·59
	270·0		·10	42·40
	267·8		0·91	3·37
	265·0		1·03	54·43
	264·5		·03	7·43
	262·7		·15	8·41
Se.	265·6	2n.	0·95	6·41
De.	261·3		·80	8·13
	262·7		·80	63·28
W. & S.	261·7	1n.	·88	72·37
	·9	,,	·95	3·36
Gl.	262·2	,,	1·1	4·54
Sp.	259·6		0·81	5·42

468   Σ. 1888.

ι BOÖTIS.

R. A.	Dec.	M.
14^h 45·8^m	19° 36'	4·7, 6·6

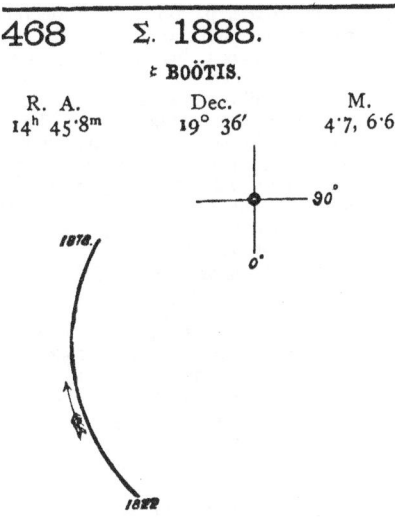

C. H₁, A, pale red; B, garnet or deeper red. Σ., A, yellow; B, reddish purple. Se., A, golden; B, red. Sm., A, orange; B, purple.

H₁ (*Phil. Trans.* 1804, p. 367). He first observed this star on the 15th of April, 1782, when the position angle was 65° 53' n.f. On the 20th of April, 1792, it was again observed, and the angle recorded was 85° 43"·5 n.p. He then discusses the observations, pointing out the changes which would be observed if the companion moved round the larger star, the plane of the orbit being coincident with the line of sight. He thus infers a retrograde orbital motion.

H₂ (*Phil. Trans.* 1824, p. 208). Discussing the observations between 1782 and 1823, he is constrained to admit that a physical connection is probable. When, however, he came to examine the measures between 1780 and 1830, he remarked, "That the motion is not rectilinear, but orbital, there seems little room, from the later observations, to doubt; but the probable errors in the positions from 1795 to 1804 prevent any certain determination of the orbit." (*Mem. R. A. S.*, vol. v., p. 36.)

In 1833, in vol. vi. of the *Memoirs*, H₂ gives the elements of the orbit. He finds, by his graphical method, the following results :—

$$a = 12"·56$$
$$e = 0·59374$$
$$\pi - 138° 24'$$
$$\lambda = 100\ 59$$
$$\gamma = 80\ 5$$
$$\text{☊} = 359\ 59$$
$$P = 117·14 \text{ tropical years}$$

Perihelion passage, Dec. 17, 1779.

Sm. (*Cycle*, p. 328). "If the relative path of the small star be really the straight line it appears to be, the angle of position will never reach 50° n.p., and the angular velocity will diminish continually from the present moment (1823). On the other hand, if the stars form a binary system, the present angular velocity of about 1° per annum will continue for some time nearly uniform, and in about fifteen or twenty years the limit of 50° n.p. will be attained or passed."

Mädler's elements are as follows:—

$$T = 1761 \cdot 71$$
$$\omega = 315° \cdot 2$$
$$\Omega = 172 \cdot 7 \text{ (Equ. } 1850 \cdot 0)$$
$$i = 52 \cdot 7$$
$$e = 0 \cdot 4540$$
$$\mu = -2° \cdot 2403$$
$$a = 5'' \cdot 591$$
$$P = 160 \cdot 695 \text{ years.}$$

And Dunér has compared these with the measures made since 1865. He finds the differences considerable and increasing.

Dr. Doberck's latest elements are—

$$\Omega = 26° \, 22'$$
$$\lambda = 117 \, 46$$
$$\gamma = 36 \, 55$$
$$e = 0 \cdot 7081$$
$$P = 127 \cdot 35 \text{ years}$$
$$T = 1770 \cdot 69$$
$$a = 4'' \cdot 86.$$

	°		″	
H₁.	24·1		3·38	1780·69
	·1		...	2·28
	...		...	91·39
	355·7		...	2·30
	354·8		...	5·22
	352·8		...	1802·25
	353·9		6·0±	4·25
H₂ & So.	342·3	In.	9·25	21·20
	340·1	,,	8·41	3·34
	337·0	4n.	7·77	5·37
	335·9	2n.	·17	8·54
	333·6	5n.	·62	30·29
	330·7	2n.	·53	3·23
Σ.	335·8		·54	22·69
	334·1	4n.	·21	9·46
	331·1	2n.	·14	32·40
	328·9	3n.	·07	5·43
	·1	4n.	·08	6·47
	327·1	2n.	6·85	8·47
Amici.	...		·66	23·30
Encke.	327·0		·79	7·41
Be.	331·2		7·30	31·40
Sm.	332·1		·3	·53
	327·4		·0	7·49
	224·8		·1	9·61
	322·9		6·9	42·42
Da.	330·3	3n.	7·54	34·44
	324·0	,,	·15	40·43

	°		″	
Da.	323·3	In.	7·26	1841·42
	322·6	2n.	·02	2·30
	318·8	,,	6·80	7·44
	317·9	,,	·71	8·50
	311·9	3n.	·26	54·46
	309·9	In.	5·90	7·42
Ga.	326·5		7·26	38·54
	325·8		·07	9·41
Ka.	·1		6·70	40·26
	322·1		·72	1·65
	·2		·64	3·68
	299·6	5n.	5·20	66·44
Mä.	324·6	4n.	7·09	41·43
	321·5		6·90	4·36
	320·8		·81	5·36
	·4	5n.	·69	6·29
	319·3	6n.	·67	7·37
	318·0	5n.	·63	8·28
	315·3	15n.	·21	52·56
	314·4	8n.	·31	3·44
	312·4	5n.	·06	4·48
	311·6	2n.	·06	5·38
	312·4	4n.	5·89	6·39
	311·2	5n.	·76	7·40
	309·8	7n.	·65	8·54
	·3	3n.	·56	9·38
O.Σ.	325·1	5n.	7·03	41·06
	319·3	3n.	6·53	7·82
	313·4	3	·22	53·54
	304·9	5	5·77	61·57
	·1	4	·58	2·47
	301·9	5	·67	3·56
	295·3	4	·09	9·02
	286·7	3	4·62	73·19
Hi.	322·3		6·12	45·37
Mo.	318·6	28	·76	·40
	319·2	20	·75	6·46
	307·8	12	5·93	58·38
D.O.	324·1		7·22	46·42
	321·5		·25	7·41
Flt.	317·4	36	6·56	51·11
Mi.	316·6	32	·51	2·30
De.	311·5	5n.	5·84	4·46
	312·6	3n.	6·15	5·22
	310·8	8n.	5·99	6·45
	308·9	2n.	·96	7·57
	·2	5n.	·82	8·36
	303·1	4n.	·65	62·55
	·0	10n.	·56	3·38
	302·0	2n.	·50	4·46
	301·1	7n.	·38	5·42
	299·0	11n.	·31	6·86
	297·4	5n.	·05	8·36
	293·8	,,	·04	70·39
	291·6	,,	4·83	1·34
	290·0	4n.	·68	2·40
	288·3	,,	·63	3·42
	287·4	,,	·62	4·39
	285·2	,,	·46	5·38
Wi.	311·7		6·00	56·55
Lu.	·8		·71	·75
Se.	310·0	12n.	·02	·88
	300·8	4n.	5·41	65·77
Po.	305·0	35	·52	1·29
Au.	303·4		·93	2·15

M.	305.9	In.	5″.68	1862.33
	295.4	,,	.50	.36
	294.7	,,	.33	8.40
	291.4	,,	.52	9.55
	292.7	,,	.21	.55
	293.0	,,	.33	.57
	.1	,,	.30	70.38
	292.9	,,	.53	.39
	.6	,,	.02	1.34
	.9	,,	4.83	.36
	286.0	,,	.93	3.39
	283.9	4n.	.92	4.36
	286.5	,,	.76	5.36
Ro.	302.4		5.79	63.28
Eng.	303.4	In.	.32	4.46
	301.4		.48	5.54
Ta.	298.5	4n.	.59	6.37
	.0	In.	.42	9.61
	295.8	,,	4.66	70.46
	296.4	,,	...	1.41
	286.6	,,	4.71	3.48
	288.6	,,	.19	4.33
	286.0	,,	...	.54
Du.	295.4	6n.	5.05	69.65
	292.9	7n.	4.79	71.94
	286.6	4n.	.44	5.51
W. & S.	291.8	5	.1	2.38
	289.3	5	.88	3.36
	.4	5	.81	.38
	288.4	5	.72	4.44
	283.0	5	.46	7.46
Lindstedt	287.0	In.	.84	3.43
Gl.	289.2	5	5.0	4.54
No.	286.3		...	5.38
Sp.	284.3		4.40	.40
Schi.	.3	In.	.40	.40
Dob.	.8	5n.	.5	6.34
	282.9	3n.	.70	7.24
Pl.	.5	8n.	.22	6.99
W.0.	284.9	In.	.59	.41
	280.6	,,	.67	.43
	284.6	,,	.65	.43
Fl.	282.7		.28	7.44

469 O.Σ. 287.

R. A.	Dec.	M.
14ʰ 47.1ᵐ	45° 25′	7.5, 7.6

C. white.

Probably a binary.

O.Σ.	97.3	2n.	0.58	1845.51
	105.4	4n.	.47	52.74
	108.4	3n.	.54	5.84
	119.0	In.	.74	68.56
De.	300.3		.64	7.23

470 O.Σ. 288.

R. A.	Dec.	M.
14ʰ 47.7ᵐ	16° 12′	6.4, 7.1

Certain change in angle and distance.

Σ.O.	228.0	3n.	0″.68	1845.35
	222.5	2n.	.53	8.96
	204.4	3n.	1.12	63.44
De.	200.4	,,	.21	6.72
W. & S.	197.8	6	.16	73.37
	196.5	4	.55	.44
	198.2	4	.26	4.44
	195.8	9	.49	7.45
	.8	6	.31	.46
Pl.	.2	6n.	.12	.24

471 Σ. 1893.

R. A.	Dec.	M.
14ʰ 50.2ᵐ	29° 58′	8.4, 10

Σ.	261.3	2n.	21.30	1831.49
	.4	In.	.82	2.29
	259.7	,,	.42	4.43
Mä.	257.9		19.36	44.41
	256.7		20.76	5.49
	255.9		.13	52.51
	256.1		.08	5.33
De.	252.3		.14	64.76

472 O.Σ. 289.

R. A.	Dec.	M.
14ʰ 51ᵐ	32° 46′	6.3, 9.8

O.Σ.	120.3	3n.	4.56	1846.34
De.	115.6	,,	.43	67.54

473 Σ. 1901.

R. A.	Dec.	M.
14ʰ 56ᵐ	31° 51′	7.7, 9.5

Σ.	203.9	In.	30.22	1831.46
	.5	,,	.47	.52
Mä.	201.0	,,	29.52	47.20
	200.6	,,	28.75	.32

474 Σ. 1909.

44 BOÖTIS.

R. A.	Dec.	M.
14ʰ 59.8ᵐ	48° 7′	5.2, 6.1

C. H_1, white. Σ., A, yellowish ; B, bluish. Sm., A, pale white ; B, lucid grey. Se., A, yellow ; B, blue.

H_1 (*Phil. Trans.*, vol. lxxii., p. 216): "Aug. 17 [1781]. Double, considerably unequal. Both W. With 227 they seem almost to touch, or at most ¼ diameter of S asunder ; with 460, ½ or ¾ diameter of S. This is a fine object to try a telescope, and a miniature of α Geminorum. Position 29° 54′ n.f."

H₂ (*Phil. Trans.* 1824, p. 218) examines the observations between 1781 and 1821, but finds them quite intractable; indeed he is unable to make quite sure that the observations relate to the same object. He thinks, however, that the positions given by Σ. in 1819·43, and that by H₂ and So. in 1821·33, "go to destroy Σ.'s idea of several revolutions having been performed in thirty-eight years." In 1830 he says, "The history of the star 44 Boötis is singularly beset with difficulties and apparent contradictions;" and it was not until the observations made in 1831, 1832, and 1833 were before him that he felt sure of the binary character of the system. He says, "Comparing the present results for 44 Boötis with the whole series of former measures, there can hardly remain a doubt of its constituting a binary system in which the orbits are very oblique to the visual ray, and the rotation performed in a period of about sixty years in the direction n.f. s.p., or direct; so that in about nine years more it will have completed a whole period in an apparent ellipse of great excentricity. This conclusion is grounded on a presumed mistake of 180° in my father's first position for 1782, and on the presumed *correctness* of his *correction* of a similar error in his second measure for 1802."

Dawes (*Mem. R. A. S.*, vol. viii.) says that the stars are most probably rapidly separating, and that the "mystery arising from the apparent contradictions in the earlier measures of Sir Wm. Herschel will, ere long, be satisfactorily solved."

Mädler, writing in 1847 (*Die Fixstern-Systeme*, p. 157) says, "Probably the connexion is physical; but the plane of the orbit passes nearly through the solar system." In 1855, however, excluding H₁'s position 62° 57' for 1802·25, he found the following elements:—

$$\text{Node } 60° \, 15'$$
$$\gamma \quad 77 \quad 36$$
$$\lambda \quad 12 \quad 36$$
$$\epsilon \quad 0·3837$$
$$P \quad 181 \text{ years}$$
$$T \quad 1784·7$$
$$a \quad 3''·10.$$

Doberck's latest elements are—

$$\text{Node } 65° \, 29'$$
$$\gamma \quad 70 \quad 5$$
$$\lambda \quad 1 \quad 18$$
$$\epsilon \quad 0·71$$
$$P \quad 261·12 \text{ years}$$
$$T \quad 1783·01$$
$$n \quad 1° \, 23'$$
$$a \quad 3''·093.$$

The proper motion of this system is $-3^s·045$ in R. A., and $-0''·03$ in N. P. D.

	°		''	
H₁.	60·1		...	1781·62
	62·9		...	1802·25
Σ.	228·0		1·5	19·43
	231·0	1n.	2·23	26·79
	233·6	2n.	·55	9·20
	234·4	3n.	·96	32·95
	235·2	6n.	3·17	5·51
	234·8	4n.	·30	6·66
	236·0	,,	·39	7·75
H₂ & So.	229·1	2n.	2·27	21·33
	234·6	1n.	·99	30·53
Da.	231·1	,,	·71	·44
	232·9	4n.	·97	1·34
	235·3	3n.	3·12	2·56
	·6	1n.	·28	3·39
	·6	4n.	·44	4·59
	·9	2n.	·76	6·58
	·7	5n.	·86	40·58
	236·0	4n.	4·00	1·48
	235·6	2n.	3·84	2·40
	8	1n.	·79	·71
	237·7	3n.	4·21	8·49
	·2	1n.	·36	9·48
	236·7	,,	·49	51·52
	237·7	,,	·58	4·74
Sm.	233·8		2·9	30·82
	235·1		3·3	4·55
	234·9		·6	6·71
	235·3		·5	9·62
	·9		·7	42·58
	236·2		4·1	7·45
O.Σ.	238·6	5n.	3·86	0·76
	236·3	3n.	4·23	8·36
	237·2	4n.	·67	56·81
	238·3	7n.	·79	63·44
Ch.	236·5	1n.	3·68	41·36
Ka.	235·2		·58	·65
	236·0		·74	3·75
Mä.	·1	1n.	·88	7·32
	237·0	2n.	·99	51·27
	238·1	9n.	4·18	·87
	237·9	15n.	·25	2·65
	·4	7n.	·25	3·64
Hi.	238·0		·26	47·09
Mit.	·3	1n.	3·74	·57
Bond.	237·0		4·5	8·55
	240·0		·3	·55
	239·6		·3	·54
Flt.	237·9	32	·3	51·47
	240·1		·6	65·60
Mi.	·2	39	·35	53·28
Ja.	238·5	9	·47	·27
Mo.	·4	20	·58	4·46
De.	239·9	11n.	·44	·55
	·1	2n.	·68	5·15
	238·8	5n.	·75	6·48
	·4	4n.	·69	8·41
	239·5	6n.	·75	63·31
	240·1	2n.	·82	6·45
	238·9	,,	·70	7·42
	240·0	,,	·72	8·36
	239·6	1n.	·93	9·61
	240·5	,,	·78	70·30
	239·6	,,	·97	1·15

Se.	238.8	7n.	4".55	1856.40
	239.3	1n.	.93	66.58
Po.	238.0	40	3.29	56.02
	239.9	9	5.38	.02
	238.8	36	.04	61.29
M.	.0	1n.	4.61	2.42
	56.9	,,	.68	9.55
	57.1	,,	.80	.57
	240.0	,,	5.06	75.28
Eng.	.6	,,	.50	64.67
	239.1		.70	5.29
Ta.	237.3	2n.	4.62	6.40
	239.5	1n.	...	9.62
	.4	,,	4.37	71.41
	236.8	,,	.71	3.48
	237.5	,,	.88	4.55
Du.	239.9	10n.	.73	69.16
	241.0	14n.	.79	71.28
	242.3	4n.	.67	5.51
Gl.	240.0	4	.69	0.32
	239.0	5	.86	1.13
	.9	5	.5	4.22
W. & S.	.8	4	5.3	1.57
	240.6	4	.3	3.25
Schi.	239.5	1n.	.16	5.41
Sp.	.6		4.90	.41
Dob.	240.3	4n.	.82	6.26
	238.5	5n.	5.04	7.29
Pl.	240.7	7n.	4.80	.18
Fl.	241.8	1n.	.61	.56

475   Σ. 1910.

R. A.	Dec.	M.
15^h 1.8^m	9° 41'	7, 7

C. yellow.

Motion probably orbital ; very slow.
Dunér has

$$1850.86. \quad \Delta = 4''.03.$$
$$P = 210°.3 + 0°.060 \ (t - 1850.0).$$

Σ.	205.5	1n.	3.98	1823.33
	209.2		.80	32.08
H$_2$ & So.	.2	3n.	4.78	23.42
H$_2$.	210.1	,,	.29	9.09
Sm.	209.7		.0	35.39
Mä.	212.2	3n.	.1	43.98
	211.5	1n.	.21	52.43
	.1	4n.	.19	5.92
	212.4	1n.	.33	61.41
Po.	209.1	12n.	3.91	45.78
Mit.	211.6	2n.	4.48	7.71
O.Σ.	210.5	6n.	.22	51.00
De.	211.3	2n.	.22	6.05
Se.	209.9	1n.	.11	.40
Mo.	211.4	2n.	.09	8.33
Du.	.1	4n.	.00	71.38
W. & S.	.4	1n.	.3	4.44
	.3		.4	5.41
Gl.	212.7	1n.	.2	4.44
Fl.	.9	,,	.27	7.43

476   Σ. 3091.

R. A.	Dec.	M.
15^h 10^m	−4° 26'	7.7, 7.7

C. yellow.

Σ.	47.3	6n.	0".50	1832.39
Mä.	35.9	2n.	.50	43.91

477   O.Σ. 294.

R. A.	Dec.	M.
15^h 10^m	56° 30'	6.8, 11.3

O.Σ.	251.2	3n.	3.26	1848.59
De.	247.8	,,	.23	67.57

478   Σ. 1926.

R. A.	Dec.	M.
15^h 10.4^m	38° 45'	6.1, 8.4

C. yellowish, blue.

Dunér gives

$$\Delta = 1''.42 - 0''.009 \ (t - 1850.0).$$
$$P = 262°.5 + 0°.10 \ (t - 1850.0).$$

Σ.	260.6	4n.	1.59	1830.60
Mä.	261.0	2n.	.46	42.69
Du.	264.9	,,	.37	71.42
	.9	,,	.17	2.51

479   O.Σ. 295.

R. A.	Dec.	M.
15^h 10.4^m	37° 16'	7.4, 9

Mä.	114.9		0.77	1843.33
	111.9		.75	6.28
	115.6		.6	7.32
O.Σ.	128.4	4n.	.74	6.38
De.	122.9	3n.	.85	66.84

480   Σ. 1925.

R. A.	Dec.	M.
15^h 10.5^m	−7° 50'	7.8, 9.3

C. A, yellowish.

Σ.	6.7	3n.	4.18	1831.69
Mit.	7.3	1n.	.19	48.49
Se.	11.1	3n.	.70	56.28
De.	10.4		.44	68.40
C.O.	9.3	2n.	.9	77.40

481　　Σ. 1930.

5 SERPENTIS.

R. A.	Dec.	M.
15ʰ 13ᵐ	2° 14′	5, 10

C. yellowish.

The stars have a rapid common proper motion. Orbital motion seems to be indicated by the slight increase in distance and diminution of the angle. (O.Σ.)

H₁.	50° to 60°		...	1783·38
Σ.	40·9	3n.	10·07	1831·69
	·6	,,	·33	6·42
O.Σ.	39·2	2n.	·52	48·38
Se.	37·1		·58	58·52

482　　Σ. 1934.

R. A.	Dec.	M.
15ʰ 13·2ᵐ	44° 14′	8·5, 8·5

C. white.

Considerable change in both coordinates.

Σ.	45·1	3n.	5·29	1830·88
H₂.	44·7		6·19	31·41
Mä.	42·8		5·94	43·59
	40·1		6·00	51·59
	41·7		5·71	3·76
	39·3		6·07	4·71
Se.	40·3	2n.	5·84	8·57
De.	38·1	4n.	6·05	64·88
O.Σ.	37·2	1n.	·23	8·52
W. & S.	35·9	4	·2	73·25
	34·7	5	5·8	·33
	35·5	6	6·2	4·43
	36·3	4	·30	7·43
	35·3	5	·33	·44
Gl.	36·2	5	·27	4·49
Dob.	33·2	4n.	·36	7·56

483　　Σ. 1932.

I (B) CORONÆ BOREALIS.

R. A.	Dec.
15ʰ 14·5ᵐ	27° 36′

Magnitudes.—Σ.,5·6,6·1. Se.,6,6·5. De., 6·9, 7·2. Σ., suspected variability.

C. white.

Certain change in both coordinates.
Dunér gives the following formulæ:

$$\Delta = 1''\cdot37 - 0''\cdot0146\,(t - 1850\cdot0).$$
$$P = 282°\cdot2 + 0°\cdot571\,(t - 1850\cdot2) + 0°\cdot0055\,(t - 1850\cdot0)^2.$$

Σ.	273·8	4n.	1·62	1830·28
H₂.	268·4	3n.	·53	·29
	267·3	1n.	·31	1·37
Da.	271·2	,,	·44	3·39
	281·0	2n.	·46	48·49
	284·0	1n.	·36	54·40
O.Σ.	279·3	2n.	·65	41·46
	280·6	1n.	·40	51·49
	295·6	2n.	·21	70·55
Mä.	278·6	,,	·50	42·42
	283·6	4n.	·45	51·88
	287·5	6n.	·32	7·27
	·1		·32	8·54
	289·4	5n.	·34	60·70
Se.	285·3	2n.	·14	56·40
De.	286·9	3n.	·2	·60
	·3	,,	·2	8·45
	290·2	4n.	·18	63·28
Kn.	288·8	1n.	·34	·78
Eng.	293·1	2n.	·57	4·48
Du.	297·0	1on.	·10	70·29
	299·2	3n.	0·99	5·51
W. & S.	296·3	7	1·02	2·49
	·8	5	·21	3·36
	298·6	6	·07	4·44
	299·4	6	·1	·49
	300·2	5	·29	5·39
	301·5	0	·07	7·47
Gl.	298·9	4	·2	4·49
Schi.	118·5	1n.	·16	5·42
Sp.	298·6		·16	·43
Dob.	303·5	3n.	·26	7·37

484　　Σ. 3093.

R. A.	Dec.	M.
15ʰ 16·4ᵐ	− 1° 6′	8, 9·2

C. A, yellowish.

The distance has diminished.

Σ.	135·5	2n.	33·28	1829·36
Mä.	137·3	,,	2·57	47·32
De.	138·6		1·15	65·35

485　　Σ. 1937.

η CORONÆ BOREALIS.

R. A.	Dec.	M.
15ʰ 18·2ᵐ	30° 43′	5·2, 5·7

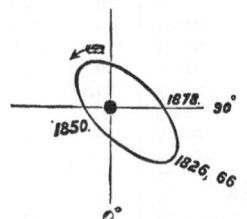

C. A, yellow ; B, certainly yellow.

H₁. "Sept. 10, 1781.—They are fairly separated so as to see the dark heaven between, but that is all. Oct. 4, 1781.—In the greatest perfection. Very near in contact. Oct. 22, 1781.—With 278 beautifully white and distinct."

H₁ (*Phil. Trans.*, vol. lxxii., p. 216). 1781. "Sept. 9.—Double. A little unequal. They are whitish stars. They seem in contact with 227, and though I can see them with this power, I should certainly not have discovered them with it ; with 460, less than ¼ diameter ; with 932, fairly separated, and the interval a little larger than with 460. I saw them also with 2010, but they are so close that this power is too much for them, at least when the altitude of the stars is not very considerable ; with 460 they are as fine a miniature of *i* Boötis as that is of *a* Geminorum. Position 59° 19′ n.f."

(*Phil. Trans.* 1804, p. 370.) "This very minute double star has undergone a great alteration in the relative situation of the two stars." "Aug. 30, 1794, they were so close that, with a 10 ft. reflector, and power of 600, a very minute division could but just be perceived." "And, May 15, 1803, I saw the separation between the two stars, with the same 7 ft. reflector, and magnifying power of 460, with which I had seen it 22 years before." He also observes that the change in angle was retrograde, and that "a parallactic motion of the largest alone" would not account for the change.

H₂ and So. (*Phil. Trans.* 1824, p. 224). H₂ thinks that the position of 1802 "is erroneous, and that the surmised motion of the stars, if any, is much less rapid than that assigned to them by H₁."

H₂ (*Mem. R. A. S.*, vol. v., p. 37). After giving the measures from 1781·69 to 1830·30, he observes that the star is very difficult, and that he does not fully rely on his recent measures. On the whole, however, he thinks there are good reasons for regarding this object as a binary. He concludes that η has made more than a revolution since 1781, and that the motion has been direct. He remarks the obvious difficulty of readily ascertaining which star precedes or follows, owing to the closeness and small difference of magnitude of the two stars. Assuming that H₁ misplaced the companion, he finds that the period has been 43·2 years, with a mean annual motion + 8°·34. This is the only star which up to that time had completed a whole revolution. Finally, he states that "as the actual motion is much less, the orbit must be elliptic, and the actual velocity, at one time or other, must have been 20° or 30° per annum, which will account for the enormous change of position which (on the above explanation of the MS. memorandum)

must have happened between 1781 and 1794."

In 1833 (*Mem. R. A. S.*, vol. viii., p. 50) he writes, "I am sure η Coronæ is closer than it used to be. The distance is below measuring. Surely not ⅓ of a second." Dawes about the same time says "not quite separated," "only elongated."

H₂ (*Mem. R. A. S.*, vol. vi., p. 154). Having obtained measures from Σ. and Da., he computed the orbit and found the following elements :—

$$a = 0''·8325$$
$$e = 0 ·26034$$
$$\lambda = 358° 38'$$
$$\tau = 1761·96 \text{ and } 1806·20$$
$$\gamma = 37° 24'$$
$$\Omega = 220 \quad 35$$
$$P = 44·242 \text{ years}$$
$$n = + 8°· 1369 ;$$

and he finds that these agree well with the observations.

With respect to the orbit of this star, he observes—

1. That the excentricity is moderate.

2. That the major axis almost coincides with the line of nodes, and that hence we see it of its natural length, the conjugate axis only being foreshortened by the effect of perspective.

3. The greatest distance in the apparent or projected orbit is 1''·049, and was attained in 1828 ; the least, about 0''·5388, in 1800 and 1812.

He is almost sure that "the distance has decreased of late," and learns with regret that Mr. Dawes has given up observing the star on account of this difficulty. He regrets these things the more because "the portion of the orbit to be passed over in the next ten or twelve years will be most important in aiding the improvement of the elements."

Σ. (*M. M.*, p. 5) says the period is about 43 years as deduced from the observations of H₁ and himself ; and he thinks that the stars will soon become so close as to defy the separating power of the largest telescopes.

Using the measures up to 1856, Dr. Winnecke made a very careful examination of the orbit of this star with the most satisfactory results. His elements are thus given in his *De Stellâ η Coronæ Borealis*, etc. :

$$a = 0''·9567$$
$$e = 0 ·2865$$
$$\Omega = 22° 18'$$
$$\lambda = 215 \quad 29$$
$$i = 60 \quad 40$$
$$P = 43·115 \text{ years}$$
$$T = 1850·329.$$

Smyth (*Cycle*, p. 340). When this observer began his measures of η, he found the work "difficult enough ;" the observations of position were mostly unsatisfactory, and those of distance were estimations. In 1842 the angular velocity was "under rapid and direct acceleration, while the distance was diminishing, so that the fine black division seen between the stars in 1832 had not only disappeared, but the object was not always elongated. "The general mean [annual motion] drawn from a comparison of my own and other observations was + 9°·41, and the period about 44 years. The excentricity, by the graphic process, is 0·3561. The connexion of the components is therefore 'fully proven.'" Da. (*Mem. R. A. S.*, vol. xxxv., p. 379). After noting the closeness, rapid motion, and the fact that two complete revolutions have been made since 1781, he says he is "sure now that H₁'s position in 1802 should be s.f. instead of n.p." He remarks, too, that the components have separated since 1854, and that it is now an easy object. "The question will be decided in between three and four years' time."

In 1841 Mädler computed the orbit, and found the following elements :—

I.

$$T = 1815\cdot20$$
$$\lambda = 263°\ 10'$$
$$\Omega = 22\ \ 35$$
$$i = 71\ \ 29$$
$$\phi - 20\ \ 43$$
$$a = 1''\cdot1912$$
$$U = 43\cdot310\ \text{years.}$$

In 1842 he published the following : —

II.

$$T = 1815\cdot230$$
$$\lambda = 261°\ 21'$$
$$\Omega = 24\ \ 19$$
$$i = 71\ \ 8$$
$$\phi = 19\ \ 44$$
$$a = 1''\cdot0879$$
$$U = 43\cdot246\ \text{years}$$

And in 1847 his last results were as follows :—

III.

$$T = 1807\cdot21$$
$$\lambda = 215°\ 11'$$
$$\Omega = 20\ \ 6$$
$$i = 59\ \ 28$$
$$\phi = 16\ \ 48$$
$$a = 0''\cdot9024$$
$$U = 42\cdot500\ \text{years.}$$

These last elements Mädler regarded as very accurate.

Villarceau published two solutions about 1852, and sought to decide between the claims of the two rival orbits, viz., those of forty-three and sixty-six years. The former (the orbit of H₂ and Mä.) was ob-

tained when the position for 1802 was reversed, and the latter when that for 1781 was so treated. Villarceau, thinking that the two orbits might be separated before 1853, solicited careful observations from O.Σ., Da., and others; and a glance at these was sufficient to show that the observations since 1847 would not agree with the orbit of forty-three years.

Here are given the three sets of elements by this eminent astronomer : in (I.) the observation in 1781 was taken as 210° 21', and that in 1802 was left intact; in (II.) the angle in 1802 was reversed, and that in 1781 taken as 30° 21'; in (III.) are exhibited his last results :—

I.

$$T = 1780\cdot124$$
$$\lambda = 194°\ 37'$$
$$\Omega = 4\ \ 25\ (1835\cdot0)$$
$$i = 58\ \ 3$$
$$\phi = 28\ \ 0$$
$$a = 1''\cdot1108$$
$$U = 66\cdot257\ \text{years.}$$

II.

$$T = 1805\cdot666$$
$$\lambda = 227°\ 10'$$
$$\Omega = 10\ \ 31\ (1835\cdot0)$$
$$i = 65\ \ 39$$
$$\phi = 28\ \ 19$$
$$a = 1''\cdot0125$$
$$U = 42\cdot501\ \text{years.}$$

III.

$$T = 1779\cdot338$$
$$\lambda = 185°\ 0'$$
$$\Omega = 9\ \ 52\ (1850\cdot0)$$
$$i = 59\ \ 19$$
$$\phi = 23\ \ 51$$
$$a = 1''\cdot2015$$
$$U = 67\cdot309\ \text{years.}$$

In conclusion, Villarceau rejects the orbit of forty-three years, and thinks that the longer one is not susceptible of being "sensibly modified by ulterior observations ;" that 66·257 years satisfied the observations anterior to 1848 ; and that the true period cannot exceed 67·309 years more than a fraction of a year.

Mr. Wilson in 1875 carefully compared the observations from 1863 to 1875 with Winnecke's orbit, and found a "systematic and increasing divergence, which is too large to be accidental." He finds on the whole that the "hypothesis that would best satisfy the observations is, that there exists in each successive revolution some shortening of the period, accompanied perhaps with a progression of the line of apsides." He thinks that the period is most probably about 41·2 years. See *Monthly Notices*, vol. xxxv.

M. Wijkander has lately computed the following elements:—

$$T = 1850\cdot26$$
$$\omega = 211°\cdot4$$
$$\text{☊} = 26\cdot7 \ (1850\cdot0)$$
$$i = 58\cdot0$$
$$e = 0\cdot2625$$
$$\mu = + 8°\cdot6588$$
$$a = 0''\cdot827$$
$$\rho = 41\cdot58 \text{ years.}$$

M. Flammarion has recently obtained these results:—

$$\text{☊} = 22°\cdot2$$
$$i = 60\cdot4$$
$$\pi = 224\cdot1$$
$$T = 1849\cdot9$$
$$e = 0\cdot287$$
$$a = 0''\cdot985$$
$$P = 40\cdot17 \text{ years.}$$

	°	n	″	
H_1.	30·6		...	1781·69
	179·6		...	1802·69
H_1 & So.	25·9	2n.	1·57	23·27
	44·4	8n.		30·30
	...	4n.	0·81	·36
	52·6	10n.	...	1·47
	57·1	9n.	...	2·50
Σ.	35·2	4n.	1·07	26·77
	43·2	2n.	0·96	9·55
	50·6	3n.	·88	31·63
	56·8	,,	·79	2·76
	74·2	6n.	·73	5·41
	88·7	,,	·56	6·52
	95·4	4n.	·38	7·47
	107·0	5n.	·36	8·44
	188·3	1n.	·60	45·64
Da.	50·7	2n.	...	31·34
	56·7	1n.	...	2·55
	63·5	3n.	...	3·39
	119·8	2n.	0·5	9·59
	135·8	,,	·5	40·62
	149·4	6n.	·49	1·65
	156·6	2n.	·5	2·58
	199·9	,,	·63	7·24
	204·4	,,	·65	8·34
	207·4	1n.	·69	·47
	218·2	2n.	·69	9·44
	238·1	,,	·55	51·42
	250·1	,,	·5	2·52
	273·3	4n.	·44	3·64
	301·4	3n.	·47	4·42
	322·4	1n.	·45	5·51
	341·7		·45	6·37
	350·8		·59	7·45
	5·5		·72	9·62
	8·4		·86	60·35
Sm.	27·5		1·07	5·44
	57·2		0·8	32·63
	61·9		·8	3·57
	68·1		·6	4·60
	75·2		·6	5·65
	89·2		·5	6·59
	102·3		·5	7·68

	°	n	″	
	109·9		0·5	1838·19
	120·1		·5	9·67
	188·5		·3	46·69
	246·8		·5	52·43
Ga.	109·8		·7	38·64
O.Σ.	132·1	2n.	[0·76]	9·82
	137·1	5n.	0·50	40·52
	149·6	4n.	·52	1·50
	159·1	2n.	·57	2·60
	179·3	6n.	·58	5·46
	195·6	3n.	·61	6·61
	203·9	5n.	·56	7·64
	209·8	2n.	·57	8·72
	220·3	3n.	·59	9·65
	230·7	,,	·49	50·52
	241·8	10n.	·47	1·56
	261·1	6n.	·43	2·62
	280·9	5n.	·32	3·56
	313·1	4n.	·33	4·66
	330·2	,,	·40	5·62
	342·5	3n.	·47	6·62
	351·8	4n.	·64	7·62
	359·5	5n.	·76	8·54
	5·8	4n.	·79	9·61
	15·8	3n.	·90	61·58
	22·5	2n.	·91	2·76
	23·6	4n.	1·10	3·54
	29·6	3n.	·13	5·35
	35·4	4n.	·13	6·66
	32·6	2n.	·24	7·47
	41·3	5n.	·04	8·55
	47·1	3n.	0·97	70·54
	55·3	5n.	·90	2·59
	57·3	4n.	·81	3·54
	64·6	,,	·83	4·61
Mä.	150·3	9n.	·59	41·54
	157·6	5n.	·55	2·26
	163·5	4n.	·55	·69
	199·2	12n.	·68	7·32
	205·3	4n.	·59	·78
	·7	3n.	·62	8·18
	228·7	,,	·41	50·70
	235·4	10n.	·36	1·68
	250·7	13n.	·27	2·65
	267·8	5n.	·27	3·35
	317·0	4n.	·26	4·73
	330·2	2n.	...	5·73
	347·2	,,	0·47	7·39
	6·2	6n.	·69	8·61
	4·9	4n.	·69	9·38
Ch.	158·6	1n.	·69	42·41
	172·5	,,	·50	3·69
Mit.	195·7	3n.	·70	6·66
Bond.	207·3		·8	8·58
	210·3		·8	·66
Flt.	235·0	2n.	·7	50·56
Ja.	257·8	7	·4	3·19
	285·3	13	·5	4·04
	355·7	3n.	·6	7·95
Se.	325·6	2n.	·32	5·40
	344·3	7n.	·47	6·59
	351·0	,,	·57	7·48
	359·1	3n.	·53	8·51
	4·5	4n.	·53	9·48

	°		″	
De.	0·8	9n.	...	1858·52
	16·9	11n.	0·7	62·56
	20·7	13n.	·82	3·42
	24·1	10n.	·7	4·43
	27·4	9n.	1·03	5·49
	30·0	,,	·04	6·44
	33·1	7n.	·04	7·50
	36·4	,,	·06	8·39
	44·0	8n.	·04	70·38
	47·7	,,	·08	1·45
	50·8	,,	·02	2·44
	56·1	,,	·00	3·44
	59·5	,,	0·97	4·42
	66·6	,,	·86	5·41
Ro.	19·7	1n.	1·06	63·56
	30·1		·59	5·52
Eng.				
Ta.	28·2	2n.	·09	4·45
	32·3	4n.	·42	6·33
	31·5	1n.	...	7·52
	44·6	,,	...	9·62
	·1	,,	1·29	70·46
	47·7	,,	...	1·41
	·8	,,	·28	2·29
	55·9	,,	...	3·48
	61·2	,,	...	4·44
	60·7	,,	...	5·38
	70·2	,,	0·83	6·45
Kn.	36·0	3n.	1·06	67·34
	46·8	1n.	·13	70·47
	45·6	5n.	·00	1·54
Du.	29·2	1n.	·12	67·69
	36·9	4n.	·14	8·65
	40·0	9n.	·02	9·53
	43·7	7n.	0·97	70·51
	47·3	9n.	·87	1·53
	51·2	7n.	·84	2·58
	55·0	2n.	1·08	3·72
	68·7	11n.	0·69	5·55
Hi.	196 5		...	
Gl.	44·6	11	1·1	70·44
	45·9	5	0·9	1·50
	47·0	4	1·0	·63
	55·0	5	·0	3·60
	54·0	7	0·85	·47
	52·2	7		·50
	55·9	10	1·16	·52
	53·5	8	...	·51
	58·8	10	1·10	4·36
	·0	10	·07	·32
	59·0	2	0·8	·49
	76·5	4n.	...	76·29
	73·8	2n.	0·8	·33
	71·6	,,	·7	·34
	75·5	,,	...	·35
W. & S.	45·2	5	1·47	41·55
	50·0	4	·38	·57
	51·0	7	·01	2·49
	58·0	5	·01	3·36
	57·6	3	·22	·38
	55·6	5	·09	·45
	58·4	8	0·93	4·44
	·1	6	...	·49
	66·7	6	·94	7·30
	68·0	7	...	·32

	°		″	
W. & S.	66·6	5	...	1847·32
	68·7	6	...	·32
	·8	8	...	·43
	82·9	9	...	·47
Fer.	51·7		0·92	2·48
Br.	64·2		·90	3·34
Schi.	66·1	1n.	·90	5·41
	72·3	,,	·79	6·51
Sp.	66·1		·91	5·42
	72·3		·79	6·51
W.O.	250·4	1n.	·76	·41
	·3	,,	·86	·43
	249·7	,,	·71	·43
	251·7	,,	·75	·44
Dob.	70·3	8n.	·84	6·38
	82·0	4n.	...	7·30
	92·6	2n.	0·61	78·40
	94·6	1n.	·61	·55

486   Σ. 1938.

μ BOÖTIS.

R. A.	Dec.	M.
15^h 20^m	37° 46′	6·7, 7·3

C. greenish white.

H_1 (*Phil. Trans.*, vol. lxxii., p. 217). 1781. "Sept. 10. Double. It is a star near μ not marked in Flamsteed's catalogue. Considerably unequal. The interval with 460 is ¾ diameter of S. The position of the small star is turned towards μ, a little following the line which joins L to μ Boötis."

In the *Phil. Trans.* for 1804, p. 372, H_1 discusses the change in angle, gives his measures in 1781, 1782, and 1802, and shows that a change of 11° had taken place in 19 years and 361 days, and that this was most probably orbital.

H_2 and So. (*Phil. Trans.* 1824, p. 227). "A very close double star. In the 5 ft. equatorial with a power of 133 it is seen elongated, but 303 shows it decidedly double. A power of 179 applied to the 7 ft. shows the discs of the two stars in contact, but 273 distinctly separates them. This double star is a severe test for a telescope, and is easily found by means of μ Boötis."

"If this double star be a binary system, of which there can be little doubt, its period is about 622 years, and the most probable mean annual motion is $0°·5783$, in the direction n.p.s.f., or retrograde."

H_2 (*Mem. R. A. S.*, vol. v., p. 38). Having the measures from 1782 to 1830 before him, he says, "It will probably, ere long, become excessively difficult or close up entirely, as both the diminution of the micrometrical distance and the rapid increase of angular velocity sufficiently indicate." "None but the finest telescopes are competent to deal with it."

Dawes (*Mem. R. A. S.*, vol. viii., p. 87). He gives measures in 1830 and 1832, and says "neatly divided; requires a superb night like this."

Smyth (*Cycle*, p. 343). "From the earliest epoch here registered [1782·68] down to my latest, an annual mean movement appears $= -0°·85$; but from Herschel junior and Sir James South's period it averages $-1°·44$, so that the period may be within 460 years; but the annual rates are as yet distressingly irregular."

Engelmann (*Ast. Nach.*, 1673—1676) writing in 1860 observes, "The period must be about 150 years; the minimum distance $0''·35$, which will be reached in 1868; the maximum about $1''·75$. Since 1857 the companion has passed through $50°$."

Dawes (*Mem. R. A. S.*, vol. xxxv., pp. 381 and 484). After noting the rapid decrease in distance and the acceleration in angle, he says that this object now requires "about a 9 inch aperture to completely separate the components."

The angle in 1852 presents an extraordinary anomaly which partly disappears when the measure is combined with those of 1851 and 1852. (O.Σ.)

Mr. Wilson in 1873 obtained by graphical processes the following elements:—

$$e = 0·51$$
$$\Omega = 172°\ 0'$$
$$D = 45\ 0$$
$$\lambda = 20\ 5$$
$$\pi = 186\ 30$$
$$P = 200·4 \text{ yrs.}$$
$$t = 1865·2.$$

In 1875 Dr. Doberck computed an orbit for this interesting binary star. The proper motion of μ^2 agreeing with that of μ Boötis "suggested the existence of a physical connexion between these two stars. Actual measures have, however, rendered such a supposition more than doubtful." Making use of Sir John Herschel's first method, the following elements were obtained (No. 2 in the table subjoined):—

No.	T	Ω	λ	γ	P	a	e	Computer.
		° '	° '	° '	years.	''		
1	1860·88	163 11	54 27	41 52	314·761	1·761	0·6832	Hind.
2	3·51	182 59	17 41	44 26	290·07	·500	·6174	Doberck.
3	·51	173 42	20 0	39 57	280·29	·47	·5974	,,
4	5·2	172 0	20 5	45 0	200·4	...	·51	Wilson.
5	5·5	169 0	23 39	46 22	198·93	...	·4957	Klinkerfues.
6	6·00	166 8	23 1	47 31	182·6	1·165	·491	Winagradskij.

Dunér says, "The connexion between u Boötis and Σ. 1938 is indubitable; otherwise the motion of μ Boötis which is considerable would cause a very great change in the relative positions of the stars. If the mass of μ Boötis were equal to the sum of the masses of Σ. 1938, the orbital motion would be only $-0°·003$ per annum, or $-0°·15$ in fifty years. In reality, an annual motion of $-0°·006$ is perhaps probable: and if this is confirmed the mass of μ Boötis would be seven times greater than that of Σ. 1938."

The common proper motion of the three stars is $-0''·16$ in R. A., and $0''·10$ in N. P. D.

	°	''	
H_1.	357·2	...	1782·68
	346·2	...	1802·66
Σ.	330·7	...	22·21
	327·0	2n.	1·38	6·77
	324·0	,,	·24	9·73

	°		''	
	319·7	3n.	1·19	1833·85
	318·6	,,	·10	5·55
	315·0	,,	·06	6·65
	·0	in.	0·90	7·70
H_2 & So.	333·7	3n.	1·65	23·41
	·5	5n.	·42	5·46
	324·1	2n.	0·85	30·24
	322·7	3n.	1·03	2·56
Da.	319·7	in.	·15	3·39
	314·8	,,	·0	7·37
	306·0	3n.	0·83	40·39
	303·2	6n.	·85	1·66
	300·9	in.	·85	2·40
	286·5	,,	·65	7·30
	280·0	,,	·65	8·52
	276·2	2n.	·68	9·44
	266·5	,,	·52	51·42
	262·2	in.	·55	2·52
	254·6	,,	·50	3·71
	249·3	3n.	·46	4·41
	190·0	,,	·48	5·46
	232·3	in.	·45	7·47

	°		″	
Sm.	319·9		1·2	1834·56
	314·8		·0	7·29
	310·6		0·9	9·32
	306·1		·8	42·52
	255·0		·5	53·60
O.Σ.	313·1	2n.	·98	40·46
	303·4	,,	·84	2·23
	287·1	4n.	·57	6·68
	272·7	2n.	·53	50·46
	262·6	3n.	·44	1·48
	268·2	,,	·48	2·65
	247·2	4n.	·53	5·11
	242·1	2n.	·59	6·57
	237·9	3n.	·57	7·65
	228·3	,,	·57	8·56
	211·2	,,	·58	60·95
	179·2	,,	·60	6·40
	167·5	2n.	·54	9·54
	158·2	4	·63	73·09
Mä.	308·7	2n.	·82	41·47
	305·1	3n.	·71	2·40
	304·9	2n.	·78	·66
	301·5		·76	3·54
	287·7	15n.	·47	7·38
	282·4	2n.	·82	8·38
	276·7	3n.	·40	50·70
	264·9	,,	·31	1·28
	263·2	4n.	·33	·78
	261·2	10n.	·41	2·61
	256·2	6n.	·35	3·49
	9	2n.	·42	5·53
	239·2	,,	·35	7·38
	236·2	4n.	·32	8·57
	226·4	3n.	·43	9·38
Ka.	303·3		...	41·67
			0·82	2·35
	295·8		...	3·67
Hi.	281·2	2n.	...	7·08
Bond.	282·0		0·5	8·53
	283·0		·5	·52
	282·8		·6	·51
	...		·7	·51
	283·8		...	·49
Ja.	265·1	9	0·45	53·23
	255·7	9	·5	4·05
Se.	234·1	2n.	·50	6·97
	180·3		...	66·54
De.	202·9	3n.	0·5?	2·55
	196·3	12n.	·5	3·38
	189·5	5n.	...	4·48
	184·6	10n.	0·5	5·45
	178·7	13n.	·5	6·94
	174·5	5n.	·5	8·38
	166·1	7n.	·62	70·39
	161·1	,,	·6	1·43
	154·9	8n.	·6	2·43
	150·9	7n.	·71	3·41
	147·8	6n.	·81	4·77
	141·9	8n.	·69	5·41
Ro.	195·8	1n.	·75	63·63
Kn.	193·6	4n.	·5	4·41
	152·0	1n.	·5	72·46
Eng.	187·5	2n.	·57	65·78
	179·3	,,	·70	7·57

	°		″	
Ta.	197·9	1n.	...	1865·72
	196·4	3n.	0·85	6·42
	170·8	1n.	...	70·65
	166·6	2n.	...	2·22
	155·7	1n.	...	3·47
	·3	,,	...	4·54
Du.	171·1	6n.	0·53	69·49
	163·9	4n.	·59	70·52
	160·8	5n.	·66	1·54
	158·0	2n.	·55	2·52
	146·7	1n.	·80	5·52
W. & S.	167·9	4	·76	1·57
	164·5	1	·4	2·33
	162·3	3	·3	·38
	151·0	4	·45	3·25
	152·0	3	...	·33
	150·1	7	0·6	·44
	149·1	7	·7	4·44
	143·5	8	...	7·43
	147·6	9	0·7	·44
	144·6	7	·84	·52
	150·6	7	...	·59
	140·4	13	0·65	·45
Gl.	152·0	10	·46	3·47
	·7	5	·5	·48
	150·6	10	·58	4·22
	·5	12	·5	·26
	164·0	5	...	0·44
	158·4	4	·5	1·65
Schi.	143·3	1n.	·63	5·46
Sp.	·3		·64	·47
W.O.	146·5	1n.	·78	6·41
	148·9	,,	·70	·43
	143·0	,,	·72	·43
	·1	,,	·73	·44
Dob.	·6	2n.	...	·35
	131·5	4n.	0·55	7·36
	137·7	4n.	·63	8·49

$$\mu^1 \text{ and } \mu^2.$$

	°		″	
H_1.	170·4	1n.	128	1781·80
Piazzi.	171·5		112	1800·00
H_2 & So.	170·4	2n.	108·9	21·35
Σ.	172·6		·7	·78
	171·9	1n.	109·1	2·67
	·9	7n.	108·4	34·64
O.Σ.	·1	1n.	·8	40·95
	·6	,,	·7	7·69
	·7	3n.	·7	51·50
	·7	,,	·6	6·77
	·6	1n.	·4	61·60
	·6	,,	...	6·22
Mä.	·8	,,	108·9	46·29
	172·0	2n.	·3	53·46
De.	171·6		·5	63·25
Kn.	·4	3n.	·1	4·40
M.	·5		·4	6·53
Du.	·7	3n.	·1	71·52
Fl.	·6	1n.	·6	7·69

487   Σ. 1944.

R. A.	Dec.	M.
15ʰ 21·8ᵐ	6° 31'	7·5, 8·1

C. white.

Probably a binary.

	°		''	
Σ.	341·6	4n.	1·34	1832·40
Mä.	339·3		·35	9·00
	·2		·30	42·42
	338·1		·34	3·33
	336·9		...	54·40
	331·4		1·33	7·39
	335·6		·00	65·52
Se.	·7	2n.	·18	56·44
W. & S.	334·9	1n.	·09	75·45

488   O.Σ. 296.

R. A.	Dec.	M.
15ʰ 22·2ᵐ	44° 26	7, 8·6

Change in angle.

	°		''	
O.Σ.	327·9	2n.	1·52	1845·53
	321·2	,,	·44	52·10
	317·3	4n.	·63	72·29
Da.	325·2	1n.	·60	48·53
De.	319·6	3n.	·51	66·70
W. & S.	316·0	6	·47	73·37
	·9	2	·61	·44
	·8	5	·33	4·50
	315·9	5	·40	5·49
	·5	5	·34	7·45

489   Σ. 1954.

δ SERPENTIS.

R. A.	Dec.
15ʰ 29·1ᵐ	10° 56'

Magnitudes.—Σ. 3, 4. Se. 4, 4·5. Sm. 3, 5. De. 3·9, 5·6. One of the stars is probably variable in its light.

C. Σ., A, yellowish white ; B, ashy. Sm., A, bright white ; B, bluish white ; " but under the very best vision both have a bluish tinge, which, in such a pair, is rather against the theory of contrast."

H₁ (*Phil. Trans.* 1803, p. 380). After observing that the position 42° 48' s.p. on September 5, 1782, was " an accurate measure," and that in 19 years and 155 days " the small star has moved, in a retrograde order, over an arch of 18° 39'," he proceeds to show that " the most natural way of accounting for the observed changes is to admit the two stars to form a binary system. In this case we calculate, with considerable probability, that the periodical time of a

revolution of the small star round δ Serpentis must be about 375 years."

H₂ and So. (*Phil. Trans.* 1824, p. 231). An examination of all the observations up to 1821 show that on the whole the distance had increased. "The angular velocity has undergone a considerable diminution, and as this corresponds with the increased distance, the orbit is probably elliptic, and so situated as to allow its ellipticity being visible without distortion. The mean annual motion is −0°·726, or retrograde."

So. (*Phil. Trans.* 1826, p. 341). Measures for 1825 are given. On these H₂ remarks : " Either there is a considerable error in these or the measures of 1821, or the result is unfavourable to the motion assigned to this star, as, instead of advancing 3° in its apparent orbit, it seems actually to have receded nearly 50'. Further observations must elucidate this difficulty." And in the *Mem. R. A. S.*, vol. v., p. 45, he writes of the measures he made in 1830, " My present observations afford no support to the evidence of motion offered by former measurements." "The present apparent fixity of δ Serpentis contrasts strongly with its former rapid motion. A considerably elongated orbit can alone account for this."

Dawes, too, with his measures in 1831 and 1833 before him, writes, " This star appears to have come to a standstill."

So also Σ., Se., Sm., and others, all note the diminished rate in the angular motion. Smyth adds that " a small movement in space has been detected in A, which, when surer known, will afford further demonstration of its physical connexion with B." Hind, however, says, " The proper motion in R. A. appears to be *nil*, but a very small one may exist in declination."

The distance by Σ. in 1852 is considerably in error. The angular change has diminished of late years, and the distance has augmented. From 1782 to 1834 the angle changed considerably ; between 1834 and 1855 but little ; since 1855 it has again been subject to change. The maximum distance, probably, has already been attained. (O.Σ.)

Dunér has the following formulæ :

$$\Delta = 3''·03 + 0''·0138\ (t-1850·0)$$
$$-0''·00015\ (t-1850·0)^2.$$

$$P = 199°·0 - 0°·273\ (t-1830·0)$$
$$+0°·0025\ (t-1830·0)^2 - 0°·0000^2$$
$$(t-1830·0)^3.$$

The common proper motion is − 0''·06 in R. A., and 0''·05 in N. P. D.

	°		''	
H₁.	227·2		...	1782·99
	208·5		...	1802·10
H₂ & So.	199·3	1n.	3·05	21·33
	198·4	6n.	·29	9·50
	·4	2n.	·04	32·31

	°		″	
Σ.	201·2	3n.	2·44	1822·68
	197·2	5n.	·66	33·07
	196·9	3n.	·56	6·30
	192·4	1n.	3·79	52·22
Sm.	196·5		2·9	31·43
	197·3		·7	8·38
	196·2		·8	42·35
	·5		3·0	51·32
Da.	188·9	3n.	2·91	32·35
	195·7	5n.	·97	41·06
	·8	1n.	·85	3·44
	194·9	3n.	3·00	8·52
	·8	2n.	·09	9·44
	·2	,,	·03	52·58
	193·1	1n.	·08	7·52
	·7	,,	·17	·56
	192·3	,,	·04	·74
	191·4	,,	·37	65·39
	·2	,,	·23	·55
Mä.	197·4	5n.	·46	41·32
	196·1	4n.	·04	2·37
	194·3	2n.	·28	52·34
	193·1	3n.	·12	4·55
	·2	4n.	·17	6·68
	·4	3n.	·22	9·38
Ch.	195·5	1n.	2·66	41·41
Ka.	196·8		·76	·65
	197·8		·92	3·66
	193·9		3·15	65·62
Hi.	194·2	2n.	·03	45·27
Mit.	193·4	1n.	2·15	7·70
De.	197·0	5n.	3·08	53·66
	194·5	,,	·23	4·54
	·1	1n.	·58	5·13
	193·0	6n.	·23	6·52
	192·4	2n.	·16	7·55
	·1	5n.	·32	8·47
	·2	,,	·19	63·43
Se.	195·5	7n.	·06	55·89
	190·4	1n.	·35	65·52
Mo.	193·1	2n.	·37	57·40
M.	190·3	1n.	2·96	62·33
	193·5	,,	3·21	7·37
	188·4	,,	·33	8·40
	186·7	,,	·37	·48
	188·5	,,	·33	9·49
	190·5	,,	·49	·61
	189·9	,,	·46	70·38
	190·2	,,	·09	·40
	193·6	,,	·24	1·34
	190·1	,,	·03	·36
	189·9	,,	·33	·48
	190·4	2n.	·18	2·39
	9·5	1n.	·30	3·39
	10·5	4n.	·64	4·41
	·9	,,	·50	5·42
Eng.	192·2	7n.	·28	65·05
Kn.	189·9	3n.	·32	5·41
Ta.	190·7	2n.	·20	5·51
	189·8	3n.	·46	6·37
	193·5	1n.	·23	9·36
	190·7	,,	2·96	71·42
	194·6	,,	·63	2·29
	188·8	,,	·44	3·48

	°		″	
Ta.	189·2	2n.	2·33	1874·36
	·0	,,	·57	5·39
O.Σ.	·0	1n.	3·11	66·49
	187·0	,,	·12	9·52
	189·7	,,	·17	74·62
Du.	191·4	6n.	·06	68·32
	189·8	4n.	·02	75·56
Gl.	193·0	5	...	1·22
W. & S.	192·1	5	3·56	·56
	191·5	5	·51	3·36
	192·9	4	·1	4·50
	191·0	2	·23	·50
	190·0	8	·41	5·51
Schi.	189·6	1n.	·28	·61
Sp.	·6		·28	·61
Dob.	186·9	6n.	·37	6·30
Pl.	188·5	4n.	·79	·97

490   Σ. 1956.

R. A.	Dec.	M.
15^h 29^m	42° 13′	8, 9·5

C. A, yellowish white.

Σ.	41·4	3n.	2·71	1831·53
Mä.	40·6	1n.	·82	45·48
Se.	37·3	2n.	·52	57·60
O.Σ.	41·4	1n.	·61	68·52
Du.	37·9	3n.	·27	70·44

491   O.Σ. 297.

R. A.	Dec.	M.
15^h 30^m	25° 25′	7·5, 11·5

Change in distance.

O.Σ.	147·3	1n.	13·56	1845·31
	·1	,,	·06	6·37
	146·1	,,	12·53	50·40
De.	147·7		10·23	67·00

492   Σ. 1957.

R. A.	Dec.	M.
15^h 30·2^m	13° 19′	7·9, 9·6

Σ.	164·6	2n.	1·47	1828·85
	161·7	,,	·35	33·35
Mä.	158·4		1·37	42·42
	157·6		·25	3·40
	156·3		...	57·39
	153·6		1·48	61·55
De.	155·7		·53	3·51
Gl.	152·5	1n.	·4	71·49
W. & S.	155·1	2n.	·24	3·38
	152·0	1n.	·5	4·50
	161·5	2n.	·4	5·51
	156·6	1n.	·5	6·48

493   Σ. 1961.

R. A. 15^h 30·3^m Dec. 43° 57′   M. 8·7, 9

Σ.	56·0	2n.	21″55	1830·65
H₂.	·2		...	31·43
Mä.	52·4		21·63	47·30
	·1		·18	51·27
	49·4		...	3·76
De.	47·8		22·23	66·77

494   O.Σ. 298.

R. A. 15^h 31·7^m Dec. 40° 13′   M. A 7, B 7·3, C 7

C. yellow.

About a quarter of a revolution has been described by A B, and the minimum distance has been attained: $\frac{A + B}{2}$ and C probably unchanged. (O.Σ.)

Dunér has the following formulæ :

$$\Delta = 0''\!·93 - 0''\!·017\,(t - 1860·0) + 0''\!·00038\,(t - 1860·0)^2.$$
$$P = 199°·5 + 1°·352\,(t - 1860·0) + 0°·0218\,(t - 1860·0)^2.$$

A B.

Mä.	179·5		1·12	1843·35
	186·5	3n.	·42	6·28
	188·6		·51	7·33
	191·8	2n.	·40	51·74
O.Σ.	181·6	3n.	·20	46·49
	195·2	4n.	·18	58·83
	212·5	1n.	0·84	68·52
	235·8	,,	·58	72·58
	264·3	,,	·53	5·52
De.	208·9	3n.	·99	66·44
	280·8	,,	...	76·46
Du.	214·1	,,	0·58	69·46
W. & S.	187·3	2	·7	74·50
	190·9	2	·5	·50
	234·0	5	...	3·40

$\frac{A + B}{2}$ and C.

O.Σ.	328·3	in.	122·53	57·68
	·0	,,	·38	61·44
	·0	,,	121·87	8·52
	327·9	,,	122·23	72·58
	328·2	,,	·38	5·52

495   O.Σ. 299.

R. A. 15^h 32^m Dec. 64° 15′   M. 7·2, 9·5

O.Σ.	20·9	3n.	3·20	1848·34
De.	23·4	,,	·24	66·81

496   Σ. 1965.

R. A. 15^h 34·9^m Dec. 37° 1′   M. 4·1, 5

C. A, greenish white ; B, greenish.

The motion has been rectilinear so far. Dunér has computed the following formulæ :

$$1849·76. \quad \Delta = 6''\!·12.$$
$$P = 301°·7 + 0°·054\,(t - 1850·0).$$

H₁.	295·8	in.	6·25	1781·27
	·5	,,	...	1802·25
Σ.	299·9		...	19·62
	·6	3n.	5·88	22·26
	300·9	5n.	6·0	9·70
So.	·9	4n.	7·17	2·30
H₂.	·0		6·0	6·00
	·5	in.	·2	32·57
Be.	·7		·18	0·68
Sm.	301·2		·4	1·61
	300·9		·5	9·50
	301·2		·1	42·57
Encke.	302·4		·53	37·44
Ga.	301·1		·21	8·59
Ka.	·5		5·92	40·26
Mä.	302·2		6·07	1·47
	301·0		...	2·40
	·8		...	3·37
	302·4		6·33	4·37
	·0		·24	7·32
	303·2		5·99	51·41
	302·1		6·01	2·47
	·6		·27	3·30
	301·3		5·90	4·65
	302·3		6·08	5·77
	·2		...	7·46
	·5		5·68	8·50
	·8		6·13	61·32
	303·1	3n.	·07	·97
Po.	300·4	6n.	·07	45·93
Mit.	301·1	in.	·16	7·70
Da.	300·8		·21	54·55
	301·3		·18	43·63
	·4	3n.	·20	7·99
	·9	1n.	·29	8·45
Mo.	299·5	30	·00	5·43
	301·3	30	·14	6·43
	·7	30	·13	52·53
O.Σ.	302·1	17n.	·05	4·58
Lu.	303·4		·66	6·17
Se.	301·7	4n.	·21	·49
M.	299·7	in.	5·93	62·23
	296·4	,,	6·31	9·57
Eng.	302·8	2n.	·19	4·48
Ta.	301·9	6n.	·30	6·32
	·7	1n.	·62	7·52
	·5	3n.	·66	9·52
Du.	302·4	,,	·03	8·60
De.	·2		·21	·70
Gl.	·0	in.	·38	71·36
Dob.	300·8	3n.	·60	6·25
Goldney	301·4	4n.	·36	8·51

497   Σ. 1967.

γ CORONÆ BOREALIS.

R. A.	Dec.
$15^h\ 37\cdot7^m$	$26°\ 41'$

Magnitudes.—Σ., 4, 7. Piazzi gives the magnitude as 6. Dawes strongly insists upon its being registered as of the 4th, and he gives the companion of the 7th.

C. Σ., A, greenish white ; B, purple. Sm., A, flushed white. Se., A, greenish ; B, purple. Da., A, light yellow ; B, "purplish."

"The star will probably become single, and after a time the companion will emerge on the opposite side." (Σ.) Ten years after its discovery he could not elongate it.

H₂, during 1832, examined it "with 320,480,600. With all, a round disc seen, but no companion."

Mädler always measured this star at sunset and sunrise, as the companion was invisible by day, and at night was hidden by the rays of the larger star.

Smyth, in 1839, on "a superb night," after much gazing, thought there was an elongation in the direction s.p. and n.f. In 1842 he found it "still a dumpy mis-shapen object, with an axis major perhaps n.p. and s.f."

Dawes, in 1843, saw it "with the companion coming out again."

The plane of the apparent orbit of this star, like that of 42 Comæ Ber., approximately coincides with the visual ray. Between 1826 and 1833 the companion was on the *following* side of the principal star ; it then passed to the north at a minimum distance, and reappeared on the *preceding* side in 1840, where it has been found up to 1873. Lately the angle has diminished considerably, and in 1874 the companion was invisible. Probably it has passed southwards to reappear on the *following* side under an angle of about 120°. The period may be about eighty years. The angle in 1840 seems erroneous. (O.Σ.)

In 1877 Dr. Doberck published the following :—

First elements (from Σ.'s observations).	Second elements.
Ω = 111°	Ω = 110° 24'
γ = 83	γ = 85 12
λ = 239	λ = 233 30
ε = 0·387	ε = 0·350
P = 95·5 yrs.	P = 95·50 yrs.
T = 1843·7	T = 1843·70
a = 0"·75.	a = 0"·70.

The common proper motion is small, −0"·007 in R. A., and − 0"·09 in N. P. D.

	°		"	
Σ.	111·0	2n.	0·72	1826·75
	110·7	3n.	·54	8·98
	103·9	5n.	·40	32·66
	105·8	2n.	·40	3·34
H₂.	single			·00
Sm.	round			4·66
	225·0		0·3 elongᵈ·	9·69 42·58
	295·0		0·5	8·37
Encke.	90·9	1n.	1·10	36·48
	95·0	2n.	0·87	8·70
O.Σ.		1n.	wedgᵈ·	40·48
	252	,,	suspected oblong	·51
	...	,,	wedgᵈ·	·57
	...	,,	oblong	1·57
	293·5	2n.	0·38	2·60
	290·4	1n.	·47	4·71
	296·0	5n.	·44	5·60
	287·7	2n.	·44	6·69
	295·5	3n.	·44	7·68
	288·5	2n.	·49	8·71
	289·1	1n.	·59	9·71
	290·0	2n.	·42	50·51
	287·6	4n.	·48	1·50
	288·8	3n.	·45	2·65
	286·6	2n.	·48	3·54
	279·0	,,	·52	5·65
	283·6	,,	·43	6·56
	288·7	,,	·45	7·63
	284·9	3n.	·44	8·56
	·3	2n.	·48	9·59
	287·7	3n.	·42	61·89
	298·9	1n.	·43	2·74
	284·2	,,	·34	3·58
	288·5	,,	·45	4·60
	286·0	2n.	·43	6·62
	264·5	1n.	·40	7·47
	255·8	,,	·33	8·56
	...	,,	wedgᵈ· prob.	9·52
	...	,,	wedgᵈ·	70·52
	...	,,	single	1·58
	...	,,	wedgᵈ· prob.	2·56
	...	,,	oblong	3·54
	...	,,	single	4·57
Mä.	332·3	1on.	0·18	41·50
	271·9	2n.	·47	2·80
	292·2	6n.	·40	51·70
	296·4	7n.	·45	2·60
	284·4	4n.	·40	3·32
	291·0	1n.	·40	4·76
	286·4	2n.	·32	7·39
	284·0	4n.	·33	8·58
	290·4	3n.	...	9·38
Da.	288·8	1n.	0·6	43·45
	285·0	,,	·57	52·07
	284·3	2n.	·69	4·40
	281·0	1n.	·5	7·52
	282·5	,,	·45	9·36
Mit.	286·8	3n.	·53	46·66
	292·5	1n.	...	·72
	293·1	,,	...	7·70

	°		″	
Hi.	300·0		...	1847·08
Bond.	294·7	In.	0·6	8·51
	290·5	,,	·3	·49
	293·0	,,	·3	·46
Ja.	294·2	10n.	·5	53·19
Wi.	295·4		·67	6·37
Se.	289·0	3n.	·45	·59
	·3	5n.	·36	7·52
	294	In.	elong^d·	64·46
	278	,,	,,	5·52
		,,	round	6·55
De.	280·5	3n.	...	58·51
	292·9	,,	...	62·56
Eng.	280·0		elong^d·	5·51
Du. companion not seen	{	In.	...	7·79
single		5n.	...	8
,,		In.	...	9·64
,,		,,	...	70·56
,,		,,	...	1·45
,,		,,	...	2·52
,,		,,	...	5·57
W.O. ,,		,,	single	
W. & S.	190	,,	...	2·45
	195	,,	...	3·36
		single	...	·40
Schi.		,,	...	5·97
Sp.		,,	...	·98
Dob.		,,	...	6
		,,	...	7

498   Σ. 1983.

R. A.	Dec.	M.
15h 46m	35° 51′	8·7, 10·8

C. A, yellow.

Σ.	77·0	3n.	17·44	1830·65
Mä.	76·7	In.	16·58	45·55

499   Σ. 1989.

R. A.	Dec.	M.
15h 46·2m	80° 22′	7·1, 8·1

C. very white.

Σ.	24·1	3n.	0·71	1832·68
	23·9	,,	·53	6·76
O.Σ.	28·1	,,	·70	40·95
Mä.	23·0		·85	1·46
Se.	21·1	2n.	·60	58·59
De.		single		65·00
		,,		70·00

500   Σ. 1984.

R. A.	Dec.	M.
15h 48·1m	53° 16′	6·2, 8·5

C. white.

Dunér gives

$$1846\cdot91. \quad \Delta = 6''\cdot42.$$
$$P = 275°\cdot3 + 0°\cdot085\,(t - 1850\cdot0).$$

	°		″	
H2.	270·3	In.	...	1830·20
Σ.	273·8	4n.	6·53	·72
Mä.	274·3	3n.	·49	43·42
	276·1		·38	51·27
	·1	3n.	·29	2·42
Se.	·6	In.	·39	7·61
Du.	·2	2n.	·42	70·90

501   Σ. 1985.

R. A.	Dec.	M.
15h 49·7m	− 1° 49′	7, 8·1

C. A, yellowish white; B, ash.

Dunér gives the following formulæ:

$$1854\cdot13. \quad \Delta = 5''\cdot70.$$
$$P = 328°\cdot3 - 0°\cdot137\,(t - 1850\cdot0).$$

H1.	316·1	In.	...	1783·32
So.	325·3	2n.	6·88	1823·42
H2.	326·5	In.	7·19	30·23
Σ.	·5	4n.	5·42	1·95
Mä.	327·0		6·20	41·47
	326·2		5·78	3·44
	327·5		·96	3·35
	328·1	In.	·74	54·47
Po.	327·0	,,	·57	46·18
Da.	·1		...	6·42
Mit.	325·5	In.	5·39	8·54
De.	328·5	4n.	·83	55·88
Se.	·5	2n.	·48	6·96
	327·7		·93	65·48
Mo.	330·1	3n.	·61	58·42
O.Σ.	·0	2n.	·70	61·44
M.	325·6	In.	·33	4·43
Eng.	330·8	3n.	·98	5·44
Ta.	329·0	In.	6·77	9·36
Du.	331·1	6n.	5·66	71·12
W. & S.	334·6	5	6·10	6·46
C.O.	331·7	2n.	5·66	7·48

502   Σ. 1993.

R. A.	Dec.	M.
15h 54m	17° 43′	8·2, 8·2

C. white.

H1.	217·9			37·85	1783·09
Σ.	37·7	3n.		33·96	1831·76
O.Σ.	38·0	,,		·93	40·79
Mä.	37·4	2n.		32·93	7·29

503   O.Σ. 303.

R. A.	Dec.	M.
15h 55·2m	13° 27′	7·4, 7·9

Certain direct motion.

O.Σ.	111·4	3n.	0·60	1846·78
	126·6	2n.	·75	65·44
	134·4	1n.	·77	75·45
Mä.	110·6		·51	43·46
	116·6		·60	7·35
	119·9		·72	51·40
Se.	·2	1n.	·4	7·57
De.	127·8	4n.	·77	67·20
Sp.	131·2		·85	76·52
Pl.	132·7	2n.	·95	·61

504   Σ. 1998.

ξ SCORPII.

R. A. Dec. M.
15^h 57·8^m −11° 2′   A 4·9, B 5·2, C 7·2

Magnitudes.—South, A 7, B 7, C 9. Σ., 4·9, 5·2, 7·2. Sm., 4½, 5, 7½. Se.; 6, 7, 8. Argelander gives A as 4·3 (decimal).

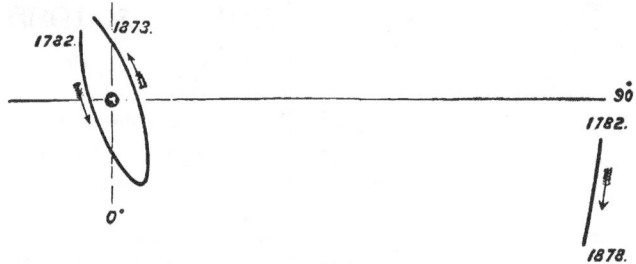

C. H$_1$, A B, fine white. South, C, "decidedly blue." Σ. A, B, yellowish white. C, bluish white. Se., A, yellow; B, white; C, blue. Sm., A, bright white; B, pale yellow; C, grey.

H$_1$ (*Phil. Trans.*, vol. lxxii., p. 218). "May 23, 1780.—Double-double. The first set very unequal. Position 1° 23′, n.f. The other set both small and obscure." In a note he adds, "In a future collection this set will be found as a treble star of the first class, the large white star, with a power of 460 and 932, appearing to be two stars. —*Orig.*"

Here the "first set" are A and C, about 6″ apart, and are H$_1$ II. 20; the "other set" are a faint pair not further alluded to. When H$_1$ discovered the duplicity of A, he registered the close pair I. 33.

H$_2$ and So. (*Phil. Trans.* 1824, p. 243). Up to 1822 these observers had not seen A double. Referring to A C, H$_2$ says, " This is perhaps a binary system with a mean annual motion of −0°·256."

So. (*Phil. Trans.* 1826, p. 343). " A and B equal. Measure of the close pair A B, June 19, 1825, 84° 43′ s.f. or n.p."

H$_2$ in 1831 saw "the division [of A B] quite well."

Da. (*Mem. R. A. S.*, vol. viii., p. 69) says, "The whole series of observations distinctly points to a direct motion, and shows that nearly a whole revolution has been completed since 1782."

Sm. (*Cycle*, p. 352) quotes Σ., who says of A B "that if H$_1$ made an error in the quadrant of the star, which the nearly equal magnitude will easily admit of, and if, from similar causes, we add 180° to South's deductions, it will show a direct motion of 182° in 55 years; giving about a century as its *annus magnus*. The stars A and C, however, are evidently retrograding at about −0°·2 per annum, which is not accountable on proper motion conditions."

Da. (*Mem. R. A. S.*, vol. xxxv., p. 386). He remarks that probably variability in the relative magnitude of the stars led to H$_1$ placing the companion in the n.f. quadrant. It was really in the 3rd quadrant. South's position in 1825 is similarly in error.

Of A B he writes that the orbital motion has been considerably accelerated, and that the distance has diminished, "and it will, no doubt, ere long require the most powerful optical means to fairly separate the components." And in 1867 he says, "The anticipated approximation of the components has come to pass, and it is now extremely difficult for any ordinary aperture to separate them, at least in this latitude."

Mädler, using the observations from 1782 to 1846, obtained the following elements for the close pair :—

$$T = 1832·611$$
$$\Omega = 4°\ 45'·2$$
$$\text{Annual motion} = +204'·688$$
$$P = 105·522 \text{ years}$$
$$i = 70°\ 13''·3$$
$$\pi' = 0''·05772$$
$$\text{Semi-axis major} = 1''·287.$$

For the distant pair, 1839·85 : 6″·801 ; 74° 29′·0 − 14′·704 t ; G = 0″·02909 ; J = 1469·0.

Dr. Doberck in 1876 published the following circular orbit of A B:—

$$\Omega = 10° 51' : i = 72° 27' : P = 95 \text{ years.}$$

Least distance 0″·38 in 1859·62.
Greatest distance 1″·25 in 1883·37.

In 1877 the following elements by the same astronomer were published:—

$$\Omega = 12° 15'$$
$$\gamma = 68\ 42$$
$$\lambda = 89\ 16$$
$$e = 0·0768$$
$$P = 95·90 \text{ years}$$
$$T = 1859·62$$
$$a = 1″·26.$$

The common proper motion of the three stars is — 0″·103 in R. A., and — 0″·105 in N. P. D.

A B.

	°		″	
H₁.	187·9		...	1782·36
Σ.	335·9	3n.	1·14	1825·47
	4·4	,,	·21	32·46
	5·8	2n.	·21	3·91
	7·7	,,	·23	5·00
	8·0	4n.	·16	6·49
	12·5	2n.	·09	7·51
H₂ & So.	351·9	8n.	·35	25·49
	1·5	4n.	·49	30·27
	9·2	2n.	·32	1·38
	10·0	8n.	·41	5·40
Da.	6·2	1n.	·15	3·39
	7·8	8n.	·16	4·50
	5·9	2n.	...	·52
	16·7	,,	1·27	9·61
	18·6	3n.	·19	40·56
	·9	,,	·19	1·58
	21·5	2n.	...	2·46
	23·5	,,	1·08	3·40
	24·5	1n.	...	·45
	30·5	3n.	1·19	48·54
	46·2	1n.	...	53·53
	156·9	2n.	1·57	65·54
Sm.	6·6		·4	34·42
	13·3		·1	8·60
	23·5		·2	42·56
	24·9		·0	6·49
Mä.	16·7	4n.	·28	1·48
	20·4	,,	·05	2·42
Mit.	23·6	3n.	0·96	6·46
	·1	,,	·96	·47
	25·5	,,	·98	·48
	26·0	1n.	1·71	7·58
	27·2	,,	0·84	8·54
Ja.	46·5	15	·93	52·98
	48·3	10	·9	4·06
De.	50·5	3n.	oblong	5·31
	57·1	5n.	,,	6·33
	318·7	3n.	wedgᵈ·	62·55
	322·0	9n.	,,	3·44
	331·9	10n.	oblong	4·50
	333·1	,,	,,	5·44
	156·6	8n.	0·5	6·46
	160·7	7n.	·82	7·45
	165·3	,,	·89	8·40

	°		″	
De.	168·7	5n.	0·88	1869·52
	170·2	7n.	·88	70·45
	173·0	,,	1·06	1·41
	·8	8n.	·11	2·46
	176·4	5n.	·19	3·42
	178·7	,,	·05	4·49
	180·5	,,	·10	5·43
Se.	53·6	4n.	0·47	55·55
	70·0	12n.	·36	6·49
	106±	6n.	single	7
		1n.	single?	8
Kn.	166·5	,,	0·99	68·48
	176·9	,,	1·12	72·46
Du.	172·5	6	0·83	69·51
	173·3	2	·88	70·54
	174·8	5	·88	1·60
	177·4	3	·96	2·53
	181·9	4	1·25	5·56
Gl.	168·2	5	...	0·21
	174·0	5	1·0	1·49
	182·8	4	·15	3·49
	176·5	2	·1	·68
W. & S.	177·3	6	0·95	2·45
	180·4	4	1·04	3·36
	183·1	4	·19	4·44
	180·0	6	·33	5·51
	184·9	8	·27	7·46
Fer.	355·7		·10	2·50
Schi.	182·0	1n.	·18	5·50
	184·1	,,	·20	6·51
C.O.	185·6	,,	·04	·44
	184·3	5n.	·26	7·46
W.O.	2·5	1n.	·23	6·46
	5·5	,,	·14	·53
	3·8	,,	·05	·54
Dob.	186·4	3n.	...	·61
	179·4	2n.	0·96	7·42

$\dfrac{A+B}{2}$ and C.

	°		″	
H₁.	88·6		6·38	1782·36
H₂ & So.	78·3	2n.	·76	1822·46
	76·6	4n.	7·07	5·46
	78·6	3n.	6·95	8·40
Σ.	·6	4n.	·75	5·48
	76·1	3n.	·70	32·46
	75·4	2n.	7·02	5·00
	74·7	3n.	·06	6·49
	·7	1n.	6·99	52·22
Sm.	76·1		7·2	34·42
	74·2		·2	8·60
	68·1·		·0	46·49
Da.	69·4	1n.	·43	0·56
Ma.	74·7		6·75	1·47
	72·8		...	2·42
	·1		6·93	61·42
Ka.	73·3		·53	43·39
	68·8		·91	65·49
	70·2		·98	6·51
Mit.	72·9	9n.	7·27	46·82
O.Σ.	...	1n.	...	52·22
	71·8	,,	7·45	6·58
	69·8	,,	·18	61·43

Ja.	68°·1	5	7''·51	1853·12
	69·2	9	·73	4·06
De.	71·6	3n.	·26	5·31
	·1	2n.	6·90	6·39
	70·1	,,	7·14	62·55
	·6	3n.	·16	3·49
	71·0	4n.	·11	5·38
	70·4	6n.	·11	6·96
	69·9	2n.	·03	8·53
	·9	,,	·21	9·50
	70·6	,,	6·99	70·35
	·1	3n.	7·08	1·36
	69·2	2n.	·19	2·48
	·5	1n.	·04	3·43
	·6	2n.	·20	4·47
	68·3	,,	·03	5·44
Se.	70·5	10n.	·50	55·54
	69·7	6n.	·10	65·45
M.	72·1	1n.	6·93	1·42
Eng.	68·9		7·41	4·48
Du.	69·8	4	·19	9·48
	72·1	1n.	·08	70·52
	·2	,,	·15	1·52
	·2	2n.	·15	2·52
Ta.	69·7	1n.	6·25	1·42
	68·7	,,	·87	4·37
W. & S.	72·0	6	·91	2·45
	1·7	3	7·	3·36
	3·4	7	·38	7·49
Fer.	243·6		7·62	2·50
Lindstedt.	65·2		·30	3·45
Sch1.	66·9	1n.	·08	5·51
	67·4	,,	·27	6·51
C.O.	66·3	1n.	·69	·44
	67·1	2n.	·32	7·51
W.0.	66·7	1n.	·27	6·46
	68·9	,,	·33	·54
Dob.	65·4	2n.	...	·46
	66·4	1n.	7·51	7·61

O.Σ. by the method of last squares finds the following :—

$$\Delta A = + 5''·342 \pm 0''·030 + (0''·0114 \pm 0''·0019)\,(T - 1850·0);$$

$$\Delta D = + 30''·280 \pm 0''·018 - (0''·0257 \pm 0''·0012)\,(T - 1850·0);$$

and these show that there is no trace of deviation from rectilinear motion.

Dunér has these formulæ :—

$$\Delta = 30''·79 - 0''·020\,(t - 1850·0).$$
$$P = 9°·84 + 0°·020\,(t - 1850·0).$$

Flamsteed	13·4	1n.	61''·7	1703·31
H1.	7·6	,,	39·98	81·82
H2 & So.	9·6	,,	31·17	1821·39
Σ.	·5	2n.	·45	2·69
	·6	3n.	·23	31·52
	·6	4n.	·01	6·33
O.Σ.	·4	3n.	·14	40·88
	·7	1n.	·03	1·60
	10·1	,,	30·72	7·69
	·1	,,	...	51·91
	·3	,,	30·66	2·55
	9·9	,,	·44	66·62
	10·9	,,	·39	8·55
	·5	,,	·12	72·57
	·7	,,	·21	·59
	·6	,,	·20	3·57
Mä.	9·7	5n.	...	41·35
	·4	,,	31·07	3·02
	·2	1n.	30·74	8·38
Se.	10·2	3n.	·41	57·20
De.	9·8	4n.	·59	8·12
Eng.	10·4	2n.	·59	63·65
	·4	,,	·15	4·36
Du.	·2	4n.	·5	9·61
W. & S.	·4	3	·75	76·46
Dob.	9·9	4n.	29·83	7·35

505   Σ. 2007

R. A. 16h 0m	Dec. 13° 39'	M. 6·5, 8

H2 & So.	328·7	2n.	31·93	1823·42
Σ.	·2		·97	30·14
Mä.	·5	1n.	33·05	43·45

506   Σ. 2010.

R. A. 16h 2·6m	Dec. 17° 22'	M. 5, 6

C. both yellow.

Σ. (P. M., p. ccxvi.) found the proper motion of the principal star was − 8''·9 and − 0''·4, that of the companion − 8''·0 and − 4''·6; hence both orbital and common proper motion

507   Σ. 2034.

R. A. 16h 3·5m	Dec. 83° 58'	M. 7, 5·8

Σ.	115·0	3n.	1·41	1831·86
O.Σ.	118·4	2n.	·60	41·14
Mä.	121·3	1n.	·64	2·72
	118·4	2n.	·34	5·61
De.	120·3	,,	·2	57·63
Du.	118·5	3n.	·44	71·25

508   ν SCORPII.

R. A. 16h 5m	Dec. − 19° 9'	M. 4, 7, 7, 8

The wide pair was discovered by H1. In 1847 Jacob detected the duplicity of B, but it was reserved for the keen eye of Burnham in 1874 to see that the principal star itself was also a close double star. A very striking group.

A B.

Bu.	357.7°	6n.	0″.45	1874.41
Newcomb.	5.2	1n.	...	.47
De.	359.1	3n.	0.84	.49
C.O.	8.9	2n.	.59	7.48

A B and C.

H_1.	334.8		38.33	1781.4
H_2 & So.	338.2		40.82	1821.4
	336.6		...	36.5
Sm.	338.5		40.00	1.5
Mit.	.9	3n.	43.00	46.54
Ja.	336.5		.57	7.7
Se.	331.3		.58	55.5
De.	336.9	2n.	.78	74.49

C D.

Mit.	39.0	2n.	1.11	46.58
Ja.	42.2		.8	7.4
	45.4		.6	8.0
Bu.	.7	6n.	...	74.41
De.	48.4	2n.	1.89	.49
C.O.	46.2	1n.	.86	7.37

509   Σ. 2021.

49 SERPENTIS.

R. A.	Dec.	M.
16ʰ 7.7ᵐ	13° 51′	6.7, 6.9

C. Σ., white. Sm., A, pale white; B, yellowish.

H_1 (*Phil. Trans.* 1804, p. 376): "In the year 1783, March 7, the position of the two stars of this double star was 21° 33′ n.p. May 20, 1802, 32° 52′; and April 2, 1804, 35° 10′; which gives a change of 13° 37′ in 21 years and 26 days. The stars are now a little farther asunder than they were formerly. A parallactic motion would account for the change of the angle, but not for the increased distance."

H_2 and So. (*Phil. Trans.* 1824, p. 247): "The motion of this star, first pointed out by Sir W. Herschel in 1804, is thus clearly established. The disagreement between our observations and M. Struve's is rather more than usual (4° 6′); but the star is close and difficult. The mean annual angular motion is about 0°.510 in the direction n.f.s.p., or direct." Measures in 1822 and 1823 are given.

So. (*Phil. Trans.* for 1826, p. 347). H_2 having the observations by So. in 1825 before him, says that the change in this star is confirmed; that the amount 6° 13′ is greater than calculation gives, viz., 1° 6′; that probably the measures in 1823 were faulty, and Σ.'s measure in 1820 (46° 33′ n.p.) worthy of more confidence.

Sm. (*Cycle*, p. 355): "A rough investigation gives above 600 years for the orbital revolution of the satellite about its primary,

—or, rather, of one sun around the other. More observations at longer epochs are, however, necessary, before it can actually be pronounced a binary system."

Later, Dawes and Secchi express their conviction that the orbital motion is certain.

These stars are transported through space by a considerable common proper motion. The distance appears to have already reached its maximum. (O.Σ.)

Dunér gives the formulæ

$$\Delta \cos P = 2''.22 + 0''.0241\,(t - 1830.0) - 0''.00010\,(t - 1830)^2.$$

$$\Delta \sin P = -2''.27 + 0''.0052\,(t - 1830.0) + 0''.000066\,(t - 1830)^2.$$

The common proper motion is $+0''.152$ in R. A., and $-0''.369$ in N. P. D.

			″	
H_1.	291.5	1n.	...	1783.18
	302.9	,,	...	1802.39
	305.2	,,	...	4.25
H_2 & So.	311.9	2n.	4.15	23.28
	318.2	4n.	3.5	5.41
	316.8	2n.	2.95	30.02
Σ.	315.5	3n.	3.19	29.48
	316.6	1n.	.03	32.53
	319.2	,,	.24	4.39
	317.0	4n.	.25	5.45
	316.8	2n.	.29	6.71
Da.	314.8	1n.	.17	1.40
	318.0	,,	.43	41.38
	321.3		...	9.44
Sm.	317.8		3.7	32.43
	318.1		.3	9.29
	323.0		.2	54.58
Mä.	319.3	1n.	...	40.84
	.2	4n.	3.62	1.45
	320.5	3n.	.40	2.38
	319.6	4n.	.39	3.34
	318.6	2n.	.36	4.38
	321.2	1n.	.34	5.11
	322.2		.39	51.40
	321.9		...	2.04
	322.4	4n.	3.52	4.62
	324.7	2n.	.37	7.39
	323.7		.58	8.42
	.7		.70	9.38
	324.7	7n.	.71	60.67
Hi.	318.9	1n.	.23	45.26
Mit.	319.1	,,	4.34	7.58
Flt.	322.9	16	3.25	51.66
De.	321.2	6n.	.67	4.63
	324.6	3n.	.53	64.80
Mo.	322.7	,,	.65	55.49
Se.	.3	6n.	.46	6.01
	325.8	1n.	.80	65.48
M.	323.7	,,	.53	2.37
	324.2	,,	4.02	8.43
	327.0	,,	3.81	70.39
	325.4	,,	.59	1.37
	324.9	,,	.88	2.45
	326.5	,,	.66	3.39
	325.9	,,	.70	4.42
	329.7	,,	.94	5.63

Eng.	325°7	4n.	3″76	1865·51

Eng.	325°7	4n.	3″76	1865·51
Ta.	...	in.	·91	6·32
	325·8	,,	·69	9·57
	327·2	,,	·93	71·42
	·6	,,	4·26	2·29
	328·4	2n.	3·76	4·46
	327·9		4·11	5·42
Br.	329·1	3	3·60	68·46
Du.	327·9	8n.	·52	70·35
Hall.	·2		·94	1·42
	·6		4·27	2·29
	328·8	3n.	3·81	6·34
W. & S.	327·7	in.	·73	2·45
	329·4	,,	·44	5·51
	·7	,,	·94	·63
	327·0	,,	·56	6·48
Sp.	·6		·69	5·48
Gl.	328·5	in.	·9	·60
Schi.	147·6	,,	·69	·47
Dob.	327·4	3n.	·90	6·25

510 O.Σ. 303.

R. A.	Dec.	M.
16h 7m	34° 43′	7·2, 8·7

Probable change in angle and distance.

O.Σ.	61·4	in.	0·40	1845·65
	58·3	,,	·40	·71
	48·0	,,	·27	6·38
	55·8	,,	·33	8·49
De.	45	obl.?		65·47
	60		,,	·52

511 Σ. 2022.

R. A.	Dec.	M.
16h 7·8m	26° 59′	6·2, 9·8

Angular change is certain.

Σ.	129·5	3n.	2·77	1830·56
Mä.	131·2		·89	44·36
Se.	136·5	2n.	·40	58·09
	·0	in.	3·26	65·52
O.Σ.	138·7	,,	2·78	8·50

512 Σ. 2023.

R. A.	Dec.	M.
16h 8·6m	5° 50′	8, 9

C. both yellowish.

Σ.	236·0	4n.	1·55	1832·41
Mä.	232·7		·51	9·74
	231·0		·50	42·42
	229·1		·41	51·40
	·5		·85	2·63
Se.	231·7	in.	·65	6·42
	229·8	2n.	·77	65·54

513 Σ. 2026.

R. A.	Dec.	M.
16h 8·9m	7° 40′	8·6, 9·1

C. yellow.

Σ.	345·9	4n.	2·54	1830·94
Mä.	342·0		·4	8·05
	337·8		·22	44·35
	334·4		·11	52·63
Se.	330·1	2n.	1·78	6·56
	325·7	in.	0·97	65·53
De.	326·1		1·50	·39
W. & S.	318·9		·4	72·45
	315·7		·5	3·46
	321·0		·4	5·51
Gl.	316·0	in.	·3	·60

514 Σ. 2032.

σ CORONÆ BOREALIS.

R. A.	Dec.	M.
16h 10·2m	34° 10′	8·6, 9·1

C. H_1, both white.

H_1 (*Phil. Trans.*, vol. lxxii., p. 215): "Aug. 7.—Treble. The two nearest pretty unequal; the third very faint, with powers lower than 460."

H_1 (*Phil. Trans.* 1804, p. 373): "This star has undergone a great change." "The great number of small stars in this neighbourhood is not favourable to a supposed connexion between any of them and σ Coronæ. As the two small stars are considerably unequal, we may suppose the larger one to be affected by a parallactic motion, which will sufficiently account for the angular changes."

H₂ and So. (*Phil. Trans.* 1824, p. 249). Measures in 1821, 1822, and 1823 are given. H₂ then discusses the entire series at considerable length. He notes the "great and almost sudden acceleration in the angular velocity of the small star," 23°·86 having been described between 1781 and 1802, 38°·6 between 1802 and 1818, and 22°·55 between 1819·6 and 1823·83, the annual rates being respectively 1°·139, 2°·298, and 6°·982. He then shows that there has also been "a very sensible diminution of distance." He explains that the phenomena by supposing that the orbit is elliptic, that its plane passes nearly through the eye, and that the star is approaching its perihelion. He then assumes the orbit to be circular, and its plane inclined 30° to the visual ray; then, taking 2°·13 as the mean annual motion, he computes the positions for the times of the recorded observations, and finds a very fair agreement between the computed and observed places.

So. (*Phil. Trans.* 1826, p. 349). Measures made in 1825 are given, and H₂ remarks that the sudden increase of angular velocity noted above is not verified; and

he thinks that "the angle 40° n.f. for 1819, on which it rests, must of necessity have been considerably in error."

H₂ (*Mem. R. A. S.*, vol. v., p. 39). After presenting the whole of the measures from 1781·79 to 1830·28, he says, "None of these angles can be depended upon, so very difficult is the star." Still, he thinks that a rapid direct motion and a great acceleration since 1800 are evident, and that the distance is still decreasing.

Sm. (*Cycle*, p. 357): "My measures afford presumptive evidence that the components are again separating; and presuming its orbit to be elliptic, with an excentricity of 0°·6988, it must occupy a period of not less than 560 years, with its motion performed in a plane passing nearly through the eye."

Σ. (*P. M.*, p. ccxxix.) shows that South's star C (magnitude 10) does not belong to the system.

O.Σ. in 1851 discovered D (O.Σ. 538) : its magnitude was about 12·5.

THE ORBIT.—The following table gives the elements obtained by several astronomers :—

Ω	π − Ω	i	e	a	P	T	Observers.
138 0			0·6112	3·679	years. 286·6	1835·60	Herschel.
20 43·9	65 54·1	40 52·2	·5899	2·3851	420·24	1825·31	Klinkerfues.
3 8	96 53	45 6	·3887	2·94	24·0	1829·7	Powell.
25 7	64 28	29 29	·69978	3·918	608·45	1826·60	Mädler.
21 3	69 24	25 39	·7256	5·194	736·88	·48	Hind.
1 57	101 57	46 47	·309	2·719	195·12	1831·17	Jacob.

In 1875 Doberck made a redetermination of the elements of this star : his results are as follow :—

$$T = 1828\cdot91$$
$$\Omega = 6° 43'$$
$$\lambda = 89 \ 17$$
$$\gamma = 29 \ 40$$
$$\rho = 843\cdot2 \text{ years}$$
$$a = 6''\cdot001$$
$$e = 0\cdot7502.$$

See also p. 131.

A B.

		″	
H₁. 347·5		...	1781·79
11·4		...	1804·74
Σ. 48·0		...	19·62
89·3	4n.	1·31	27·02
104·9	3n.	·22	30·11
118·8	,,	·29	2·99
130·4	5n.	·30	5·50
134·7	6n.	·43	6·59
139·9	5n.	·41	7·55
H₂ & So. 71·5	2n.	·44	22·83
77·5	6n.	·48	5·44
92·1	,,	...	8·50

H₂ & So. 105·0	9n.	1·22	1830·28
108·7	3n.	·38	31·36
113·5	6n.	·07	2·52
119·9	3n.	·33	3·26
Sm. 107·6		·3	0·76
114·9		·4	2·37
120·7		·2	3·58
130·9		·4	5·50
145·1		·6	9·67
155·9		·8	43·35
162·4		2·0	6·60
176·8		·2	52·25
Da. 115·4	3n.	...	32·55
120·6	4n.	1·30	3·36
125·6	3n.	...	4·55
136·8	1n.	...	7·47
144·3	,,	1·60	9·53
147·8	3n.	·65	40·57
150·3	,,	·65	1·48
153·2	1n.	...	2·37
156·5	,,	1·77	3·47
166·0	2n.	·88	7·44
168·6	3n.	·99	8·53
170·1	1n.	2·09	9·45
173·8	,,	·26	51·42

	°		″	
Da.	177.8	4n.	2.38	1853.63
	178.4	3n.	.25	4.56
	180.1	1n.	.43	5.48
	185.5	2n.	.71	60.36
	191.4	1n.	3.07	5.38
Ga.	147.8		1.55	39.52
O.Σ.	149.3	5n.	.53	40.63
	153.7	1n.	.56	1.60
	168.2	2n.	.76	6.68
	169.6	1n.	.69	7.69
	170.8	,,	.91	8.74
	172.4	3n.	.96	9.74
	168.9	,,	.99	50.52
	173.4	6n.	2.05	1.63
	174.3	5n.	.06	2.63
	175.6	6n.	.17	3.65
	179.0	2n.	.24	4.66
	.1	4n.	.29	5.61
	.9	,,	.46	6.57
	181.6	3n.	.50	7.63
	182.3	2n.	.51	8.57
	185.9	3n.	.58	9.68
	186.8	1n.	.72	60.74
	187.4	5n.	.69	1.57
	189.5	1n.	.77	2.74
	188.7	,,	.77	.79
	.2	4n.	.77	3.60
	190.2		.89	4.60
	192.7	2n.	.96	5.74
	...	1n.	...	.74
	192.0	,,	2.92	6.49
	193.5		...	.49
	.2	5n.	3.01	.65
	192.8		2.97	8.55
	196.6		.99	.61
	195.6		...	.61
	.3	3n.	3.26	72.57
	198.7	1n.	.17	3.54
	196.6		.12	.57
	199.8	4n.	.41	4.61
Ka.	148.8		.57	41.56
	156.3		.66	3.68
	193.9		2.86	66.68
Mä.	152.2	7n.	1.59	41.56
	156.3	4n.	.81	2.31
	157.5	,,	.86	.73
	165.0	11n.	2.07	6.46
	166.5	14n.	.16	7.44
	168.3	2n.	.39	8.41
	173.0	,,	.23	50.70
	174.5	6n.	.34	1.25
	176.2	9n.	.43	.76
	177.5	11n.	.39	2.60
	.7	6n.	.46	3.38
	178.7	2n.	.65	.77
	179.4	5n.	.51	4.70
Ch.	148.2	1n.	1.56	41.54
	153.5	,,	.35	2.41
	160.4	,,	.63	3.68
G.O.	157.2		.53	4.44
Mit.	166.7	1n.	.33	7.70
Bond.	171.2		2.2	8.42
	172.5		.2	.42
Flt.	174.4	43	.32	51.22
Mi.	176.4	24	.38	2.31

	°		″	
Mo.	178.6	20	2.22	1854.67
	184.9	20	.70	9.34
Ja.	177.9	2n.	.18	3.14
	.8	3n.	.24	4.05
	181.2	,,	.52	6.73
	183.0	,,	.52	7.66
	.9		.56	8.20
De.	179.7	5n.	.38	4.66
	.9	3n.	.40	5.18
	181.7	6n.	.68	6.42
	180.0	2n.	.52	7.66
	184.7	6n.	.73	8.49
	189.3	5n.	.84	62.53
	190.4	9n.	.74	3.39
	.9	6n.	.75	4.48
	191.7	,,	.83	5.41
	193.2	11n.	.88	6.92
	195.2	5n.	.94	8.36
	196.4	4n.	3.04	9.53
	.9	6n.	.09	70.43
	197.1	,,	.07	1.46
	.8	,,	.16	2.83
	198.3	,,	.27	3.48
	.8	,,	.30	4.48
	199.0	5n.	.25	5.39
Wi.	181.6		.49	55.54
	182.8		.52	6.39
Se.	180.8	4n.	.30	5.61
	182.3	2n.	.45	6.43
	183.5	,,	.42	7.61
Eng.	190.5		.11	64.45
Ro.	189.1		...	5.72
Ta.	.2	2n.	3.73	6.43
	193.7	1n.	.62	8.29
	195.1	,,	.59	9.57
	196.7	,,	.30	71.42
	197.9	,,	.34	2.29
	200.5	,,	.63	3.55
	199.2	2n.	.16	4.46
	.8	1n.	.55	5.42
	.8	2n.	.44	6.56
M.	12.0	1n.	.00	67.37
	200.5	,,	.55	74.44
Du.	195.5	1n.	2.79	67.72
	194.7	4n.	3.14	8.59
	195.1	5n.	.01	.62
	196.5	4n.	.15	71.35
	199.5	5n.	.28	5.54
Kn.	194.3	3n.	2.97	67.34
	195.3	4n.	3.22	71.53
Br.	194.0	4	.11	68.55
W. & S.	.3	3	.51	71.51
	197.7	8	.25	2.53
	198.4	5	.14	3.42
	200.6	5	.47	5.50
	202.2	6	.68	7.46
Schi.	198.6	1n.	.34	5.46
Dob.	199.4	4n.	.89	6.28
	.5	5n.	.58	7.32
	201.3	4n.	.14	8.53
Pl.	199.2	3n.	.78	6.52
	202.3	5n.	.62	7.65
W.O.	201.1	1n.	.44	6.44
	200.0	,,	.58	.44
	199.0	,,	.46	.45

A C.

H₁.	244.9		20″	1832.60
O.Σ.	231.6	5n.	20.41	54.40
Hall.	221.7		15.92	76.40

A D.

H₁.	65		24	1781.00
So.	90.6		42.2	1825.53
H₂.	89.2		44.2	8.40
Sm.	90.0		43.3	30.76
	88.7		44.1	2.37
	89.3		.0	6.50
	88.9		.2	9.67
	90.0		46.3	52.00
Σ.	88.8		43.75	36.69
	.6		44.17	7.66
	.6		.88	40.58
O.Σ.	87.9		47.52	50.26
	.7		.98	1.69
Mä.	90.1		.83	3.32
De.	88.4		51.0	62.00
W. & S.	.2	In.	52.6	72.53
	87.9	,,	...	3.42
	.9	,,	54.18	6.48
Fl.	.6	,,	.15	7.46

515   O.Σ. 309.

R. A.	Dec.	M.
16ʰ 15ᵐ	41° 56′	7.5, 7.8

The relative brightness of the two stars is probably variable.

O.Σ.	234.4	In.	0.50	1842.71
	55.3	,,	.66	5.65
	239.4	,,	.54	7.55
	56.7	,,	.40	51.67
De.	231.5	3n.	...	67.98

516   Σ. 2044.

R. A.	Dec.	M.
16ʰ 20ᵐ	37° 19′	7.8, 8

C. white.

Dunér gives

$$1854.98. \quad \Delta = 8″.46.$$
$$P = 345°.4 - 0°.065 (t - 1850.0).$$

H₂ & So.	346.4	2n.	10.15	1823.41
Σ.	.9	3n.	8.54	30.03
Mä.	.7	In.	.68	43.61
Se.	344.6	2n.	.54	57.56
De.	.5	3n.	.09	8.22
Du.	.5	5n.	.49	69.89

517   O.Σ. 310.

R. A.	Dec.	M.
16ʰ 21ᵐ	38° 13′	7.6, 10.2

O.Σ.	221.5	In.	3.15	1845.35
	.9	,,	2.88	.68
	217.5	,,	.97	51.67
	224.4	,,	.96	74.67
De.	225.7	3n.	3.15	67.43

518   O.Σ. 312.

R. A.	Dec.	M.
16ʰ 21ᵐ	61° 47′	2.1, 8.1

Probable increase in the distance.

O.Σ.	143.9	5n.	4.66	1843.71
	142.2	4n.	.86	51.21
	.1	2n.	.99	60.11
	145.5	,,	5.20	73.67
Da.	141.4	In.	4.71	47.41
Mä.	144.8	2n.	.01	52.69
De.	142.0	4	.90	66.50
Du.	.4	3n.	.9	70.52
W. & S.	.5	In.	.9	3.42

519   Σ. 2054.

R. A.	Dec.	M.
16ʰ 22ᵐ	61° 58′	5.7, 6.9

H₂.	351.5		...	1830.24
Σ.	7.4	6n.	0.90	2.22
	6.1		.96	5.76
O.Σ.	6.9	5n.	1.08	41.44
	1.9	3n.	.01	55.63
	0.7	In.	.07	72.61
Mä.	3.4		.06	43.53
	2.0		.16	52.33
	.8		0.92	9.40
Se.	.2	2n.	.94	7.74
De.	.9		1.12	67.85

520   α SCORPII (Antares).

R. A.	Dec.	M.
16ʰ 22ᵐ	− 26° 10′	1.5, 7.7

C. Da., A, red; B, "blue," "purple," "very blue," "green."

Discovered to be double by Professor O. M. Mitchell with a refractor of 11¼ in. aperture, in 1846. Dawes could see and measure this object with his 6½-inch refractor.

The proper motion of Antares is − 0″.006 in R. A. and + 0″.034 in N. P. D., and in this the companion most probably partakes.

Burg.	270		...	1819.28
Mit.	270	1on.	2.52	46.59
	...	4n.	.8	7.50
	273.0	2n.	3.11	8.59

Da.	273°·9	2	...	1847·29
	·6	9	3·47	·29
	270·0	5	·64	8·55
	271·6	5	·41	·59
	275·9	5	·24	9·40
	·7	5	·67	64·43
Bond.	·3	1n.	·8	48·28
	277·0	,,	·8	·28
	272·7	,,	·6	·49
	273·1	,,	·6	·58
	272·0	,,	·4	·55
	270·4	,,	·4	9·52
Mä.	276·2	80	·69	·68
Ja.	272·8	15	2·94	52·63
	273·5	3n.	3·20	6·22
	275·0	,,	·40	7·18
Se.	273·8	6n.	·07	5·56
	·3	4n.	·25	6·55
	·2	,,	2·69	7·54
	272·9		·92	66·17
Sm.	270·0		3·5	57·40
Mo.	275·8	4n.	·30	8·35
Po.	271·9		...	61·09
Kn.	275·7	4n.	3·37	4·44
De.	270·4	3n.	2·99	5·56
W. & S.	268·7	4	3·41	73·42
Gl.	·4	10	·29	·62
	267·6	8	·32	·63
Schi.	273·9	1n.	·22	5·81
Sp.	274·0		·22	·81
C.O.	273·3	4n.	2·85	7·42

O.Σ.	217°·6	1n.	1·06	1847·47
	213·7	,,	·36	68·67
	211·3	,,	·22	74·58
Mä.	217·6	3n.	·00	42·63
	216·7	1n.	·09	6·39
Se.	213·2	2n.	·10	56·50
De.	214·9	,,	·1	7·05
Du.	210·2	11n.	·19	70·29
Dob.	208·5	4n.	0·95	7·47

523　Σ. 2052.

R. A.	Dec.	M.
16h 23·6m	18° 40'	7·5, 7·5

The common proper motion is $-0''·33$ in R. A. and $-0''·36$ in N. P. D.

Dunér gives

$$1855·85. \quad \Delta = 2''·85.$$
$$P = 106°·2 - 0·150\,(t - 1850·0).$$

Σ.	109·3	1n.	2·66	1822·69
	·7	3n.	·98	9·52
So.	·2	1n.	3·24	3·43
H₂.	107·9	2n.	2·86	30·27
Mä.	109·8	1n.	·80	42·45
De.	105·4	5n.	3·14	54·69
	103·2		2·99	65·55
Se.	104·2	2n.	·95	56·49
	103·1	3n.	·75	68·99
Mo.	104·8	,,	·62	58·44
M.	96·2		·75	64·75
Eng.	104·1	3n.	3·16	5·53
Du.	103·0	,,	·46	70·46
Gl.	·0	2n.	·65	4·50
W. & S.	·3	,,	·63	·60
Sp.	101·0		·61	6·31

521　O.Σ. 311.

R. A.	Dec.	M.
16h 22·6m	21° 10'	7·5, 10·3

The distance has diminished about 4″.

O.Σ.	183°·5	1n.	13″·57	1845·35
	·5	,,	·55	6·37
	·5	,,	12·81	52·46
	189·0	,,	10·40	68·67
	188·1	,,	·15	73·47
Mä.	183·5		13·94	50·45
	·4		...	2·61
De.	186·6	3n.	10·73	66·60
W. & S.	189·0		8·00	77·46

522　Σ. 2049.

R. A.	Dec.	M.
16h 23m	26° 15'	6·5, 7·5

Dunér gives the following formulæ:

$$1854·84. \quad \Delta = 1''·12.$$
$$P = 215°·1 - 0°·180\,(t - 1850·0).$$

Σ.	215·6	6n.	1·04	1833·08
	216·1	3n.	·03	6·54
H₂.	220·0	1n.	...	0·20
O.Σ.	223·8	,,	1·25	40·69
	·7	2n.	·03	1·52

524　Σ. 2055.

λ OPHIUCHI.

R. A.	Dec.	M.
16h 24·9m	2° 15'	4, 6·1

C. A, yellow; B, bluish.

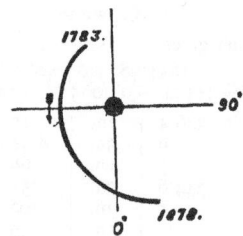

H₁ (*Phil. Trans.* 1804, p. 375): "The position, March 9, 1783, was 14° 30', n.f. May 20, 1802, it was 20° 41'. The difference in 19 years and 72 days is 60° 11'. March 9, 1783, the distance, with 460, was

¼ or ⅓ diameter of the small star. May 1 and 2, 1802, I could not perceive the small star, though the last of the two evenings was very fine. May 20, 1802, with 527, I saw it well, but with great difficulty. The object is uncommonly beautiful; but it requires a most excellent telescope to see it well, and the focus ought to be adjusted upon ε of the same constellation, so as to make that perfectly round. The appearance of the two stars is much like that of a planet with a large satellite or small companion, and strongly suggests the idea of a connexion between the two bodies, especially as they are much insulated. The change of the angle of position might be explained by a parallactic motion of the large star; but the observations on the distance of the two stars can hardly agree with an increase of it, which would have been the consequence of that motion."

H₂ (*Mem. R. A. Soc.*, vol. viii., p. 53). His measures were made in 1831, 1832, and 1833: his notes are "a very good and measurable elongation and notched disc,"—"a distinct notch in the wedge."

Da. also wedged it in these years.

Σ. (*M. M.*, p. 6) "gives his measures from 1825 to 1834. He notes the probable error of 183° in H₁'s measure in 1783 and 1802. The measure in 1783 taken as given by H₁, when compared with Σ's in 1834, shows an angular change of 275°·1 in 51·24 years. This indicates a period of revolution of about sixty-six years."

Smyth (*Cycle*, p. 365): "My observations are not indicative of the acceleration which has been spoken of by other astronomers." "From the shown course and velocity, it is evidently making an elliptical and rapid orbit, of which the *annus magnus* may be between eighty and ninety years."

Subjoined are the elements obtained by Mädler and Hind :—

	Mädler.	Hind.
Perihelion passage ...	1798	1791·214
Position at perihelion	...	177° 50'
Ascending node ...	184°	30 23
Inclination	45°—50°	49 40
Angle between π and ☊ on orbit	...	135 24
Excentricity	0·37	0·4772
Semi-axis major ...	1″·1	0″·847
Period	88 yrs.	95·88 yrs.

Dr. Doberck gives the following (*Ast. Nachr.*, No. 2126) :—

$$\begin{aligned}
☊ &= 157° \ 21' \\
\lambda &= 94 \ 16 \\
\gamma &= 44 \ 44 \\
e &= 0·4930 \\
P &= 233·89 \text{ years} \\
T &= 1803·91.
\end{aligned}$$

O.Σ. (in 1876) writes : "My father has

already remarked (*M. M.*, p. 6) that one of the two observations of H₁ appears to be gravely in error, and he has suggested that the direction in 1802 should be changed 180°. It appears to us, however, it would be quite as admissible, and more in accordance with the most recent observations, to suppose that in his observation of 1802 H₁ was mistaken in the designation of the quadrant in which the companion was seen. In his memoir of 1804 he admits that in 1802 the companion was found b.s.q., as in 1783. If we write a.s.q. for b.s.q., we shall have for 1802·39 the angle 110°·68, and in the interval between 1802 and 1825 the companion described an arc of 220° in passing its apparent periastre at a very small distance from the principal star. The continued increase of the distance since 1825 indicates that the position of the principal star in the apparent orbit is very excentric.

	°			″	
H₁.	75·5			...	1783·18
	69·3			...	1802·39
Σ.	331·8	3n.	0·83	25·51
	342·1	,,	·81	8·51
	349·4	2n.	1·04	31·90
	350·6	,,	0·98	4·42
	352·4	5n.	·99	5·55
	353·3	,,	1·01	6·50
	356·8	1n.	·03	7·59
H₂.	337·7	4n.	...	1·37
	347·5	,,	1·07	2·57
	·8	3n.	·00	3·33
Sm.	351·2		·0	4·48
	352·9		·1	6·51
	356·5		·0	9·67
	1·4		·1	42·50
	15·5		·2	53·25
Da.	349·5	6n.	0·93	34·55
	354·8	1n.	1·17	7·68
	358·3	3n.	·07	40·54
	359·4	4n.	·13	1·54
	·4	3n.	·10	2·58
	1·6	2n.	...	3·47
	0·3	1n.	1·09	·41
	8·8	2n.	·24	8·47
	·9	7n.	·26	·85
	14·0	4n.	·31	54·14
	19·5	1n.	...	60·36
O.Σ.	2·7	6n.	0·97	40·57
	4·4	2n.	1·03	1·58
	6·3	1n.	0·98	2·60
	8·1	3n.	1·09	5·63
	9·3	2n.	·12	6·69
	·9	,,	0·97	7·67
	12·3	1n.	1·30	52·55
	11·6	,,	·29	3·55
	·5	,,	·21	·57
	17·2	,,	·28	6·58
	·1	,,	·51	7·59
	19·7	,,	·30	8·56
	18·6	,,	·35	61·63
	29·5	,,	·61	6·62

	°		″	
O.Σ.	30·1	1n.	1·43	1868·56
	29·8	2n.	·48	72·58
	33·9	,,	·49	4·57
Ch.	5·2	,,	0·97	41·52
	357·4	1n.	1·05	2·43
	2·4	,,	·10	3·40
Mä.	·8	7n.	·29	41·59
	1·6	5n.	·11	2·38
	10·3	4n.	·17	7·43
	12·9	,,	·05	50·58
	14·8	1n.	·25	1·40
	15·9	8n.	·06	2·57
	17·8	4n.	·09	4·63
	·5	3n.	·29	6·73
	19·8	5n.	·27	8·62
Bond.	9·8		1·4	48·56
Ja.	12·6	15	·21	52·67
	15·2	15	·35	4·06
	·5	6n.	·37	6·44
	16·8	3n.	·39	8·12
De.	15·3	,,	1″+	5·30
	·5	4n.	1·2	·35
	14·4	6n.	·2	6·56
	15·9	4n.	·2	7·58
	·9	5n.	·24	8·50
	18·2	7n.	·45	62·55
	21·5	5n.	·65	3·44
	25·2	7n.	·51	5·49
	26·5	11n.	·51	6·95
	27·4	5n.	·45	8·46
	28·2	4n.	·59	9·55
	·6	,,	·51	70·45
	·8	5n.	·57	1·41
	30·1	4n.	·64	2·44
	·5	3n.	·68	3·43
	32·2	4n.	·58	4·56
	33·9	,,	·62	5·44
Se.	17·9	3n.	·36	55·58
	18·2	2n.	·37	6·59
	19·8	3n.	·33	7·51
Mo.	15·6	2n.	·29	7·58
Ro.	20·7	1n.	·16	63·57
	21·0	2n.	·10	5·52
Ta.	22·7	,,	...	6·41
	26·7	3n.	1·41	70·50
	·7	1n.	·56	1·49
	·3	,,	·86	3·55
	25·8	3n.	·53	4·44
	23·6	1n.	·83	5·42
	26·8	,,	·98	6·61
M.	23·9	,,	·34	67·61
	28·3	,,	·42	8·46
	26·2	,,	·47	·41
	...	,,	·41	71·48
	32·0	,,	·49	·50
Du.	26·3	3n.	·41	68·68
	·7	7n.	·50	9·62
	·5	3n.	·53	70·57
	28·9	,,	·45	1·62
	33·0	2n.	·48	5·55
Gl.	28·2	7	·6	0·44
	27·9	4	·7	1·32
	29·0	5	·5	·63
	30·0	4	·5	3·68
	33·5	10	·26	4·62

	°		″	
W. & S.	28·9	5	1·56	1872·45
	30·3	8	·62	3·42
	33·6	4	·31	4·62
	·4	6	·54	5·52
W.O.	·7	1n.	·74	4·74
	36·1	,,	·45	6·44
	33·7	,,	·48	·45
	·7	,,	·52	·45
	31·7	,,	...	·46
	32·1	,,	·64	·53
Schi.	·8	,,	·44	5·49
	·8	,,	·43	6·58
Dob.	30·5	5n.	·44	·50
	32·3	2n.	·56	7·42
	33·5	,,	·59	8·47
Pl.	31·2	4n.	·82	6·53

525   O.Σ. 313.

R. A.	Dec.	M.
16ʰ 29ᵐ	40° 21′	7·2, 7·8

Probable change.

	°		″	
O.Σ.	162·1	1n.	0·81	1842·71
	161·2	,,	·81	5·71
	162·7	,,	·71	7·55
	164·0	,,	·89	9·71
	160·8	,,	·77	51·67
Mä.	155·8	4n.	·8	46·30
	156·7	5n.	·9	52·03
Da.	159·6	1n.	·94	47·41
De.	153·8	4n.	·92	66·76
Du.	152·6	,,	·93	9·51

526   Σ. 2084.

ζ HERCULIS.

R. A.	Dec.	M.
16ʰ 36·8ᵐ	31° 49′	3, 6·5

C. A, yellowish ; B, reddish.

H_1 : "July 21, 1782.—20° 42′.

"Aug. 30, 1782.—Saw it better than ever I did. I could plainly distinguish that the small star is ash-coloured, and the large fine blue-white.

"Sept. 20, 1802.—I cannot see the star double. A conjunction of the two stars may have taken place." Although looked at every evening, it was not seen "lengthened" or "wedge-formed" till the 29th of September.

H_1 (*Phil. Trans.* 1803, p. 378) : "My observations of this star furnish us with a phenomenon which is new in astronomy ; it is the occultation of one star by another. This phenomenon, whatever be the cause of it, will be equally remarkable, whether owing to solar parallax, proper motion, or motion in an orbit whose plane is nearly coincident with the visual ray."

H₁ discovered this object July 18, 1782, and measured the angle and distance. He gives his measures up to 1803, and then examines the several ways in which the phenomena may be explained, observing that "the observations I have made on this star are not sufficient to direct us in the investigation of the nature of the motion by which this change is occasioned."

H₂ and So. (*Phil. Trans.* 1824, p. 267): "April 27, 1821.—Decidedly single, with powers 133 and 303. The evening exceedingly favourable, and the star perfectly round and well-defined."

June 19, 1822.—Fine evening. Powers 133 and 381 would not separate it.

Oct. 17, 1823.—Gave the same result.

South, on July 28, 1825, failed to divide the star with powers 181, 327, 413, 512, and 787 ; nor was there any trace of elongation. "With 787 it was exquisitely defined, and as round as possible."

Σ. (*M. M.*, p. 6). He gives his measures from 1826 to 1834. In 1828·71, he found the distance 0″·65 and the angle 349°·5 ; but in 1828·76 he was "uncertain whether the point of light seen" was the companion or a spurious image. In 1828·77 he writes, "not double : fine air : just after sunset." Up to 1831 his remarks are "not double" —"certainly not double," "no *comes*," and so on. But in 1832·75 he says, "no doubt the comes is seen," and gives the distance 0″·81 and the angle 220°·5. In 1833·27, "no comes seen;" 1833·41, "a red point suspected in the direction 105°;" 1834·45, "air very fine : ξ Libræ examined with altitude 20° bears 1000 : λ Ophiuchi was then examined, and finely seen double ; then ʃ Herculis was examined, and the companion detected at once." Distance, 0″·92, 0″·90 : angle, 203°·5, 202°·5. This sudden reappearance of the comes at first led him to suspect that variability was the cause of the difficulty he had so long experienced, and he was much surprised to find the companion in almost the same straight line as that in which he saw it eight years before. Great was his delight, on copying the observation into the book containing mean results, when he found that the *comes* was in the same line nearly, but in the opposite direction. There had been an occultation, and all the difficulty which had so often vexed him for eight years was completely explained.

Σ. says this star "offers the astounding velocity of an apparent and very elliptical orbit revolving in little more than fourteen years."

Sm. (*Cycle*, p. 369). This observer elongated the star in 1835, but could not 'notch' it till 1838. He prefers an orbit "with an excentricity of 0·4186 and a period of about thirty-five years."

The following are the principal orbits hitherto computed :—

	Mädler.	Villarceau.	Dunér.
Perihelion passage	1829·50	1830·481	1830·01
Node	39°26'	214° 21'	45° 56'
Perihelion from node	262 4	284 55	
Inclination ..	50 53	±136 17	34 52
Excentricity	0·45454	0·4482	0·4239
Semi-axis major ..	1‴·189	..	1″·223
Mean annual motion	−730″·45		
Period	31ʸʳˢ·4678	36·357	34·221

The common proper motion of the system is − 0ˢ·034 in R. A., and − 0″·45 in N. P. D.

	°		″	
H₁.	69·3		...	1782·55
Σ.	23·4	5n.	0·91	1826·63

single from 1828 to 1831

	°		″	
	220·5	1n.	0·81	1832·75
	203·5	2n.	·91	4·45
	169·9	5n.	1·09	5·45
	186·2	,,	·09	6·60
	175·4	4n.	·09	7·47
	168·6	2n.	·03	8·44
	160·4	1n.	·16	9·67
	159·9	4n.	·29	40·66

H₂ and So. : "Decidedly single, with powers 133 and 303," in 1821·320. "Perfectly round with 381″ in 1822·465. "No elongation with 578 on the 5-feet equatorial" in 1823·794. "Perfectly round," with powers up to 787″, in 1825·57.

	°		″	
Sm.	190·0		0·5	1835·68
	176·3		·7	6·73
	169·0		·2	8·65
	136·9		·2	42·57
	108·5		·0	8·59
	83·8		·3	52·53
Encke.	190·6	2	·24	36·59
	168·4	4	·37	8·70
	170·3	2	·97	9·50
O.Σ.	157·4		1·20	·67
	·1	5n.	·24	40·66
	147·7	3n.	·23	1·60
	146·0	,,	·21	2·64
	125·4	2n.	·12	4·71
	121·3	3n.	·24	5·63
	110·5	2n.	·33	6·69
	111·3	,,	·42	7·68
	104·2	,,	·53	8·76
	98·5	,,	·49	9·73
	93·8	1n.	·52	50·53
	88·4	5n.	·47	1·62
	84·1	,,	·51	2·63
	79·9	4n.	·48	3·59
	76·8	3n.	·56	4·66
	70·8	4n.	·44	5·62
	64·7	3n.	·49	6·62
	58·4	4n.	·49	7·64
	51·0	,,	·63	8·62
	42·3	,,	·29	9·63
	32·5	1n.	·38	60·74

O.Σ.	17°.1	4n.	1″.05	1861.57
	341.2	1n.	.00	2.74
	228.6	2n.	0.97	6.74
	204.6	1n.	1.13	8.56
	202.7	,,	.34	.61
	179.6	,,	.34	71.52
	168.8	3n.	.14	2.60
	171.5	2n.	.10	3.50
	167.2	1n.	.37	.52
	162.9	4n.	.40	4.62
Da.	161.9	,,	.22	39.76
	150.6	8n.	.23	40.66
	142.9	4n.	.23	1.65
	138.4	3n.	.06	2.58
	129.8	,,	.29	3.64
	112.1	5n.	...	6.89
	107.9	1n.	.62	7.53
	102.0	6n.	.56	8.61
	.3	3n.	.50	.61
	99.2	1n.	.70	9.48
	86.9	3n.	.59	51.80
	80.0	7n.	.66	3.40
	69.5	1n.	.59	5.68
	45.7	6n.	.33	9.61
	235.1	3n.	0.85	66.70
	229.2	2n.	.82	.81
	225.0	,,	.98	.99
Mä.	149.2	9n.	1.11	41.44
	141.5	3n.	0.91	2.40
	.4	4n.	.98	.75
	130.3		.91	3.58
	104.6	17n.	1.30	7.47
	98.7	3n.	.08	8.41
	91.4	,,	.26	50.55
	84.8	,,	.29	1.23
	.6	8n.	.18	.88
	77.2	,,	.23	3.39
	74.3	3n.	.19	.82
	72.3	5n.	.33	4.68
	60.0	4n.	.07	7.39
	48.6	7n.	.20	8.66
	41.2	4n.	.10	9.39
Ka.	140.1		.42	42.55
Ch.	138.6	1n.	.19	43.61
Mit.	109.2	,,	.09	7.71
Flt.	91.7	10	.4	50.54
	89.3	48	.3	1.51
	84.0	32	.24	2.64
Ja.	81.1	10	.52	3.15
	78.0	14	.52	4.06
	66.2		.60	6.25
	57.0		.46	7.86
Se.	69.7	3n.	.52	5.53
	64.1	6n.	.41	6.53
	59.5	,,	.29	7.59
	54.6	2n.	.06	8.48
	43.2	,,	.06	9.52
	80	1n.	round	65.54
	86	,,	,,	.54
	...	,,	,,	.55
De.	68.5	8n.	...	4.80
	70.8	4n.	...	5.23
	63.7	15n.	1.2	6.52
	59.0	5n.	.25	7.75
	49.9	8n.	.0	8.55

De.	1°.7	9n.	...	1852.53
	342.4	4n.	...	3.49
	244.6	5n.	0.5	6.46
	225.6	7n.	.8	7.52
	210.6	,,	.94	8.42
	200.9	,,	1.08	9.58
	190.8	11n.	.09	70.49
	180.7	12n.	.27	1.50
	173.8	,,	.34	2.48
	162.4	11n.	.39	3.52
	157.0	10n.	.36	4.53
	149.0	8n.	.41	5.52
Mo.	73.1	4n.	.45	55.66
	60.2	3n.	.60	7.46
Du.	221.4	2n.	.03	67.72
	213.2	5n.	.05	8.67
	203.1	11n.	.05	9.62
	193.6	6n.	.21	70.59
	183.7	12n.	.18	1.60
	177.2	6n.	.22	2.58
	166.3	4n.	.40	3.70
	154.9	1n.	.35	4.65
	147.3	12n.	.27	5.61
Br.	212.4	2	.19	68.73
Knott.	206.5	4n.	0.99	8.48
	183.3	5n.	1.02	71.54
W. & S.	165.4	3	.2	3.45
	166.8	3	...	.46
	168.0	8	0.76	.47
	153.5	4	1.06	4.55
	159.3	4	.1	.62
	149.4	5	...	5.57
	151.2	5	...	.58
Gl.	153.8	5	0.75	4.51
	157.3	8	.81	.63
Sp.	147.2		1.21	5.55
	138.1		.17	6.54
Schi.	147.1	1n.	.21	5.55
	138.0	,,	.17	6.54
W.O.	144.0	,,	.29	.45
	142.6	,,	.33	.55
Dob.	129.0	3n.	.40	8.45
	124.9	,,	.28	.61

527　Σ. 2094.

R. A.	Dec.	M.
16ʰ 39ᵐ	23° 44′	7.3, 7.6, 11

In A B a retrograde movement is very probable. A C unchanged.

A B.

Σ.	82.8	5n.	1.62	1831.41
O.Σ.	79.6	3n.	.71	41.52
	77.6	1n.	.59	61.44
Mä.	81.7	5n.	.23	43.45
Se.	79.9	3n.	.51	56.47
De.	80.5	1n.	.3	7.50
Du.	81.0	4n.	.26	9.49
Dob.	80.6	3n.	.39	77.48

A C.

De.	312.6	1n.	24.97	61.44

528

R. A.	Dec.	M.
16ʰ 40·3ᵐ	43° 42′	8, 8

C. white.

Discovered by Baron Dembowski in 1869. Rapid change in angle.

De.	131·7		0·93	1870·44
	117·5		·67	6·68
8p.	114·0		·5	7·44

529   Σ. 2097.

R. A.	Dec.	M.
16ʰ 40·5ᵐ	35° 57′	8·5, 8·7

Σ.	89·9	3n.	2·14	1829·63
Mä.	86·9		·14	43·36
	85·6		1·93	4·35
Se.	84·3	2n.	2·15	65·54

530   O.Σ. 315.

R. A.	Dec.	M.
16ʰ 45·3ᵐ	1° 25′	6·2, 8·1

Slow retrograde motion.

O.Σ.	173·3	2n.	0·87	1844·49
	164·6	4n.	·71	54·46
	162·1	1n.	1·01	73·47
Mä.	175·8		0·63	45·51
	168·6		·6	7·35
Se.	160·0	2n.	elongᵈ·	57·50
Da.	167·6		1·33	65·60
De.	162·6	5n.	0·86	67·38

531   Σ. 2106.

R. A.	Dec.	M.
16ʰ 45·4ᵐ	9° 37′	6·7, 8·4

C. white.

Change both in angle and distance.

Σ.	339·0	2n.	1·08	1825·52
	337·2	1n.	0·86	9·53
	336·5	,,	1·00	32·52
	335·8	,,	·05	3·45
Mä.	·9		0·80	42·42
	331·1		·8	3·55
	·8		·9	4·35
Se.	328·4	3n.	·84	56·45
De.	321·3	2n.	·5	63·53
O.Σ.	323·3	1n.	·71	8·50
	320·4	,,	·70	71·50
	324·7	,,	·59	5·48
W. & S.	310·7	3	·5	3·45
	316·6	3	...	·46
	311·9	6	0·6	·46
	329·8	4	·4	5·57
Gl.	310·5	4	·5	4·40

532   Σ. 2107.

R. A.	Dec.	M.
16ʰ 47·1ᵐ	28° 52′	6·5, 8

C. A, yellowish ; B, bluish.

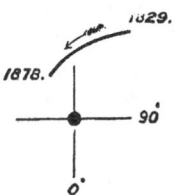

Discovered by Σ. and measured by him from 1828 to 1836 : at the latter date he wrote, "There can be no doubt about the increase in the angle." Dembowski says, "Very difficult owing to the sombre colour of B."

O.Σ., in 1877, says the angular movement has increased considerably in the last twenty-five years, and this augmentation has been accompanied by a notable diminution of distance.

Dunér has published the following formulæ:

$$\Delta \cos P = -1'''04 + 0''00475 (t - 1825\cdot5) + 0''00059 (t - 1852\cdot5)^2.$$
$$\Delta \sin P = +0''12 - 0''02000 (t - 1852\cdot5) - 0''00010 (t - 1852\cdot5)^2.$$

Σ.	148·6	3n.	1·12	1829·91
	156·4	,,	·25	36·54
	159·2	,,	·05	7·74
O.Σ.	160·5	2n.	·06	40·54
	164·3	,,	·08	1·54
	165·7	1n.	0·85	8·46
	169·0	2n.	·73	9·74
	·9	,,	·85	51·58
	170·1	2n.	·80	2·63
	172·4	1n.	·79	3·57
	170·4	,,	·97	5·50
	175·2	,,	·92	6·58
	174·7	2n.	·93	7·63
	175·7	1n.	·89	8·56
	·3	2n.	1·02	61·53
	185·5	,,	0·78	6·67
	194·0	1n.	·91	8·67
	203·4	,,	·75	72·58
	208·9	,,	·72	4·67
	218·7	,,	·72	5·48
Da.	162·0	4n.	1·26	40·95
	170·0	1n.	·23	8·43
	175·7	,,	0·93	54·40
	178·7	,,	1·12	·52
Mä.	163·2	2n.	·0	41·55
	162·7	3n.	·03	2·40
	166·9	6n.	0·88	6·41
	168·3	3n.	·83	7·35

Mä.	174.7	3n.	0.88	1851.20
	176.0	6n.	.87	.77
	.6	10n.	.80	2.61
	180.1	4n.	.74	3.48
	178.2	3n.	.86	4.72
Ka.	...	,,	1.27	41.68
	162.9	11in.	...	2.81
De.	176.8	5n.	1.0	56.54
	189.7	4n.	...	62.55
	188.9	5n.	0.93	3.36
	185.7	,,	.6	4.47
	190.3	7n.	.8	5.50
	189.5	8n.	1.08	7.11
	193.8	5n.	0.93	8.48
	195.8	3n.	.85	9.55
	198.2	6n.	.78	70.46
	202.0	3n.	.80	1.59
	203.6	5n.	.90	2.44
	208.0	4n.	.8	3.50
	.9	,,	.77	4.35
	212.3	,,	.72	5.45
Se.	175.4	2n.	.97	56.60
	191.2	,,	.42	65.59
Eng.	190.0		.6	4.43
	193.8	5n.	.94	5.48
Gl.	200.3	5	.5	70.55
	203.0	5	.5	1.70
	208.4	10	.78	4.63
	.1	8	.84	.66
W. & S.	210.0	5	.77	2.49
	207.5	6	.7	3.48
	208.4	3	.7	4.65
	216.3	3	...	5.58
	215.5	5	...	.59
Sp.	207.4		0.84	.55
Schi.	.3	1n.	.84	.55
Du.	212.3	9n.	.99	.61

533   O.Σ. 317.

R. A.	Dec.	M.
16h 49m	44° 36'	7.2, 11.8

Considerable change in both angle and distance. O.Σ. in 1874 observed a third star of the 8th magnitude. (*See* Measures.)

A B.

O.Σ.	235.6	1n.	15.86	1845.73
	.0	,,	.61	7.69
	232.9	,,	...	52.69
	226.4	,,	16.91	74.74

A C.

O.Σ.	318.1	1n.	13.67	4.74

534   O.Σ. 318.

R. A.	Dec.	M.
16h 51m	14° 18'	6.7, 9.3

C. A, yellow.

Mä.	254.0	2n.	2.48	1845.44
	253.7	1n.	.49	52.61

O.Σ.	250.9	3n.	2.74	1847.74
De.	253.1	,,	.51	66.22
Du.	255.8	2n.	.68	70.96

535   Σ. 3107.

R. A.	Dec.	M.
16h 52.5m	4° 10'	8.5, 8.5

Σ.	112.3	3n.	1.6	1831.87
Ka.	...		.27	41.68
	162.9		...	2.81
Mä.	92.0		...	2.43
Se.	104.0	1n.	0.13	59.51
De.	.4		1.32	64.53
O.Σ.	109.0	1n.	.58	8.50
	106.4	,,	.44	75.48
W. & S.	104.7		.5	3.47
	102.0		.6	5.58
Gl.	104.0		.4	.70

536   O.Σ. 321.

R. A.	Dec.	M.
16h 54m	14° 29'	7.7, 8.7

Mä.	14.9		0.37	1843.29
	18.0		.3	52.61
O.Σ.	1.7	3n.	.51	48.82
Se.	4.6	2n.	elongd.	57.59
De.	5.2		,,	66.45

537   Σ. 2118.

R. A.	Dec.	M.
16h 55.8m	65° 13'	6.4, 6.9

The distance has diminished from 0".85 to 0".27; in the angle there has been but little change.

H1.	245		...	1781.76
	251.5	1n.	...	3.26
H2.	242.6		0.63	1830.32
	246.1		.70	1.37
Σ.	.4	5n.	.84	2.30
	247.0	3n.	.70	6.75
Sm.	245.0		.8	2.41
	243.7		.7	9.72
Da.	252.0		...	4.57
	242.9		...	40.77
	241.6	3n.	0.61	54.81
O.Σ.	245.3	,,	.77	41.24
	235.7	2n.	.58	59.67
	238.0	1n.	.27	72.42
Mä.	248.4		.8	43.32
	244.6	2n.	.65	7.97
Mit.	243.3		...	.71
	244.6		0.65	.97
De.	241.0		.61	54.81
	single			62.50
	,,			3.49
Se.	240.1	3n.	...	57.35
W. & S.	303.2	2	0.4	74.70
Du.	single			6.68

538 Σ. 2114.

R. A.	Dec.	M.
16^h 56.2^m	8° 37′	6.2, 7.4

C. white.

H$_2$ in 1831 found it "extremely difficult." Da. thought it was not a binary, and noted the fact that Σ.'s measures differ largely *inter se*. A small change in angle with sensible change in distance. (O. Σ.)

Dunér has computed the following formulæ:

$$1850.90. \quad \Delta = 1''.25.$$
$$P = 143°.9 + 0°.3125\ (t - 1850.0).$$

Σ.	137.9	In.	1.51	1825.53
	134.5	,,	.26	9.53
	128.1	2n.	.32	31.61
	140.0	,,	.32	2.52
	139.0	In.	.28	3.45
H$_2$.	132.9		.91	1.37
	134.0		0.84	2.41
Sm.	137.0		1.5	.41
O.Σ.	145.1	4n.	.19	41.81
	.4	In.	.26	2.60
	150.7	,,	.27	61.52
	156.5	,,	.58	8.52
	153.9	,,	.29	75.48
Mä.	143.5	4n.	.23	42.41
	142.8	2n.	.28	52.22

Mä.	142.7		1.25	1853.35
	.9		.18	5.63
	148.2	In.	.10	61.56
Da.	143.3	,,	.4±	47.63
Mit.	139.7	,,	0.83	.57
De.	147.1	3n.	1.2	55.52
	145.7	4n.	.07	6.49
	147.5	3n.	.30	63.39
Se.	145.2	,,	.25	56.84
Mo.	147.3	In.	.31	9.30
Du.	149.6	5n.	.28	69.79
W. & S.	151.9	9	.6	73.46
	152.2	6	.39	.47
	151.7	4	0.9	4.62
	146.5	4	1.33	2.49
Gl.	152.2	10	.6	4.66
	153.2	6	.3	.69
Schi.	151.4	In.	.16	5.57
Sp.	.5		.17	.58
Dob.	153.3	3n.	.10	7.48

539 Σ. 2120.

R. A.	Dec.	M.
17^h 0^m	28° 15′	6.4, 9.2

Magnitudes.—De. has 6.8, 9.7, and suspects variability in B.

C. Σ., A, yellow or red; B, fine blue.
De. gives A white, B blue.

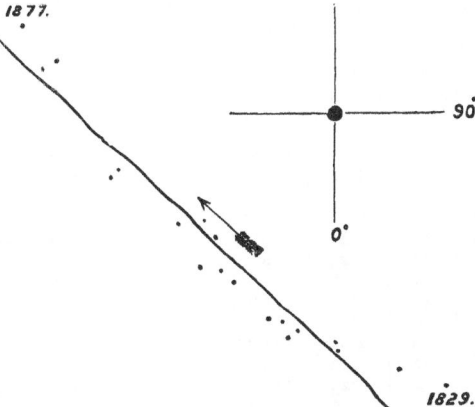

A beautiful double star discovered by Σ. In the *M. M.* he gives his measures from 1829 to 1835, and from them alone he inferred change both in distance and angle. Treating the distances by the method of least squares, he finds the formula $x = 3''.445 - 0''.112\ (t - 1833.25)$. Computing the distances for the epochs of the observations, he finds the agreement between the observed and computed values all but perfect. The motion he found was retrograde. He notes the great discrepancies between the angles made in autumn and those in spring.

At p. 293 he gives his measures in 1835 and 1836, and pronounces the change in angle and distance beyond doubt.

O.Σ., in 1877, after having reduced the observations to 1850, found that the results were not well represented by uniform rectilinear motion: a term depending on the

336 DOUBLE STARS.

square of the time then being introduced, he was led to the following formulæ, which on the whole are satisfactory:—

$$\Delta A = -1''\!\cdot\!619 \pm 0''\!\cdot\!0203 - (0''\!\cdot\!1122 \pm 0''\!\cdot\!00130)(t-1850\!\cdot\!0) + (0''\!\cdot\!00042 - 0''\!\cdot\!000035)(t-1850\!\cdot\!0)^2.$$

$$\Delta D = +1''\!\cdot\!647 \pm 0''\!\cdot\!0195 - (0''\!\cdot\!1018 \pm 0''\!\cdot\!00125)(t-1850\!\cdot\!0) + (0''\!\cdot\!0010 + 0''\!\cdot\!000033(t-1850\!\cdot\!0)^2.$$

Dunér has made a special investigation of the movement of this star (see *Ofversigt af Svenska Vetenskaps-Akademiens Förhandlingar*, 1873). A similar special treatment by O.Σ. is found in vol. v. of the *Mélanges Mathématiques et Astronomiques de St. Petersbourg*. The former astronomer found the following two systems of equations:

$$\Delta \cos P = +1''\!\cdot\!7493 - 0''\!\cdot\!10298(t-1850\!\cdot\!0);$$
$$\Delta \sin P = -1''\!\cdot\!4902 - 0''\!\cdot\!10662(t-1850\!\cdot\!0).$$

$$\Delta \cos P = +1''\!\cdot\!7827 - 0''\!\cdot\!10242(t-1850\!\cdot\!0) - 0''\!\cdot\!000200(t-1850\!\cdot\!0)^2;$$
$$\Delta \sin P = -1''\!\cdot\!5252 - 0''\!\cdot\!10724(t-1850\!\cdot\!0) + 0''\!\cdot\!000209(t-1850\!\cdot\!0)^2.$$

In this investigation all observations (except those by O.Σ.) from 1828 to 1872 were employed.

According to the first formula the motion is rectilinear and uniform; the second represents a curve of which the centres of curvature are on the same side of the principal star.

An extract from Du.'s table showing the results of his latest formula compared with the observed angles and distances is subjoined.

1783·20	11″·88	42°·2	+1″·68	+8°·9	H₁.
1837·00	3 ·06	359 ·8	+0 ·01	+3 ·0	Σ.
1847·57	2 ·19	324 ·6	−0 ·13	−3 ·1	O.Σ.
1851·97	2 ·19	306 ·9	−0 ·09	−4 ·5	,,
1857·60	...	292 ·7	...	+0 ·8	De.
1865·09	2 ·98	272 ·9	−0 ·12	−0 ·2	,,
1871·10	3 ·73	263 ·5	−0 ·04	+0 ·8	Gl.
1873·53	3 ·88	261 ·0	−0 ·18	+0 ·2	W. & S.
1874·34	4 ·03	258 ·8	−0 ·13	−0 ·5	Gl.
·99	4 ·21	258 ·1	−0 ·04	−0 ·5	De.
1875·53	4 ·37	259 ·1	+0 ·07	+1· 0	Du.

On the whole, Du. is inclined to think that the movement is rectilinear and uniform. and that therefore the observation of H₁ is erroneous. A physical relation is, however, possible, but more good measures must be obtained before the question can be decided.

	°		″	
H₁.	42·0	1n.	11·88	1783·20
Σ.	11·4	2n.	3·83	1829·60
	8·2	1n.	·54	32·52
	2·5	4n.	·45	3·47
	4·9	2n.	·25	4·67
	358·4	,,	·18	5·38
	0·1	4n.	·10	6·55
	1·9	3n.	·20	5·65
O.Σ.	345·8	2n.	2·83	41·12
	324·6	3n.	·19	7·57
	314·5	,,	·25	50·00
	306·8	,,	·19	1·97
	296·5	2	·29	5·10
	290·7	2	·49	7·12
	281·5	1	·97	61·63
	269·2	2	3·48	7·70
	260·7	2	4·21	73·66
Mä.	347·0	3n.	2·8	41·44
	345·9	5n.	·76	2·49
	342·2	58n.	·62	3·57
	339·4	6n.	·52	4·38
	336·0	15n.	·52	5·42
	328·9	27n.	·46	7·15

	°		″	
Mä.	315·0	6n.	2·37	1851·09
	308·3	42n.	·35	3·00
	288·1	11n.	·37	8·28
	283·9	15n.	·88	60·97
Ka.	...	5n.	·56	42·55
	342·8	7n.	...	2·93
De.	297·8	3n.	2·85	55·51
	294·0	2n.	·6	6·34
	289·2	5n.	·5	8·47
	278·4	4n.	3·03	62·51
	275·9	6n.	2·99	3·39
	274·0	4n.	·87	4·48
	272·1	6n.	3·05	5·49
	269·2	9n.	·26	7·16
	265·7	4n.	·45	8·45
	·7	3n.	·51	9·55
	264·0	4n.	·79	70·45
	262·5	,,	·75	1·44
	261·2	,,	·96	2·42
	259·8	,,	4·08	3·50
	258·8	,,	·16	4·52
	257·5	,,	·26	5·45
Mo.	298·1	2n.	2·58	55·62
Se.	290·4	3n.	·49	6·76
	269·8	,,	3·60	65·59
	272·9		·56	·40
Eng.	268·7	3n.	·41	·36
Ro.	270·6		...	·51
Ta.	272·9	2n.	3·56	6·41
	273·0	,,	·58	8·41
	270·4	1n.	2·97	71·51

Ta.	255°·9	1n.	3″·89	1873·55
	258·3	2n.	·13	4·48
	259·9	1n.	·38	5·42
	257·6	,,	...	6·61
M.	261·2	6n.	4·05	0·38
	258·9	,,	3·80	·41
	253·4	,,	4·25	4·47
Gl.	264·0	7	3·8	0·60
	263·2	5	·6	1·21
	·4	5	·8	·50
	259·2	4	4·1	3·68
	258·6	10	·0	4·66
	·5	10	·0	·69
Du.	265·1	3n.	3·64	1·14
	259·1	7n.	4·49	5·53
Kn.	263·3	3n.	3·86	1·51
W. & S.	262·9	4	·78	2·48
	261·7	6	·65	3·50
	258·5	4	4·2	4·62
Schi.	257·9	1n.	·16	5·54
Sp.	·9		·16	·54
W.O.	255·9	1n.	·66	6·45
	257·2	,,	·55	·53
	256·7	,,	·58	·54
Dob.	257·3	3n.	·19	·51
Pl.	258·1	4n.	·30	·53
	256·5	5n.	·39	·65
Fl.	·5	1n.	·32	7·64

540   O.Σ. 323.

R. A.	Dec.	M.
17ʰ 1·6ᵐ	47° 8′	7·4, 10·5

Change in angle and distance.

O.Σ.	112·3	1n.	6·98	1845·73
	111·4	,,	·73	6·69
	112·8	,,	7·08	9·71
	108·6	,,	6·85	51·61
Mä.	284·9		5·50	45·71
	281·6		·50	9·40
	278·4		·5	52·69
De.	103·4	1n.	7·52	66·88

541   Σ. 2130.

μ DRACONIS.

R. A.	Dec.	M.
17ʰ 2·9ᵐ	54° 38′	5, 5·1

C. white.

Discovered by H₁ October 19, 1779.

H₁ (*Phil. Trans.* 1804, p. 364): "The change in the relative situation of the two stars of this double star is pretty considerable."

"The two stars being nearly of an equal magnitude, we can have no inducement to suppose them to be at very different distances from us. This makes it not probable

that the difference of their parallactic motion should be the cause of the change in the angle of position ; otherwise, the direction of that motion would be sufficiently favourable." (H₂ and So., *Phil. Trans.* 1824, p. 271.)

H₂ finds that their recent measures confirm the motion announced by H₁, the average amount per annum being — 0°·5792. "There can be little doubt of its being a binary system—a miniature of α Geminorum."

Having the measures made by So. in 1825 before him, H₂ finds the change in 4·55 years has been — 0°·44 instead of — 2° 36′ "which a computation founded on a mean motion of — 0°·5792 per annum would give." He thinks that the position for 1820 is not very reliable. (*Phil. Trans.* 1826.)

Σ. (*M. M.*, p. 51.) His own measures from 1826 to 1835 show that "the distance has diminished steadily." But his observations in 1836 "do not favour the opinion before expressed as to decrease of distance."

Smyth (*Cycle*, p. 380): "A geometrical rough-cast of the whole [of the observations] yields a period of about 600 years for the orbital revolution ; since the velocity has appeared to decrease to — 0°·3 per annum, and then to accelerate to — 0°·7, during this small S.W. portion of its orbit."

Certain change in both angle and distance. The diminution in the distance will probably soon be followed by a much more rapid angular change : hitherto the angular change has been very uniform. (O.Σ.)

Dr. Doberck has the following formulæ for this star :—

$$\Delta = 3''·44 - 0''·019\,(t - 1830·0).$$
$$P = 205°·32 - 0°·6274\,(t - 1830·0) - 0°·001532\,(t - 1830·0)^2.$$

"The latter formula represents very nearly the five positions on which it is based," and the comparison with the measures is very satisfactory.

The proper motion of μ is — 0″·12 in R. A., and — 0″·07 in N. P. D.

H₁.	23·2	1n.	4·35	1781·73
	219·5	,,	...	1802·17
	221·0	,,	...	4·09
	215·9	,,	...	·10
H₂ & So.	208·4	6	3·90	21·38
	·9	35	4·33	5·52
Σ.	210·8		...	1·80
	207·2	1n.	3·61	6·89
	·8	,,	·15	8·73
	209·2	,,	·25	9·94
	204·4	2n.	·20	32·43
	203·6	1n.	·24	3·37
	·0	3n.	·23	5·39
	202·8	4n.	·35	6·78

Sm.	206°·7		3"·6	1830·79
	200·3		·3	9·53
	191·6		·0	47·51
	190·7		·0	54·48
O.Σ.	199·5	3n.	3·20	40·83
	196·1	1n.	·06	2·73
	191·1	,,	2·82	51·74
	182·0	2n.	·75	61·47
	181·5	1n.	·70	6·73
	177·9	,,	·66	72·42
	176·2	,,	·50	4·73
G.O.	199·2	26	3·13	40·59
Ka.	198·4		2·98	1·65
	199·3		·95	3·88
	179·7		·66	66·75
Mä.	197·1	2n.	3·17	43·36
	195·9	,,	·06	5·47
Mit.	190·9	1n.	2·90	7·63
Fl.	·1	28	3·09	51·75
Mi.	188·0	48	2·97	2·25
De.	·2	5n.	2·89	3·71
	·1	6n.	3·13	4·71
	187·3	4n.	2·92	6·63
	182·7	2n.	·73	62·80
	·2	3n.	·56	3·39
	181·3	12n.	·59	4·76
	178·3	1n.	·49	9·60
	176·3	2n.	·49	71·56
	175·1	,,	·49	2·52
	173·8	3n.	·51	3·67
	172·7	,,	·39	4·59
	171·8	2n.	·49	5·75
Mo.	188·4	30	·93	54·68
Se.	·4	4n.	·75	7·51
	181·0	3n.	·72	65·99
M.	...	1n.	·84	2·41
	355·5	,,	·91	7·37
	173·2	,,	·81	9·76
	177·8	2n.	·82	72·49
Ro.	...	1n.	3·03	65·72
Ta.	175·1	,,	·00	6·40
	177·8	3n.	·26	8·51
	176·2	1n.	2·62	71·51
	180·6	,,	...	5·52
Du.	179·6	3n.	·69	68·36
	176·8	8n.	·65	71·80
	172·5	2n.	·62	5·65
W. & S.	177·5	4	2·44	1·56
	174·0	8	·7	2·50
	·4	4	·5	·78
	173·7	4	·8	3·50
	172·3	4	·85	4·62
	171·8	4	·64	5·57
	·9	6	·53	6·58
Gl.	172·2	9	·9	4·66
	·6	7	·9	·69
	·4	3	·9	·70
Dob.	171·0	4n.	·68	6·54
	·0	2n.	·49	7·64

542   O.Σ. 324.

R. A.	Dec.	M.
$17^h\ 3^m$	$31°\ 23'$	6·3, 10·8

C. A, yellow.

O.Σ.	221°·4	1n.	3"·79	1845·47
	219·8	,,	·99	6·69
	212·8	.,	4·03	53·40
	217·7	,,	3·73	68·61
De.	219·9	3n.	·98	7·12

543   Σ. 2135.

R. A.	Dec.	M.
$17^h\ 7^m$	$21°\ 22'$	7·1, 8·4

C. A, yellowish ; B, bluish.

Dunér gives the formulæ

$$1852·51. \quad \Delta = 6''·78.$$
$$P = 168°·9 + 0°·133\ (t - 1850·0).$$

Σ.	166·1	4n.	6·70	1829·45
Mä.	167·2	3n.	·60	43·12
	168·9	1n.	·82	50·69
Mo.	171·1	2n.	·95	5·61
Se.	170·5	,,	·86	6·98
	169·5	1n.	7·03	65·38
Du.	171·1	5n.	6·79	71·34

544   36 OPHIUCHI.

R. A.	Dec.	M.
$17^h\ 8^m$	$-26°\ 25'$	A 4½, B 6½, C 7½

C. A, ruddy ; B, pale yellow (Sm.)

It is probable that one of the two brighter stars is variable in its light.

H_2 and So. (*Phil. Trans.* 1824, p. 272). Measures in 1822 and 1823 are given. On the 10th April, 1823, "the measure of a distant small star of the 10th magnitude was 19° 5' n.p., and 3' 0"·735," and this "will serve to verify the proper motion of A (36), which has been supposed in some way connected with the star 30 Scorpii, though at a great distance (12') from it, by reason of an observation of Bessel, that they have a common proper motion."

In 1825, however, more measures were made, and the distance of C from B as given above was found "decidedly wrong." Then follow many measures connecting A with 30 Scorpii ; a diagram is given, and the proper motions and their effects examined at great length ; and he shows that 36 Ophiuchi and 30 Scorpii are "journeying together through space."

Smyth (*Cycle*, p. 381): "Mayer made the two stars to be exactly on the same meridian [in 1780], with a difference of declination = 13" : this accidental statement was the cause of considerable error ;" for this position, combined with those of H_1 and So., seemed to indicate direct motion. Smyth's measures, however, "show a motion exactly contrary," and an observation made at his request by the Astronomer Royal in 1843 confirms this. Sm.

gives the following measures of a small sta^r in the neighbourhood of B :—

A C 289°·9, 193"·8.

This small star is double having "a most minute *comes* near the s.f. vertical." (Sm.)

The proper motion of A is — 0^s·029 in R. A., and + 1"·20 in N. P. D.

Bessel first pointed out the fact that a common proper motion animates 36 Ophiuchi and 30 Scorpii (see his *Fundamenta Astronomiæ*). The differences between these stars observed since Flamsteed's time are as follows according to Flammarion :—

Diff. in R. A.	Diff. in Dec.	Observer.
1690 + 13′ 32"·4	+ 2′ 56"·0	Flamsteed.
1755 ...	+ 3 2 ·7	Bradley.
1756 + 13 13 ·1	...	T. Mayer.
1800 + 13 7 ·0	+ 3 4 ·2	Piazzi.
1831 + 13 11 ·4	+ 3 3 ·6	Smyth.
1839 + 13 10 ·6	+ 3 4 ·4	,,
1860 + 13 7 ·8	+ 3 6 ·05	Greenwich.
1864 + 13 7 ·0	+ 3 7 ·24	,,

The proper motions of A, B and 30 are respectively

$$
\begin{array}{ll}
- 0^s\cdot029 \text{ in R. A.,} & \text{and} + 1''\cdot20 \text{ in N. P. D.} \\
- 0\ \cdot043 & + 1\ \cdot05 \\
- 0\ \cdot042 & + 1\ \cdot17
\end{array}
$$

Several small stars * are seen in the neighbourhood of this remarkable system : C, of the 10th magnitude, distant about 200" from A ; D of the 10th magnitude, and E of the 12th. From the investigations of Flammarion it appears probable that C and D are fixed, and that E partakes of the common proper motion of the system.

A B.

H$_2$ & So.	227·3	12	5·50	1822·52
	228·5	15	·2	4·86
Sm.	226·1		·2	31·57
	221·4		·0	5·33
	219·5		·3	9·28
	216·6		4·9	42·46
	213·8		·6	57·30
Da.	219·3		·78	41·59
Airy.	213·3	in.	5·32	3·52
Ja.	216·1		4·66	6·21
	214·9		·49	50·62
	·4	10	·23	4·07
Mit.	215·8	In.	·27	47·62
Bond.	·5	,,	·34	8·55
Mo.	213·0	20	·45	54·69
	·3	24	·40	8·42
Se.	212·9	2n.	·59	6·58
	211·3	In.	·25	7·56
	208·6		·2	66·72
Po.	210·0		·62	1·06
De.	212·4		·22	2·40

* Challis in 1839 detected four small stars, in addition to those seen by Sm.

M.	218·8	In.	4·2	1862·43
	209·0	,,	·41	8·49
	208·2	,,	·69	9·51
	206·0	,,	·21	72·49
	209·1	,,	3·93	3·73
W.O.	205·8	7	4·99	63
	202·2	In.	·47	76·55
Ta.	210·6		5·00	1·51
W. & S.	204·2	4	4·6	2·51
	·5	4	...	·52
	·1	2	...	·53
Schi.	203·5	In.	4·25	5·58
Sp.	·6		·25	·58
C.O.	204·3	2n.	5·17	6·54
Dob.	200·1	In.	3·98	7·44
Fl.	204·1	,,	4·28	·50
Pl.	203·3	3n.	·16	6·57

A C.

Sm.	289·9			193·8	31·57
Ja.	298·3		4	180·0	54·07

B C.

Ja.	296·8		2	...	54·07

545 Σ. 2140.

α HERCULIS.

R. A.	Dec.	M.
17^h 9·1^m	14° 32′	3, 6·1

H$_1$: "Aug. 29, 1779. Double. On May 2, 1781, Dr. Maskelyne very politely offered to show me a double star which he mentioned having discovered about four years ago." This was α Herculis.

"Not the slightest change of relative position since 1779." (O.Σ.)

Dunér's formulæ are

$$1851\cdot83. \quad \Delta = 4''\cdot58.$$
$$P = 118°\cdot0 - 0°\cdot080 (t - 1850\cdot0).$$

H$_1$.	117·2	In.	5·04	1782·69
	121·9	,,	...	1803·40
H$_2$ & So.	119·5	3n.	·26	21·74
Σ.	118·4	12n.	4·64	9·63
	119·4	13n.	·63	35·74
Be.	118·5	6n.	·99	0·92
Da.	119·7	4n.	·90	1·25
	118·5	3n.	·65	48·52
Sm.	119·4		·6	32·51
O.Σ.	·6	3n.	·76	40·73
	121·2	2n.	·77	1·62
	118·0	In.	·69	2·60
	117·6	,,	·68	5·65
	118·0	,,	·69	51·91
	116·9	,,	·62	2·67
	117·3	,,	·59	3·57
	116·8	,,	·70	8·59
	118·0	,,	·69	61·63
	·0	,,	·68	5·72

Mä.	118°.8	11n.	4.42	1842.67
	117.5	2n.	.82	6.97
	.9	,,	.51	52.65
	116.6	14n.	.44	6.67
	117.3	11n.	.57	61.71
Hi.	119.8	2n.	.69	45.69
Mit.	117.6	1n.	.93	7.61
Po.	118.0	2n.	.68	6.66
Mo.	116.6	1n.	.92	52.62
	118.1	3n.	.57	7.62
Ja.	117.9	36	.51	3.30
	.7	70	.56	7.00
De.	118.2	5n.	.62	3.63
Se.	117.7	6n.	.74	6.32
Eng.	.9	1n.	.86	64.52
Ka.	115.4	7n.	.64	5.69
Du.	.7	5n.	.44	9.07
W. & S.	.3	3n.	.68	72.86
Gl.	.6	,,	.60	4.68
Dob.	.2	4n.	.76	6.54

546   Σ. 3127.

δ HERCULIS.

R. A.	Dec.	M.
17ʰ 10.1ᵐ	24° 59'	3, 8.1

C. Σ., A, green; B, ashy white. De., A, clear yellow; B, blue.

H₂ and So. (*Phil. Trans.* 1824, p. 276): "There can be no doubt of a material change both in position and distance having taken place in this star : + 9° 42' in the one, and − 5″·349 in the other, are quantities too large to leave any room for doubt. The proper motion of δ, if correctly stated in Piazzi's catalogue, should have carried it in forty years, − 8″ in R. A. and − 5″·6 in declination, in the direction s.p., at an angle of 37° with the parallel. Had the small star then remained at rest, the angle of position, instead of 82°, would now have been only 54° s.f., and the distance 32″·3."

So. (*Phil. Trans.* 1826, p. 364). After recording the measures made by So. in 1825, H₂ says, "The change stated to have taken place in this star is confirmed by the present observations ; according to which, compared with those of 1821, a motion of + 1° 23' in angle and − 2″·175 in distance has taken place since our former measures. This is a remarkable verification of the relative motion both in position and distance ; and as the change is contrary to what the presumed proper motion of the large star would alone produce, this star merits particular attention."

Σ. (*M. M.*, p. 195). He gives his own measures from 1829 to 1835, and adds, "A notable decrease of distance, conjoined with a small increase of angle, is shown by these

measures." He finds that the distances computed from the formulæ $25''\cdot422 - 0''\cdot1766\ (t - 1833\cdot49)$ agree well with the observations.

Sm. (*Cycle*, p. 387): "My last epoch [1839·62] was under the very best atmospheric and instrumental circumstances : and on the whole I am led to infer that if all the series could be depended on, B had lately passed its apastron in the S.E. portion of its orbit, and that it is slackening its march as it recedes from the extremity of the ellipse, now barely moving a degree in ten years."

O.Σ. finds that the following formulæ represent the observations quite well, and hence that there has been no deviation from uniform rectilinear motion :—

$$\Delta A = +\ 1''\cdot233 \pm 0''\cdot027 - (0''\cdot0833 \pm 0''\cdot0020)\ (t - 1850\cdot0).$$

$$\Delta D = -\ 22''\cdot539 \pm 0''\cdot016 + (0''\cdot1618 \pm 0''\cdot0011)\ (t - 1850\cdot0).$$

Assuming that the relative change is entirely due to difference of proper motion, the minimum distance, 9″·2, will be attained in 1963. If, on the contrary, the stars form a binary system, the distance will continue to diminish for a shorter period.

The proper motion of δ is −0″·10 in R. A., and + 0″·12 in N. P. D.

Dunér gives
$$\Delta \cos P = -\ 22''\cdot65 + 0''\cdot1605\ (t - 1850\cdot0).$$
$$\Delta \sin P = +\ 1\ \cdot29 - 0\ \cdot0808\ (t - 1850\cdot0).$$

	°			
H₁.	...	1n.	34.69	1779.76
	...	,,	33.75	80.53
	162.5	,,	34.22	1.80
H₂ & So.	172.2	8	28.86	1821.36
	173.5	28	26.69	5.50
Σ.	...	2n.	27.84	1.85
	173.7	1n.	26.11	9.77
	174.1	3n.	25.63	31.67
	.0	2n.	.37	2.78
	.2	5n.	24.98	5.62
	.8	3n.	.88	6.58
	.3	,,	.58	7.74
Sm.	173.9		26.0	0.71
	174.9		24.7	7.49
	175.1		.5	9.62
O.Σ.	177.4	1n.	.06	40.83
	175.2	2n.	.27	1.61
	.8	1n.	23.95	2.60
	.7	,,	.40	5.68
	176.0	,,	.23	6.71
	177.7	,,	22.49	9.73
	.5	,,	21.98	53.83
	.2	,,	.73	5.64
	178.3	,,	.12	8.56
	179.2	3n.	.71	61.48
	180.7	1n.	19.28	8.67
	183.0	,,	18.52	74.58
	182.5	2n.	.61	5.48

	°		''	
Mä.	175·1	In.	24·17	1841·53
	177·1		23·24	7·32
	176·9	In.	22·03	54·69
	178·1	,,	21·33	8·61
	180·8	,,	20·98	62·74
Ka.	174·9	6n.	23·89	41·67
	175·9	7n.	·42	3·97
Ja.	177·0	11	22·31	52·73
	·6	10	21·99	3·15
	·2	18	·86	4·08
	·1		·55	6·22
	178·2		·28	7·94
De.	·5	4n.	·97	4·79
	176·0	2n.	·79	5·22
	178·6	In.	·64	·80
	·0	3n.	·71	6·47
	177·7	In.	·08	7·54
	178·6	3n.	·18	8·39
	179·4	4n.	20·55	62·75
	·2	5n.	·46	3·43
	·6	8n.	·18	5·48
	180·1	7n.	19·95	6·94
	·9	4n.	·68	8·49
	·9	,,	·48	9·54
	181·4	,,	·38	70·45
	·5	,,	·11	1·49
	·4	,,	·33	2·49
	·6	,,	18·82	3·50
	·5	3n.	·67	4·54
	·8	5n.	·59	5·54
Se.	178·1	3n.	21·63	57·22
M.	·0	In.	20·02	62·38
	181·0	,,	·07	74·44
Eng.	179·9	2n.	·34	64·42
Kn.	·6	,,	·19	6·74
	180·2	3n.	19·33	71·60
Du.	·9	4n.	·17	0·80
W. & S.	181·2	6	20·0	1·48
	180·1	7	19·3	2·48
	181·2	4	·2	·52
	·5	10	·3	·53
	·7	3	18·8	3·50
Gl.	·1	In.	19·0	·68
	·0	,,	·3	5·60
Dob.	·6	2n.	...	6·62
Fl.	·6	In.	18·43	7·00

547 Σ. 2145.

R. A. 17h 11·8m	Dec. 26° 43′		M. 8, 9·5	
Σ.	174·4	In.	9·72	1829·68
	·1	,,	·87	32·30
Mä.	177·0	3n.	10·61	43·65
	·2	4n.	·74	5·17
	176·9		11·29	51·69
	178·0		·26	2·33
	176·8		·12	4·78
	178·8		...	8·72
De.	·2		11·32	63·41
W. & S.	·8	In.	12·93	76·47
Fl.	179·1	,,	·64	7·45

548 Σ. 2153.

R. A. 17h 14·8m	Dec. 49° 26′	M. 8·6, 9·1

C. yellowish.

Dunér's formulæ are

$$1847·98. \quad \Delta = 1''·90.$$
$$P = 276°·7 - 0°·214 (t - 1850·0).$$

	°		''	
Σ.	282·3	2n.	1·67	1828·74
	280·2	In.	2·22	32·93
	282·0	,,	1·98	4·91
Mä.	277·3	2n.	2·06	43·70
	275·0	In.	·18	5·60
	·2	2n.	...	51·74
	·2	In.	1·91	4·78
Se.	276·5	2n.	·59	8·89
	270·0	In.	·96	66·84
De.	271·1	3n.	·92	·48
Du.	274·0	,,	·85	71·20

549 Σ. 2161.

ρ HERCULIS.

R. A. 17h 19·5m	Dec. 37° 15′	M. 4, 5 1

C. both white (H_1). A, greenish white ; B, greenish (Σ.)

This was one of the double stars known to Mayer and other astronomers before H_1 began his survey.

Piazzi enters it "Double, the smaller precedes."

It was first examined by H_1, Aug. 29, 1779.

H_2 and So. (*Phil. Trans.* 1824, p. 277). Measures from 1871 to 1822 are given. "It seems extremely probable that this elegant double star has undergone a sensible alteration in its position. The distance has increased materially."

Sm. (*Cycle*, p. 390). All the observations subsequent to 1824 "tend to prove its fixity."

Da. (*Mem. R. A. S.*, vol. xxxv., p. 399). H_1's distance in 1781 is probably much too small, and that of H_2 and So. considerably too large, "as is frequently the case." He thinks that the binary character of the star is doubtful.

Dunér gives

$$1853·02. \quad \Delta = 3''·80.$$
$$P = 309°·4 + 0°·114 (t - 1850·0).$$

	°		''	
H_1.	300·3		2·96	1781·79
	301·2		...	1802·17
H_2 & So.	307·0	8	4·46	21·38

Σ.	306·2	In.	3·68	1876·89
	·6	,,	·60	8·71
	308·0	2n.	·56	32·89
Da.	·4		·86	0·63
	·9	3	·77	40·83
	·3	5	·85	7·48
	309·7	5	·75	8·51
	308·9	5	·78	53·76
	·7	3n.	·86	9·72
Sm.	·5		·6	31·60
	·9		·7	9·74
	309·1		·8	47·61
	310·5		·5	53·79
O.Σ.	309·9	In.	·77	39·88
	310·1	2n.	·83	40·83
	311·7		·82	1·62
	310·1	In.	·62	2·73
	·0	,,	·62	5·65
	311·0	,,	·75	51·67
	310·4	,,	·66	9·62
	309·8	3n.	·70	61·52
Mä.	310·5	,,	·87	41·44
	309·7	,,	...	2·38
	310·4	2n.	3·72	3·49
	309·2	In.	·79	4·43
	·6	5n.	·65	5·36
	310·2		·74	6·51
	·4		·74	7·67
	·5	4n.	·68	8·45
	311·1	3n.	·62	52·13
	310·3		...	7·39
	·3	10n.	3·63	60·88
Ja.	·6		4·05	41·46
	309·0	20	3·72	52·78
Mo.	·2	36	·87	45·51
	307·0	30	·78	6·55
	309·7	60	·84	55·65
Po.	307·9	7n.	·81	46·09
D.O.	·1		4·14	·48
De.	309·9	5n.	3·50	53·66
	·3	In.	...	5·13
	·1	,,	3·91	·77
	·5		·61	62·30
Mit.	·4	In.	4·91	47·71
Se.	·7	2n.	3·83	56·60
Br.	·9	2	·75	68·59
M.	·1	In.	·61	2·37
	310·4	,,	4·02	8·46
	305·7	·,	3·23	9·61
	306·6	,,	·97	·76
	311·2	,,	·90	70·38
	308·9	,,	·93	·78
	311·2	,,	·74	1·37
	309·9	2n.	·79	2·49
	·0	,,	·93	3·72
	312·9	In.	·94	4·70
	311·2	2n.	4·05	5·73
Eng.	·5	4n.	3·81	65·51
Ta.	...	In.	·81	6·32
	306·6	,,	·85	·46
	308·2	,,	4·01	8·54
Ka.	311·0		3·67	6·68
Du.	310·6	2n.	·66	7·71
	311·9	,,	·69	8·66
	313·3	In.	·85	71·72

W. & S.	311·3	4	4·28	1871·52
	·9	3	·48	·53
	·6	4	3·96	2·51
	·2	6	4·04	·52
	312·7	4	·17	·52
	311·7	10	...	3·46
	·9	4	3·8	·63
	312·6	11	·85	6·48
Gl.	·0	9	4·00	4·70
Pl.	310·8	4n.	3·71	6·52
Schi.	312·6	In.	·65	5·54
Sp.	·6		·66	·54
Dob.	311·2	4n.	.87	6·54

550   Σ. 2165.

281 (B) HERCULIS.

R. A.	Dec.	M.
17^h 21·6^m	29° 34'	7, 8·5

A gradual increase in distance; motion rectilinear hitherto.

Dunér has

$$\Delta = 6''\!\cdot\!91 + 0''\!\cdot\!0125 \; (t - 1850\!\cdot\!0).$$
$$P = 48°\!\cdot\!7 + 0°\!\cdot\!182 \; (t - 1850\!\cdot\!0).$$

Σ.	45·6	In.	6·94	1829·68
	·5	2n.	·64	32·78
	46·2	In.	·62	3·43
Mä.	·4	,,	·75	40·61
	47·4	,,	·81	2·72
	48·7	,,	·68	3·41
	46·9	6n.	·80	5·42
	48·4	,,	·80	51·94
	47·3	4n.	7·09	9·75
Mo.	49·6	12	·0	7·48
Se.	50·0	4n.	·20	·62
De.	51·2	6n.	·10	64·57
Du.	52·9	4n.	·19	72·23
W. & S.	51·5	In.	·7	3·50
	53·7	,,	·6	4·63
	52·5	,,	·42	5·57
	·6	,,	·54	6·58
Gl.	53·5	3n.	·68	4·71
Fl.	·8	In.	·71	7·77
Dob.	50·7	4n.	·59	·54

551   Σ. 2171.

R. A.	Dec.	M.
17^h 22·6^m	−9° 54'	7·5, 7·6

Σ.	75·6	4n.	1·61	1830·53
O.Σ.	71·9	In.	·68	41·55
Mä.	72·1		·59	37·35
	73·7	In.	...	42·41
	70·1	,,	1·54	3·42
	65·0	2n.	·55	5·43
Mit.	68·1	In.	·41	8·58
Se.	70·0	2n.	·52	56·47

552 Σ. 2173.

221 (B) OPHIUCHI.

R. A.   Dec.
17^h $24\cdot7^m$ $- 0°$ $58'$

Magnitudes.—5·8, 6·1. O.Σ. suspects a variability in the light of these stars, and De.'s observations confirm it.

C. yellow.

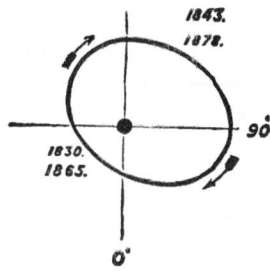

Discovered by Σ., and often measured by him without artificial light. He measured it easily in 1829 and 1832, but in 1836 with the finest sky it was single. "We have therefore a new example of occultation or very close conjunction, such as γ Coronæ, and ω Leonis, and others have presented. This star is worthy of the most careful attention."

Da., who regarded this object as a binary, could just discern a slight elongation in 1840. From that time he found that the distance increased.

Owing to the equality in magnitude of the two stars, and to a probable variability in one, it is difficult to determine whether the occultations in 1836 and 1864 embrace a revolution or merely represent two periastron passages, the smaller star having been alternately on the N. and S. side of the principal star. If the companion has already been on the N. side, the period is about 28 years. The measures of 1874, however, seem to favour the hypothesis that the period is about 46 years. The two passages through the apparent periastre thus divide the elongated orbit into two branches, one of which is passed over in about 28 years, and the other in 18 years. The distance has diminished since 1872, and 1875·65 gives a relation between the stars, the sign of the direction being changed, identical with that of 1829. (O.Σ.)

Dunér has computed the following elements :—

$$T = 1874\cdot35$$
$$\omega = 1°\cdot84$$
$$\mathcal{Q} = 152\cdot56$$
$$i = 80\cdot01$$
$$e = 0\cdot0839$$
$$\mu = - 7°\cdot630$$
$$a = 1''\cdot051.$$

Not satisfied with these, and the small excentricity of the true orbit rendering the graphical method uncertain, Dunér, with the aid of the method of least squares, has sought the general equation of the second degree which best represents the rectilinear coordinates deduced from the normal places: he finds the following :—

$$- 29\cdot8609\,x^2 - 8\cdot7106\,y^2 - 29\cdot8333\,xy$$
$$+ 0\cdot0622\,x + 0\cdot3363\,y + 1\cdot0000 = 0.$$

This equation gives the elements ω, $\mathcal{Q}$, i, e, a of the orbit: μ and T are found by another method. The results are—

$$T = 1872\cdot91$$
$$\omega = 7°\cdot26$$
$$\mathcal{Q} = 152\cdot65 \text{ (Equ. 1850·0)}$$
$$i = 80\cdot53$$
$$e = 0\cdot1349$$
$$\mu = - 7°\cdot9248$$
$$a = 1''\cdot009$$
$$P = 45\cdot43 \text{ years.}$$

With these elements the observations are compared, and the following extract will exhibit the resulting errors :—

1829·57	0''·62	327°·2	−0''·23	−3°·0	Σ.
31·68	·68	318·6	−0·05	−7·3	,,
40·64	·55	355·7	+0·13	+0·8	O.Σ.
51·60	1·07	330·7	−0·07	−1·7	,,
61·63	0·51	315·2	+0·09	+4·7	,,
67·79	·65	174·5	+0·17	+6·3	Du.
70·35	·85	336·5	+0·12	−2·5	Gl.
74·57	·85	151·3	+0·01	+0·3	,,
74·59	·90	331·7	+0·04	+0·4	W. & S.
76·63	·72	148·6	−0·04	+1·6	Du.

To aid those observers who wish to watch the star through its next minimum, Dunér supplies the following short ephemeris:—

1876·43	0''·78	147°·4
8·43	·62	142·1
1880·43	·40	131·8
2·43	·20	97·1
4·43	·25	21·8
6·43	·46	353·1
8·43	·68	344·8
1890·43	·87	340·4

Σ.	327·2	2n.	0''·62	1829·56
	321·6	1n.	·52	30·86
	318·6	,,	·68	1·68
	324·4	,,	·67	2·52
	single	3n.	...	6·69
Da.	167·0	1n.	0·5	40·47
	·4	2n.	·71	1·64
	163·3	3n.	·75	2·67
	161·2	6u.	·9	3·54
	159·4	1n.	1·10	8·45

O.Σ.	355·5	3n.	0·61	1840·64
	352·2	,,	·67	1·61
	344·7	2n.	·75	2·60
	339·8	1n.	·72	4·71
	334·0	2n.	·83	5·63
	338·5	,,	·93	6·69
	335·6	,,	·73	7·70
	331·5	1n.	1·07	51·60
	333·0	,,	·08	2·66
	149·2	,,	0·82	3·57
	151·3	,,	1·21	4·63
	146·6	,,	0·90	5·66
	326·0	,,	·85	6·58
	325·7	2n.	1·00	7·67
	·7	1n.	0·81	8·71
	323·4	2n.	·65	9·64
	315·2	1n.	·48	61·63
	190·8	,,	·53	5·72
	355·7	2n.	·44	6·62
	164·5	1n.	·58	7·47
	159·7	2n.	·68	8·53
	157·0	,,	1·02	71·51
	334·9	3n.	0·81	2·58
	329·3	2n.	·77	4·62
Mä.	172·4	6n.	·55	41·35
	169·8	3n.	·70	2·50
	168·2	8n.	·76	3·48
	165·0	3n.	·80	4·35
	162·9	9n.	·89	5·47
	159·4	5n.	1·07	6·46
	·2	2n.	·16	7·47
	154·1	4n.	·26	51·32
	·8	,,	·21	2·24
	150·5	3n.	·37	4·66
	148·3	2n.	0·88	8·62
Ka.	174·9	5n.	...	42·46
	165·2	10n.	...	3·65
	...	2n.	0·68	·71
Mit.	160·4	1n.	1·23	8·58
De.	330·0	5n.	·0	56·51
	single?			64
	,,			5
	161·1	2n.	0·5	8·60
	157·2	5n.	·58	9·57
	·0	6n.	·82	70·44
	155·0	4n.	·99	1·44
	152·3	5n.	·89	2·54
	150·8	4n.	·78	3·50
	·0	,,	·91	4·46
	147·5	,,	·74	5·53
Se.	145·9	2n.	·84	58·56
Mo.	·0	1n.	·25	·61
Eng.	160·0		·6	64·45
Du.	174·5	1n.	·65	7·79
	161·3	3n.	·65	8·66
	169·1	6n.	·66	9·63
	159·7	4n.	·77	70·67
	156·5	6n.	·84	1·64
	152·6	1n.	1·06	3·67
	148·7	5n.	0·87	5·67
	·6	4n.	·72	6·63
Gl.	336·2	5	·9	0·24
	·7	4	·8	·46
	330·9	10	·7	4·51
	331·7	7	·99	·63

W. & S.	334·1	7	1·10	1873·50
	·0	9	0·91	·51
	331·6	6	·90	4·55
	330·9	4	...	·63
	327·8	7	1·0	5·57
W.O.	·2	1n.	·07	4·66
	330·4	,,	·10	·72
	148·3	,,	0·66	6·45
	152·6	,,	·90	·53
	146·3	,,	·76	·54
Schi.	146·5	1n.	·82	5·57
	143·8	,,	·83	6·59
Sp.	146·5		·83	5·58
	143·8		·83	6·60
Dob.	331·4	3n.	...	·67
	333·4	,,	0·66	7·68
	322·5	1n.	·52	8·40
C.O.	141·6	2n.	·55	·49

553   O.Σ. 331.

R. A.	Dec.	M.
17^h 26^m	2° 55′	7·5, 9

Probable change in angle.

O.Σ.	324·1	1n.	0·88	1845·62
	330·3	,,	·88	6·69
	324·4	,,	·78	52·67
De.	332·6		·89	66·22

554   Σ. 2185.

R. A.	Dec.	M,
17^h 29^m	6° 6′	7, 10

Σ.	5·3	2n.	27·50	1830·49
Mä.	·1	1n.	·97	47·36

555   Σ. 2190.

R. A.	Dec.	M.
17^h 30·9^m	21° 4′	6, 9·5

Σ.	33·2	2n.	10·17	1829·66
H.	18·4		10	31·50
Mä.	33·6	1n.	·16	43·74
De.	28·0		·18	63·39
Kn.	23·3	4n.	9·63	72·40
W. & S.	24·9	1n.	10·58	6·53

556   Σ. 2192.

R. A.	Dec.	M.
17^h 35·4^m	29° 18′	7·5, 9·9

Σ.	88·4	3n.	10·34	1832·63
	88·2	2n.	·52	4·67
Mä.	85·7	,,	·09	43·72
	·5	3n.	·45	5·68
De.	76·2		·23	64·72
Gl.	73·4	1n.	·42	74·60
W. & S.	72·4	2n.	·4	·60
	73·5	1n.	·54	5·58

557   Σ. 2199.

R. A.	Dec.	M.
17^h 36·4^m	55° 49′	7·2, 7·8

Distance perhaps unchanged : a gradual diminution in angle.
Dunér's formulæ are

1858·50. Δ = 1″·52.
P = 108°·8 − 0°·378 (t − 1850·0).

Σ.	116·3	3n.	1″·66	1830·94
Mä.	111·4	4n.	·5	43·50
	·5	1n.	·5	5·09
	107·2	3n.	·52	52·37
	106·6	,,	·53	7·13
	103·7	2n.	·62	9·39
	102·8	,,	·62	60·87
	103·9	,,	·56	1·89
O.Σ.	111·2	3n.	·67	48·73
	101·2	2n.	·60	72·52
Se.	106·8	,,	·56	57·64
De.	101·4	3n.	·65	63·06
Du.	102·0	5n.	·38	71·38
W. & S.	99·6	10	·51	2·50
	98·4	8	·54	3·33
	100·7	7	·44	4·70
	·4	9	·73	5·58
Gl.	100·8	10	·51	4·73
	·7	6	·47	·79
	·9	5	·5	·80

558   Σ. 2203.

R. A.	Dec.	M.
17^h 37·5^m	41° 43′	7·5, 7·8

Probable change in angle.
Dunér gives

1849·66. Δ = 0″·71.
P = 332°·2 − 0°·140 (t − 1850·0).

Σ.	333·4	3n.	0·72	1830·13
Mä.	336·6	3n.	·78	·13
	334·9	,,	·78	43·31
	328·3		...	55·29
O.Σ.	335·9	4n.	0·79	41·13
	332·2	2n.	·85	54·51
	327·6	1n.	·77	72·61
De.	328·3	2n.	...	55·29
Se.	329·3	,,	0·63	7·18
Du.	330·3	3n.	·70	70·52
W.O.	325·8		·89	4·73
Sp.	327·8		·75	5·59

559   Σ. 2202.

R. A.	Dec.	M.
17^h 39^m	2° 38′	5·5, 5·8

Dunér's formulæ are

1851·60. Δ = 20″·47.
P = 93°·9 − 0°·020 (t − 1850·0).

H$_1$.	90	1n.	...	1781·55
H$_2$ & So.	93·5	2n.	20·52	1821·77
Σ.	94·1	4n.	·54	7·37
	·4	2n.	·25	36·61
Mä.	·1	3n.	·01	43·27
	·5	2n.	·10	52·63
	93·7	3n.	·25	5·68
	·0	1n.	21·06	61·56
De.	·8	,,	20·48	57·64
Eng.	·6	5n.	21·03	63·63
Du.	·4	3n.	20·38	9·30
W. & S.	·3	2n.	·60	74·55
Gl.	·5	,,	·84	·57

560   Σ. 2218.

R. A.	Dec.	M.
17^h 39·5^m	63° 44′	6·5, 7·7

Dunér gives

Δ = 2″·38 − 0″·0080 (t − 1850·0).
P = 354°·0 − 0°·084 (t − 1850·0).

H$_2$.	359·9		2·78	1831·36
Σ.	336·7	3n.	·50	2·72
	355·1	,,	·47	6·78
Sm.	356·7		·5	5·70
Mä.	334·5	3n.	·53	44·63
Se.	353·3	,,	·30	57·53
	358·5	1n.	·60	66·85
De.	353·0	2n.	·5	58·00
	351·6		·22	68·41
Mo.	353·1	2n.	·29	59·35
Du.	352·1	4n.	·08	72·34

561   Σ. 2213.

R. A.	Dec.	M.
17^h 40·3^m	31° 11′	7·5, 8

Dunér has

1851·48. Δ = 4″·43.
P = 332°·5 − 0°·064 (t − 1850·0).

So.	335·5	2n.	5·03	1825·47
Σ.	332·3	3n.	4·29	9·43
	333·3	,,	·45	36·60
Mä.	·3	3n.	·55	43·60
Se.	·1	3n.	·61	56·82
De.	331·2	2n.	·59	7·83
Mo.	332·2	12	·23	8·52
Du.	331·2	5n.	·40	69·49

562   Σ. 2205.

R. A.	Dec.	M.
17^h 40·4^m	17° 46′	8·3, 8 ·7

Dunér's formulæ are

$$1849\cdot56. \quad \Delta = 2''\cdot36.$$
$$P = 296°\cdot1 + 0°\cdot270\ (t - 1850\cdot0).$$

Σ.	291·4	2n.	2·52	1829·58
	290·1		·53	33·45
Mä.	293·4		·49	9·28
	294·1	2n.	·28	42·71
	295·5	3n.	·62	5·59
Se.	297·3	,,	·16	57·23
De.	301·6		·19	69·31
Du.	·6	3n.	·19	·31

563　　Σ. 2215.

R. A.	Dec.	M.
$17^h\ 40\cdot6^m$	$17°\ 45'$	5·9, 7·9

Change in angle uncertain.

Σ.	310·6	7n.	0·74	1831·53
	307·8	3n.	·81	5·99
O.Σ.	311·6	4n.	·85	41·56
	304·6	2n.	·77	70·54
Mä.	311·4	8n.	·75	42·22
Se.	304·6	3n.	·66	55·92
De.	·6		·67	·92
	306·0	1n.	...	6·52
Du.	307·0	4n.	0·74	68·45
W.O.	296·5		·91	74·66
Sp.	300·7		·70	5·54
Dob.	306·6	4n.	...	7·54

564　　Σ. 2220.

μ^1 HERCULIS.

R. A.	Dec.	M.
$17^h\ 41\cdot8^m$	$27°\ 48'$	A 3·8, B 9·5, C 10·5

C. Σ., A, yellow. Sm., B, cerulean blue.
Da., B and C, white.

H$_1$ IV. 41 forms the double star whose components are designated μ_2 and μ_1, the latter preceding and being the smaller or companion star. This object was measured by H$_1$, Σ., etc., the distance being about 30″. In 1856, however, Mr. Alvan Clark, with an aperture of 7¾ in., discovered that μ_1 was itself double. Σ. with apertures of 9·6 in. and 15 in. had overlooked it; and Mädler did not notice the duplicity of the companion.*

A B. Σ. measured it from 1829 to 1836, and thought there was no sign of motion.

Smyth found it "difficult to measure, especially in distance, from its bearing illumination badly."

He also gives as "the assigned values" of its proper motion

* Dawes in 1859 writes, " Seen double with A.C.'s 8-in. O.G." "Seen double with powers 312 and 697. Best measured with 697," and "very difficult in distance, as the small star bears but little illumination."

P	R.A.	− 0″·29	Dec.	− 0″·84
B		− 0 ·36		− 0 ·72
A		− 0 ·39		− 0 ·72.

Σ., on the ground that the two stars have a common proper motion, thought "it very probable that they are physically connected."

B C. Dawes says "probably binary," but in the notes to his observations in 1864 he writes, "*undoubtedly binary.* Annual motion = + 3°·12 ±."

O.Σ. The recent measures of B C indicate a very rapid revolution; this would produce changes in the relative situation of B and C in the period preceding the discovery of C by Clark which may perhaps explain why the star was not seen earlier. In 1860 the measures of B C presented no difficulty, but in 1873 their separation could not be effected on any but the finest nights. In the case of Σ. the fact that his measures were made with a bright field sufficiently explains how it was that he never detected this object. In A B the distance increased from 1830 till 1860, and then began to diminish. The angle seems to have increased up to 1850, and to have been nearly stationary since. The discrepancies in the distances are in part removed when the measures made before Clark's discovery are referred to $\dfrac{B + C}{2}$. Since 1859 the optical centre between B and C has remained fixed with reference to A.

A B.

	°		″	
H$_1$.	240		18	1781·77
So.	·8		29·30	1825·50
Σ.	241·0	1n.	·83	9·68
	·5	2n.	·91	32·55
	·8	3n.	30·25	6·51
Sm.	·8		·1	7·67
	242·9		·8	57·73
Mä.	241·6	3n.	·2	43·74
	·7	1n.	·14	4·43
	242·7	,,	·69	6·39
	·9		·27	51·89
	·8		·29	6·36
O.Σ.	243·8	1n.	·33	1·88
	244·6	,,	32·42	60·30
	·5	,,	31·62	2·83
	243·9	,,	·37	6·62
	244·6	,,	·04	·73
Se.	243·4	,,	·19	57·85
Da.	·2		·35	9·76
	242·6		...	60·87
	·5		...	1·42
M.	244·0	1n.	31·78	·46
Eng.	·4		·50	4·49
Kn.	243·8	2n.	·32	5·43
De.	·6		·32	6·86
W. & S.	244·8	3	·2	71·51
	245·2	2	·0	3·50
Fl.	·1	1n.	·15	7·75

B C.

	°		"	
Da.	59.2	2n.	1.81	1857.50
	60.3	3n.	2.05	9.70
	77.5	In.	1.80	64.43
Se.	71.7	,,	.74	57.85
O.Σ.	67.7	,,	.64	60.30
	78.5	,,	.50	2.83
	91.0	,,	.18	6.62
	88.0	,,	.02	.73
	98.7	,,	0.88	8.50
	156.8	,,	.62	71.52
	185.5	,,	.63	3.50
Eng.	67.5		1.7	64.49
Kn.	79.6	2n.	.84	5.43
De.	82.0		.2	.44
	98.7		0.88	8.50
		single		73.67
	216.0		0.83	6.68
	229.7		.87	7.54
W. & S.	100.0	In.	.6	1.51
	90.0	,,	.6	3.50
Gl.	100.9		.6	1.51
	.0	10	.4	4.63
	101.0	6	.4	.66
W.0.	202.4		.76	.48

A and $\dfrac{B+C}{2}$.

	°		"	
M.	244.0	In.	31.78	61.46
O.Σ.	.2	,,	.23	8.50
	.5	,,	.09	71.52

---

565   O.Σ. 337.

R. A.	Dec.	M.
17h 44.8m	7° 17'	7.5, 8

Probable small change in angle and distance.

	°		"	
Mä.	307.6		0.47	1843.37
O.Σ.	305.0	In.	.68	5.62
	304.3	.,	.67	.73
	306.0	,,	.52	51.67
	123.0	,,	.56	5.66
Se.	298.1	2n.	.42	7.05
De.	114.8	4n.	oblong	67.32
Du.	293.8	5n.	0.39	70.44

---

566   O.Σ. 338.

R. A.	Dec.	M.
17h 46.5m	15° 21'	6.6, 6.9

C. golden.

Considerable retrograde motion : a slight increase probably in the distance.

	°		"	
Mä.	43.0		0.58	1843.37
O.Σ.	44.2	4n.	.68	5.21
	38.9	3n.	.65	52.30
	36.1	,,	.70	5.63
	27.8	,,	.82	72.56

	°		"	
Se.	33.0	2n.	0.60	1857.05
De.	25.9	4n.	...	67.33
W.0.	.3		0.85	74.72
W. & S.	.4	7	.83	7.45
	26.6	7	.86	.46

---

567   Σ. 2262.

τ OPHIUCHI.

R. A.	Dec.	M.
17h 56.5m	−8° 11'	5, 5.7

C. H1, both pale red or white red; Sm., both pale white; Σ., yellowish.

H1. "April 28, 1783.—The closest of all my double stars; can only be suspected with 460; but 932 confirms it to be a double star. It is wedge-formed with 460; with 932, one-half of the small star, if not three-quarters, seem to be behind the large star. The morning is so fine that I can hardly doubt the reality; but according to custom I shall put it down as a phenomenon that may be a deception."

Σ. examined this object three times by day in 1825 without being able to see the companion. Nor was he more successful in 1827. In 1835.66 and 1835.67, however, Σ., his son, and the amanuensis all agreed that it was oblong; and a few days later "two stars of the 5th and 6th magnitude" were seen in contact. The powers used were 480 and 600. On trying 1000, "in moments of best definition," Σ. saw the stars separated. He notes that the motion is direct, and that the period may be from 80 to 90 years.

Sm. failed to elongate it in 1832; but in 1838 he found it measurable.

Da. says that the low altitude and oblique position render it difficult, although its distance has increased.

Sm. also observed a small star of the 10th magnitude :—

	°	"
1832	114.5	83.1
38	115 .0	82 .7.

THE ORBIT.—Hind in 1852 found that the period was about 120 years, and the excentricity 0.575.

Dr. Doberck in 1875 obtained the following definitive elements :—

$$T = 1818.50$$
$$\Omega = 67° 1'$$
$$\lambda = 36 \ 26$$
$$\gamma = 46 \ 8$$
$$e = 0.6055$$
$$P = 217.87 \text{ years}$$
$$a = 1''.193;$$

and a comparison of the elements with the observations from 1783 to 1871 shows very satisfactory agreement.

	°			
H₁.	331·6 ᶜ		elong^d·	1783·34
	360		,,	1802·74
	360		,,	4·44
	...		single	25·67
	146·0		elong^d·	7·28
Σ.	...	2n.	single	5·65
	356	In.	wedg'd	·71
	326	,,	,,	7·28
	196·7	,,	oblong	35·65
	193·5	,,	,,	·66
	190·3	,,	0·35	·67
	·2	,,	in cont^ct·	·68
	192·4	,,	,,	·71
	196·6	,,	0·35 separat^d·	·71
	197·3	,,	0·35	6·42
	203·3	,,	·46	·64
	200·0	,,	·46	·68
	199·2	,,	·45	·70
	·7	,,	·46	·71
Sm.	...		round	2·55
	214·0		0·5	8·58
	227·0		·9	42·52
	238·8		1·1	55·34
O.Σ.	223·1	In.	0·94	40·51
	228·1	3n.	·86	1·61
	232·4	In.	·87	5·65
	230·5	2n.	·96	6·69
	233·9	In.	·97	7·82
	238·2	,,	1·19	51·67
	237·4	,,	·29	2·64
	241·9	,,	·18	·67
	236·1	,,	·20	4·70
	240·3	2n.	·30	5·67
	239·9	In.	·47	7·67
	240·5	,,	·42	·67
	·9	,,	·47	8·71
	242·7	,,	·64	9·63
	·9	,,	·43	61·63
	244·1	,,	·51	5·72
	243·3	,,	·75	6·62
	248·1	,,	·69	72·58
	251·1	,,	·63	4·67
Da.	221·5	4n.	0·88	40·68
	225·7	5n.	·79	1·66
	226·9	In.	...	2·64
	228·9	2n.	0·95	3·61
	232·7	In.	1·01	8·66
	238·0	,,	·22	54·67
Mä.	217·3		0·75	41·53
	225·6		·77	2·57
	228·8		·80	3·54
	229·8		·79	4·34
	238·6		1·27	52·66
	·3		·17	3·79
	·2		·09	4·71
	240·0		·44	8·64
	244·0		·29	61·60
Ka.	224·6		0·80	43·11
	228·9		·95	·61
	249·4		1·40	65·52
Ch.	218·1	In.	0·79	44·74
Ja	239·5		1·0	6·20
	230·7		0·96	6·69
	234·0	21	·0	50·77

	°		''	
Ja.	239·5	II	1·10	1852·65
	243·6		·41	8·20
Mit.	229·4	8n.	0·78	46·51
	·7	2n.	1·18	8·10
De.	238·1	3n.	·26	5·48
	240·5	6n.	·2	6·58
	241·3	4n.	·25	7·62
	·8	6n.	·16	8·52
	244·3	7n.	·36	62·60
	·8	6n.	·43	3·55
	245·6	8n.	·41	5·47
	246·1	Ion.	·43	7·06
	·6	4n.	·37	8·53
	247·4	,,	·43	9·57
	·8	,,	·42	70·50
	·3	,,	·55	1·50
	248·2	,,	·56	2·52
	·6	,,	·63	3·59
	·5	,,	·56	4·57
	249·0	5n.	·64	5·55
Se.	236·9	2n.	·27	55·55
	240·7	4n.	·20	6·24
	239·6	3n.	·26	·55
	245·8	In.	·30	60·77
	247·6	3n.	·60	6·72
Kn.	246·8	4n.	·20	3·57
Ro.	243·1	In.	·17	5·57
	...	,,	·29	·65
Ta.	246·2	,,	·65	6·43
	·3	,,	...	8·59
	248·4	,,	...	9·56
	·9	,,	2·11	73·55
	250·7	,,	1·48	4·57
Du.	248·1	6n.	·40	69·64
	250·9	3n.	·29	71·35
Gl.	247·6	5	·70	0·32
	·0	8	·50	1·73
	249·6	4	·60	3·60
	248·5	10	·66	4·80
M.	...		·74	0·48
	250·1	In.	·56	·49
	·3	,,	·60	·50
	253·5	2n.	·45	3·73
W. & S.	248·9	4	·67	2·49
	250·3	4	·70	·52
	·8	10	·71	3·52
	249·4	4	·64	4·63
	·9	5	·71	6·60
Schi.	248·9	In.	·61	5·60
	247·6	,,	·73	6·59
Sp.	248·9		·61	5·60
	247·6		·73	6·60
C.O.	250·4	In.	2·05	·2
	·5	8n.	1·90	7·61
Dob.	248·9	4n.	·70	6·61
	251·3	3n.	·47	7·53
Pl.	249·8	2n.	·64	6·55

568   Σ. 2271.

R. A.	Dec.	M.
17ʰ 57·7ᵐ	52° 51'	7·3, 8·3

Dunér's formulæ are

$$1851·27. \quad \Delta = 2''·07.$$
$$P = 264°·6 + 0°·127 \, (t - 1850·0).$$

H₂.	259·5	2n.	1·70	1829·71
Σ.	262·2	3n.	·87	31·48
Mä.	263·6	1n.	2·36	42·72
Se.	266·1	2n.	·53	59·72
	271·1	1n.	·43	66·85
Du.	266·9	4n.	·06	9·45
O.Σ.	·6	1n.	·37	70·87
	265·2	,,	·41	4·73

569   Σ. 2267.

R. A.	Dec.	M.
17ʰ 57·8ᵐ	40° 11′	8, 8

Σ.	234·1	3n.	1·41	1830·68
Mä.	236·8		·48	9·07
	239·5		·48	42·68
	237·0		·08	3·35
	238·1		·32	51·72
	237·2		·42	2·33
	241·7		·25	3·38
	243·5		·40	9·88
O.Σ.	·6	1n.	·65	40·84
	62·5	3n.	·54	1·55
	56·2	2n.	·47	54·51
	59·5	1n.	·33	70·87
Mä.	239·5	2n.	·48	42·67
	236·9	1n.	...	4·36
	...	2n.	1·59	5·46
Se.	240·0	,,	·46	57·64
Sp.	242·6		·00	75·63
W. & S.	241·1	6	·09	6·60

570   Σ. 2268.

R. A.	Dec.	M.
17ʰ 58ᵐ	25° 22′	8, 9

Σ.	218·2	2n.	18·12	1829·70
Mä.	214·7	,,	·87	43·75
	215·1	,,	·76	5·47

571   Σ. 2272.
70 OPHIUCHI.

R. A.	Dec.	M.
17ʰ 59·4ᵐ	2° 33′	4·1, 6·1

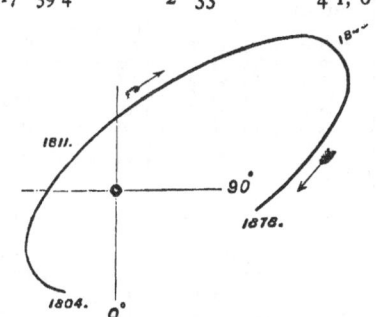

C. Σ., A, yellow ; B, purple.

H₁ (*Phil. Trans.*, vol. lxxii., p. 217):
"Aug. 29, 1779.—Double. Considerably unequal. With 227, 1⅜ diameters of L ; with 460, much above 2 diameters of L."

H₁ (*Phil. Trans.* 1804, p. 374): "The alteration of the angle of position that has taken place in the situation of this double star is very remarkable." The change amounted to 131° 59′ in 24 years and 234 days. "This cannot be owing to the effect of systematical parallax, which could never bring the small star to the preceding side of the large one."

H₂ and So. (*Phil. Trans.* 1824, p. 288). H₂ says that the angles of 1779 and 1781 contradict each other ; that that of 1779 is preferable ; that the motion seems exceedingly capricious, the diminution of angular velocity since 1821 being so great and sudden as almost to throw a doubt on the observations. He is unable to say which observation is in fault.

Having numerous observations by So. made in 1825 before him, H₂ finds the angular velocity greatly below that indicated by the observations up to 1820. An examination of the observations of distance leads him to put the distance in 1780 at 3″·5, in 1804 at 2″·5625, and hence to regard a decrease as established for that period. With the decrease of angular velocity there has also been an increase of distance.

An examination of the measures from 1779 to 1830 shows "the extreme uncertainty which must attend any determination of the elements of the orbit of a double star on principles which include the measured distances among the data."—(*Mem. R.A.S.*, vol. v.)

Σ. (*M. M.*, p. 98) : "A fine series of measures were made by Σ. between 1825 and 1835. He often measured the star during the early twilight.

Sm. (*Cycle*, p. 404) : "70 Ophiuchi was designated by the letter φ in the British catalogue ; but as there is no such letter in Bayer's map, Mr. Baily has properly rejected it in his late edition of Flamsteed."

"It may be stated in round numbers that 70 Ophiuchi describes its ellipse in a period of about eighty years."

H₂'s investigations led him to think that Σ.'s distances in 1818, 1819, 1825, 1826, 1827, and 1828 were "the only irreconcilable contradictions to the curve," owing to their being too small. Σ., on comparing his measures with the small telescope used in 1819 with those by the great Fraunfoper equatorial, found that the latter instrument gave smaller results than the other. He at last appealed to Bessel and the Königsberg heliometer : "A comparison of the distances of 39 stars, taken by both [observers], shows that those of Dorpat are, on an average, 0"·19 smaller than those of Königsberg."

Da. says, "one of the most interesting and beautiful of the binary systems."

The proper motion of A is thus given :—

	R. A.	Dec.
Piazzi	+ 0"·30	− 1"·17
Bessel	+ 0 ·26	− 1 ·09
Argelander	+ 0 ·22	− 1 ·10.

THE ORBIT.—Encke was the first to compute the orbit of this splendid object ; his elements are as follow :—

Perihelion passage ...	1806·877
Position of perihelion	283° 3'
Node	147 12
Inclination	46 25
Angle of excentricity	25 28
Semi-axis major ...	4"·3284

Mean annual motion   − 292'·43
Period ... ...   73·76 years.

In 1832 appeared a set of elements by H_2, obtained by means of his graphical method ; they are—

$$a = 4"·392$$
$$e = 0 ·4667$$
$$\pi = 292° 25'$$
$$\Omega = 137 \quad 2$$
$$\gamma = 48 \quad 5$$
$$\lambda = 145 \quad 46$$
$$n = − 4°·4812$$
$$P = 80·34 \text{ years.}$$
$$\tau = 1807·06 \text{ A.D. ;}$$

and the following selection from his table of comparison of elements and observations will exhibit the nature of the agreement :—

Date.	Angle.		Distance.		Observer.
	Observed.	Computed.	Observed.	Computed.	
	° ′	° ′			
1779·77	90 0	90 13	...	...	H_1.
1819·63	168 42	163 8	4"·66	3"·5	Σ.
21·51	156 50	157 3	...	...	H_2 and So.
22·54	154 30	155 0	...	...	,,
22·60	...	...	4·40	4·36	Σ.
25·56	148 12	147 12	...	...	H_2 and So.
27·40	143 54	144 8	4·51	5·20	Σ.
28·75	..	...	4·79	5·42	,,
30·50	137 28	138 13	5·65	5·61	Be., H_2, Da.

Mädler in 1835 published the following elements :—

Perihelion passage ...	1806·746
Position of perihelion	287° 14'
Node	133 47
Inclination	42 52
Angle of excentricity	28 30
Semi-axis major ...	4"·3159
Mean annual motion	− 267'·957
Period	80·61 years.

Discussing the measures made up to 1841, this eminent astronomer found that the law of gravity does not hold good in this system ; he found that the elements which were on the whole most satisfactory gave angles widely different from those observed between 1804 and 1823. (See *Ast. Nachr.*, No. 444.)

Hind and Jacob have arrived at the following results for this star :—

	Jacob.	Hind.
T	1807·60	1807·48
π	293° 17'	294° 6'
Ω	128 33	122 14
i	51 30	47 20
e	0·4820	0·4973
P	87·52 years	88·48 years
a	4·675	...

In 1868 M. Schur published the elements which follow (see *Ast. Nachr.*, No. 1681).

$$T = 1808·79$$
$$\omega = 155°·7$$
$$\Omega = 125·4 \ (1850·0)$$
$$i = 57·9$$
$$e = 0·49149$$
$$\mu = − 3°·8148$$
$$a = 4"·704$$
$$P = 94·37 \text{ years.}$$

From the ephemeris Dunér has constructed a table of comparison of elements and observations from which the following is extracted :—

			Differences.		
1868·72	4"·84	97°·6	− 0"·53	− 0°·6	Du.
70·51	4 ·45	94 ·0	− 0 ·67	− 1 ·7	Gl.
71·49	4 ·60	92 ·6	− 0 ·38	− 1 ·5	W. & S.
71·53	4 ·27	92 ·6	− 0 ·70	− 1 ·5	De.
73·57	4 ·01	88 ·6	− 0 ·66	− 2 ·1	W. & S.
73·51	3 ·89	88 ·8	− 0 ·78	− 2 ·0	De.
74·12	3 ·91	88 ·5	− 0 ·66	− 1 ·2	Gl.
76·62	3 ·15	81 ·5	− 1 ·01	− 3 ·0	Du.

	°		"	
H_1.	90·0		3·6	1779·76
	...		4·4	80·49
	99·2		...	1·73
	336·1		...	1802·25
	319·0		...	4·41
Σ.	168·5	5n.	...	19·64
	160·2	2	...	20·77

	°		"	
Σ.	157·6	5	...	1821·74
	153·9	3		2·64
	148·2	14n.	3·98	5·57
	145·1	2	4·37	7·02
	140·2	4	·78	8·71
	138·0	6	5·08	9·59
	135·7	2	·31	30·84
	134·7	5	·41	1·68
	133·9	3	·55	2·75
	131·1	4	·85	4·47
	130·7	5	6·10	5·60
	129·5	8	·13	6·66
H₂ & So.	156·2		3·68	21·31
	154·8		4·85	2·42
	153·6		...	3·35
	148·2		3·98	5·57
	142		5	7·51
	138·1		·95	30·36
	136·1		·97	1·52
	135·5		·49	2·57
	129·3		6·97	6·65
	120·8		·77	45·43
Be.	135·6	6n.	5·45	30·41
	·7	2n.	·51	·69
Da.	137·3	6n.	·53	·57
	132·7	5n.	·71	2·55
	·8	3n.	6·14	3·42
	130·6	7n.	·12	4·57
	125·8	2n.	·55	9·65
	124·8	4n.	·62	40·59
	123·4	,,	·63	1·68
	·3	2n.	·72	2·53
	118·8	3n.	·80	8·12
	114·6	7n.	·48	53·60
	113·7	4n.	·33	4·73
	112·6	1n.	·51	5·66
	113·2	2n.	·46	·69
	110·0	1n.	·38	7·57
	·2	2n.	·52	·58
	109·3	5n.	·24	9·72
Sm.	136·4		5·43	30·76
	132·5		6·0	3·59
	130·6		5·97	5·56
	·8		6·11	·60
	128·6		·19	6·81
	127·5		·26	7·64
	126·5		·25	8·51
	122·4		·64	42·55
	119·7		·8	7·48
	114·9		·5	52·44
Encke.	128·0	3n.	·46	36·50
	·2	5n.	·72	7·46
	126·6	7n.	·63	8·56
	125·2	2n.	·78	9·51
O.Σ.	·7	4n.	·34	·87
	127·1	7n.	·59	40·75
	125·8	3n.	·65	1·65
	124·7	5n.	·62	2·70
	121·5	1n.	·62	4·71
	120·9	4n.	·55	5·68
	121·3	2n.	·61	6·73
	119·7	,,	·50	7·76
	118·5	,,	·78	8·79
	117·8	3n.	·54	9·78
	115·4	5n.	·52	51·67

	°		"	
O.Σ.	115·0	5n.	6·55	1852·67
	113·6	3n.	·47	3·78
	112·8	,,	·54	4·69
	111·9	3n.	·49	5·66
	·7	2n.	·37	6·73
	110·1	4n.	·40	7·69
	109·8	2n.	·21	8·72
	108·5	3n.	·19	9·68
	106·4	1n.	5·88	61·63
	105·2	2n.	·85	2·77
	102·7	1n.	·32	5·72
	101·2	,,	·26	·80
	100·4	4n.	·29	6·66
	99·1	2n.	4·69	8·71
	93·6	4n.	·08	72·60
	87·4	3n.	3·78	4·69
Ka.	127·9	1n.	6·00	40·35
	123·4	,,	·53	1·74
	122·6	2n.	·48	2·59
	100·6	,,	5·31	65·62
Mä.	125·4	8n.	6·45	41·44
	124·4	2n.	...	2·39
	·7	4n.	6·25	·66
	123·3	14n.	·42	3·53
	122·5	3n.	·43	·72
	·0	5n.	·48	4·57
	120·8	17n.	·58	5·54
	119·8	1on.	·64	6·57
	118·2		·83	8·50
	116·7		·94	50·64
	115·4		·67	1·47
	·5		·67	·74
	114·7		·56	2·73
	113·3		·56	3·76
	·3		·31	4·68
	111·5		·32	6·50
	108·9		5·94	8·63
	106·0		·85	61·97
	105·2		·70	2·72
Ch.	125·0	2n.	·86	42·62
	123·5	3n.	6·69	3·56
	121·5	1n.	5·96	4·59
Hi.	120·1		6·14	6·46
D.O.	117·1		7·43	·56
	·2		·19	7·45
Mit.	120·3	1n.	5·53	·59
Ja.	·5		6·83	6·01
	115·1	10	·86	50·48
	114·0	15	·73	2·74
	113·6	21	·36	4·08
	111·9		·45	6·12
	·4		·39	·37
	110·6		·46	7·13
	109·7		·10	8·12
Bond.	118·1	1n.	·9	48·52
	117·7	2n.	·8	·52
De.	116·4	6n.	·45	53·54
	113·5	11n.	·26	4·58
	·1	1n.	·52	5·21
	111·7	6n.	·40	6·63
	109·7	4n.	·31	7·63
	·3	4n.	·09	8·44
	105·5	9n.	5·73	62·62
	104·2	,,	·60	3·51
	103·5	11n.	·46	4·60

	°		"	
De.	102·3	9n.	5·37	1865·51
	101·0	15n.	·17	7·01
	99·0	7n.	4·85	8·46
	96·5	8n.	·71	9·60
	94·5	,,	·56	70·51
	92·6	,,	·27	1·53
	90·7	9n.	·08	2·49
	88·8	8n.	3·89	3·51
	86·1	,,	·66	4·56
	83·7	9n.	·48	5·52
Se.	111·6	13n.	6·25	55·45
	106·3	3n.	·07	60·61
	101·1	,,	5·27	6·61
Flt.	112·3		6·36	57·42
Mo.	110·1	12	·15	·67
	108·6	12	6·08	8·39
Au.	109·0		·61	9·75
	107·9		·49	60·74
M.	·0	In.	5·89	1·46
	278·1	,,	·33	7·41
	277·8	,,	·27	8·46
	281·8	,,	4·96	·69
	275·1	,,	·83	70·45
	274·2	,,	·80	·47
	276·4	,,	·84	·50
	·6	,,	·65	·79
	270·8	2n.	·28	2·49
	268·9	In.	·20	3·41
	·7	2n.	·23	·73
	88·7	4n.	·00	4·47
	84·8	,,	3·84	5·65
W.O.	106·2	5	5·19	63
	·9	6	·74	63
	·3	6	6·04	63
	105·9	6	5·87	63
	·0	6	·72	63
	104·8	6	·74	63
Ro.	·5	In.	·76	63·55
	·5	,,	·30	5·53
	103·9	2n.	·24	·55
Eng.	104·8		·42	4·48
Ta.	101·8	6n.	·26	6·49
	100·5	In.	...	7·52
	101·1	2n.	5·49	8·64
	100·2	In.	·31	9·61
	94·4	2n.	4·62	70·60
	96·7	In.	·34	1·55
	84·7	,,	3·95	3·55
	88·6	,,	·67	4·58
Kn.	99·8	2n.	5·22	67·44
	98·5	,,	4·97	8·56
	94·5	3n.	·30	71·59
Br.	98·0	5	·92	68·90
Du.	97·5	4n.	·83	·72
	96·9	3n.	·58	9·69
	92·6	In.	·20	71·72
Gl.	94·7	8	4·6	0·30
	93·2	7	·3	·72
	92·9	5	·4	1·50
	93·1	5	·4	·63
	92·1	5	·08	·80
	89·5	6	3·90	3·51
	87·5	10	·92	4·73
W. & S.	92·8	3	4·6	1·48
	·5	4	·60	·49

	°		"	
W. & S.	...		4·61	1872·49
	90·7	9	·32	·50
	91·9	4	·17	·52
	·8	6	·07	·54
	88·8	5	·1	3·51
	·5	4	3·93	·63
	95·1	4	·49	5·62
	80·5	4	...	6·61
	·4	6	3·37	·61
Fer.	91·5		4·30	2·49
Schi.	84·1	In.	3·44	5·62
	81·2	,,	·34	6·59
Dob.	78·9	3n.	·46	·52
	77·6	,,	·46	7·52
	75·4	4n.	·03	8·54
Pl.	80·2	4n.	·55	6·54
	78·5	5n.	·39	7·65
C.O.	·5	4n.	·12	·68

572 Σ. 2275.

R. A.	Dec.	M.
$17^h 59·3^m$	39° 21′	9, 9·2

Σ.	127·9	3n.	1·08	1832·20
Mä.	126·8	In.	0·80	44·37

573 Σ. 2277.

401 (B) HERCULIS.

R. A.	Dec.	M.
$18^h 0^m$	48° 27′	6·3, 8·2

Dunér's formulæ are

$$1850·91.\quad \Delta = 27''·29.$$
$$P = 119°·09 + 0°·0630\,(t - 1850·0).$$

Σ.	117·9	3n.	27·59	1830·06
Mä.	118·6	In.	26·94	44·90
De.	119·9	,,	27·71	57·70
Du.	120·3	3n.	·03	71·49

574 Σ. 2278.

R. A.	Dec.	M.
$18^h 1^m$	56° 26′	6·8, 7·3, 7·8

C. white.

Dunér has the following formulæ:

$$\Delta = 38''·62 - 0''·0171\,(t - 1850·0).$$
$$P = 22°·99 + 0°·0348\,(t - 1850·0).$$

A B.

H_2.	21·8	3n.	38·78	1830·34
Σ.	22·5	,,	·92	1·56
Se.	23·0	2n.	·66	57·00
De.	·3	In.	·63	8·11
Du.	·7	4n.	·18	69·61

B C.

H₁.	147·1	2n.	6"65	1829·64
	146·5	1n.	5·60	31·73
Σ.	147·8	3n.	·97	1·56
Mä.	146·1	1n.	·87	44·90
Se.	147·1	2n.	6·24	57·00
De.	·3	,,	·12	8·11
Du.	·6	5n.	5·99	69·41
Ta.	·0	1n.	6·05	74·59
Dob.	145·8	2n.	...	6·61

575   O.Σ. 342.

72 OPHIUCHI.

R. A.	Dec.	M.
18ʰ 1·6ᵐ	9° 33'	4, 8

A very difficult object, if really double. O.Σ. has examined it with very great care on twelve nights since 1841. Sometimes the companion was readily seen, sometimes there was no trace of it, and on other occasions there was some appearance of the principal star being oblong. He suspects the companion of rapid variability.

O.Σ.	156·6	1n.	1·3	1842·72
	162·4	,,	·61	7·59
	168·1	,,	·6	·70
	169·6	,,	·49	51·67
	...	'simple'		2·63
Mä.	208·7		0·35	45·74
	206·6		·27	7·75
Da.	...	single		48
	...	,,		54
Se.	...	,,		6·53
	345·9	elongated		7·57
De.	...	single		64
W. & S.	...	round		73·51
	...	elongated		5·62
Sp.	...	single		6

576   Σ. 2281.

73 OPHIUCHI.

R. A.	Dec.	M.
18ʰ 3·6ᵐ	3° 58'	5·7, 7·2

Dunér's formulæ are

$$\Delta = 1''\cdot33 - 0''\cdot0112\ (t-1850\cdot0).$$
$$P = 255°\cdot5 - 0°\cdot181\ (t-1850\cdot0).$$

H₁.	265·7	2n.	...	1783·35
	264·3	1n.	...	1802·38
H₂.	257·6	2n.	...	22·46
So.	...	4n.	1·99	·93
Σ.	259·7	7n.	·48	34·86
Sm.	260·5		·7	·60
	259·9		·5	8·74
	255·0		·4	42·39

Mä.	257·8		1·35	1840·81
	256·1		·43	2·70
	258·9		·06	4·35
	254·5	4n.	·30	51·21
	·6		·28	·71
	252·0		·32	4·68
	254·1		·20	5·66
	253·7	14n.	·25	6·81
	252·7		·27	8·65
	248·7		·20	9·81
	252·8	11n.	·17	61·89
O.Σ.	254·2	5n.	·61	41·26
	251·6	4n.	·35	9·72
	247·5	5n.	·13	70·12
Da.	256·2	4n.	·47	44·58
	255·0	1n.	·38	9·48
Mit.	253·3	,,	·27	7·59
Flt.	256·2	24	·5	51·37
Mo.	253·9	1n.	·47	4·60
De.	·9	4n.	·2	5·69
	255·1	2n.	·35	62·69
	251·6	1n.	·24	5·34
	253·4	,,	·09	6·68
	252·4	,,	·05	70·66
	251·7	,,	·17	7·74
Se.	254·6	7n.	·33	56·27
	252·3	3n.	·47	65·92
Eng.	247·8	2n.	·91	4·48
Ro.	250·8	3n.	·50	5·54
Ta.	249·2	2n.	·61	6·46
	245·3	1n.	·64	74·61
Du.	253·9	6n.	0·92	1·07
W. & S.	252·8	6	1·45	·50
	253·7	4	·37	2·54
	·6	3	...	·54
	255·9	6	1·35	·55
	·6	3	·29	3·53
	254·2	4	·06	4·64
	·7	5	·12	5·62
	252·9	4	·19	6·55
	253·9	5	·14	·61
Gl.	254·4	8	·06	4·73
Sp.	250·4		0·93	5·61
Dob.	·0	3n.	1·03	6·61

577   O.Σ. 344.

R. A.	Dec.	M.
18ʰ 4ᵐ	49° 41'	6·7, 10·8

O.Σ.	156·2	1n.	2·22	1842·67
	·1	,,	·08	8·73
	153·1	,,	·20	51·67
De.	147·9	3n.	·21	66·55

578   Σ. 2289.

417 (B) HERCULIS.

R. A.	Dec.	M.
18ʰ 4·8ᵐ	16° 27'	6, 7·1

Very slow angular change.

Dunér's formulæ are

$$1853 \cdot 46. \quad \Delta = 1'' \cdot 06.$$
$$P = 238° \cdot 2 - 0° \cdot 212 \ (t - 1850 \cdot 0).$$

	°		″	
Σ.	243·1	4n.	1·20	1829·96
H₂.	241·6	1n.	·26	30·67
Mä.	240·1		·01	40·11
	239·1	12n.	0·95	3·49
	·0	2n.	1·15	52·63
	235·0	8n.	·03	6·79
	232·4		·05	9·81
	234·2		·07	60·86
	232·8	8n.	·08	1·99
O.Σ.	239·2	5n.	·33	41·11
	237·3	3n.	·16	54·69
Se.	236·1	4n.	·05	7·08
	235·3	1n.	·24	65·63
De.	234·3	3n.	·24	2·95
Eng.	235·8	,,·	·67	5·66
Du.	234·9	9n.	·05	70·59
W. & S.	235·7	4	·97	2·50
	236·3	5	·25	·50
	235·8	7	·03	3·54
	238·1	5	·14	4·64
	237·2	4	·12	5·65
	235·3	4	·39	6·55
	234·5	4	·33	·61
Gl.	237·6	7	·15	4·69
Dob.	235·9	4n.	0·12	7·50

---

579   O.Σ. 345.

R. A.	Dec.	M.
18ʰ 7ᵐ	5° 47′	7·3, 10·3

	°		″	
O.Σ.	65·8	1n.	1·02	1842·71
	64·7	,,	...	5·73
	·3	,,	1·07	7·59
De.	·8	3n.	·32	66·61

---

580   Σ. 2292.

R. A.	Dec.	M.
18ʰ 7·3ᵐ	27° 37′	8, 8·1

Dunér's formulæ are

$$1859 \cdot 06. \quad \Delta = 1'' \cdot 23.$$
$$P = 261° \cdot 3 + 0° \cdot 14 \ (t - 1850 \cdot 0).$$

	°		″	
Σ.	261·2	4n.	1·39	1830·40
O.Σ.	260·4	2n.	·60	40·78
	·3	,,	·44	1·57
	258·4	1n.	·62	56·56
	259·8	,,	·38	68·50
	263·6	,,	·71	73·53
Mä.	260·2	1n.	·22	42·63
Se.	·0	2n.	·25	56·55
De.	258·9	1n.	·5	6·78
Du.	264·2	8n.	·11	71·94
	265·7	2n.	·22	6·71
Lindstedt.	264·6	1n.	...	·75

---

581   Σ. 2294.

R. A.	Dec.	M.
18ʰ 8·4ᵐ	0° 9′	7·4, 7·7

Probable change in angle and distance.

	°		″	
Σ.	91·8	4n.	1·06	1831·00
O.Σ.	87·8	1n.	·04	41·55
	84·2	,,	0·85	65·80
	·9	,,	·66	72·56
Mä.	90·3		·80	43·60
Se.	93·3		·62	56·83
Sp.	89·2		·37	76·11

---

582   O.Σ. 349.

R. A.	Dec.	M.
18ʰ 12·2ᵐ	83° 54′	7·5, 8

	°		″	
O.Σ.	99·3	1n.	0·66	1842·73
	94·4	,,	·71	5·71
	92·3	,,	·48	51·73
De.	103·5	4n.	...	67·62

---

583   Σ. 2303.

R. A.	Dec.	M.
18ʰ 13·6ᵐ	− 8° 2′	6·7 9·2

	°		″	
So.	213·0		...	1825·20
Σ.	216·4	5n.	3·22	31·20
Mä.	219·4		·31	7·15
	220·6		·52	43·64
	225·7		...	50·75
	223·3		·50	1·77
Mit.	219·9	1n.	·22	48·62
Mo.	222·9		·56	55·64
	227·0		...	8·60
Se.	222·1	2n.	2·79	7·59
De.	223·1	1n.	3·23	64·59
	225·0	,,	2·94	6·68
	221·3	,,	3·15	7·61
	225·3	,,	2·70	72·52
C.O.	222·4	2n.	·79	7·51

---

584   Σ. 2311.

R. A.	Dec.	M.
18ʰ 17ᵐ	11° 23′	8·9, 9·9

	°		″	
Σ.	170·7	4n.	8·65	1830·30
Mä.	168·9	1n.	...	43·19

---

585   O.Σ. 347.

R. A.	Dec.	M.
18ʰ 19ᵐ	7° 10′	7·2, 11

	°		″	
O.Σ.	342·2	1n.	3·12	1847·59
	337·5	,,	·59	51·82
De.	347·6	3n.	·07	66·27

586   Σ. 2315.

R. A. 18h 20·2m	Dec. 27° 20'	M. 7, 8

Change in angle.

	°		"	
Σ.	281·1	4n.	0·60	1830·74
O.Σ.	271·4	,,	·68	41·12
	257·6	3n.	·52	63·73
Mä.	276·9		·54	41·71
	272·2		·45	3·43
	270·9		·30	5·45
	267·4		·26	52·02
	255·9		elong^d·	4·71
	257·6		0·3	9·84
Se.	260·7	2n.	elong^d·	7·05
	267·0	1n.	,,	65·63
Sp.	247·0		0·38	75·90

587   Σ. 2318.

R. A. 18h 20·6m	Dec. 25° 56'	M. 8, 10·2

Σ.	257·2	2n.	12·85	1829·74
Mä.	256·8		20·03	43·74
De.	255·7		·66	65·07
Bu.	254·9		.81	78·40

588   Σ. 2316.

R. A. 18h 21m	Dec. 0° 7'	M. 5·5, 7·8

C. A, yellow ; B, blue.

Dunér has these formulæ :

$$1851·69. \quad \Delta = 3''·88.$$
$$P = 315°·0 + 0°·05 \, (t - 1850·0).$$

H1.	314·5	1n.	...	1781·79
	312·4	,,	...	1802·34
H2 & So.	·3	4n.	...	22·72
	318·6	5n.	4·46	5·54
	319·8	1n.	5·10	9·0
Σ.	318·1	3n.	4·15	2·95
	314·2	6n.	3·94	8·62
	315·0	3n.	·83	36·22
Da.	314·2	,,	4·34	0·61
	313·8	1n.	3·94	40·69
Sm.	314·3		4·3	33·53
	·2		3·9	42·53
Mä.	316·5	1n.	·73	5·61
Mit.	317·0	,,	·54	7·61
Ja.	314·4	5n.	·66	52·76
De.	313·7	6n.	·96	4·70
Mo.	314·4	3n.	4·00	5·61
Se.	317·6	,,	3·79	7·89
Ro.	311·3	1n.	·81	65·57
Ta.	·4	,,	·95	6·53
	310·4	·,	...	9·55
	305·6	,,	3·92	70·65
Du.	317·6	7n.	·80	·76
W. & S.	316·1	4n.	4·08	3·06
Gl.	315·6	2n.	3·58	4·80

589   O.Σ. 351.

R. A. 18h 22·1m	Dec. 48° 42'	M. 7·3, 8

	°		"	
O.Σ.	31·7	1n.	0·49	1842·67
	22·3	,,	·51	4·85
	20·9	,,	·48	51·67
Mä.	44·5		·33	43·65
	42·1		·35	6·68
De.	30·6	5n.	...	67·69

590   Σ. 2323.

39 DRACONIS.

R. A. 18h 22·1m	Dec. 58° 44'	M. 4·7, 7·7

Very slow changes. Distinct retrograde movement.

A B.

H1.	9·6	2n.	...	1791·80
So.	4·0		3·6	1823·63
	5·3		·59	5·55
H2.	359·8		...	9·99
	0·2		...	30·26
Σ.	5·9	7n.	3·14	3·20
Sm.	5·5		·3	6·39
Mä.	4·1		·13	43·38
	·4		·3	7·42
	3·2		2·21	51·85
	4·9		3·45	2·34
	7·2		·36	3·39
	4·6		·03	8·80
	0·6		·33	61·45
Mo.	0·2	2n.	·22	45·71
	359·6	12	·45	56·58
O.Σ.	·3	1n.	·26	1·90
	2·0	,,	·86	66·73
	·1	,,	·70	70·87
Se.	·4	3n.	·35	57·59
M.	3·9		·76	65·44
W. & S.	3·4	3	...	75·74
	1·2	3	...	3·55
	0·7	4	3·97	·58
No.	·5		·85	6·80

A C.

H1.	26·6		...	1780·77
Σ.	22·2		...	1819·52
	21·7		88·96	34·27
So.	·9		89·61	23·46
	·4		88·94	5·55
Sm.	·7		89·2	36·39
O.Σ.	·4		·14	51·90
M.	·3	1n.	·69	66·48
W. & S.	20·7	2	...	73·58
	21·6	3	...	5·74
No.	·5		90·58	6·80

B C.

No.	22·4		86·85	76·80

591 O.Σ. 353.
φ DRACONIS.

R. A. 18ʰ 24ᵐ	Dec. 71° 17′		M. 4·8, 6·5
O.Σ. 63·6	6n.	0·56	1856·13
De. 62·9	5n.	...	67·73

592 Σ. 2330.

R. A. 18ʰ 25·7ᵐ	Dec. 13° 6′		M. 7·3, 9
H₂. 177·2.		20·94	1828·65
Σ. ·0	1n.	·19	·71
176·9	2n.	·38	9·58
Mä. 174·4		·09	47·57
175·0		·18	52·62
De. 174·5		19·26	64·90
W. & S. 173·4	2n.	·0	74·68
Gl. ·6	1n.	...	4·79

593 Σ. 2342.

R. A. 18ʰ 29·6ᵐ	Dec. 4° 50′		M. 5·7, 8·5
Σ. 12·0	1n.	26·60	1828·71
·1	,,	·80	9·62
·1	,,	27·17	31·69
11·7	,,	·06	2·81
Mä. ·9	2n.	·89	43·61
10·6	,,	28·32	5·46

594 O.Σ. 357.

R. A. 18ʰ 30ᵐ Dec. 11° 37′ M. 7·5, 76

Certain retrograde motion.

O.Σ. 275·5	2n.	0·48	1845·15
264·4	3n.	·53	55·67
256·9	2n.	·53	72·58

595 Σ. 2345.

R. A. 18ʰ 30·4ᵐ	Dec. 20° 59′		M. 8·4, 10·1
Σ. 182·6	1n.	7·43	1829·75
186·0		7·33	1·84
·0		·44	5·59
H₂. ·6		10	30·50
Mä. 188·0		·51	43·77
·1		·45	7·48
De. 194·1		·70	65·21

596 O.Σ. 358.

R. A. 18ʰ 30·5ᵐ Dec. 16° 54′ M. 6·8, 7·2

Considerable change in both angle and distance.

Dunér has

$$\Delta = 1''\cdot46 + 0''\cdot022\,(t - 1860\cdot0).$$
$$P = 209°\cdot9 - 0°\cdot55\,(t - 1860\cdot0) + 0°\cdot00822\,(t - 1860)^2.$$

Mä. 223·8		0·82	1843·54
O.Σ. 227·0	3n.	1·23	5·41
207·9	2n.	·73	63·16
203·9	,,	·83	72·58
Da. 218·6	1n.	·18	48·56
Se. 214·2	,,	·35	57·71
207·2		2·15	65·63
De. ·5	4n.	1·72	6·68
Du. 205·4	6n.	·67	9·58
203·3	,,	·65	71·02
·2	2n.	·81	5·69
Sp. 202·5		·67	·62
·1		·73	6·59
Pl. 200·8	3n.	·72	·57

597 Σ. 2346.

R. A. 18ʰ 30·5ᵐ	Dec. 7° 26′		M. 7·5, 9
Σ. 282·9	4n.	15·41	1829·64
Mä. 286·6		16·41	43·69
·1		17·04	52·62
287·2		·14	9·84
De. 288·2		18·06	64·83
W. & S. 289·4	2n.	19·0	74·65
·2	1n.	...	7·62
Gl. ·7	,,	19·34	4·79
Fl. ·5	,,	·26	7·80

598 O.Σ. 359.

R. A. 18ʰ 31ᵐ	Dec. 23° 31′		M. 6·6, 6·9
Da. 356·9	2n.	0·73	1848·56
358·6	,,	·60	54·16
O.Σ. 354·1	6n.	·66	49·54
Mä. 358·0	3n.	·69	51·77
Se. 357·6	2n.	·58	7·19
De. 173·3	4n.	...	67·91
Du. 352·6	8n.	0·57	70·80

599 α LYRÆ.

R. A. 18ʰ 32·8ᵐ Dec. 38° 40′

Magnitudes.—Σ., 1, 10·5; Sm., 1, 11.

C. Σ., A, bluish white. Sm., A, pale sapphire; B, smalt blue. H₁, A, fine brilliant white.

H₁ (*Phil. Trans.*, vol. lxxii., p. 223). "Sept. 24, 1781.—Double. Excessively unequal. By moonlight I could not see the small star with 278, and saw it with great difficulty with 460; but in the absence of the moon I have seen it very well with

227. L fine brilliant white ; S dusky. Distance 37″ 13‴. Position 26° 46′ s.f.″

On the 22nd of October H$_1$ applied a power of 6450 to his telescope and examined this fine star for fifteen minutes. He found the image "perfectly round, and occasionally separated from rays that were flashing about it." He was led to think that this star has light enough to bear a power of 100,000 with 6 inches of aperture. The diameter of the disc of α, taken with his new micrometer, was 0″·3553.

H$_2$ and So. measured this double star in 1822. From a consideration of the change indicated by the measures and the proper motion of α given by Piazzi, H$_2$ was led to believe, "First, that the proximity of the large and small stars is merely apparent and accidental, no connexion existing between them ; and, secondly, that the proper motions assigned to α are not very remote from truth."

So. measured it again in 1825, and Dawes has measures in 1830. Both H$_2$ and Dawes observe that the small star bears illumination well.

Smyth measured this pair in 1830, 1837, and 1843. He says (*Cycle*, p. 423), "α Lyræ is one of the insulated bodies, and is worthy of ranking with Sirius, Canopus, and Capella. Yet, by the experiments of Dr. Wollaston, it appeared that the light afforded us by this star is about $\frac{1}{180,000,000,000}$th part of the Sun's light, or only about one-ninth part of the light of Sirius, but still it offers a glorious blaze." Brinkley found its parallax between 2″ and 2″·52, Struve 0″·125, while Airy "has pronounced that its annual parallax is too small to be sensible to our best instruments."

Brünnow remarks that the proper motion of this fine star appears to have decreased : the movement as deduced by three eminent astronomers is here given :—

	R. A.	N. P. D.	Years.
Bradley	0″·2839	0″·2908	1750—1830
Argelander	·2661	·2675	1837—1852
Brünnow	·2414	·2643	1852—1869

The following are some of the values of the parallax which have been found :

Σ.	0″·261
O.Σ.	·147
Br.	·212
Pe.	·103.

H$_1$.	116°·8	38″	1782·36
	·2	43	92·31
H$_2$.	132·1	41·52	1822·87
	137·9	40	30·00
So.	133·5	41·13	25·56
Sm.	135·2	43·1	30·84
	137·9	42·7	7·51
	140·3	43·4	43·34

	°		″	
O.Σ.	144·5		44·16	1851·85
Se.	147·7	5n.	45·24	7·47
De.	150·2	6n.	46·16	65·63*
Br.	150° 58″·55		·23	69 †
W. & S.	151·8	4	57·64	71·58
	152·5	6	49·09	2·72
	·2	6	47·50	3·45
	154·0	3	48·4	4·70
	·5	5	...	·62
Gl.	·0	6	...	3·79
	153·8	10	48·5	4·79
	·8	9	·61	·80
	·7	4	·5	·85
	·4	4n.	...	75·71
	154·7	3n.	...	8·60
Fl.	155·1	1n.	48·14	7·00

600   O.Σ. 360.

R. A.	Dec.	M.
18^h 33^m	4° 45′	6·5, 10

	°		″	
O.Σ.	292·6	3n.	1·11	1849·67
De.	291·3	4n.	·40	67·16

601   Σ. 2356.

R. A.	Dec.	M.
18^h 33·6^m	28° 36′	8, 9

Dunér gives

$$1858·69. \quad \Delta = 0″·96.$$
$$P = 53°·2 + 0°·205\ (t - 1850·0).$$

	°		″	
Σ.	47·1	3n.	1·03	1831·42
Mä.	52·6	4n.	0·86	43·94
	53·2	2n.	·85	51·88
	57·9	3n.	·88	7·26
	58·7		·80	9·40
	66·3	3n.	·55	62·73
De.	single			3·53
	,,			5·31
Eng.	,,			·54
Se.	54·3	1n.	0·96	6·63
Du.	58·2	9n.	·99	70·65
W. & S.	54·9	1n.	1·13	4·69
	56·8	,,	0·98	5·62
Gl.	55·4	,,	1·18	4·80

602   Σ. 2367.

R. A.	Dec.	M.
18^h 35·9^m	30° 11′	7, 7·5, 8·4

Probable change in angle (A B).

A B.

	°			
Σ.	...	1n.	single	1829·75
	...	,,	,,	32·31
	58·7	,,	elong^d·	·80
	72·2	,,	,,	·86
	64·5	,,	,,	4·91

* "Distance corrected for refraction = 46″·171."
† Position and distance determined from a very large number of observations extending over fourteen months.

O.Σ.	79.3	In.	0.53	1841.65
	77.3	,,	.54	7.59
	76.7	,,	.51	51.62
	62.9	,,	.53	66.68
Mä.	70.4	elong^d.		43.77
De.	73.5	,,		64.67
Sp.	242.3		0.32	75.65

$$\frac{A+B}{2} \text{ and } C.$$

Σ.	193.9	5n.	14.13	32.53
O.Σ.	194.4	In.	.40	41.65
	193.7	,,	.23	7.59
	.3	,,	.28	51.62
	194.8		.20	66.68
G. Bond.	193.3	In.	...	51.62

603   Σ. 2384.

R. A.	Dec.	M.
18h 38.4m	67° 0′	8, 8.5

C. yellow.

A small double star discovered by Σ. He says "there is scarcely any doubt that change has here taken place."

In Mr. Bishop's volume, p. 141, Hind says, "Notwithstanding the strange coincidence of Σ.'s individual measures in 1836, t seems very doubtful whether any alteration has taken place since the star was first measured."

Dawes (*Mem. R. A. S.*, vol. xxxv., p. 408) writes, "Notwithstanding the strong doubts expressed by Mr. Hind, I cannot but regard this star as constituting a highly interesting binary system ; though its orbital movement is far less rapid than was indicated by the comparison of Σ.'s measures in 1832 and 1836.

O.Σ. (1876). The angle has not changed since 1836, but the distance has probably slowly diminished since 1832.

Σ.	307.1	3n.	0.82	1832.34
	318.3	5n.	.65	6.87
O.Σ.	315.1	In.	...	40.57
	319.0	,,	0.85	.61
	313.6	,,	.53	6.69
	321.3	,,	.46	51.67

604   Σ. 2375.

R. A.	Dec.	M.
18h 40m	5° 23′	6.2, 6.6

Dunér's formulæ are

$$1851.40. \quad \Delta = 2''.22.$$
$$P = 111°.2 + 0°.153 \, (t - 1850.0).$$

Σ.	107.8	2n.	2.22	1825.60
	.8	In.	.34	8.66
	108.6	2n.	.20	32.88
H₂.	103.1		1.88	28.65
	109.4		2.48	9.80
	108.8		.14	31.60
Da.	109.8		.27	2.56
	.5		.27	4.50
	111.5	In.	.07	48.56
Mä.	.4		.05	1.62
	.6		.27	2.54
	.4	6n.	.33	3.42
	112.9		.37	50.86
	111.9	4n.	.27	1.72
	.5		.24	2.01
	112.1	I In.	.28	7.26
	113.4		.30	9.81
De.	110.1	5n.	.28	4.59
Se.	111.6	In.	.23	6.92
	113.5	,,	.02	66.56
Mo.	.o	3n.	.33	58.68
M.	112.0	In.	.50	65.67
Du.	114.6	5n.	1.85	71.47

605   Σ. 2382.

ε¹ LYRÆ.

R. A.	Dec.	M.
18h 40.4m	39° 33′	4.6, 6.3

C. Σ., A, greenish white ; B, bluish white. H₁, A, very white ; B, a little inclining to red ; Sm., A, yellow ; B, ruddy.

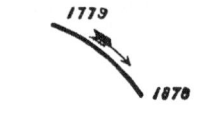

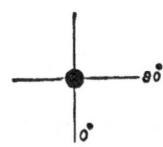

H₁ (*Phil. Trans.*, vol. lxxii., p. 217) : "Aug. 29, 1779. A very curious double-double star. At first sight it appears double at some considerable distance, and by attending a little we see that each of the stars is a very delicate double star."

(*Phil. Trans.* 1804, p. 373.) "This remarkable double-double star has undergone a change of situation in each double star separately, which is not very considerable, but deserves our notice on account of a certain similarity in the directions of the alteration. The position of II. 5, Nov. 2, 1779 was 56° 5′ n.f; and, by a mean of three observations, taken Sept. 20, 1802, May 26 and 29, 1804, it was 59° 14′, which gives a change of 3° 9′, the motion of the angle being retrograde." He then states that this change could not be due to "the position of the apex of the translation of the solar system;" that the parallax thus arising would explain the change of the preceding, but not of the following star; and adds, "The situation of both, however, is in a part of the heavens which is so rich in scattered small stars, that a variety of casual and merely apparent combinations may be expected."

H₂ and So. (*Phil. Trans.* 1824, p. 311) made measures in 1822. "The measures on the whole are favourable to a slow variation in the angle of position." But as the change is so small, "it must be regarded as still open to further inquiry."

H₂, in the *Mem. R. A. S.*, vol. v., p. 42, says, "The strong suspicion that these two elegant double stars, so very similar to each other in appearance, distance, and velocity of motion, form in reality a twin system, and have a combined rotation about their common centre of gravity, is increased by the fact that their rotations are performed in the same direction, which, from the analogy of the planetary system, and from that of ζ Cancri, the only ternary star hitherto suspected on any grounds of observation to exist, might be expected."

Σ. (*M. M.*, p. 52): "A small indirect angular motion appears indubitable."

Sm. (*Cycle*, p. 429): "The proper motions in space assigned to this quadruple related system form a link in the chain of evidence which proves the connexion. Indeed, it may be roundly stated that B will revolve round A in about a couple of thousand years;" "and possibly both double systems may move about the central ones in something less than a million of years."

O.Σ. . "The indirect motion still continues. The distance appears to have changed very little. Our measures of distance in 1841 and 1856 are no doubt too large."

Dunér's formulæ are

$$1854·06. \quad \Delta = 3''·12.$$
$$P = 21°·6 - 0°·185 (t - 1850·0).$$

The coefficient of $t - 1850$ is very sure.

A B.

	°		"	
H₁	33·9	In.	...	1779·53
	...	,,	(3·44)	·93
	30·8	3n.	...	1803·84
H₂ & So.	25·9	26	4·01	22·12
	·3	In.	3·62	31·18
Σ.	24·2	,,	2·95	28·72
	27·0	2n.	3·14	30·94

	°		"	
Σ.	25·5	In.	3·09	1831·96
	26·0	3n.	2·98	2·50
	20·5		3·12	42·71
Da.	23·7	2n.	·57	30·53
	25·2	In.	...	2·57
	24·3	2n.	3·32	4·52
	21·9	5n.	·25	41·38
	·7	In.	·16	6·95
	·4	3n.	·14	8·59
	20·4	2n.	·20	53·71
	19·6	4n.	·12	4·71
	·5	2n.	·03	·84
	·3	In.	·06	9·73
Be.	25·2		·31	30·72
Sm.	·3		·5	·73
	23·9		·2	6·45
	21·9		·3	9·78
	20·6		·2	42·59
	19·7		·0	53·71
Encke.	24·0		·42	37·59
Ga.	22·3		·50	8·72
Ka.	23·8		·34	9·99
	21·1		·06	41·67
	24·9		·19	3·03
	·7		·11	·94
	19·9	6n.	·14	31·65
	18·5		2·95	65·80
O.Σ.	22·5	4n.	3·52	40·74
	23·6	In.	·33	2·71
	19·1	,,	·31	51·88
	18·8	,,	·19	2·73
	21·1	,,	·16	7·51
	18·8	,,	·37	7·76
	·5	3n.	·22	8·57
	19·5	,,	·06	61·50
	16·1	,,	·26	2·83
	17·3	2n.	·12	3·53
	19·2	In.	...	6·48
	17·4	,,	...	·49
Mä.	22·1		3·19	42·47
	23·2	12n.	·12	3·68
	24·0		·23	4·67
	21·7	7n.	·20	7·53
	·9		·22	50·54
	·9		·26	1·84
	·3		·10	2·58
	20·6		·20	3·75
	·9		·02	4·70
	·5	11n.	·13	6·43
	·2		·16	8·47
	19·4	13n.	·05	61·35
Po.	20·7	4n.	·09	45·64
Ja.	21·9		·95	6·45
Mit.	20·4	In.	2·46	7·60
Bond.	21·0	,,	3·1	8·47
Flt.	19·7	43	·15	51·65
Mi.	21·4	35	·18	1·82
De.	...	5n.	·33	3·63
	20·7	4n.	·35	4·70
	19·4	In.	·11	5·21
	20·0	3n.	·18	6·49
	19·3	4n.	·03	62·64
	·4	,,	·06	3·53
Se.	22·4	3n.	·07	55·90
	19·0	In.	·28	66·72

	°		"	
Mo.	20.7	34	3.09	1845.64
	18.7	18	.20	58.43
	.8	12	.16	9.00
M.	20.4	1n.	.06	61.45
	13.8	,,	.29	7.41
	14.0	,,	.00	8.79
	13.6	,,	.22	9.76
	17.4	,,	.37	70.38
	15.1	,,	.18	.41
	14.9	,,	.28	.43
	15.3	,,	.29	.79
	19.0	,,	.18	2.45
	17.2	,,	.24	.47
	19.0	2n.	.25	3.43
	14.2	3n.	.19	4.59
	.2	2n.	.10	5.69
Eng.	19.8	7n.	.29	64.45
Ta.	23.0	2n.	.41	6.51
	.5	1n.	.60	7.54
	18.3	3n.	.14	.73
	17.4	2n.	.15	8.73
	18.3	3n.	2.96	9.73
	.0	4n.	3.09	73.84
Br.	19.7	2	.33	68.59
W. & S.	14.3	4	.18	71.57
	17.1	4	.13	.52
	16.0	3	...	.53
	.6	8	3.02	3.55
	17.4	11	.22	6.59
Gl.	.0	6	...	3.48
	18.4	8	...	.81
	.5	5	3.1	.50
	16.1	2n.	...	5.18
	.7	6n.	...	.56
	.2	2n.	...	8.63
Schi.	.9	1n.	3.07	5.63
Sp.	.9		.08	.63
Dob.	17.4	3n.	...	.93
	15.3	,,	3.24	7.43
	16.3	,,	.04	8.56
Goldney.	18.6	4n.	.18	.71

606   Σ. 2383.

ε² LYRÆ.

R. A.	Dec.	M.
18h 40.4m	39° 29'	4.9, 5.2

90°
0°
1779.
1878.

Magnitudes.—"The difference in brightness seems variable." (Σ.) The magnitudes of E, F, G, are 9.5, 12, 12, respectively.

C. H_1, white; Σ., very white; Sm., white.

H_1 (*Phil. Trans.*, vol. lxxii., p. 217): "The stars of the second set are equal, or the preceding of them rather larger than the following. The interval between the equal set with a power of 227 is almost 1½ diameter of either; with 460, full 1¾ diameter; with 932, two diameters; with with 2010, 2½ diameters. These estimations are a mean of two years' observations. Position of the equal set 72° 57' s.f."

H_1 (*Phil. Trans.* 1804, p. 373) found the position on Sept. 20, 1802, May 26 and 29, 1804, was 83° 28' and 75° 35' s.f.; this gives a difference of 7° 58', the motion being retrograde.

H_2 and So. (*Phil. Trans.* 1824, p. 314): "The change surmised by Sir Wm. Herschel in 1804 seems to be well borne out by subsequent observations, the total alteration in the angle being no less than 13° 51', averaging 0°.325 per annum in the direction n.p.s.f., or retrograde."

H_2 (*Phil. Trans.* 1826, p. 375). After giving his observations in 1825.53, he remarks: "The change of position in 3.11 years amounts to − 0°.45. Calculating on the presumed angular motion − 0°.325, it should have been −1° 0'. The difference is nearly insensible."

Σ. (*M. M.*, p. 52): "Indirect motion is beyond doubt."

The angular movement in this pair is in the same direction, and very nearly of the same amount, as in the preceding pair. Both the distance and the angular change have been very constant: between 1779 and 1831 it was − 0°.355; between 1831 and 1862, −0°.335 per annum. (O.Σ.)

Dunér gives

$$1853. \quad \Delta = 2''.58.$$
$$P = 149°.2 − 0°.360 (t − 1850.0).$$

C D.

	°		"	
H_1.	173.5	1n.	...	1779.83
	167.3	5	...	1804.08
H_2 & So.	159.9	8	3.80	22.42
	.2	28	.34	5.53
	156.4		...	8.72
	157.0		3.26	32.57
Σ.	159.6	1n.	2.50	28.72
	154.1	2n.	.68	30.94
	156.3	1n.	.69	1.96
	154.0	3n.	.48	2.50
Da.	157.3	5n.	.92	0.57
	156.2	1n.	.62	2.57
	154.2	2n.	.86	4.52
	152.8	,,	.57	40.72
	151.8	,,	.65	1.66
	149.9	1n.	.44	6.95

	$^\circ$		$''$	
Da.	148.9	2n.	2.59	1848.54
	147.4	,,	.45	53.71
	146.2	4n.	.42	4.72
	145.9	2n.	.47	.84
	.8	1n.	.54	9.73
	144.8	,,	.55	65.75
Be.	156.1		.82	30.72
Sm.	157.1		.8	.73
	154.3		.5	4.52
	150.9		.6	42.59
	148.1		.5	53.71
Encke.	153.7		.95	37.59
Ga.	155.4		3.35	8.72
	152.8		2.5	9.78
Ka.	151.0		.71	.99
	152.7		.49	41.67
	.0		.61	3.03
	151.3		.69	.94
	149.0	6n.	.52	51.65
	143.0		.34	65.80
	142.6		.37	6.84
O.Σ.	151.1	3n.	.73	40.83
	150.1	1n.	.50	2.71
	147.6	,,	.78	51.88
	.9	,,	.51	2.73
	144.6	2n.	.55	7.63
	143.9	1n.	.42	8.50
	144.6	2n.	.49	.61
	146.2	3n.	.52	61.50
	145.6	2n.	.54	2.83
	.5		.45	3.53
	142.6	1n.	...	6.48
Mä.	153.6		2.50	41.49
	.2		.85	2.47
	151.9	12n.	.72	3.68
	.5		.79	4.67
	150.2	7n.	.74	7.53
	149.6		.83	50.54
	148.5		.81	1.84
	149.2		.48	2.58
	147.4		.93	3.75
	146.8		.66	4.70
	.6	12n.	.70	6.43
	.3		.65	84.7
	144.5		.21	9.40
	145.2	13n.	.62	61.35
Ja.	150.7		3.10	46.14
Po.	.8	3n.	2.76	.63
Mit.	149.2	1n.	.55	7.60
Bond.	.0	,,	.5	8.47
Flt.	147.0	45	.42	51.57
Mi.	146.7	24	.49	.82
De.	147.0	5n.	.60	3.63
	146.7	4n.	.70	4.69
	147.5	2n.	.68	6.46
	143.8	4n.	.50	62.64
	144.2	,,	.45	3.53
Se.	148.4	4n.	.57	56.06
	...	1n.	.49	65.54
Mo.	150.8	30	.75	45.64
	146.7	3n.	.48	58.42
M.	152.6	1n.	.43	61.45
	318.2	,,	.10	7.41
	139.4	,,	.57	8.79
	137.6	,,	.44	9.76

	$^\circ$		$''$	
M.	137.2	1n.	2.37	1869.76
	138.4	,,	...	.48
	139.6	,,	2.52	70.43
	.8	,,	.40	.45
	.5	,,	.47	.79
	137.2	2n.	.50	2.45
	138.3	1n.	.38	3.42
	.2	2n.	.35	4.61
Eng.	143.4	7n.	.55	65.39
Ro.	...	1n.	.49	.54
Ta.	141.0	2n.	...	6.51
	142.3	1n.	2.53	7.54
	139.4	,,	...	74.61
Du.	143.2	2n.	2.41	67.71
	139.9	,,	.29	8.73
	141.4	3n.	.35	9.73
	.5	4n.	.43	73.84
Br.	.8	2	.64	68.69
W. & S.	146.3	4	.48	71.57
	142.0	4	.7	.52
	144.2	1	...	.53
	141.2	4	2.7	.55
	139.4	12	.62	6.59
Gl.	141.2	7	...	3.48
	142.1	6	2.5	.81
	138.6	2n.	...	5.28
	136.3	6n.	...	6.56
	138.2	5n.	2.40	8.63
Schi.	139.2	1n.	.39	5.63
Sp.	.3		.40	.63
Dob.	.3	3n.	...	.93
	137.3	,,	2.35	7.43
	139.1	,,	.24	8.56
Goldney.	.3	4n.	.43	.71

Small stars between ϵ^1 and ϵ^2.

A C.

	$^\circ$			
W. & S.	173.0	4	207.0	1871.57
	.0	3	206.2	.52
	.8	1	...	.53

A E.

W. & S.	135.2	1	...	2.52
	134.7	2	144	.52
	.5	1	...	.53

A F.

W. & S.	180.5	1	139	.52
	.0	1	...	.53

A G.

W. & S.	167.5	1	108	.52
	164.6	2	...	.53

C E.

W. & S.	36.6	1	129	.52
	35.9	1	...	.53
	37.9	1	...	.53

C F.

W. & S.	338.4	1	71	.52
	339.3	1	...	.53

C G.

W. & S.	1·2	I	101″	1872·52
	·4	I	...	·53

E F.

W. & S.	248·9	I	111	·52
	250·0	I	...	·53

E G.

W. & S.	268·5	I	77	·52
	·5	I	...	·53

G F.*

So.	220		53	23
H₂.	221		48	31
W. & S.	219·2	In.	44	72
Bu.	38·4		46·7	78

E H.

Bu.	357·0		25·0	78·36

ϵ^1 and ϵ^2.

Sm.	172·5		207·3	30·73
	·9		206·8	6·45
Σ.	·2		207·1	5·23
De.	·9		·7	63·15
Eng.	·8		206·3	4·45
M.	·4	In.	209·7	6·47
W. & S.	173·0	,,	206·6	72·05
Fl.	172·8	,,	207·1	7·52

* The values for A E, A F, A G, C E, C F, C G, E F, E G, F G, given by W. and S., were calculated, not measured.

607   Σ. 2398.

R. A.	Dec.	M.
18h 40·9m	59° 25′	8·2, 8·7

C. A yellowish, B bluish.

Σ.	134·4		12·42	1832·17
Mä.	137·4		·97	44·91
	138·6		13·27	7·32
De.	142·8		15·56	65·04
Fl.	144·7	In.	16·52	77·88

608   Σ. 2396.

R. A.	Dec.	M.
18h 42·8m	10° 38′	7·7, 11·2

Σ. (P. M., p. ccxxx.) shows that the smaller star does not partake in the considerable proper motion of the larger star; and O.Σ. finds that the following formulæ represent the observations quite well :

$$\Delta A = -11''\cdot790 \pm 0''\cdot051 - (0''\cdot1222 \pm 0''\cdot0044)\,(t-1850\cdot0).$$

$$\Delta D = +2''\cdot204 \pm 0''\cdot051 + (0''\cdot4579 \pm 0''\cdot0044)\,(t-1850\cdot0).$$

Σ.	232·6	3n.	11·74	1829·60
	285·3	2n.	12·28	51·90
Mä.	267·4		10·31	43·71

Mä.	275·7		10·32	1846·73
	276·5		11·75	7·77
	284·7		...	50·75
	286·3		12·57	1·73
	287·6		·40	2·78
O.Σ.	278·2	3n.	11·71	49·09
	292·4	2n.	13·96	57·14
	304·7	,,	16·64	65·76
De.	·4		·39	·44
W. & S.	313·5	4	19·9	74·65
	311·9	4	·92	5·63
Fl.	314·6	In.	21·02	7·75

609   O.Σ. 363.

R. A.	Dec.	M.
18h 43m	77° 33′	7·5, 7·7

O.Σ.	199·5	In.	0·57	1842·73
	13·8	,,	·63	4·85
	20·2	,,	·50	6·69
	26·5	,,	·49	75·34
De.	19·7	3n.	...	67·61

610   Σ. 2400.

R. A.	Dec.	M.
18h 43·5m	16° 7′	8·1, 10·6

At first glance the observations seem to indicate an occultation produced by orbital motion : this view, however, is not confirmed on a more careful scrutiny. On discussing the observations, it was found that the changes have been caused by the proper motion of the larger star, and that in 1871 the minimum distance would be reached. Some uncertainty still remains for future measures to remove. (O.Σ.)

Σ.	305·2	2n.	2·96	1829·18
	303·2	,,	·74	33·16
Mä.	300·1		1·99	43·70
	301·0		2·20	6·47
	299·9		1·80	8·45
O.Σ.	275·3	In.	·77	51·62
	...	,,	1 ±	4·63
	246·?	,,	0·8?	5·67
	...	,,	1 ?	7·61
	...	,,	single	·67
	...	,,	,,	8·53
	...	,,		·59
	238	,,	1 ±	65·72
	236·3	,,	1·02	72·61

611   Σ. 2402.

R. A.	Dec.	M.
18h 44·1m	10° 32′	8, 8·4

The measures by O.Σ. and Secchi seem to indicate a slight angular change.

Σ.	196°3	2n.	0″68	1828·68
	·2	1n.	·76	9·64
	201·8	,,	·85	33·77
Mä.	204·3		·70	8·83
	208·4		·68	43·65
	212·0		·68	52·63
	229·4		...	61·66
O.Σ.	218·5	1n.	0·91	40·51
	215·6	,,	·81	1·55
	218·4	,,	·95	·66
	208·8	,,	1·04	72·61
Se.	213·4	2n.	0·89	56·64
	203·8	1n.	·84	65·63
W. & S.	206·9	7	·98	72·51
	205·9	2	1·0	·55
	202·1	3	·0	·56
	·2	4	·0	4·65
	207·1	4	0·8	5·63
Gl.	203·2	6	·97	4·73
W.O.	205·7	1n.	1·09	·68
	204·2	,,	·13	·68
Schi.	206·6	1n.	0·85	5·66
Sp.	·6		·85	·67

H₁.	306°0		26″65	1780·76
Σ.	350·8		...	1814·13
	346·5	3n.	30·26	32·60
	345·6	6n.	·38	6·39
So.	349·2		29·95	22·14
Sm.	347·6		30·4	30·78
	345·5		·3	7·89
O.Σ.	·1	1n.	·52	9·85
	·1	,,	·64	40·84
	342·7	,,	·82	51·67
	338·9	,,	31·52	70·87
	·9	,,	·66	4·73
Mä.	344·8		32·10	41·48
	·2		30·59	3·32
	·0		·48	7·81
Ka.	345·0		·27	1·74
De.	341·5	3n.	·93	58·21
	340·6	4n.	31·01	63·14
M.	341·0	1n.	32·10	5·43
Fi.	339·4	,,	31·87	77·76

612 Σ. 2409.

R. A.	Dec.	M.
18ʰ 46ᵐ	13° 23′	8, 9·3

The amount of angular change, if any, is still uncertain.

Σ.	32·5	6n.	0·98	1832·76
O.Σ.	48·0	1n.	1·00	40·57
	49·7	,,	0·82	1·55
	45·3	,,	1·03	2·72
	42·4	,,	·18	7·59
	38·5	,,	...	8·73
	43·1	,,	1·06	·74
	37·1	,,	·12	52·67
	38·5	,,	·01	72·61
Se.	31·7	2n.	·05	56·65

613 O.Σ. 364.

R. A.	Dec.	M.
18ʰ 48ᵐ	25° 12′	7·5, 10·5

O.Σ. could not see the companion in the years 1845, 1847, and 1852. De. in 1865 found it "not round."

O.Σ.	162·8	1n.	0·74	1842·67

614 Σ. 2420.

R. A.	Dec.	M.
18ʰ 49·4ᵐ	59° 14′	4·6, 7·6

Probably a case of rectilinear motion. The proper motion of A is + 0ˢ·005 in R. A., and − 0″·01 in N. P. D.

615 O.Σ. 525.

R. A.	Dec.	M.
18ʰ 50ᵐ	33° 48′	A 5·1, B 10·3, C 7·1

C. A yellow, C blue.

O.Σ. observes that it is very remarkable the stars A C should have been measured three times without B being detected.

De. has estimated the magnitude of B as the 8th and 9th. O.Σ. has twice entered it as the 11th.

A B.

O.Σ.	128·0	7n.	1·55	1849·70
De.	132·8	2n.	·36	69·77

A C.

O.Σ.	350·5	10n.	45·50	46·98
De.	·6	2n.	·43	69·77

616 Σ. 3130.

O.Σ. 365.

R. A.	Dec.	M.
18ʰ 52ᵐ	44° 4′	A 7·4, B 8·5, C 11·1

A B.—In August 1841 the star A was readily seen to be double, but in the September following it was twice examined without the companion being detected. It was oblong till 1851, quite round in 1852 and 1854, and again readily separated in 1857, the relative position being the same as in 1841. Hence the period is probably sixteen years: possibly, however, the phenomena may be explained by the variability of the companion. (O.Σ.)

A C, probably unchanged.

A B.

O.Σ.	168°.1	1n.	0".50	1841.65
	212	,,	oblong	4.85
	232	,,	,,	.85
	235	,,	obl.?	5.65
	212	,,	,,	6.69
	226	,,	oblong	7.59
	242	,,	,,	8.74
	250	,,	obl.?	9.82
	276	,,	0.20	51.60
	273	,,	oblong	.75
	...	,,	single	2.63
	...	,,	,,	4.64
	...	,,	,,	.69
	166.1	,,	0.50	7.67

A C.

Σ.	262.9	6n.	2.69	33.37
O.Σ.	266.2	1n.	.94	41.65
	260.3	,,	.59	4.85
	265.7	,,	.78	7.59
	261.6	,,	.78	8.74
	262.6	,,	.36	9.82
	264.2	,,	.74	52.63
	256.5	,,	.64	4.64
	264.6	,,	.99	.69
	.4	,,	.96	7.67

# 617	Σ. 2422.

R. A.	Dec.	M.
18^h 52.2^m	25° 56'	7.6, 7.7

Probable change in angle.
Dunér gives

$$1849.08. \quad \Delta = 0''.79.$$
$$P = 104.4 - 0°.1 \ (t - 1850.0).$$

Σ.	106.0	6n.	0.85	1832.10
O.Σ.	.5	2n.	.98	40.69
	101.1	1n.	.74	52.67
	100.3	,,	.87	7.61
	96.8	,,	.85	72.61
Mä.	105.1	7n.	.77	43.08
Se.	106.8	3n.	.83	56.88
De.	100.6	1n.	.7	63.53
	99.0	,,	.79	5.73
	100.5	,,	.84	8.66
	98.0	,,	.8	73.49
	100.6	,,	.74	4.49
Du.	.5	6n.	.72	69.16
W.O.	97.4	,,	.97	74.66
Sp.	98.5	,,	.76	5.68

# 618	Σ. 2424.

R. A.	Dec.	M.
18^h 53.5^m	13° 28'	5.7, 9.2

Considerable changes (see *P.M.*, p. ccxxi.), probably due to the proper motion of the brighter star.
Smyth's magnitude of A is from Piazzi: his own estimate was that "it certainly appeared bright enough to be rated among the 6th, on careful comparison." He gives B as of the 10th magnitude, A as of the 7th. Dunér's estimates are 5, 10.
Dunér gives

$$\Delta \sin P = -16''.55.$$
$$\Delta \cos P = - 6''.91 + 0''.111 \ (t - 1850.0).$$

H₁.	238°.6	1n.	7"	1802.76
Σ.	236.4		20.06	20.64
	241.6	3n.	18.66	31.31
	248.6	2n.	17.82	51.90
So.	240.5	5n.	19.66	25.11
Sm.	.9		.1	32.61
Mä.	244.9	3n.	17.07	44.22
	247.5	,,	.45	52.05
	.5	2n.	16.50	6.82
	248.1	1n.	.13	62.72
De.	252.1	4n.	17.43	3.48
O.Σ.	254.2	1n.	16.87	8.75
Du.	.0	2n.	17.23	9.77
W. & S.	255.8	4	16.9	74.65
	258.5	5	17.7	.65
	255.9	4	.29	5.63
Gl.	257.4	4	.5	4.73
Dob.	256.7	2n.	16.74	7.52

# 619	Σ. 2429.

R. A.	Dec.	M.
18^h 54^m	36° 16'	8.3, 9.8

H₁.	285.0		...	1783.21
So.	290.3	2n.	5.47	1825.57
Σ.	289.5	3n.	.32	9.83
Mä.	288.8	1n.	.67	43.40
Ta.	287.9	,,	.96	66.47

# 620	Σ. 2438.

R. A.	Dec.	M.
18^h 55.5^m	58° 4'	7, 7.6

Certain change in angle and distance. The periastre was probably passed either between 1842 and 1870 or since the latter year. (O.Σ.)

H₁.	355		1	1782.68
	358.4		...	3.26
H₂.	337.8		0.7	1830.00
Σ.	340.6	4n.	.72	2.53
Sm.	341.0		.7	4.53
O.Σ.	348.7	2n.	.69	40.57
	341.8	1n.	.52	6.69
	306	,,	.53	70.87
	single			.92
Mä.	338.0		0.6	41.48
	346.6		...	3.32
Da.	335.1	1n.	0.65	1.80
Se.	333.2	4n.	.4	57.54
De.	330.0	elongated		63.62
W. & S.	not elongated			73.58

621   Σ. 2434.

R. A.	Dec.	M.
18h 56·5m	− 0° 53′	A 7·9, B 8·4, C 10·3

C. Sm., A and B, white; C, blue.

A B is H₁ IV. 127. B C is Σ. 2434.

Between 1831 and 1864 the distance of A B seems to have diminished about 1″, and the angle to have decreased about 10°. The change in the angle of B C also amounted to about 10° in the same time, the distance remaining nearly as when first measured.

In A B the motion is rectilinear. B C form a physical system in rapid motion.

A B.

	°		″	
H₁.	159·9		17·7	1783·60
H₂ & So.	148·8	10	26·01	1823·48
So.	·6		25·8	31·48
	146·8		·6	8·59
Σ.	147·0	4n.	·56	1·57
Mä.	145·8		·45	5·53
	142·9		·66	4·45
Mit.	141·3	1n.	·77	8·65
	139·8		24·24	51·75
Se.	138·9		·48	6·93
	136·8	2n.	·23	66·65
M.	137·1	1n.	23·28	1·47
W.O.	133·4	6	24·73	63
De.	136·8	2n.	·29	4·66
W. & S.	132·7	2	23·5	72·56
	·2	4	24·1	·62
	134·1	7	·0	3·55
	133·4	6	...	·55
	132·9	4	24·0	4·67
	133·4	3	·18	5·66
Gl.	·2	6	23·8	4·73
U.O.	·6	2n.	24·31	7·67
Fl.	·8	1n.	23·89	·76

B C.

Σ.	80·5	3n.	1·93	1831·58
Sm.	85·0		2·0	8·59
Mä.	80·6		·20	44·45
Mit.	72·4	2n.	1·54	8·12
Se.	68·7	,,	·73	57·12
	72·0	1n.	·0	66·57
De.	69·6	2n.	·79	4·66
W. & S.	70·4	2	·5	72·56
	67·5	4	·09	3·55
C.O.	63·2	2n.	·51	7·68

622   Σ. 2437.

R. A.	Dec.	M.
18h 56·6m	19° 0′	7·8, 8

Certain but slow movement.

Dunér, excluding the observations of Mädler, finds the following:

$$1854·21.\quad \Delta = 0''·92.$$
$$P = 74°·8 − 0°·312\,(t − 1850·0).$$

Σ.	80·8	5n.	1·08	1830·79
H₂.	82·9		·0	1·00
Mä.	76·7		0·99	9·60
	74·0	8n.	·93	44·26
	70·5	2n.	·98	52·22
	65·8	,,	1·01	5·26
	53·8	1n.	0·45	8·81
	62·3	,,	·6	9·74
	63·0	,,	·75	62·70
O.Σ.	74·1	2n.	1·36	40·76
	72·5	1n.	·04	2·76
	74·4	,,	·02	6·69
	71·4	,,	0·96	7·59
	74·7	,,	1·05	9·73
	66·4	,,	·02	72·61
Se.	71·6	2n.	0·94	57·10
	68·5	1n.	·5	66·73
De.	71·4	4n.	·8	3·06
Eng.	65·5	1n.	1·06	4·49
Du.	72·7	5n.	0·80	70·09
W. & S.	62·8	4	·86	2·50
	68·5	6	1·0	3·55
	67·0	3	0·86	·56
	68·3	4	·8	4·67
	·0	4	·8	6·52
Gl.	67·9	11	1·02	4·73
Sp.	70·7		·02	5·61

623   Σ. 2441.

R. A.	Dec.	M.
18h 58·1m	31° 13′	7·7, 9·3

Indirect motion.

Σ.	291·9	3n.	5·22	1830·34
Mä.	290·1		·19	7·33
	288·4		·11	43·77
	·5		·20	5·41
	287·4		4·69	51·88
	283·8		5·03	2·68
	285·5		·18	7·99
Se.	·3	2n.	·64	·13
	284·0	1n.	·77	65·63
O.Σ.	281·7	,,	·49	8·61
	280·4	,,	·05	74·72
De.	285·1	,,	·36	63·47
	284·4	,,	·22	6·54
	283·4	,,	·26	9·50
	·1	,,	·14	74·70
	282·4	,,	·27	7·78

624   Σ. 2442.

R. A.	Dec.	M.
18h 58m	16° 48′	8, 9·5

Σ.	207·6	2n.	23·05	1828·77
Mä.	·2	1n.	22·55	43·70

625 Σ. 2454.

R. A. 19ʰ 1·5ᵐ　Dec. 30° 15′　M. 8, 9·2

Considerable angular change.
While setting for this star in 1840, O.Σ. detected another pair, of which he gives the following measures:

Angle.	Distance.		Magnitudes.
160°	0″·68	1840·51	8, 9
155°·8	·43	2·72	

In 1866 a careful search failed to reveal the existence of this object.

Σ.	203·9	3n.	0″·75	1831·50
O.Σ.	223·7	In.	1·00	0·84
	214·9	,,	0·80	6·68
	233·9	,,	·95	66·68
	236·3	,,	·82	72·61
Mä.	208·1		·6	43·76
Se.	217·0	In.	·45	57·57
De.	225·9	,,	1·26	65·32
Du.	235·2	5n.	0·81	72·45
W. & S.	·1	In.	1·0	3·55
	230·0	,,	...	·56
	232·4	,,	0·79	6·63
W.O.	231·6		·88	4·69
Sp.	235·0		·87	5·89

626 Σ. 2456.

R. A. 19ʰ 1·6ᵐ　Dec. 38° 21′　M. 8·2, 8·2

Σ.	13·6	3n.	29·07	1829·43
Mä.	12·0		...	44·90
	11·0		27·50	7·85
	·4		28·00	50·81
	·2		27·34	2·02
	10·7		...	9·86
Eng.	9·0		26·76	64·46
De.	·3		·48	·82
Dob.	7·1	In.	...	76·63
Fl.	·5	,,	25·59	7·86

627 Σ. 2455.

R. A. 19ʰ 1·8ᵐ　Dec. 21° 59′　M. 7·2, 8·3

Dunér has computed the following formulæ:

$$\Delta \sin P = + 3''\cdot11 + 0''\cdot011\,(t - 1850\cdot0) - 0''\cdot00031\,(t - 1850\cdot0)^2.$$
$$\Delta \cos P = - 2''\cdot48 + 0''\cdot058\,(t - 1850\cdot0) + 0''\cdot00025\,(t - 1850\cdot0)^2.$$

The measures appear to indicate a physical relation, but the curvature is so slight that the movement may possibly be rectilinear and uniform.

H₂.	140°		4″	1827·64
Σ.	144·5	3n.	4·93	8·77
Mä.	136·6		·42	39·29
	132·6	6n.	·16	44·57
	125·2	In.	3·74	50·74
	124·9	2n.	4·07	2·64
	·2	5n.	3·66	3·99
	122·2	2n.	·65	8·73
	120·5	In.	4·04	61·74
	116·1	2n.	3·73	2·84
Mo.	124·3	3n.	·98	55·66
Se.	123·0	,,	·70	7·29
	113·4	In.	·77	65·64
De.	115·6	4n.	·52	4·60
	·3	2n.	·55	5·66
	114·4	4n.	·39	7·20
	111·8	In.	·48	9·63
	109·9	,,	·22	71·65
	·7	,,	·48	2·67
	106·5	,,	·40	3·71
	·5	,,	·08	4·67
	105·3	,,	·35	5·67
Eng.	114·2	3n.	·51	65·58
Du.	110·8	6n.	·41	9·95
W. & S.	109·7	6	·7	72·64
	107·2	3	...	·55
	109·2	4	3·42	3·55
	·7	8	·37	4·67
Gl.	·5	10	·30	4·73
Schi.	104·8	In.	·48	5·61
Sp.	·8		·49	·61

628 Σ. 2464.

R. A. 19ʰ 3·6ᵐ　Dec. 11° 41′　M. 8·2, 10·5

Σ.	19·2	3n.	1·36	1830·36
Mä.	16·2		...	42·36
	24·1		1·10	3·67
	33·5		...	51·73
Se.	21·0	2n.	1·25	7·12
W. & S.	26·5	In.	·0	76·32

629 Σ. 2471.

R. A. 19ʰ 5ᵐ　Dec. 7° 56′　M. 7·9, 10·7

Certain change in angle and distance.

Σ.	121·8	4n.	7·63	1830·18
O.Σ.	126·6	In.	8·24	72·56
	127·9	,,	·10	4·72

630 Σ. 2479.

R. A. 19ʰ 5·9ᵐ　Dec. 55° 8′　M. A 7·1, B 8, C 9·4

Dunér has

1847·57. Δ = 6″·63.
$$P = 36°\cdot4 - 0°\cdot10\,(t - 1850\cdot0).$$

A B.

De.	40°9		elong^d	1863·87
	36·8		,,	70·83
	20·2		0·6	4·09

A C.

Σ.	38·0	4n.	6·65	32·61
Mä.	37·0	1n.	·40	44·37
Se.	34·0	,,	7·23	59·80
De.	·9	,,	6·72	63·49
Du.	35·6	2n.	·66	71·13

631   Σ. 2481.

R. A.	Dec.	M.
19^h 7·1^m	38° 35'	8, 8, 9

The common proper motion of A B is
− 0"·29 in R. A., and + 0"·097 in N. P. D.
C is Secchi No. 3.
Dunér gives the following formulæ :

$$1853·72. \quad \Delta = 3''·90.$$
$$P = 231°·6 - 0°·23 (t - 1850·0).$$

A and $\dfrac{B+C}{2}$.

H₁.	261·6		...	1783·33
Σ.	234·3	3n.	3·83	1830·45
Mä.	235·3	1n.	4·10	43·74
Po.	231·5	,,	3·73	8·53
O.Σ.	49·3	4n.	4·02	55·56
Mo.	230·3	2n.	3·96	6·59
Se.	·8	3n.	·99	7·80
De.	·1	1n.	4·73	8·44
M.	223·7	,,	·31	65·44
	226·0	,,	·32	9·77
Sp.	225·5		3·95	75·65
Du.	224·5	3n.	...	6·75
Lindstedt.	·5	1n.	...	·75

B C.

Se.	93·4		0·4	56·83
	98·2		·4	9·61
O.Σ.	88·7	1n.	·59	66·74
Sp.	86·5		·52	75·64

632   Σ. 2484.

R. A.	Dec.	M.
19^h 9^m	18° 52'	7·4, 8·9

Σ.	218·4	5n.	2·50	1831·76
Mä.	220·3		·47	6·46
	224·3		·43	43·72
	221·9		·43	51·80
	223·3		·64	2·63
	229·9		...	61·70
Se.	224·0	3n.	2·49	57·26
De.	·5	1n.	·58	63·48
	227·7	,,	·68	6·70
	·8	,,	·68	7·63
	228·4	,,	·29	72·75

O.Σ.	224·8	3n.	2"·61	1871·34
W. & S.	232·3	1n.	·5	4·73
	224·6	,,	·78	6·31
Gl.	232·5	,,	·51	4·84

633   Σ. 2486.

R. A.	Dec.	M.
19^h 9^m	49° 37'	6, 6·5

The distance has diminished.
The common proper motion is − 0"·22
in R. A., and − 0"·647 in N. P. D. (Σ.)
Dunér's formulæ are

$$\Delta = 10''·28 - 0''·012 (t - 1850·0).$$
$$P = 223°·2 - 0·08 (t - 1850·0).$$

Σ.	224·7	6n.	10·49	1834·62
Mä.	223·6		·23	43·84
	222·8	2n.	·38	5·85
	·1		·53	7·85
	·5	3n.	·17	51·90
	·3	2n.	·55	60·65
Po.	·8	4n.	·22	47·36
O.Σ.	221·8	2n.	·28	51·85
	222·3	1n.	·18	70·92
	220·9	,,	9·90	4·73
Mo.	222·6	6n.	10·17	56·70
De.	·8	5n.	·21	4·76
	221·9		9·97	67·64
Se.	223·0	3n.	10·16	57·52
M.	221·8		9·99	63·31
Du.	·8	4n.	10·07	9·73
W. & S.	·4	4	9·8	74·72
	222·1	7	10·1	·78
	221·3	4	·26	5·66
Gl.	·4	10	·2	4·84
Dob.	220·3	4n.	9·81	7·45

634   O.Σ. 368.

R. A.	Dec.	M.
19^h 10·5^m	15° 57'	7·3, 8·5

O.Σ.	217·5	6n.	0·81	1850·40
Mä.	219·0		·60	43·39
	·3		·67	51·73
Se.	...	single		6·67
	217·1		0·80	7·71
De.	214·5	4n.	·93	67·13
W.O.	235·6		1·84	74·72

635   Σ. 2491.

R. A.	Dec.	M.
19^h 11·4^m	28° 4'	7·9, 9·2

Direct motion,

Σ.	206˚6	4n.	1″08	1828·77
Da.	211·5	3n.	·23	41·41
Mä.	204.4		0·96	2·56
	·4		1·20	8·45
Se.	211·5	2n.	·05	56·68
O.Σ.	222·1	1n.	·19	72·61
W. & S.	208·8		·03	5·67
Dob.	215·8	1n.	...	6·71

636   Σ. 2509.

P. XIX. 108 DRACONIS.

R. A.	Dec.	M.
19ʰ 15·6ᵐ	62° 59′	7, 8·1

Σ. says (*M. M.*, p. 296): "Angular motion is very probable."

H₂ (*Mem. R. A. S.*, vol. vi., p. 53) writes in 1830, "perfectly divided with 480 and the whole aperture: 320 gave a sensible elongation, but would not reduce the discs small enough, and separate them from the flare surrounding them."

Probable diminution in angle and increase in distance. (O.Σ.)

Σ.	356·7	1n.	0·42	1831·96
	347·0	,,	·45	2·34
	353·6	,,	·66	·45
	354·7	,,	·55	·45
	345·8	3n.	·57	6·93
Sm.	349·0		·5	8·78
O.Σ.	352·4	1n.	·76	40·61
	347·7	,,	·63	5·69
	345·2	,,	·64	6·69
Mä.	343·8		...	1·47
	339·3		0·55	3·40
	346·5		·67	8·20
Se.	340·3	3n.	·69	57·43
De.	339·7	,,	1	8·53
	345·1	,,	0·8	62·70
	341·7	2n.	·9	3·39
W. & S.	·6	5	·81	72·64
	·4	4	·81	3·58
	342·3	7	·66	4·73
Gl.	341·8	10	1·0	·84

637   Σ. 2514.

R. A.	Dec.	M.
19ʰ 16·8ᵐ	67° 28′	9, 11·3

Σ.	277·0	3n.	7·39	1832·67
De.	306·8		8·12	66·58

638   Σ. 2515.

R. A.	Dec.	M.
19ʰ 20ᵐ	21° 17′	8, 9

Σ.	18˚3	2n.	18″74	1829·20
Mä.	19·8		17·07	47·69
	20·4		16·49	50·74
	·1		·42	1·85
	21·1		·78	2·64
De.	22·5		14·60	65·04

639   O.Σ. 372.

R. A.	Dec.	M.
19ʰ 19·9ᵐ	46° 59′	A 7, B 8·8, C 10·5

B C form a binary system, most probably.

A B.

O.Σ.	57·2	2n.	79·44	1849·67
De.	·2	3n.	·53	67·93

B C.

O.Σ.	293·6	4n.	3·38	47·46
Mä.	298·0		2·95	3·65
	286·9		3·20	5·68
De.	296·1	3n.	·74	67·93

640   Σ. 2521.

R. A.	Dec.	M.
19ʰ 21ᵐ	19° 39′	5·5, 10·3

Certain change in angle and distance.

H₂.	45		15	1827·64
	46·4		20	30·00
Σ.	43·5	3n.	22·64	29·40
	40·9	2n.	23·36	51·89
Sm.	44·8		25·0	33·58
	43·5		22·26	48·06
O.Σ.	40·2	1n.	23·96	66·74
	39·6	,,	·81	8·75

641   Σ. 2524.

R. A.	Dec.	M.
19ʰ 21·6ᵐ	25° 15′	8·3, 8·5

Dunér has

$$1854·23. \quad \Delta = 6''·80.$$
$$P = 102°·8 - 0''·09 (t - 1850·0).$$

Σ.	104·6	3n.	7·16	1829·76
Mä.	103·5	1n.	6·31	43·63
Se.	105·6	,,	·67	56·59
	101·7	,,	7·18	7·65
Mo.	·1	2n.	6·93	·64
Du.	·3	5n.	·57	68·51

642   Σ. 2525.

R. A.	Dec.	M.
19ʰ 21·6ᵐ	27° 5′	7·4, 7·6

The angle and distance have diminished.

Σ.	255·9°	5n.	1″33	1830·43
	·5	2n.	·30	6·14
O.Σ.	251·8	1n.	·04	40·56
	253·1	,,	·52	·84
	246·8	,,	·05	54·63
	240·6	,,	0·75	65·72
	242·4	,,	·73	·80
	234·0	,,	·66	72·61
Da.	255·5		1·25	40·62
Mä.	251·0		0·82	2·41
	254·0		·95	3·69
Se.	247·1		·85	56·61
	239·9	1n.	·40	65·64
De.	240·8	7n.	·60	·22
W. & S.	232·6	4	...	72·64
	225·8	2	0·5	3·57
	237·8	2	·5	4·75
	3·6	3	...	5·66
Gl.	234·2	10	0·48	4·84
Sp.	232·2		·43	6·00

643 Σ. 2538.

R. A.	Dec.	M.
19h 27m	36° 27'	A 8·2, B 8·3, C 8·7

A B.

So.	245·2	2n.	53·23	1825·57
H₂.	242·1	1n.	55·06	30·60
Σ.	245·2	2n.	53·04	·85
Se.	244·7	,,	51·78	57·90
Du.	246·1	3n.	52·95	69·68
O.Σ.	248·1	1n.	·81	70·92
W. & S.	243·2	,,	·85	4·75
Gl.	·4	,·	...	·85

B C.

So.	56·5	2n.	6·30	25·57
H₂.	53·9	3n.	7·04	30·08
Σ.	52·5	,,	6·07	·87
Mä.	54·2	2n.	5·95	43·75
Se.	·8	,,	6·02	57·90
Du.	52·8	5n.	5·92	69·58
O.Σ.	54·5		·96	70·92
W. & S.	·6	1n.	·9	4·75
Gl.	53·9	,,	6·04	·85

644 Σ. 2541.

R. A.	Dec.	M.
19h 30·2m	− 10° 42'	8·2, 9·8

Σ.	339·9	3n.	2·84	1831·01
	338·4	2n.	3·61	51·85
Sm.	·4		·2	35·58
Mä.	336·9		2·86	43·63
Mit.	·2	2n.	3·02	8·19
O.Σ.	337·6		·36	51·91
Se.	·6	2n.	·47	7·17
Da.	340		2·5	9·80
M.	323·0	1n.	3·25	61·74

De.	329·1°	,,	3″29	1864·59
	332·0	,,	·48	6·67
	·3	,,	·49	7·61
	·3	,,	·42	72·71
W. & S.	330·2		2·76	6·61
C.O.	319·8	3n.	3·69	7·56

645 O.Σ. 376.

R. A.	Dec.	M.
19h 31m	33° 57'	7·1, 9·8

O.Σ.	228·6	6n.	2·60	1848·52
De.	233·5	3n.	·72	66·65

646 Σ. 2544.

R. A.	Dec.	M.
19h 31·3m	8° 3'	7·8, 9·5

A C unchanged.

A B.

Σ.	218·4	3n.	1·14	1828·99
H₂.	221·2		·5	30·00
Mä.	217·8		·2	42·71
De.	208·9		·2	64·21
W. & S.	207·7	1n.	0·5	74·17
Gl.	205·2	,,	·42	·84

A C.

239·2	3n.	16·1	28·99

647 O.Σ. 377.

R. A.	Dec.	M.
19h 32m	35° 22'	A 8·4, B 8·5, C 9·2

Direct motion in A B.

A B.

O.Σ.	51·2	2n.	0·88	1842·68
	45·0	,,	·86	53·20
De.	38·3	3n.	·7	67·64
Du.	45·0	5n.	·85	71·07

$\frac{A + B}{2}$ and C.

De.	154·7	3n.	25·14	67·64
Du.	·7	4n.	·36	70·70

648 O.Σ. 378.

R. A.	Dec.	M.
19h 32m	40° 44'	7·2, 9

O.Σ.	283·8	3n.	1·29	1846·05
De.	287·4	,,	·25	66·44

649   Σ. 2556.

R. A. 19^h 34·3^m Dec. 21° 59'   M. 7·3, 7·8

Variable? Certain indirect motion.

Σ.	188·4	3n.	0·56	1829 83
O.Σ.	183·4	1n.	·73	40·84
	176·1	,,	·59	50·77
	345·9	,,	·57	6·58
	163·4		·55	72·64
Mä.	191·1		·5	41·56
	·7		·45	2·67
	188·1	7n.	·98	3·04
	189·9		·55	4·44
Se.	179·1	3n.	·49	56·88
De.	167·7		...	64·91
Du.	175·0	4n.	0·52	8·96
W. & S.	...	round		73·57
W.O.	167·7	1n.	0·63	4·68
Sp.	126·2		·45	5·61

650   O.Σ. 380.

R. A. 19^h 36·9^m Dec. 11° 33'   M. 6, 7·2

Slow retrograde motion in A B. In 1842 O.Σ. discovered the small star C, but it was invisible in 1844, 1849, 1851, and 1872. Dawes never saw it, but De. suspected its existence in 1865 and 1869.

A B.

Mä.	80·8		0·73	1843·53
	67·8		·33	51·73
	72·4		·20	2·72
H₂.	74·6		·59	45·53
Da.	73·0		·49	8·65
	74·9		·47	53·71
O.Σ.	·7	8n.	·62	0·72
Se.	·0	1n.	·54	6·83
	77·0	,,	elong^d·	7·71
	79·7	,,	,,	9·61
De.	69·6	5n.	...	67·82
W.O.	77·4		0·51	74·69

A C

O.Σ.	160·3	1n.	1·21	42·72
		C invisible {		4
				9
				51
	69·0	1n.	1·33	6·57
		C invisible		72
Mä.	342·7		1·30	43·54
		C invisible		51·73
	349·3		...	2·72
Se.	346·9	1n.	...	9·61
De.	355·9	,,	1·7	65·46
		C invisible {		·74
				8·65
	363·3	1n.	1·71	9·74
	359 ?		...	·74
		C suspected		·78

651   O.Σ. 383.

R. A. 19^h 38·8^m Dec. 40° 26'   M. 7, 8·5

Mä.	25·4		0·62	1843·39
O.Σ.	27·4	3n.	·91	5·07
Da.	23·7		·85	53·75
Se.	21·7	2n.	·76	8·22
De.	25·1	4n.	·81	67·64
W.O.	24·2		·85	74·70

652   Σ. 2574.

R. A. 19^h 39·1^m Dec. 62° 23'   M. 8, 8

The angle has increased and the distance diminished.

Σ.	131·3	5n.	0·90	1834·10
O.Σ.	139·4	1n.	·92	40·61
	137·0	,,	·81	5·69
	134·7	,,	·75	6·69
Mä.	136·9		1·05	1·47
	131·7		0·75	3·39
	132·5		·65	4·90
De.	208·9		1·20	64·21

653   Σ. 2576.

R. A. 19^h 41^m Dec. 33° 20'   M. 7·8, 7·8

Certain change in angle; probable diminution in distance.

Dunér has

$$1861·44. \quad \Delta = 3''·27.$$
$$P = 313°·6 - 0°·295 (t - 1850·0).$$

So.	326·2		...	1823·65
H₂.	322·8		4·33	9·65
Σ.	318·8	3n.	3·59	31·80
	141·9		·55	51·80
Mä.	316·9		·53	37·89
	315·1	3n.	·46	43·99
	313·0	2n.	·35	51·80
	311·1	3n.	·37	7·18
	·4	1n.	·21	9·86
O.Σ.	131·3	3n.	·40	1·82
	124·4	1n.	·42	66·76
	125·3	,,	·24	70·92
Mo.	312·2	2n.	·47	56·56
Se.	132·0	4n.	·49	7·15
	308·6		·31	6·60
De.	310·8	2n.	·49	8·02
	308·8		·27	63·35
Eng.	·4	2n.	·19	5·64
M.	296·8	1n.	·46	6·46
Du.	307·3	12n.	2·97	71·61

W. & S.	304°2	5	3˝32	1872·63
	306·4	7	2·93	·75
	304·2	4	3·23	3·59
	7·6	6	·51	4·73
	126·7	4	·52	6·68
	125·0	2	·14	·61
Gl.	305·7	9	·48	4·85
	·3	5	·25	·85
Sp.	304·9		·14	5·65
Fl.	·5	In.	·26	7·64

654   Σ. 2579.

δ CYGNI.

R. A.	Dec.	M.
19ʰ 41·2ᵐ	44° 50′	3, 7·9

Magnitudes.—Σ., 3, 7·9. Sm., 3·5, 9.
Σ. gives the magnitude of B as 6·5 on
one occasion ; Da. always 8 or 9. De.
thinks that B varies both in colour
and magnitude. Dawes, on the other
hand, never "suspected its brightness to
be variable." O.Σ. confirms Σ.'s suspicion that B is variable.

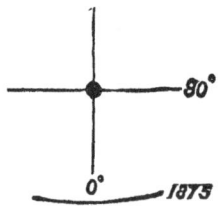

C. Σ., A, greenish ; B, ash. Sm., A, pale
yellow ; B, sea-green.

H₁ (*Phil. Trans.* 1804, p. 377) : "This
double star, I believe, has furnished us
with a second instance of a conjunction re-
sembling that of ζ Herculis. The position,
September 22, 1783, was 18° 21′ n.f.
January 3, 10, and 11, 1802, I could no
longer perceive the small star, which must
have been at least so near the large one as
to be lost in its brightness. January 29,
1804, I examined this star with powers
from 527 to 1500, and saw it as a lengthened
star, but not with sufficient clearness to
take a measure of its position. May 22,
1804, in a very clear evening, I tried 527
and 1500, with the 10-ft. reflector, which
acted remarkably well on the double stars,
but I could not perceive the small star of
δ Cygni." He then tried the 20-ft. reflector
with powers 157 and 360, with the same
result. He then adds : "A parallactic
motion of δ will perfectly account for this
occultation, for the situation of the two
stars, in 1783, was such, that this motion

must have carried the large star, by this
time, nearly upon the small one."

H₂ and So. (*Phil. Trans.* 1824, p. 339).
These observers, using the 5-ft. refractor,
examined δ carefully in 1823, but could
not see the least appearance of elongation.
"The star perfectly round and admirably
defined ; the night beautiful."

Σ. (*M. M.*, p. 25). In 1826 Σ. turned
the Fraunhofer equatorial on this object,
and saw it double on the first examination.
He says, "It is very probable that between
1783 and 1826 the small star performed a
whole revolution + 34° in an orbit very
elliptical, so that the period may be less
than forty years." At p. 297 he thinks
the above remarks need correction, the
period certainly not being forty years.

H₁'s inability to see the companion in
1802 he thinks inexplicable, unless due to
variability or periastron passage, the latter
being perhaps the more probable.

Dawes (*Mem. R. A. S.*, vol. xxxv., p. 416).
Speaking of the difficulty of this star, he
observes that this is a case "in which
great perfection of telescope is of far more
importance than large aperture beyond
about six inches." In the *Astronomical
Register*, 1865, p. 225, he expresses his
opinion that Behrmann's elements are far
from correct if his own measures are not
"egregiously in error." "According to
my own measures, the distance has scarcely
varied for the last twenty-five years."
Behrmann's ephemeris gave 320°± 0˝·4
for 1865, while Dawes's measures in
1865·58 were 349°·62 and 1˝·675.

THE ORBIT.—Mr. Hind was probably
the first to publish elements of this system.
Making use of the observations from 1783
to 1842, he obtained the following results :—

$$T = 1862, \text{ Nov. } 14.$$
$$\pi - \Omega = 243° \ 24'$$
$$\Omega = \ 24 \ \ 54$$
$$i = \ 46 \ \ 23$$
$$e = 0·6067$$
$$P = 178 \text{ years and } 256 \text{ days}$$
$$a = 1˝·811.$$

About 50° of the apparent orbit had then
been described.

Behrmann in 1865, using Klinkerfues'
method (see *Astr. Nach.*, No. 1127),
deduced the elements which follow :—

$$T = 1866·3512$$
$$\pi - \Omega = 280° \ 20'·6$$
$$\Omega = 166 \ \ 26 \ ·4$$
$$i = \ 64 \ \ 38 \ ·4$$
$$e = 0·8470$$
$$\mu = - 1°·283$$
$$P = 280·56 \text{ years}$$
$$a = 3˝·165.$$

His ephemeris gives the following quantities:—

1826	40°.9	1″.816
30	37.3	.772
40	27.2	.635
50	14.8	.432
60	355.6	.000
70	176.8	0.702
78	157.0	1.518
79	155.9	.585
80	154.7	.646

Behrmann used the measures made from 1783 to 1856. A comparison of the computed with the observed quantities shows that the elements require corrections. In 1866, having received the careful measures by Dawes, Dembowski, etc., Behrmann computed a fresh set of elements: they are as follows:

$$a = 2″.30974$$
$$e = 0.28583$$
$$\pi = 289° 42'$$
$$\gamma = 37 \ 46$$
$$\Omega = 91 \ 8$$
$$\lambda = 203 \ 2$$
$$P = 415.11486 \text{ years}$$
$$n = 0°.86723$$
$$T = 1904.1023.$$

Behrmann also compares the observation from 1783 to 1865 with the elements, and a very satisfactory agreement is found. Finally, he gives the ephemeris from 1826 to 1878: the following extract will be of interest:

1826	41°.5	1″.93	1860	358°.7	1″.49
30	37.6	.87	65	350.7	.47
35	32.3	.79	70	342.5	.454
40	26.6	.72	75	334.1	.452
45	20.3	.65	76	332.5	.453
50	13.6	.59	77	330.8	.454
55	6.4	.54	78	329.1	.456

Engelmann gives the following simple formula for the angles of position:

$$P = 20°.4 - 1″.410 \ (t - 1845.0).$$
$$\Delta = 1″.68.$$

On this Dunér remarks that it fairly represents the modern observations, but makes that of H₁ in error to the extent of 36°; and that if, instead of 18° 21′ n.f., we read 18° 21′ s.f., perfect agreement is produced. Dr. Doberck's formulæ are

$$\theta = 12°.48 - 1°.402 \ (\tau - 1850) + 0°.0006 \ (\tau - 1850)^2.$$
$$P = 1″.64 - 0″.0067 \ (\tau - 1850).$$

	°		″	
H₁.	71.6		2.50	1783.72
	single			1802
				4
H₂ & So.	"perfectly round"			23
	"round"			5
Σ.	32.5		1.5	32.72
	40.6	2n.	.91	26.55
	36.9	1n.	.91	8.80
	.7	,,	.57	31.73
	.2	2n.	.70	3.81
	34.7	1n.	.68	5.66
	31.9	4n.	.80	6.52
Mä.	.3		.61	7.27
	26.6		...	41.50
	21.6		1.46	2.77
	22.7		.28	3.45
	23.9		.47	4.36
	21.9		.32	5.65
	20.2		.33	6.35
	19.0		...	7.18
	13.8		1.19	52.44
Sm.	30.9		.5	37.78
	25.6		.8	42.56
	14.7		.5	52.69
Da.	27.4	2n.	.5	39.66
	25.1	,,	...	40.67
	23.7	4n.	1.66	1.89
	16.7	3n.	...	7.39
	14.5	,,	1.76	8.75
	11.5	2n.	.65	51.51
	10.7	,,	.68	2.74
	7.3	3n.	.76	3.73
	4.3	1n.	.68	4.56
	357.7	3n.	.67	9.58
	349.6	2n.	.67	65.38
Ka.	25.7		.72	41.94
	.8		.71	3.12
O.Σ.	19.6	3n.	.68	4.78
	8.3	,,	.51	52.70
	3.3	,,	.65	8.71
	353.8	,,	.60	63.74
	341.7	2n.	.47	72.81
Flt	10.3	30	.75	51.68
Mo.	1.0	26	.11	4.79
	0.4	30	.27	5.74
Se.	...	single		55
	3.2		1.41	6.84
	350.4	3n.	.23	66.08
De.	355.4	5n.	...	2.75
	.5	7n.	1.58	3.61
	351.3	4n.	.68	4.72
	350.5	1n.	.55	5.64
	348.9	15n.	.51	7.06
	347 2	5n.	.56	8.61
	346.4	,,	.58	9.60
	343.4	7n.	.72	70.56
	342.3	5n.	.59	1.50
	339.3	7n.	.51	2.60
	336.2	6n.	.55	3.56
	.3	7n.	.61	4.62
	333.7	5n.	.58	5.58
Eng.	354.4		2.30	64.74
Kn.	349.0	2n.	1.70	5.43
	348.3	,,	.70	6.68

Kn.	337.9	4n.	1".69	1871.74
	330.2	1n.	.69	2.67
Ro.	340.7	,,	..	65.73
Ta.	344.7	,,	1.50	6.59
Br.	349.0		.52	8.69
Du.	343.7	13n.	.53	70.85
	336.5	4n.	.52	5.69
W. & S.	.8	5	.70	2.78
	9.1	7	...	4.70
	5.8	7	...	5.70
Dob.	Companion not seen			.77

655 O.Σ. 385.

R. A. 19h 42m	Dec. 40° 16'		M. 7.5, 9.8

O.Σ.	55.0	3n.	1.31	1845.07
De.	51.1	,,	.38	66.62

656 Σ. 2583.

R. A. 19h 43m	Dec. 11° 31'		M. 6, 6.8

Dunér has

$$1856.19. \quad \Delta = 1".43.$$
$$P = 123°.1 - 0°.088 \ (t - 1850.0).$$

H$_1$.	124.4	1n.	...	1783.65
	127.5	,,	...	1802.72
H$_2$ & So.	135.5	,,	1.96	23.70
	123.5	4n.	.55	5.61
	127.4	1n.	.68	32.56
Σ.	120.7	6n.	.50	29.96
Da.	124.5	2n.	.83	30.56
	120.8	1n.	.63	65.74
Sm.	122.0		.5	31.70
	121.3		.7	6.81
Mä.	123.2	4n.	.39	42.17
	120.8	2n.	.40	7.96
O.Σ.	122.3	4n.	.50	8.24
Mo.	120.8	2n.	.49	55.88
Kn.	121.1	4n.	.45	65.67
Ka.	117.6	6n.	.34	5.84
Du.	119.0	1on.	.27	72.17
W. & S.	.7	3n.	.48	3.37
Gl.	121.4	2n.	.48	4.85
Dob.	119.9	4n.	.27	7.70

657 O.Σ. 386.

R. A. 19h 44m	Dec. 36° 51'		M. 7.7, 8

O.Σ.	77.5	3n.	0.97	1846.63
Mä.	83.8	1n.	.75	47.73
Da.	78.6	,,	.97	8.55
Se.	79.9	3n.	.84	58.68
De.	77.8	,,	.92	67.27
Du.	82.2	7n.	.80	70.69

658 O.Σ. 387.

R. A. 19h 44.2m	Dec. 35° 0'		M. 7.2, 8.2

Rapid direct motion.

Mä.	118.3		0.52	1843.39
	119.4		.60	7.73
	94.7		.52	53.78
O.Σ.	129.4	2n.	.50	44.18
	103.8	3n.	.47	51.97
	90.7	2n.	.57	5.63
	78.4	,,	.60	61.22
Da.	89.7	3n.	.53	53.75
Se.	91.9	1n.	.3	6.83
	198.2	,,	.25	9.61
De.	52.6	3n.	...	68.25
	.5	1n.	oval	70.56
	39.6	,,	oblong	1.57
	26.6	,,	wedg^d.	2.55
	23.0	6n.	oblong	3.73
	20.7	5n.	,,	4.57
	22.0	3n.	0.34	5.40
	17.7	2n.	.46	7.67
Sp.	23.2	...	.38	5.57

659 a AQUILÆ.

R. A. 19h 44.9m	Dec. 8° 33'		M. 1.5, 10.2

The proper motion of Altair is + 0".56 in R.A., and + 0".39 in Dec.

H$_1$.	334.7		143	1781.55
Σ.	326.2		153.5	1821.16
	324.7		152.9	5.53
	322.1		.3	36.29
Sm.	323.1		.6	4.81
O.Σ.	318.5		153.3	51.81
De.	314.9		154.	65.07
W. & S.	313.9	1n.	156 4	71.58
	312.7	2n.	.4	3.60
	.9	1n.	...	5.68
Gl.	313.0	,,	157.0	4.70
Fl.	311.9	,,	156.1	7.82

660 O.Σ. 388.

R. A. 19h 47m	Dec. 25° 33'		M. A 7.6, B 7.6, C 8.8

A B.

Mä.	156.9	1n.	3.89	1847.73
O.Σ.	140.5	5n.	.70	8.51
Da.	139.8	1n.	.85	.51
	140.0	,,	...	53.73
	139.4	,,	3.89	4.72
De.	.5	3n.	.70	65.89
Du.	.3	4n.	.63	9.59

B C.

De.	138°·4	3n.	26″·8	1865·89
Du.	·2	4n.	·84	9·53

661 Σ. 2596.

R. A.	Dec.	M.
19ʰ 48·5ᵐ	14° 59′	7·2, 8·6

Σ.	353·1	4n.	2·12	1831·26
Mä.	351·8		·05	42·71
De.	343·4		·02	64·52

662 Σ. 2603.

R. A.	Dec.	M.
19ʰ 48·6ᵐ	69° 58′	4, 7·6

Is B variable?

The proper motion of A is $+ 0^s\cdot015$ in R. A., and $+ 0''\cdot01$ in N. P. D.

Dunér gives

$$1853\cdot19. \quad \Delta = 2''\cdot94.$$
$$P = 357°\cdot1 + 0°\cdot152 \, (t - 1850\cdot0).$$

H_1.	333·2	1n.	2·5	1781·81
	354·5	,,	...	1804·39
H_2.	355·3	,,	2·59	23·58
	348·5	,,	3·27	8·64
	353·2	,,	·27	30·67
Da.	354·8		·09	·67
	353·2	1n.	2·84	43·78
	356·5	,,	·83	8·87
Σ.	354·5	6n.	·79	32·44
Sm.	·6		3·1	3·68
	356·3		·0	46·77
Mä.	355·7		2·69	1·54
	358·8	4n.	·81	3·88
De.	357·8	1n.	·92	56·53
Se.	358·0	2n.	·74	·75
Mo.	360·4	4n.	3·06	9·75
M.	353·1	1n.	2·65	61·82
O.Σ.	359·3	2n.	3·04	3·66
Ro.	349·0	,,	·02	5·65
Ta.	..	1n.	·01	9·49
	355·9	,,	·03	74·61
Du	360·5	7n.	2·99	0·79
W. & S.	·1	2n.	3·20	4·72
Gl.	·9	,,	·15	·84
Dob.	359·8	3n.	2·89	6·61
C.O.	360·0	4n.	·93	7·97

663 O.Σ. 532.

β AQUILÆ.

R. A.	Dec.	M.
19ʰ 49ᵐ	6° 6′	3·4, 11·3

O.Σ.	17·1	4n.	12·36	1852·44
	·0	,,	·60	8·12
	15·5	2n.	·63	63·70
	·6	,,	·67	74·74

664 Σ. 2606.

R. A.	Dec.	M
19ʰ 53·9ᵐ	32° 57′	7·5, 8·2

Dunér gives

$$1849\cdot77. \quad \Delta = 1''\cdot16.$$
$$P = 133°\cdot3 + 0°\cdot09 \, (t - 1850\cdot0).$$

Σ.	132·0	6n.	1″·19	1834·39
Mä.	133·7		·22	42·42
	·2	1n.	·13	4·34
O.Σ.	134·9	4n.	·33	5·48
Da.	137·8	2n.	·19	51·24
De.	134·8	1n.	·1	6·65
Se.	113·9	2n.	·01	·80
Du.	134·5	4n.	·17	68·25

665 Σ. 2607.

O.Σ. 392.

R. A.	Dec.	M.
19ʰ 54ᵐ	41° 56′	A 7·2, B 9, C 9·2

A B. The angle has diminished. No change in A C.

A B.

O.Σ.	330·0	1n.	0·50	1842·67
	317·9	2n.	·41	5·65
	316·6	1n.	·38	51·62
	310·2	,,	·40	4·69
	323·5	,,	·41	7·67
	318·0	,,	·53	8·59
De.	306·9	4n.	...	67·45

A C.

Σ.	293·4	3n.	3·22	31·52
O.Σ.	292·0	1n.	·29	42·67
	290·3	2n.	·20	5·65
	289·6	1n.	·07	51·62
	·5	,,	·29	4·69
	290·1	,,	·33	7·67
	·9	,,	·41	8·59

666 O.Σ. 393.

R. A.	Dec.	M.
19ʰ 54ᵐ	44° 4′	7·5, 8·4

The distance has diminished.

O.Σ.	225·7	3n.	21·75	1847·44
	226·7	,,	·12	71·48
De.	·1	2n.	·21	65·96
Du.	·3	,,	20·78	9·78

667 O.Σ. 395.

R. A. 19h 56·9m Dec. 24° 36′ M. 5·8, 6·2

Probable direct motion.

Mä.	89·2		0·50	1843·53
	67·4	In.	·45	5·79
	74·2	,,	·45	7·73
	·5		·52	·86
	78·3	In.	·40	51·72
	91·0	,,	·6	2·85
O.Σ.	79·3	2n.	·64	44·16
	80·0	,,	·57	50·22
	96·3	In.	·68	74·76
Da.	80·6	2n.	·57	52·60
Se.	93·1	In.	elongd.	9·61
Eng.	82·9	2n.	0·64	65·60
De.	89·1	4n.	...	7·41
W.O.	91·4		0·67	74·72
Sp.	92·7		·69	5·63
Du.	91·7	2n.	·63	·69

668 Σ. 2619.

R. A. 19h 57m Dec. 47° 56′ M. A 8·1, B 8·1, C 12

Probable change in A C.
C was discovered by O.Σ. in 1851·82.

A B.

Σ.	244·9	4n.	4·29	1831·91
Mä.	245·8	In.	3·99	43·80
Se.	246·3	3n.	4·37	57·54
Mo.	244·5	2n.	·21	·71
De.	·0	In.	·42	8·39
O.Σ.	63·9	3n.	·25	9·14
Du.	245·9	8n.	3·91	70·37

A C.

O.Σ.	296·6	In.	17·79	54·69
	302·7	,,	16·88	70·92

669 Σ. 2627.

R. A. 20h 2m Dec. 4° 26′ M. 9, 11·5

Σ.	23·2	3n.	1·96	1829·37
Mä.	25·3	In.	2·21	42·72

670 Σ. 2640.

R. A. 20h 3·2m Dec. 63° 33′ M. 6, 9·9

Σ.	30·5	In.	4·73	1831·87
	28·8	,,	·95	2·02
	24·9	2n.	5·01	3·38
Da.	23·3		4·99	41·80
Ro.	...	In.	·72	65·65
	25·3	,,	5·07	·74
Br.	22·8		·13	8·73
	23·8		·02	9·69
Ta.	26·9	In.	·51	·49
W. & S.	21·0	,,	4·77	73·73
	23·6	,,	5·36	5·70
Gl.	20·3	,,	...	3·79
Dob.	22·7	3n.	5·23	7·74

671 Σ. 2637.

θ SAGITTÆ.

R. A. 20h 4·6m Dec. 20° 33′ M. 6, 8·3

Probable increase of distance in A B : certain increase in A C. The proper motion of A is probably the cause of these changes.

The common proper motion of A B is given by Σ. as $+0''·061$ in R. A., and $-0''·147$ in N. P. D.

A B.

H1.	...		11·07	1781·64
So.	328·0		·78	1824·98
H2.	320·0		10	7·64
	325·9		18·0	30·00
Σ.	326·7	8n.	11·40	2·82
Sm.	327·1		·4	4·77
Mä.	328·1		·17	42·71
	325·7		·37	3·62
	326·3		·21	7·98
	·5		·72	51·80
O.Σ.	·3	In.	·73	·80
	328·3	,,	·54	74·72
	·7	,,	·74	·79
Mo.	326·7	60	·40	53·68
M.	·2	In.	·62	61·79
	·8	,,	12·33	6·46
	428·5	,,	11·50	75·63

A C.

H1.	...		58	1781·64
So.	226·8		69·66	1824·98
Σ.	·8	4n.	70·22	8·73
	·6	5n.	·98	35·28
Sm.	·6		·1	4·77
O.Σ.	·0	In.	72·86	51·80
	225·1	,,	75·83	74·72
	·1	,,	·65	·79
Mo.	·7	26	72·91	53·70
M.	·0	In.	73·09	61·79
	224·8	,,	·93	6·46
	·7	,,	74·73	75·65

672   Σ. 2636.

R. A. 20ʰ 5·3ᵐ	Dec. − 4° 56′		M. 8·2, 9·2
Σ. 201·8	2n.	12·51	1827·24
Mä. 204·0		13·27	43·75
·3		·63	52·39
·2		...	3·79
Mit. 202·3	1n.	12·23	48·66
C.O. 203·7	1n.	·73	77·73

673   Σ. 2641.

R. A. 20ʰ 5·9ᵐ	Dec. 3° 27′		M. 7·5, 11·2
Σ. 170·1	2n.	20·34	1827·76
Mä. 169·5	1n.	·78	43·70

674   O.Σ. 400.

R. A. 20ʰ 6·2ᵐ	Dec. 43° 37′		M. 7·2, 8·2

A binary system in retrograde motion. The distance is sensibly constant.

O.Σ. 334·9	3n.	0·64	1845·73
324·6	2n.	·59	53·23
319·3	,,	·62	60·10
Mä. 326·7		·50	43·39
Da. 320·5	1n.	·65	53·89
311·2	,,	...	65·51
130·3	,,	separatᵈ	·94
De. 307·8	3n.	...	6·67
121·9	1n.	oblong	8·55
W. & S.		not divided	73·62
		elongated ?	6·77
Schi. 267·9	1n.	0·33	5·67
Sp. ·9		·33	·67

675   Σ. 2652.

R. A. 20ʰ 7ᵐ	Dec. 61° 43′		M. 7·3, 7·6
Σ. 281·3	5n.	0·43	1834·18
O.Σ. 292·5	1n.	wedged	40·61
289·4	,,	oblong	1·62
282·2	,,	0·4	54·69

676   Σ. 2649.

R. A. 20ʰ 7·6ᵐ	Dec. 31° 43′		M. 7·7, 8·8
Σ. 152·3	3n.	26·08	1832·19
Mä. 151·1	1n.	25·94	47·79

677   Σ. 2646.

R. A. 20ʰ 8ᵐ	Dec. − 6° 25′		M. 7, 8·8
So. 50·6	3n.	25·11	1825·69
Σ. 51·6	,,	24·70	9·42
Mä. ·2		·30	43·81

678   O.Σ. 403.

R. A. 20ʰ 10ᵐ	Dec. 41ᵛ 45′	M. A 7, B 7·2, C 9·5

Dembowski's distance is probably too great.

A B.

O.Σ. 173·0	5n.	0·59	1848·10
De. ·4	3n.	1·00	66·85
Du. 171·6	2n.	0·79	75·68

$$\frac{A + B}{2} \text{ and } C.$$

O.Σ. 33·1	5n.	11·83	48·10
De. 34·0	3n.	·70	66·85

679   Σ. 2658.

R. A. 20ʰ 10·5ᵐ	Dec. 52° 45′	M. A 7, B 9, C 10

C. A yellowish white, B blue.

A B.

H_2. 126·6		5·36	1829·70
Σ. ·9	4n.	·49	31·62
Mä. 124·8	2n.	·34	43·58
122·4		4·80	7·85
127·9		·58	51·87
122·2		·29	3·38
De. 123·0		5·45	63·51
W.O. 125·1		4·89	76·77

A C.

H_2. 220·2		33·48	30·76
Σ. 216·8	3n.	32·07	2·14
De. 213·2		37·80	63·51

680   O.Σ. 405.

R. A. 20ʰ 14ᵐ	Dec. 32° 52′		M. 7·7, 8·7
O.Σ. 152·6	3n.	0·61	1846·43
De. 144·7	,,		67·72

681   Σ. 2690.

R. A. 20ʰ 25·5ᵐ	Dec. 10° 51′	M. A 7, B 7·5, C 7·6

In A and $\dfrac{B+C}{2}$ the distance has gradually increased.

In 1840 Dawes detected the duplicity of B. In this pair there is probably a slow retrograde motion, with decrease of distance. The star was also detected by O.Σ. independently in 1842.

A B.

Σ.	256°·3	4n.	14"·19	1831·26
Mä.	255·2		·20	53·76
Ro.	257·2	3n.	·73	65·68

B C.

Da.	211·4	1in.	0·65	41·95
	·2	4n.	·58	50·19
Sm.	210·5		·7	42·58
	215·0		·5	57·71
O.Σ.	210·7	1n.	·63	42·67
	34·5	,,	·60	5·75
	37·1	,,	·57	7·71
	207·0	,,	·49	51·67
Ro.	...	2n.	·49	65·64
De.	202·2	3n.	...	6·30
W. & S.	220·0	est$^{n.}$	...	71·53
	227·3	1	...	3·68

A C.

125		20	fixed

A and $\dfrac{B+C}{2}$.

Σ.	256·0	2n.	14·05	1829·64
	·5	,,	·33	32·88
Da.	·4	4n.	·56	41·37
O.Σ.	·4	1n.	·87	2·67
	257·0	,,	·61	5·75
	256·1	,,	·61	7·71
	257·1	,,	·77	51·67
De.	256·1	4n.	·88	65·15
W. & S.	255·3	5	15·2	71·53
	256·3	4	·39	3·68
	·5	2	·09	5·74
Gl.	255·0	4	·1	3·91
Pl.	254·7	4n.	·13	7·07

682 Σ. 2695.

R. A.	Dec.	M.
20h 26·8	25° 24'	6·2, 8

Σ.	76·5	5n.	0·79	1831·78
O.Σ.	75·0	4n.	1·04	50·84

683 Σ. 2696.

R. A.	Dec.	M.
20h 27·5m	5° 2'	8, 8·4

O.Σ.'s measure in 1872 shows that the angle is probably unchanged. The more recent measures seem to indicate a slight increase of angle and decrease of distance.

Σ.	300·0	1n.	1"·27	1825·71
	298·5	3n.	·32	32·84
H$_2$.	290·1		·0	1·00
Mä.	302·8		0·99	8·27
	308·5		·90	42·72
	302·6		·90	3·69
	309·7		·90	61·76
Se.	310·3	1n.	·72	56·62
W. & S.	303·4	5	·66	72·64
	305·2	4	·85	3·69
	306·0	5	...	5·75
Gl.	304·4	8	0·8	3·91
	3·7	7	·80	·91
Fer.	...	in contact		4·54

684 Σ. 2703.

R. A.	Dec.	M.
20h 31·2m	14° 19'	7·6, 7·6

Increase of distance in B C and A C.

Dunér's formulæ for the motion of C with reference to $\dfrac{A+B}{2}$ are

$$\Delta \sin P = 46''\cdot11 - 0''\cdot0250\,(t - 1853\cdot0).$$
$$\Delta \cos P = -39''\cdot80 - 0''\cdot0582\,(t - 1853\cdot0).$$

A B.

H$_1$.	288·5	1n.	26	1783·65
Σ.	290·5		...	1822·14
	291·1	4n.	25·28	9·52
So.	290·0	2n.	·08	4·81
Mä.	291·4		24·89	42·13
	290·6	1n.	25·06	7·69
Se.	·9	2n.	·06	57·26
De.	291·0	,,	·11	8·17
	290·9	4n.	·15	64·60
Du.	·5	5n.	·23	8·53

A C.

Σ.	238·6		...	21·85
	239·4	3n.	66·72	9·40
Se.	238·5	1n.	...	56·69
De.	·2	2n.	67·27	8·17
	·2	4n.	68·66	64·60
Du.	·0	3n.	·75	8·53

B C.

Σ.	216·8		...	21·88
	217·9	3n.	54·38	9·42
So.	·1		·30	4·78
Se.	·7	1n.	56·07	56·69
De.	·3	2n.	55·92	8·17
	·2	4n.	57·02	64·60
Du.	·4	,,	·03	8·53

685   Σ. 2704.

β DELPHINI.

R. A.   Dec.   M.
20ʰ 31·9ᵐ   14° 11'   A 3·5, B 4·5, C 11

C. A green.

A B.—In 1873 Mr. Burnham discovered that A was a close double star, and Dembowski's measure in 1875 seems to indicate rapid angular change.

According to Σ. (*P. M.*, ccxxxi.) the star C does not partake in the proper motion of the system.

A B.

Burnham	355°		0·7	1873·60
	15·5	5n.	·65	4·66
	20·1	4n.	·54	5·65
	25·8	,,	·48	6·63
	29·7	5n.	·51	7·71
O.Σ.	8·0	1n.	·69	4·73

$$\frac{A + B}{2} \text{ and } C.$$

H₁.	348·0		27·4	1781·58
Σ.	343·8		32·48	1829·40
	339·9	1n.	33·77	51·88
Sm.	341·8		30·0	34·79
Mä.	342·1		...	43·63
	340·6		...	51·80
O.Σ.	338·6	1n.	33·71	·81
De.	336·6		34·64	64·94
W. & S.	338·9	3	...	75·74

686   O.Σ. 533.

R. A.   Dec.   M.
20ʰ 33·3ᵐ   9° 40'   4·7, 11·3

The proper motion of A is + 0ˢ·0227 in R. A., and + 0"·340 in Dec.

This star, owing to the great difference in the brightness of the components, is very difficult to measure. The distance is probably unchanged since 1851. O.Σ. has deduced the following formulæ :

$$\Delta A = -\,0''\cdot262 \pm 0''\cdot036 - (0''\cdot2851 \pm 0\cdot0040)\,(t - 1860\cdot0).$$

$$\Delta D = +\,10\cdot021 \pm 0\cdot036 - (0\cdot0155 \pm 0\cdot0040)\,(t - 1860\cdot0).$$

It is probable that the changes are due to the proper motion of κ. There is a third star following κ about 3'·5 distant which most probably forms with it a binary system.

O.Σ.	11·9	2n.	10·26	1851·71
	9·8	1n.	·25	2·63
	·8	,,	·63	3·82
	5·4	,,	·22	6·57

O.Σ.	359·4	1n.	10·32	1859·62
	348·3	,,	·12	65·78
	338·2	,,	·34	72·64
	335·1	,,	·88	4·79

687   Σ. 2708.

R. A.   Dec.   M.
20ʰ 34·1ᵐ   38° 13'   7, 8·7

C. A yellow, B blue.

The formulæ for rectlinear motion deduced by Σ. (see *P. M.*, p. ccxxxii.) still fairly represent the observations. O.Σ. obtains the following :

$$\Delta A = -\,4''\cdot800 \pm 0''\cdot040 - (0''\cdot1786 \pm 0''\cdot0029)\,(t - 1850\cdot0);$$

$$\Delta D = +\,14''\cdot528 \pm 0''\cdot040 + (0\cdot1939 \pm 0''\cdot0029)\,(t - 1850\cdot0);$$

and the differences show that there has yet been no departure from rectilinear motion. Dunér finds the following formulæ :

$$\Delta \cos P = +\,14''\cdot47 + 0''\cdot1875\,(t - 1850\cdot0),$$
$$\Delta \sin P = -\,4''\cdot69 - 0''\cdot1745\,(t - 1850\cdot0);$$
and
$$\Delta \cos P = +\,14''\cdot34 + 0''\cdot1865\,(t - 1850\cdot0),$$
$$\Delta \sin P = -\,4''\cdot76 - 0''\cdot1693\,(t - 1850\cdot0).$$

So.	2·3	1n.	9·65	1823·68
H₂.	355·1	,,	10·45	8·76
	352·3	,,	11·32	32·34
Σ.	354·6	2n.	10·82	29·86
	351·2	,,	·96	32·36
	349·3	,,	11·97	5·78
	348·1	4n.	12·23	6·89
Da.	352·8		11·24	0·71
	·0		...	2·56
	351·1	1n.	11·46	3·87
	350·5	5n.	·70	4·55
	347·4	2n.	12·61	7·75
	346·9	,,	·91	9·79
	345·9	,,	13·16	40·67
	·6	,,	·46	1·63
	·0	,,	...	2·65
	342·2	,,	13·69	3·86
	340·5	1n.	16·01	53·82
O.Σ.	346·0	2n.	13·06	39·86
	343·2	3n.	14·58	46·69
	340·2	2n.	16·52	54·82
	336·2	,,	19·52	67·74
Ka.	345·9		13·04	41·83
	337·5	6n.	18·66	65·85
Mä.	343·4	,,	14·34	44·76
Flt.	342·0	37	15·76	51·79
Mo.	340·4	3n.	16·10	4·66
De.	339·1		15·91	5·13
	338·6		16·69	7·38
	337·1	5n.	18·31	63·02
Se.	338·3	1n.	17·26	57·91
Po.	337·7	15	18·10	9·85
M.	336·7	1n.	17·93	62·48

Eng.	336·2	In.	18″·80	1865·58
Bo.	335·3	2n.	17·77	·71
Du.	336·2		19·83	9·40
Ta.	·9	5	...	71·54
W. & S.	335·5	4	21·0	3·69
	·0	5	·0	·69
	334·7	3	·2	·72
	336·6	5	·28	5·74
	333·2	3	22·08	6·77
Gl.	335·0	8	21·4	3·91
Dob.	333·6	2n.	...	6·62
	·6	,,	21·81	7·69
Pl.	·3	4n.	·86	·20
Fl.	·9	In.	·67	·78

688   OΣ. 410.

R. A. 20^h 35·1^m Dec. 40° 9′   M. A 6·4, B 6·7, C 7·7

A B.

Mä.	23·1		0·52	1843·42
	28·9		·42	6·93
O.Σ.	23·3	7n.	·63	50·60
Da.	27·7		·53	3·82
	24·1		·5	67·35
De.	·1	4n.	·40	57·35

$$\frac{A+B}{2} \text{ and } C.$$

O.Σ.	69·8	4n.	68·69	51·45

689   OΣ. 411.

R. A. 20^h 38·3^m Dec. 45° 25′   M. 7·4, 10·2

Direct motion: change in distance.

O.Σ.	273·7	2n.	15·26	1845·36
	278·7	3n.	14·80	52·11
	291·5	In.	15·02	70·92
Mä.	273·7		14·28	46·04
De.	288·9	3n.	·62	66·91

690   Σ. 2725.

R. A. 20^h 40·6^m Dec. 15° 28′   M. 7·3, 8

Change in angle and distance.
Dunér has

$$\Delta = 4''{\cdot}56 + 0''{\cdot}0154\,(t - 1850{\cdot}0).$$
$$P = 358°{\cdot}1 + 0°{\cdot}110\,(t - 1850{\cdot}0).$$

H₁.	348·7	In.	...	1783·29
Bo.	355·0	5n.	...	1825·08
	...	,,	4·98	·40
Σ.	357·6	4n.	·28	31·78

O.Σ.	353·9	3n.	4″·58	1839·86
	355·8	2n.	·61	44·62
	358·4	,,	·95	59·66
	359·8	In.	·94	72·64
Da.	355·7	4n.	·65	41·16
	356·8	3n.	·74	54·32
Mä.	·4		·78	41·57
	357·0		·54	2·74
	·0	Ion.	·78	3·30
	358·3	3n.	5·00	9·79
Mo.	357·9	,,	4·61	54·75
De.	358·9	5n.	·71	·68
	359·4		·71	67·39
Se.	·9	4n.	·77	56·85
Ro.	·5	In.	·60	65·71
Ka.	·9		·78	6·75
Du.	·8	5n.	·67	8·55
Ta.	·2	In.	...	·69
	358·7	6	5·05	71·68
Fer.	I·0		4·73	2·59
Sp.	0·4		·85	6·11
Dob.	358·6	3n.	5·6	·78
	·6	2n.	·06	7·69
	359·7	3n.	4·44	8·58
Fl.	0·9	In.	·81	·82

691   Σ. 2726.

R. A. 20^h 41^m Dec. 30° 17′   M. 4, 9 2

C. A very yellow.

Σ.	57·2	4n.	6·61	1830·82
Se.	59·6	3n.	·33	57·35
O.Σ.	61·2	2n.	·44	62·26
Dob.	60·5	,,	·28	77·70

692   Σ. 2727.

γ DELPHINI.

R. A. 20^h 41·8^m Dec. 15° 42′   M. 4, 5

C. A, golden : B, bluish green. B appears to vary in colour : it is given as yellow, green, and blue, by different observers.

The common proper motion is − 0ˢ·004 in R. A., and + 0″·19 in N. P. D.
Dunér's formulæ are

$$\Delta = 11''{\cdot}91 - 0''{\cdot}0170\,(t - 1850{\cdot}0).$$
$$P = 273°{\cdot}5 - 0°{\cdot}035\,(t - 1850{\cdot}0).$$

H₁.	...	IIn.	1·01	1780·17
	274·5	4n.	...	·65
	273·3	In.	...	1804·44
Σ.	·6	,,	11·83	23·34
	·8	5n.	·90	30·89
So.	·7	In.	12·32	23·68
Du.	·4		·07	31·59
	272·1		11·52	59·05

Sm.	273°6		12"1	1831·60
	·4		·3	4·52
	·3		11·8	9·71
H₂.	·1	2n.	12·05	2·57
Mä.	272·8	6n.	·03	5·84
	273·5	,,	11·46	42·52
	·2	1n.	·44	5·72
	·3	3n.	·39	9·64
	272·5	2n.	·36	54·81
Po.	·3	3n.	·70	45·70
Mo.	·4	2n.	·28	56·51
Se.	·5	7n.	·69	7·03
De.	271·4	3n.	·42	8·23
	272·4	1n.	·45	63·89
	·1	,,	·52	5·88
	271·5	2n.	·29	7·53
M.	270·8	1n.	·40	3·71
Ro.	·9	,,	·54	5·74
Kn.	272·7	,,	·73	5·78
Ka.	271·7	6n.	·18	6·74
Du.	272·4	,,	·42	8·28
W. & S.	271·7	1n.	·1	73·69
Gl.	272·2	,,	·36	·91
Fl.	270·8	,,	·25	7·82
Dob.	272·1	3n.	·16	8·74
Goldney.	·4	,,	·12	·75

O.Σ.	122°3	4n.	0"65	1842·66
	118·1	3n.	·60	5·18
	109·5	4n.	·56	8·80
	106·8	,,	·55	52·02
	95·5	3n.	·67	6·98
	93·4	,,	·66	60·97
	86·3	2n.	·70	71·75
Mä.	114·3		·55	43·53
	36·8		·3	7·82
Sm.	130·0		·7	3·74
Da.	108·8	1n.	·7	51·99
	103·4	5n.	·64	4·07
	96·5	1n.	·71	60·81
	92·5	,,	·68	6·99
Se.	100·2	3n.	·64	58·76
De.	92·6	6n.	...	66·39
	85·1	3n.	oblong	55·88
	93·8	4n.	0·4	65·73
	91·2	1n.	·5	6·84
	89·8	,,	...	8·55
	88·7	5n.	0·6	71·41
	83·9	,,	·51	6·71
Du.	92·5	7n.	·62	69·68
	90·5	4n.	·71	71·26
	87·4	3n.	·68	5·70
W. & S.	88·5	4	·45	2·65
	93·7	2	...	3·69
Schi.	82·5	1n.	0·72	5·59
Sp.	·5		·72	·60
Dob.	86·7	1n.	·58	·76

693   O.Σ. 413.

λ CYGNI.

R. A.	Dec.	M.
20ʰ 42·5ᵐ	36° 3′	5, 6·3

C. Da., pale yellow ; Sm., all bluish.

H₁ VI. 32, is λ Cygni, a wide double star, the components being of 5th and 12th magnitude, according to South, and the distance about 1¼ minutes. The measures indicate fixity. In 1843 the larger star was first seen double by O.Σ.

Dawes (*Mem. R. A. S.*, vol. xxxv., p. 427) writes, "A close and beautiful binary, discovered by Mr. Otto Struve." "On one night Mädler observed an object which he has called O.Σ. 413 ; but the angle recorded is so far from the true one, that the star cannot have been seen really elongated, though it might reasonably be expected that the Dorpat telescope of 9·6 inches aperture would be capable of even separating such an object." At p. 498 he adds, "The retrograde movement in the position of this close double star continues so as satisfactorily to prove its binary character." He notes the great discrepancies in the measures, the great difficulty of the object, and the absence of change in the distance.

A C unchanged.

The angular motion has probably slackened. (1878.)

694   O.Σ. 414.

R. A.	Dec.	M.
20ʰ 43ᵐ	41° 59′	7·2, 8·3

The distance appears to have increased.

Mä.	94·2	1n.	9·25	1847·82
O.Σ.	95·9	6n.	·88	8·30
De.	·5	3n.	·92	66·80
Du.	·6	6n.	10·02	9·72

695   Σ. 2729.

4 AQUARII.

R. A.	Dec.	M.
20ʰ 45·1ᵐ	− 6° 4′	5·9, 7·2

H₁ (*Phil. Trans.* 1804, p. 371) : "The position of the two stars, July 23, 1783, was 81° 30′ n.p. ; and, by a mean of two observations, August 28 and 29, 1802, it was 61° 5′ n.f. Both the last measures are positive with regard to the position being following, and not preceding, as it certainly was in the year 1783. This proves a change of 37° 25′ in 19 years and 37 days. The distance is perhaps a little increased.

September 5, 1782, it was ⅓ diameter of s. August 29, 1802, less than ½ diameter of s." He infers "a real motion, the nature of which cannot remain many years unknown; its velocity, hitherto, having been at the rate of nearly two degrees, per year, of angular change."

Σ. (*M. M.*, p. 8) began his measures of this star in 1825, but was unable to separate the components till 1833. His measures in 1829·76 led him to think that H_1's measure in 1783·56 was erroneous, an entire revolution between 1802 and 1829 being at variance with his observations from 1825 to 1833. He found it a very difficult object even in 1836, but suspected a decrease of distance and a direct angular motion.

Sm. (*Cycle*, p. 488) found this object excessively difficult; "but after succeeding in making it wedge-shaped in a direction towards a 14th magnitude star in the n.f. quadrant, long gazing brought up a bright point of light in the same direction."

Da. (*Mem. R. A. S.*, vol. xxxv., p. 427) says that the distance has diminished, and that there has been an acceleration of the direct motion since Σ. measured this star. A careful examination of the measures led him to think that it was H_1's result in 1802 that was in error rather than that in 1782; and he further notes that Σ.'s positions differ 18° *inter se*, and that the mean result is too small.

The common proper motion is $+0''·061$ in R. A., and $-0''·043$ in N. P. D. (Σ.)

	°		''	
H_1	351·5		0·5	1783·36
	28·9		...	1802·66
Σ.	25·0	1n.	0·81	25·59
	30·0	,,	·80	·61
	13·4	,,	·69	30·92
	23·0	,,	...	2·90
	31·2	,,	·67	3·77
H_2.	46·6		·67	2·73
Sm.	45·0		·5	4·69
Da.	62·1	2n.	...	9·68
	65·5	,,	0·6	40·72
	72·7	1n.	...	1·80
	81·7	,,	...	3·76
	95·9	,,	0·5	53·70
	101·7	,,	·3	4·75
Mä.	24·6		·6	41·49
	27·2		·45	2·82
	31·8		·52	3·70
Ka.	81·7		...	3·76
Se.	107·9	1n.	0·3	56·81
	125	,,	elong^d.	65·71
Ro.	143·6	,,	,,	·74
De.	140		,,	6·12
Sp.	157·0		0·42	75·62
Schi.	·0	1n.	·42	·62
W. & S.	not separated			6·86
C.O.	158·5	1n.	0·5	77·70

696 OΣ. 416.

R. A.	D ec.	M.
$20^h 47·7^m$	43° 19'	7·8, 8·1

Dunér gives

$$1859·79. \quad \Delta = 6''·98.$$
$$P = 143°·1 - 0°·2 \, (t - 1860·0).$$

	°		''	
O.Σ.	146·7	3n.	6·97	1846·13
	142·4	2n.	7·17	63·23
Mä.	145·9	1n.	·31	43·56
	143·8	,,	6·80	7·82
Da.	·6	1n.	7·05	53·89
De.	141·7	3n.	·01	66·10
Du.	·3	,,	6·99	9·80
W. & S.	139·8	1n.	7·35	76·78

697 OΣ. 417.

R. A.	Dec.	M.
$20^h 48^m$	28° 41'	A 7·5, B 8·1, C 9·4

A B.

	°		''	
O.Σ.	39·4	5n.	0·57	1847·98
De.	35·4	1n.	...	69·78

$\frac{A+B}{2}$ and C.

	°		''	
O.Σ.	109·0	5n.	30·49	47·98
De.	·3	3n.	·87	66·86

698 Σ. 2734.

R. A.	Dec.	M
$20^h 48·3^m$	12° 39'	8·2, 8·7

	°		''	
Σ.	181·7	3n.	28·50	1829·79
De.	187·9		26·72	63·54
W. & S.	191·7	1n.	27	76·77

699 OΣ. 418.

R. A.	Dec.	M.
$20^h 49·9^m$	32° 15'	7·3, 7·4

Gradual decrease in angle and increase in distance.

	°		''	
O.Σ.	301·8	2n.	0·56	1842·67
	292·8	,,	·67	8·81
	287·9	,,	·74	53·20
	291·2	,,	·88	60·64
	293·0	1n.	·96	8·77
Se.	112·6	2n.	·75	58·57
De.	292·4	3n.	1·01	66·90
W.O.	110·4		·04	74·72

700   O.Σ. 420.

R. A. 20^h 50^m	Dec. 40° 15′		M. 7, 11·2
O.Σ.	0·6	3n.	5·79	1848·30
De.	4·7	,,	·54	67·00

701   O.Σ. 422.

R. A. 20^h 51^m	Dec. 44° 42′		M. 7·4, 9·1
O.Σ.	331·9	5n.	2·72	1851·35
De.	334·8	4n.	·60	67·43

702   Σ. 2737.

ε EQUULEI.

R. A.   Dec.
20^h 53·1^m 3° 50′

Magnitudes.—Σ., A 5·7, B 6·2, C 7·1.
De., 6·2, 7·1, 7·5.

A C.—H$_1$ (*Phil. Trans.*, vol. lxxii., p. 219): "Aug. 2, 1780.—Double. Considerably unequal, L. W; s. much inclining to R. Distance 9″·375 mean measure. Position 5° 39′ n.f."
This is A C. It was measured by H$_2$ and So. in 1823, and they noted the increase in distance.
Σ. also measured this star as a double from 1825 to 1832.
A B.—In 1832, however, Σ. discovered that A was double. "1835·62: power 480; elongated; 800 gave 0″·4, 300°·5; in contact." He could not separate the pair in 1835.
Smyth says (*Cycle*, p. 490), "It is clear that A and B are binary."
Da. (*Mem. R. A. S.*, vol. xxxv., p. 429) says "an increase of distance has certainly occurred in this close pair; and a very small diminution of angle is probable in both sets."
O.Σ. (1877). In A B the distance has increased: in $\dfrac{A + B}{2}$ and C there has been but very little change. The plane of the apparent orbit of the former coincides very nearly with the visual ray.
The common proper motion of the system is − 0^s·011 in R. A., and + 0″·13 in N. P. D.
For A B, Dunér gives

$$\Delta = 0''\cdot87 + 0''\cdot0165\ (t - 1855\cdot0)$$
$$-0''\cdot00035\ (t - 1850\cdot0)^2.$$
$$P = 288°\cdot17 - 0°\cdot186\ (t - 1855\cdot0)$$
$$+ 0°\cdot00415\ (t - 1850\cdot0)^2.$$

For $\dfrac{A + B}{2}$ and C,

1853·19. Δ = 10″·68.
$$P = 76°\cdot8 - 0°\cdot0833\ (t - 1850\cdot0).$$

A B.

Σ.	300·5	In.	0′·4	1835·62
	287·3	,,	·35	·64
	293·9	,,	...	·68
	295·4	,,	...	·69
	293·1	,,	0·3	·70
Sm.	290·0		·5	8·83
Da.	286·3	In.	·7	9·69
	285·7	2n.	·7	40·66
	290·9	,,	·59	1·82
	287·8	In.	·7	2·82
	285·2	2n.	·66	3·77
	287·0	In.	·73	7·63
	·2	2n.	·87	8·67
	285·1	3n.	·97	53·85
	·5	In.	1·04	9·67
	·1	,,	0·86	63·85
Mä.	297·5		·65	41·53
	293·4	9n.	·60	2·57
	290·1		·6	3·68
	296·0		·6	4·88
	290·4	In.	·93	50·75
	288·9	3n.	·94	1·20
	291·0	Ion.	·89	5·48
	292·4		·91	6·79
	·6		·97	8·81
	290·8	6n.	1·02	61·43
O.Σ.	287·5	2n.	0·68	43·64
	281·7	,,	·80	52·26
	285·3	3n.	1·02	9·63
	283·8	2n.	·15	70·32
Mit.	288·1	,,	0·57	47·63
Ja.	286·8	9	·8	53·88
De.	280·2	5n.	...	4·62
	·3	2n.	...	5·84
	285·1	6n.	1·0	6·66
	·3	3n.	·0	8·49
	283·8	4n.	0·6	62·64
Se.	287·4	5n.	·81	55·88
	·5	3n.	1·02	66·72
Kn.	287·1	2n.	·01	3·66
	290·5	,,	·11	4·74
	288·1	In.	·07	5·70
Ro.	285·6	,,	0·79	·68
Ka.	283·0	6n.	1·02	·85
Du.	288·6	4n.	0·99	9·69
	289·0	In.	1·00	70·73
	288·8	2n.	·06	5·68
W.O.	296·6	In.	·16	...
W. & S.	287·2	5	0·86	72·65
	285·2	8	1·09	·85
	286·9	6	·15	3·70
	289·2	4	·12	5·79
Gl.	287·0	4	·10	3·91
Sp.	286·4		0·97	6·44
Schi.	·4	In.	·97	·43
Dob.	288·1	3n.	·83	7·74
	·4	2n.	...	8·78

$\dfrac{A\,B}{2}$ and C.

	°		''	
H₁,	84.3	In.	9.37	1781.81
Σ.	80.4		...	1821.25
	77.5	In.	10.56	5.62
	79.4	,,	.98	9.90
	78.0	,,	.77	31.57
	79.1	,,	.81	2.88
	78.0	,,	.92	...
	77.8	6n.	.75	35.65
H₂ & So.	79.3	10	12.37	23.57
	70	11		7.79
Sm.	77.6		10.7	33.77
	78.1		11.2	8.83
Da.	77.0	In.	10.76	9.68
	.2	,,	...	40.80
	76.8	,,	10.74	1.84
	.7	,,	11.25	3.79
	77.3	,,	.22	8.66
	76.2	,,	10.33	53.75
O.Σ.	78.8	5n.	11.20	41.39
	76.5	,,	.11	56.68
	.8	2n.	.19	70.32
Mä.	77.3		10.52	41.54
	78.0		.50	2.76
	77.4		.30	3.68
	78.1		.48	4.88
	76.9		.86	9.26
	.4		.24	51.73
	.5		.44	3.80
	77.3		...	5.79
	76.5		10.43	8.83
	.4		.47	61.39
Mit.	.4	In.	11.08	47.63
Ja.	75.3	9	10.0	53.88
De.	76.4	5n.	.60	4.89
	.1	2n.	.51	5.84
	75.9	5n.	.58	6.61
	76.3	2n.	.56	8.46
	.1	4n.	.83	62.64
Se.	74.6	5n.	.90	55.88
Po.	76.8	15	.58	.89
Kn.	75.6	In.	.39	63.66*
	77.6	2n.	11.10	4.74
	75.8	3n.	10.59	5.67*
Ro.	76.0	,,	.25	.68
M.	77.5	In.	.74	7.61
	76.5	,,	.51	.65
	74.5	,,	.62	8.62
	75.6	,,	.81	.62
	72.8	,,	.68	9.54
	74.1	,,	.53	.67
	.6	,,	...	.61
	75.0	,,	...	.67
	.3	,,	10.89	70.78
	.6	3n.	.62	2.79
	74.9	4n.	.99	3.76
	.2	3n.	11.36	4.70
	75.0	4n.	.38	5.66
Du.	74.3	2n.	10.66	69.66
	.3	In.	.57	70.73
W. & S.	76.2	4	9.3	2.65
	75.9	4	...	.85

* A C.

	°		''	
W. & S.	73.7	5	10.10	1873.70
	77.5	3	9.65	5.79
Ta.	76.5	In.	...	3.74
Schi.	75.2	,,	10.68	5.90
Sp.	.3		.69	.90
Fl.	73.4	In.	.82	7.76
Dob.	75.3	,,	.77	.81

703   O.Σ. 424.

R. A.	Dec.	M.
20ʰ 54ᵐ	15° 6'	7.5, 8.7

O.Σ.	325.4	3n.	0.42	1848.34
De.	330		oblong	65.74

704   Σ. 2741.

R. A.	Dec.	M.
20ʰ 54.6ᵐ	50° 0'	6, 7.3

The observations are very discordant.

H₁.	43.6	In.	1.15	1783.73
H₂.	36.2		.41	1828.55
	34.2		2.89	9.61
	33.7		1.81	30.63
	.2	In.	.76	1.62
Da.	32.8	,,	2.42	0.57
	.1	,,	...	4.50
	.3	,,	2.04	41.80
	.2		...	6.98
	.3		1.71	7.91
Σ.	35.8	3n.	.93	31.49
Sm	34.6		2.1	3.69
Mä.	35.3		.06	4.27
	33.6		.19	41.48
	.8	7n.	.11	2.68
	34.9	In.	1.71	51.85
O.Σ.	33.0	2n.	2.07	41.22
Mo.	31.3	4n.	1.88	55.75
De.	32.4	2n.	2.0	6.01
Se.	30.2	3n.	1.94	7.16
M.	29.6	In.	.85	62.49
Eng.	30.9	4n.	2.25	5.49
Ro.	33.9	2n.	.15	.69
Du.	31.4	7n.	1.99	72.95
Ta.	27.7	In.	.74	3.74
Fer.	31.3		.39	4.55
W. & S.	.4	In.	2.05	.85
	32.2	,,	1.97	5.79
Gl.	33.7	,,	.89	4.91
Pl.	30.3	2n.	2.12	6.86
Dob.	28.6	,,	1.91	.77

705   O.Σ. 425.

R. A.	Dec.	M.
20ʰ 56ᵐ	48° 13'	A 7, B 10.5, C 11

A B.

O.Σ.	27.6	3n.	12.32	1847.49
De.	29.9	,,	.71	67.60

A C.

O.Σ.	46°0	In.	11″·59	51·70

B C.

O.Σ.	135·0	In.	4·11	51·70

706 Σ. 2744.

R. A. 20h 57m	Dec. 1° 4′	M. 6·3, 7

Retrograde motion.

Σ.	190·5	5n.	1·52	1830·16
H₂.	188·3		2·0	1·00
Ma.	·5		1·75	41·63
	187·8		·60	2·73
	189·2		·71	3·75
Da.	185·8		·43	8·68
De.	177·5	6n.	·50	63·24
O.Σ.	184·5	2n.	·93	43·24
	170·6	1n.	·77	74·84
Se.	184·3	4n.	·57	56·46
W. & S.	175·2	4	·27	72·65
	·7	6	·60	3·72
	174·5	6	·31	5·79
Gl.	176·2	4	·5	3·91
Sp.	172·9		·52	6·63

707 Σ. 2746.

R. A. 20h 57m	Dec. 38° 47′	M. 8, 8·6

Direct motion.

Σ.	276·2	5n.	0·87	1830·82
	279·3	2n.	·98	5·63
O.Σ.	270·9	,,	·98	40·72
	279·2	,,	1·03	58·22
S3.	281·2	,,	0·88	6·86
De.	283·7	,,	·80	63·33
Sp.	282·9		·96	75·67
W. & S.	290·3	5	1·09	6·78

708 Σ. 2749.

R. A. 20h 58·7m	Dec. 3° 3′	M. 7·7, 8·9

Probably a ternary system.

A B.

So.	149·5		3·61	1825·60
Σ.	148·7	5n.	·51	30·10
H₂.	150·0		·5	1·00
Ma.	149·4		·74	43·70
De.	151·0		·47	56·70
Da.	·0		·54	63·90

B C.

Se.	127°·0	0″·6	56·64
De.	141·7	·8	63·71
Bu.	150·0	1·2	74·82

709 Σ. 2758.

61 CYGNI.

R. A. 21h 1·4m	Dec. 38° 7′	M. 5·3, 5·9

C. golden.

H₁ (*Phil. Trans.*, vol. lxxii., p. 221): "Sept. 20, 1780.—Double. It is a star preceding τ. Pretty unequal. L. pale R; or L. R; S. garnet. Distance 16″ 7‴. Position 36° 28′ n.f."

H₂ and So. (*Phil. Trans.* 1824, p. 365). These observers give a complete list of measures from 1753·8 to 1819·9, and after observing that the proper motion given by Piazzi is + 5″·38 in R. A. and + 3″·30 in Dec., H₂ goes on to say, "This affords indisputable proof of their connection in a binary system, otherwise the lapse of nearly seventy years, during which they have been observed, one of them would doubtless have left the other behind, without supposing a coincidence too extraordinary to have resulted from accident."

From the measures he finds a mean annual motion in angle amounting to +0°·730, and then, computing the positions for the dates of the observations, he finds a very fair agreement. "The mean angular motion of these stars then about their common centre of gravity is not far short of that of the two stars of Castor, while their apparent mutual distance is at least three times as great. This circumstance, taken in connection with the rapidity of their apparent proper motion, affords a presumption of their being much nearer to us, and renders 61 Cygni a fit object for the investigation of parallax."

Σ. (*M. M.*, p. 169) gives his measures from 1821 to 1835, and from them infers an increase in angle and distance. Treating the distances by the method of least squares, he arrives at the formula

$$15''·727 + 0''·0749 \, (t - 1832·58) \, ;$$

and the computed and observed distances then agree very well. In the P. M., Σ. states that the motion up to 1851 had been rectilinear, and gives the following formulæ:

$$\Delta \sin P = + 16''·163 + 0''·0620$$
$$(t - 1840·02).$$

$$\Delta \cos P = - 1''·959 - 0''·1890$$
$$(t - 1840·02).$$

Smyth (*Cycle*, p. 494) says: "It affords a positive instance of a double star which, besides the individuals revolving round each other, or about their common centre of gravity, has a progressive uniform motion towards some determinate region. This path is relatively spiral, but still so vast as to appear rectilinear; but too little is yet known of its amount and direction to refer it to definite laws."

"The difference between the proper motions of the components here shown would produce a change in R. A. of $7''\cdot2$ since Bradley's time, and an alteration of declination amounting to $18''\cdot9$, corresponding to a change in distance of $19''\cdot7$. Bessel considered the series of positions and distances very inadequate to afford a trustworthy set of elements. He concluded that the annual angular motion is somewhere about $0°\cdot67$, and that the distance at the beginning of the present century reached a minimum of about $15''$. Hence, he remarks, we are enabled to conclude that the period of revolution must be more than 540 years, and that we see the semi-major axis of the orbit under an angle of more than $15°$." (Hind.)

In making his observations for the determination of the parallax of this star, "Bessel chose two stars cf about the 9·10 magnitudes, one being nearly in the direction of the line joining the double star, and the other perpendicular to this direction. The distance of each of these stars from the point which bisects the distance between the two stars of 61 Cygni, was measured sixteen times every night of observation." The resulting parallax was $0''\cdot3136$, equivalent to a distance from the sun 657,700 times the length of the semi-axis of the earth's

orbit. (See Mr. Bishop's volume of Observations.)

After reducing the angles to the equinox f 1850·0, O. Σ. finds the following formulæ :

$$\Delta A = + 16''\cdot659 \pm 0''\cdot036 + (0'\cdot0464 \pm 0''\cdot0028)\,(t - 1850\cdot0).$$

$$\Delta D = - 3''\cdot783 \pm 0''\cdot013 - (0''\cdot1906 \pm 0''\cdot0010)\,(t - 1850\cdot0).$$

An examination of the differences resulting from a comparison of the measures with the calculated quantities shows that the formulæ are not satisfactory. Hence it is evident that the traces of orbital motion may soon be very distinctly recognized.

Dunér has found that Σ.'s formulæ do not represent the observations between 1866 and 1876, and he gives the following :

$$\Delta \sin P = + 15''\cdot09 + 0''\cdot0788\,(t - 1825\cdot0) - 0''\cdot00062\,(t - 1825\cdot0)^2.$$

$$\Delta \cos P = + 0''\cdot89 - 0''\cdot1858\,(t - 1825\cdot0).$$

He observes that the deviation from a straight line is already apparent; that his formulæ give very considerable differences in the early observations; and that probably no formulæ would agree well with both the early and recent measures.

The proper motion of this object has been carefully determined. Argelander's values are—

First Star.		Second Star.	
R. A.	Dec.	R. A.	Dec.
$+ 5''\cdot11$	$+ 3''\cdot23$	$+ 5''\cdot19$	$+ 3''\cdot00$

H_2 (*Phil. Trans.* 1824, p. 367) gives the following list of the early observations :

Date.	Position. (n.f.)	No. of Obs.	Distance.	No. of Obs.	Δ R.A.	No. of Obs.	Δ Decl.	No. of Obs.	Authority
1753·8	54° 36′		19″·628		14·40	2	16·0	1	Bradley, cited by Bessel.
1778·0	39 2		15 ·244		15·00	6	9·6	5	C. Mayer, ditto.
1781·9	36 11	2	16 ·333	3					H_1, Catalogue and MS.
1784·4					22·50	1	6·9	1	Dagelet, cited by Bessel.
1793·6	37 14		14 ·873		15·00	1	9·0	1	Lalande, ditto.
1800·0	19 43		19 ·267		21·60	17	6·5	13	Piazzi, Catalogue for 1800.
1805·0	11 32		14 ·502		18·00	6	2·9	8	,, cited by Bessel, Fund·
1812·3	10 53		16 ·741		19·80		3·1		Bessel, Fund· Astronomia.
1813·8					19·60	37			Lindenau, cited by Bessel.
1814·5					20·32	2			Struve, Catalogus primus.
1819·9	6 58	5	15 ·20		19·10	14	1·85		,, Additamenta, p. 180.
1822·9	5 19	35	·425	33					Herschel and South, mean result.

	°		″			°		″	
Bradley.	35·4		19·63	1753·80	Bessel.	79·1		16·74	1812·30
C. Mayer.	50·9		15·24	78·00	Lindenau.	69·1		·56	13·80
H_1.	53·8		16·33	81·90	Σ.	68·9		17·20	14·50
Lalande.	52·7		14·87	93·60		83·5		15·11	20·51
Piazzi.	70·2		19·27	1800·00		85·8		14·93	2·72
	78·5		14·50	5·00		89·4		15·31	8·72

	°		″	
Σ.	91·1	4n.	15·63	1831·70
	92·0	1n.	·79	2·77
	93·8	6n.	·97	5·65
	94·4		16·08	6·57
	95·4		15·93	7·71
So..	86·9	63	·44	25·70
H₂.	89·3		·43	8·52
	·9		·43	9·47
	90·8		·61	30·56
	·7		·45	1·74
Da.	·3	2n.	·69	0·66
	92·4	1n.	·88	3·80
	93·3	2n.	16·12	4·62
	94·8	,,	·20	7·56
	96·0	,,	·57	9·75
	97·2	,,	·40	40·73
	·9	1n.	·55	1·87
	98·9	,,	·76	3·76
	99·6	2n.	...	·98
Sm.	90·5		15·6	30·81
	92·3		·4	2·65
	93·2		16·2	4·76
	·6		15·8	5·59
	95·1		·9	7·65
	96·3		16·3	9·69
	99·8		·4	48·07
	103·7		17·0	53·80
Mä.	94·1		15·59	35·54
	98·5		16·49	41·49
	99·0		·86	2·62
	98·9		·78	3·76
	100·1		·35	4·48
	103·1	4n.	·80	50·95
	104·1	19n.	·83	2·44
	·4		·90	3·13
	105·0		17·63	4·55
	106·9	12n.	·48	7·55
	107·6		·56	9·54
	108·7	22n.	·93	61·03
Encke.	95·2		16·27	37·63
Galle.	·4		15·91	·71
	96·1		16·70	8·73
Ka.	97·1	6n.	·0	40·05
	·6	,,	·1	1·81
	111·2	,,	18·47	65·89
O.Σ.	99·0	3n.	16·67	43·53
	100·9	,,	17·02	7·46
	102·4	2n.	·18	50·30
	103·6	3n.	·34	1·81
	104·5	2n.	·46	2·67
	105·2	,,	·57	4·25
	106·5	,,	18·02	7·20
	108·7	3n.	·22	60·80
	112·5	4n.	·81	8·54
	116·1	2n.	19·42	74·74
Ja.	99·3		16·02	45·87
	90·7		17·12	6·70
	100·8		16·81	7·96
	102·9	11	17·43	50·62
	103·3	20	·48	1·77
	104·3	10	·40	2·75
	·7	10	·68	3·89
	106·4		·9	6·81
	107·2		18·0	7·82
	·3		17·9	8·27

	°		″	
D.O.	99·3		17·12	1846·71
Mit.	101·1	1n.	·85	7·54
Flt.	102·9	19	16·96	50·90
	103·9	27	17·20	2·72
	107·5		·7	6·67
Mo.	104·3	3n.	·28	2·76
	105·4	,,	·45	4·83
	108·3	2n.	·88	9·91
Mi.	103·9		·17	2·93
De.	105·5	7n.	·29	4·73
	106·1	1n.	·34	5·84
	·4	4n.	·45	6·62
	107·3	3n.	·73	7·61
	·8	4n.	·73	8·53
	109·4	8n.	18·36	62·76
	·6	4n.	·37	3·39
	110·4	10n.	·53	4·74
	·9	8n.	·57	5·65
	111·7	16n.	·72	7·16
	112·8	5n.	·83	8·68
	113·5	7n.	·96	9·70
	·9	,,	19·16	70·58
	114·2	,,	·23	1·55
	·3	8n.	·33	2·60
	·8	7n.	·44	3·55
	115·3	6n.	·50	4·53
	·9	7n.	·58	5·56
Se.	105·7	2n.	17·56	55·55
	·2	,,	·89	6·63
	111·8	1n.	18·81	60·84
Po.	106·6	40	17·88	55·92
	108·6	7n.	18·2	9·89
M.	109·9	1n.	17·64	61·79
	108·9	,,	18·13	·84
	107·7	,,	17·89	·85
	108·4	,,	·66	2·57
	110·4	4n.	18·63	5·56
	111·8	1n.	·89	8·60
	112·7	,,	·71	·60
	·5	,,	·51	·61
	111·8	,,	19·06	9·54
	112·8	,,	18·24	·62
	·6	,,	19·48	70·48
	·1	,,	·07	·50
	114·1	,,	·77	2·78
	115·1	18n.	·55	5·51
Ro.	109·7	1n.	...	65·75
	111·4	,,	18·76	·76
Kn.	·6	2n.	·76	6·72
	113·4	3n.	19·16	71·60
	114·4	2n.	·60	3·00
Ta.	112·8	1n.	18·84	66·74
	113·6	,,	·93	72·70
	·6	,,	·68	3·71
	117·4	,,	...	5·34
	116·1	,,	19·43	·38
	115·9	,,	20·03	·59
	·9	,,	...	·81
Du.	112·0	4n.	18·49	67·89
	·2	6n.	·62	8·82
	113·0	8n.	·82	9·89
	·5	2n.	·91	70·90
	115·1	3n.	19·41	3·87
	·7	,,	·39	5·95
Gl.	113·8	6	·12	0·20

Gl.	113°9	6	19″38	1870·65
	114·0	6	·00	·80
	·1	6	·18	1·32
	·0	6	·32	·73
	113·9	6	·20	3·73
	116·1	4	...	·68
	115·9	3	...	·81
	·0	4	19·5	·70
	·2	10	·46	4·91
	·6	8	·60	·91
	116·3	4n.	19·61	5·71
	·3	12n.	·68	6·46
	·7	7n.	...	8·66
W. & S.	113·1	5	·1	1·59
	115·6	6	·0	2·65
	114·2	4	...	·72
	·4	4	...	·73
	·0	7	...	·74
	·5	5	...	·74
	·7	4	18·7	·75
	·7	9	...	3·72
	117·0	7	19·40	5·79
	115·9	6	·59	6·59
	116·2	10	·68	·60
	115·8	8	·96	·61
	116·6	4	·66	·61
	·2	5	·72	·62
Pl.	·4	2n.	20·03	·75
	·3	7n.	19·78	7·77
Fl.	·2	1n.	·76	·79
Dob.	·7	4n.	·52	8·60
Goldney.	·7	,,	20·02	·72

Mä.	222°7		12″42	1841·50
	223·3	3n.	11·75	5·68
	·8		·70	7·76
	·7	3n.	·40	50·94
	·9	,,	10·90	1·93
	·6	2n.	·79	6·92
	221·4	,,	·78	9·86
Da.	223·9	1n.	12·15	41·67
	·4	3n.	11·98	3·82
Ka.	·9	6n.	·92	1·81
	225·4	,,	9·64	65·91
Mo.	224·3	3n.	10·91	54·72
	·5	1n.	·57	5·86
De.	225·0		·52	7·08
	224·7		·12	63·02
	·9		9·84	5·22
	225·0		·68	7·12
	·5		·35	9·19
	·3		·24	70·96
	·2		8·99	3·05
	·2		·79	5·03
Du.	224·5	4n.	9·58	68·55
	·7	6n.	·54	9·83
	225·2	3n.	8·29	74·81
Ta.	222·6	6	7·35	2·70
	223·2		8·58	3·74
W. & S.	224·4	5	9·5	·72
Gl.	225·3	4	·42	4·70
Pl.	224·2	2n.	8·72	6·79
Fl.	225·4		·22	7·83

710 Σ. 2760.

R. A. 21^h 1·9^m Dec. 33° 39′ M. 7·3, 8·1

The distance has diminished considerably, but the angle has probably not changed at all since 1830. O.Σ. finds that the distances are represented by

$$e = 11''·555 \pm 0''·017 - (0''·0991 \pm 0''·00137)\,(t - 1850·0) ;$$

and the differences between the observed and computed values are very small.

Dunér gives

$$\Delta = 11''·37 - 0''·102\,(t - 1850·0).$$
$$P = 224°·1 + 0°·047\,(t - 1850·0) + 0°·00043\,(t - 1850·0)^2.$$

So.	222·8	2n.	14·32	1825·61
H$_2$.	223·9	1n.	13·49	9·84
Σ.	·2	2n.	·66	·87
	222·9	,,	·38	32·40
	223·5	3n.	12·95	5·63
	224·1	,,	·70	6·67
	·0	,,	·76	7·77
O.Σ.	·9	2n.	13·02	9·86
	223·5	4n.	11·67	47·90
	·8	2n.	·02	55·81
	224·3	,,	9·70	68·76
	·2	,,	·23	73·72

711 Σ. 2762.

R. A. 21^h 4^m Dec. 29° 43′ M. 6, 8

C. A greenish white, B bluish.

Dunér's formulæ are

$$1855·0. \quad \Delta = 3''·43.$$
$$P = 314°·3 - 0°·047\,(t - 1850·0).$$

H$_1$.	315·3	1n.	...	1783·70
So.	·2	2n.	3·57	1824·70
Σ.	·6	3n.	·54	9·75
Mä.	314·7	1n.	·62	47·70
De.	310·7	5n.	·42	54·62
	315·1	1n.	·60	6·45
Se.	313·7	3n.	·53	7·06
Du.	·1	5n.	·19	68·59
W. & S.	·6	1n.	·43	73·72
Gl.	·0	,,	·40	4·80

712 Σ. 2777.

R. A. 21^h 8·6^m Dec. 9° 28′ M. A 4·1, B 10, C 10

A B. The orbit has a great resemblance to that of 42 Comæ Berenicis. The period is still uncertain: it may be six or seven years, or about double that time.

$\dfrac{A + B}{2}$ and C. The observations being reduced to 1850, and weights being assigned, M. Doubiago finds the following formulæ for uniform rectilinear motion :

$$\Delta A = +\ 16''\cdot136 \pm 0'\cdot030 - (0''\cdot0600 \\ \pm\ 0'\cdot0024)\ (t - 1850\cdot0);$$

$$\Delta D = +\ 26''\cdot267 \pm 0'\cdot031 + (0''\cdot2943 \\ \pm\ 0'\cdot0024)\ (t - 1850\cdot0);$$

and the differences indicate no deviation from such movement. (O.Σ.)

The proper motion of δ is + 0''·08 in R. A., and − 0''·28 in N. P. D. In this C has no share.

For $\dfrac{A + B}{2}$ and C, Schiaparelli gives

$$\Delta \sin P = +\ 16''\cdot90 - 0''\cdot0632 \\ (t - 1839\cdot0).$$

$$\Delta \cos P = +\ 22''\cdot98 + 0''\cdot2873 \\ (t - 1839\cdot0).$$

A B.

O.Σ.	22·5	In.	...	1852·64
	18·8	,,	...	·67
	191·9	,,	...	3·91
	...	single		4·69
	...	,,		6·57
	207·6	In.	0·21	·67
	211·5	,,	·23	·67
	16·8	,,	·40	8·59
	13·5	,,	·39	9·65
	236	,,		61·57
	203·3	,,	·50 ±	5·91
	24·0	oblong		74·67
	...	wedged		·73
	...	oblong		...
	221·2	In.	0·33	74·75
Du.	8·0	2n.	·25	0·73

$\dfrac{A + B}{2}$ and C.

H₁.	78·4		19·53	1781·80
So.	41·9		25·81	1825·26
H₂.	40		20·	7·63
	39·5		27·83	30·35
Σ.	41·4	3n.	26·64	28·80
	39·7	,,	27·48	32·10
	37·8	2n.	·56	4·90
	·8	4n.	·63	5·64
	·4		28·07	6·65
	36·7		·26	7·77
Sm.	38·8		27·1	0·67
	37·6		·9	6·78
	36·8		28·2	8·59
Mä.	34·5		...	41·49
	·9		29·88	3·63
O.Σ.	·8	In.	28·82	1·65
	32·2	,,	30·48	7·82
	30·9	,,	31·07	51·84
	·9	,,	·38	2·64
	29·2	,,	·57	3·91
	·7	,,	·65	4·69
	·4	,,	32·36	6·58

	28·7	In.	32·59	1857·67
	·4	,,	·84	8·59
	·2	,,	·87	9·65
	26·1	,,	34·70	65·91
Ka.	34·0		28·5	42·64
	33·9		29·2	4·17
De.	27·0		33·76	63·14
	25·0		·70	74·80
Kn.	27·5		34·46	65·72
Du.	25·5	2n.	35·80	9·67
W. & S.	24·2	In.	37·66	76·81
Fl.	·0	,,	·57	7·82

713   Σ. 2779.

R. A.	Dec.	M.
1ʰ 9·3ᵐ	28° 35′	8·5, 8·5

C. yellowish.

Change in both angle and distance. Dunér gives

$$\Delta = 18''\cdot61 - 0''\cdot029\ (t - 1850\cdot0).$$
$$P = 187°\cdot1 - 0°\cdot125\ (t - 1850\cdot0).$$

Σ.	189·4	2n.	19·22	1828·81
H₂.	185·4	In.	23·8	9·80
Mä.	188·0	,,	18·30	43·71
	186·6		·40	8·02
	187·7	3n.	·70	50·93
O.Σ.	·2	In.	·77	39·83
	184·7	,,	17·99	68·76
	185·1	,,	·90	·77
	·1	,,	·71	74·84
De.	·2	5n.	18·13	64·59
Du.	184·2	4n.	·07	9·81
Fl.	182·8	In.	17·86	77·82

714   Σ. 2778.

R. A.	Dec.	M.
21ʰ 9·5ᵐ	− 1° 44′	8·4, 10·6

Σ.	267·0	4n.	21·19	1828·24
Mä.	268·4	In.	20·10	43·70

715   O.Σ. 432.

R. A.	Dec.	M.
21ʰ 9·7ᵐ	40° 39′	6·8, 7·2

Mä.	130·4		1·04	1843·65
O.Σ.	·4	4n.	·19	7·94
Da.	129·7	2n.	·07	53·86
	126·4	In.	·03	9·68
De.	128·3	4n.	·16	67·01
Du.	·3	2n.	·23	9·72
W. & S.	·4	In.	·17	73·73
	·7	,,	·17	5·80
Sp.	126·4		·21	·63
Pl.	124·6	3n.	·32	6·86

716 τ CYGNI.

R. A.	Dec.	M.
21ʰ 10ᵐ	37° 32′	5·6, 7·9

C. yellow, blue.

Rapid change in angle. Discovered by Alvan Clark in 1874. The position of Holden's third star in 1876·9 was 260°·3; distance 15″·68.

W.0.	162·6	1″·10	1874·83
	161·5	·25	6·79
	160·2	·04	·90
De.	174·8	1n.	·06	4·90
	·5	2n.	·24	5·12
	170·5	3n.	·32	·69
	161·5	2n.	·24	6·79
	155·3	8n.	·26	7·70
Bu.	150·0		·06	8·41

717 A. C. 19.

R. A.	Dec.	M.
21ᵇ 11·4ᵐ	63° 57′	7, 7

This double star was discovered by Alvan Clark in Dawes's observatory, on July 8, 1859.

Dawes says, "a neat star, sharply defined and pretty steady;" and in his notes he observes, "if there is no error of identity, it must have rapidly separated in the interval [since the date when O.Σ. frequently examined it and entered it as single], and may now perhaps have arrived nearly at its maximum distance; the plane of the orbit lying nearly in the line of sight."

O.Σ.		single		1842·00
Da.	246·2		0·88	59·73
	·4		·93	60·70
	244·5		·98	6·83
W. & S.	247·4	4	·93	72·78
	251·2	4	·99	3·81
De.	247·6		·95	·11

718 O.Σ. 435.

R. A.	Dec.	M.
21ʰ 15·4ᵐ	2° 23′	7·5, 8

Variable ?

Mä.	24·2		0·45	1843·65
	17·0		·52	5·88
	16·0		·45	51·71
O.Σ.	23·8	3n.	·59	48·13
Da.	201·8	4n.	·55	52·43
De.	196·2	3n.	...	66·03

719 O.Σ. 437.

R. A.	Dec.	M.
21ʰ 15·9ᵐ	31° 56′	7, 10·5

Indirect motion.

O.Σ.	67·7	4n.	1·37	1845·43
	58·1	3n.	·51	58·74
Mä.	63·7	1n.	...	45·63
	61·8	,,	1·29	51·76
Da.	·2	4n.	·29	·98
	60·2		·29	3·50
De.	54·6	4n.	·40	66·57
Du.	53·2	7n.	·35	71·15
W. & S.	51·4	5	·32	3·73
	·8		·55	6·30

720 Σ. 2797.

R. A.	Dec.	M.
21ʰ 20·9ᵐ	13° 10′	6·7, 8·2

H₂.	213·3	1n.	4·65	1828·64
Σ.	·3	3n.	3·17	31·26
Mä.	215·5		·20	42·71
	216·4		·18	5·52
	215·9		·23	51·75
	216·5		·23	2·72
	218·1		·31	5·81
	222·2	1n.	·63	8·81
Mo.	214·5	2n.	·53	6·82
De.	216·0	7n.	·0	7·64
Du.	·3	5n.	·18	68·94

721 Σ. 2801.

R. A.	Dec.	M.
21ʰ 22·1ᵐ	79° 50′	7·3, 8

The common proper motion of this pair is + 0ˢ·108 in R. A., and −0″·106 in N. P. D.

Σ.	273·1	3n.	1·42	1832·38
	271·6	2n.	·45	7·07
O.Σ.	269·9	,,	·52	46·74
	265·7	1n.	·73	75·46
Se.	279·4	,,	·16	59·92

722 Σ. 2799.

20 (B) PEGASI.

R. A.	Dec.	M.
21ʰ 23ᵐ	10° 34′	6·6, 6·6

Dawes first measured it in 1832, and says (*Mem. R. A. S.*, vol. xxxv., p. 436), "there is sufficient evidence of a retrograde orbital motion since the first observation."

O.Σ. (1877). Indirect motion : distance not sensibly changed.

So.	338°·1		1'·20	1825·68
Σ.	·9	In.	·26	8·76
	335·8	,,	·44	9·64
	333·4	3n.	·37	32·91
	332·2	In.	·30	3·77
H₂.	334·6		·0	0·64
Mä.	332·9		·39	5·81
	330·1		·58	41·63
	328·9		·34	2·76
	324·4		·70	50·75
	·7		·48	1·74
	322·7		·33	3·80
	323·7		·36	4·79
	·0		·36	8·80
	318·5		·37	9·88
	321·6		·47	61·81
	317·8		·38	2·76
Da.	327·4	9n.	·26	40·72
	322·6	3n.	·26	51·96
	320·3	5n.	·17	4·74
O.Σ.	329·8	2n.	·48	43·75
	320·8	,,	·44	53·20
De.	142·3	4n.	·2	5·52
	137·8	5n.	·2	6·57
	317·9	4n.	·45	62·67
	·1	3n.	·43	3·64
Se.	320·7	4n.	·23	56·28
	317·8	In.	·24	66·89
Mo.	320·8	16	·38	59·81
Ro.	315·2	2n.	·29	65·67
Ta.	317·1	In.	...	6·74
	315·4	,,	1·47	71·70
	309·9	,,	·00	3·71
	312·4	,,	·45	·74
Br.	325·8		2·98	68·70
W. & S.	314·0	8	1·36	72·66
	312·5	5	·28	3·73
	·4	4	·26	5·81
Gl.	313·7	4	·41	3·80
Schi.	128·7	In.	·21	6·44
Sp.	308·8		·22	·45
Pl.	310·0	5n.	·33	7·54
Dob.	·8	3n.	·33	8·62

723 Σ. 2802.

R. A.	Dec.	M.
21ʰ 27ᵐ	33° 17'	8, 8

Dunér gives

$$1852\cdot79. \quad \Delta = 3''\cdot90.$$
$$P = 10\cdot0 - 0°\cdot059 \,(t - 1850\cdot0).$$

So.	10·6	2n.	4·32	1825·65
Σ.	11·3	3n.	3·84	30·48
Mä.	12·3	In.	4·25	43·78
	10·0	3n.	3·94	8·72
Se.	9·9	2n.	·93	56·82
De.	·1	In.	·75	7·64
Du.	8·7	4n.	·86	71·61

724 Σ. 2804.

29 (B) PEGASI.

R. A.	Dec.	M.
21ʰ 27·1ᵐ	20° 11'	7·3, 8

This object was discovered by Σ. and South. The former (M. M., p. 55,) says that the angle had increased between the years 1828 and 1834, and that the motion was almost beyond doubt.

Dawes (Mem. R. A. S., vol. xxxv., p. 436) writes : "There can be no doubt of the binary character of this object."

The direct angular movement suspected by my father is perfectly confirmed by the later observations, the distance remaining unchanged. (O.Σ.)

Dunér gives

$$1852\cdot90. \quad \Delta = 2''\cdot90.$$
$$P = 320\cdot9 + 0°\cdot2685 \,(t - 1850\cdot0).$$

So.	311°·7	2n.	2'·58	1825·70
H₂.	313·3	In.	·71	8·64
	318·0	,,	3·38	30·66
	316·1	,,	·22	1·73
Σ.	314·4	2n.	2·93	28·75
	317·5	7n.	·79	35·35
Da.	314·2		3·12	2·87
	317·0		·18	5·44
	·8	4n.	2·93	41·44
	319·3	2n.	·83	6·18
	321·3	4n.	·88	54·00
Mä.	318·1		·86	38·24
	320·8		·80	42·76
	319·9	7n.	·90	3·85
	320·8		·89	51·01
	321·7		·77	·87
	322·8		·90	4·45
	·7	8n.	·92	6·67
	·4		·68	8·86
	324·8	4n.	3·13	62·26
Hind.	322·4	In.	2·62	45·45
	323·0	,,	...	6·54
Mo.	322·9	30	3·02	54·77
Se.	321·5	3n.	2·87	6·47
	328·0	2n.	3·02	66·81
M.	327·0	In.	2·85	4·67
De.	324·5	6n.	·75	·87
Ro.	...	2n.	·65	5·67
Ta.	323·7	In.	·60	6·78
	324·8	,,	3·16	73·74
O.Σ.	327·7	,,	2·94	68·67
	331·1	,,	·92	74·67
Du.	327·6	9n.	·76	0·68
Kn.	325·0	2n.	·96	1·78
W. & S.	·0	4	...	·99
	327·3	4	2·91.	2·66
	·5	5	·8	3·73
	326·7	4	...	5·81
Gl.	328·5	6	3·02	4·79
	325·3	4	...	·80
Sp.	326·9		2·91	5·96
Pl.	325·6	6n.	·96	7·60

725   Σ. 2822.

μ CYGNI.

R. A.	Dec.	M.
21h 38·9m	28° 12'	4, 5

C. A white, B bluish white.

This bright object was first seen double by C. Mayer in 1777.

Considerable diminution of distance combined with a very slow increase in the angle. It is strange that the former change has not been accompanied by a proportionate augmentation of the angular motion. The distance will no doubt be very small when the apparent periastre is reached. The apparent orbit is probably very elongated. (O.Σ.)

The common proper motion of this object is about $+ 0^s\cdot 016$ in R. A., and $+ 0''\cdot 26$ in N. P. D.

The small star C does not belong to the system.

Dunér gives the following formulæ:

$$\Delta = 5''\cdot 60 - 0''\cdot 0330\,(t - 1830\cdot 0) - 0''\cdot 00021\,(t - 1830\cdot 0)^2.$$

$$P = 113^\circ\cdot 5 + 0^\circ\cdot 06707\,(t - 1830\cdot 0) + 0^\circ\cdot 000555\,(t - 1830\cdot 0)^2 + 0\cdot 0000052\,(t - 1830\cdot 0)^3.$$

	Angle °	n	Dist "	Epoch
H$_1$.	109·2	2n.	...	1780·84
So.	113·1	,,	5·74	1823·69
H$_2$.	112·7	1n.	·67	30·77
Σ.	114·5	4n.	·55	1·63
Sm.	113·8		·6	2·79
	114·3		·4	9·62
Mä.	115·0		·10	41·60
	114·9		4·63	2·77
	117·1		5·32	3·96
	115·8		·51	4·88
	·1		·41	7·92
	116·1		4·76	50·83
	115·3		·37	1·87
	·3		·25	3·54
	·8		·22	7·83
	·4		·52	8·06
	116·9	8n.	·18	62·16
Da.	114·3	6n.	5·46	42·08
Mit.	112·1	1n.	4·89	7·61
Po.	·2	3n.	·94	·69
O.Σ.	·4	1n.	·78	51·84
	115·7	,,	·66	4·69
	116·0	,,	·44	61·63
	·4	,,	·34	6·72
Mo.	·6		·88	54·84
	115·2	6n.	·50	5·84
	116·7	2n.	·41	9·75
De.	·4		·66	5·62
	·4	6n.	·40	62·98
	117·0	5n.	·14	6·18
	·8	,,	3·94	72·85
Se.	116·7	3n.	4·63	1856·94
	117·0		·39	66·90
M.	114·8	1n.	·46	2·53
Eng.	115·8	3n.	·12	5·37
Ro.	·7	1n.	...	·68
Ta.	116·3	,,	4·64	6·78
	113·4	,,	·82	8·75
	117·2	,,	3·56	73·74
Ka.	115·5	5n.	4·00	66·84
Du.	116·9	,,	3·89	8·17
	118·3	6n.	·51	75·46
Br.	117·3		4·18	68·75
W. & S.	...		·12	71·68
	118·5	5	3·91	3·73
	117·8	3	·83	·73
	·6	3	...	·78
	118·2	4	3·65	5·81
	119·4	3	4·05	6·78
	·4	4	3·84	·83
Fer.	116·8		4·08	2·62
Gl.	117·7	5	...	3·80
	·5	8	3·89	4·91
	·6	2n.	·76	6·22
	120·5	,,	...	8·75
Pl.	119·3	5n.	3·79	7·42
Fl.	118·4	1n.	·76	·57
Dob.	119·2	2n.	·65	8·82
Goldney.	116·5	4n.	·76	·79

726   Σ. 2824.

R. A.	Dec.	M.
21h 39·2m	25° 6'	3·9, 10·8

	Angle °	n	Dist "	Epoch
Σ.	308·5	5n.	11·01	1831·56
Sm.	310·0		12·0	5·66
Mä.	307·3		11·49	44·89
	306·5		·20	8·01
	·3		·60	51·00
	302·8		9·82	60·82
De.	303·9		11·6	4·84
W. & S.	301·7	1n.	·6	71·51
	·8	,,	·8	3·73
	303·3	,,	...	5·89
Gl.	302·5	,,	12·1	4·80

727   Σ. 2825.

R. A.	Dec.	M.
21h 40·9m	0° 18'	8, 8·2

	Angle °	n	Dist "	Epoch
Σ.	100·2	3n.	1·09	1827·72
H$_2$.	·1		·0	31
Mä.	·3		0·95	42·69
	·0		·95	4·33
	118·8		·97	51·78
Se.	107·7	1n.	·95	5·78
	105·5	2n.	·87	7·34
	106·5	1n.	·7	66·85
De.	107·9		1·08	·06
W. & S	110·7	1n.	·06	76·86

728   Σ. 2837.

R. A.	Dec.	M.
21ʰ 42·9ᵐ	82° 23′	8·5, 9

Σ.	321·3	3n.	2″·16	1832·30
Mä.	311·0		·49	44·44
De.	306·3		·31	66·24
W. & S.	302·2	In.	·57	76·94

729   Σ. 2828.

R. A.	Dec.	M.
21ʰ 43·5ᵐ	2° 50′	8, 9

A B. — The angle is unchanged, but the distance has increased considerably. Secchi's angle in 1856 is probably 10° in error. In B C the angle may have increased about 3°.
Dunér gives

$$\Delta = 24''\cdot 87 + 0''\cdot 051\ (t - 1850\cdot 0).$$
$$1855\cdot 26. \quad P = 142°\cdot 6.$$

A B.

Σ.	142·4	3n.	23·79	1829·09
Se.	141·9	2n.	25·17	56·64
De.	142·4	3n.	·61	64·68
Du.	143·4	2n.	·86	8·80
	142·6	In.	26·26	75·68
O.Σ.	·3	,,	·26	4·84
W. & S.	143·4	2	·15	5·89
Fl.	142·7	In.	·52	7·88

B C.

Σ.	36·9	3n.	3·64	29·09
Mä.	37·3		·93	42·73
De.	40·0	3n.	·91	64·68
O.Σ.	·3	In.	·77	74·84
Du.	·9	5n.	4·06	0·45
W. & S.	38·5	2	3·93	5·89
Fl.	41·0		·89	7·86

A C.

Du.	133·4	In.	25·32	75·68
W. & S.	134·5	,,	...	·81
	·2	,,	25·78	·89

730   Σ. 2840.

R. A.	Dec.	M.
21ʰ 48ᵐ	55° 14′	6, 7

C. A greenish white, B bluish white.

Dunér has

$$\Delta = 20''\cdot 14 - 0''\cdot 0184\ (t - 1830\cdot 0).$$
$$1850\cdot 63. \quad P = 194°\cdot 6.$$

	°	In.	21″·22	1782·97
H₁.	...	In.	21″·22	1782·97
	192·2	,,	...	3·62
H₂ & So.	193·8	2n.	20·31	1823·74
	195·3		·31	30·76
Σ.	193·8	2n.	·08	2·46
	194·2	,,	19·94	3·46
Richardⁿ·	195·1		·59	40·50
Mä.	194·8	In.	...	2·81
	196·1	,,	20·15	4·90
	195·0	2n.	19·69	5·58
	194·8	,,	·20	52·20
De.	·4	3n.	·60	7·70
Eng.	·5		·68	64·43
Du.	·5	3n.	·28	8·49
W. & S.	195·4	In.	·9	74·85
Gl.	194·9	,,	20·16	·91

731   O.Σ. 456.

R. A.	Dec.	M.
21ʰ 51ᵐ	51° 59′	7·8, 8

O.Σ.	25·7	3n.	1·35	1847·73
De.	30·6	4n.	·50	66·64

732   Σ. 2842.

R. A.	Dec.	M.
21ʰ 52·1ᵐ	19° 40′	8·2, 10·7

Σ.	274·7	2n.	1·14	1829·17
	267·9	In.	·00	32·90
	264·7	,,	·30	42·72
De.	·0		·4	63·91

733   O.Σ. 458.

R. A.	Dec.	M.
21ʰ 52·6ᵐ	59° 14′	7·1, 8·6

In 1873 Mr. Burnham detected a distant star, position angle 40°, distance 25″.

O.Σ.	348·8	7n.	0·71	1851·75
Mä.	44·6		elongᵈ·	45·74
De.	353·7	3n.	0·80	66·94

734   Σ. 2860.

R. A.	Dec.	M.
21ʰ 59·4ᵐ	60° 16′	7·7, 9·3

C. A very yellow, B blue.

Σ.	250·8	3n.	3·32	1832·30
Mä.	252·4		4·45	44·43
	254·0		·50	7·95
De.	·6		5·15	64·94

735 Σ. 2863.

R. A.	Dec.	M.
22h 0·3m	64° 2'	4·7, 6·5

C. A yellowish, B blue.

Dunér gives

$$\Delta = 5''{\cdot}96 + 0''{\cdot}0226\ (t - 1850{\cdot}0).$$
$$P = 287°{\cdot}8 - 0°{\cdot}128\ (t - 1850{\cdot}0).$$

	°		"	
H₁.	...	1n.	5·0	1780·37
	290·3	,,	...	81·96
	293·8	,,	...	1803·22
So.	·2	,,	5·82	23·62
H₂.	290·8	,,	6·37	30·67
Sm.	289·5		5·6	·91
	288·8		·8	9·65
Σ.	·9	3n.	·60	1·77
Mä.	287·2	7n.	·7	45·57
	288·5	1n.	·87	50·72
	287·6	3n.	·71	2·66
	·3	,,	6·22	61·80
Po.	·5	,,	5·88	45·90
Mit.	·4	1n.	6·17	7·69
De.	·5	7n.	·00	54·83
	286·0	8n.	·30	64·84
Mo.	·4	2n.	5·85	58·66
Du.	285·5	,,	6·48	70·95
W. & S.	284·4	,,	·45	1·86
Gl.	·8	1n.	·6	4·91

736 Σ. 2865.

R. A.	Dec.	M.
22h 1·1m	69° 38'	8·5, 9

So.	173·7		16·61	1825·27
H₂.	172·3		16	31·40
Σ.	175·1	2n.	·36	3·38
Mä.	177·6		17·50	47·90
De.	181·3		18·08	63·56

737 O.Σ. 463.

R. A.	Dec.	M.
22h 4m	13° 9'	7·5, 11·4

O.Σ.	346·8	4n.	4·53	1848·08
De.	352·7	3n.	·51	66·55

738 Σ. 2872.

R. A.	Dec.	M.
22h 4·5m	58° 42'	A 7·2, B 8, C 8

The relative brightness of B and C is variable.

A B.

Σ.	316·5	8n.	21·35	1834·42
O.Σ.	·4	4n.	·75	51·24

B C.

Σ.	334·5	2n.	0"·54	1833·63
	335·6	3n.	·45	6·17
Sm.	330·0		·5	9·77
Mä.	332·1		·52	41·54
O.Σ.	333·2	5n.	·62	9·11
Se.	328·3	2n.	·40	56·95
De.	325·9		·5	67·65

739 O.Σ. 465.

R. A.	Dec.	M.
22h 7m	49° 37'	7·2, 10·7

O.Σ.	324·3	3n.	15·31	1848·10
De.	323·5	,,	14·94	66·66

740 Σ. 2878.

R. A.	Dec.	M.
22h 8·5m	7° 23'	6·5, 8

Σ.	130·8	4n.	1·36	1830·31
Mä.	132·6		·34	9·70
	135·2		·34	42·72
	134·8		·33	50·99
O.Σ.	137·8	3n.	·26	46·32
Mä.	134·7		·38	51·82
	132·7		·27	9·88
Se.	130·1	2n.	·44	6·82
Mo.	132·2	16	·26	9·84
Fer.	125·6		0·98	73·71
Gl.	128·9	1n.	1·26	·80
W. & S.	132·2	,,	·45	5·92
Sp.	125·6		·20	6·90

741 Σ. 2877.

P. XXII. 33 PEGASI.

R. A.	Dec.	M.
22h 8·5m	16° 36'	6·4, 9·6

C. Σ., A, yellow ; B, blue. Sm., A, lucid yellow ; B, sea-green.

H₁ (*Phil. Trans.*, vol. lxxv., p. 649) : "Fl. 33 Pegasi. Double. 89° 12' n.f.")

H₂ and So. (*Phil. Trans.* 1824, p. 379) : "The proper motions assigned by Piazzi to this star are + 0"·40 in R.A., equivalent to 0"·38 on the parallel, and − 0"·01 in declination. In forty years, therefore, it should have moved 15"·2 from its place in a direction almost exactly coincident with the parallel ; and supposing the small star at rest and the position of 1783 correct, the angle at present should be 75° 38'. coineiding exactly with the observed. The proper motion of this star appears therefore to be well established in fact, and correct in quantity."

The proper motion of A is −0″·20 in R. A., and + 0″·117 in N. P. D.
Dunér gives

$$\Delta \cos P = + 7''{\cdot}53 + 0''{\cdot}0925\ (t-1850{\cdot}0).$$
$$\Delta \sin P = - 3\ {\cdot}71 + 0\ {\cdot}0730\ (t-1850{\cdot}0).$$

The movement is rectilinear and uniform.

	°		″	
H₂.	310		...	1827·65
Σ.	316·8	3n.	7·59	8·72
	315·2	1n.	·75	9·64
	334·9	,,	8·73	51·91
Sm.	315·4		6·5	33·63
Mä.	322·4		7·83	6·57
	328·8		8·36	43·64
	335·5	3n.	·49	50·99
	336·4	2n.	·71	1·85
	339·4	7n.	·57	6·99
	341·0	6n.	9·21	61·63
Mit.	331·5	1n.	8·04	47·61
O.Σ.	332·5	,,	·51	51·90
	350·1	,,	10·26	74·73
Se.	337·4	3n.	8·56	57·35
	345·2	1n.	9·53	66·90
De.	342·2	3n.	8·99	3·67
Du.	345·6	4n.	9·48	8·53
M.	·8	1n.	·31	·70
	348·6	,,	·55	9·77
	347·9	,,	·26	73·78
W. & S.	349·0	4	·8	2·66
	348·2	7	·8	3·73
	349·1	6	·7	4·84
Gl.	347·9	5	...	3·74
	348·5	4	10·0	·79
	351·0	3	...	·79

742   Σ. 2895.

R. A.	Dec.	M.
22ʰ 15·1ᵐ	24° 21′	8·5, 10

Dunér's formulæ are

$$\Delta = 5''{\cdot}43 + 0''{\cdot}0290\ (t - 1850{\cdot}0).$$
$$P = 17\ {\cdot}5 + 0°{\cdot}5125\ (t - 1850{\cdot}0) - 0°{\cdot}00288\ (t - 1850{\cdot}0)^2.$$

Σ.	6·1	3n.	4·85	1830·09
Mä.	15·0	4n.	5·61	44·39
	20·3	3n.	·33	52·36
	18·2	1n.	·15	6·79
	25·8	,,	·73	61·76
De.	23·1	3n.	·86	3·73
Du.	27·5	6n.	6·00	9·93
W. & S.	28·5	2n.	·5	74·85
Gl.	·8	1n.	·62	·91

743   O.Σ. 469.

R. A.	Dec.	M.
22ʰ 15ᵐ	34° 31′	7·2, 8·8

O.Σ.	280·5	3n.	31·80	1846·79
	282·2	1n.	·05	74·72
	·4	,,	30·89	6·74
De.	281·4	3n.	31·01	66·71

744   Σ. 2900.

R. A.	Dec.	M.
22ʰ 17·9ᵐ	20° 14′	A 6, B 9·2, C 7·9

In A B there has been no sensible change. Σ. (see *P. M.*, ccxxxii.) shows that C is fixed : his formulæ are

$$P = - (18''{\cdot}299 \mp 0''{\cdot}035) - (0''{\cdot}3482 \mp 0''{\cdot}0053)\ (T - 1838{\cdot}0).$$
$$p' = + (54''{\cdot}250 \mp 0''{\cdot}035) + (0''{\cdot}0218 \mp 0''{\cdot}0053)\ (T - 1838{\cdot}0).$$

O.Σ. finds that the following formulæ represent the observations well :

$$\Delta A = -22''{\cdot}307 \pm 0''{\cdot}049 - (0''{\cdot}3266 \pm 0''{\cdot}0033)\ (t - 1850{\cdot}0).$$
$$\Delta D = + 54''{\cdot}321 \pm 0''{\cdot}049 + (0''{\cdot}0094 \pm 0''{\cdot}0033\ (t - 1850{\cdot}0).$$

A B are probably a physical pair, while the changes in A C are due to the proper motion of A.

A B.

	°		″	
Da.	181·3		3·12	1830·75
	·7		·04	2·86
	179·1,		2·83	40·15
	·2		...	51·58
	180·8		2·40	4·78
Sm.	181·6		·7	31·74
	180·2		·5	8·88
	178·9		·7	9·69
Σ.	180·4	1in.	·45	5·06
O.Σ.	181·8	in.	·83	9·88
	175·5	,,	·64	45·74
	180·7	,,	·60	7·87
	175·8	,,	·40	51·71
	182·6	,,	·46	65·91
	178·4	,,	·25	74·66
	176·3	,,	·81	·72
Mä.	179·2		·57	42·78
	180·2		·07	5·51
	179·5		·24	7·97
	182·5		...	51·01
	181·3		2·50	2·26
	180·9		3·80	60·82
Mit.	178·9	in.	1·74	47·61
Se.	177·7	,,	...	56·76
De.	·6		2·25	63·25
Ta.	...		·46	70·59
	184·8	6	·40	·60
W. & S.	175·8	4	...	1·93
	176·5	2	...	3·73
	177·0	4	...	4·87
	175·5	4	1·8	·85
Gl.	·1	7	·83	·91
	176·2	5	·88	·91

A C.

	°		″	
H₁.	360·8		45·05	1783·62
So.	345·7		56·04	1823·71
Da.	343·4		·61	30·66
	340·7		57·9	40·15

Sm.	344·0		56·9	1831·74
	341·0		·6	8·38
Σ.	343·2	5n.	...	2·25
	341·8	6n.	...	6·07
O.Σ.	340·7	2n.	57·79	9·88
	337·8	3n.	58·60	48·44
	331·7	,,	61·90	71·76
Mo.	336·7	26	59·16	54·85
Se.	335·6	3n.	60·06	6·85
De.	334·1		·5	63·25
Ro.	...	1n.	·47	5·68
Ta.	332·2	,,	61·85	70·60
	331·7	,,	62·28	3·71
W. & S.	·7	4	·7	1·61
	330·0	4	·4	·93
	331·2	2	...	3·73
	·4	3	60·0	4·85
	330·2	1	68·5	6·83
Gl.	·4	5	63·3	3·80
	332·1	4		·80
	331·9	5	63·3	4·91
Fl.	330·2	1n.	·5	7·84

745   Σ. 2909.

ƺ AQUARII.

R. A.	Dec.	M.
22^h 22·6^m	−0° 38′	4, 4·1

C. Mayer saw this star double in 1777; distance about 3″, angle about 18°.

H$_1$ (*Phil. Trans.*, vol. lxxii., p. 217): "Sept. 12, 1779.—Double.* Equal, or the preceding rather the larger. Both W. With 229, 1¼ diameter; with 449, 1½ diameter; with 460, 2 diameters; with 932, 2½ diameters; with 2010, pretty distinct, but too tremulous to estimate. With my 20-ft. reflector, power 600, full 2 diameters, very distinct. Position 71° 39′ n.f. Distance 4″·56, mean of two years' observations."

H$_1$ (*Phil. Trans.* 1804, p. 367). He finds an angular change of 6° 58′ in 22 years and 38 days. The equality of the stars, and the insulated situation they occupy, lead him to think a physical connexion highly probable.

H$_2$ in 1825 and 1829 discussed the measures, and was led to conclude that the indirect motion was fully confirmed.

Σ. (*M. M.*, p. 55): "There can be no doubt concerning the indirect angular motion. The distance has probably diminished, as it should if the angular velocity has increased."

Sm. (*Cycle*, p. 518): "By roundly assuming a mean of ¼° yearly, there may be a period of 750 years."

Da. (*Mem. R. A. S.*, vol. xxxv., p. 440). He says that some of the earlier measures

* Known to Mayer, etc

were enormously too large, and that Σ, early pointed this out.

Dr. Doberck's elements are

$$T = 1924·15$$
$$☊ = 140° 51′$$
$$\lambda = 134 \ 40$$
$$\gamma = 44 \ 42$$
$$a = 7″·64$$
$$e = 0·6518$$
$$P = 1578·33 \text{ years.}$$

Dunér's formulæ are

$$1854·46. \quad \Delta = 3″·49.$$
$$P = 346°·4 − 0°·4945 \ (t − 1850·0).$$

The proper motion of the system is + 0^s·010 in R. A., and − 0″·04 in N. P. D.

H$_1$.	18·9	1n.	4·22	1779·73
	...	,,	5·31	·94
	...	,,	·62	80·48
	...	,,	4·38	·60
	18·4	,,	...	1·73
	17·9	,,	...	2·38
	12·0	,,	...	1802·00
H$_2$ & So.	360·5	22	4·98	22·27
	361·1	70	·01	5·73
	356·2	1n.	5·22	8·56
	·4	,,	4·73	9·60
	·2	2n.	3·84	31·64
	352·4	1n.	·55	5·55
	·0	,,	·91	6·43
Σ.	359·8	2n.	·6	25·73
	355·2	5n.	·46	32·81
	353·0	1n.	·50	9·83
	349·5	,,	·77	51·89
Be.	355·7		·52	30·98
	354·3		·69	4·77
Sm.	356·0		·16	1·83
	355·3		4·1	2·71
	353·8		3·8	4·90
	352·4		·5	8·04
	348·9		2·7	42·59
Encke.	352·0		4·05	36·52
	351·8		3·8	·60
	350·6		·78	7·61
Da.	·7	2n.	·57	·38
	·1	6n.	·73	9·77
	348·4	2n.	·47	41·86
	·0		·54	2·67
	349·1		·43	·89
	348·1		·53	3·72
	·2		...	4·00
	347·5	3n.	3·48	6·95
	346·8		·38	7·93
	·4		·27	8·05
	345·2	9n.	·43	53·53
	343·6	2n.	·32	4·91
	340·3	4n.	·44	9·69
	336·3	1n.	·33	66·99
Galle.	350·4		·85	38·67
O.Σ.	351·6	1n.	·67	9·83
	349·3	,,	·84	·85
	350·2	,,	4·05	·85
	339·2		3·36	65·91

	°		″	
Ka.	353.7	7n.	3.49	1840.01
	350.0	4n.	.29	1.82
	338.9	5n.	.17	66.89
Mä.	352.2		4.12	41.48
	350.3		3.47	2.76
	348.7	1n.	.27	7.86
	345.8	,,	.58	52.80
	346.5		.60	3.85
	345.0		.77	4.86
	343.3		.47	5.78
	344.4		.89	6.79
	343.8		.65	8.01
	342.2		.51	.80
	340.4	9n.	.63	61.81
Hind.	348.2	1n.	.53	43.79
	.6	,,	.57	5.63
	.1	,,	...	6.53
Ja.	.1		3.2	5.87
	347.8		.82	6.48
	.6		.94	.80
	.0	10	.59	51.73
	346.8	,,	.78	2.73
	342.3	3n.	.28	7.87
D.O.	347.4		.83	46.75
	348.1		...	.85
	...		3.92	.86
Mit.	346.7	1n.	.95	7.57
	345.5	,,	.6	8.72
Flt.	348.1	40	.35	50.88
	347.0	16	.21	2.91
Mi.	346.7	32	.34	1.72
	345.7	16	.53	2.94
Po.	.8	35	...	3.77
	342.3	30	3.66	5.79
Mo.	345.6	3n.	.61	3.94
	.9	,,	.57	5.83
De.	344.8	6n.	.74	4.88
	343.9	5n.	.61	5.90
	342.6	,,	.59	6.74
	341.9	,,	.54	7.82
	.0	4n.	.67	8.74
	339.3	9n.	.53	62.76
	338.5	6n.	.50	3.69
	337.0	7n.	.34	7.18
	336.9	2n.	.34	8.72
	335.3	1n.	.38	9.85
	336.7	,,	.39	70.50
	.4	,,	.31	1.59
	334.6	,,	.54	2.52
	.9	,,	...	3.72
	.4	,,	3.45	4.73
	.4	3n.	.33	5.71
Se.	345.1	,,	.47	55.77
	343.0	2n.	.32	6.76
	337.8	,,	.50	66.77
Lu.	349.4		4.01	56.19
	342.1		3.32	62.85
Au.	341.6		.58	1.45
M.	340.7	1n.	.28	.78
	333.2	,,	.56	7.65
	335.7	,,	.44	8.62
	333.5	,,	.49	.82
	335.5	,,	.63	9.66
	.2	,,	.58	.77
	332.6	,,	.68	70.81

	°		″	
M.	334.5	1n.	3.56	1870.83
	333.9	,,	.60	2.72
	335.1	,,	.47	.77
	333.6	6n.	.37	3.77
	332.6	7n.	.41	5.74
Ro.	339.4	1n.	.20	63.18
	338.8	5n.	.27	5.69
Kn.	337.0	3n.	.64	6.71
	333.6	,,	.34	71.61
Ta.	339.8	2n.	4.24	66.74
	.1	,,	.42	8.76
	333.3	1n.	...	70.63
	334.0	2n.	3.8	3.67
Du.	337.2	,,	.34	67.69
	336.6	,,	.22	8.84
	337.4	,,	.17	9.48
	.2	1n.	.35	70.99
	336.1	,,	.52	5.71
Br.	.6		.42	68.76
Gl.	335.9	5	.27	70.63
	336.8	5	.40	1.70
	.2	4	...	3.73
	335.9	5	3.6	.74
	336.8	3	...	.80
	335.2	6	3.44	.87
	334.9	9	.6	4.91
	.9	8	.59	.91
	333.2	2n.	.3	5.69
	.2	4n.	.6	6.60
	334.0	2n.	...	8.77
W. & S.	...		3.79	1.61
	335.2	4	.81	.75
	336.2	4	.7	2.67
	335.7	4	.9	.71
	336.6	5	...	.74
	334.3	6	3.52	3.73
	335.3	3	...	.78
	.1	4	3.48	.81
	.7	2	.78	.83
	.0	5	...	5.92
	336.1	4	3.48	6.85
Schi.	334.4	1n.	.40	5.65
	154.8	,,	.38	6.87
Sp.	334.5		.40	5.65
	.9		.39	6.88
C.O.	335.3	9n.	.73	.77
	334.5	6n.	.38	7.67
Pl.	338.0	1n.	...	.32
	...	9n.	3.52	.54
Dob.	333.9	6n.	.26	8.80
Goldney.	332.9	5n.	.35	.80

746   Σ. 2910.

R. A.	Dec.	M.
22ʰ 22.5ᵐ	22° 55′	8.3, 8.8

Dunér gives

$$1856.12. \quad \Delta = 5''.33.$$
$$P = 345°.1 - 0°.05 \, (t - 1850.0).$$

	°		″	
Σ.	347.2	3n.	5.30	1832.14
Mä.	345	1n.	.49	45.65
	.5		.58	51.01
	344.7	3n.	.28	2.10

Mo. 344.2	2n.	5˙45	1856·75
Du. 343.8	3n.	·27	68·44
W. & S. ·7	2n.	·30	74·85
Gl. ·8	1n.	·38	·91

747   Σ. 2912.

37 PEGASI.

R. A.	Dec.	M.
22ʰ 23·9ᵐ	3° 49′	5·8, 7·2

Σ. thought there was no evidence of orbital motion.

H_2 (*Mem. R. A. S.*, vol. vi., p. 67) writes: "Divided with 320 and 6 inches aperture."

Sm. (*Cycle*, p. 518) writes: "It is clear that the angle is undergoing a rapid change direct, already indicative of a period of about five centuries."

Da. (*Mem. R. A. S.*, vol. xxxv., pp. 442, 502) says that the angle is probably increasing and the distance possibly diminishing, but that the question of binarity is still unsettled. And Secchi was of opinion that the angular motion was then doubtful, but that the distance had certainly diminished.

O.Σ.'s measure in 1852 shows that the angular change is very slow.

The common proper motion is −0″·063 in R. A., and +0″·123 in N. P. D.

Σ. 114·5	1n.	1·16	1825·69
109·3	,,	·24	32·82
114·1	,,	·08	4·84
·1	,,	·31	51·89
Sm. 116·8		·3	35·81
118·9		·1	9·66
Mä. 117·8		0·91	·70
106·2		·65	41·64
121·1		·85	2·80
120·2		·83	3·63
119·8		·82	6·74
126·3		·67	51·85
118·5		·81	4·36
O.Σ. 123·5	1n.	1·13	1·65
116·4		0·83	2·67
Da. ·2		1·10	43·87
118·5		0·91	54·44
119·8		...	60·70
Mit. 121·8	1n.	0·98	47·57
Se. 117·6	,,	·74	57·09
single			66·71
Ja. 116·3		0·7	57·87
W. & S. elongated			72·71
119·3	4	0·5	3·78
122·3	1	·5	1·92
Gl. 119·6	6	·5	3·87

748   Σ. 2915.

R. A.	Dec.	M.
22ʰ 26·5ᵐ	6° 48′	8·5, 8·7

So. 169·7		12″·90	1825·74
Σ. ·0	3n.	·27	7·76
H₂. 171·9		15	31·00
Mä. 166·5		13·03	43·71
De. 158·9		12·20	63·75

749   Σ. 2919.

R. A.	Dec.	M.
22ʰ 27·4ᵐ	20° 33′	9, 10·5

Σ. 273·8	4n.	14·30	1829·75
Mä. 270·8		...	43·79
De. 267·8		15·54	65·24

750   Σ. 2924.

R. A.	Dec.	M.
22ʰ 29·5ᵐ	69° 17′	6·8, 7·3

Σ. 257·3	3n.	0·84	1831·76
259·1	,,	·72	6·69
H₂. 258·3		1·0	1·80
O.Σ. 254·8	2n.	0·97	41·12
Mä. 263·7		·84	51·66
Se. ·5	1n.	·84	9·54
W. & S. 265·3	,,	1·24	73·83
266·5	,,	·23	4·84
Gl. ·5	,,	·14	·91

751   Σ. 2928.

R. A.	Dec.	M.
22ʰ 33·1ᵐ	− 13° 14′	8, 8

The angle has diminished.

So. 326·8		6·01	1825·29
Σ. 327·7	3n.	4·69	30·82
Mä. 325·9		5·34	43·71
Mit. 324·2	1n.	4·09	8·74
Se. 322·0	2n.	·39	57·40
De. 319·3		·38	63·11
Fer. 321·4		·55	7·91
W. & S. 318·8	3	·42	72·71
316·0	5	3·86	·75
319·5	4	...	5·97
Gl. 317·5	5	4·07	3·82
C.O. 316·8	3n.	·43	7·75

752   Σ. 2934.

R. A.	Dec.	M.
22ʰ 36·1ᵐ	20° 48′	8·2, 9·2

Secchi thought that there was perhaps some ground for suspecting variability.

Σ. 186·3	1n.	1·31	1828·86
191·6	,,	·26	9·72
185·6	,,	·10	33·77

Mä.	182·2		1·20	1838·19
	177·9		·20	42·77
	176·0		·10	3·77
	181·1		·25	5·64
	169·0		...	53·99
	172·7		...	6·80
Se.	168·2	2n.	1·10	·87
De.	164·7	3n.	·21	63·82
Fer.	168·5		·00	7·91
W. & S.	164·1	1	...	72·71
	163·8	6	1·22	3·78
	162·0	5	·15	4·84
	158·0	4	...	5·97
Gl.	163·6	6	1·16	3·87

753　　O.Σ. 477.

R. A.	Dec.	M.
22^h 38^m	45° 22′	7·2, 11·1

Rapid change in angle and distance. O.Σ. finds the following formulæ :

$$\Delta A = + 5''·687 - 0''·1795 \ (t - 1860·0);$$
$$D \Delta = - 4''·972 + 0''·0167 \ (t-1860·0);$$

and the differences are very small.

O.Σ.	122·7	3n.	9·60	1846·06
	148·2	2n.	5·54	75·74
De.	138·5	3n.	6·48	67·06

754　　Σ. 2941.

R. A.	Dec.	M.
22^h 40·1^m	18° 37′	7·5, 10·2

Σ.	270·5	3n.	8·73	1830·07
Mä.	269·5		9·27	43·70
	267·1		·67	7·83
	265·1		...	51·81
	267·6		9·40	3·03
De.	·0		·71	64·58

755　　Σ. 2942.

R. A.	Dec.	M.
22^h 41^m	38° 51′	7, 9·2

The distance has increased a little.

Σ.	282·4	4n.	2·65	1831·61
O.Σ.	278·2	3n.	·83	46·76
	279·3	2n.	3·04	69·86
Se.	276·3	3n.	2·96	56·93

756　　Σ. 2943.

R. A.	Dec.	M.
22^h 41·3^m	— 14° 41′	6, 9·2

Σ.	112·2	2n.	30·7	1831·80
Mä.	·5	1n.	...	47·71
C.O.	114·9	3n.	28·32	77·80

757　　Σ. 2944.

R. A.	Dec.	M.
22^h 41·6^m	— 4° 51′	A 7, B 7·5, C 8·2

In A B the angle has increased and the distance diminished.

Σ. (see *P. M.*, ccxxxiii.) showed that the changes in A C are produced in a straight line : his formulæ are

$$P = + (22''·210) \mp 0''·045) + (0''·2038$$
$$\mp 0''·0060) \ (T - 1837·0).$$

$$p' = - (50''·110) \mp 0''·045 + (0''·3102$$
$$\mp 0''·0060) \ (T - 1837·0).$$

O.Σ. finds that the following formulæ represent the observations well :

$$\Delta A = + 24''·862 \pm 0''·044 + (0''·1997$$
$$\pm 0''·0034) \ (t - 1850·0).$$

$$\Delta D = - 46''·149 \pm 0''·044 + (0''·3036$$
$$\pm 0''·0034) \ (t - 1850·0).$$

Dunér has the following formulæ :
For A B,

$$\Delta = 4''·25 - 0''·0189 \ (t - 1830·0).$$
$$P = 256°·7 + 0°·1257 \ (t - 1830)$$
$$+ 0°·00057 \ (t - 1830·0)^2$$

For A C,

$$\Delta \cos P = - 46''·20 + 0''·3000 \ (t - 1850·0).$$
$$\Delta \sin P = + 24''·95 + 0''·2318 \ (t - 1850·0).$$

A B.

H$_1$.	243·1	2n.	...	1792·72
So.	245·6	1n.	4·35	1822·90
Σ.	246·9	8n.	·12	32·98
	247·5	3n.	·19	6·33
Sm.	·4		·2	5·88
Mä.	·8	3n.	·00	44·99
	251·1	1n.	3·99	58·86
Da.	248·6		·98	46·06
O.Σ.	247·1	2n.	4·20	·78
	251·5	,,	·28	58·16
Se.	249·5	,,	3·91	5·84
Mo.	248·9	1n.	·69	7·91
De.	250·4	3n.	·68	62·68
Ro.	...	,,	·69	5·73
Ta.	252·2	2n.	·92	6·74
	253·0	1n.	...	8·75
	254·3	,,	4·32	70·60
	251·5	1n.	·86	3·64
Du.	253·8	3n.	3·46	68·82
W. & S.	254·1	3	·4	73·78
Gl.	·0	1n.	·38	·82
C.O.	·4	2n.	·73	5·83
Pl.	255·1	3n.	·68	6·94

A C.

So.	162°.5	In.	57".38	1822.90
Σ.	157.7	4n.	56.03	31.84
	156.7	3n.	55.11	4.57
	.3	4n.	54.93	6.41
Sm.	158.0		55.1	5.88
Da.	154.8		...	45.63
O.Σ.	150.9	2n.	52.12	51.90
	148.8	,,	51.09	8.16
Se.	150.2	,,	.65	6.34
	145.1	In.	49.89	66.96
De.	148.4		51.48	57.90
	146.7		50.67	62.68
Mo.	148.7	2n.	51.28	57.91
Du.	144.5	3n.	49.96	68.82
W. & S.	142.5	4	48.2	72.78
	141.7		...	.77
	.5		...	3.78
Gl.	.5		49.08	.82

B C.

Ro.	325.6	3n.	54.95	65.70
Ta.	326.0	2n.	...	6.74
	318.9	In.	...	70.63
	323.7	,,	...	3.68
W. & S.	317.2	I	...	.77

758 O.Σ. 481.

R. A.	Dec.	M.
22^h 42^m	77° 53′	7.5, 9.3

There is a third star (8.9) about 1′.5 distant. (O.Σ.)

O.Σ.	267.7	6n.	2.43	1855.18
De.	269.2	3n.	.37	66.61

759 Σ. 2947.

R. A.	Dec.	M.
22^h 44.9^m	67° 56′	7.2, 7.2

H$_1$.	86.4		...	1782.74
H$_2$.	78.6		3.53	1828.64
	74.4		.56	30.63
	73.9		.37	1.73
Σ.	76.0	3n.	2.98	2.45
Mä.	74.6		3.72	44.42
Mo.	70.5	12	.24	58.72
M.	72.8		.25	64.69
De.	69.9		.31	5.27

760 O.Σ. 482.

R. A.	Dec.	M.
22^h 48^m	82° 31′	5.2, 9.9

O.Σ.	30.2	6n.	3.46	1850.59
De.	33.0	3n.	.70	66.61

761 Σ. 2959.

R. A.	Dec.	M.
22^h 50.9^m	−3° 53′	6.5, 10.5

Σ.	96.7	4n.	15".66	1832.10
Mä.	97.8		14.88	43.64
	98.4		.05	8.00
De.	101.7		.21	64.78
C.O.	.5	2n.	.14	77.80

762 O.Σ. 536.

R. A.	Dec.	M.
22^h 52.5^m	8° 43′	7, 7.5

The common proper motion is + 0".43 in R. A., and + 0".24 in N. P. D.

O.Σ.	338.4	2n.	0.36	1852.67
	343.7		.46	3.91
			'simple'	9.65
	261		oblong?	61.66

763 O.Σ. 484.

R. A.	Dec.	M.
22^h 52^m	72° 12′	A 7.1, B 8, C 11

In A B there is rapid retrograde motion.

A B.

O.Σ.	117.7	2n.	0.36	1846.42
	99.3	,,	.46	55.56
De.	89.5	,,	...	67.66

$$\frac{A+B}{2} \text{ and } C.$$

O.Σ.	255.4	2n.	30.72	55.56

764 O.Σ. 483.

52 PEGASI.

R. A.	Dec.	M.
22^h 53.2^m	11° 5′	6.2, 7.7

Change in angle and distance.

O.Σ.	180.8	2n.	0.94	1845.28
	187.9	3n.	.94	52.78
	191.8	2n.	1.24	9.66
Mä.	186.3		0.73	45.70
Da.	...		.91	7.86
	190.9		1.23	53.88
Se.	203.4	2n.	0.96	7.85
De.	198.5	4n.	1.14	65.24
W. & S.	202.2	4	.4	73.83
	204.2	15	.21	.82
	203.9	4	.43	4.84
Pl.	.5	2n.	...	6.92

765　　Σ. 2976.

R. A.	Dec.	M.
23ʰ 1·6ᵐ	5° 57′	A 8·3, B 10·2, C 8·8

In A B the distance has changed considerably, the angle very little; while in A C there is decided change in both angle and distance.

A B.

	°		″	
Σ.	262·0	3n.	7·94	1828·43
H₂.	263·3		6·5	31·00
Mä.	262·8		7·08	43·74
Se.	263·0	2n.	5·09	57·38
O.Σ.	265·0	1n.	8·06	65·99
	264·2	,,	7·22	75·90
W. & S.	265·1		...	·97

A C.

	°		″	
Σ.	177·6	3n.	15·88	28·43
H₂.	·8		·0	31·00
Sm.	179·7		16·0	7·72
Mä.	180·6		15·86	43·74
	182·4		16·81	52·93
Se.	183·2	2n.	·31	7·42
O.Σ.	185·1	1n.	·40	65·99
	187·4	,,	·72	...
W. & S.	·2	4	·5	73·78
	186·7	2n.	·8	5·97
Gl.	·0	4	·45	3·82

766　　O.Σ. 489.

π CEPHEI.

R. A.	Dec.	M.
23ʰ 4·1ᵐ	74° 44′	5·2, 7·5

C. A yellow, B purple.

The wide pair, h. 1852, was measured by Sm., who gives the magnitudes as A 5, the small companion 10, B 12. His measure of the wide pair in 1838 was 241°·5, and difference of R. A. 11ˢ·8.

O.Σ., however, detected the duplicity of the principal star, the close companion being of the 8·9 magnitude, and the distance about 1½″.

On being apprised of this discovery, Sm. examined the object at Hartwell in 1843, and was able to see the companion and estimate its position and distance. Mr. Lassell and Mr. Dawes in the same year saw it with the 9-in. Newtonian, power 400.

Rapid direct motion: the distance has probably increased.

The proper motion of π is + 0ˢ·002 in R. A., and + 0″·04 in N. P. D.

O.Σ.	° 351·4	2n.	″ 1·15	1846·48
	358·5	,,	·21	51·42
	23·9	1n.	·32	76·25
Sm.	330·0		·8	43·77
De.	14·2	2n.	·24	65·88
	17·0	1n.	...	6·43
	10·8	,,	1·42	7·60
	17·6	,,	·16	9·78
W. & S.	21·4	,,	·38	73·81
	19·3	,,	·26	4·84
Gl.	·9	,,	·24	·91
	16·0	2n.	...	5·18

767　　O.Σ. 490.

R. A.	Dec.	M.
23ʰ 5ᵐ	56° 47′	7·2, 9·2

O.Σ.	308·5	3n.	1·36	1846·80
De.	301·9	,,	·56	66·95

768　　Σ. 2998.

R. A.	Dec.	M.
23ʰ 12·8ᵐ	− 14° 7′	5·2, 7·2

C. Σ., A yellow, B ash.

H₁.	...	1n.	13·75	1781·63
	342·8	,,	...	1802·67
So.	346·7	,,	14·99	22·87
Σ.	345·2	3n.	13·37	30·90
	344·7	,,	·82	6·65
Sm.	·9		·5	1·87
	345·4		14·0	8·91
Mä.	347·0	2n.	13·83	44·69
De.	344·8	,,	·71	58·12
Du.	346·1	,,	14·02	68·34
C.O.	348·6	3n.	13·30	75·91
	346·8	2n.	·70	7·71
W. & S.	·6	1n.	·83	5·97
Fl.	348·8	,,	·74	7·92
Goldney.	345·2	3n.	·82	8·89

769　　Σ. 3001.

o CEPHEI.

R. A.	Dec.	M.
23ʰ 13·7ᵐ	67° 27′	5·2, 7·8

C. A very yellow, B very blue.

Discovered by Σ., and measured by him in 1832 and 1833.

Sm. says, "Little can be said upon the dates, until a longer lapse of time has intervened, when it may very probably prove to be a physical object."

Secchi (p. 60) writes, "motion certain."

The common proper motion is − 0ˢ·019 in R. A., and − 0″·02 in N. P. D.

Sm.	173.8		2.5	1831.00
	.8		.5	4.95
Σ.	174.9	3n.	.35	2.84
Da.	175.1		.30	4.02
Mä.	138.2		.39	9.55
	179.8		.54	42.80
	180.1		.28	4.43
	184.9		.20	52.65
Mit.	183.0	1n.	.81	47.69
O.Σ.	187.0	,,	.65	51.87
	196.9	,,	.88	70.18
De.	183.9	5n.	.57	54.82
	187.2	1n.	.73	5.81
	184.5	,,	.46	6.49
	185.4	2n.	.52	8.57
	187.0	6n.	.47	64.68
Po.	182.6	...		55.92
	184.3	...		61.01
Ja.	186.8		2.53	56.92
Se.	187.1	.2n.	.47	8.44
	185.7	1n.	.56	66.97
Mo.	186.9		.60	58.62
M.	182.3	1n.	.28	62.57
	178.7	,,	...	.57
	0.6	,,	2.56	9.67
W. & S.	188.0	4	.57	72.80
	181.0	4	.69	3.82
	191.0	4	.56	4.84
	189.6	4	.81	.85
Gl.	.8	7	...	3.80
	191.0	5	...	.81
	190.7	8	2.7	4.91
	189.9	8	.6	.91
	191.2	2n.	...	5.69
	189.9	1n.	...	6.71

770   Σ. 3006.

R. A.   Dec.   M.
23h 15.4m   34° 47'   8.5, 9

C. white.

Dunér's formulæ are
$$1858.25. \quad \Delta = 4''.97.$$
$$P = 176°.9 - 0°.25 \ (t - 1850.0).$$

So.	183.8	2n.	5.12	1825.70
H₂.	188		.5	7.88
	182.9		...	9.67
	176.2		5.05	30.76
Da.	178.3	1n.	.48	.81
	176.9	,,	4.93	41.30
	177.0	,,	5.22	3.78
	176.4		...	5.51
Σ.	182.8	3n.	4.65	31.55
Mä.	177.1		.32	43.80
	183.5	1n.	5.22	4.90
Mo.	174.7	2n.	4.98	56.88
De.	173.5	7n.	.94	64.92
Ro.	172.9	2n.	.95	5.71
Du.	171.6	3n.	.98	70.20
Ta.	168.7	1n.	5.05	3.64
W. & S.	170.4	2n.	4.92	.80
	173.0	1n.	.94	4.85
Gl.	169.9	,,	5.0	3.87
	172.9	,,	4.9	4.91

771   Σ. 3007.

R. A.   Dec.   M.
23h 17m   19° 54'   6.5, 9.5

Σ.	79.2	3n.	5.68	1829.83
	82.9	1n.	6.13	51.80
O.Σ.	83.3		.12	69.79

772   Σ. 3008.

P. XXIII. 69 AQUARII.

R. A.   Dec.   M.
23h 17.5m   −9° 7'   7, 8

The proper motion of A is − 0".11 in R. A., and +0".09 in N. P. D.

So.	274.1	10	7.98	1824.80
De.	264.7	3n.	5.95	57.84
	.8	1n.	.58	8.57
	263.1	4n.	.60	62.74
	262.4	2n.	.58	3.74
	.0	6n.	.32	6.42
	260.6	1n.	.18	8.55
	259.9	,,	.07	70.52
	.4	,,	.05	1.59
	258.9	,,	.11	2.56
	.4	,,	4.88	3.64
	.5	,,	.83	.88
	256.6	2n.	.88	4.86
	.7	1n.	.75	5.60
Ro.	260.7	2n.	5.81	65.72
W. & S.	248.1	3	4.8	72.75
	249.0	4	5.02	.75
	248.5	4	.17	.85
Gl.	259.6	5	.16	3.82
C.O.	258.0	4n.	...	5.90
	255.9	3n.	4.90	7.75
Pl.	257.0	,,	.72	6.91

773   O.Σ. 495.

R. A.   Dec.   M.
23h 19m   56° 52'   7.3, 7.5

O.Σ.	310.4	3n.	0.56	1846.57
De.	140.3	4n.	...	67.89

774   O.Σ. 496.

R. A.   Dec.
23h 24m   57° 53

M.
A 5.4, B 7.4, C 8.9, D 10.

C. A white, B reddish, C red

Of the existence of D there can be no doubt. O.Σ. saw it in 1845, but failed to detect it in 1853. De. has not seen it.

26

Da. discovered the duplicity of B independently. The companions given by H_2 in his quarto catalogue probably do not exist. (O.Σ.)

A B.

O.Σ.	269°·2	5n.	76"·1	1849·64
De.	·3	3n.	75·63	68·68

B C.

O.Σ.	224·2	5n.	1·38	49·64
De.	223·8	4n.	·39	67·94

A D.

O.Σ.	336·8	1n.	1·51	51·76

---

775   O.Σ. 500.

R. A.	Dec.	M.
23ʰ 31·8ᵐ	43° 46'	6·1, 7

C. A white, B blue.

Probable direct motion.

Mä.	92·9		0·3	1843·90
	113·3		·35	51·75
O.Σ.	229·4	2n.	·45	45·24
	308·5	,,	·45	52·82
De.	313·6	4n.	...	67·21
W. & S.	140·8	6	0·66	74·86
Gl.	141·2	1n.	·7	·91

---

776   So. 356.

R. A.	Dec.	M.
23ʰ 39·8ᵐ	− 19° 21'	6, 7·5

Σ.	143·4		5·12	1821·91
So.	·5		·96	3·79
H_2.	141·3		6·12	30·67
	145·6		8·0	1·70
Sm.	141·8		5·5	2·80
Se.	140·7		·68	55·93
De.	139·9		·63	66·40
C.O.	140·7	2n.	...	75·90
	139·4	,,	6·34	6·75

---

777   Σ. 3037.

R. A.	Dec.	M.
23ʰ 40ᵐ	59° 48'	7, 8·5

C. A very yellow, B blue.

Σ.	214·3	3n.	2·72	1831·73
	213·0	1n.	·67	3·47
Mä.	229·1	3n.	·15	45·95

---

778   Σ. 3038.

R. A.	Dec.	M.
23ʰ 40·4	61° 59'	9, 9·5

C. white.

Σ.	275·0	3n.	4"·36	1833·83
Mä.	277·8		·30	45·75
	278·2		...	7·95
	·6		4·14	8·13
	279·4		·47	52·65

---

779   Σ. 3039.

R. A.	Dec.	M.
23ʰ 41ᵐ	27° 45'	7·3, 9·7

C. A very yellow.

H_1.	39·5	1n.	...	1782·89
So.	36·5	,,	...	1824·81
Σ.	·3	2n.	30·1	30·52
	·6	1n.	·6	1·73
Mä.	·0	,,	31·57	42·77
	35·4	2n.	·58	5·32

---

780   O.Σ. 507.

R. A.	Dec.	M.
23ʰ 43ᵐ	64° 13'	A 6·8, B 7·5, C 7·8

A B.

O.Σ.	224·4	2n.	0·56	1847·01
De.	240·9	3n.	...	67·96

$$\frac{A+B}{2} \text{ and } C.$$

O.Σ.	353·8	2n.	48·83	47·01
De.	·9	3n.	·89	67·96

---

781   O.Σ. 510.

R. A.	Dec.	M.
23ʰ 45ᵐ	41° 25'	A 7·5, B 7·8, C 9

C is probably variable. O.Σ. has estimated it as 9 and 10·11; De. as 11 and 12.

A B.

O.Σ.	347·8	3n.	0·40	1848·43
De.	163·6	,,	...	67·63

$$\frac{A+B}{2} \text{ and } C.$$

O.Σ.	344·0	1n.	20·78	47·91
De.	345·1	4n.	21·03	67·21

782 Σ. 3046.

R. A.	Dec.	M.
23ʰ 50·2ᵐ	− 10° 10′	8, 8·5

Direct motion.
The common proper motion is − 0ˢ·380 in R. A., and + 0″·097 in N. P. D.

	°		″	
Σ.	232·2	4n.	2·51	1830·15
	239·4	1n.	·81	51·88
Mä.	234·8		·46	43·80
Se.	241·2	2n.	·78	57·42
	238·0		3·02	67·91
De.	241·0	2n.	2·90	3·92
W. & S.	240·5	5	...	72·89
Gl.	·5	2	3·2	3·87
C.O.	243·0	3n.	·03	7·75

783 Σ. 3050.

R. A.	Dec.	M.
23ʰ 53·4ᵐ	33° 4′	6, 6

Probable change in angle and distance. Dunér gives

$$\Delta = 3''\!\cdot\!60 - 0''\!\cdot\!0196\,(t - 1850\!\cdot\!0).$$
$$P = 194°\!\cdot\!8 + 0°\!\cdot\!278\,(t - 1860\!\cdot\!0) + 0°\!\cdot\!00151\,(t - 1850\!\cdot\!0)^2.$$

	°		″	
H₁.	180		...	1790·91
	180		...	94·71
H₂.	188·4	2n.	5·26	1821·92
	195		·5	7·88
	189·7	3n.	4·37	30·04
Da.	·0	4n.	·07	·73
	191·7		·14	7·03
	193·0	1n.	3·65	43·81
	196·0		·47	54·81
	·4		·60	5·47
Σ.	191·0	3n.	·78	32·65
Mä.	190·8		·60	6·49
	193·6		·43	45·65
	196·0	2n.	·87	50·85
	198·2	8n.	·34	62·20
De.	196·5		·66	54·68
	199·5	10n.	·18	64·84
Mo.	196·6	3n.	·50	55·98
Se.	·4	,,	·44	7·51
	200·2		·63	66·97
Eng.	·1	2n.	·49	5·38
Ro.	199·4	,,	·40	·73
Du.	200·7	3n.	·09	8·56
	203·0	,,	2·96	75·70
W. & S.	202·2	5	·94	1·94
	201·0	6	·93	2·89
	200·7	5	3·21	·80
	201·0	4	·13	·80
	202·1	9	·01	·80
	·1	8	·20	3·81
	203·6	4	·09	6·06
Gl.	201·0	6	...	3·79
	·2	6·	...	·81
	200·7	5	...	·82
	202·5	7	3·16	4·91

784 B. A. C. 8350.

R. A.	Dec.	M.
23ʰ 55·9	26° 27′	6, 9

Probably variable in magnitude.
The rapid changes are due to the proper motion of A, which is + 0ˢ·064 in R. A., and + 0″·97 in N. P. D.
Brünnow has found a parallax of 0″·054 for this star.

	°		″	
Br.	77·0		16·0	1870·00
Fl.	49·8	1n.	14·0	7·94

785 Σ. 3056.

R. A.	Dec.	M.
23ʰ 58·5ᵐ	33° 36′	A 7·4, B 7·4, C 9

Slight change in the angle of A B. In $\dfrac{A+B}{2}$ and C the distance has increased considerably.

A B form a binary system, most probably.

A B.

	°		″	
Σ.	159·6	2n.	0·60	1828·84
	156·7	,,	·60	33·81
O.Σ.	155·0	,,	·57	41·11
	146·8	,,	·57	53·70
Mä.	156·9		·44	41·56
	159·0		·40	2·76
Se.	154·2	2n.	...	56·87
De.	152·3	5n.	0·6	64·84

$\dfrac{A+B}{2}$ and **C.**

	°		″	
H₂.	352		15	1827·88
Σ.	355·5	2n.	20·25	8·84
	·3	3n.	·63	33·51
O.Σ.	356·1	2n.	21·04	53·70
De.	·4	5n.	·45	64·84

786 Σ. 3062.

R. A.	Dec.	M.
23ʰ 59·9ᵐ	57° 46′	6·9, 8

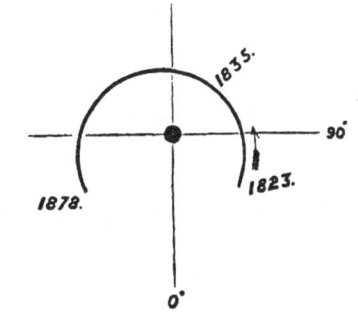

H₁ (*Phil. Trans.*, vol. lxxv., p. 645) : "Double. 50° 42′ n. prec."

Σ. (*M. M.*, p. 9). Measures in 1831 and 1833 are given. Σ. notes the great angular change in two years with diminished distance. In 1822 he saw the stars separated in the 5-ft. meridian instrument, and presenting no difficulty. In 1824 he could see them with difficulty. Hence he puts the distance in 1823·81 at from 1″ to 1″·25, and this places the continued decrease of distance from 1822 to 1834 beyond doubt. Thus, 147°·87 have been passed over in 51·06 years, in a direct sense. The apparent orbit must thus be very elliptical.

Mädler (*Die Fixst. Syst.*) discusses the observations made up to 1846. He finds the observations in 1782·65 and 1823 intractable; but by giving a double weight to the later observations he finds the following elements :

$$T = 1834 \cdot 01$$
$$U = 146 \cdot 83 \text{ years}$$
$$\phi = 35° \ 7' \cdot 5$$
$$\lambda = 42 \ 10 \ \cdot 3$$
$$\Omega = 77 \ 21 \ \cdot 2$$
$$i = 38 \ 35 \ \cdot 9$$
$$a = 0'' \cdot 9982$$
$$\pi' = 0 \ \cdot 03456.$$

He is of opinion that another fifty years' observations are needed to enable the orbit to be fairly dealt with.

M. Schur's elements are

$$T = 1835 \cdot 196$$
$$\omega = 97° \cdot 5$$
$$\Omega = 32 \ \cdot 2$$
$$i = 29 \ \cdot 9$$
$$e = 0 \cdot 5009$$
$$\mu = + 3° \cdot 1959$$
$$a = 1'' \cdot 310$$
$$P = 112 \cdot 644 \text{ years.}$$

Dunér's comparison of these elements with the observations exhibits considerable differences.

M. Schur gives an ephemeris from which the following have been taken :

1870	278°·2	1″·551
72	281 ·6	·576
74	284 ·9	·600
76	288 ·1	·620
78	291 ·1	·639
80	294 ·2	·655
82	297 ·2	·670
84	300 ·1	·680
86	302 ·9	·690

D₁ Doberck's provisional elements are
$\Omega = 38° 35'$, $\lambda = 92° 7'$, $\gamma = 32° 11'$, $c = 0·4612$, $P = 104·415$ years, $T = 1834·88$, $a = 1''·27$.

The proper motion of this system is $+ 0''·346$ in R. A., and $- 0'·020$ in N. P. D.

	°		″	
H₁.	319·4		...	1782·65
			·1		3·05
Σ.	36·7±		1·2±	1823·81
	87·5	2n.	0·82	31·71
	108·5	3n.	·56	3·71
O.Σ.	186·5	4n.	·65	40·32
	220·3	2n.	·97	6·42
	229·7	,,	1·14	8·22
	232·5	3n.	·09	9·19
	233·8	,,	·16	50·04
	235·7	2n.	·35	1·16
	238·3	3n.	·23	2·49
	243·4	4n.	·47	4·11
	242·6	3n.	·38	5·05
	247·8	2n.	·40	6·66
	250·4	3n.	·49	7·37
	255·3	,,	·45	9·16
	261·7	2n.	·54	62·18
	270·4	,,	·47	6·20
	276·5	,,	·59	8·98
	279·2	,,	·48	70·18
Da.	187·0		0·8	40·78
	193·4		·95	1·86
	210·0		·94	3·80
	228·8		1·16	8·87
	235·2		...	50·93
	244·3		1·28	4·32
	265·6		·40	63·86
Mä.	193·6		0·89	41·58
	207·3		·87	2·80
	213·8		·85	4·50
	216·8		·96	5·54
	220·7		1·05	6·53
	225·1		·12	7·53
	232·3		·28	50·71
	237·0		·16	1·18
	234·6		·27	·76
	241·3		·02	2·21
	238·0		·25	·72
	248·8		·43	6·81
De.	249·8	3n.	...	4·91
	·9	,,	...	5·05
	·2	6n.	1·2	·69
	250·2	5n.	·22	6·37
	252·2	4n.	·2	7·71
	·4	2n.	·2	8·54
	263·6	9n.	·48	62·73
	265·4	,,	·43	3·57
	266·1	2n.	·37	4·10
	268·7	6n.	·40	·67
	270·7	7n.	·35	5·61
	274·1	11n.	·40	7·25
	277·8	5n.	·42	8·55
	280·0	,,	·47	9·68
	282·2	7n.	·44	70·52
	283·9	,,	·39	1·55
	285·7	6n.	·47	2·63
	287·5	9n.	·44	3·63
	289·6	6n.	·40	4·63
	291·9	5n.	·46	5·60
	293·0	1n.	·54	6·00
Mo.	247·9		·33	55·91
Se.	253·4		·25	7·60
	270·0		·34	66·97

Kn.	265.7°	In.	1".40	1863.52	Gl.	291.1°	8	1".4	1874.91
	269.9	3n.	.43	5.71		.0	9	.3	.91
	282.7	2n.	.38	72.60		293.2	In.	...	5.18
Ro.	271.9	3n.	.14	65.71		.0	3n.	...	6.41
Ta.	270.3	,,	.46	6.64		295.6	2n.	...	7.95
	268.3	,,	.66	8.76		293.5	In.	...	8.73
	280.6	In.	.63	70.64	W. & S.	286.3	4	1.45	2.80
	297.8	,,	0.91	3.80		287.8	8	.45	3.82
	299.1	,,	1.08	4.72		291.2	8	.37	4.86
Gl.	281.0	5	.5	0.44		298.8	6	.44	6.93
	284.0	4	.6	1.60	Du.	292.9	5n.	.43	5.69
	289.4	6	.5	3.81	Pl.	294.5	,,	.46	6.99
	286.7	4	.6	.87	Dob.	302.1	,,	.38	8.89

SUPPLEMENTARY LIST OF MEASURES.

Ref. No.	Σ.				Ref. No.	Σ.			
787	**35.**				**795**	**371.**			
Σ.	268.3°	8".69	1830.1		Σ.	74.7°	3".35	1831.2	
De.	267.5	7.88	68.2		De.	81.7	.32	67.4	
788	**51.**				**796**				
Σ.	131.5	4.16	30.8		H_2.	199.3	9.06	35.9	
Se.	127.6	.05	57.9		Ja.	202	7	57	
789	**149.**				**797**	**3114.**			
Σ.	118.2	1.35	33.2		Σ.	190.1	1.92	32.4	
De.	108.0	.33	67.7		De.	179.4	2.26	64.6	
790	**171.**				**798**	**[78.]**			
Σ.	157.6	27.9	29.9		Mä.	241.7	2.25	46.1	
De.	159.1	29.14	65.4		De.	247.7	.72	66.9	
791	**[37.]**				**799**	**531.**			
Mä.	223.1	1.39	43.3		Σ.	291.9	0.8	30.5	
De.	214.6	.41	67.7		O.Σ.	395.2	1.07	71.2	
792	**254.**				**800**	**536.**			
Σ.	334.2	13.33	31.7		Σ.	149.5	1.75	31.0	
W.	343.8	12.9	74.0		W.	160.2	.40	74.1	
793	**325.**				**801**	**579.**			
Σ.	253.4	11.70	30.9		Σ.	30.1	16.48	31.5	
Gl.	228.0	9.2	74.0		W.	35.6	.21	75.1	
794	**360.**				**802**	**596.**			
Σ.	146.4	1.34	31.2		Σ.	280.8	11.12	31.1	
De.	139.1	.67	69.3		W.	287.5	10.33	74.1	

Ref. No.	Σ.		
803	**620.**		
Σ.	225°·9	3″·70	1828·2
W.	232·0	·63	75·1
804	**704.**		
Σ.	8·5	26·53	31·3
De.	10·4	23·77	65·1
805	**Bu. 320.**		
Bu.	269·0	2·8	74·9
De.	292·3	3·06	77·1
806	**782.**		
Σ.	309·2	35·96	30·1
De.	308·6	38·66	66·6
807	**787.**		
Σ.	81·1	1·37	30·3
Gl.	75·3	·3	79·1
808	**[122.]**		
Mä.	117·8	0·22	43·2
De.	single.		65
809	**826.**		
Σ.	115·5	1·84	32·4
W.	128·6	·76	75·2
810	**879.**		
Σ.	67·7	8·40	27·3
De.	71·6	7·46	67·9
811	**[139.]**		
Mä.	132·5	0·77	43·2
De.	317·2	elong^d.	73·5
812	**910.**		
Σ.	170·9	0·67	29·5
De..	165·6	·7	69·1
813	**[152.]**		
O.Σ.	40·2	0·86	50·0
De.	33·9	·85	
814	**974.**		
H$_1$.	216·9	23·5	1782·8
Fl.	224·1	22·2	1877·8
815	**991.**		
Σ.	173·2	3·72	28·2
Gl.	167·0	·7	74·2

Ref. No.	Σ.		
816	**1047.**		
Σ.	19°·5	20″·66	1828·5
De.	22·3	21·51	68·4
817	**1046.**		
Σ.	231·0	12·07	29·4
De.	234·2	10·92	67·7
818	**1051.**		
Σ.	268·5	1·23	31·3
Mä.	278·4	·22	58·1

This is A B.   A C is fixed at 81° 31″.

Ref. No.	Σ.		
819	**1171.**		
Σ.	338·6	2·8	28·9
De.	330·1	·5	64·8
820	**1213.**		
Σ.	327·7	8·43	30·9
De.	324·0	7·26	67·7
821	**1230.**		
Σ.	194·1	28·0	29·2
Fl.	192·8	30·82	77·8
822	**1234.**		
Σ.	71·3	20·77	31·0
Fl.	69·0	21·75	77·8
823	**1243.**		
Σ.	221·3	1·99	33·9
De.	225·3	1·84	67·2
824	**1285.**		
Σ.	339·2	27·57	28·3
Fl.	337·9	25·45	77·3
825	**1343.**		
Σ.	271·1	10·22	36·2
Ferguson.	269·4	6·79	63·3
826	**1402.**		
H$_2$.	96·0	20·42	30·2
Fl.	·98·7	23·12	77·8
827	**1476.**		
H$_2$.	351·9	2·27	31·0
W.	359·1	·57	75·3
828	**So. 621.**		
So.	25·5	43·43	25·2
Fl.	38·8	57·84	77·5

Ref. No.	Σ.		
829	**1549.**		
Σ.	115°·9	14″·03	1828·7
Fl.	·2	12·98	77·4
830	**1594.**		
Σ.	165·0	16·95	31·9
W.	161·2	13·31	75·8
831	**1621.**		
Σ.	124·0	3·44	30·3
De.	140·0	·13.	67·9
832	**γ CRUCIS.**		
H₂.	28	120	35
Po.	36·5	99	60
833	**1682.**		
Σ.	308·8	33·65	31·6
Bu.	306·2	31·98	78·3
834	**[267.]**		
O.Σ.	300·8	0·25	49·6
De.	single ?		72·4
835	**h. 4649.**		
H₂.	64·4	12·0	35·4
	69·4	·0	37·5
836	**1804.**		
H₁.	27·6	...	1796·6
W.	19·9	4·20	1874·3
837	**[276.]**		
O.Σ.	202·7	0·5	42·0
De.	194·3	elong^d.	69·4

This is A B. A C seems unchanged: 74°·2, 9″·57, 1869 De.

838	**3124.**		
Σ.	150	elong^d.	36·2
W.	135	,,	74·4
839	**1846.**		
Σ.	108·8	3·69	26·8
W.	112·6	4·27	75·4
840	**So. 184.**		
H₁.	128·2	11·88	1783·0
St.	129·8	9·68	1876·4
841			
So.	270·1	10·82	23·3
Bu.	291·3	15·62	78·3

Ref. No.	Σ.		
842	**π LUPI.**		
H₂.	112°·8	0″·8	1835·3
Ja.	106.2	1·2	48·1
843	**1908.**		
Σ.	137·2	1·46	32·5
W.	143·8	·26	74·5
844	**3095.**		
Σ.	349·7	2·85	31·3
De.	337·5	·84	69·3
845	**1972.**		
H₁.	86·8	28·0	1783·5
De.	82·1	30·6	1865·8
846	**1988.**		
Σ.	263·3	2·91	30·0
W.	262·3	3·04	76·4
847	**2006.**		
Σ.	203·5	1·69	28 7*
De.	197·4	·65	68·9*
Σ.	224·0	43·72	28·7†
De.	221·0	44·31	68·5†
848	**2017.**		
Σ.	249·7	25·03	31·4
De.	251·2	·95	67·6
849	**2041.**		
Σ.	4·4	3·06	31·4
De.	1·5	2·58	68·0
850	**3105.**		
Σ.	59·4	0·41	30·9
De.	53·2	·50	70·0
851	**2080.**		
Σ.	29·3	5·61	30·4
De.	28·2	4·42	68·4
852	**2089.**		
Σ.	61·0	2·30	30·6
De.	67·3	·41	68·3
853	**2096.**		
H₁.	93·1	20·45	1783·2
Gl.	90·5	23·0	1874·5

* A B.   † A C.

Ref. No.	Σ.				Ref. No.	Σ.			
854	**2156.**				**867**	**2488.**			
Σ.	31°8	3"54	1829·5		Σ.	318°5	1"29	1829·0	
W.	35·4	·31	74·6		W.	329·5	·28	76·6	
855	**2160.**				**868**	**[371.]**			
Σ.	61·9	4·15	29·6		Mä.	149·3	0·72	43·4	
Gl.	66·7	3·97	74·7		Newcomb.	154·0	96	74·7	
856	**2163.**				**869**	**h. 5113.**			
Σ.	103·5	1·51	30·0		H₂.	129·1	25	37·5	
Bu.	97·2	·58	78·3		St.	169·8	16·72	77·6	
857	**A. C. 9.**				**870**	**h. 5114.**			
Da.	231·2	1·12	57·5		H₂.	131·0	1·75	37·6*	
Sp.	237·4	0·91	75·6		Ja.	260·0	·5	57·8*	
858	**h. 5014.**				H₂.	270·7	69·43	37·6†	
H₂.	69·2	0·75	36·7		Ja.	266·0	66·37	57·2†	
Ja.	135·1	·55	57·3		**871**	**[375.]**			
859	**[524.]**				Ma.	119·5	0·55	43·5	
O.Σ.	86·5	0·37	53·3		Newcomb.	144·4	·67	74·7	
De.	68·8	elong^d·	70·8		**872**	**2553.**			
860	**2286.**				H₂.	78·0	0·7	30·0	
Σ.	322·0	2·42	31·7		De.	91·6	·98	74·1	
De.	315·1	·53	65·6		**873**	**2564.**			
861	**η SERPENTIS.**				Σ.	184·0	10·78	32·3	
H₁.	99·1	81	1781·8		Bu.	175·1	·04	78·4	
Fl.	67·1	142·8	1877·5		**874**	**Da. 10.**			
862	**2310.**				Da.	314·4	0·53	59·6	
Σ.	233·7	5·07	29·7		Sp.	308·2	·60	75·8	
W.	237·6	·25	75·6		**875**	**2585**			
863	**A. C. 11.**				H₁.	304·2	8·83	1781·9	
Da.	178·1	0·42	54·7		Dob.	312·4	·64	1877·7	
Sp.	172·0	·33	75·6		**876**	**h. 2904.**			
864	**Bu. 134.**				H₂.	173·5	20	31·0	
O.Σ.	141·0	1·11	51·8		St.	141·4	18·32	77·7	
De.	134·0	·07	75·0		**877**	**A. C. 16.**			
865	**γ Cor. Aust.**				Da.	234·3	0·35	59·6	
H₂.	37·1	1·23	34·5		Newcomb.	241·4	·45	74·7	
St.	253·1	·67	76·6		**878**	**[392.]**			
866	**2461.**				Mä.	324·3	0·2	43·8*	
Σ.	330·6	3·72	30·7		De.	304·3	elong^d·	69·5*	
W.	321·9	·86	76·6		Σ.	293·4	3·23	31·5†	
					De.	291·8	·08	70·1†	

* A B.   † A C.

Ref. No.	Σ.		

879   2612.

Σ.	52°.8	36″.6	1827.7
Bu.	53.3	38.0	78.5

880   Σ. 2624.

H₁.	179.3	...	1783.7*
W.	176.8	2.06	1875.7*
H₁.	320.0	...	1783.1†
W.	328.8	42.3	1875.7†

881   2626.

Σ.	121.7	1.17	31.1
W.	130.1	0.98	75.7

882   2662.

Σ.	38.9	1.72	31.0
De.	41.8	.6	67.4

883   2666.

Σ.	239.9	2.59	28.8
W.	246.2	.53	75.7

884   [406.]

O.Σ.	136.3	0.54	45.8
Sp.	112.9	.45	76.8

885   2668.

Σ.	293.6	3.30	31.1
W.	288.3	.05	75.7

886   2673.

Σ.	335.1	2.53	30.7
W.	329.4	1.92	74.7

887   2674.

Σ.	1.3	15.51	29.6
W.	3.2	14.8	75.7

888   2723.

Σ.	85.6	1.49	31.7
De.	92.2	.22	66.7

889   2728.

Σ.	24.7	4.22	31.8
De.	.5	5.17	67.4

890   Bu. 269.

H₁.	234.8	1	1783.4
Bu.	252.6	1.08	1876.2

* A B.   † A C.

Ref. No.	Σ.		

891   2751.

H₂.	345°.9	2″.42	1828.6
De.	349.3	1.62	69.6

892   Bu. 368.

De.	99.4	0.5	75.8
	93.8	.64	7.8

893   [527.]

O.Σ.	306.2	0.4	46.8
Bu.	99.4	.64	77.7

894   H₁ I. 47.

H₁.	354.8	...	1783.5
St.	321.8	3.0	1877.7

895   θ INDI.

H₂.	306.7	3.68	34.5
Russell.	292.0	...	71.0

896   [445.]

O.Σ.	113.1	0.78	47.5
De.	107.7	.80	72.2

897

H₂.	13.3	30.11	36.6
Ja.	7.2	33.91	56.8

898   2833.

Σ.	341.7	8.73	25.7
De.	337.2	9.29	66.6

899   2842.

Σ.	99.4	3.08	31.9
De.	110.7	.41	67.3

900   2846.

Σ.	264.6	3.19	31.8
De.	268.5	.28	66.3

901   2847.

Σ.	293.8	1.35	30.7
W.	305.1	.38	76.9

902   2881.

So.	111.3	1.79	25.7
W.	103.2	.52	75.9

903

H₂.	300.0	10.0	26.0
St.	305.9	8.42	76.7

Ref. No. 904	Σ. [476.]		
Mä.	335°·6	0″·65	1843·5
Bu.	elong^d·	·3	74·8

Ref. No. 905	2950.		
Σ.	319·1	2·04	32·2
Fer.	312·1	·48	74·8

906	2977.		
Σ.	335·1	2·19	33·2
De.	344·1	·38	66·9

907	2989.		
Σ.	144·1	1·47	28·7
De.	138·9	·62	67·9

Ref. No. 908	Σ. 3041.		
Σ.	183°·4	3″·27	1832·2*
De.	180·5	·21	66·3*
Σ.	347·6	71·1	32·2†
De.	349·4	69·0	66·2†

909	3047.		
Σ.	64·7	1·12	31·7
Gl.	72·7	1·02	74·9

910	3060.		
Σ.	109·7	3·79	28·7
De.	114·6	·49	76·5

* B C.   † A and $\dfrac{B+C}{2}$.

APPENDIX.

WHILE the last sheets of the Measures were going through the press, the following list of double and multiple stars was most kindly placed at our service by their discoverer, Mr. Burnham. Most of them are very interesting objects; many are quite recent discoveries, and as yet unpublished; many, too, are naked-eye stars; and several are well-known Struvian pairs which Mr. Burnham has found triple; and, lastly, the measures have been largely supplied by that most excellent observer Dembowski. It is therefore with much pleasure and gratitude that we give this valuable list as an Appendix to our book.

As before, Bu. = Burnham; De. = Dembowski; Hl. = Hall; C.O. = the observers at Cincinnati Observatory.

Ref. No.	Burnham's No.	Name or Catalogue.		R. A. 1880.	Decl. 1880.	P.	D.	Mags.	1870.	Observer.
				h. m.	° ′	°	″		+	
1001	483	.		0 2·8	40 11	44·7	2·37	7·5, 11·8	8·7	Bu.
1002	391	B. A. C. 10		3·2	−28 39	97·2	0·78	6, 6	6·8	C.O.
1003	253			4·1	57 51	49·9	0·39	8·3, 8·5	5·9	De.
1004	255			5·6	27 45	99·0	0·38	7·5, 7·9	5·8	De.
1005	492	B. A. C. 201		38·5	54 34	152·6	1·91	6, 12	8·7	Bu.
1006	495			42·4	18 2	230·9	0·58	7·8, 7·5	8·7	Bu.
1007	232		A B	43·6	49 59	288·4	0·44	8·0, 8·5	6·2	De.
			A C			292·8	28·70	10·2	6·0	De.
1008	1		A B	45·6	55 58	81·0	1·42	8·1, 10·1	5·3	De.
			A C			133·3	3·70	8·9	5·3	De.
			A D			192·9	8·82	9·5	5·3	De.
			A E			360·±	15·±			De.
1009[a]	235		A B	1 3·5	50 22	74·0	0·45	7·0, 7·4	5·6	De.
1010	258			5·5	61 4	260·4	0·79	6·2, 8·9	5·2	De.
1011[b]	110			14·1	−16 26	24·6	1·50	7·2, 7·3	6·2	De.
1012	4			16·6	10 44	81·0	0·37	7·0, 7·5	7·7	Bu.
1013	399	Ceti 211		21·8	−11 31	302·8	1·39	6·5, 10	6·5	De.
1014	506	η Piscium		25·1	14 44	12·9	1·02	4, 11	8·7	Bu.
1015	5	103 Piscium		32·8	16 1	289·4	1·34	7·0, 9·0	5·5	De.
1016	6			38·7	−7 22	167·1	2·59	6·4, 9·2	5·5	De.
1017[c]	510		A B	42·1	15 43	337·4	1·59	8, 12	8·1	Bu.
			A C			326·4	53·56		8·1	Bu.
1018	260			46·7	14 51	228·0	0·56	8·3, 9·0	5·8	De.
1019	7	58 Ceti		51·9	−2 39	12·1	2·86	7·0, 11·8	5·5	De.

[a] This is the principal star of the wide triple, O.Σ. 24. The distant companions have minute attendants.
[b] Identical with H₂ 2036.   [c] The wide pair is H₁ V. 92.

Ref. No.	Burnham's No.	Name or Catalogue	R. A. 1880.	Decl. 1880.	P.	D.	Mags.	1870.	Observer.
			h. m.	° ′	°	″		+	
1020	513	48 Cassiopeæ	52·1	70 19	264·4	1·04	5, 7	8·7	Bu.
1021	8		2 15·0	8 20	200·4	0·96	8·1, 9·4	5·3	De.
1022^d			16·8	32 58	233·0	1·30	7·5, 13	8·9	Bu.
1023	518	Ceti 389	23·2	9 2	139·0	1·57	7, 11	7·9	Bu.
1024	521	Persei 67	35·0	47 45	153·7	5·86	6, 11	8·7	Bu.
1025	306	Arietis 107	36·9	25 8	17·3	2·93	6·4, 11·0	6·8	De.
1026	9		39·6	35 3	160·6	1·52	6·3, 8·4	5·9	De.
1027^e	524	20 Persei	46·1	37 51	158·7	0·34	5·5, 5·5	8·7	Bu.
1028	11	ρ² Eridani	56·8	-8 9	87·2	2·72	5·4, 9·6	5·6	De.
1029	526	β Persei A B	3 0·3	40 30	155·3	58·79	13	8·0	Bu.
		A C			144·7	67·72	13	8·2	Bu.
		A D			192·6	81·92	10	8·7	Bu.
		D E			115·1	10·64	12½	7·7	Bu.
1030	84		10·1	-6 22	10·3	0·50	7·2, 7·4	5·8	De.
1031	533	B. A. C. 1101	28·1	31 17	149·3	0·43	7, 7	8·7	Bu.
1032	535	38 Persei	36·8	31 54	60·5	0·96	4, 8·5	7·8	Bu.
1033^f		A B	37·0	31 47		0·5±	8·5, 8·5		
		A C			37·9	23·87	8·7	8·9	Bu.
1034^g	536	A B	39·1	23 49	336·4	0·44	8, 9·5	8·7	Bu.
		A C			302·4	36·72	8·0	8·7	Bu.
1035	537		39·9	23 28	185·9	0·60	8·5, 11	7·9	Bu.
1036	263		48·8	32 50	71·6	0·67	8·2, 8·5	5·9	De.
1037	545		59·4	37 42	310·0	1·02	8, 11	8·2	Bu.
1038	547	47 Tauri	4 7·4	8 58	359·7	0·89	5, 7	7·8	Bu.
1039	311	Eridani 315	21·9	-24 21	147·5	0·89	6·5, 7·0	7·1	C.O.
1040	550	Aldebaran	29·0	16 16	109·0	30·45	1, 14	7·9	Bu.
1041		46 Eridani	29·1	-7 0	57·1	1·28	6, 10·5	9·0	Bu.
1042		A B	44·6	10 52	17·5	0·35	7, 7	9·0	Bu.
		A C			148·5	18·35	14	9·0	Bu.
1043	314	Leporis 3	53·6	-16 34	149·9	0·43	6·6, 6·9	6·7	De.
1044	188	τ Orionis A B	5 11·8	-6 58	250·1	35·98	4, 14	6·2	Hl.
		A C			59·8	35·97	12	6·2	Hl.
		B B			49·3	3·77	16	6·2	Hl.
1045	189	Orionis 81	14·5	-5 29	283·6	4·27	6·8, 11·5	5·9	De.
1046^h	190	Orionis 82	14·6	-8 9	355·3	0·61	7·9, 8·7	6·1	De.
1047	556		18·7	-2 36	228·2	0·79	6·5, 11·8	8·2	Bu.
1048^i		A B	20·2	34 19	223·5	1·11	8·5, 10	8·9	Bu.
		A C			131·4	18·04	10	8·9	Bu.
		A D			200·7	20·77	11·5	8·9	Bu.
1049	320	β Leporis	23·1	-20 51	288·3	2·68	4, 11	7·9	Bu.
1050^j	557	A B	23·3	3 3	149·8	24·32	7	8·1	Bu.
		B C			142·4	0·46	9·5, 9·5	8·1	Bu.
1051	558	δ Orionis	25·9	-0 23	227·0	33·79	2, 14	8·9	Bu.
1052	321	Leponis 45 A B	34·0	-17 55	144·5	0·65	6·8, 8·3	7·3	De.
		C D			357·7	1·26	9·3, 9·7	7·3	De.
		A C			136·0	89·46		6·6	De.
1053^k	89		31·5	-1 30	344·2	0·55	7·9, 8·5	5·7	De.
1054	16	3 Monocerotis	56·2	-10 36	354·8	1·62	6·0, 9·7	5·6	De.
1055^l	17	4 Monocerotis A B	6 2·8	-11 8	178·0	3·16	6·8, 10·5	5·9	De.
		A C			244·5	8·95	11·5	6·8	De.
1056^m		A B	12·4	28 29	133·3	0·27	7·5, 7·5	9·2	B.
		A C			250·3	2·83	9·5	9·2	B.

d The principal star of Σ. 258.   A C 145°·4 : 68″·70 (1878·9) ; C D 280°·8 : 6″·16 (1878·9).
e As a wide pair this is Σ. 318, 236°·8 : 14″·08 (1829·1).   f A and C = Σ. 439.
g A and C = So. 437.   This and No. 38 in the Pleiades.
h The wide pair, A C, is Σ. 692.   Σ. gives, 4°·2 : 34″·86 (1831·5).   No change since.
i The wide pair, A C, is Σ. 707.
j Struve's companion (Σ. 721) found to be a close pair.
k Moving? Bu. 361°·7 : 0″·71 (1879·1).   A mean of three measures in each case.
l C discovered by Mr. George Knott.   m A and C make Σ. 888.

Ref. No.	Burnham's No.	Name or Catalogue.	R. A. 1880.	Decl. 1880.	P.	D.	Mags.	1870.	Observer.
			h. m.	° ′	°	″		+	
1057		Monocerotis 97	6 44·7	− 0 23	30·9	5·60	6·5, 12	9·1	Bu.
1058	327	A B	52·5	− 2 52	100·1	0·96	7·5, 8·0	6·8	De.
		A C			102·6	13·22	11·5	6·8	De.
1059[n]	328	Canis Maj. 139 A B	7 1·0	−11 7	128·4	0·3	6·3, 7·5	5·9	De.
		A C			360·±	20·±	10		
1060[o]	575	A B	9·3	−15 16	199·0	0·68	8, 8	8·2	Bu.
		A C			2·1	15·87	9·8	8·1	Bu.
1061		65 Aurigæ A B	14·0	36 59	8·3	10·36	6, 12·5	8·9	Bu.
		A C			26·8	36·10	13	8·9	Bu.
1062[p]	577		14·5	0 38	136·8	0·90	7·5, 7·7	8·2	Bu.
	21				9·9	14·54	13	8·2	Bu.
1063		η Canis Maj.	21·6	7 11	27·4	4·09	5·5, 11·3	5·6	De.
1064[q]	332	A B	22·2	−11 19	166·3	0·80	6·3, 8·2	5·5	De.
		A C			312·1	20·20	8·7	(32·1)	Σ.
		A D			157·2	23·41	9·8	8·1	Bu.
		A E			41·4	31·06	13	8·1	Bu.
1065[r]	579	A B	26·7	33 23	219·1	0·84	8·0, 11·5	8·2	Bu.
		A C			232·2	16·60	12	8·2	Bu.
1066	580	Pollux A B	38·0	28 19	274·9	43	2, 13-14	7·9	Bu.
		A C			70·8	174·52		8·2	Bu.
		C D			132·6	1·40	9, 12	8·2	Bu.
1067	101	9 Argus	46·2	−13 35	289·4	0·46	5·6, 6·7	5·7	De.
1068	581	A B	57·7	12 38	176·9	0·40	8, 8	8·1	Bu.
		A C			185·3	4·76	11·0	8·1	Bu.
1069[s]	582	A B	58·1	12 25	204·5	19·75	8½, 8½	8·1	Bu.
		B C			59·8	3·76	12	8·1	Bu.
1070	204		8 7·0	10 45	302·1	1·06	7·1, 10·1	5·9	De.
1071[t]	584	P. VIII. 124	33·0	19 58	291·1	1·61	8, 12	8·0	Bu.
1072	585	Cancri 109	34·3	20 54	106·4	0·40	7·5, 9·0	8·1	Bu.
1073[u]	587	15 Hydræ	45·7	−6 44	159·9	0·46	6, 9	8·2	Bu.
1074	211	Hydræ 68	55·7	3 9	257·8	1·11	7·5, 10·0	5·2	De.
1075	105	κ Leonis	9 17·7	26 42	203·8	3·05	4·9, 10·7	6·2	De.
1076	590	29 Hydræ	21·3	−8 42	176·8	10·80	7, 12	8·2	Bu.
1077	591		23·5	−2 36	35·8	0·78	7·7, 8·5	8·2	Bu.
1078	215		48·7	−27 26	337·8	1·78	7·5, 10·5	7·1	C.O.
1079	592		49·3	−15 38	192·8	9·80	6·5, 12·13	8·2	Bu.
1080	594	Leonis 150	10 16·3	15 58	143·4	1·58	6·5, 11·0	8·2	Bu.
1081	596	Leonis 222	43·0	17 47	277·3	2·39	6·5, 13	8·3	Bu.
1082	597		49·3	24 14	46·9	0·88	8·5, 11	8·2	Bu.
1083	599	65 Leonis	11 1·8	2 30	82·4	1·78	5·5, 11·5	8·2	Bu.
1084[v]	600	A B	10·9	−6 29	226·4	1.25	6·5, 12	8·1	Bu.
		A C			97·3	61·53	8	8·1	Bu.
1085[w]	601	A B	22·9	−16 40	331·5	28·16	7·5	8·3	Bu.
		B C			226·9	0·81	8, 9	8·3	Bu.
1086	456		30·7	−11 41	68·2	0·65	9, 9	7·3	Hl.
1087		Corvi 17	12 9·6	−22 41	232·4	0·77	6·5, 7·0	9·4	Bu.
1088	605	B. A. C. 4149	14·0	−21 30	144·2	1·25	6, 8	8·2	Bu.
1089	28	B. A. C. 4213	23·9	−12 44	353·7	1·95	6·4, 10·2	5·3	De.
1090		31 Virginis	35·9	7 28	28·7	3.56	6, 12	9·3	Bu.
1091	341	Hydræ 348	57·3	−19 56	136·4	0·78	6·5, 7·0	6·3	De.
1092		48 Virginis	57·7	−2 55	229·4	0·48	6, 6	9·4	Bu.
1093		B. A. C. 4389	13 0·5	45 54	109·2	2·68	6, 12	9·3	Bu.
1094	608	15 Canes Ven.	4·2	39 10	284·9	1·22	5·5, 10·5	8·3	Bu.
1095	609		4·5	−4 18	356·1	0·89	7, 11	8·3	Bu.

[n] A and C make Σ. 1026 *rej.*
[o] A and C make Σ. 1057.
[p] A and C make Σ. 1074.
[q] A and C make Σ. 1097.
[r] A and C make O.Σ. 173.
[s] A and B make Σ. 1179.
[t] The principal star of a very wide pair (So. 571).
[u] The large star and two distant companions make H₁ V. 120.
[v] The wide pair is H₁ N. 26.
[w] The wide pair is H₁ IV. 112.

Ref. No.	Burnham's No.	Name or Catalogue.	R. A. 1880.	Decl. 1880.	P.	D	Mags.	1870.	Observers.
			h. m.	° ′		″		+	
1096		Virginis 454	13 4·9	13 57	205·1	5·06	6·7, 12	9·3	Bu.
1097	610	Virginis 504	17·6	−20 19	18·3	4·02	6·8, 10·5	8·2	Bu.
1098	113		23·1	12 6	188·8	1·57	8·4, 11·1	5·3	De.
1099	114		28·0	−8 0	137·1	1·49	7·6, 8·6	5·3	De.
1100ˣ		Virginis 550	28·5	−12 36	81·2	0·47	6, 6·5	9·4	Bu.
1101ʸ		A B	29·1	33 45	30·4	1·98	8·5, 9·0	9·3	Bu.
		A C			21·6	34·43	12	9·3	Bu.
1102	612	B. A. C. 4559	33·7	11 21	56·1	0·23	6, 6	8·3	Bu.
1103ᶻ		86 Virginis A B	39·5	−11 49	298·4	1·61	5·5, 10·5	9·3	Bu.
		C D			274·2	1·72	11·5, 13	9·3	Bu.
		A C			164·7	26·94		9·3	Bu.
1104	614		48·0	10 44	268·4	0·60	8, 12	8·4	Bu.
1105	224		14 7·6	13 8	71·0	0·67	8·9, 9·3	5·7	De.
1106			7·8	−7 58	160·5	0·69	8, 8	9·4	Bu.
1107ᵃ	225	A B	18·8	−19 26	295·5	35·03	6	5·7	De.
		B C			101·9	1·40	7·3, 8·2	5·7	De.
1108		52 Hydræ	21·1	−28 57	276·8	4·00	5, 11·3	9·4	Bu.
1109	616	γ Boötis	27·3	38 50	98·6	26·18	3, 13	8·2	Bu.
1110	346	Libræ 23	41·8	−16 50	235·7	1·18	7·2, 8·0	6·3	De.
1111ᵇ	617		42·4	−23 45	219·6	59·44	6½, 8	8·3	Bu.
					336·6	2·71	9	8·3	Bu.
1112	106	μ Libræ	42·7	−13 39	335·0	1·38	5·4, 6·3	5·6	De.
1113	239	59 Hydræ	51·6	−27 10	129·5	0·93	6, 6	8·4	Bu.
1114ᶜ	618	ι¹ Libræ A B	15 5·4	−19 20	110·5	57·46	6	8·3	Bu.
		B C			24·3	1·86	10, 10	8·3	Bu.
1115			25·6	48 8	128·5	10·74	6·5, 12·5	9·3	Bu.
1116	122		33·0	−19 23	204·0	1·76	7·0, 7·3	5·4	De.
1117	619	Serpentis 55	37·6	14 3	359·7	0·57	6·5, 7·0	8·3	Bu.
1118ᵈ	620	A B	38·9	−27 41	166·8	0·86	7·5, 7·5	8·4	Bu.
		A C			214·1	50·25	9·0	8·4	Bu.
1119		B. A. C. 5248	44·7	55 45	152·0	1·31	5·2, 11·0	9·3	Bu.
1120ᵉ		β Scorpii	58·5	−19 58	87·0	0·73	2, 10	9·4	Bu.
1121ᶠ		Libræ 213 A B	59·3	−5 58	150·3	1·51	6·5, 9·5	9·4	Bu.
		A C			233·7	28·54	10·4	9·4	Bu.
		A D			192·7	52·28	10·7	9·4	Bu.
1122	39	11 Scorpii	16 0·9	−12 25	256·5	3·35	6·1, 10·4	5·7	De.
1123	355		4·2	45 42	279·3	0·34	7·3, 8·0	6·3	De.
1124	120	ν Scorpii A B	5·0	−19 9	359·9	0·73	4·3, 6·7	5·9	De.
		C D			47·9	1·89	7·0, 8·0	5·4	De.
		A C			336·5	40·77		5·4	De.
1125	625	ω Herculis A B	19·9	14 19	175·3	1·90	5, 12	8·6	Bu.
		A C			103·2	33·80	11·5	8·6	Bu.
1126	627	52 Herculis	45·7	46 12	309·4	1·83	5, 10	8·4	Bu.
1127		54 Herculis	51·0	18 38	175·4	2·56	5, 12	9·4	Bu.
1128	282		17 8·5	−14 27	154·1	4·23	6·7, 11·8	5·4	De.
1129	126	P. XVII. 43 A B	12·9	−17 38	261·3	1·74	6·4, 7·5	5·1	De.
		A C			150· ±	20· ±			
1130	242	A B	17·3	−11 35	68·9	0·96	8·2, 8·9	5·9	De.
		A C			63·4	8·90	10·4	6·0	De.
		A D			63·8	47·46		6·0	De.
1131	129	B. A. C. 5896	21·2	−25 25	100·3	0·89	7·5, 8·0	7·4	C.O.

ˣ A variable star, discovered in 1866 by Schmidt. There is a distant faint companion, 156°·5 : 23″·88 (1879·4).
ʸ A and C make H₂ 3261.
ᶻ As a wide pair this is Σ. 1780 *rej.* Both components are double, and the smaller very difficult.
ᵃ The wide pair is H₁ N. 80.   ᵇ The wide pair is So. 663.
ᶜ The wide pair is H₁ IV. 44 (= Str. 376).   ᵈ The wide pair is H₂ 4803.
ᵉ Just as this page is sent to press a letter from Mr. Burnham announces that he hàs succeeded in dividing the principal star of β Scorpii. The old companion is thus given by Struve :
A C, 22°·4 ; 13″·10 : 1823·3. Mag. 2, 4.
ᶠ The large star and two distant companions is Σ: 2005 *rej.*

Ref. No.	Burnham's No.	Name or Catalogue.	R.A. 1880. h. m.	Decl. 1880. ° '	P. °	D. ''	Mags.	1870. +	Observer.
1132		26 Draconis	17 33·7	61 58	149·1	1·36	5·5, 10·5	9·3	Bu.
1133	631	Ophiuchi 255	33·8	—0 35	70·2	0·36	7·7	8·6	Bu.
1134	130	90 Herculis	49·4	40 3	123·0	1·82	5·8, 9·1	5·5	De.
1135	633	γ Draconis	53·8	51 30	152·1	20·88	2·13	8·4	Bu.
1136	132	B. A. C. 6158	18 4·1	—19 52	240·1	0·78	6·8, 7·2	5·0	De.
1137g	638	A B	4·3	2 34	152·0	22·33	8·9, 8·9	8·1	Bu.
		B C			10·5	1·71	11·8	8·6	Bu.
1138	286	16 Sagittarii	8·1	—20 25	218·5	5·67	6, 13	8·5	Bu.
1139h	639	A B	11·7	—18 40	155·3	0·57	7·5, 7·5	8·6	Bu.
		C D			330·±	4·±	8, 1·4	8·6	Bu.
		A C			51·7	17·30		8·6	Bu.
1140	133	B. A. C. 6261	20·3	—26 42	265·3	1·80	7·5, 7·5	5·7	Schi.
1141	135	Scuti Sob. 45	31·3	—14 6	184·0	2·40	6·7, 11·5	5·1	De.
1142		Draconis 205	44·4	49 18	353·6	0·59	6·5, 8·5	9·3	Bu.
1143	265		44·6	11 23	235·9	1·46	7·1, 9·1	5·3	De.
1144	137	A B	49·8	37 14	123·8	1·15	8·2, 8·7	5·6	De.
1145	648	B. A. C. 6480	52·5	32 45	312·5	0·60	6, 9·5	8·5	Bu.
1146	287	ξ Aquilæ	59·9	13 41	59·6	4·92	3, 12	8·5	Bu.
1147	139	Aquilæ 59	19 7·2	16 39	139·5	0·72	6·7, 8·0	5·9	De.
1148	248	2 Vulpeculæ	12·6	22 49	125·0	1·83	5·7, 9·5	5·6	De.
1149j	141	A B	16·8	22 17	80·6	0·71	7·5, 9·1	6·0	De.
		A C			155·2	26·53	11·5	5·3	De.
		A D			90·5	50·75	11·0	7·8	
1150i	652	A B	27·3	28 1	331·4	4·78	7·9, 13	8·5	Bu.
		A C			3·7	5·60	9·7	8·5	Bu.
1151	654	52 Sagittarii	29·4	—25 10	160·8	2·93	5·11	8·6	Bu.
1152	658	B. A. C. 6762	39·0	26 50	295·2	0·57	6·5, 10·0	8·5	Bu.
1153	148		45·4	—10 40	333·2	0·93	7·9, 8·3	5·3	De.
1154k		η Cygni	51·8	34 46	211·7	7·54	5, 12·5	9·4	Bu.
1155	428		20 1·1	12 36	343·7	0·52	7·2, 8·5	6·5	De.
1156l	430	A B	6·8	35 28	17·7	1·12	9·5, 10·0	6·6	De.
		A C			51·3	16·91	9·7	6·6	De.
1157	661	Cygni 166	12·6	40 0	67·0	12·60	6·5, 13	8·4	Bu.
1158	431		15·4	35 53	221·4	0·55	8·5, 8·8	7·1	De.
1159	60	π Capricorni	20·4	—18 36	145·2	3·27	5·1, 8·7	5·0	De.
1160	62		23·1	29 44	135·5	1·20	8·5, 9·4	5·5	De.
1161	668	B. A. C. 7080	25·8	—10 16	29·0	4·64	6, 12	8·6	Bu.
1162	670		27·3	13 32	58·3	0·75	8·5, 8·8	7·7	De.
1163m	151	β Delphini	31·9	14 11	15·6	0·61	4·3, 5·7	4·7	De.
1164	675	51 Cygni	38·5	49 54	101·5	2·78	5, 12	8·2	Bu.
1165n	64		39·3	12 17	172·4	0·64	8·7, 9·0	6·2	De.
1166	152	Cephei 55	39·3	56 57	111·0	0·44	7·2, 8·0	6·0	De.
1167	268		43·2	41 38	221·4	0·38	7·4, 8·2	5·9	De.
1168	367	A B	49·9	27 38	115·7	0·55	7·5, 7·9	6·4	De.
		A C			28·2	30·88	12	5·6	De.
1169	156		57·6	46 6	241·6	1·05	7·1, 9·4	5·4	De.
1170	368	Aquarii 45	21 1·0	—8 43	97·1	0·55	7·2, 7·6	6·8	De.
1171o	159	A B	6·3	47 12	318·7	1·32	6·1, 9·2	6·4	De.
		A C			189·5	134·10	6·8	4·1	De.
1172p	270	A B	7·5	6 43	354·6	0·62	7·4, 9·7	5·8	De.
		A C			30·±	20·±			
		A D			173·0	183·2	7·8		
1173	289	A B	13·4	34 35	137·8	0·90	8·2, 12	8·5	Bu.
		A C			262·1	5·39	13	8·5	Bu.

g The wide pair is Σ. 2287 *rej.*
i A and C make H₉. 2867.
k H. noted two very distant and larger stars (= H₂ 1455).
l A and C make H₂ 1489.
m Binary in rapid motion. Bu. 53°·7 : 0''·24 (1878·6). It is apparently single now (1879).
n The wide pair is O.Σ. (App.) 209.
p The wide pair, A D, is So. 781.

h The wide pair, A C, is Sh. 264.
j A and C make Σ. 2539.

o The wide pair is O.Σ. (App.) 215.

Ref. No.	Burnham's No.	Name or Catalogue.	R. A. 1880. (h. m.)	Decl. 1880.	P.	D.	Mags.	1870.	Observer.
1174^q	164	A B	21 19·2	8 52	241·6	0·57	8·0, 8·5	5·5 +	De.
		A C			242·2	26·51	8·7		
1175^r	167	Cygni 363	31·0	29 31	89·2	2·08	7·0, 11·4	6·5	De.
1176	686	A B	33·7	55 14	127·9	0·38	7·7, 8·0	7·7	De.
		A C			11·0	41·67	8·0	8·1	
1177^s	449	A B	34·7	41 11	19·1	6·78	7, 12	6·8	Bu.
		A C			170·5	13·71	10·8		De.
		A D			248·2	17·94	7·7	6·8	De.
1178	688		37·7	40 30	209·2	0·38	7·5, 7·5	8·1	Bu.
1179	690	μ Cephei	39·8	58 14	259·8	19·28	5, 12	8·5	Bu.
1180	276	η Piscis Austr.	53·9	−29 2	117·4	1·71	5, 6	6·6	C.O.
1181	694	Lacertæ 4	59·1	43 54	352·2	0·50	6·0, 8·5	8·7	Bu.
1182	696		58·7	15 19	353·8	0·65	8·5, 9·0	8·2	Bu.
1183	172	51 Aquarii	22 17·9	−5 27	21·9	0·44	6·7, 6·7	5·7	De.
1184	290	34 Pegasi	20·5	3 47	223·3	2·61	6, 12	5·7	Hl.
1185	291		21·6	3 55	157·8	0·32	8·4, 8·4	5·8	De.
1186	76		23·4	−0 49	335·3	1·47	8·2, 10·1	6·2	De.
1187	277		34·2	40 45	199·4	0·50	8·2, 8·4	5·3	De.
1188	480		35·3	4 6	65·8	0·86	9·0, 9·8	7·5	De.
1189^t	382	B. A. C. 7983 A B	48·3	44 7	205·7	1·07	6, 8	6·4	De.
		A C			353·6	26·43	10·7	6·2	De.
1190	178	Aquarii 252	49·0	−5 38	324·6	0·5±	6·2, 8·0	5·0	De.
1191	384	Aquarii 265	56·2	−19 11	72·2	1·27	7·2, 9·2	7·1	De.
1192	180		23 2·2	60 11	176·8	0·57	7·5, 8·0	5·1	De.
					106·3	34·50	10·5	5·5	De.
1193^u	385	A B	4·5	31 50	135·8	0·42	7·1, 7·9	6·4	De.
		A C			77·1	58·05	8·7	6·7	De.
1194	181	Aquarii 286 A B	7·5	−14 3	309·2	1·51	7·1, 10·4	6·3	De.
		A C			234·9	18·78	12	7·7	Bu.
1195	714	B. A. C. 8084	7·9	−3 17	145·5	0·57	7, 10	8·6	Bu.
1196	717	8 Andromedæ	12·2	48 22	161·4	7·61	5, 13	8·2	Bu.
1197	718	64 Pegasi	16·0	31 9	86·9	0·46	6, 8	8·7	Bu.
1198	720	72 Pegasi	28·0	30 40	127·7	0·40	6, 6	8·7	Bu.
1199	730	27 Piscium	52·5	−4 13	265·8	1·42	6, 11	8·4	Bu.
1200^v	733	85 Pegasi	55·9	26 27	274·0	0·67	6, 12·5	8·7	Bu.

q Measures of A C from Struve (= Σ. 2793).
r A and C make O.Σ. (App.) 220.
s A and C make O.Σ. 447 (= H₁ III. 110).
t A and C = H₂ 1828.
u A and C = H₂ 5532.
v The principal star has a large proper motion and sensible parallax.   There is a 9 m. companion, 33°·6 : 14″·40 (1878·5). This does not partake of the proper motion of A. There is also a fourth, faint star, 227°·1 : 61‴·73 (1879·0). Mr. Burnham's latest measures are as follows :—

	°		″	
A B,	276·0	4 n.	0·75	1878·75
	287·2	1 n.	·73	9·46
A C,	33·0	7 n.	14·55	8·70
	29·0	3 n.	15·00	9·32
A D,	277·1	1 n.	61·73	8·96

C is of 9th and D of the 13th mag. A B evidently form a physical system.

ADDITIONAL NOTES TO MEASURES.

No. 86 (o Ceti). H₁ on Oct. 20, 1777, wrote, "Double. Very unequal. Large, garnet; small, dusky. Distance, mean of some very accurate measures 1′ 44″·218 ; mean of other very accurate measures 1′ 53″·032." The earliest measure of this star was probably made by Cassini, about 1863, with a telescope 34 feet in length.

No. 159 (α Tauri). H₁ in 1781, Dec. 19, wrote : "Double. Extremely unequal. Large, red ; small, dusky. Distance, 1′ 27″ 45‴ ; position, 52° 58′ nf. With 460, the apparent diameter of this star, when on the meridian, measured 1″ 46‴, a mean of two very complete observations ; they agreed to 6‴ ; with 932, it measured 1″ 12‴, also a mean of two excellent observations ; they agreed to 8‴. The apparent disc was perfectly well defined with both powers."

No. 274, p. 244, line 4. The duplicity was detected by Bird in 1864.

No. 300 (Σ. 1273). Hall discovered a faint companion in 1875 : Position, 190°± ; distance, 12″ ± ; and Mr. Burnham gives a measure in 1878, 192°·2, 14″·74.

No. 456 (α Centauri). Mr. Ellery's latest measures of this star are, 1879·252 : 174°·4 : 3″·41 : 13 observations. (See *Observatory*, No. 27.)

No. 520 (Antares). Burg, of Vienna, was the first to see the companion of this star : he was watching an occultation of Antares by the Moon in April 1819. Mr. Grant detected it in 1844 in India.

No. 1120 (β Scorpii). "Aug. 19, 1761. —Found the little star, which is 14″ north of β Scorpii, to precede it one second of time, by my parallactic wires, with my watch, which makes four beats to a second of time. If anything, the difference was something more than a second of time ; the little star may therefore be supposed to precede β Scorpii 17″ in R.A." The difference of Dec. was 13″·97.—Maskelyne (*Phil. Trans.*, vol. liv.) Powell, in 1859, found the difference of R. A. + 6″·3. Smyth's magnitudes are A 2, B 5½, C 5 Powell's, A 2½, B 5, C 7.

No. 708 (Σ. 2749). The magnitude of C is 9.

No. 725 (μ Cygni, A C, 7·5).

1800	62°·3	216″·5	Piazzi.	
77	57 ·1	209 ·9	Fl.	

CLASSIFICATION.

FOR a very exhaustive and interesting classification of double stars we must refer our readers to M. Flammarion's Catalogue of these objects. A few general remarks and a much more simple classification are all that we can here present.

Sir John Herschel's great Catalogue gives the places of 10,320 double stars. Adding 700 of Mr. Burnham's and a few hundreds for the discoveries of other astronomers we may take 12,000 as a rough total of the number of known double stars. Unfortunately, observers of these objects have confined their attention till lately too much to the Herschelian and Struvian pairs, and hence at present in our attempts to ascertain the number of physical double stars we deal almost exclusively with the discoveries of those great observers.* A very extensive examination of nearly all known measures of these and many of Sir John Herschel's stars leads us to believe that the number in which orbital motion has already shown itself since discovery may be put at about 600. If to these be added the relatively fixed pairs which are known to possess a common proper motion, we get at least 700. But this is not all: Mr. Burnham's discoveries will in all probability yield a large number of binary systems. Hence it does not appear too much to say that if this branch of astronomy continue to command the attention which has been given to it during the last ten years, the number of known physical systems will soon rise to at least 1000. The careful examination of Herschel and South's and Burnham's stars by Dembowski, Mr. Burnham, and the Cincinnati observers, is almost weekly adding to this important and interesting class.

---

* It is with much pleasure that we find Mr. Burnham and Mr. O. Stone energetically protesting by word and deed against this narrow circle of observation. The extremely clear, compact, and complete form in which Σ. published his double-star work no doubt led to this custom.

SYSTEMS FOR WHICH ORBITS HAVE BEEN COMPUTED.

	Name.	a	e	P	Computer.
		$''$		years.	
1	42 Comæ Ber.	0·65	0·48	25	O.Σ.
2	ζ Herculis	1·22	·42	34	Du.
3	Σ. 3121	0·71	·26	37	Dob.
4	η Cor. Bor.	0·82	·26	41	Wijkr·
5	Σ. 2173	1·01	·13	45	Du.
6	γ Cor. Anstr.	2·40	·69	55	Schi.
7	ζ Cancri	0·90	·33	58	Dob.
8	ξ Ursæ Maj.	2·55	·39	60	Du.
9	a Centauri	18·45	·53	88	Dob.
10	70 Ophiuchi	4·70	·49	94	Schur.
11	γ Cor. Bor.	0·70	·35	95	Dob.
12	ξ Scorpii	1·26	·07	95	,,
13	Σ. 3062	1·27	·46	104	,,
14	ω Leonis	0·85	·55	114	,,
15	ϸ Eridani	3·82	·37	117	,,
16	Σ. 1768	0·75	·66	124	,,
17	ξ Boötis	4·86	·71	127	,,
18	γ Virginis	3·97	·89	185	Thiele.
19	τ Ophiuchi	1·40	·60	217	Dob.
20	η Cass.	9·83	·56	222	,,
21	λ Ophiuchi	1·19	·49	233	,,
22	44 Boötis	3·09	·71	261	,,
23	μ² Boötis	1·47	·59	280	,,
24	36 Andromedæ	1·54	·65	349	,,
25	γ Leonis	2·00	·74	402	,,
26	δ Cygni*	2·31	·28	415	Behrmann.
27	σ Cor. Bor.	5·88	·75	845	Dob.
28	Castor	7·43	·33	1001	,,
29	ζ Aquarii	7·64	·65	1578	,,

The period of O.Σ. 365 (No. 616) may be about sixteen years, while that of O.Σ. 535 (No. 712) is either seven or fourteen years.

TERNARY SYSTEMS.

Under this head may probably be placed all the following systems: γ Andromedæ, Σ. 183, 719, 948, 1001, 1110, 1196, 1938, 1998, 1426, 2006, 2220, 2479, 2481, 2607, 2737, 2749, 2607; and O.Σ. 276, 380, 392.

QUATERNARY SYSTEMS.

ϵ^1 and ϵ^2 Lyræ, ν Scorpii, Σ. 2576 and 17, χ Cygni (see Flammarion's Catalogue), θ Orionis (?).

* These elements are quoted rather to give completeness to the table than because of their intrinsic value. They depend mainly on H_1's angle; it should have been taken as sf., and not nf. Dr. Doberck's formulæ give better results in every way (see p. 372).

Lastly, if we tabulate the most important binaries according to the arc described since discovery, we find the following approximate numbers:—

One or more revolutions			14
Between 270° and 360°			5
,,	180 ,,	270	10
,,	90 ,,	180	15
,,	45 ,,	90	20
,,	20 ,,	45	100
,,	0 ,,	20	200

NOTE

On systematic Errors in the Measures of Angle and Distance of Double Stars.

FROM our remarks on p. 418, it might be inferred that we wished to discourage the study of the Struvian stars. This is far from being the case. There are scores of Σ.'s and $O.\Sigma$.'s stars which need careful remeasurement in order to determine the amount of change, if any; and where change has taken place, to find out its nature. More than this: no one with even a slight acquaintance with the distressing discrepancies and difficulties which are met with in attempts to deal in a satisfactory way with the orbits of H_1's and Σ.'s binaries would desire that attention to them should be relaxed. And it is quite certain that there are some difficulties which numerous and careful series of measures (especially at the critical times) would considerably diminish or altogether remove. We have said *careful series* of measures. By this expression we mean series of measures by practised observers on a uniform plan, and supplemented by a rigorous determination of systematic error. Mistakes and accidental errors are not serious matters, but constant personal errors (if the observations are so made that the constant error cannot be ascertained and the correction applied) may render worthless the honest work of long years under the most favourable circumstances. To ascertain his systematic errors in the measurement of the position angles and distances of double stars, Struve made extensive series of measures of artificial double stars. His distinguished son O. Struve employed the same method; and to ascertain any change in the errors he repeated the observations about every ten years. Dawes as we have seen got rid of the error, or some portion of it, as regards the angles, by the use of his prism ; and the Cincinnati observers keep the line joining the two eyes parallel to or normal to that joining the two stars measured, and then from the results deduce the necessary corrections. Dembowski, in order to ascertain the corrections to be applied to his angles and distances, has undertaken a most laborious series of measures of twenty-four double stars in which the changes are so small that they may be disregarded ; and other eminent observers have promised to measure the same objects. For the convenience of those who may wish to join in such an investigation, the names, places for 1875, and magnitudes of these selected pairs are subjoined.*

		R. A.		Dec.		Mag.
		h.	m.	°	′	
Σ.	170	1	43·9	+75	38	6·7, 7·8
	191		52·4	73	16	6 , 8·9
	1169	7	57·5	79	52	7·8, 8
	1321	9	5·8	53	14	7·8, 7·8
	1350		22·0	67	19	7 , 7·8
	1603	12	1·9	56	10	7 , 7·8
	1685		45·7	19	51	7 , 7·8
	2034	16	3·7	83	58	7·8, 8
	2326	18	17·5	81	27	7·8, 8·9
O.Σ.	353		22·6	71	16	5 , 7
	363		43·5	77	34	7 , 7
Σ.	2452		57·4	75	37	6·7, 7·8
	2571	19	34·7	77	59	7·8, 8
	2603		48·6	69	57	4 , 7·8
	2675	20	13·1	77	18	4 , 8
	2796	21	16·9	78	4	7·8, 9
	2801		22·2	79	49	7·8, 8
	2806		27·0	70	0	3 , 8
	2893	22	10·6	72	43	5·6, 7·8
	2924		29·4	69	15	7 , 7·8
	2923		29·7	69	44	7 , 9
O.Σ.	481		41·8	77	52	7 , 9
	489	23	3·9	74	43	5 , 7·8
	3051		55·6	79	36	7·8, 9·10

* See *Observatory*, vol. ii., p. 214, for some valuable remarks by Dr. Doberck on this subject.

Just as this sheet was ready for the printer an excellent paper on systematic errors was received from the author, M. Thiele. The subjoined results are taken from it :—

	Mean error of one night's work.	
	Distance.	Angle.
Brünnow	0″·149	2°·44
Dawes	·095	0 ·41
Dembowski	·116	·69
Doberck	·153	1 ·00
Dunér	·099	0 ·94
Gledhill	·062	·89
Herschel (Sir J.)	·460	·91
Herschel (Sir W.)	·39	4 ·4
Knott	·109	0 ·61
Mädler	·141	·71
Main	·171	2 ·97
Plummer (W. E.)	·123	1 ·30
O.Σ.	·082	0 ·95
Σ.	·095	·86
Talmage	·173	1 ·05
Wilson, Seabroke, and others	·145	...

The numbers given above for Dawes, Dembowski, Main, and O.Σ. are the arithmetical means of the values at different periods. And as an example of the way in which these systematic errors change in the course of series of measures extending over many years, the case of one of the most experienced and skilful observers, Dembowski, is here given more fully :—

Mean error of one day, 1853 to 1856 0″·079 0°·59
  ,,   ,,   1862 ,, 1871 0 ·109 0 ·46
  ,,   ,,   Since 1871   0 ·16   1 ·04

The above results were obtained by comparing the several observers' measures of Castor with the computed position angles and distances. For a full explanation of the process, see Thiele's " CASTOR. *Calcul du mouvement relatif et critique des observations de cette étoile double. Copenhagen,* 1879."

PART IV.

BIBLIOGRAPHY.

LIST A.

PAPERS ON DOUBLE STARS, ETC.

ADOLPH.

Measures of 70 Ophiuchi.—*Ast. Nachr.*, vol. lxxi., p. 155.

AIRY.

On the Parallax of *a* Lyræ.—*Memoirs of R. A. S.*, vol. x., p. 265.

AMICI.

A few of his Double Star Measures (1815 to 1823).— *Ast. Nachr.*, vol. xc., p. 304.

ARGELANDER.

On some Double Stars.—*Ast. Nachr.*, vol. lxii., p. 253.

AUWERS (Dr. A.)

"Nachtrag zu den Untersuchungen über die Veränderliche Eigenbewegung des Procyon."—Berlin, 1873.

"Untersuchungen über veränderliche eigenbewegungen Von G. F. T. Arthur Auwers."—Erster Theil.

"Inaugural-Dissertation zur Erlangung der Philosophischen Doctorwürde."—Königsberg, 1862. It contains investigations of the proper motion of *a* Virginis, *β* Orionis, *a* Hydræ, the elements of the orbit of Procyon, etc.

On the Companion of Sirius.—*Ast. Nachr.*, No. 1371.

On variable Proper Motions. The Proper Motion of

Procyon. The Parallax of Procyon.—*Ast. Nachr.*, vol. lviii., pp. 33, 35, etc.

Measures of Double Stars with the Heliometer.—*Ast. Nachr.*, vol. lix., p. 1. 1859 to 1862.

On 61 Cygni, Procyon, ζ Ursæ Majoris, γ Lalande. —No. 21258, *Ast. Nachr.*, vol. lix. On the general Method of Observation ; On the Errors in Measures of Distance ; On the Micrometer Screw, etc.—See same paper.

On the Orbit of Sirius.—*Ast. Nachr.*, vol. lxiii., p. 273.

BARCLAY.

" Leyton Astronomical Observations," 4 vols.

BEHRMAN.

On the Orbit of δ Cygni.—*Ast. Nachr.*, vol. lxvi., pp. 1, 141.

BESSEL.

Ast. Nachr., No. 514, 515, 516 : on the Companion of Sirius.

" Astron. Untersuchungen" (Königsberg, 1841), On the Double Star 70 Ophiuchi.

" Fundamenta Astronomiæ, pro MDCCLV." Measures of some Double Stars ; On the common Proper Motion of Groups and distant Pairs.

" Abhandlungen von F. W. Bessel," edited by Engelman. 3 vols. Leipsig, 1876. Papers on the Parallax of 61 Cygni.

" Verzeichniss Von 257 Doppelsternen," etc.—*Ast. Nachr.*, vol. iv., p. 301.

"Von Doppelsternen und Vergleichung mit Struves," vol. x., p. 389 ; see also p. 317.

" On 70 Ophiuchi," vol. xiii., p. 11 ; vol. xv., p. 105.

" On 61 Cygni," vol. xvi., p. 65 ; vol. xvii., p. 257.

" Ueber Veränderlichkeit der eigenen Bewegungen der Fixsterne," vol. xxii., pp. 145, 169, 185.

"On the Change in the Proper Motion of Sirius," vol. xxii., p. 172.

See also his "Fundamenta Bradleianis."

BISHOP.

"Astronomical Observations," 1852.

BODE.

See Bode's Jahrbuch for 1784 for a list of Double Stars known before Sir William Herschel's time.

BOND.

On the Companion of Sirius.—*Ast. Nachr.*, No. 1353, and No. 1374 ; see also the *American Journal of Science* for March 1862.

Measures of Mizar.—*Ast. Nachr.*, vol. xlviii.

Stellar Photography.—*Ast. Nachr.*, vol. xlix., p. 84.

"On the relative Precision of Measures of Double Stars taken photographically and by direct vision."—*Monthly Notices*, vol. xviii.

BRINKLEY.

Parallax of *a* Lyræ.—*Phil. Trans.*, vol. c. ; *Mem. of R. A. S.*, vol. i., p. 329 ; *Phil. Trans.*, 1824, pt. ii., p. 471.

BROTHERS (A.)

Catalogue of Binary Stars.—*Mem. of the Lit. and Phil. Soc. of Manchester*, vol. iii., 3rd series. See also *Astron. Register*, 1868.

BRÜNNOW (Dr.)

"Astronomical Notices," No. 28. On the Companion of Sirius. See also Dunsink Observations.

BURNHAM (S. W.)

Catalogues of New Double Stars.—*Monthly Notices of R. A. S.*, vols. xxxiii., xxxiv., xxxv., xxxvi., xxxvii., xxxviii.

On Σ. 2344.—*Ast. Nachr.*, vol. lxxxviii., p. 285. See also *Ast. Nachr.*, vols. lxxxv., lxxxvi., lxxxviii.

CALANDRELLI (J.)

"Atti dell' Accademia Pontificia de Nuovi Lincei, 5 Aprile, 1853."

"On the Companion of Sirius."

"On the Proper Motion of Sirius." See *Ast. Nachr.*, vol. li., p. 224.

CASSINI.

"A New Double Star."—*Histoire de l'Académie Royale des Sciences*, tome i. 1678.

"Occultation of γ Virginis Star seen Double."—*Histoire de l'Académie Royale des Sciences*, 1720.

CHACORNAC.

"On the Companion of Sirius."—*Ast. Nachr.*, No. 1355. Also No. 1368 (vol. lvii.)

CINCINNATI OBSERVATORY Volumes :—

Mitchell's Measures ; New Double Stars ; Measures in 1875-76-77.

CLARK (A. C.)

"New Double Stars discovered by Mr. Alvan Clark." —*Monthly Notices of R. A. S.*, vol. xvii., p. 257 ; vol. xx., p. 55.

"Discovery of Companion of Sirius."—*Ast. Nachr.*, vol. lvii., p. 131.

COOPER.

"Double Star Measures in 1832 and 1833."—*Ast. Nachr.*, vol. xc., p. 303.

DARBY (Rev. W. A.)

The *Astronomical Observer*.

D'ARREST.

"On the Influence of Aberration on the Position Angle and Distance."—*Ast. Nachr.*, vol. lix., p. 231.

DARQUIER.

"Mémoire sur les étoiles doubles et le mouvement des fixes. Par M. Darquier."—"*Histoire et Mémoires de l'Acad. Royale des Sciences de Toulouse*, tome ii., 1784.

DAWES.

"Double Stars discovered by Dawes."—*Monthly Notices of R. A. S.*, vol. xxiv., p. 117.

Measures, in vols. ii., iii., x., xv., xxvii.

On Measuring Angles of Position, vols. xviii., xxvi., xxvii.

On Star Magnitudes, vols. xi., xiii.

On μ Herculis, 70 Ophiuchi, σ Orionis, Sirius, vols. xv., xx., xxiii., xxiv.

On Eyepieces, vols. xxiii., xxv.

On H. I. 13, vol. xxiii.

On δ Cygni.—*Ast. Nachr.*, vol. lxv., p. 251.

The following papers are in the *Memoirs of the Royal Astronomical Society* :—

"Observations of the Triple Star ζ Cancri," vol. v., p. 135.

"Observations of Double Stars," vol. v., p. 139.

"Micrometrical Measurements of the Positions and Distances of 121 Double Stars, taken at Ormskirk during the years 1830, 1831, 1832, and 1833," vol. viii., p. 61.

"Micrometrical Measures of Double Stars, made at Ormskirk between 1834 and 1839-40," vol. xix., p. 191.

"Catalogue of Micrometrical Measurements of Double Stars," vol. xxxv., p. 137.

DEMBOWSKI (Baron).

The following papers are in the *Astronomische Nachrichten* :—

His method of observing Double Stars, Instruments, etc., vol. xiii., p. 231. Naples, 1855.

Measures of the Dorpat Double Stars (lucidæ), vol. xlii., pp. 47, 77, 109, 285, 359, 375.

Measures of Double Stars, vol. xliv., p. 57.

Introductory Remarks on his Second Series of Measures, vol. xlvi., p. 267.

Measures, vol. xlvi., p. 317.

Measures, vol. xlvii., pp. 79 to 333 ; vol. l., pp. 129 to 317 ; vol. li., pp. 55 to 139.

Mean Places of Fifty-two Double Stars, vol. liii., p. 113.

Measures, vols. lxxii., lxxv., lxxvi., lxxvii., lxxix., lxxxvii., xcii.

New Double Stars, vol. lxxiii.; also No. 1475, vol. lxii.

New Double Stars, vol. lxxxiii., p. 170.

Measures in 1872-3, vol. lxxxiii., p. 161.

On the Value of his Micrometer, vol. lxxxi.

DOBERCK (Dr. William).

The following papers are in the *Astronomische Nachrichten* :—

Elements of μ Bootis, σ Cor. Bor., τ Ophiuchi,—vol. lxxxv.

Elements of τ Ophiuchi, γ Leonis, etc.,—vol. lxxxvi.

Elements of η Cass., σ Cor. Bor., μ Drac., ι and ω Leonis, ξ Libræ, λ Ophiuchi, Σ. 175 and 1819, μ^2 Boötis 44 Boötis, η Cass., γ Leonis, τ Ophiuchi,—vol. lxxxviii., pp. 45—297 ; and vol. lxxxviii., p. 199.

Elements of ξ Boötis, 1st Elements of γ Cor. Bor. Elliptical Elements of ζ Libræ, 2nd Elements of γ Cor. Bor., Elements of λ and τ Ophiuchi, and ξ Boötis,—vol. lxxxix., pp. 95—259.

On Double Star Calculation,—vol. xc., p. 57.

On δ Cygni,—vol. xc., p. 153.

Elements of p Eridani,—vol. xc., p. 191.

On Cooper's and Amici's Double Star Measures,—vol. xc., p. 303.

Provisional Elements of Σ. 1768 and Σ. 3121,—p. 313.

Elements of Σ. 3062,—p. 319.

Provisional Elements of Castor ; On Double Star Calculations ; On Double Star Orbits,—vol. cxi.

On Double Star Observations ; Elements of μ^2 Boötis; Double Star Measures made at Markrie Observatory ;

On the Correction of Approximate Double Star Orbits,—vol. xcii.

In the *Monthly Notices* are the following :—Elements of μ^2 Boötis, σ Cor. Bor., etc., vol. xxxv. See also the *Transactions of the Royal Irish Academy*, and *The Observatory*.

DUNÉR.

Elements of ζ Herculis and η Coronæ.—*Ast. Nachr.*, vol. lxxviii., p. 315.

" Measures Micrometriques," etc. 1 vol., 1876.

DUNLOP.

Catalogue of 253 Double and Triple Stars observed at Paramatta, N. S. W.—*Mem. of R. A. S.*, vol. iii., 1829.

DURHAM OBSERVATIONS (Measures in 1848).

ENCKE.

" Über die Berechnung der Bahnen der Doppelsterne." —*Ast. Jahrbuch*, 1832.

On γ Virginis.—*Ast. Nachr.*, vol. xv., p. 22.

See also the Berlin volumes.

ENGELMANN (Dr. R.)

"Messungen von 90 Doppelsternen. Leipzig, 1865." —It consists of historical introduction, description of telescope and micrometer, measures and notes.

Measures in 1864 ; the Measures of Eng., De., Se., compared.—*Ast. Nachr.*, vol. lxiv., p. 81.

Measures and Discussion,—vol. lxx., p. 257.

Mean places of many of Σ.'s Double Stars: 1873.—*Ast. Nachr.*, vol. lxxxiv., p. 177.

EULER.

Photometric Formula.—*Berlin Memoirs*, vol. vi.

FERRARI.

Terza serie delle misure micrometriche della stelle doppie."—Roma, 1875.

FLAMMARION (C.)

Periods of Four Double Stars.—*Ast. Nachr.*, vol. lxxxiv., p. 95.

See also *Monthly Notices,* vol. xxxvi ; Comptes Rendus, vol. lxxvii.

"Catalogue des étoiles doubles et multiples."   1878.

FLETCHER (I.)

"Results of Micrometrical Measures of Double Stars, made at Tarn Bank, Cumberland, from 1850-52 to 1853-4."—*Mem. R. A. S.,* vol. xxii., p. 167.

Measures of 70 Ophiuchi.—*Ast. Nachr.,* vol. xxxiii., p. 53.

FORBES (Prof. J. S.)

"On the alleged Evidence for a Physical Connexion between Stars forming Binary or Multiple Groups arising from their Proximity alone."—*Phil. Mag.,* 1849, 1850.

FRITZCHE (H.)

"Untersuchungen über dem Doppelstern ; Σ. 3121."—*Bulletin de l'Académie Impériale des Sciences de St. Pétersbourg,* tome x.

FUSS (V.)

"Untersuchungen über die Bahn der Doppelsterns Σ. 3062."—*Bulletin de l'Académie Impériale des Sciences de St. Pétersbourg,*—tome xi., 1867.

FUSS (Prof.)

"Betrachtungen über die Fixstern Trabauten von Hrn. *Prof. Fuss in St. Petersburg."—Ast. Jahrbuch für das Jahr,* 1785.   Berlin, 1782.

"Reflexions sur les étoiles."   1780.

GASPARIS.

Formulæ for the Computation of Orbits.—*Ast. Nachr.,* vol. lxxviii., p. 333.   See also *Comptes Rendus,* vol. lxxiii., and the *Trans. of the Acad. of Science of Naples.*

GAUSS.

On the Proper Motion of the Star P. XIII. 194.—*Ast. Nachr.,* vol. xxii., p. 191.

GAUTIER.

"Notice sur les travaux astronomiques les plus recents

relatifs aux étoiles doubles, par M. A. Gautier."—*Archives des Sciences Physiques et Naturelles.* Genève, 1850.

GILLISS (Lieut. James M.)

Appendix I. 'A Catalogue of 1963 Stars, and of 290 Double Stars, observed by the U. S. Naval Astronomical Expedition to the Southern Hemisphere during the years 1850-51-52.—Washington, Government Printing Office, 1870.

Catalogue of 290 Double Stars observed by the U. S. Naval Astronomical Expedition in 1850-51-52.—Washington.

GLEDHILL (J.)

"Measures of 484 Double Stars, made at Mr. Edward Crossley's Observatory, Bermerside, Halifax."—*Mem. R. A. S.*, vol. xlii., p. 101.

List of Binary Stars.—*Monthly Notices*, vol. xxxvii.

GOLDSCHMIDT (H.)

On the close Companions of Sirius.—*Ast. Nachr.*, vol. lx., p. 109. See also *Monthly Notices,* vol. xxiii.; *Comptes Rendus*, vol. lvi.

GORE (J. E.)

"Southern Stellar Objects for Small Telescopes." 1877.

GREENWICH OBSERVATIONS (Measures).

GRUBER (Dr. L.)

On the Motion of η Cass.—*Ast. Nachr.*, vol. lxxxviii., p. 361.

HALL (Professor A.)

On the Companion of Sirius.—*Ast. Nachr.*, vol. lxxxiv., p. 28 ; vol. lxxxviii., p. 137 ; vol. xc., p. 163 ; vol. lxxix., p. 247.

"Double Star Measurements."—*Ast. Nachr.*, vol. xc., p. 163.

HALLEY (Dr.)

"The Parallax of the Stars."—*Phil. Trans.*, vol. xxxi. 1720.

HENDERSON (Thomas).

"On the Parallax of *a* Centauri."—*Mem. R. A. S.*, vol. xi., p. 61.

"On the Parallax of Sirius,"—vol. xi., p. 239.

"The Parallax of *a* Centauri," etc.,—vol. xii., p. 329.

HERSCHEL (Sir John).

The following papers occur in the *Memoirs of the Royal Astronomical Society* :—

"Descriptions and approximate Places of 321 New Double and Triple Stars," made with a 20-feet Reflecting Telescope,—vol. ii., p. 459.

"Micrometrical Measures of 364 Double Stars," with a 7-feet Equatorial Achromatic Telescope, taken at Slough in the years 1828, 1829, and 1830,—vol. v., p. 13.

"On the Investigation of the Orbits of Revolving Double Stars,"—vol. v., p. 171.

"Approximate Places and Descriptions of 295 New Double and Triple Stars," etc.,—vol. iv., p. 47. Made with a 20-feet Reflecting Telescope.

"A Catalogue of 384 New Double and Multiple Stars," etc.,—vol. iii., p. 177. Made with a 20-feet Reflecting Telescope.

"Fourth Series of Observations with a 20-ft. Reflector; containing the mean places and other particulars of 1236 Double Stars, as determined at Slough, in the years 1828 and 1829,"—vol. iv., p. 331.

"Fifth Catalogue of Double Stars observed at Slough in the years 1830 and 1831, with the 20-ft. Reflector," etc., —vol. vi., p. 1.

"Remarks on the Fifth Catalogue,"—vol. vi., p. 74.

"Notice of the Elliptic Orbit of ξ Boötis, with a second approximation to the Orbit of γ Virginis. A Notice of the Elliptic Orbit of η Coronæ,"—vol. vi., p. 149.

"A List of Test Objects, principally Double Stars,

arranged in classes, for the trial of Telescopes in various respects, as to light, distinctness," etc.,—vol. viii., p. 21.

"A Second Series of Micrometrical Measures of Double Stars," etc.,—vol. viii., p. 37. Made at Slough, with the 7-feet Equatorial.

"Sixth Catalogue of Double Stars," etc. : at Slough, in 1831 and 1832, with a 20-feet Reflector,—vol. ix., p. 193.

"On the determination of the most probable Orbit of a Binary Star,"—vol. xviii., p. 47.

"Seventh Catalogue of Double Stars, observed at Slough in the years 1823—1828, inclusive, with the 20-feet Reflector ; 84 of which have not been previously described,"—vol. xxxviii., p. 1. 1871.

"A Catalogue of 10,300 Multiple and Double Stars," etc.,—vol. xl.

"A Synopsis of all Sir William Herschel's Micrometrical Measurements and estimated Positions and Distances of the Double Stars described by him," etc.,— vol. xxxv., p. 21.

"On the Parallax of the Fixed Stars."—*Phil. Trans.* 1826, pt. iii. ; and 1827, pt. i.

"Measures of 380 Double and Triple Stars made in 1821, 1822, and 1823, etc., by Herschel and South."— *Phil. Trans.* 1824, pt. iii.

"Results of Astronomical Observations made during the years 1834, 5, 6, 7, 8, at the Cape of Good Hope."— Published in 1847.

In the *Monthly Notices* are the following papers :— Catalogues of Double Stars, vols. i., ii., iii. ; On the Orbits of Double Stars, vol. ii. ; List of Objects, vol. iii. ; Measures, vol. iii. ; On γ Virginis, vol. iii.

HERSCHEL (Sir William).

"On the Places of 145 New Double Stars."—*Memoirs of R. A. S.*, vol. i., p. 166.

Papers in the *Philosophical Transactions:*—

On the Parallax of the Fixed Stars. By Mr. Herschel, F.R.S.,—vol. lxxii., 1782.

Catalogue of Double Stars,—vol. lxxii., 1782.

Catalogue of Double Stars,—vol. lxxv., 1785.

On his Forty-feet Telescope,—vol. lxxxv., 1795.

"Account of Changes that have happened during the last Twenty-five Years, in the relative Situation of Double Stars," etc. 1803. Part ii.

Continuation of "Account," etc. 1804. Part ii.

"Experiments for ascertaining how far Telescopes will enable us to determine very small Angles," etc.,—vol. xcv., p. 31.

"On the Power of Penetrating into Space by Telescopes," etc.,—vol. xc., p. 49.

HIND (J. R.)

"On the Double Stars δ Cygni and γ Leonis."—*Memoirs of R. A. S.*, vol. xvi., p. 291.

Elements of the Binary Star γ Virginis,—vol. xvi., p. 461.

The following are in the *Astronomische Nachrichten:*—

Elements of γ Virginis, μ^2 Boötis, a Geminorum, σ Cor. Bor.,—vol. xxiii., pp. 225, 351, 379.

Elements of δ Cygni,—vol. xxiv., p. 209.

Elements of Σ. 1938, λ Ophiuchi, τ Ophiuchi,—vol. xxvi., p. 319.

In the *Monthly Notices* are the following papers:—On Castor, σ Cor. Bor., δ Cygni, γ Leonis, vol. vii.; On μ^2 Boötis, vols. viii., xxxii.; Elements of Binary Stars, vol. ix.; On γ Virginis, vol. xi.; On a Centauri, vols. xv., xxxiii.; On ξ Boötis, vol. xxxii.; On Castor and ξ Ursæ, vol. xxxiii.; On a Centauri, vol. xxxvii.

HOUZEAU.

"On a novel Effect of the Aberration of Light peculiar to the Double Stars which have proper motion; On

61 Cygni and 70 Ophiuchi."—*Ast. Nachr.*, vol. xxi., pp. 241, 273 ; also vol. xxii., p. 249.

Annual Parallax, etc., of 70 Ophiuchi.—*Ast. Nachr.*, vol. xxi., p. 278.

HOWE (H. A.)

"Catalogue of 50 New Double Stars, discovered with the 11-inch Refractor of the Cincinnati Observatory." 1876. These are Stars of S. declination from 8° to 38°.

HUTH.

"An Doppelsternen."—*Ast. Jahrbuch*, 1807. These are measures of the distances of Double Stars made in Frankfort in 1804 : the distances are in diameters.

JACOB (Captain).

"Relative Path of 61 Cygni." —*Edinburgh New Phil. Journal*, 1858.

"Double Stars observed at Poonah in 1845-6."—*Mem. R. A. S.*, vol. xvi., p. 311.

"Catalogue of Double Stars, deduced from Observations made at Poonah from Nov. 1845 to Feb. 1848."— *Mem. R. A. S.*, vol. xvii., p. 79.

"Micrometrical Measures of 120 Double or Multiple Stars," etc. Madras, 1856-58.—*Mem. R. A. S.*, vol. xxviii., p. 13.

"On the Orbit of *a* Centauri."—*Ast. Nachr.*, vol. xliv., p. 41.

In the *Monthly Notices :*—Orbits of Binary Stars, vols. vii., x., xv., xvi. ; Measures, vols. vii., viii., xiv., xvii., xix. ; On *v* Scorpii, vols. viii., xix. ; On 51 Libræ, vol. xviii. See also *Brit. Assoc. Report*, 1855, on 70 Ophiuchi.

KAISER (Dr. F.)

"Doppelsternmessungen mit Airy's Doppelbild-micrometer und mit dem Fadenmicrometer."—*Annalen der Sternwarte in Leiden.* Dritter band. Haag, 1872.

"Double Star Measures made at Leiden."—*Ast. Nachr.*, vol. xviii., p. 1.

"On 70 Ophiuchi."—*Ast. Nachr.*, vol. xix., p. 204; also vol. xx., pp. 187, 262.

"Measures of Double Stars,"—vol. xx., p. 112.

"On Airy's Micrometer."—*Ast. Nachr.*, xlv., p. 209.

"Measures from 1840 to 1844."—*Ast. Nachr.*, vol. lxxiv., p. 97.

KIRKWOOD (Prof.)

"On the High Excentricity of the Orbits of Binaries." —*Proceedings of the American Assoc. for the Advancement of Science.* See also *Silliman's American Journal*, 2nd series, vol. xxxvii.

KLINKERFUES (Dr.)

"On the Computation of the Orbits of Double Stars." —*Ast. Nachr.*, vol. xlii., p. 81.

His general method of computing the Orbits of Double Stars; with an example.—*Ast. Nachr.*, vol. xlvii., p. 353.

"On the Orbit of 70 Ophiuchi,"—vol. xlviii., p. 101.

KNOTT.

Micrometrical Measures of Double Stars.—*Mem. R. A. S.*, vol. xliii.

KRÜGER (Dr. A.)

"On the Parallax of 70 Ophiuchi," etc.—*Ast. Nachr.*, vol. li., p. 145. This paper deals with the Bonn heliometer, the micrometer screw, influence of temperature, etc., on it, etc.

"On the Parallax of 70 Ophiuchi."—*Ast. Nachr.*, vol. lix., p. 161.

LALANDE.

"A Catalogue of 195 Double Stars," etc.—*Mem. R. A. S.*, vol. iv., p. 165.

LAMBERT.

"Lettres Cosmologiques." 1761.

LAMONT (Dr.)

Measures of Double Stars.—*Annalen der Königlichen Sternwarte bei Munchen*, vol. xvii., 1869.

LASSELL.

"Companion of Sirius."—*Ast. Nach.*, No. 1360 ; also see *Mem. R. A. S.*, vol. xxxvi., p. 38.

"Description of a 20-feet Newtonian," etc.—*Mem. R. A. S.*, vol. xviii., p. 1.

"Measures of Double Stars at Malta."—*Ast. Nachr.*, vol. xxxvi., p. 287. See also *Monthly Notices*, vols. xvii., xxiii., xxiv.

LAUGIER.

"On the Proper Motion of Sirius."—*Ast. Nachr.*, vol. xlviii., p. 209.

LINDENAU.

"On the Parallax of 61 Cygni."—*Bohnenberger und Lindenau's Zeitschrift*, vol. ii., 1815.

"On the Proper Motion of 61 Cygni."—*Ast. Jahrbuch*, 1818.

LINDSAY (Lord).

"Summary of the *Mensuræ Micrometricæ*."—*Dun Echt Observations*, vol. i., 1877.

LUTHER (Dr. E.)

"Measures of Binaries in 1862, 1863."—*Ast. Beobachtungen*, Königsberg, 1870. See also the vol. for 1865 ; also *Ast. Nachr.*, vol. xlvi., p. 355, 1857.

MACLEAR (Sir Thomas).

"Determination of the Parallax of a^1 and a^2 Centauri," etc.—*Mem. R. A. S.*, vol. xx., p. 70 ; also *Ast. Nachr.*, vol. xxxii., p. 243. See also *Monthly Notices*, vols. xi., xvi.

MÄDLER.

The following papers are in the *Astronomische Nachrichten* :—

440 DOUBLE STARS.

"Dopplestern Messungen,"—vol. xii., p. 265. Made at Berlin.

"Elements of the Orbit of σ Coronæ,"—vol. xii., p. 399.

"On ξ Ursæ Majoris,"—vol. xii., p. 268.

"Double Star Measures, at Berlin,"—vol. xiii., pp. 183, 247, 259.

"On 70 Ophiuchi,"—vol. xiii., p. 9.

"On the Orbit of Castor,"—vol. xiv., p. 75.

"On ξ Ursæ Majoris,"—vol. xiv., p. 109.

"Double Star Measures for 1836,"—vol. xiv., p. 183.

"On Σ. 3062, Orbit of,"—vol. xv., p. 151.

"Orbit of η Coronæ,"—vol. xv., p. 303.

"Orbit of γ Virginis and ζ Herculis,"—vol. xvi., pp. 33 and 42.

"On ε Boötis, ζ Boötis, ξ Boötis,"—vol. xviii., p. 364; ζ Cancri, p. 320; φ² Cancri, p. 363; Castor, pp. 79, 364; 42 Comæ Ber., p. 364; η Coronæ, p. 364; γ Coronæ, p. 266; σ Coronæ, p. 363," etc., etc.

"On 70 Ophiuchi,"—vol. xix., pp. 201, 349.

"The Orbit of ξ Ursæ,"—vol. xxi., p. 93.

"List of 504 Dorpat Double Stars which show no change of place since the earliest Measures,"—vol. xxi., p. 147.

"Observations of η Coronæ, ζ Herculis, 70 Ophiuchi, and Σ. 1938, at Dorpat, in 1843,"—vol. xxi., pp. 151, 152.

"Second List of Dorpat Double Stars which appear unchanged since the earliest Measures,"—vol. xxii., p. 27.

"On the Proper Motion of Procyon,"—vol. xxxii., p. 81.

See also *Untersuchungen über die Fixstern-Systeme*, 1847; *Comptes Rendus*, vol. vi., On the Direction of the Orbits of the Multiple Systems of Stars; Popular Astronomy; *Schumacher's Jahrbuch*, 1839; and the Dorpat volumes.

MAIN (Rev. R.)

See the Radcliffe Observations for Measures.

MARTH (A.)

On α Centauri.—*Monthly Notices,* vol. xxxvii.

MASKELYNE (Dr.)

"Annual Parallax of Sirius."—*Phil. Trans.,* vol. li., 1760.

MAYER (C.)

" De novis in Cœlo sidereo Phænomenis."—*Acta Academiæ Theodoro Palatinæ,* vol. iv.

" Gründliche Vertheidigung neuer beobachtungen von Fixsterntrabanten."—*Christian Mayer, Mannheim,* 1778.

MESSIER.

"A New Double Star."—*Connoissance des Temps,* 1783.

"Occultation of γ Virginis by the Moon in 1775."—*Histoire de l'Académie Royale des Sciences,* 1774.

MICHELL (Rev. John).

"On the Means of Discovering the Distance, Magnitude, etc., of the Fixed Stars," etc.—*Phil. Trans.,* vol. lxxiv., 1784.

" Parallax of the Fixed Stars."—*Phil. Trans.,* vol. lvii., 1767.

MILLER (J. F.), Whitehaven.

The following are in the *Astronomische Nachrichten :*—

" Measures of Binary Stars (Σ. 2708, η Cass., Castor),"—vol. xxxiii., p. 367.

" Measures of ε Lyræ, ξ Ursæ, Σ. 1263, μ Draconis,"—vol. xxxiv., p. 213.

" Measures of γ Virginis, ξ Boötis, Aquarii,"—vol. xxxvi., pp. 129, 361. See also *Monthly Notices,* vols. xii., xiii., and *Mem. R.A.S.,* vol. xxii.

MITCHELL (Prof. O. M.)

Measures of 176 Double and Triple Stars, 1846 to 1848.—*Cincinnati Observatory* publications.

MITCHELL (Miss).

American Journal of Science and Astronomy, vol. xxxvi. Measures, 1859 to 1863.

NOBILE (A.)

" Sulle due stelle multiple, Σ. 1263 e σ Cor. Bor. nota per A. Nobile."—Accad. delle Scienza, Naples.

" Saggio di un nuovo metodo per l'osservazione delle distanze seambievoli delle stelle multiple."—Naples.

"Misuri di Angoli di Posizione di Stelle Multiple."— Firenze, 1875.

OELTZEN.

" New Double Star."—*Ast. Nachr.*, vol. xxxvii., p. 395.

PEARSON (Dr.)

"On a doubly-refracting property of Rock Crystal, considered as a principle of Micrometrical Measurements, when applied to a Telescope,"—vol. i., p. 67.

On the Construction and Use of a Micrometrical Eyepiece of a Telescope,"—vol. i., p. 82.

" On the Construction of a new Position Micrometer, depending on the doubly-refractive power of Rock Crystal,"—vol. i., p. 103.

PETERS (Dr. C. H. F.)

The following papers are in the *Astronomische Nach-richten :*—

" On the Parallax of Polaris,"—vol. xxi., pp. 84, 87.

" On the Proper Motion of Sirius,"—vol. xxxii., pp. 1, 17, 33, 49 ; and vol. xxxi., p. 219.

" The Elements of the Orbit of Sirius,"—vol. xxxi., p. 239.

" Measures of Binaries,"—vol xliv., p. 158.

" On the Companion of Sirius,"—vol. lvii., p. 176.

" On the Proper Motion of Σ. 1300, —(Dr. C. H. F. Peters), vol. lxxi., p. 240.

PIERCE.

On γ Virginis.—*Gould's Astron. Journal,* vol. xviii.

PIGOTT.

" Double Stars discovered in 1779."—*Phil. Trans.,* vol. lxxi., 1781.

POGSON.

Measures.—*Brit. Assoc. Report*, 1858.

POND.

"On the Changes which have taken place in the Declination of some of the principal Fixed Stars." By J. Pond, Esq., Astronomer Royal.—*Phil. Trans.*, 1823, pt. i. ; see also 1823, p. 529.

"On the Parallax of *a* Lyræ." By J. Pond, Esq.— *Phil. Trans.*, vol. cxiii.

"On the Parallax of *a* Aquilæ."—*Phil. Trans.*, vol. cviii.

POWELL (E. B.)

"Observations of Double Stars taken at Madras in 1853, 1854, 1855, and the beginning of 1856."—*Mem. R. A. S.*, vol. xxv., p. 55.

"On the Orbit of *a* Centauri."—*Mem. R. A. S.*, vol. xxiv., p. 91.

"Second Series of Observations of Double Stars," etc. Madras, 1859 to 1862.—*Mem. R. A. S.*, vol. xxxii., p. 75.

The following are in the *Monthly Notices :*—

On Orbits, vols. xv., xxi., xxiv. Measures, vol. xvi.

RÜMKER (Ch.)

" Positionen von Doppelsternen."—*Ast. Nachr.*, vol. xvi., p. 31.

SAFFORD (T. H.)

"The observed Motions of the Companion of Sirius, considered with reference to the disturbing body indicated by theory." By. T. H. Safford.—*Proceedings of the American Academy of Arts and Sciences*, vol. v., 1863.

See also *Monthly Notices*, vols. xxii., xxiii.

SAVARY.

"On the Orbit of ξ Ursæ Majoris."— See *Connoissance des Tems*, 1822 and 1830.

SCHIAPARELLI.

"Measures of Double Stars."—*Ast. Nachr.*, vol. lxxxix., p. 317. "Orbit of γ Cor. Anst.,"—vol. lxxxvii.

SCHJELLERUP (Dr.)

"Einfacher Beweis des A. N., No. 1227 angeführten geometrischen Satzes zur Berechnung von Doppelstern-Bahnen."—*Ast. Nachr.*, vol. lv., p. 230.

"On some New Double Stars."—*Ast. Nachr.*, vol. lxxii., p. 331.

SCHLUTER.

"Measures of Double Stars with the Königsberg Heliometer."—*Ast. Beobachtungen*, edited by Busch. 1838.

SCHMIDT (J. F. J.)

"On the Colour of Arcturus."—*Ast. Nachr.*, vol. xlii., p. 226.

"Measures of Double Stars."—*Ast. Nachr.*, vol. lxv., p. 104.

SCHUBERT.

"On the Companion of Sirius."—*Astronomical Journal*, vol. i. ; see also *Gould's Astronomical Journal*, 16.

SCHULTZ.

"On H. VIII. 20. Measures of 104 Stars."—*Ast. Nachr.*, vol. lxxx.

SCHUR (W.)

"Orbit of Σ. 3062."—*Ast. Nachr.*, vol. lxix., p. 49.

"Orbit of 70 Ophiuchi,"—vol. lxxi., p. 1 ; also vol. lxxiii., p. 301.

SCHWAUS.

"On 61 Cygni."—*Ast. Nachr.*, vol. xvi.

SECCHI (A.)

The following papers are in the *Astronomische Nachrichten*:—

"On some Double Stars, Colours," etc.,—vol. xli., p. 109.

"Measures of Binaries," etc.,—vol. xli., p. 238.

"On the Companion of Antares,"—vol. xli., p. 238.

"Measures of Σ.'s Double Stars,"—vol. xliii., pp. 139, 141 ; vol. xlv., p. 251.

"On δ Cygni,"—vol. lxvi., p. 62.

"Measures of Double Stars,"—vol. lxviii., p. 87 ; also vol. lxiv., p. 84.

"Descrizione del Nuovo Osservatorio del Collegio Romano, D.C.D.G." Roma, 1856.

"Catalogo di 1321 stelle Doppie." 1860.

SEELIGER (H.)

"Zur Theorie des Doppelsternbewegungen."—Leipsig, 1872. Inaugural Dissertation. He gives historical sketch and new formulæ for computing an orbit.

SIRIUS.

Papers in the *Astronomische Nachrichten*, vols. lxi. to lxxx.:—vols. lxii., lxiii., lxiv., lxvi., lxvii., lxx., lxxi., lxxiv., lxxvi., lxxvii., lxxviii., lxxix. ; by Auwers, Bruhns, Dunér, Engelmann, Eastman, Foerster, Gylden, Hall, Newcomb, Pechüle, Tempel, Tietjen, Vogel, Goldschmidt. See also vols. lxxxiv., lxxxviii., xc.

In the *Monthly Noticee*, see vols. xviii., xx., xxii., xxiii., xxiv., xxv., xxvi., xxvii., xxviii., xxix.

SMYTH (Admiral).

"A Cycle of Celestial Objects," etc. 2 vols. London, 1844.

"Sidereal Chromatics." 1864.

"Speculum Hartwellianum." 1860.

"Observations of γ Virginis," etc.—*Mem. R. A. S.*, vol. xvi., p. 19.

SOUTH (Sir James).

"Observations on the best mode of Examining the Double or Compound Stars ; together with a Catalogue of those whose Places have been identified."—*Mem. R. A. S.*, vol. i., p. 109.

"Measures of 458 Double and Triple Stars," etc. Also, "Re-examination of 36 Double and Triple Stars," etc., 1823, 1824, 1825.—*Phil. Trans.* 1826, pt. i. See also *Phil. Trans.* 1824, pt. iii.—*Edinburgh Journal of Science*, vols. vii., viii.

STEINHEIL.

"On the Separation of bright Double Stars."—*Ast. Nachr.*, vol. xiv., p. 205.

STRUVE (F. W.)

The following paper is in the *Memoirs of the Royal Astronomical Society :—*

"A Comparison of Observations made on Double Stars," vol. ii., p. 443.

The following papers are in the *Astronomische Nachrichten :—*

Papers on his Review of the Heavens, New Double Stars, etc.,—vol. iv., pp. 50, 62, 65, 474.

"On γ Virginis,"—vol. xii., p. 271.

"Double Star Measures at Dorpat,"—vol. xiii., p. 249.

"On 40 Eridani,"—vol. xiv., p. 315.

"On the Parallax of *a* Lyræ,"—vol. xvii., p. 177.

"On 70 Ophiuchi,"—vol. xix., p. 203.

"Rapport fait à la classe physico-mathématique, sur un nouvel ouvrage rélatif aux étoiles doubles et multiples,"—vol. xxii., p. 49.

"An Account of the Instruments at Pulkowa,"—vol. xviii., p. 33.

"Description de l'observatoire Astronomique Central de Poulkova." Par F. G. W. Struve, St. Petersbourg, 1845.

"Catalogus 795 stellarum duplicium." 1822 :—"Ueber die Doppelsterne nach einer mit dem grossen Refractor von Fraunhofer," etc. 1832 :—"Memoire sur let étoiles doubles." 1832 : — "Ueber doppelsterne." 1837 :— "Catalogus novus stellarum duplicium," etc. 1827 :— "Stellarum fixarum positiones mediæ." 1852 :— "Mensuræ micrometricæ." 1837 :—"Additamentum in mensuras," etc. 1840 :—Report on Double Stars (*Edinburgh Journal of Science*, vol. ix.) 1828 :—"Stellar Astronomy."

"Catalogue de 514 étoiles doubles et multiples." St. Petersbourg, 1843.

"On Waldbeck's Computation of the Angle and Distance of γ Virginis in 1720."—See *Brewster's Edinburgh Journal*, vol. i.

STRUVE (O.)

"Catalogue de 256 étoiles doubles principles," etc. St. Petersbourg, 1843.

"Catalogue revu et corrigé," etc.—See the *Recueil de Mémoires des Astronomes de l'Observatoire central de Russie*, vol. i. St. Petersbourg, 1853.

"Mém. Acad." St. Petersbourg, vii., 1853.

"Mém. de Poulk," i., 1853.

The following papers are in the *Bulletin de l'Aladémie Impériale des Sciences de St. Petersbourg :*—

"Bullet. Scient. Acad., St. Petersbourg,—x., 1842 ; xiii., 1855 ; xvii., 1859.

"On the Companion of Procyon,"—tome xxii., 1876.

"On the Orbit of 42 Comæ Ber.,"—tome xxi., 1875, and tome x., 1866.

"On Σ. 2120,"—tome xxi., 1876.

"On Σ. 634,"—tome xix.

"Observation du Procyon, comme étoile double,"—tome xviii.

"Results of some Supplementary Observations made on Artificial Double Stars,"—tome xii., and tome iv., 1866.

"On the Companion of Sirius,"—tome x., also vii.

"Observations of some Double Stars recently discovered,"—tome i.

"Nouvelle détermination de la parallaxe des étoiles a Lyræ et 61 Cygni."—*Mémoires de l'Académie*, vii. série, tome i. St. Petersbourg, 1859.

On the Orbit of Σ. 1728 and Σ. 2120.—*Bulletin de l'Académie de St. Petersbourg*, tome xxi.

"On ζ Cancri, ξ Ursæ, γ Virg., η Cor. Bor., ω Leonis, and Σ. 2173."—*Ast. Nachr.*, vol. xviii., p. 43.

" Notice sur une révision de l'hemisphère céleste boréal," etc.—*Ast. Nachr.*, vol. xix., p. 283.

On New Double Stars.—*Monthly Notices*, vol. xx. On Sirius,—vols. xxiii., xxvi.

" Observations de Poulkova,"—vol. ix., 1878.

TEMPEL (W.)

"On the Companion of Sirius."—*Ast. Nachr.*, vol. lxii., p. 119.

"On the Stars in the Trapezium of Orion."—*Ast. Nachr.*, vol. lxxx.

THIELE (Th. N.)

"On the Orbit of ξ Libræ."—*Ast. Nachr.*, vol. l., p. 353.

"On the Orbit of Castor."—*Ast. Nachr.*, vol. lii., p. 39.

TISSERAND.

"On 70 Ophiuchi."—*Comptes Rendus*, vol. lxxxii.; and *Acad. of Sciences of Toulouse*, 1876.

VILLARCEAU.

"Elements of ζ Herculis."—*Ast. Nachr.*, vol. xxvi., p. 305.

"On η Cor. Bor."—*Ast. Nachr.*, vol. xxxvii., p. 57.

" Méthode pour Calculer les orbites relatives des étoiles doubles."—*Conn. des Temps*, 1852 and 1877.

See also *Comptes Rendus:* vols. xxviii., xxxvi., on η Cor. Bor.; vols. xxviii., xxxviii., on ζ Hercules; vol. xxix., formulæ for the case of an orbit whose plane coincides with the line of sight; vol. xxxiv., method of computing an orbit in general terms, and on the effect of the velocity of light on the form of the orbit of a double star.

WALDO (L.)

Double Star Measures made at Harvard Observatory in 1876.—*Ast. Nachr.*, vol. xcii.

WICHMANN (Dr.)

"On the Königsberg Heliometer Observations."—*Ast. Nachr.*, vol. xliii., p. 17.

"On the Influence of Temperature on the Observations," p. 20; also on other cognate matters. Comparison between the Oxford and Königsberg heliometers.

See also *Enganzungs-Heft zu den Ast. Nachr.* Altona, 1849.

WILSON (J. M.)

On Castor, ξ Ursæ, ζ Cancri.—*Monthly Notices*, xxxii.

On Castor, ξ Ursæ, ζ Çancri, μ₂ Boötis,—vol. xxxiii.

On ε Lyræ and Sirius,—vol. xxxiv.; On 61 Cygni and η Cor. Bor.,—vol. xxxv.

WILSON (J. M.) AND SEABROKE.

Catalogue of Micrometrical Measurements of Double Stars, made at the Temple Observatory.—*Mem. R. A. S.*, vol. xlii., p. 61.

Second Catalogue of Measures.—*Mem. R. A. S.*, vol. xliii.

WINNECKE (A.)

"The Orbit of ζ Cancri and η Cor. Bor."—*Ast. Nachr.*, vol. xli., pp. 102, 107.

"Measures of Double Stars."—*Ast. Nachr.*, vol. lxxiii., p. 145.

WROTTESLEY (Lord).

"A Catalogue of the Positions and Distances of 398, Double Stars."—*Mem. R. A. S.*, vol. xxix., p. 85.

LIST B.

PAPERS ON THE MICROMETER.

AIRY.

"On a New Construction of the Divided Eye-glass Double-Image Micrometer. By G. B. Airy, Esq., Astronomer Royal."—*Monthly Notices of R. A. S.*, vol. vi., p. 229; also vol. x., p. 160. *Mem. R. A. S.*, vol. xv.,

p. 199.   *Cambridge Trans.*, vol. iii.; and "Account of Northumberland Equatorial," p. 34.

ARAGO.

"Popular Astronomy," vol. i., p. 382.

AUZOUT.

Phil. Trans., No. 21.   1666.   He used two silk threads, one fixed and the other moveable by means of a fine screw.

BABBAGE.

"On a new Zenith Micrometer.   By Chas. Babbage, Esq., F.R.S., etc."—*Mem. R. A. S.*, vol. ii., p. 101.   1825.

BARLOW (P.)

"On the Principle of Construction and general Application of the Negative Achromatic Lens to Telescopes and Eyepieces of every description."—*Phil. Trans.* 1834, pt. i.

BIDDER.

"On a new Form of Position Micrometer.   By G. P. Bidder, Esq., Q.C."—*Monthly Notices of R. A. S.*, vol. xxxiv., p. 394.

BOGUSLAWSKI.

"On the Use of a New Micrometer," etc.—*Monthly Notices of R. A. S.*, vol. vi., p. 219.   A single web is placed in the focus of the object-glass as a diameter across the field of view.   It is so arranged that it can be turned round the centre in every direction, and make with the declination circle any given angle.   It was specially designed for the "observations of Mars and neighbouring stars for the purpose of determining his parallax."   Also *Mem. R. A. S.*, vol. xv., p. 193.

BOSCOVICH.

"Of a new Micrometer and Megameter.   By the Abbé Boscovich, etc."   1777.—*Phil. Trans.*, vol. lxvii.

BRADLEY.

"Directions for using the common Micrometer."—*Phil. Trans.*, vol. lxii.   1772.

BREWSTER (Sir David).

"Treatise on New Instruments." 1813.

BRÜNNOW (Dr. F.)

"Spherical Astronomy." A valuable chapter on the Heliometer and Micrometer. London, 1865.

CASELLA.

"On a Micrometric Diaphragm." By L. P. Casella.—*Monthly Notices of R. A. S.*, vol. xxi., p. 178.

CAVALLO (Mr. T.), F.R.S.

"Description of a simple Micrometer for Measuring Small Angles with the Telescope."—*Phil. Trans.*, vol. lxxxi. 1791. This was a thin and narrow slip of mother-of-pearl finely divided, and placed in the focus of the telescope.

CHAUVENET.

"A Manual of Spherical and Practical Astronomy," etc. By Prof. Chauvenet. Trübner and Co.: 1868. A chapter on the Micrometer.

CLARK.

"Mr. Alvan Clark's New Micrometer for Measuring Large Distances."—*Monthly Notices of R. A. S.*, vol. xix., p. 324. 1859.

CLAUSEN (Th.)

"Beschreibung eines neuen Micrometer."—*Ast. Nachr.*, vol. xviii., p. 96.

DAWES (Rev. W. R.)

"On an Improvement in Mr. Dollond's Micrometer."—*Monthly Notices of R. A. S.*, vol. ii., p. 180.

"New Arrangement of two Solar Prisms for use with the Micrometer."—*Monthly Notices of R. A. S.*, vol. xxv., p. 218.

Valuable hints and information may be obtained from the following papers by this eminent observer:—

Mem. R. A. S., vol. viii., p. 62; vol. xix., p. 191; vol. xxxv. Also *Monthly Notices*, vol. xviii., p. 58.

DEMBOWSKI.

"D etermination de la valeur en arc des Revolutions du Micromètre."—*Ast. Nachr.*, vol. lxxxi., p. 247.

DOBERCK.

"On Amici's Double-Image Micrometer and Graham's Square-Bar Micrometer."—*Ast. Nachr.*, vol. cxii.

Examination of the Merz Micrometer.—*Ast. Nachr.*, vol. cxii.

DOLLOND (John).

Contrivance for Measuring Small Angles.—*Phil. Trans.*, vol. x., pp. 364, 462 ; also vol. x., p. 409.

"Account of a Micrometer made of Rock Crystal." By G. Dollond, F.R.S.—*Phil. Trans.*, 1821, pt. i.

"An Account of a Concave Achromatic Glass Lens, as adapted to the wired Micrometer when applied to a Telescope, which has the power of increasing the magnifying power of the Telescope without increasing the diameter of the Micrometer Wires." By George Dollond, F.R.S., etc.—*Phil. Trans.*, 1834, pt. i.

ENCYCLOPÆDIA BRITANNICA.

Article "Micrometer,"—vol. xiv., p. 742.

EPPS.

"Formulæ for reducing Micrometric Observations."—*Monthly Notices of R. A. S.*, vol. iii., p. 198.

FRAUNHOFER.

"Mikrometer, über eine neue Art."—*Ast. Nachr.*, vol. ii., p. 51 ; vol. ii., p. 361.

GASCOIGNE.

"On Mr. Gascoigne's Micrometer. By Mr. Richard Townley." 1667.—*Phil. Trans.*, vol. ii., p. 161.

"Description of Mr. Gascoigne's Micrometer. By Mr. Hook." 1667.—*Phil. Trans.*, vol. ii., p. 195.

"An Account of Mr. Gascoigne's Micrometer by Bevis." —*Phil. Trans.*, vol. x., p. 369.

See also Costard's *History of Astronomy*, and Baily's *Account of the Rev. John Flamsteed.* 1835.

HALLEY (Dr.)

"On Cassini's Micrometer."—*Phil. Trans.*, No. 363.

HERSCHEL (Sir William).

"Description of a Micrometer for taking the Angle of Position. By Mr. Wm. Herschel, of Bath." 1781. *Phil. Trans.*, vol. lxxi. 1781.

"Description of a Lamp Micrometer, and the Method of using it. By Mr. Wm. Herschel, F.R.S." 1782.— *Phil. Trans.*, vol. lxxii.

"A Description of the Dark and Lucid Disc and Periphery Micrometers. By Wm. Herschel, Esq., F.R.S." 1782.—*Phil. Trans.*, vol. lxxiii.

"On Fixing Spider Lines." Capt. John Herschel.— *Monthly Notices*, vol. xxxiv., p. 396.

HIRE (De la).

"On Huyghens's Micrometer."—*Royal Academy of Sciences*, 1717.

HOOKE.

See Sprat's "History of the Royal Society," Hooke's "Micrographia," and Hooke's "Posthumous Works."

HUSSEY.

"Description of a Lamp Micrometer."—*Ast. Nachr.*, vol. x., p. 385.

HUYGHENS.

See his *Systema Saturnium*, published in 1659. By placing a strip of metal at the focus of the telescope, and making its breadth equal to that of the object observed, the apparent diameter could be deduced.

KAISER (Dr. F.)

"On Airy's Micrometer."—*Ast. Nachr.*, vol. xlv., p. 209.

"Erste Onderzoekingen med den Micrometer van Airy."—*Ast. Nachr.*, vol. xlviii., p. 109.

"Measures of Double Stars with Airy's Double-Image Micrometer and the Parallel-Wire Micrometer."

"Annalen der Sternwarte in Leiden." Dritterband Haag, 1872.

See also *Monthly Notices of R. A. S.*, vol. xxvi., p. 305, and vol. xxvii., p. 11 ; *Ast. Nachr.*, vol. xlv., p. 209 ; vol. xviii., p. 1 ; and vol. lx.

KLINKERFUES.

"Mikrometer von Repsold, für den Göttinger Meridian Kreis ueber dasselbe von Klinkerfues."—*Ast. Nachr.*, vol. xlii., p. 107, and vol. lx.

LAMP (Dr.)

On Bessel's Formula for the Correction of Micrometer Screws.—*Ast. Nachr.*, vol. lxxxviii.

MALVASIA (Marquis).

See his "Ephemerides," published at Bologna in 1662. This micrometer was a network of fine silver wires, forming small squares in the focus of the telescope.

See *Mém. Acad. des Sciences*, 1717.

MASKELYNE.

"Description of a Method of Measuring Differences of R.A. and Dec., with Dollond's Micrometer; with other New Applications of the same. By the Rev. N. Maskelyne, B.D., F.R.S." 1771.—*Phil. Trans.*, vol. lxi.

"Of a New Instrument for Measuring Small Angles, called the Prismatic Micrometer. By the Rev. Nevil Maskelyne, D.D., F.R.S., etc." 1777.—*Phil. Trans.*, vol. lxvii.

MÖSTA.

"Mikrometer, Untersuchungen über periodische Ungleichheiten derselben, von Mösta."—*Ast. Nachr.*, vol. lix., p. 257.

PAPE.

"Untersuchung der Microscop Mikrometer des Altoner Meridiankreises, von Dr. C. F. Pape." 1859.—*Ast. Nachr.*, vol. l., p. 337.

PEARSON (Rev. W., LL.D., F.R.S., etc.)

"An Introduction to Practical Astronomy," etc., 2 vols. London, 1829. In this work will be found "An Historical Account of the different Methods of Measuring small

Celestial Arcs," and also full descriptions of reticles, parallel wire, angular, circular, and double-image Micrometers, etc., etc.*

POGSON.

"On the Ocular Crystal Micrometer."—*Brit. Assoc. Report*, 1858.

PORRO.

"Nuovo micrometro per mezzo di linee luminose ad uso dell' astronomia." J. Parro.—*Ast. Nachr.*, vol. xlviii., p. 65. 1858.

POWELL (Rev. B.)

"On a New Double-Image Micrometer."—*Monthly Notices of R. A. S.*, vol. vii., p. 24.

RAMSDEN.

"The Description of Two New Micrometers. By Mr. Ramsden, Optician." 1779.—*Phil. Trans.*, vol. lxix.

REES'S CYCLOPÆDIA.

Dr. Smith on Micrometers.

RESLHUBER.

"Mikrometer mit Lichtpunkten im Refractor zu Kremsmünster, über dasselbe von Reslhuber." 1858.—*Ast. Nachr.*, vol. xlviii., p. 149.

ROGERS.

"Modification of the Micrometer Head, devised by Mr. Joseph A. Rogers, Aid U. S. Naval Observatory."—*Ast. Nachr.*, vol. lxiii., p. 78. 1864.

SAVERY.

"On a Double-Image Micrometer."—*Phil. Trans.*, vol. xlviii. 1783.

SECCHI.

"Doppel-bild Mikrometer, über dasselbe von Secchi."—*Ast. Nachr.*, vol. xlviii., p. 309. 1858.

SEELIGER (H.)

Theory of the Heliometer. Leipsig, 1877.

* See also *Mem. R. A. S.*, vol. i.

SIMMS.

"Notice of an Improvement in the Double-Image Position Micrometer. By William Simms, Jun., Esq."— *Monthly Notices of R. A. S.*, vol. xviii., p. 64 ; also vol. xxvii., p. 11.

SMEATON.

"An Equatorial Micrometer," invented by Mr. John Smeaton, F.R.S., with Observations. 1786.—*Phil. Trans.*, vol. lxxvii.

STEINHEIL.

"Mikrometer-Ocular mit leuchtenden Faden." Remarks on, by Steinheil. 1859.—*Ast. Nachr.*, vol. li., p. 353.

VALZ.

" On the Divided Eyepiece Micrometer."— See *Monthly Notices of R. A. S.*, vol. x., p. 160 ; also *Zach's Correspondance Astronomique*, tome i., p. 353.

WILSON (Dr.)

"An Improvement proposed in the Cross Wires of Telescopes."—*Phil. Trans.*, vol. lxiv. 1774.

WINNECKE (A.)

Ueber ein neues Hülfsmittel, die periodischen Fehler von Mikrometer-Schrauben zu bestimmen.—*Ast. Nachr.*, vol. cxi. (See *The Observatory*, No. 12.)

WOLLASTON.

" Of a Single-lens Micrometer," by Dr. Wollaston.— *Phil. Trans.*, vol. ciii.

" On a Method of Cutting Rock Crystals for Micrometers," by Dr. Wollaston.—*Phil. Trans.*, vol. cx.

"A Description of a New System of Wires in the Focus of a Telescope," etc.—*Phil. Trans.*, vol. lxxv. 1785.

WREN (Sir Christopher).

See Sprat's " History of the Royal Society."

ZENGER.

" The Stereo-Micrometer. By Professor Chas. V.

Zenger."—*Monthly Notices of R. A. S.*, vol. xxxvi., p. 252.

See also Σ.'s *Mensuræ Micrometricæ, O.Σ.'s* last vol. of the *Poulkova Observations* (vol. ix.) ; Σ.'s *Description de l'Observatoire Astronomique Central de Poulkova.* 1845.

LIST C.

SOME PAPERS ON THE COLOURS OF STARS.

ANDRE AND BORROSCH.
  Ueber das farbige Licht der Doppelsterne. Prag., 1842.

ARAGO.
  See his *Astronomie Populaire.*

BALLOT (Dr.)
  Papers in *Poggendorff's Annalen* 1845, and vol. lxviii.

BOLZANO (Dr.)
  See *Pogg. Annalen* 1843.

DOPPLER (Dr.)
  His papers are to be found as follows :—In vol. viii. of *Sitzungsberichte der Mathematisch-Naturwissenschaftichen Classe der Kaiserlichen Akademie der Wissenschaften.* Wien, 1852. *Weitere Mittheilungen meine Theorie des farbigen Lichtes der Doppelsterne betreffend*, in *Pogg. Annalen* 1850 and 1851. *Ueber das farbige Licht der Doppelsterne und einiger auderer Gestirne des Himmels*, in *Abhandlungen der Königlichen Böhmischen Gesellschaft der Wissenschaften.* Prag., 1843.

KLEIN.
  " On the Changes of Colour of the Fixed Stars."—*Ast. Nachr.*, vol. lxx.

KREIL.
  "On Doppler's Theory," in *Ein Astronomisch-Meteorologischen Jahrbuche für Prag.* 1844.

NIESTEN.

"On the Colours of Double Stars."—*Bulletin de l'Académie royale de Belgique.* 1879.

SCHMIDL (Dr.)

In den österreichischen Blättern für Literatur und Kunst.

SCHMIDT (Dr.)

"On the Colours of the Stars."—*Ast. Nachr.*, vol. lxxx.

"On the Colour of Arcturus."—*Ast. Nachr.*, vol. xlii.

SESTINI.

Memoria sopra i colori delle stelle del Catalogo di Baily.
Roma, 1845 and 1847. See *Gould's Astron. Journal,* 1850.

SMYTH (Admiral).

"Sidereal Chromatics."   London, 1864.

SMYTH (Prof. Piazzi).   See his "Teneriffe."

ZENKER (Dr.)

"On Doppler's Theory."—*Ast. Nachr.*, vol. lxxxv.

ZOLLNER (Dr.)

Ueber Farbenbestimmung der Gestirne, vol. lxxi.

See also *Monthly Notices*, vols. x., xi., xx., xxvi., xxvii.

ADDITIONAL NOTES.

176. Σ. 13.   102°·5 in. ...   1879·66
177. O.Σ. 4.   155°·3   0″·56   78·6   Bu.
186. O.Σ. 515.   273°·0   0″·2   77·7   ,,
                      272°·3   0″·28   8·8   ,,
      O.Σ. 28.   314°·0   0″·85   78·7   ,,
187. Σ. 93 (Polaris).

"Some three years ago the discovery by M. De Böe, of Belgium, of two small companions about 4″ distant was announced, and several observers subsequently claimed to have seen them in the described places. . . . . I have no hesitation in saying these supposed stars do not exist."—(Mr. Burnham's *Double Star Observations*, p. 87.)

193. Σ. 196.   52°·0   2″·49   1877·7   Bu.
199. o Ceti.   82°·8   115″·62   77·8   ,,

Mr. Burnham thus describes his new companion :

                      91°·5   74″·11   Mag. 13   1877·8.

201. Σ. 293.   75°·5   8″·30   1878·6   Bu.
209. Σ. 453.

Mr. Burnham, with 6 in. and 18½ in. apertures, finds no trace of duplicity. (1874 to 1879.)

211. O.Σ. 531.   138°·5   2″·60   1878·6   Bu.
      Σ. 518.   A B 125°·4   3″·66   79·0   ,,
               A D 145°·6   37″·10   77·8   ,,
                   137°·3   35″·99   79·0   ,,
212. Σ. 547.   Companion not seen.   1865   De.
                      ,,   ,,   1873·7   Bu.
                      9°·8   2″·46   77·9   ,,
                      14°·0   ·25   79·0   ,,
213. Σ. 554.   7°·0   0″·60   78·0   ,,
216. O.Σ. 92.   247°·4   2″·78   78·0   ,,
220. Σ. 728.   196°·2   0″·45   78·0   ,,
221. Σ. 748.
      A E (5th star) 354°·3   4″·02   77·9   ,,
      C F (6th star) 119°·8   3″·74   77·9   ,,

Bu. has never suspected the existence of any other star either within the trapezium or near the principal stars. Nor does he think the 5th and 6th stars variable.

224. So. 503.

Bu. has discovered a faint star nearer than C.

                      157°·3   28″·09   Mag. 13.   1878.

227. Σ. 943.   147°·4   19″·65   1878·2   Bu.
233. O.Σ. 159.   361°·0   0″·48   78·1   ,,

Bu.'s third star C is thus given :

                      31°·4   23″·64   Mag. 12-13.   1878·1.   Bu.

PAGE					
235.	O.Σ. 165.	78°·1	2″·45	1878·1	Bu.
236.	Σ. 1074.	140°·0	0″·71	78·0	,,
237.	Σ. 1081.	226°·1	1″·36	78·1	,,
243.	Procyon and D.				
		262°·3	56″·59	36·7 *	
		311°·8	44″·59	74	Newcomb.
		317°·3	·62	77	Bu.
245.	Σ. 1142.	254°·5	22″·84	78·2	Bu.
246.	Σ. 1175.	222°·5	1″·81	78·1	,,
252.	Σ. 1216.	158°·8	0″·61	78·2	,,
256.	Σ. 1329.	248°·4	22″·84	78·1	,,
256.	Σ. 3121.				

Dr. Doberck's provisional elements are:

☊ 16° 0′, λ 149° 30′, γ 74° 15′, e 0·2600, P 37·03 yrs., T 1842·78, a [0″·71].

260.	Σ. 1356.				

Dr. Doberck's latest elements are:

☊ 149° 15′, λ 122° 19′, γ 64° 5′, e 0·5510, P 114·55 yrs., T 1841·57, a 0″·85.

262.	δ Sextantis.	117°·4	0″·2 ±	78·2	Stone.
		161°·0	0″·2 ±	78·2	Bu.
	O.Σ. 523.	298°·7	6″·67	78·1	,,
	Σ. 1423.	69°·7	1″·28	78·2	,,
267.	O.Σ. 230.	13°·4	8″·36	78·2	,,
277.	O.Σ. 234.	282°	obl.	70·5	O.Σ.
		187°·1	0″·35 †	78·2	Bu.
		151°·7	0″·27	78·3	,,
	O.Σ. 235.				

Dr. Doberck's elements are:

☊ 96° 17′, λ 129° 55′, γ 60° 13′, e 0·5870, P 94·406 yrs., T 1839·10, a 1″·066.

278.	O.Σ. 243.	14°·5	0″·94	78·3	,,	
279.	Σ. 1607 = H_2 202 = H_3 516.					
		A B 358°·1	30″·85	Mag. 7·8, 8·3	1878·2	Bu.
		B C 309°·8	21″·04		78·2	,,
	O.Σ. 249.					
		A B 307°·1	0″·46	1878·3	Bu.	
		A C 148°·3	12″·68	78·3	,,	
	Σ. 1641.	39°·2	8″·96	78·3	,,	
280.	Σ. 1643.	46°·6	1″·76	78·3	,,	
287.	Σ. 1703.	281°·5	18″·79	78·3	,,	
	Σ. 1707.	35°·5	9″·84	78·3	,,	
290.	Σ. 1757.	Doberck's formulæ are:				

P = 46°·61 + 1°·094 (t — 1850) — 0°·0153 (t − 1850)².
Δ = 1″·83 + 0″·016 (t — 1850).

293.	O.Σ. 270.	351°·9	8″·71	1878·3	Bu.
295.	Σ. 1820.	67°·4	2″·32	78·3	,,
	Σ. 1819.	Doberck finds:			

P = 51°·16 − 1°·491 (t − 1850) + 0°·0138 (t — 1850)².
Δ = 1″·09 + 0″·010 (t − 1850).

* Annalen König. Stern. München., xvii.
† "Too large."

297.	O.Σ. 283.	141°·3	5″·27	1875·5	O.Σ.
		131°·1	5″·00	78·3	Bu.
303.	Σ. 1879.	39°·1	0″·37	77·4	Schi.
		42°·9	·34	77·5	De.
		217°·1	·42	78·4	Bu.
305.	Σ. 1888.	279°·4	4″·22	79·40	Dob.
307.	Σ. 3091.	225°·0	0″·25 ±	78·4	Bu.
	O.Σ. 294.	248°·7	2″·75	78·3	,,
	O.Σ. 295.	128°·7	0″·75	78·4	,,
	Σ. 1909.	241°·4	4″·93	79·34	Dob.
308.	Σ. 3093.	139°·9	30″·79	78·3	Bu.

313. Doberck's second elements were obtained from equations of condition.

317.	Σ. 1961.	44°·6	22″·45	1878·3	Bu.
	O.Σ. 298.	295°·2	0″·3	77·4	De.
		310°·7	0″·27	78·3	Bu.

O.Σ. 298.

Dr. Doberck gives the following :

        ☉ 14° 38′, λ 342° 31′, γ 56° 10′, e 0·4872, P 68·802 yrs.
        T 1812·96, a 0″·886.

322. ν Scorpii, Bu. 120.

	360°·6	0″·79	77·5	De.
	363°·7	1″·04	78·3	Bu.
333.	Σ. 2106.	323″·7	0″·64	78·4	,,
346.	A. C. 7.	234°·9	1″·05	78·5	,,
355.	Σ. 2315.	251°·7	0″·31	78·4	,,
357.	a Lyræ. A B 154°·9	48″·01	78·4	,,
	·9	·11	78·5	,,

A faint companion of the 12-13 mag. was discovered by Winnecke in 1864.

A C 298°·8	46″·87	1864·8	Wi.
289°·9	51″·66	78·3	Bu.
292°·9	·97	78·4	,,
293°·1	·93	78·4	,,

Mr. Burnham has never seen the slightest trace of the faint stars supposed to have been seen by Mr. Buckingham and others.

363. O.Σ. 364.

        Mr. Burnham has hitherto failed to see the companion.

365. Σ. 2434.   64°·1   1″·53   1878·6   Bu.

367. O.Σ. 368.

        Bu. gives the following measure of a new small star :

A C 98°·2	17″·37	1878·6	Bu.
A B 216°·2	1″·03	78·7	,,

368.	Σ. 2514.	316°·3	8″·51	78·4	,,
	Σ. 2515.	23°·8	12 ″·98	78·6	,,

370. O.Σ. 380.

        Neither Burnham nor Newcomb can see the star C.

370.	Σ. 2574.	145°·0	0″·63	1878·5	Bu.
374.	O.Σ. 532.	16°·5	12″·00	78·6	,,

Σ. 2607 A B = O.Σ. 392.

        317°·0   0″·31   78·5   ,,

375. Σ. 2619.

        C was first observed by H₂.

        294°·2   18″·40   78·4   ,,

PAGE
376.   O.Σ. 400.

Single in 1878 (Bu.)

Σ. 2658.

A B 119°·9	5″·21	78·6	Bu.
A C 212°·4	40″·50	78·6	,,

378.   O.Σ. 533.   329°·2   10″·59   78·2   ,,
381.   Σ. 2734.   191°·2   25″·82   78·6   ,,
       O.Σ. 418.   294°·8   1″·2±   77·7   ,,
383.   O.Σ. 424.   327°·7   0″·45   78·6   ,,
384.   Σ. 2749.

Secchi discovered C in 1856.

                   148°·9   1″·13   77·8   ,,
389.   τ Cygni.   150°·0   1″·06   78·4   ,,
392.   Σ. 2860.   255°·6   5″·96   78·5   ,,
397.   Σ. 2912.   130°·0   0″·32   78·6   ,,
       Σ. 2915.   155°·1   12″·29   77·8   ,,
398.   O.Σ. 477.   154°·1   4″·62   78·6   ,,
399.   Σ. 2959.   102°·2   13″·77   77·8   ,,

Mr. Burnham's new star C is thus given :

                   95°·9   8″·31   1877·9   Mag. 12-13.
399.   O.Σ. 536.   Round   ...   74·8   Bu.
                   161°·5   0″·47   77·8   ,,
406.   No. 804.   12°·3   22″·63   1879·21 Dob.
       No. 818.   275°·8   0″·99   79·15   ,,

Σ. 1058.   Mag. 8·2, 11·7.

                   282°·7   23″·78   32   Σ.
                   283°·0   22″·47   44   Mä.
                   Companion not seen 65   De.
                        ,,   ,,   74   Bu.
                        ,,   ,,   75   ,,
                   281°·0   22″·84   79   ,,
                     ·0   ·32   79   ,,

Is one of the components variable? (Bu.)
The place for 1880 is
         R. A. 7ʰ 10·3ᵐ.   Dec. 9° 37′.

INDEX.

Hazell, Watson, and Viney, Printers, London and Aylesbury.

CORRECTIONS, NOTES, &c.

TO THE

HANDBOOK

OF

DOUBLE STARS.

"So long as no presumption à *priori* can be adduced why the most minute star in the heavens should not give us that very information respecting parallax, proper motion, and an infinity of other interesting points, which we are in search of and yet may never obtain from its brighter rivals, the minuteness of an object is no reason for neglecting its examination. . . . . But if small double stars are to be watched, it is first necessary that they should be known ; nor need we fear that the list will become overwhelming. It will be curtailed at one end, by the rejection of uninteresting and uninstructive objects, at least as fast as it is increased on the other by the accession of new candidates for examination ;—and if in the result, but one individual of such a nature as above hinted at should at length be found, the toils of whole generations of astronomers would be amply repaid ;— it would be the diamond remaining after searching and throwing away the rubbish." —*Sir John Herschel in Mem. R.A.S.*, vol. ii., pp. 472, 473.

London :

MACMILLAN AND CO.,

1880.

LONDON :

R. CLAY, SONS, AND TAYLOR,
BREAD STREET HILL.

INTRODUCTION.

A CAREFUL examination of the Handbook has led to the detection of a very considerable number of errors, some of them serious, but by far the larger portion of no great importance. The sources of these errors are not difficult to find : they may be thus enumerated—hasty reading of the proofs, errors made in copying results from books in the London libraries, slips or failure of type during printing, misprints in the works mostly consulted, and those arising from the printing of sheets before the final revise had been received. By far the most fruitful of these sources of error was the too implicit reliance on the lists of measures in well-known catalogues ; the original works ought to have been consulted in every case, and, where possible, even the original MSS.

The corrections have been thrown into two classes : the first contains those which from their importance demand immediate attention in order to save waste of time. These the reader is requested to insert at once. In the second list will be found a large number of corrections which may be entered as the stars are observed or read.

A very copious set of additional notes has also been drawn up embodying, so far as we know them, the most recent and important orbits, measures, and discoveries.

It will readily be understood that the construction of these lists has involved very considerable labour. Indeed it was soon found that it would not be possible to make them without very frequent reference to works and MSS. in London. Fortunately, however, several observers who have kindly taken much interest in the success of the Handbook very kindly undertook to give any help in their power in order to facilitate the work of correction. Foremost among these must be mentioned Mr. Herbert Sadler, whose acquaintance with double-star literature is second to none. To him our best thanks are due for suggestions, original observations, critical remarks, and extracts from MSS. in the Royal Astronomical Society's library. We are also much indebted to Dr. Wm. Doberck, Director of Markree Observatory, Collooney, Ireland, for very numerous and valuable formulæ, and his most recent elements of interesting binaries. Short lists of errata, suggestions, and recent discoveries were also very kindly contributed by Mr. Burnham. The following observers have also sent us useful notes, measures, or lists of errata: Dr. Dunér, Mr. Seabroke, Prof. O. Stone, Prof. T. N. Thiele, and the Rev. T. W. Webb, M.A. To these gentlemen we beg to express our gratitude. Nor must we forget to acknowledge the suggestions and corrections of misprints by the gentlemen who have so favourably reviewed the Handbook.

BERMERSIDE, HALIFAX,
October, 1880.

CORRECTIONS.

PARTS I. AND II.

Page 1. Line 12, for Kirsch, read Kirch.
4. Line 23, for 400, read 460 ; line 28, for ε, read i.
8. Last line but one, for Mitchell, read Mitchel ; last line, *dele* Spörer.
24. Line 20, for screen, read screw.
25. Line 10, after 1¾ in., add a comma ; line 14, read 13½ lbs.
36. Line 13, 5 in. aperture, *i.e.* 5 French inches.
40. Line 5, for Anwers, read Auwers.
43. Line 9, for 9·5 in., read 8·6 in.
45. Line 10, read 7,000 lbs.
47. Line 11, 11³ should be 11¾.
56. In the map of the Pleiades the dots for Maia and Asterope have failed to print : the place of Maia is less than $\frac{1}{10}$ inch to left of lower part of M ; that for Asterope $\frac{1}{10}$ inch to right of the final e.
78. The diagram should be so placed that the straight lines are horizontal.
90. W's angle, last line but 1, should be 285°·5.
99. Last line but three : in the formula, read $\dfrac{\Delta t}{1} - \dfrac{\Delta^2 t}{2}$, &c.
106. Last line but two : *dele* the words *the number of nights ; and, lastly, in column* 7.

PART III.

THE CATALOGUE.

Page 152. No. 2. The Dec. should be 28° 26′. In No. 9, the Dec. should be 37° 34′.
153. No. 85 is not 65 Arietis (B). It is incorrectly so called in the *M. M.*, p. 89. See Σ.'s *P. M.*, pp. 156, 174. The star 65 Arietis (B) is a single star. (Sadler.)
154. No. 100. The Dec. should be 28° 55′.
155. No. 142. The Dec. should be 39° 11′.

Page 157. No. 258. H₂'s No. should be 3012.
164. No. 563. The R. A. should be 17 h. 41·8 m.
166. No. 687. H₂'s No. should be 8692.
167. No. 712. The Dec. should be 9° 31'. For No. 722, Σ.'s No. is 2799, and for No. 723, Σ.'s No. is 2802.
168. No. 757. H₂'s No. should be 9742 : for No. 766, it should be 9930.
169. No. 799. The R. A. should be 4 h. 17 m.
170. No. 849. The R. A. should be 16 h. 15·7 m., and the Dec. 1° 26'.

The errors in Dec. and R. A. should also be corrected in the Measures.

The following omissions in the Catalogue may be entered at the reader's convenience.

No. 2. H₂'s No. is 10317; No. 9. H₂'s No. is 87; No. 86 is 875 in H₂; No. 159 is H₂'s No. 1684; No. 223 is 2473 in H₂; No. 508 is H₂ 6621; No. 659 is H₂ 8180, and the magnitudes are 1·5, 10·2; No. 778 is H₂ 10171; No. 810 is H₂ 2517; No. 840 is H₂ 6091.

In some copies the Dec. column has 1800 for 1880 on pp. 156, 157, 158, 170 (the R. A. also).

No. 893. The arc is given 207° : but if we add 180° to Bu's angle the change in 30 years is about 20°.

It will be found that some stars are common to the two lists A, B : *e.g.* No. 805 is 1049; No. 892 is 1170; No. 933 is 1054; No. 994 is 1186. We were unwilling to break the order and completeness of Mr. Burnham's list by omitting such stars, and those in list A were already beyond our control when list B arrived. In Mr. Burnham's list his most recent measures and discoveries are given.

THE MEASURES.

A.—THE MORE IMPORTANT CORRECTIONS.

Page 174. In the "Abbreviations, &c." for Spörer (Sp.) read Schiaparelli (Schi.) : and in the "Measures" for Sp. read Schi, and *dele* Sp.

In *some* copies all the following errata will be found.

175. No. 1. The Dec. should be — 5° 12'. No. 2. The Dec. should be 28° 26' : and H₁'s angle should be 259°·4. It is similarly misprinted in the Mem. R. A. S. Vol. 35, p. 77.
176. No. 5. Dob's angle should be 101°·7.
177. No. 9. The Dec. should be 37° 34'.
178. No. 12. Smyth's distance should be 35″

Page 185. No. 38. The Dec. should be 14° 44'. No. 39. The proper motion is secular.

186. No. 40. Secchi's first distance is 3"·57.

187. No. 48. The Dec. should be − 1° 9'.

188. No. 49. Line 4, read 1830 for 1870.

190. No. 57. The proper motion given is secular. No. 59, col. 2, last line but three, read A D.

192. No. 66. Secchi's distance is 1"·78 : De's second distance 2"·15 : and Mä's last distance is 1"·93.
   No. 67. H₁'s distance is 4"— 8".

195. No. 75. In fifth line of footnote run the pen through "On the other hand, the discoverer" and put "Dawes" : and in last line but six, put smalt for small.

198. No. 81. The last two distances should be 16"·0, 15"·86.

199. No. 86. The Dec. should be − 3° 32'.

202. No. 99. Prefix 2 to Mä's last and Flt's distance.
   The Dec. of No. 100 is 28° 55'.

203. No. 102. A B. Gl.'s last angle is 122°·5 and his first epoch 1874·01.

205. No. 110. Σ.'s angle should be 8°·5. No. 112. O.Σ.'s last distance is 1"·11.

208. No. 130. Smyth's second angle should be 235°·9.

210. No. 142. The Dec. should be 39° 11'.

211. No. 145. The first angle should be 328°·1.

212. No. 149. A B. H₁'s distance should be 81"·78, and the first distance of B C should be 4"– 8". In No. 151, erase the first Mag. 7·9.

214. No. 159. H₁'s distance is so given by Σ. but it should be 87"·75.

215. No. 163. Prefix 8 to the last two angles of W. and S.

217. No. 178. The last two distances are 14"·53, 13"·80. In No. 182 Mä.'s distances are 14"·91, &c.

218. No. 189. H₁'s distance should be 46"·70.

219. The angles in No. 191 should be 304°·1, 306°·4, 304°·2, 307°·6.

222. Col. 2, last line, read indirect motion.

223. No. 206. O.Σ.'s measures are 149°·9, 2"·70. In A C, H₁'s angle is 6°·6.

224. No. 211. Last line, col. 2, add "and A C."
   No. 212. The proper motion (secular) of A is + 5"·7 in R. A. and − 11"·6 in Dec. (Σ.) And in A C, H₁'s distance, *not* angle, is 150" ±. C is of the 10th mag.
   No. 214. Mä.'s distance should probably be 1"·3 ; his angle 353°·6. It is 253°·6 in the *Dorpat Obs.* vol. xiii.

225. No. 217. O.Σ.'s distances are 1"·30, 1"·65.

233. No. 248. The measures under W. O., C. O., Bu., belong to Sirius.

234. Col. 1. H₁'s second angle is 183°·9. Po.'s distance in 1861·12 is 6"·07. In No. 253, for A C read B C.

235. No. 258. The first seven distances are O. Σ.'s, the next three are Σ.'s, and O.Σ.'s first angle is 150°·7. The first three angles and dates are Σ.'s.

Page 239.   Col. 1, in line 13, read H$_2$, line 17, Hi. ; col. 2, line 23, read Hi. for H$_1$.

245.   No. 278.   Mädler's distance is 5″·238.

247.   Col. 2, last line but one, read ξ for ζ.

249.   Col. 1.   Sm.'s and Ka.'s distances are 1″·3, &c. ;   O.Σ.'s first is 0″·99.

          Col. 2.   Ch.'s measures belong to A C.

250.   Col. 1.   The first date is 1871·19.   Col. 2.   Ch.'s measures belong to A B.

254.   No. 301.   The magnitudes should be 7·8, 8·9.

256.   No. 310.   The last measure points to decrease in angle.

260.   Line 11.   The elements here referred to are T = 1842·77, ω = 122°·9, ☊ = 151°·6, i = 65°·4, e = 0·5028, μ = + 3°·3451, P = 107·62 years.

262.   No. 331.   Line 6, read 1852.

263.   Col. 2.   In lines 27, 28, 31, place *sf* after the angles.

266.   No. 349 is O.Σ. 222.

269.   Col. 2.   Put Σ. one line lower.

270.   No. 368.   The magnitudes should be 4, 5.

275.   No. 370.   The angles by W. and S. are 335°·8, 333°·6, 334°·5. No. 371, line 18, read R. A.

276.   No. 372.   So.'s distance is 6″·29.

279.   No. 388.   De.'s angle is 346°·8.

282.   Col. 1, lines 12 and 16, read other for others, and in line 33, for almost, read almost always.   In Col. 2, last line but 9, read, "In fact the orbit," &c.

287.   No. 407.   A C.   The second measure by Ta. belongs to A B, and the last two distances are 28″·69, 28″·68.

289.   Col. 2.   The expression in brackets should be $\left(\dfrac{0\cdot127}{0\cdot072}\right)^{x}$.

290.   Col. 1, line 4.   The distance is 14″·04.

291.   No. 425.   Challis's distance is 0″·6.

292.   No. 428.   Secchi's angle is 198°·5.

294.   No. 442 is Σ. 1816.

299.   No. 458, lines 6 and 11, read 1779.

304.   Col. 1, first paragraph, H$_2$ wrote it ; last line but four, the angle is 324°·8.

306.   Col. 1, last line, read — 0^{s}·045.

308.   No. 484.   The distances are 33″·38, 32″·57, 31″·15.

312.   Col. 1.   The date of Hind's measure is 1847·08.

314.   Col. 2.   The angle by H$_2$ and So. is 171°·8.

315.   Col. 2, line 5, read shows for show ; last line but nine, read 0°·00002.

318.   Col. 2.   Encke's angles are 290°·9, 295°·0.

320.   Col. 1.   In H$_1$'s remarks, the "other set" is Σ. 1999.

322.   No. 506.   The proper motion given is secular.

324.   No. 514.   The magnitudes should be 5, 6·1, 10, 12·5.

Page 325. Col. 2, line 16, read 0·6988; line 23, read, re-discovered C; and in the table, Powell's period is 240 years. In col. 2, for D, in line 23, read C. O.Σ. re-discovered C in 1851.

332. Col. 2, last line. For De. read O.Σ.

334. No. 533. The distance of A C is 113″·67.

337. Col. 2. H_1's first angle is 232°·4 and the last 219°·9.

339. Col. 1, line 30. The mag. of C should be 7½.
Col. 2. A C, Jacob's distance should be 150″ (estimated).

344. No. 555. De.'s angle is 23°·0.

346. No. 562. *Dele* De.'s measure.

349. Col. 2. In the seventh paragraph, for φ read *p*, and in the last, read Fraunhofer.

354. No. 578. The last distance is 1″·12.

362. No. 606. GF. H_2's distance is 45″.

363. Col 2. The first angle is 360°·0. This star is *o* Draconis.

373. No. 654. Col. 1. The last date is 1875—77.

374. No. 664. Secchi's angle is 134°·9.

377. No. 681. Col 1. For A C, read A D.

378. No. 685. Col. 1. The first distance of $\frac{A + B}{2}$ and C is 25″·9.

379. No. 692. The first distance is 11″·01.

387. No. 712. The Dec. is 9° 31′.

391. No. 725. O.Σ.'s first angle is 114°·4.

393. No. 741. The remarks of H_1, H_2, and So. refer to No. 744, *i.e.* to Flamsteed's 33 Pegasi, not to Piazzi's XXII. 33 Pegasi.

398. No. 757. Ta.'s last distance is 3″·86.

402. No. 775. O.Σ.'s first angle is 299°·4.

404. Col. 1, last line, read — 0″·020.

405. No. 799. O.Σ.'s angle is 295°·2.

422. In the second and twenty-sixth lines, for systematic read accidental, or, average amount of the accidental errors.

B.—The Less Important Corrections.

178. No. 13. The date of the last measure is 1876·93.

179. No. 17. The date of H_2 and So.'s measures should be 1823·87.

181. Col. 1, first line, read H_1, and H_1's angles are 62°·1, 70°·8.
Col. 2, Peters's angle is 112°·7.

182. Col. 1. Du.'s third distance is 6″·02. In No. 27, read — 0″·01 in Du.'s formula: in the measures Du.'s last date should be 1872·09. The corrected mean distance is 3″·08.

183. No. 32. Read " Discovered by Σ." ; in last line but four, read H_2.

185. No. 36. For He. read De.

DOUBLE STARS.

Page 186. No. 39. Du.'s corrected means are 159°·8, 30‴·07, 1869·05.

No. 44. Expunge Mädler's second measure of angle and distance.

189. No. 54. H₁'s distance, more correctly, 15″·82.

191. Col. 2. Σ. s second angle is 88°·2.

192. No. 68. Mädler's third and last distances are 8‴·53, 8‴·70.

194. Col. 1, last line, read 1802·08.

Col. 2. Dele Be.'s second measure.

195. Col. 1. Du.'s corrected means are 325°·8, 2‴·84, 1871·17. W. and S.'s last date is 1873·01.

196. Du.'s corrected means are, A B. 62°·8, 10‴·16, 1871·33 ; B C. 110°·6, 0″·61, 1871·01.

198. No. 80. In Du.'s formula for 4″·01243, read 0″·01243.

199. No. 87. The first angle should be 194°·7, and Mä. should be one line lower. In No. 85, Du.'s formula should read 1854·50 in the second line.

200. No. 91. O.Σ.'s angle should be 341°·1.

201. No. 92. For Da. read De.

No. 93. H₁'s date should be 1783·6. In No. 97, for O.Σ., read Σ. and O.Σ. In No. 98, last line, read 0‴·104 ; and the date of H₁'s measure is 1783·66.

203. No. 101. W. and S.'s last measure belongs to Gl.

204. No. 107. Near the end of the fourth paragraph, read "from ε in the micrometric." Du.'s corrected means are 199°·3, 1‴·04, 1872·13 ; and W. and S.'s fourth, sixth, and last angles are 197°·5, 201°·2, 200°·7.

208. Col. 2. The second date is 1822·08.

209. No. 136. In Σ.'s last measure read, "Wedge-shaped, stars not separated."

211. No. 145. The first distance is O.Σ.'s.

212. No. 149. B C. The last distance should be 4″±.

213. No. 157. In O.Σ.'s measures for "in contact," read "not distinctly seen." The last date is 1874·07.

214. Col. 1, line 18, read Baily for Bessel. In No. 159, De.'s angle should be 34°·9.

216. No. 173. The third angle and distance are O.Σ.'s. In No. 176, Du.'s means are 219°·2, 1‴·40, 1871·68 ; and in No. 175, his first distance should be 0‴·95, and his second 1‴·22.

220. No. 195. H₁'s distance is 16″·08.

221. No. 201. A B. H₁'s first date is 1780·13, and similarly for A C, D C ; for D B it is 1780·5 ; for O.Σ.'s fourth angle in A E read 350°·2.

226. No. 230. For Σ. read O.Σ.

228. No. 240. In A C, O.Σ.'s angle is 13°·6 ; col. 2, sixth line from bottom, read p. 319. In No. 241, Σ.'s angle is 93°·2.

230. Col. 2, eighth line from bottom, for Bessel read Baily.

232. No. 246. A B. Du.'s corrected means are 68°·6 and 11‴·26, 64°·0 and 11‴·21, 59°·8 and 11‴·14, 60°·9 and 10‴·65, 57°·1 and 10‴·81.

Page 236. No. 262. Σ.'s distance should be 0″·51.

238. Col. 1, line 9, for S read L.

240. Col. 2. Sm.'s distances are 4″·7, &c., and Encke and Galle's 5″·28, &c.

243. Line 32, for Bessel read Baily.

248. Col. 2, last line but one, read minimum.

250. Col. 1. A C. Lower H₂ and So. one line.

252. No. 294. Secchi's angle should be 147°·99.

253. Col. 1. The first angle should probably be 33°·8 (*see* Mem. R. A. S. vol. 35, p. 46).

254. No. 300. The last measure of W. and S. belongs to Gl.

257. No. 313. The last measure belongs to De.
No. 315. The first date should be 1782·4.

260. Col. 1. The first date should be 1782·8.

263. Col. 2, line 13, read Baily for Bessel.

265. Col. 1, first distance, read 111″·38.

266. No. 348. For O.Σ. read Σ.

267. No. 355. The decimal figures of O.Σ.'s last two distances should be ·53, ·69.

269. Col. 1, last line but five, read "due to the application."

276. No. 372. In Duner's formulæ, read 1851·37 and 1830·0.
No. 373. H₁'s distance is 82″·7.

279. No. 387. In the last line of Du.'s formulæ, read + 0″·0995.

280. No. 396. The last measure is by O.Σ. Col. 2. Σ.'s angle is 198°·1. In No. 399, A C, the distance should be 102″·88.

282. Col. 1, last line but three, read H₂.

286. No. 406. In Du.'s formulæ, read − 0°·24 ; Σ.'s distances are 32″·73, 32″·45.

291. Col. 2. No. 424. O.Σ.'s second distance is 0″·71, and his last date 1859–72. In No. 425, sixth line, read H₂'s 20 ft.

299. Col. 2. First date should be 1780·1.

300. Col. 2, last line, read + 0°·26.

301. Col. 2, line 11, read 1¾.

302. Col. 1. Last distance should be 2″·86.

303. No. 467. The fourth date should be 1830·37.

304. No. 468. Dele De.'s eighth measure.

316. Col. 2. The fourth angle should be 189°·9.

319. No. 501. In Du.'s formulæ, read + 0°·137, and in No. 502, O.Σ.'s distance should be 33″·39.

322. No. 507. The magnitudes should be 7·5, 8.
No. 508. A B and C. The date of H₁'s angle (339°·4) is 1782·29.
No. 508. A B and C, Ja.'s distance is 40″·57.

327. Col. 1, line 2, read H₂. No. 520, line 7, Mitchell should have but one l.

333. No. 532. In the first line of Du.'s formulæ, read 1852·5.

335. Col. 1, last line but 10, read H₁ for Σ.

336. Col. 1. The first angle should be 42°·2.

Page 337. Col. 2, line 5, for H₂ and So., *Phil. Trans.* 1824, p. 271, read *Phil. Trans.* 1804, p. 365.

343. No. 552. Dele 221 (B.)

345. No. 559. H₁'s distance is 19″·67, and the date of the estimated angle is 1782·65.

No. 560. Σ.'s first angle should be 356°·7, and Mä.'s 354°·5.

347. No. 565. O.Σ.'s corrected distances are 0‴·63, 0″·62, 0‴·47, 0″·51.

348. Col. 1. Raise Σ two lines.

350. Col. 1, line 7, ·read Baily for Bessel.

353. No. 576. The first angle should be 267°·2. Sm.'s third angle 259°·5.

358. Col. 1, line 4. The distance should be 0″·33.

359. Col. 1. The first date should be 1779·83.

360. Col. 2. In Du.'s formulæ, read 1853·64.

364. Col. 1, last line but four, read ccxxxi. This star is 11 Aquilæ.

368. No. 636. Sm.'s epoch should be 1833·78.

No. 640. The last distance should be 23″·88, and Sm.'s second measure should be Mä.'s.

No. 641, line 6, – 0°·09.

370. No. 652. De.'s measures should be 141°·4, 0‴·6, 1866·3.

373. No. 656. In Du.'s formulæ read 1830·0 for 1850·0.

No. 659. H₁'s distance 143″·3, 1781·82.

376. No. 679. A B. For W. O. read W. and S.

379. No. 688. The last distance should be 68″·99.

380. No. 693, line 10, read 1842.

382. No. 702. In Du.'s formulæ, for A B, read 1855 throughout.

384. No. 708. In A B, De. should be Se., Da. should be De., and the last epoch 1864·79.

387. No. 712. Mag. of B. is 5.

388. No. 713. The R. A. should be 21 h. 9·3 m.

389. No. 720. De.'s measures were on one night.

391. No. 726. The second date should be 1836·66.

401. Col. 1, line 5, read 178°·2.

402. Col. 1, line 3, read fourth, for quarto.

404. Col. 1, last line but five, read Dr. for D₁. In first line of col. 2, read 320°·7.

405. No. 802. For W. read S., also in Nos. 800, 809, 827, 830, 839, 862, 880, 883, 885, 887, 901, 902 ; and in Nos. 803, 836, 843, 846, 866, 886, read W. and S.

412. No. 1044. The measures are by Holden.

413. No. 1062*p*. A B and A C should be added to 577, and the 21 (Bu.'s number) should be lowered a line.

414. Note c. For H₁ IV. 44, read H₁ VI. 44 ; for (= Str. 376) read (= Sh. 376).

417. Col. 1, line 9, read 1663 ; line 22, add "and independently by Bu. and De."

418. Line 7, for 10,320, read 10,317.

PART IV.

Page 426. Line 5. Expunge γ, and read Lalande No. 21258—*Ast. Nachr.*, &c.

Line 13. For Behrman read Behrmann.

429. Line 6 from bottom. For vol. xlii. read vol. xliii.

430. Line 5. For vol. lxxii. read vol. lxxiii.

Line 18, read Σ. 1757 for Σ. 175 ; in last line, read Markree.

Line 3 from bottom. For vol. cxi. read vol. xci.

432. Line 2. For vol. lxxvii. read vol. lxxviii.

Line 23, read Trabanten for Trabauten.

438. Line 7. For vol. lxxiv. read vol. lxiv.

Line 4 from bottom add, edited by Labaume.

440. Last line. Add vols. xx.-xxxvi.

441. Line 6 from bottom. Dele one l in Mitchell.

443. Line 8 from bottom. For vol. v., 1863, read vol. vi., 1862–1865, p. 143.

Last line, for Anst., read Aust.

457. Line 18, for Naturwissenschaftichen, read Naturwissenschaftlichen ; line 23, for auderer, read anderer ; last line, the last syllable of Meteorologischen is -es not -en ; *dele* final e in Jahrbuche.

459. The first measure is by Doberck ; for p. 212, read 213 ; next line, for 1873·77, read 1873 to 77.

NOTES, CRITICISMS, &c.

Page 24. From the list of necessary and useful books *Chambers's Handbook of Astronomy* (Clarendon Press Series), was accidentally omitted. For clear, full, and accurate information on almost all astronomical matters it is without a rival.

38. Sir Wm. Herschel used several telescopes while making his famous "sweeps" and "reviews" in search of missing stars, nebulæ, double stars, &c. Those most used were as follows;—the 7 ft. Newtonian reflector, aperture $4\frac{1}{2}$ in., power 222; the 7 ft. $1\cdot2$ in. reflector, aperture $6\cdot2$ in., powers 227 and 460; the 20 ft. Newtonian reflector, aperture $8\frac{7}{10}$ in.; the 20 ft. reflector, aperture 12 in.

39. Sir John Herschel's Cape Reflector was a "Front-view."

176. No. 4. Doberck has
$$\theta = 55^\circ\cdot78 - 0^\circ\cdot502\ (t - 1850). \quad \rho = 0''\cdot660.$$
No. 5. Doberck's formulæ are
$$\theta = 112^\circ\cdot82 - 0^\circ\cdot637\ (t - 1850). \quad \rho = 0''\cdot516.$$

177. No. 6. $162^\circ\cdot4$ $0''\cdot44$ $1878\cdot91$ Bu.
No. 6. Doberck gives
$$\theta = 201^\circ\cdot75 - 1^\circ\cdot198\ (t - 1850). \quad \rho = 0''\cdot447.$$

178. No. 12 is 42 Piscium; No. 13 is Cass. 49; No. 15 is λ Cass.
No. 16.

	$^\circ$		$''$		
A B	$131\cdot6$	2n.	6.36	$1878\cdot8$	Bu.
A C	$163\cdot0$	,,	$29\cdot06$	,, $\cdot8$	,,
A D	$180\cdot0$	,,	$41\cdot25$	,, $\cdot8$	,,

The C in these measures is one of Bu.'s new stars.

179. No. 17 is 51 Piscium.

No. 21. Doberck has
$$\theta = 94''\cdot10 + 0^\circ\cdot720\ (t - 1850). \quad \rho = 1'''\cdot420.$$
No. 22. $114^\circ\cdot4$ 2n. $9''\cdot58$ $1878\cdot8$ Bu.

182. No. 29. This star is Brisbane 114—116 (where it is called λ¹ Tucanæ); it is 2 Dunlop; Rumker 29; H₂ $3407\frac{1}{2}$. According to the *Uranometria Argentina* it is λ¹ Toucanæ: Houzeau in his *Uranométrie Générale* styles it λ¹; according to the B.A.C. it is neither λ¹ nor λ². Gilliss gives the following:
$$75''\cdot5\ ;\ 21''\cdot2\ ;\ \text{mag. } 7\cdot5,\ 8\ ;\ 1851\cdot85.$$
(the angle and distance computed graphically by E. B. Merriman).

183. No. 31 is 66 Piscium; No. 32 is 36 Andr.

Page 183. No. 31. Doberck has

$$\theta = 74°·17 - 1°·537 \ (t - 1850). \quad \rho = 0''·630.$$

184. No. 34. So. 390. A comparison of the observations of this star as given by Secchi in the *Descrizione del Nuovo Osservatorio* and the *Catalogo delle Stelle Doppie* shows that the author has transferred the angle due to So. 390 to χ¹ Ceti. The correct observation is as follows: 33°·17, 6''·359, 1855·95. Subtracting 180° from the angle by the C.O. we have 34°·2. Hence probably no change in this star (Sadler).

185. No. 36. 35°·2 2n. 0''98 1879·1 Bu.
Bu. can find no other pair at or near the place given for this star by O.Σ.
No. 37. The date of H₁'s distance is 1783·64.
No. 39 = H₁ iv. 116 = H₁ iv. 9, and H₁ has 158°·4, 28''·98, 1783·03.

186. No. 40. Doberck's formulae are

$$\theta = 324°·55 - 0°·147 \ (t - 1850) + 0°·00036 \ (t - 1850)^2.$$
$$\rho = 3''·694 + 0'''·0090 \ (t - 1850).$$

No. 41 is φ Andr. 272°·3, 0''·28, 1878·8, Bu.
No. 42. The proper motion is secular; the date of H₁'s distance is 1780·65.

187. No. 48 is 42 Ceti.
Doberck's formulae are

$$\theta = 339°·35 + 0°·444 \ (t - 1850) - 0°·00207 \ (t - 1850)^2.$$
$$\rho = 1''·225 + 0'''·0057 \ (t - 1850).$$

188. No. 50 is ψ Cass. The date of H₁'s distance is 1783·26.

190. No. 57. H₁'s angle is probably 5° in error.
No. 59. The small star C, mag. 14, was discovered by the Washington observers in 1875. There is also a faint star D measured by Jacob, mag. 11½ of H₂'s scale:

69°·3 74''·44 1845·91

For A C in the text, read A D, last line but three in col. 2, p. 190.
Doberck's formulae are, for A B,

$$\theta = 25°·65 + 0°·295 \ (t - 1850). \quad \rho = 1''·493.$$

191. No. 62 is ψ' Ceti.
192. No. 65. Doberck gives

$$\theta = 113°·86 + 0°·398 \ (t - 1850). \quad \rho = 0''·500.$$

No. 66. Doberck's formulae are

$$\theta = 251°·56 + 0°·288 \ (t - 1850). \quad \rho = 2''·056.$$

193. No. 70. Bu. has 349°·0, 0''·31, 1878·87. No. 72. Sm.'s dist. of A D probably erroneous.
No. 71. Doberck's formulae are

$$\theta = 33°·60 - 0°·160 \ (t - 1850). \quad \rho = 1''298·.$$

195. No. 75 is γ Andr. H₁ has 62°·4 for the angle in 1804·1.
197. No. 76 is 10 Arietis. No. 78. C was discovered by Dawes.
198. No. 80 is Andr. 259; No. 81 is 66 Ceti. For No. 83 Doberck's formulae are $\theta = 233°·44 - 0°·312 \ (t - 1850). \quad \rho = 0''·733.$

Page 199.　No. 88.　　$57^n\cdot5$　　$0''\cdot62$　　$1878\cdot9$　　Bu.

No. 90.　Doberck gives

$$\theta = 269°\cdot55 - 0°\cdot225 \ (t - 1850).$$
$$\rho = 1''\cdot984.$$

200.　The date of H_1's angle of A C is $1782\cdot5$.

No. 91.　H_1's angle is an estimation.

201.　No. 95.　Du.'s corrected distance $1''\cdot28$.

202.　No. 99 is γ Ceti ; No. 101 is Arietis 114.

No. 100.　Doberck has

$$\theta = 300°\cdot67 + 0°\cdot151 \ (t - 1850).$$
$$\rho = 2''\cdot992.$$

203.　No. 102 is π Arietis.　Probably a confusion between A B and A C
by H_1.　His measure of A B probably belongs to A C.　For A B,
H_1 has $122°\cdot7$, $1803\cdot45$.　It is H_1 I. 64.　(See *Mem. R. A. S.*, vol.
xxxv., p. 38.)

No. 104.　Doberck gives

$$\theta = 296°\cdot86 + 0°\cdot172 \ (t - 1850) - 0°\cdot00052 \ (t - 1850)^2.$$
$$\rho = 1''\cdot440 + 0''\cdot0043 \ (t - 1850).$$

205.　No. 111 is 52 Arietis.

No. 112.　Doberck gives

$$\theta = 230°\cdot00 - 0°\cdot745 \ (t - 1850). \quad \rho = 0''\cdot980.$$

No. 116　　　$136°\cdot2$　　$0''\cdot44$　　$1878\cdot8$　　Bu.

Doberck gives

$$\theta = 151°\cdot65 - 0°\cdot645 \ (t - 1850). \quad \rho = 0''\cdot490.$$

206.　No. 117.　Doberck's formulæ are

$$\theta = 272°\cdot00 - 0°790 \ (t - 1850) - 0°\cdot01610 \ (t - 1850)^2.$$
$$\rho = 0''\cdot710 - 0''\cdot0145 \ (t - 1850).$$

No. 118.　Doberck's formulæ are

$$\theta = 272°\cdot83 - 0°\cdot763 \ (t - 1850) + 0°\cdot00734 \ (t - 1850)^2.$$
$$\rho = 0''\cdot697 + 0''\cdot0067 \ (t - 1850).$$

No. 119.　There is a third star, mag. $11\cdot5$; $223°\cdot3$, $25''\cdot55$, $1829\cdot9$
Σ.

No. 120.　Doberck's formulæ are

$$\theta = 84°\cdot51 - 0°\cdot522 \ (t - 1850).$$
$$\rho = 1''\cdot210.$$

No. 122.　Mädler's angle is thus given in his great work, *Die
Fixstern-Systeme*, part I., p. 20.　Probably it should be $108°\cdot9$.

No. 124.　Mädler's angle should be $174°\cdot6 + 180°$.

207.　No. 128.　Doberck's formulæ are

$$\theta = 257° \ 80 - 0°\cdot687 \ (t - 1850) - 0°\cdot00481 \ (t - 1850)^2.$$
$$\rho = 0''\cdot500 - 0''\cdot0035 \ (t - 1850).$$

Probably the instrument used by H_2 and So. was too small to show A
double.　The common proper motion (secular) of A B is $+ 3''\cdot5$
and $- 4''\cdot7 \ (\Sigma.)$.　The third paragraph relates to A C.

208.　No. 128.　$\dfrac{A\ B}{2}$ and C.　The angle by H_2 and So. in 1821 is probably
$10°$ in error : the distance was estimated.　(See *P. M.* p. ccxxvi.)

Page 208. No. 129. This is H_1 III. 78. H_1 has $357°\cdot9$, $7''\cdot17$, $1783\cdot05$.

210. No. 139. H_1 has $347°\cdot3$, $5''\cdot8$, $1804\cdot099$. This is 32 Eridani.

No. 143. Bu. has $203°\cdot8$, $1''\cdot08$, $1878\cdot9$. For No. 141, Doberck's formulæ are $\theta = 7°\cdot66 + 0°\cdot814\,(t - 1850)$, and $\rho = 0''\cdot819$.

211. No. 145. Mädler's angle is correctly quoted from the *Dorpat Obs.*, vol. xiii., p. 67, but he adds, "ungewiss" (uncertain). Probably it was $328°\cdot1$.

No. 148. Bu. has $288°\cdot4$, $0''\cdot49$, $1878\cdot8$.

No. 149. Doberck is at present engaged on the orbit of this star. On calculating the circular elements from four angles he found the resulting orbit would not represent the observations at all. The orbit is therefore very eccentric and not circular, as he had been led to suppose from a provisional graphical construction.

212. No. 151. This is 55 Tauri. Doberck gives

$$\theta = 27°\cdot27 + 1°\cdot077\,(t - 1850). \quad \rho = 0''\cdot620.$$

No. 153 is Tauri 230.

213. No. 154 is 56 Persei.

No. 155. Doberck has

$$\theta = 226°\cdot32 - 1°\cdot758\,(t - 1850). \quad \rho = 0''\cdot980.$$

No. 156. $11°\cdot6$, $4''\cdot2$, $1879\cdot10$. C. O. The angle was "computed by means of Σ.'s distance ; computing the personal equation by means of β's distance we have $P = 11°\cdot1$." (*Publications of the Cincinnati Observatory*, 1878–79.)

No. 157. Bu. has $7°\cdot4$, $0''\cdot57$, $1879\cdot0$.

214. No. 160. Doberck's formulæ are

$$\theta = 309°\cdot40 + 0°\cdot335\,(t - 1850) - 0°\cdot00175\,(t - 1850)^2.$$
$$\rho = 1''\cdot585 + 0''\cdot0083\,(t - 1850).$$

No. 161. Doberck has

$$\theta = 304°\cdot32 - 0°\cdot326\,(t - 1850). \quad \rho = 1''\cdot620.$$

No. 162. Du.'s corrected distance is $3''\cdot29$.

215. No. 163. Doberck's formulæ are

$$\theta = 268°\cdot80 - 0°\cdot360\,(t - 1850). \quad \rho = 1''\cdot563.$$

No. 168. South quotes H_1's angle as $5°\cdot1$ and his own as $173°\cdot8$ or $185°\cdot1$. In the *Phil. Trans.* vol. 75, H_1's angle is given $174°\cdot9$. Hence probably no change had taken place (*Mem. R. A. S.*, vol. 35, p. 321). Mr. Sadler has kindly examined H_1's MSS. and finds that they confirm the latter angle.

Doberck has

$$\theta = 176°\cdot93 - 0°\cdot161\,(t - 1850). \quad \rho = 2''\cdot520.$$

216. No. 171 is 5 Aurigæ. No. 173. De. says "white," O.Σ. "reddish." Doberck has

$$\theta = 230°\cdot56 + 0°\cdot581\,(t - 1850). \quad \rho = 2''\cdot755.$$

No. 172. Doberck has

$$\theta = 68°\cdot04 - 0°\cdot377\,(t - 1850). \quad \rho = 1''\cdot080.$$

No. 173. Doberck has

$$\theta = 348°\cdot63 - 0°\cdot607\,(t - 1850). \quad \rho = 0''\cdot610.$$

B

Page 216. No. 176 is H$_1$ I. 45.   H$_1$ 222°·7, 1782·87.

217. No. 181 is ι Leporis.   "Probably an error of 1 rev. (24°) in H$_1$'s reading : hence 335°·3.   The second angle is merely an estimation."—
(Sadler).

No 182 is 14 Aurigæ.   Of A B Du.'s corrected distance is 14"·74.

218. No. 185.   Doberck's formulæ are

$$\theta = 271°·51 — 0°·433 \, (t — 1850).$$
$$\rho = 1"·758.$$

β Orionis (Rigel) may here be added.   Its place for 1880 is R.A. 5h.
8·8m., Dec. - 8° 20', A B.   The principal star forms with the 8th
mag. companion the well known pair H$_1$ II. 33 = Σ. 668. Σ. has
199°·77, 9"'·137, in 1831·53.

In 1871, with the 6-in. refractor, Burnham suspected an elongation of
B.   Mr. Sadler had also been led to think that B was a close double
before 1876.   After numerous trials with the 6 in. and 18$\frac{1}{2}$ in. Mr.
Burnham at last satisfied himself of the duplicity of this object.   His
mean result is

B C   179°·0, 0"·35, 1878·16.

Mitchel detected a faint and distant star, mag. 12-13.   For this
we have

A D   1"·5, 44"'·47, 1877·82.   Bu.

219. No. 192.   Doberck gives

$$\theta = 52°·43 + 0°·149 \, (t - 1850). \quad \rho = 2"·992.$$

No. 194 is 118 Tauri.

220. No. 195.   A C, date of H$_1$'s angle 1783·74.   No. 197 is 32 Orionis.
Bu. has 197°·1, 0"'·52, 1879·0.

221. No. 200.   Doberck has

$$\theta = 250°·88 + 0°·187 \, (t - 1850) - 0°·00032 \, (t - 1850)^2.$$
$$\rho = 3"·330 + 0"·0057 \, (t - 1850).$$

No. 201.   The following notes on this remarkable system will be read
with interest by the amateur.

Huyghens in 1656 gave a figure of these stars, but only three are
shown.   Mairan in 1731 and Messier in 1771, gave the four as we
now see them.   (See *Mem. R. A. S.*, vol. iii., pp. 187 to 189 and
213 ; also vol. ii., pp. 487-495.)

THE FIFTH STAR.—Σ. discovered it in 1826.   He thought it was
not variable, and that it was seen or not according to the state of the
air.   Sometimes he could not see it at midnight and yet it was quite
an easy object at other times in the strong morning twilight (*M. M.*, p.
242.)*   H$_2$ examined the four stars and their immediate neighbourhood
with great care seven months before Σ. discovered this star, but saw
no trace of any small stars.   Found it quite easy in 1827 and 1828.
Thought it was a new star and certainly variable in light (*Mem.
R. A. S.*, vol. iii.)   Dawes always found it of uniform brightness.
Thought it neither new, nor variable to any considerable degree.

* It has been seen after sunrise in the 15 in refractor at Cambridge, U.S.

Page 221. Found it a somewhat difficult object and of a "reddish (?) yellow."—*Mem. R. A. S.*, vol. xxxv.

Lassell observed that the fifth was less bright or smaller than the sixth in 1856.—*M. N.*, vol. xvii.

O. Σ.'s observations also show that the fifth was equal in brightness to the sixth prior to 1857, that the fifth then became the brighter, and that one of them is probably a variable.—*Poulkova Observations*, vol. ix., *M. N.*, vol. xvii.

Magnitude.—Σ. 11·3 ; O.Σ. 10 ; Dawes, 10¾ ; H₂ 17 in 1828 and 12 in 1834.

Colour.—Dr. Copeland says it is "of an intense red colour" in the Parsonstown 6-ft. reflector. Mr. Sadler's observations confirm this.

THE SIXTH STAR.—This difficult object was discovered by H₂ in 1830 while looking at the nebula with Sir James South's 11¾-in. refractor.—*M. N.*, vol. xvii.

Lassell discovered it independently in 1841.—*Astronomical Register*, 1865.

Dawes usually found it not very difficult to see with an object-glass of 6⅛-in. aperture, in a fine state of the air ; suspected variability to the extent of 1 or 1½ mag. (*Mem. R. A. S.*, vol. 35). O.Σ. thinks it is a variable and that it equalled the fifth in brightness prior to 1857. Lassell found it brighter than the fifth in 1856, and De la Rue has at times seen it much brighter that the fifth.

Σ. could never see it at Dorpat, where its meridian altitude was 26°.

Magnitude.—Dawes, 12 ; O.Σ. 11·3 ; H₂, 14 in 1834.

Aperture required to show these Stars.—Sir John H. at the Cape of Good Hope often examined them ; he says, "Small as this star is, if the state of the atmosphere be favourable, it does not require the full aperture of 18 inches to render it visible." De la Rue has seen it with a 4⅛-in. refractor, and Burnham with 3¾-in.* Our experience at Bermerside shows that the visibility of these objects depends much more on the state of the air than on aperture. If they are visible at all any aperture down to five or six inches will show them.

Other faint Stars within or near the Trapezium.—Porro in 1857 records a minute star between A and B ; Dumouchel two near C.—*M. N.* vol. xvii. (Dumouchel was De Vico's assistant.)

Huggins, in 1866, saw nine stars in all,. viz., the four bright ones, the fifth and sixth, one between A and B, one between B and D and a little within the line joining them, and one near the centre of the trapezium.

---

* The following results are given by Captain Noble in the *Astronomical Register* for April 1880. *The 5th star* : 1857—steadily seen in the autumn; got fainter towards the end of the year. In the beginning of 1858 it was visible in bright moonlight. In January 1859, it was seen in an illuminated field : in November it was much fainter. It was very bright and conspicuous in December 1863. Early in 1875 it was fainter, but bright early in 1876. *The 6th star :* It was nearly as bright as the 5th in January 1874, while in December 1876 it was brighter than the 5th. The instrument used was a fine 4·2-in. refractor by Ross and Dalmeyer.

Page 221.   Bond, De Vico, Lassell, and others, have also seen faint stars within
          or on the borders of the trapezium.   De Vico records three within it
          in 1839.   The instrument used by De Vico and his assistant was a
          refractor of 6¼ in. by Cauchoix, power 68.

             On these statements O.Σ. remarks : (1) That he could never see
          the stars with his large telescope when studying the nebulæ. (2) That
          neither H₂ nor Lamont when working for years on the same object
          ever suspected their existence.   (3) That De Vico could not see
          the 6th star.

             Lastly, Mr. Burnham has most carefully, on several fine nights,
          examined this object with apertures ranging from 6-in. to 18½-in.,
          without ever suspecting the existence of any other than the six well-
          known stars.   Nor do his observations favour the supposed variability
          of either the fifth or sixth.

             See *Nature*, vol. xxi., pp. 117, 286, for an interesting account
          of the small stars seen within and near the trapezium, by Bond,
          &c.

             The last measure under A B belongs to D C.

222.   No. 202.   Doberck's formulæ are

             $\theta = 193°\cdot60 — 0°\cdot470\ (t — 1850)$.

             $\rho = 0''\cdot625$.

223.   No. 203.   Doberck has

             $\theta = 86°\cdot75 - 0°\cdot362\ (t - 1850)$.

             $\rho = 0''\cdot620$.

          No. 206.   A C.   The angle by W. and S. should probably be 9°·7 or
             8°·7.   Copeland has 8°·84, 59″·6, 1874·1.

          No. 206 is ζ Orionis = H₁ iv. 21.   " B was discovered by Kunowsky
             in 1819 with a 4·3-in. refractor."—(Sadler).

224.   No. 210.   De.'s distance is 0″·5.

          Doberck has

             $\theta = 304°\cdot72 + 0°\cdot568\ (t - 1850)$.

             $\rho = 0''\cdot595$.

          No. 212.   The star has been found double by O.Σ., No. 545.

°	″		
5·5	2·15	1871·4	O.Σ.
3·2	·05	5·9	De.
3·8	1·90	8·0	Bu.

          Bu.'s recent measure of A B is

             292°·7   45″·51   1878·0

          No. 214.   Mädler's angle is taken from the *Dorpat Observations*, vol.
             xiii.   Probably a misprint for 353°·6.

225.   No. 219.   Doberck has

             $\theta = 315°\cdot30 + 0°\cdot241\ (t - 1850)$.

             $\rho = 1''\cdot642$.

226.   No. 226 is 5 Lyncis.

          No. 229.   This is Σ. 919, 11 Monocerotis.   Three other companions
             have been seen.

Page 226.

AD	$56 \cdot 1$	$25 \cdot 79$	5, 12–13	$1878 \cdot 0$ Bu.
AE	$190 \pm$	$85 \pm$	,, $11 \cdot 5$	$1875 \cdot 3$ Sadler.
BF	$336 \cdot 5$	251	$5 \cdot 5$, 8	$1824 \cdot 1$ H_2 & So.

No. 230. The first measure is by O.Σ.

227. No. 234. Doberck's formulæ are

$$\theta = 336° \cdot 38 - 0° \cdot 196 \, (t - 1850). \quad \rho = 2'' \cdot 386.$$

No. 237. Doberck's formulæ are

$$\theta = 78° \cdot 81 + 0° \cdot 043 \, (t - 1850).$$

$$\rho = 1'' \cdot 935.$$

228. No. 240 is 15 Monoc. Dawes wrote, "The very minute distant star I have not measured, though it was seen at an angle of about $12° \cdot 6$ and distance $11''$ ± [this is A C], and also a fourth in the np quadrant, whose position is about 308°, and distance $40''$ ±."—*Mem. R. A. S.*, vol. xxxv. p. 330.

This is a nebulous star in the centre of a fine cluster. Cf. D'Arrest, "Siderum Nebulosorum Obs. Hav." and Lord Rosse, "Scientific Trans. Royal Dublin Society," ii. [N.S.] p. 53.

A small star of the $11 \cdot 5$ mag. has been measured : Da. $307° \cdot 8$, $40'' \pm$ $1841 \cdot 2$, H_2 240°, $40'' \pm$ $1828 \cdot 05$. Also a variable mag. $8 \cdot 5$, $139° \cdot 2$, $75'' \cdot 7$, $1863 \cdot 16$ (Main) ; of the 12 mag. $1875 \cdot 3$ (Sadler).

No. 242. C.O. $272° \cdot 3$, $1'' \cdot 01$, $1878 \cdot 16$.

No. 243. Doberck has

$$\theta = 144° \cdot 45 - 0° \cdot 515 \, (t - 1850).$$

$$\rho = 1'' \cdot 654.$$

230. No. 246. Recent measures of AB. Bu. $52° \cdot 4$, $10'' \cdot 83$, $1878 \cdot 01$; $50° \cdot 7$, $10'' \cdot 44$, $1879 \cdot 13$; $48° \cdot 3$, $10'' \cdot 00$, $1880 \cdot 11$. W.O. $51° \cdot 7$, $10'' \cdot 76$, $1878 \cdot 24$; $50° \cdot 1$, $10'' \cdot 57$, $1879 \cdot 20$; $47° \cdot 8$, $10'' \cdot 30$, $1880 \cdot 25$. C.O. $50° \cdot 0$, $10'' \cdot 61$, $1878 \cdot 70$.

232. No. 246. Sm. considerably underrated the distance of the small companion he saw; its distance is really more than $1000''$. Mr. Hunt in Feb. 1880 found the distance $1062''$. The star seen by Secchi is probably Marth's star, the distance being erroneous.

233. No. 248. Doberck has

$$\theta = 6° \cdot 10 - 0° \cdot 666 \, (t - 1850). \quad \rho = 0'' \cdot 610.$$

No. 250. De. thus records his observation :—" $1868 \cdot 259$. Wonderful spectacle ! I see the clear yellow disc of A covering a part— about ⅓ of the diameter—of the ashy disc of B. Not a trace of rings." He was greatly astonished when he beheld this phenomenon on placing the images in the centre of the field, and was quite sure that it was not an optical illusion. He adds, however, that the phenomenon may be explained by the peculiar augmentation which the diameter of the disc of a star often receives in warm and moist weather.—*Astr. Nachr.*, No. 1810. Mr. Burnham's third star C is a very faint object, mag. 12—13. Angle $341° \cdot 4$.

Page 233. Doberck has
$$\theta = 330°\cdot57 + 1°\cdot173 \ (t - 1850). \quad \rho = 0''\cdot516.$$

234. No. 252. Gl.'s date is 1874·13, distance 3''·28.

236. No. 259. Doberck's formulæ are
$$\theta = 39°\cdot94 + 0°\cdot212 \ (t — 1850). \quad \rho = 3''\cdot565.$$
C.O. 40°·2, 3''·37, 1879·22.

No. 262. Doberck's formulæ are
$$\theta = 128°\cdot70 + 0°\cdot590 \ (t - 1850) - 0°\cdot00601 \ (t - 1850)^2.$$
$$\rho = 0''\cdot560 + 0'''\cdot0057 \ (t - 1850).$$

237. No. 265. Doberck's formulæ are
$$\theta = 222°\cdot55 + 0°\cdot127 \ (t — 1850). \quad \rho = 1''\cdot363.$$

241. Col. 2. Du.'s corrected distances are 5'''·28, 5''·39, 5''·45, 5'''·50.

242   No. 274. In all probability no star ever existed where Sm. thought
he had seen one of the eighth mag. The object seen by Sm. was
probably Secchi's companion, *i.e.*, E on p. 244. (See *M.N.*, vol.
xxxix., p. 188.)

243. Procyon. Secchi, p. 116, gives 83°·6 and 33''·162. This distance
should be 333''·162. Hence Secchi's companion, Smyth's companion, and E, are probably the same star.

244. No. 275. Doberck gives
$$\theta = 136°\cdot70 + 0°\cdot230 \ (t - 1850). \quad \rho = 1''\cdot393.$$

No. 276. Doberck gives
$$\theta = 145°\cdot04 - 0°\cdot913 \ (t - 1850). \quad \rho = 0''\cdot510.$$

No. 277. The last two distances are probably 1'' in error.
C.O. 237°·0, 19''·84, 1879·18.

245. No. 279. C was probably beyond the power of the 5-in. telescope
used by H$_2$ and So.

No. 284. O.Σ. writes, "Il faut croire que, depuis ce temps [1855] les
deux étoiles se sont éclipsées mutuellement."

246. No. 287. Doberck gives
$$\theta = 300°\cdot49 - 0°\cdot766 \ (t - 1850). \quad \rho = 0''\cdot436.$$

No. 291. Doberck's latest elements are : ☊ = 358° 3', λ = 182° 33',
γ = 18° 31', e = 0·3318, T = 1870·82, P = 59 486 years,
a = 0''·886.

255. No. 305. Doberck gives
$$\theta' = 351°\cdot97 + 0°\cdot236 \ (t - 1850). \quad \rho = 10''\cdot161.$$

No. 308 is σ^2 Ursæ Majoris.

256. No. 310. A comparison of Σ.'s with O.Σ.'s angle shows an increase.
De.'s later measure indicates considerable decrease.

257. No. 313. Bu. has 193°·0, 0''·39, 1879·2 ; No. 315 is 38 Lyncis, p. 258,
No. 317 is Lyncis 157, and No. 316 is 37 Lyncis.

259. No. 321. Doberck's formulæ are
$$\theta = 331°\cdot60 - 0°\cdot220 \ (t - 1850) - 0°\cdot00293 \ (t - 1850)^2.$$
$$\rho = 0''\cdot856 - 0''\cdot0114 \ (t - 1850).$$

260. No. 322, col. 1, line 11. The elements referred to are as follows :—
T = 1842·77, ω = 122°·9, ☊ = 151°·6, i = 65°·4, e = 0·5028,
μ = + 3°·3451, P = 107·62 years.

Page 262. No. 337. 69°·7, 1″·28, 1878·2, Bu. Secchi's distance is probably 1″ in error.

265. No. 339. Doberck's formulæ are
$\theta = 264°·04 + 0°·319 (t - 1850)$. $\rho = 0″·708$.

No. 340. Doberck gives
$\theta = 161°·96 - 0°·669 (t - 1850)$. $\rho = 1″·660$.

No. 342. This is H_1 I. 71. H_1 has 87°·9, 4″ ±, 1782·88. The note on p. 38 of vol. xxxv. of the *Memoirs of the R. A. S.* is incorrect, owing to a mistake of Σ.'s in the *Dorpat Cat.* (See *P.M.* p. C.)

266. No. 344. Doberck has
$\theta = 60°·33 + 0°·302 (t - 1850)$. $\rho = 1″·079$.

No. 346. Doberck's formulæ are
$\theta = 127°·00 - 0°·213 (t - 1850)$. $\rho = 1″·930$.

268. No. 360. H_1 6″·41, 1781·14 ; 7″·80, 1782·27.

No. 361. Doberck's formulæ are
$\theta = 318°·13 - 0°·179 (t - 1850) + 0°·00185 (t - 1850)^2$.
$\rho = 1″·137 + 0″·0118 (t - 1850)$.

No. 362. Doberck's formulæ are
$\theta = 280°·18 + 0°·200 (t - 1850)$. $\rho = 1″·158$.

No. 364. Du.'s mag for C is 12.

275. No. 370. Doberck's formulæ are
$\theta = 336°·07 - 0°·238 (t - 1850)$. $\rho = 4″·805$.

276. No. 373. Secchi's angle clearly should be — 90°.

277. No. 375. Doberck's latest elements are
$☋ = 99° 35', \lambda = 134° 55', \gamma = 54° 27', e = 0·5000, P = 94·406$ years, $T = 1839·10. a = 0″· 80$

No. 376 is 90 Leonis.

278. No. 379. A C. Add

	°	″	
Eng.	142·4	13·42	1865·27
De.	145·4	21·1	71·31

No. 380. Mädler's angle should be + 180°.

No. 380. Doberck's formulæ are
$\theta = 285°·34 - 0°·859 (t - 1850) + 0°·01460 (t - 1850)^2$.
$\rho = 0″·735 + 0″·0125 (t - 1850)$.

No. 381. 14°·5   0″·94   1878·3   Bu.
No. 383. 18°·2, 1″·36, 1879·27, C.O.

No. 384 is 2 Comæ.

279. No. 387. C.O. 93°·5, 43″·97, 1879·35.

No. 388. Doberck's formulæ are
$\theta = 346°·50 - 0°·150 (t - 1850)$.
$\rho = 1″·237$.

No. 391. Doberck's formulæ are
$\theta = 288°·90 - 0°·390 (t - 1850) - 0°·00798 (t - 1850)^2$.
$\rho = 0″·865 - 0″·0177 (t - 1850)$.

280. No. 395. Doberck's formulæ are
$\theta = 66°·68 - 0°·466 (t - 1850)$. $\rho = 1″·877$.

Page 280. No. 398. Doberck's formulæ are

$$\theta = 209°\text{·}25 + 0°\text{·}352\ (t - 1850) - 0°\text{·}00075\ (t - 1850)^2.$$
$$\rho = 1''\text{·}265 + 0''\text{·}0027\ (t - 1850).$$

No. 399. Doberck's formulæ are

$$\theta = 345°\text{·}85 + 0°\text{·}225\ (t - 1850). \quad \rho = 2''\text{·}135.$$

281. No. 401. Doberck's formulæ are

$$\theta = 115°\text{·}15 - 0°\text{·}438\ (t - 1850). \quad \rho = 0''\text{·}792.$$

No. 402. There are two distant stars of about the eleventh mag.

282. No. 405. For a critical examination of the occultation of γ Virginis observed by Cassini, see Prof. Newcomb's "Researches on the Motion of the Moon" in the *Washington Observations* for 1875, vol. 22.

286. No. 407. Doberck's formulæ are

$$\theta = 40°\text{·}46 + 0°\text{·}754\ (t - 1850). \quad \rho = 1'''\text{·}368.$$

The remarks attributed to A. C. I am unable to trace. Probably a mistake.

No. 407. A C is H_1 V. 130 : 126°·8, 31'''·28, 1783·15.

287. No. 408. Doberck has

$$\theta = 230°\text{·}40 + 0°\text{·}585\ (t - 1850). \quad \rho = 0''\text{·}572.$$

No. 414. Doberck's formulæ are

$$\theta = 340°\text{·}80 - 0°\text{·}110\ (t - 1850). \quad \rho = 3'\text{·}410.$$

289. No. 417. Doberck gives

$$\theta = 198°\text{·}71 - 0°\text{·}335\ (t - 1850) + 0°\text{·}00273\ (t - 1850)^2.$$
$$\rho = 0''\text{·}945 + 0''\text{·}0077\ (t - 1850).$$

No. 418. 350°·7, 1'''·29, 1879·4, C.O.

No. 419. For Mizar and Alcor, Powell has 72°·19, 710''·1, 1856. Snyth gives 71°·7, 690'', 1839. Probably Sm.'s distance is 10'' too small.

290. No. 421. Doberck's formulæ are

$$\theta = 326°\text{·}60 + 0°\text{·}385\ (t - 1850) - 0°\text{·}00027\ (t - 1850)^2.$$
$$\rho = 1''\text{·}125 + 0''\text{·}0008\ (t - 1850).$$

291. No. 425. 193°·8 　 2''·68 　 1878·2 　 Bu.

"The only Double Star ever discovered by Sm." (Bu.)

H_2 has 190°·0, $1\frac{1}{2}''\pm$, 1835. The distances of H_2 and Da. are estimations.

No. 427 is 1 Boötis.

292. No. 429 is 84 Virginis.

293. No. 431 is τ Boötis.

°	''		
345·±	20 to 25	1825·00	H_2
349·4	10·10	48·42	Da.
·3	9·36	54·28	,,
351·9	8·71	78·3	Bu.

The stars have a common proper motion. The companion is a variable of short period (Sadler). The colour of the smaller star always appears decidedly red to Mr. Sadler.

Page 293. No. 432. Doberck has

$$\theta = 179°\cdot15 + 0°\cdot7715\ (t - 1850) + 0°\cdot00585\ (t - 1850)^2.$$
$$+ 0°\cdot000176\ (t - 1850)^3,$$
$$\rho = 3''\cdot230 - 0''\cdot0245\ (t - 1850) - 0''\cdot00055\ (t - 1850)^2.$$

No. 433. Doberck's formulæ are
$$\theta = 60°\cdot80 + 0°\cdot390\ (t — 1850).$$
$$\rho = 2''\cdot514.$$

294. No. 435. Doberck's formulæ are
$$\theta = 107°\cdot60 + 0°\cdot213\ (t - 1850) - 0°\cdot00242\ (t - 1850)^2.$$
$$\rho = 0''\cdot775 + 0''\cdot0088\ (t - 1850).$$

No. 436. Doberck's formulæ are
$$\theta = 71°\cdot11 + 0°\cdot130\ (t — 1850).$$
$$\rho = 2''\cdot755.$$

No. 438. Doberck gives
$$\theta = 142°\cdot60 - 0°\cdot904\ (t - 1850) + 0°\cdot00646\ (t - 1850)^2.$$
$$\rho = 0''\cdot420 + 0''\cdot0030\ (t - 1850).$$

295. No. 443. Doberck's formulæ are
$$\theta = 53°\cdot62 + 0°\cdot429\ (t - 1850) + 0°\cdot00077\ (t - 1850)^2.$$
$$\rho = 2''\cdot336 - 0''\cdot0042\ (t - 1850).$$

No. 444 is κ Boötis.

296. No. 446. Doberck's formulæ are
$$\theta = 182°\cdot50 - 0°\cdot172\ (t - 1850) + 0°\cdot00017\ (t - 1850)^2.$$
$$\rho = 3''\cdot825 + 0''\cdot0037\ (t - 1850).$$

No. 448. There is a third star (*h.* 1251) mag. 14, distance 19''·16, angle 69°·6, 1873·3 (Bu.)

For A B Bu. has 131°·4, 0''·65, 1878·3.

No. 450. Enormous errors in the early observations led to the belief that rapid changes had taken place. (See *Mem. R. A. S.*, vol. xxxv. p. 372 ; also vol. vi., p. 81.)

No. 451. Doberck's formulæ are
$$\theta = 319°\cdot71 — 0°\cdot361\ (t — 1850). \quad \rho = 1''\cdot428.$$

300. No. 461. Mädler's angles are probably largely in error.

303. No. 467. Doberck's formulæ are
$$\theta = 266°\cdot95 - 0°\cdot252\ (t - 1850) - 0°\cdot00250\ (t - 1850)^2.$$
$$\rho = 1''\cdot047 - 0''\cdot0104\ (t - 1850).$$

No. 468. H_1's MSS. show that the discovery of this star was made on the 9th of April, 1780 ; the angle was first measured 1782·285. His first distance should be 3''·23. (Sadler.)

305. No. 469. Doberck's formulæ are
$$\theta = 282°\cdot40 + 1°\cdot842\ (t - 1850) - 0°\cdot04831\ (t - 1850)^2.$$
$$\rho = 0''\cdot407 + 0''\cdot0107\ (t - 1850).$$

308. No. 481. Σ. in 1836 frequently suspected A of being oblong : to O.Σ. it has always appeared round.

No. 482. Doberck's formulæ are
$$\theta = 40°\cdot88 - 0°\cdot205\ (t - 1850) + 0°\cdot00055\ (t - 1850)^2.$$
$$\rho = 5''\cdot704 + 0''\cdot0154\ (t - 1850).$$

No. 484. A, a close double ? 240°, 1''·0 (Howe).

Page 308.　No. 485. Lamont made the following obs. : 1836, June 28, 266°·7, 0″·92; and for the distant star, 277°·0, 15″·79 : the angle "uncertain."

There is a minute star of the fifteenth magnitude (H_2's scale) :—

$\overset{\circ}{}$	$\overset{''}{}$		
33·6	30±	1832	H_2.
26·3	48·58	1856	Winnecke.

Doberck has recently computed the following provisional elements of η Cor. Bor. :—$\Omega = 25°\ 43'$, $\lambda = 218°\ 36'$, $\gamma = 59°\ 41'$, $e = 0·2667$, $P = 41·562$ years, $T = 1850·792$, $a = 0″·918$.

315. No. 487. Doberck has
$$\theta = 338°·36 - 0°·150\ (t - 1850) - 0°·00089\ (t - 1850)^2.$$
$$\rho = 1″·240 - 0″·0080\ (t - 1850).$$

No. 488. Doberck gives
$$\theta = 325°·20 - 0°·351\ (t - 1850).$$
$$\rho = 1″·560.$$

317. No. 496. This is ζ Coronæ. The date of H_1's distance is 1780·48.

319. No. 499 is π^2 Ursæ Minoris.

No. 503. Doberck's formulæ are
$$\theta = 116°·11 + 0°·816\ (t - 1850) - 0°·00942\ (t - 1850)^2$$
$$\rho = 0″·627 + 0″·0072\ (t - 1850).$$

320. ρ Coronæ Borealis may be added ; the changes are due to proper motion. It is H_1, vi. 93.

R. A. $15^h\ 56·4^m$, dec. $+ 33°\ 41'$, mag. A 6, B 11 (Smyth).

$\overset{\circ}{}$	$\overset{''}{}$		
144·4	87·73	1783·6	H_1.
125·1	79·20	1825·5	So.
92·4	76·1	79·4	C. O.

A is creamy white ; B blue (Smyth).

322. Col. 2. The angle 7°·6 by H_1 is the original measure ; in H_1's printed Catalogue it is 10°·38. (See *Mem. R. A. S.*, vol. xxxv. p. 75.)

This star is κ Herculis.

No. 507. The last two angles are reduced by Du. to the equinox 1850.

No. 508. Mitchel also discovered the duplicity of B.

327. No. 518 is η Draconis.

330. No. 525. Doberck gives
$$\theta = 158°·79 - 0°277\ (t - 1850).\quad \rho = 0″·850.$$

331. The following very accurate elements of ζ Herculis have just been received from Dr. Doberck :—$\Omega = 44°\ 6'$, $\lambda = 251^v\ 47'$, $\gamma = 44°\ 32'$, $e = 0·4666$, $P = 34·411$ years, $T = 1864·78$, $a = 1″·345$.

333. No. 530 is 21 Ophiuchi.

Doberck's formulæ are
$$\theta = 170°·20 - 0°·306\ (t - 1850) + 0°·00345\ (t - 1850)^2$$
$$\rho = 0″·763 + 0″·0086\ (t - 1850).$$

No. 531. Doberck's formulæ are
$$\theta = 331°·68 - 0°·337\ (t - 1850) - 0°·00481\ (t - 1850)^2.$$
$$\rho = 0″·785 - 0″·0112\ (t - 1850).$$

Page 334. No. 535. Kaiser's measure, taken from Fl., really belongs to Σ. 2107. Secchi's distance is so given by himself, but it is no doubt a misprint for 1″·13.

No. 537. Of this star H₁ wrote in 1781 : "One of the most minute of all the double stars I have hitherto found. It is in vain to look for them if every circumstance is not favourable. The observer as well as the instrument must have been long enough out in the open air to acquire the same temperature. In very cold weather, an hour at least will be required ; but in a moderate temperature, half an hour will be sufficient." (*Phil. Trans.* vol. 72, p. 216.) This star is 20 Draconis.

335. No. 538 is P. XVI. 270 ; No. 539 is Herculis 210.

337. No. 540. Mädler's angles should no doubt be — 180°.

°	″		
104·6	7·91	1871·6	De.
106·9	7·85	4·7	,,

339. No. 544. Smyth's small star is H₂ and So.'s 10ᵐ star at 19° 5′ *np.*

No. 545. O.Σ.'s fifth measure is really Σ.'s.

344. No. 555. As Burnham points out (*Mem. R. A. S.*, vol. xliv.) Σ.'s angle is probably 10° in error.

345. No. 559. H₁'s angle is a "mere estimation." This star is 61 Ophiuchi.

346. No. 564. H₁'s angle and distance are estimations.

347. No. 565. Doberck's formulæ are

$$\theta = 304°·20 - 0°·388\,(t - 1850) - 0°·00610\,(t - 1850)^2.$$
$$\rho = 0″·553 - 0″·0087\,(t - 1850).$$

No. 566. Doberck's formulæ are

$$\theta = 39°·90 - 0°·759\,(t - 1850) + 0°·00560\,(t - 1850)^2.$$
$$\rho = 0″·637 + 0″·0047\,(t - 1850).$$

349. No. 571. There are two faint stars C and D : C is of the 12th mag. and is in the *sf* quadrant (Sm.). Se. 67°·2 in 1856·6. "There is also a little *comes* in the sp. preceding A by 5ˢ·5."— (Sm.). Se. 11ᵐ 215°·1, 87″·6, 1856·6. This is D.

355. No. 586. Doberck's formulæ are

$$\theta = 268°·91 - 0°·785\,(t - 1850) - 0°·00726\,(t - 1850)^2.$$
$$\rho = 0″·490 - 0″·0045\,(t - 1850).$$

The following are recent measures

251°·7	0″·31	1878·4	Bu.

No. 587. Burnham suggested that Σ.'s distance should be 19″·85 (*Mem. R. A. S.*, vol. xliv. ; *Astr. Nachr.* 2210). An examination of the original Observations has shown that the true distance was 20‴·5.

No. 590. The mag. of C is 7·1.

Doberck has for A B

$$\theta = 3°·11 - 0°·140\,(t - 1850) + 0°·00048\,(t - 1850)^2.$$
$$\rho = 3″·325 + 0″·0115\,(t - 1850).$$

Page 356.   No. 591.   Doberck gives

$\theta = 66°·76 - 0°·367 (t - 1850).$ $\rho = 0'''·532.$

No. 598 is P. XVIII. 132.

Doberck's formulæ are

$\theta = 358°·10 - 0°·211 (t - 1850) - 0°·00245 (t - 1850)^2.$

$\rho = 0''·690 - 0''·0080 (t - 1850).$

No. 599.   The companion appears to be brightening: De. measured this star ¼ hour after sunset and estimated the mag. of the companion as 8·8, and in 1874, Mr. Sadler measured the angle with ease ¾ hour after sunset.

1837.   Nov. 24, 137°·1, 41'''·38.   Lamont.

For the third faint star, mag. 13—14, we have

$\overset{\circ}{}$	$''$		
361·7	—	1837·9	Lamont.
298·8	46·87	1864·8	Winnecke.
292·0	51·85	78·4	Bu.

358.   No. 602.   O.Σ.'s fourth measure ("observation uncertain") excluded from the mean ; distance 0''·33.

No. 603.   Recent observations are

	Single	1865	De.
	,,	1878	Bu.

361.   AC.   172°·8   207'''·1   1848·47   Bond.

362.   No. 606.   Lamont in 1839 made the following measures :

CF. 339·3 ; CG. 1°·9 ; CE. 35°·6.

The ε Lyræ group may be thus described :—The two binaries ε¹, ε² ; the *debilissima* pair discovered by H_1, and so named by H_2 ; the 9½ mag. star following, discovered by H_1 ; three small stars preceding the southern star of the *debilissima* pair, and a faint star to the north, discovered by Dawes in 1833 ; one to the north of the northern star of the *debilissima*, discovered by Lassell ; and one following the *debilissima* to the south-east, discovered by Ward and Sadler.   In H_1's MSS. we find the following remarks: "Nov. 29, 1782.   Inter ε and *e* Lyræ (Mayer's *e*) the 20 ft. shows me two small stars I have never before taken notice of with the 7 ft.   That which is below [mag. 9·5] I have often noticed." H_1 therefore discovered these faint objects. H_2 first saw the *debilissima* on the 27th of Oct. 1823, with the 20-ft. reflector, aperture 18 in. ; the existence of the pair could not be "even suspected" with the refractors of 5-in. and 3·8-in. aperture ; the 6-in. and 9-in. reflectors also failed to show them.   Nor did H_1 ever notice them with his 6-in. reflector (see *Phil. Trans.* for 1824, part II., p. 313, and H_1's MSS.).   Mr. Pratt sees 24 stars in this group (see the *Observatory* for Sept. and Nov. 1880).

363.   No. 615 is σ. 596, Lyræ 91.   Σ. appears to have seen B in 1823 in the Reichenbach circle of 4·3 in. aperture.   Variable ? (Sadler).

364.   No. 620.   Doberck gives

$\theta = 334°·67 - 0°·328 (t - 1850).$

$\rho = 0''·593 - 0''·0056 (t - 1850).$

Page 365. No. 621. Burnham and Sadler after a careful examination of the stars and Herschel's MSS. have shown that H_1 iv. 127 $= \Sigma$. 2447 (see *M. N.*, vol. xxxiv.) B C is P. XVIII. 274. Flammarion says A B $=$ P. XVIII. 274, 275. H_1's measures belong to Σ. 2447.

No. 623. Here again Bu. and Sadler have cleared up a doubtful point.

Σ. 2441 is H_1 I. 60. H_1 286°·8 in 1783·1. In the *Mem. R. A. S.*, vol. xxxv., p. 38, in the note, for "with a change of 118°," read "with a change of 5°."

366. No. 625. Doberck's formulæ are

$\theta = 216°·94 + 0°·699 (t - 1850).$

$\rho = 0''·856.$

No. 629. 120°·8   7'''·49   1879·4   C.O.

367. No. 631. H_1 observed a third star, angle 244°·05, in 1783.

No. 633 is Cygni 6.

No. 634. Doberck's formulæ are

$\theta = 218°·57 - 0°·353 (t - 1850) + 0°·00667 (t - 1850)^2.$

$\rho = 0''·715 + 0''·0135 (t - 1850).$

368. No. 636. Doberck's formulæ are

$\theta = 344°·66 - 0°·212 (t - 1850) + 0°·00323 (t - 1850)^2.$

$\rho = 0''·690 + 0''·0105 (t - 1850).$

No. 640 is P. XIX. 128.

No. 642 is Cygni 22.

Doberck's formulæ are

$\theta = 251°·04 - 0°·480 (t - 1850) - 0°·01038 (t - 1850)^2.$

$\rho = 0''·985 - 0''·0213 (t - 1850).$

369. No. 643. The measures by Se., W. and S., and Gl. are computed by Du. from measures of A C.

A C, Se. 247°·8, 47''·07, 1857·89.

No. 644 is P. XIX. 185.

The C.O. Obs. belongs to Σ. 2545. The stars Σ. 2541, 4545, 2547, and a new pair of Bu.'s are all close together and have been the cause of much confusion. (Sadler.)

No. 646. The mag. of C is 8·5.

The measure of A C is by Σ.

No. 647. Doberck gives

$\theta = 48°·06 - 0°·519 (t - 1850).$

$\rho = 0''·860.$

370. No. 650 is χ Aquilæ.

No. 652. Doberck's formulæ are

$\theta = 136°·78 + 0°·289 (t - 1850).$

$\rho = 0''·750.$

I am unable to trace the measures assigned to De.; his results for 1866·3 are 141°·37, 0''·6.

No. 653. Doberck has

$\theta = 313°·38 - 0°·306 (t - 1850) - 0°00087 (t - 1850)^2.$

$\rho = 3''·425 - 0''·0097 (t - 1850).$

Page 373. No. 656 is π Aquilæ.
           No. 658. Doberck gives
$$\theta = 110°\cdot98 - 3°\cdot463\ (t - 1850).$$
$$\rho = 0''\cdot478.$$
374.   No. 662, is ε Draconis.
           No. 663 was discovered by Lamont; his measures are 23°·8,
           11'''·81, 1838·8 (Sadler.) O.Σ. observes that the companion shares
           in the large proper motion of A, and that traces of orbital motion
           have already manifested themselves.—*Poulkova Observations*, ix.,
           407.
           No. 665 is Cygni 116.
375.   No. 667 is 16(*h*) Vulpeculæ.
           Doberck has
$$\theta = 79°\cdot47 + 0°\cdot488\ (t - 1850).\quad \rho = 0''\cdot633.$$
           No. 670. Doberck's formulæ are
$$\theta = 25°\cdot92 - 0°\cdot122\ (t - 1850).\quad \rho = 4''\cdot952.$$
           No. 671. H₁'s distance of A C is really 57″·82.
376.   No. 678 = Σ. 2657 *rej.*

The System (?) of α Capricorni.

Of α² H₂ says "the latter is one of the most beautiful and delicate
objects in the heavens."
For α¹ and α² we have

291°·4	372″·99	1822·58	H₂ and So.
291·1	376·5	1865	De.
290·6	376·3	1876	Schur.

Dr. Schur has recently discussed all known observations of α¹ and α²,
from Flamsteed's to his own, and finds a relative proper motion
which increases the difference of R.A. by + 0″·05, and of N.P.D.
by − 0″·02 ; but the observations do not suffice to decide the
question of physical connexion.
α¹ is triple.

α¹ B	225°·0	30″ ±		H₁.
α C	200 ±	30 ±	1874	Bu.

Magnitudes :—α¹, 4 ; B, 9-10 (H₂) ; C, very minute.
α² is quadruple.

α² A	150°±	8″±	1826±	H₂.
	150·2	7·41	1878	Bu.

A was detected by H₂, who says, "the small star is brighter than the
18th magnitude . . . . a very beautiful and delicate object," &c.
A was found double in 1862, by Mr. A. G. Clark ; we have

A B	57°·6	1″·72	1874	Holden.
	58·6	1·24	1874	Newcomb.
	65·2	1·14	75	Hall.
	61·2	1·06	78	Bu.

α² C 145°·7, 198″·0, 1838, Sm.

Page 376. Magnitudes :—a^2, 3 ; A, 12 ; B, 12 ; C, 9 (Sm.)

Of A Mr. Burnham writes, "Under favourable conditions a 6-inch refractor will show it fairly. Mr. Clark discovered its duplicity with the 18-inch refractor now at the Dearborn Observatory. Professor Young was able to see it with the 9·4-inch refractor of the Dartmouth College Observatory, when observing at Sherman, Colorado, from an altitude of more than 8,000 feet above the sealevel " (Burnham's " Double Stars discovered by Alvan G. Clark."

The place of a^2 for 1880 is

R.A. 20h. 11·4m. Dec. − 12° 55′.

No. 681 is Σ. 2690 = H_1 III. 16 = P. XX. 177, 178.—*Mem. R. A. S.*, vol. xxxv. p. 423.

For A D, not A C, we have

121°·5	20″±	—	H_2
105	—	1841	Da.
108 ·4	23 ·4	1878·2	Bu.

D was detected by H_2, who assigned to it a magnitude of 16

B E, 255°±, 20″±, mag. 14 of Σ.'s scale, 1877·8. This small star was discovered by Mr. Sadler in Aug. 1877.

The measure of A D is by Sm. H_2's angle (and also Sm.'s) probably wrong.

B C is Da. No. 1 = O.Σ. 407 = P. XX. 177. Σ. did not see any elongation in B in 1829 and 1832 ; nor did Mädler on sixteen nights from 1841 onwards.—*Mem. R. A. S.*, vol. xxxv. p. 423.

It is the pair A and $\dfrac{B\ C}{2}$ which form H_1 III. 16 ; H_2 260°·30, 12″·50, 1781·82.

377. No. 682 is Vulpeculæ 94.

378. No. 685. The binary A B is Burnham No. 151. " It is now (1878) excessively difficult." — " It has been gradually closing up since 1873." (Bu.)

	°	″		
A B	53·7	0·24	1878·6	Bu.
A D	115·7	27·37	8·6	,,
A C	335·6	35·06	7·7	,,

D was first measured by H_2, angle 107°·7 distance 18″ ±, 1830. Bu. has 115°·7, 27″·37, 1878·6. Mag. 11·5 ; mag. of C, 10·5.

No. 686 is κ Delphini.

327°·8	11‴·10	1877·7	De.
329°·2	10″·59	8·2	Bu.

379. No. 691 is 52 Cygni.

H_1 II. 25, H_1 58°·95, 1781·65.

380. No. 693 Doberck has

$\theta = 108°·70 − 0°·937 (t − 1850).$

$\rho = 0″·637.$

381. No. 698.   191°·2   25·82   1878·6   Bu.

No. 699.   294°·8   1″·2±   77·7   ,,

Page 381.   Doberck's formulæ are

$$\theta = 292°\!\cdot\!76 - 0°\!\cdot\!183 \, (t - 1850) + 0\!\cdot\!00413 \, (t - 1850)^2.$$
$$\rho = 0''\!\cdot\!672 + 0''\!\cdot\!0152 \, (t - 1850).$$

383.   No. 704. H_1's distance, $1''\!\cdot\!15$, is from Flammarion ; no distance is given in the Synoptical Catalogue.

No. 705 is P. XX. 440.

384.   No. 706. Doberck's formulæ are

$$\theta = 183°\!\cdot\!10 - 0°\!\cdot\!380 \, (t - 1850). \quad \rho = 1'''\!\cdot\!497.$$

No. 708. Secchi detected the duplicity of B in 1856.

B C	148°·9	1''·13	1877·8	Bu.

387.   No. 712 is δ Equulei = O.Σ. 535.

A B	156°·4	0''·2±	1877·7	Bu.
	No certain elongation		78·6	,,
	Elongated in 180° ?		78·6	,,

388.   No. 715 = P. XXI. 50.

389.   No. 716. Discovered by Alvan G. Clark, the son of Alvan Clark (A. C.) It is A. G. C. 13.

No. 717 = H_1 I. 48. It was re-discovered by A. C.

259°·9	1783·18	H_1.

No. 721.   269°·7   1''·66   1878·8   C.O.

Secchi's angle is probably 10° in error.

No. 722. Doberck's formulæ are

$$\theta = 324°\!\cdot\!47 - 0°\!\cdot\!524 \, (t - 1850).$$
$$\rho = 1'''\!\cdot\!331.$$

391.   No. 725. H_1's distance of A B, $6''\,64$ in 1780·84.

	°	''		
A B	116·2	3·88	1878·4	Bu.
A C	62·3	216·5	1800·0	Piazzi.
	58·9	214·1	55·8	De.
	57·1	209·9	77·6	Fl.
	56·3	208·5	78·4	Bu.

Mr. Burnham gives a nearer star of the twelfth mag.

262°·8	35''·28	1878·0	Bu.

392.   No. 729. The mag. of C is 9·2. The angle by Se. is given by him 132° 82, the distance 24'''·69. Flammarion gives them 141°·9, 25''·17. Secchi's angle is probably 10° in error. The formulæ given are for A B. For B C Dunér gives

$$1851·74 \; \Delta = 3''\cdot87.$$
$$P = 8°\cdot6 + 0°\cdot1058 \, (t - 1850·0).$$

No. 733. Mädler's angle is as given (see *Dorpat Observations*, vol. xiii., p. 76). Perhaps it should be 344°·6. He adds, "ungewiss, ob länglich" (uncertain, oblong ?).

No. 734. Bu. has   255°·6   5''·96   1878·5.

393.   No. 735 is ξ Cephei.

No. 738 = P. XX. 11.12.

No. 740 = Pegasi 148.

No. 741. Bu. has   351°·9   10'''·51   1878·6.

Page 394. No. 744 is 33 Pegasi. The date of Sm.'s second measure is 1838·88.

395. No. 745. Doberck's latest corrected elements are

$$\Omega = 141° 8', \lambda = 134° 40', \gamma = 44° 50', e = 0\cdot6000.$$
$$P = 1624\cdot8 \text{ years, } T = 1927\cdot74, a = 7''\cdot64.$$

397. No. 747. Bu. has   130°·0   0''·32   1878·6.

No. 748. Bu. has   155°·1   12''·29   1878·8.

No. 752. Doberck's formulæ are

$$\theta = 175°\cdot25 - 0°\cdot627\ (t - 1850).$$
$$\rho = 1''\cdot197.$$

398. No. 756 is H_1 V. 80. We have H_1 109°·9, 36''·78, 1783·60 ; and 34''·50 in 1783·00.

No. 757 is P. XXII. 219.

399. No. 759. Doberck gives

$$\theta = 72°\cdot43 - 0°\cdot167\ (t - 1850) + 0°\cdot00055\ (t - 1850)^2.$$
$$\rho = 3''\cdot180 + 0''\cdot0105\ (t - 1850).$$

No. 760 is Cephei 34 H.

No. 763 is Σ. 2966, *rej.*

No. 764. Doberck has

$$\theta = 186°\cdot40 + 0°\cdot932\ (t - 1850) - 0°\cdot01009\ (t - 1850)^2.$$
$$\rho = 0''\cdot942 + 0''\cdot0102\ (t - 1850).$$

400. No. 765. Probably no real change.

No. 766. Doberck gives

$$\theta = 356°\cdot01 + 1°\cdot010\ (t - 1850).$$
$$\rho = 1''\cdot257.$$

No. 767. Doberck's formulæ are

$$\theta = 308°\cdot19 - 0°\cdot341\ (t - 1850) + 0°\cdot00338\ (t - 1850)^2.$$
$$\rho = 1''\cdot322 + 0''\cdot0131\ (t - 1850).$$

No. 768. Doberck has

$$\theta = 345°\cdot88 + 0°\cdot044\ (t - 1850).$$
$$\rho = 13''\cdot721.$$

401. No. 770. Doberck's formulæ are

$$\theta = 176°\cdot64 - 0°\cdot218\ (t - 1850).$$
$$\rho = 4''\cdot935.$$

No. 772. The angles by W. and S. probably 10° in error.

No. 774. According to O.Σ. this is P. XXIII. 100, 101 (see *Poulkova Observations*, vol. ix. p. 395). Dawes's remarks in the *Mem. R. A. S.*, vol. xxxv. p. 445, may be thus summarised :—O.Σ. 496 = Dawes 2 = P. XXIII. 100 [*i.e.* the pair formed by P. XXIII. 100 and the minute companion discovered by Dawes and O.Σ.]. Smyth thought he had seen a very faint star 20'' from P. XXIII. 100 and in the *np* quadrant (constituting *h* 1886 mag. 7 and 13, P. = 340°·5, D. = 18''), and it was while verifying this that Dawes detected the close companion. P. XXIII. 100 must not be confounded with Σ. 3022, which is 9' further south. H_2 failed to pick up the close companion detected by Dawes, although his observations were made with the 20-ft. reflector.

C

Page 401.   O.Σ. says, "As to the companions, forming with A and B the Double Stars introduced by him [H₂] in his fourth Catalogue under the numbers 1886 and 1888, they have not been seen by any other observer since. Their existence appears very doubtful."—*Poulkova Observations*, vol. ix. p. 395. The stars there referred to are thus given by H₂ in the *Mem. R. A. S.*, vol. iv. p. 376 :—"h. 1886, R.A. 23ʰ 22ᵐ 2ˢ·6, Dec. 57° 36′ 47″ (1830). Position 340°·5, distance 18″, mag. 7, 13. Precedes A.C. [Catalogue of the Astronomical Society of 2881 stars] 2807. The large star of this, with A.C. 2807, make up the Double Star, *sh.* 355."

	h. m. s.		angle.	dist.	mag.
"h. 1888 (*a*)	R. A. 23 22 12	Dec. 57° 36′ 37″			
(*b*)	,, ,, 13·1	,, ,, 47	120°	20″	6, 12
(*c*)	,, ,, 13·0	,, ,, 34			

(*a*) is A. C. 2807 ⎱ Place observed.
(*b*) ,,   ,, 2807· ⎰ Has a companion *sf*, and a small close double
(*c*) ,,   ,, 2807 ⎰ star *np*."

Mr. Sadler, who has paid much attention to this object, thus enumerates the various components of it :—

A a. mag. 5·4, 10; 336°·8, 1″·51 ; O.Σ. 1851·76. Burnham 342°·6, 1″·26, 1880·62; 6, 10·5.

A B. mag. 5·4, 7·4 ; 269°·3, 75″·63 ; De. 1868·9. These are the stars P. XXIII. 100 and 101 of the Catalogues.

B C. mag. 7·4, 8·9 ; 223°·8, 1″·39 ; De. 1868·9. C was discovered by Da. and O.Σ., and the pair is O.Σ. 496 = Da. 2.

A F. mag. 5·4, 9·7 ; 113°·6, 43″·36 ; De. 1868·9. This is *h.* 1888.

B G. mag. 7·4, var.; 341°·8, 25″ ± ; Da. 1841·8. This is *h.* 1886. The mag. of G is thus given :—H₂ 13, Sm. 14, Piazzi Smyth at Teneriffe 16, Sadler 10·5. The first three estimates are in H₂'s scale.

A D. mag. 5·4, 9·2 ; 338°·3, 67″·1 ; De. 1868·9.

D E. mag. 9·2, 9·6 ; 73°·5, 10″·29 ; De. 1868″·9.

From this it appears (1) that O.Σ. is mistaken with regard to *h.* 1886 and 1888 ; (2) that the star is really octuple.

At Bermerside, G was carefully looked for on several fairly good nights in March last, but no trace of it was seen. On the other hand, it was visible to Mr. Knott on the 15th of March, with the aperture reduced to 6 in. Mr. Knott estimated the mag. at 12½ or 13 of Argelander's scale and 15 or 16 of Smyth's. During the present summer, at Bermerside, G was always readily seen with 6 in. Aa requires the very finest nights.

402.   No. 775.   Doberck gives
$$\theta = 303°·33 + 0°·675 (t - 1850).$$
$$\rho = 0″·463.$$

No. 776.   This is Sh. 356. South's distance is 5″·06.

No. 777.   211°·3, 2″·89, 1873·8, W. & S.

No. 779.   H₁'s distance is 26″·20.

Page 402. No. 781 is h. 1911.

403. No. 782. Doberck's formulæ are

$\theta = 238°\cdot14 + 0°\cdot265\ (t - 1850) - 0°\cdot00107\ (t - 1850)^2.$

$\rho = 2''\cdot720 + 0''\cdot0110\ (t - 1850).$

No. 784. Bu. detected a closer companion in 1878, and his measures already show the binary character of the pair. The mag. of B is

12·5.	274°·0	0''·67	1878·7	Bu.
	284°·6	0''·75	79·4	,,

See p. 416, No. 1200.

405. No. 787. Bu. has 8'''·62 in 1877·9.

St. = Prof. O. Stone.

No. 796. Dunlop 202°·2, 7''est., 1826·11 ; H₂ 200°·3, 8'''·29, 1836·91 ; Ja. 199°·9, 7'''17, 1845·86 ; 202°·8, 7'''·07, 1856·18.

406. No. 805. β Leporis.

281·5, 3'''·02, 1880·009, Hall. De.'s angle is 293°·6.

No. 813. The date of De.'s measures is 1867·9, and his angle is 38°·9.

No. 814. H₁ 219°·95, 23'''·50, 1783·895. The angle given on p. 71 of vol. xxxv. of the *Mem. R. A. S.* is a misprint ; 50° 3' *np.* for 50° 3' *sp.*

No. 818. Mädler's angle probably 10° in error ; O.Σ. has 269°·0, 1'''·19 in 1855·2.

407. No. 829. The C.O. have 13'''·61 in 1879·3.

No. 830. The second distance was an estimation only.

No. 833. Bu. 305°·8, 32'''·35, 1878·3.

No. 835. The measures in 1837·5 were also by H₂.

No. 836. Σ. has 18°·3, 4'''·37, 1829·6.

No. 840 is Sh. 184 = H₁ III. 97 = 54 [10] Hydræ. H₁'s distance is 11''·28.

No. 841. This is P. XIV. 212. There are several other companions. " A system not unlike 61 Cygni, having large proper motion " (Burnham and Flammarion).

The proper motion of A is very rapid : it is + 0ˢ·066 in R. A. and + 1''·72 in P. D.

No. 845. Σ. has 83°·96, 30''·15, 1832·6.

No. 846. De has 266°·0, 3'''·1, 1857·5. Σ.'s angle is 266°·3.

No. 849 = O.Σ. 308.

No. 853 is 19 Ophiuchi. Σ. has 92°·6, 22'''·25, 1832·14.

408. No. 864 = O.Σ. 543. The first measure was by Σ.

No. 866. O.Σ. has 330°·4, 3'''·68, 1851·7.

No. 868. O.Σ. has 154°·15, 0''·80, 1846·5, and De. 153°·7. 0''·96, 1867·4.

No. 870. In the Cape Observations, H₂ gives for A B, 130°·9, 1¾'', 1837·663. "Some doubt." In his MSS. he has 139°·95, 1'''·75, 1837·663. In the Cape Observations he has the following note :— " Excessively difficult, yet I feel convinced it *is* double. [N.B. After all, there remains some degree of suspicion attached to this

Page 408.   star, the quadrant being *sf*, in which an illusory appendage has been
              frequently observed.]"
              Of C he says, "distance by a single difference of R.A. posi-
              tion by a single measure." Ja. has for A B, 248°·58, 1"·3 est.,
              1856·55; 260°·0, 1"·5 est., 1857·85; for A C, 265°·79, 66"·37,
              1856·55.

              No. 871. Mädler's angle probably wrong. O.Σ. has 145°·4, 0"·61,
              1842·7.

              No. 875. In 1802·4 H₁ found the angle 310°·7. A is a very close
              double and is A. G. C. 11. In 1878 Bu. found 155°·4, 0"·35.

              No. 877. Bu. has 237°·7, 0"·45, 1878·7.

              No. 878. Bu. has 317°·0, 0"·31, 1878·5.

409.   No. 880. A C. misprint in *Mem. R. A. S.*, vol. xxxv., p. 40; 320°·0
              should be 326°·0. The epoch is 1783·7.

              No. 881. Du. has 125°·3, 1"·09, 1868·99.

              No. 882. Bu. has 36°·6, 1"·65, 1878·5.

              No. 890 is H₁ I. 62.

              No. 891. H₂'s distance is probably much too large; Dawes has
              347°·7, 1"·52, 1837·68.

              No. 893. Bu. has 285°·0 in 1878·3; O.Σ., 290°·6 in 1859.

              No. 894. H₁ gave 336°·8 in 1802·7, and H₂ 328°·0, 2"·12, 1836·64.

              No. 903. H₂ and So. give 303°·1, 10"·03 in 1823·26.

              No. 903 is 53 Aquarii = H₁ No. 41.

416.   No. 1177.

              H₁ III. 110 = O.Σ. 447 = Bu. 449.

              The following are the chief measures of this quintuple star :

	°	″			
A B	19·1	6·78	7, 12	1876·8	Bu.
A C	157·6	13·90	,, 10·8	1783·81	H₁.
	169·4	13·96	,, ,,	1848·30	O.Σ.
	170·5	13·71	,, ,,	1866·58	De.
A D	248·2	17·94	,, 13–14	1876·8	Bu.
A E	49·5	25·97	,, 7·7	1783·81	H₁.
	45·4	29·00	,, ,,	1848·30	O.Σ.
	45·7	29·13	,, ,,	1866·58	De.

Mr. Burnham says : " B and D are very minute, and might be easily overlooked
with even a large aperture."

BIBLIOGRAPHY.

PAPERS ON DOUBLE STARS.

ABBÉ (Prof. C.)
"On O. M. Mitchel's Observations at Cincinnati."—*M. N.*, vol. xxxvii., 121.

AMICI.
Double Stars (Measures of Distances only).—*Zach's Correspondance Astronomique*, vol. viii., pp. 73 and 216.

BALL (Dr.)
"On the Parallax of the preceding Star of 61 Cygni." —*Dunsink Observations*, part III. 1879.

BOND (W. C. and G. P.)
Proceedings of the American Academy, vol. ii., p. 144, 135. "Double Stars observed at Cambridge Observatory, 1848-49."

BREEN.
"On the Orbit of ξ Ursæ Majoris," cf. *Wochenschrift für Astronomie*, &c., 1862, p. 358.

BRÜNNOW (Dr. F.)
"On the Parallax of *a* Lyræ, 85 Pegasi," &c.—*Dunsink Observations*, part II. 1873.
"Observations of Double Stars."—*Dunsink Observations*, part III. 1879.

BURNHAM (S. W., M.A.)

"Double Star Observations, made in 1877, 1878, at Chicago, with the 18½ inch Refractor of the Dearborn Observatory, comprising I. a Catalogue of 251 new Double Stars with Measures ; II. Micrometrical Measures of 500 Double Stars."—*Memoirs of R.A.S.*, vol xliv.

New Double Stars.—*Monthly Notices*, vol. xl.

New Double Stars, Catalogue vi., *Astr. Nachr.*, No. 2062 ; vii., *Astr. Nachr.*, No. 2103 ; viii., *American Journal of Science and Arts*, July 1877.

On "Errors and Omissions in the Catalogues of Sir W. Herschel's Double Stars."—*M. N.*, vols. xxxiii., p. 151, xxxiv., p. 98.

"Catalogue of Red Double Stars."—*M. N.*, vol. xxxvi. p. 332.

"An Examination of the Double-Star Measures of the Bedford Catalogue. By S. W. Burnham, Esq." *M. N.*, vol. xl., No. 8.

BURTON (C. E.)

"Observations of Double Stars."—*M. N.*, vol. xxxiv., p. 125.

CINCINNATI OBSERVATORY.

Publications of the Cincinnati Observatory ; Micrometrical Measurements of Double Stars, 1878-1879.

CLARK (Alvan G.)

"List of 14 Double Stars discovered by him ; collected Measures of the same, by Mr. Burnham."—*American Journal*, 1879, April.

COPELAND.

"Ueber die Bahnbewegung von a Centauri."—*Vierteljahrsschrift der A. G.*, vol. v., p. 312.

CRULS.

" Southern Multiple Stars, by M. Cruls, Director of the Imperial Observatory of Rio Janeiro."—*Comptes Rendus*, vol. lxxxix., No. 8.

DOBERCK.

Measures of Double Stars.—*Astr. Nachr.*, vol. xciv.

Elements of 36 Andromedæ, *a* Centauri, and γ Leonis. —*Astr. Nachr.*, vol. xciv.

On the Brightness, Parallax, and Distribution of Double Stars. On the Brightness of the Components of Revolving Double Stars. Elements of Σ. 3062 and O.Σ. 298.—*Astr. Nachr.*, vol. xcv.

Elements of 4 Aquarii and μ^2 Herculis.—*Astr. Nachr.*, vol. xcvi.

Elements of O.Σ. 235.—*Astr. Nachr.*, vol. xcvi.

ELLERY.

" Observations of *a* Centauri."—*M. N.*, xxxvii., p. 435.

English Mechanic.

Nearly every volume contains papers, &c., on Double Stars.

GILL (D.)

" Heliometer Measures of *a* Centauri."—*M. N.*, vol. xxxix., p. 123.

GOLDNEY (G. A.)

Double Star Measures.—*Astr. Nachr.*, vol. xcv.

HARVARD COLLEGE ZONES.

Of the double stars in these Observations some are new, some are Σ.'s and H_2's; the distances are estimated ; the position angles are not given, the quadrant of the smaller star alone being observed.

HOWE (H. A.)

On the Determination of the Zero of a Position Circle. —*Astr. Nachr.*, vol. xcvi.

JACOB (Captain).

Madras Obs., 1847-52.

KNOBEL (E. B.)

Notes on a Paper entitled "An Examination of the Double-Star Measures of the Bedford Catalogue. By S. W. Burnham, Esq."—*M. N.*, vol. xl., No. 8.

KNOTT (G.)

"On ξ Ursæ Majoris.—*M. N.*, vol. xxxiii., p. 101.

MITCHEL (O. M.)

Sidereal Messenger, 1846-48.

OXFORD (University Observatory.)

Astronomical Observations (Measures of Double Stars, Computed Orbits, &c.) No. 1. 1878.

RUSSELL (H. C.)

Measures of a Centauri.—*M. N.*, vol. xxxvii. p. 462.

SADLER (H.)

Notes on the late Admiral Smyth's "Cycle of Celestial Objects."—*Monthly Notices*, vol. xxxix.

Notes on "A Catalogue of 10,300 Multiple and Double Stars," &c., forming vol. xl. of the *Memoirs of the Royal Astronomical Society*. Hours 0 to vi.—*Monthly Notices*, vol. xl.

SAFFORD.

"On the Proper Motion of Sirius in Declination."—*Brünnow's Astronomical Notices*, No. 28.

SANTARELLI.

" Memoria intorno a parecchie osservazioni fatte nelle specola del Collegio Romano nel 1842." Rome, 1842.

SCHJELLERUP.

"Stjernefortegnelse indeholdende 10,000 Positioner af teleskopiske Fixstjerner − 15° og + 15° Graders Deklination."

SCHUR (Dr. W.)

Double Star Measures.—*Astr. Nachr.*, vol. xliv.

SEELIGER (Dr. H.)

Ueber Mädler's Doppelsternmessungen.—*Astr. Nachr.*, vol. xcvi.

" On the Orbit of λ Ophiuchi."—*Zur Theorie der Doppelsternbewegungen.* 1872.

SMYTH (Admiral).

Speculum Hartwellianum.

SMYTH (Prof. P.)

"Edinburgh Obs. 1855-59."—*M. N.*, vol. xxiii.

THIELE (T. N.)

" Castor. Calcul du Mouvement relatif et critique des observations de cette étoile double. Par T. N. Thiele." Copenhagen, 1879.

Sur la compensation de quelques erreurs quasi-systématiques par la méthode des moindres carrés. Par T. N. Thiele. Copenhagen, 1880.

"Undersögelse af Omlöbsbevægelsen i Dobbelstjernesystemet Gamma Virginis udfört tildels efter nye Methoder af T. N. Thiele." Copenhagen, 1866.

VINOGRADSKIJ.

"On the Orbits of ξ Boötis, μ² Boötis." — *Mémoire Scientif. Univ. de Kasan.* 1872.

PAPERS ON THE MICROMETER.

BOWDEN (A.)

A Self-registering Micrometer. — *Monthly Notices*, vol. xl.

DOBERCK.

On the Adjustments of a large Equatorial.—*Ast. Nachr.*, vol. xcvi. (The effects of the errors on the measures of double stars, &c.)

F. R. A. S.

An excellent article on Micrometers, by " F.R.A.S." —See *English Mechanic*, vol. xix.*

KNORRE (Dr. V.)

" Ueber ein neues Mikrometer zum Registriren von Declinationsdifferenzen."—*Astr. Nachr.*, vol. xciii.

PETERS (Dr. C. H. F.)

On the Determination of the Radius of a Ring Micrometer.—*Brünnow's Astronomical Notices*, No. 13.

PAPERS ON THE COLOURS OF STARS.

BIRMINGHAM (J.)

The Red Stars, Observations and Catalogue.—*Trans. R. I. A.*, vol. xxxvi.

DOBERCK.

On the Colour of Revolving Double Stars. — *Astr. Nachr.*, vol. xcv.

FEARNLEY.

Coloured Stars in zone + 64° 50′ to + 70° 10′.—*Astr. Nachr.*, No. 2121.

FRANKS (W. S.)

" A Catalogue of the Colours and Magnitudes of 3,890 Stars between the North Pole and 25° South Declination." (Presented to the Royal Astronomical Society by the Author. 1878.)

---

* In the volumes of the *English Mechanic* will also he found a large amount of useful matter relating to telescopes, double stars, &c., &c.

HOLDEN (Prof. E. S.)

Note on a relation between the colours and magnitudes of the components of Binary Stars.—*The American Journal of Science*, vol. xix.

PEIRCE (C. S.)

On the Colours of Double Stars.—*Nature*, vol. xxii p. 291.

PICKERING (Prof.)

For some valuable and interesting observations on the magnitudes and colours of 180 bright double stars, the methods of observing them, &c., see *Annals of Harvard College Observatory*, vol. xi., parts I. and II. 1877-79.

SADLER (H.)

Colours of some Double Stars.—*Astr. Reg.*, vol. xiv.

SCHJELLERUP.

" Catalog der rothen isolirten. Sterne."—*Astr. Nachr.*, Nos. 1591 and 1613.

" Zweiter Catalog der rothen isolirten. Sterne."— *Vierteljahrsschrift der Ast. Ges.*, vol. ix., p. 252.

SECCHI.

Sugli spettri Prismatici delle stelle Fisse.—*Atti del' Academia Pontificia.* 1872.

Le Stelle, pp. 359-389.

Prodromo di un Catalog fisico delle stelle colorate.— *Memorie della Società degli spettroscopisti Italiani*, vol. vi. 1876. See also *Memorie della Società Italiana del* xl. third series, vol. i., and *Memoria* ii., Sugli spettri, &c.

SMYTH (Prof. P.)

Edinburgh Obs., 1855-59, and Prof. Smyth's "Teneriffe" Obs.

TUPMAN (Captain).

Colours and Magnitudes of Southern Stars.—*M. N.*, vol. xxxiii.

INDEX.